# MOLECULAR SIGNALS IN PLANT-MICROBE COMMUNICATIONS

Editor

**Desh Pal S. Verma, Ph.D., FRSC**
Associate Director
Ohio State Biotechnology Center
and
Professor
Department of Molecular Genetics
The Ohio State University
Columbus, Ohio



CRC

**CRC Press**
**Boca Raton Ann Arbor London**

**Library of Congress Cataloging-in-Publication Data**

Molecular signals in plant-microbe communications / edited by Desh Pal S. Verma.

p. cm.

Includes bibliographical references (p. ) and index.

ISBN 0-8493-5905-8

1. Plant-microbe relationships—Molecular aspects. I. Verma, D. P. S. (Desh Pal S.), 1944- .

QR351.M65 1991

632'.3—dc20

91-19662

CIP

This book represents information obtained from authentic and highly regarded sources. Reprinted material is quoted with permission, and sources are indicated. A wide variety of references are listed. Every reasonable effort has been made to give reliable data and information, but the authors and the publisher cannot assume responsibility for the validity of all materials or for the consequences of their use.



Direct all inquiries to CRC Press, Inc., 2000 Corporate Blvd., N.W., Boca Raton, Florida 33431.



International Standard Book Number 0-8493-5905-8

Library of Congress Card Number 91-19662

Printed in the United States of America 2 3 4 5 6 7 8 9 0

**DATE DUE**

| | |
|---|---|
| | |
| | |
| | |
| | |
| | |
| | |
| | |
| | |
| | |
| | |
| | |
| | |
| | |
| | |
| | |
| | |

# MOLECULAR SIGNALS IN PLANT-MICROBE COMMUNICATIONS

# PREFACE

All species are interdependent for their biotropic needs. Microbes interact with higher plants primarily for their carbon and nitrogen supplies since plants can fix carbon from the atmosphere utilizing solar energy and assimilate nitrogen from soil. In the quest for these primary nutrients, some microbes have evolved highly sophisticated mechanisms to exploit plants for developing unique ecological niches. At the same time, plants have evolved a variety of mechanisms to defend themselves against invading organisms and to allow development of associations that are beneficial to them. Some of the metabolites in the host defense pathways are often used by microorganisms as signals for controlling their specific genes for initial interaction with the plant. Whether an association between a microbe and a plant is beneficial or harmful depends upon the net balance of the carbon and nitrogen exchange and how much of its resources each organism allows the other to take.

Due to heavy losses inflicted by certain microbes upon important agricultural crops, chemical protection has been successfully employed to improve crop yield. The strategy used in chemical protection of crop plants varies depending upon the pathogen and the crop species. However, the enormous negative effect caused by heavy application of chemicals in modern agriculture is forcing scientists to develop alternative approaches to control plant-microbe interactions. Therefore, an understanding of the molecular signals involved in communication of plants and microbes is essential in order to manipulate both beneficial and harmful associations. This book summarizes some of the signals and mechanisms used by various organisms, including parasitic angiosperms, to interact with their hosts. That various parasitic angiosperms also use a signal from their host for germination and initiation of haustoria suggests that chemical communication may be universal for interaction between species. Understanding this ''language'' of communication will not only allow one to manipulate plant-microbe interactions, but the basic knowledge gained will also help understand the regulation and function of genes involved in various developmental pathways in plants and organ-specific metabolisms. Altering the metabolic pathways in plants by genetic engineering may allow us to increase the production of compounds that act as signals for beneficial organisms while limiting harmful associations. Such germ plasm can be easily controlled and maintained.

**Desh Pal S. Verma**

# THE EDITOR

**Desh Pal S. Verma, Ph.D., FRSC,** is Associate Director of the Ohio State Biotechnology Center, Columbus, Ohio and Professor of Molecular Genetics at The Ohio State University, Columbus, Ohio.

Dr. Verma received his M.Sc. degree in 1964 from Agra University and his Ph.D. degree from the University of Western Ontario in 1970. After doing postdoctoral work at the Institute for Cancer Research, Philadelphia, he joined McGill University, Montreal in 1972 as a Research Associate, was appointed Assistant Professor in 1974, Associate Professor in 1978, and Professor in 1983. In 1985, he became the Director of the Centre for Plant Molecular Biology at McGill University. In 1988 he assumed his present position.

Dr. Verma is currently President of the International Society of Molecular Plant-Microbe Interactions, elected Fellow of the Royal Society of Canada and a member of the Academy of Sciences, Senior Editor for *Molecular Plant-Microbe Interactions,* and member of the Editorial Boards for *Plant Science* and *Plant Molecular Biology*. He has also served on the Natural Sciences and Engineering Research Council (NSERC) of Canada and National Science Foundation (NSF) Advisory Panels. He is included in the *International Directory of Distinguished Leadership,* ABI, U.S.A. (1986), is cited in *Who's Who in Frontiers of Science and Technology in North America* (1986), served as Canadian Pacific Scholar (1985-1986), received the C.D. Nelson Award of the Canadian Society of Plant Physiologists (1984), and was Steacie Fellow, NSERC (1981-1983).

Dr. Verma has contributed extensively to the area of plant molecular biology in general and molecular plant-microbe interactions in particular. He has presented numerous lectures in international meetings in the field. His research program has been well supported by many national and international agencies (NSERC, NSF, USDA, and Rockefeller Foundation). He has published more than 120 papers and edited several books. His current major research interests include molecular communications between legume plants and *Rhizobium* leading to the establishment of endosymbiosis and the biogenesis of the subcellular compartment housing the bacteria in root nodules. Major emphasis is on the function and regulation of host genes.

# CONTRIBUTORS

**Paola Bonfante-Fasolo, Ph.D.**
Dipartimento di Biologia Vegetale dell' Università
Torino, Italy

**Vitaly Citovsky, Ph.D.**
Department of Plant Biology
University of California
Berkeley, California

**Steve Daubert, Ph.D.**
Department of Plant Pathology
University of California
Davis, California

**Keith Davis, Ph.D.**
Biotechnology Center and Department of Plant Biology
The Ohio State University
Columbus, Ohio

**Ralph A. Dean, Ph.D.**
Department of Plant Pathology and Physiology
Clemson University
Clemson, South Carolina

**Ruud A. de Maagd, Ph.D.**
Department of Plant Molecular Biology
Leiden University
Leiden, The Netherlands

**Jean Dénarié, Ph.D.**
Laboratoire de Biologie Moléculaire des Relations Plantes-Microorganismes
Centre National de la Recherche Scientifique (CNRS)
and
Institut National de la Recherche Agronomique (INRA)
Castanet Tolosan, France

**Yves Dessaux, Ph.D.**
Institut des Sciences Végétales
Centre National de la Recherche Scientifique (CNRS)
Gif-sur-Yvette, France

**Richard A. Dixon Ph.D.**
Plant Biology Division
Samuel Roberts Noble Foundation
Ardmore, Oklahoma

**Dominique Expert, Ph.D.**
Laboratoire de Pathologie Végétale
Institut National Superieur d'Agronomie
Paris, France

**Stanton B. Gelvin, Ph.D.**
Department of Biological Sciences
Purdue University
West Lafayette, Indiana

**Paul Gill, Ph.D.**
Laboratoire de Pathologie Végétale
Institut National Superieur d'Agronomie
Paris, France

**James X. Gray, Ph.D.**
Plant-Microbe Interactions Group
Australian National University
Canberra, Australia

**Elizabeth Greene**
Department of Plant Biology
University of California
Berkeley, California

**Lee A. Hadwiger, Ph.D.**
Molecular Biology of Disease Resistance Laboratory
Department of Plant Pathology
Washington State University
Pullman, Washington

**Elizabeth Howard, Ph.D.**
Department of Plant Biology
University of California
Berkeley, California

**Nicola Sante Iacobellis, Ph.D.**
Istituto Tossine e Micotossine da Parassiti Vegetali
Consiglio Nazionale Richerche
Bari, Italy

**Andrew W. B. Johnston, Ph.D.**
Department of Genetics
John Innes Institute
Norwich, England

**Clarence I. Kado, Ph.D.**
Department of Plant Pathology
Davis Crown Gall Group
University of California
Davis, California

**Jan Kijne, Ph.D.**
Department of Plant Molecular Biology
Leiden University
Leiden, The Netherlands

**Thomas J. Knight, Ph.D.**
Department of Biological Sciences
University of Southern Maine
Portland, Maine

**P. E. Kolattukudy, Ph.D.**
Biotechnology Center
The Ohio State University
Columbus, Ohio

**Adam Kondorosi, Ph.D.**
Institut des Sciences Végétales
Centre National de la Recherche
Scientifique (CNRS)
Gif-sur-Yvette, France
and
Institute of Genetics
Biological Research Center
Hungarian Academy of Science
Szeged, Hungary

**Gretchen Kuldau, Ph.D.**
Department of Plant Biology
University of California
Berkeley, California

**Christopher J. Lamb, Ph.D.**
Plant Biology Laboratory
Salk Institute for Biological Studies
La Jolla, California

**Pat J. Langston-Unkefer**
Isotope and Nuclear Chemistry Division
Los Alamos National Laboratory
Los Alamos, New Mexico

**Ben J. J. Lugtenberg, Ph.D.**
Department of Plant Molecular Biology
Leiden University
Leiden, The Netherlands

**B. Gail McLean, Ph.D.**
Department of Plant Biology
University of California
Berkeley, California

**Peter J. Murphy, Ph.D.**
Department of Crop Protection
University of Adelaide
Glen Osmond, Australia

**Dale Noel, Ph.D.**
Department of Biology
Marquette University
Milwaukee, Wisconsin

**Hachiro Oku, Ph.D.**
Laboratory of Plant Pathology and
Genetic Engineering
College of Agriculture
Okayama University
Okayama, Japan

**Richard P. Oliver, Ph.D.**
Norwich Molecular Plant Pathology
Group
School of Biological Sciences
University of East Anglia
Norwich, England

**Silvia Perotto, Ph.D.**
IPSR—John Innes Institute
Norwich, England

**Annik Petit, Ph.D.**
Institut des Sciences Végétales
Centre National de la Recherche
Scientifique (CNRS)
Gif-sur-Yvette, France

**P. J. Punt, Ph.D.**
TNO Medical Biological Laboratory
Rijswijk, The Netherlands

**A. Quispel, Ph.D.**
Department of Plant Molecular Biology
Leiden University
Leiden, The Netherlands

**James L. Riopel, Ph.D.**
Department of Biology
University of Virginia
Charlottesville, Virginia

**Philippe Roche**
Centre de Recherches de Biochimie et de Génétique Cellulaire
Centre National de la Recherche Scientifique (CNRS)
Toulouse, France

**Barry G. Rolfe, Ph.D.**
Plant-Microbe Interactions Group
Australian National University
Canberra, Australia

**C. P. Saint, Ph.D.**
Department of Crop Protection
University of Adelaide
Glen Osmond, Australia

**Gerrit Smit, Ph.D.**
Department of Plant Molecular Biology
Leiden University
Leiden, The Netherlands

**Giuseppe Surico, Ph.D.**
Istituto di Patologia e Zoologia Forestale e Agraria
Università di Firenze
Firenze, Italy

**Jon Y. Takemoto, Ph.D.**
Department of Biology and Molecular Biology Program
Utah State University
Logan, Utah

**Jacques Tempé, Ph.D.**
Institut des Sciences Végétales
Centre National de la Recherche Scientifique (CNRS)
Gif-sur-Yvette, France
and
Institut National Agronomique
Paris, France

**Yvonne Thorstenson, M.S.**
Department of Plant Biology
University of California
Berkeley, California

**M. P. Timko, Ph.D.**
Department of Biology
University of Virginia
Charlottesville, Virginia

**C. A. M. J. J. van den Hondel, Ph.D.**
TNO Medical Biology Laboratory
Rijswijk, The Netherlands

**Keith Wycoff, Ph.D.**
Plant Biology Division
Samuel Roberts Noble Foundation
Ardmore, Oklahoma

**Patricia Zambryski, Ph.D.**
Department of Plant Biology
University of California
Berkeley, California

**John Zupan, Ph.D.**
Department of Plant Biology
University of California
Berkeley, California

# TABLE OF CONTENTS

## SECTION III: BACTERIAL-PLANT INTERACTIONS (SYMBIOTIC)



## SECTION IV: INTERACTION OF OTHER MICROORGANISMS WITH PLANTS

# *Section I: Fungal Development and Interactions with the Host*

# REGULATION OF GENE EXPRESSION AND SIGNALS IN FUNGAL DEVELOPMENT

**Ralph A. Dean**

## TABLE OF CONTENTS

# I. INTRODUCTION

Fungi have enormous impacts on our everyday lives which range from the production of antibiotics, such as penicillin, to devastation and famine caused by fungal diseases of crop plants. The study of fungal development has been an exciting area of research for the better part of this century. Fungi also offer excellent model systems for studying eukaryotic developmental processes. Advantages include limited cell type complexity, ease and rapid rate of growth, and in some fungi, well-defined genetic systems. The advent of DNA-mediated transformation systems and related technology has resulted in extraordinary advances in our understanding of development. Three different aspects of fungal development, sexual reproduction, asexual reproduction, and formation of infection structures, are discussed in this chapter. For each aspect, one particular organism has been chosen to highlight the advances that have been made. This approach should serve to encourage research with fungal systems that are only now becoming amenable to molecular genetics and to dissect various signals involved in regulating specific genes involved in growth and development. The reader should note that the lack of consistency in gene symbols in this review is deliberate and represents those commonly accepted by investigators working in these particular fields.

Extensive studies of budding yeast (*Saccharomyces cerevisiae*) have contributed most significantly to our understanding of the mechanisms that control growth and development. Although yeast is a unicellular organism, haploid and diploid forms exist. Haploids mate to produce diploids, and the diploids undergo meiosis and produce (sporulate) haploid spores; however, certain strains exhibit a homothallic life cycle (a life cycle in which a single haploid cell can give rise to diploid cells) whereas other strains are heterothallic and have a life cycle in which a single haploid cell is unable to produce diploid cells. Studies have revealed the molecular basis which explains why only haploids mate, but diploids sporulate and do not mate. These studies also explain the basis of homothallism versus heterothallism. All of these processes are governed by a master regulatory locus, the mating type locus (*MAT*). Much of the molecular basis of how *MAT* controls mating and sporulation, and the mechanism by which *MAT* is controlled (to create homothallic behavior) is now understood.[1,2,3]

Filamentous fungi, most notably *Neurospora crassa* and *Aspergillus nidulans,* have been exploited for physiological, biochemical, and genetic studies of developmental processes. *Aspergillus nidulans* has proved valuable for elucidating the genetic control of asexual reproduction. Much information has accumulated over recent years following the development of a sophisticated genetic system including DNA transformation. This chapter summarizes the knowledge accumulated about genes that regulate the asexual reproductive pathway leading to the formation of the multicellular reproductive apparatus, the conidiophore with mature conidia.[4]

The final aspect of fungal development reviewed is the formation of specialized structures by *Magnaporthe grisea* essential for infection of rice, the appressorium and penetration peg. Comparatively this process is the least understood. However, a few fungal pathogens including *M. grisea* have been selected as model systems and are being developed by a growing number of researchers. Fungal pathogens form infection structures of varying complexity. Spores from some fungi, such as fusarium species, germinate and directly penetrate their hosts without forming specialized structures. For others, such as rust fungi, the elaboration of these structures is highly complex. How fungi respond to their environment and develop infection structures usually only on host tissues and not when grown in culture remains an intriguing mystery. The fact that many pathogenic fungi, including *M. grisea,* can be "tricked" into forming infection structures on a number of artificial surfaces should prove to be useful tools for gaining access to the underlying molecular mechanisms.

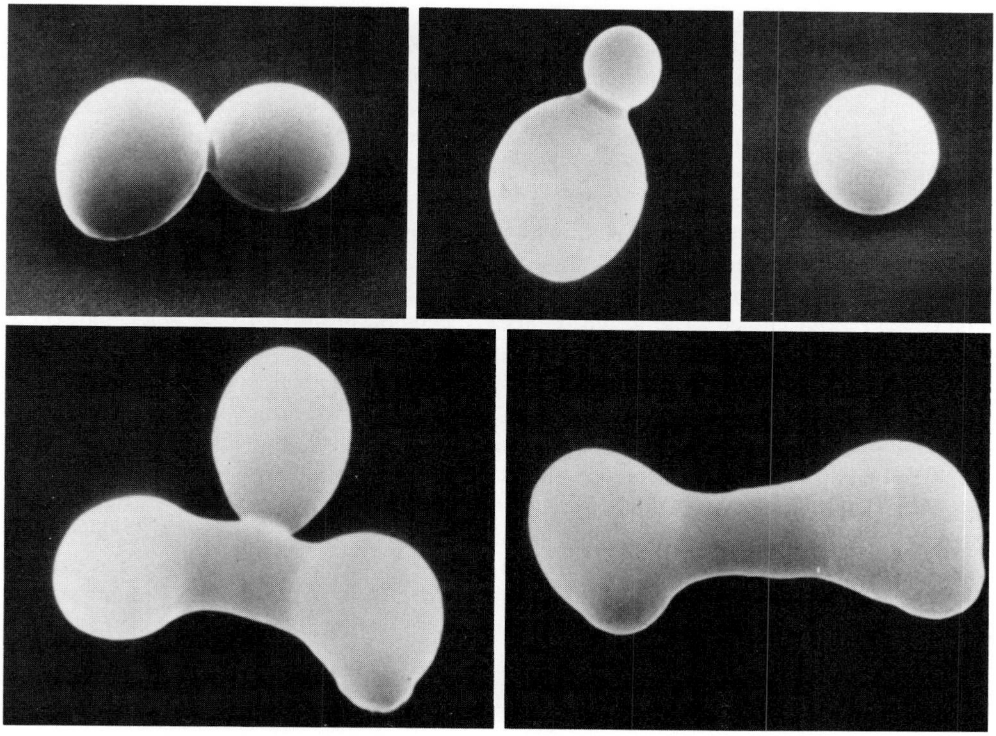

FIGURE 1. Morphology of budding in *S. cerevisiae* and zygote formation. Upper row shows cells with different bud sizes. The lower row shows zygote formation. The zygote was formed by mating between an **a** and an α cell. Daughter **a**/α cells often bud from the neck of the zygote (as shown on lower right). Magnification approximately × 5200. Reproduced with permission.[1]

## A. *SACCHAROMYCES CEREVISIAE*

The term "yeast" can be applied to many organisms, but in this chapter will be used only to refer to *Saccharomyces cerevisiae.* Yeast exists in nature in two distinct forms, both of which are unicellular and elliptical, but differ considerably in size. The predominant and most stable in nature is the larger diploid form, 6.0 × 5.0 μm, compared with the smaller haploid form of 4.75 × 4.2 μm. The cell volume of the haploid is about one half that of the diploid. Both forms are capable of mitotic division, usually described as budding. The original "mother" cell gives rise to a daughter with entirely new surface material. A bud begins on the spore wall over a thickened region called a plaque. As this region begins to enlarge, the chromosomes (17 in *S. cerevisiae*) attach to the spindle pole body and are replicated (S phase) along with the spindle pole body. During mitosis (M phase), the spindle becomes oriented so that the duplicated chromosomes partition between the two cells. The daughter cell is slightly smaller than the mother cell and must increase in size before it can undergo division (Figure 1). With adequate nutrition, cells can double approximately every 100 min, but when nutrients or other growth requirements are withheld, cell growth ceases, and they arrest, unbudded, in the $G_1$ phase of the cell cycle. Diploid cells then initiate meiosis and produce four haploid spores. It is not clear how starvation is monitored, but cAMP levels may be involved. Haploid cells also arrest in $G_1$ when starved, where they survive and resume growth upon the availability of nutrients. In addition, haploid cells arrest in $G_1$ when other haploid cells with which they can mate are in close proximity. During mating, haploid cells elongate and become pear-shaped (schmoos), eventually fusing and forming diploids.[5] Details of the complexities of mating are reviewed in some detail in the following sections.

### 1. Environmental Signals Induce Mating

In most wild-type strains of yeast, the four haploid products of meiosis are two of each of the mating types **a** and α. When an **a** and α cell are placed in close proximity, they will mate with nearly 100% certainty (Figure 1). Conjugation and mating require communication between the cells.[6] Both **a** and α cells produce and secrete specific signaling molecules, or pheromones (12 and 13 amino acid peptides, respectively), which exert their effects through receptors specific to each mating type.[7,8] There are two specific consequences of this cell-cell communication. First, growth is inhibited, and the cells come to rest in $G_1$ of the cell cycle, so that each cell has only one copy of each chromosome. Next, the proteins necessary for conjugation and diploidization are synthesized. The genes that encode the specific receptors have been cloned. The **a** and α factor receptors are products of *STE3* and *STE2* genes, respectively.[9] These genes, as well as many of the other genes necessary for mating, were isolated by functional complementation of strains that were defective in mating. Pheromone receptors are likely to be coupled to G-proteins that transduce the mating signal across the plasma membrane. The receptors are membrane proteins and have seven potential membrane-spanning regions. This structure is typical for receptors involved in other transduction signalling systems such as rhodopsin, the vision receptor, and the β-adrenergic receptor for epinephrine.[10,11] The components of the mating signal transducer have been identified genetically and the genes cloned. On the basis of the predicted amino acid sequence and loss of function, phenotypes *GPA1, STE4*, and *STE18* are believed to code for the Gα, Gβ, and Gγ subunits of the heterotrimeric nucleotide-binding proteins (G-proteins). How this leads to growth arrest and induction of mating is uncertain. The current model of G-protein function in mammalian cells is that the binding of the signal to the receptor leads to guanine nucleotide exchange (GDP for GTP) on the α-subunit and the activated α-subunit dissociates from the βγ-subunit. Which of these free subunits acts as effector of the pheromone response is currently unresolved.[12,13,14] Genetic evidence and the results from constitutively expressed *GPA1, STE4*, and *STE18* strongly support the notion that Gβγ mediates the mating signal and not Gα as is the more widely held view in other systems. Both subunits are capable of interacting with a diverse group of intracellular effectors in different systems.[15] The activated G-protein subunit may interact with *STE7* and *STE11* gene products. These gene products show sequence similarity with protein kinases and, in turn, may phosphorylate *STE12*. The *STE12* product has recently been shown to bind to upstream regions of **a**- and α-specific genes.[16]

### 2. *MAT* Locus Controls a, α, and aα Cell Types

Cells that have the *MAT*α allele exhibit the phenotype of α cells, cells with *MAT***a** are **a** cells, and cells with both alleles act as **a**α cells. This is a result of the different mating-type alleles directing the expression of different genes.[17,18] (Figure 2). *MAT***a** and *MAT*α direct the transcription of **a**1 and **a**2, and α1 and α2, respectively. In α cells, α1 is required for the expression of the α-factor and **a**-factor receptor genes, referred to as the α-specific genes (α*sg*). Expression of the inappropriate **a**-specific genes (**a***sg*) is repressed by α2.[19] **a** cells, however, produce their specific mating factor and receptor because they lack the specific repressor α2. Neither **a***sg* nor α*sg* are expressed in the **a**α diploid. **a***sg* genes are repressed by α2. In addition, α2 combines with **a**1 to form **a**1α2, which represses the synthesis of α1 and prevents the activation of α*sg*.[20] **a**1α2 also represses the expression of haploid-specific genes (*STE4, STE5, STE12*) and other genes required for conjugation and diploidization, such as, *FUS1*[21] (required for cell fusion) and *KAR1*[22] (required for karygamy). The operator site has been identified as a consensus sequence within many **a**1α2 genes. The *MAT* gene products do not exert their effects on their own, but interact with a common transcriptional factor, which has been called Pheromone and Receptor Transcription Factor (PRTF),[23] encoded by *MCM1* gene.[24] In α cells, PRFT and α1 act synergistically to induce

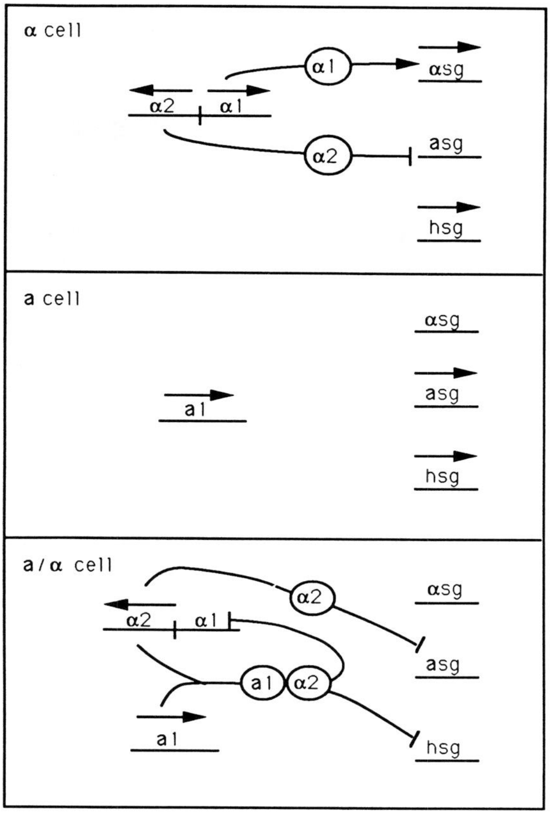


FIGURE 2. Mating locus controls expression of cell-type specific gene sets. The mating-type locus (shown on left) codes for three regulatory activities; α1, α2 and **a**1-α2. These activities regulate the expression of the cell type specific gene sets (α*sg*, α-specific genes; **a***sg*, **a**-specific genes; and *hsg*, haploid specific genes) drawn on the right. Arrowheads indicate stimulation of gene expression and lines with blunt ends indicate inhibition of gene expression. Redrawn with permission.[1]

α*sg*. Transcription of **a***sg* appears to be repressed by α2 binding to PRFT. **a***sg* repression may be due to α2 covering the activation domain of PRFT so that it cannot interact with the transcription machinery. In **a** cells, PRFT binds with **a***sg* and stimulates transcription. Since α1 is absent, PRFT cannot bind and activate α*sg*.[25,26] This combinatorial gene regulation with **a**- and α-specific factors is one of the most intriguing examples of pleiotropic control through repression in eukaryotic systems. **a**1α2 signals mating is complete and not only turns off genes required for mating and mating type switching controlled by *HO*, the homothallism locus, but initiates meiosis and sporulation.

### 3. Entry into Meiosis

The *RME1* gene is a key negative regulator of meiosis. It has been cloned, is expressed in haploid cells, and is transcriptionally repressed in diploid cells.[27] As predicted, overexpression in normal diploid cells prevents meiosis. A null mutation has no effect in **a**1α2 diploids, indicating that *RME1* transcript is not required for meiosis.[28] A number of genes that have been isolated act as inducers of meiosis. *IME1* and *IME2* were isolated by expression on high copy number plasmids that could circumvent *RME1* expression and induce sporu-

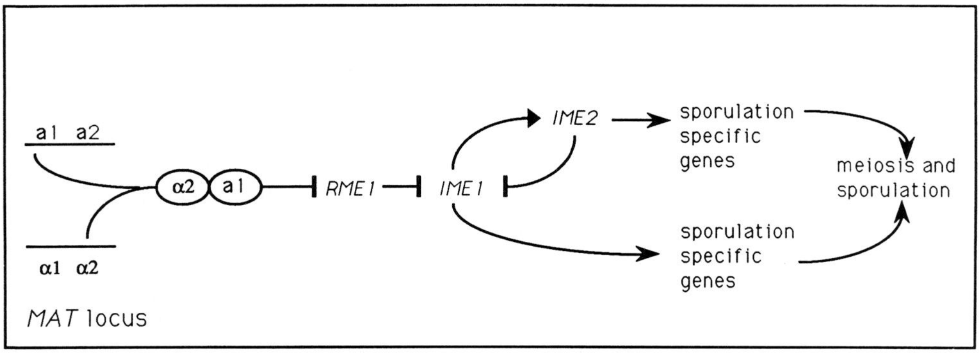


FIGURE 3. Entry into meiosis regulated by mating-type locus. **a**1α2 is only present in **a**/α diploids and is required to turn off the expression of the *RME1* (regulator of meiosis) gene. *RME1* is normally expressed in haploids where it represses the expression of the *IME1* (inducer of meiosis) gene. In **a**/α diploids, *IME1* is not repressed and stimulates *IME2*. Both *IME1* and *IME2* are required for the correct expression of *SPO* (sporulation specific) genes.

lation when cells were subjected to appropriate starvation.[29] As expected of positive regulators, null mutations in either gene cannot induce sporulation, even in the absence of *RME1*. *IME1* and *IME2* are downstream of *RME1* in the meiotic regulation pathway. The *IME1* product is required for *IME2* expression, indicating *IME2* is located downstream of *IME1* in the regulatory cascade; however, because overexpression of *IME2* in a null *ime1* mutant only weakly induced sporulation, it is believed that *IME1* may have additional roles in meiosis. Interestingly, *IME2* apparently controls *IME1* expression. During meiosis, *IME1* expression normally peaks early during meiosis and then declines. In a null *ime2* mutant, *IME1* transcript levels remain high throughout meiosis. In addition, several genes such as *SPO11* and *SPO13* that are expressed during sporulation, are also expressed in the null *ime2* mutant. However, they are not shut off in late meiosis as normally occurs[30] (Figure 3).

How *IME1* and *IME2* regulate sporulation-specific genes is unclear. Several putative negative regulators of *SPO11* and *SPO13* have been identified.[31] These genes, referred to as *UME* (unscheduled meiotic gene expression), were identified through the isolation of mutants that allow mitotic expression of *SPO* genes. Epistasis experiments suggest that *IME1* and *IME2* may activate *SPO11* and *SPO13* indirectly through relieving expression by *UME* products, or that *UME* and *IME* products act independently. As previously mentioned, the environmental signal that induces meiosis is starvation. Exactly how this happens is unclear, but the signal transduction pathway may be mediated by *IME1*. Starvation reduces the intracellular levels of cAMP.[32] The adenylate cyclase gene *CYR1* is posttranslationally controlled in yeast by *RAS1* and *RAS2* products.[33] Loss of function mutations in *RAS2*, like null mutations in *CYR1*, are insensitive to the environmental signal, and are able to sporulate in rich media. The environmental signal may be transferred to *RAS2* by *CDC25* and *IRA1*. *CDC25* is required to activate *RAS2*, probably by assisting the binding of GTP.[34] *IRA1* appears to be involved in negative regulation of *RAS2*.[35] How sporulation-specific genes respond to cAMP levels is unresolved. At least one factor for entry into meiosis is regulated by phosphorylation. Mutations in genes for the active subunits of cAMP-dependent kinase that result in the loss of kinase activity allow sporulation without starvation. Conversely, mutations in the regulatory kinase subunit (encoded by *BCY1*) which leave the kinase active, regardless of cAMP, create cells that fail to sporulate. The target(s) of the kinase is unknown, but it is postulated that phosphorylated protein(s) direct entry into meiosis, perhaps by negative control of *IME1*.[36]

**4. Regulation of Mating Type Switching**

The regeneration of the diploid state in yeast is remarkable. Mating generally only occurs between two haploid cell types, designated **a** and α. However, most naturally occurring yeast strains are "self-fertile," that is, a singly isolated haploid cell, when allowed to mature and form a colony, is able to form diploids. Thus, one haploid cell type has the ability to produce both cell types during mitotic cell divisions. The evidence for two mating types, rather than a universal mating type, was provided by early genetic studies which showed that certain single-spored cultures were unable to form diploids. When two "self-sterile" (heterothallic) haploid colonies of complementing mating types were combined, a diploid zygote formed which always segregated to give 2**a**:2α in each ascus.[37] Subsequently, the **a** and α mating types were defined by a pair of allelles (*MAT***a** and *MAT*α) located at the mating type locus (*MAT*) on the right arm of chromosome III.[38] The different behaviors of the homothallic versus the heterothallic strains are due to the presence of the *HO* gene. Homothallic strains have a functional copy (*HO*), whereas heterothallic strains have a defective version (*ho*). In strains carrying *HO*, cells can switch their mating type as frequently as every cell division. The basis for switching is the presence of silent copies of the **a**- and α-specific sequences. Direct genetic evidence for such copies came from observations that mutant alleles of *MAT* can be healed by two successive mating type switches (for example, α− to **a**+ to α+).[39] The identity and location of these silent copies was revealed after extensive genetic analysis. Generally, the silent *MAT*α, situated near the left end of chromosome III, is referred to as *HML*α and is necessary for **a** to α conversions. The silent *MAT***a** (*HMR***a**) is situated near the other end of chromosome III. In some strains the **a**- and α-specific information is reversed; **a** information is at *HML* (*HML***a**) and α at *HMR* (*HMR*α). These three loci, *MAT, HML,* and *HMR,* have been cloned by functional complementation and structurally analyzed (Figure 4). All three share regions of homology called "X" and "Z1," which flank the allele-specific region "Y" (Y**a** or Yα).[40,41] *HML* and *MAT* share additional homologous regions, "W" and "Z2." W is upstream and adjacent to X. Interconversion of mating type during vegetative growth of homothallic strains represents the alternate transposition of the **a**- and α-specific sequence ("cassettes") to *MAT* (the "playback locus") from *HML* or *HMR* (the "storage loci"). This kind of analogy should not be taken too far, since the information at the *HML* and *HMR* loci remains unchanged. Thus, mating interconversion is nonreciprocal and is a gene conversion event. Interconversion is catalyzed by the *HO* gene product. The *HO* protein is a site-specific endonuclease that recognizes an 18-bp region at the junction of Y and Z regions of the *MAT* locus and generates a double strand break with a 4-base 3′ overhang.[42] The DNA segment at *MAT* is deleted and repaired with information from *HML* or *HMR*. Other genes, such as the *RAD52* product, participate in strand repair.[43] It is not clear why *HML* and *HMR* loci are insensitive to *HO* cleavage. The junction may be sequestered by chromatin structure or the DNA may be modified.[3] It is also a mystery as to why there is no transcription from *HML* and *HMR,* since they contain functional promoters and in certain mutants are activated *in situ*. However, it appears that transcription from *HML* and *HMR* is controlled by "silencer" DNA regions to the left (upstream) of these loci.[44] These silencers also determine the mating type information that will be switched to *MAT* by apparently blocking access to the *HO* cleavage sites at either *HML* or *HMR*. This region has been functionally analyzed and is composed of multiple domains. Repression of *HML* and *HMR,* in addition to the silencer, require at least four *trans*-acting genes called *SIR1, SIR2, SIR3*, and *SIR4*.[45] The products of these genes act as negative regulators. The molecular mechanisms by which these regulators act is still being resolved.

Mating type switching in homothallic yeast strains has been shown to follow specific rules. These were deduced from examining lineages in *HO* cells. Switching always occurs in pairs, and only in cells that have undergone one cell division cycle (mother cells).[46]

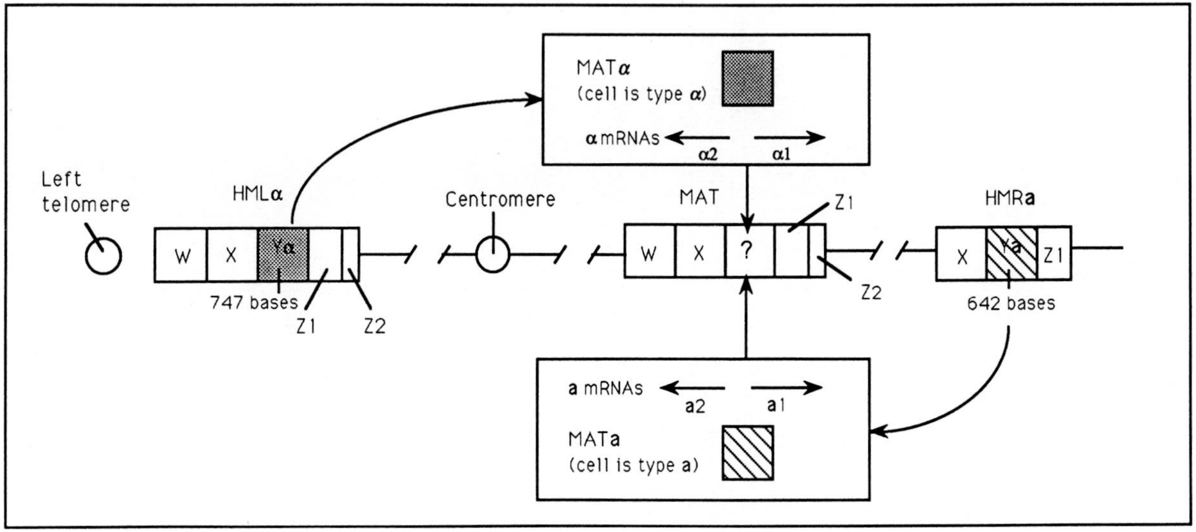


FIGURE 4. Cassette mechanism for mating-type interconversion. Three copies of the mating type information are present on chromosome III. The active copy is at the *MAT* locus and can be **a** or α. Two unexpressed copies of each mating type also exist, one at *HML* and the other at *HMR*. Mating type interconversion is initiated by the *HO* gene product, an endonuclease, which produces a double stranded break at *MAT*. This results in the deletion of the information at *MAT* and a duplication of *HML* or *HMR* information to the *MAT* locus.

Switching appears to be exclusively controlled by *HO* expression. *HO* is unlinked to *MAT* and is located on chromosome IV. It is activated transiently only as mother cells undergo START (commitment to mitotic cycle) during late $G_1$ phase of the cell cycle. The switch must be initiated before *MAT* replication to ensure that both progeny of the mother cell change mating type. The *HO* promoter is large; at least 1400 bp of upstream DNA are required for correct *HO* expression.[3] The URS1 region and the TATA box are required for mother-daughter regulation, and URS2 is required for cell-cycle control. At least six genes, *SWI1* to *SWI6,* are required for *HO* activation.[47] Other genes, such as *SDI1* and *SDI2,* act as repressors.[48] Repression in diploids likely occurs through **a**1α2, since ten recognition sites for this product are found throughout the *HO* promoter.[49] *SWI4* and *SWI6* are required for cell cycle START-dependent activation probably via CACGAAAA repeats within URS2, whereas *SWI5* seems to be the major determinant for mother cell specificity. The *SWI5* gene encodes an 85-kDa DNA binding protein containing three repeated "zinc fingers" with specificity for UAS1.[50] *SWI5* is cell cycle regulated and is expressed only during the latter half of the cell cycle. Consequently, during *HO* expression no SWI5 protein is being synthesized. The presence of the SWI5 protein is sufficient to cause *HO* expression. Ectopic expression of *SWI5* throughout the cell cycle causes daughter cells to switch.[51] Therefore, because in normal circumstances switching must rely on SWI5 protein from the previous cell cycle, a spore cell undergoes a complete cell cycle during which it fails to express *HO* and then turns it on during the next. The reason why the daughter following this first division does not express *HO* remains unexplained. *SWI5* does not appear to be asymmetrically distributed at the end of mitosis, therefore, *SWI5* partitioning appears unlikely as the answer. The SWI5 protein appears equally concentrated in mother and daughter cells, although its cytological distribution is cell cycle-dependent.[52]

Before leaving this subject, a few comments on the biological significance of mating type switching are appropriate. First, diploidization is beneficial. All diploid cells possess unique biological properties that accommodate recessive mutations because they have an additional gene copy, but this does not resolve the issue of interconvertability. Since yeast mates so efficiently, **a**1α2 diploids form between sibling spores, resulting in inbreeding. Situations may exist where only a single product of meiosis is viable. In an isolated environment, switching will allow mitotically produced cells to diploidize. The pattern of switching is also significant, because it ensures the original cell types remain in the population. Daughter cells are produced at every division and are unable to switch, therefore, they comprise a stem-cell lineage.

### B. *ASPERGILLUS NIDULANS*

Unlike reproduction in yeast, asexual reproduction in *A. nidulans* involves the orderly differentiation and spatial arrangement of several well-defined cell types that make up a complex reproductive apparatus, the conidiophore. Conidia, the dispersal units, are derived from specialized cells on top of the conidiophore (Figure 5). Conidia of phytopathogenic fungi have a particularly important biological role because they represent the dispersal units that lead to spread of disease. In phytopathogenic fungi this process is poorly understood. Current insights into the role of asexual reproduction in filamentous fungi results from classical and molecular genetic studies with *A. nidulans*.[4,53,54]

#### 1. Development of the Asexual Reproductive Apparatus

Conidial formation at the ultrastructural level for several fungi has been reviewed extensively.[55] A detailed description for *A. nidulans* was recently published.[56] The first indication of asexual reproduction is the formation of thick-walled foot cells. The environmental signals that trigger the differentiation of hyphal segments into foot cells is not clearly understood, but, unlike yeast development, it is not induced by starvation. Even continued

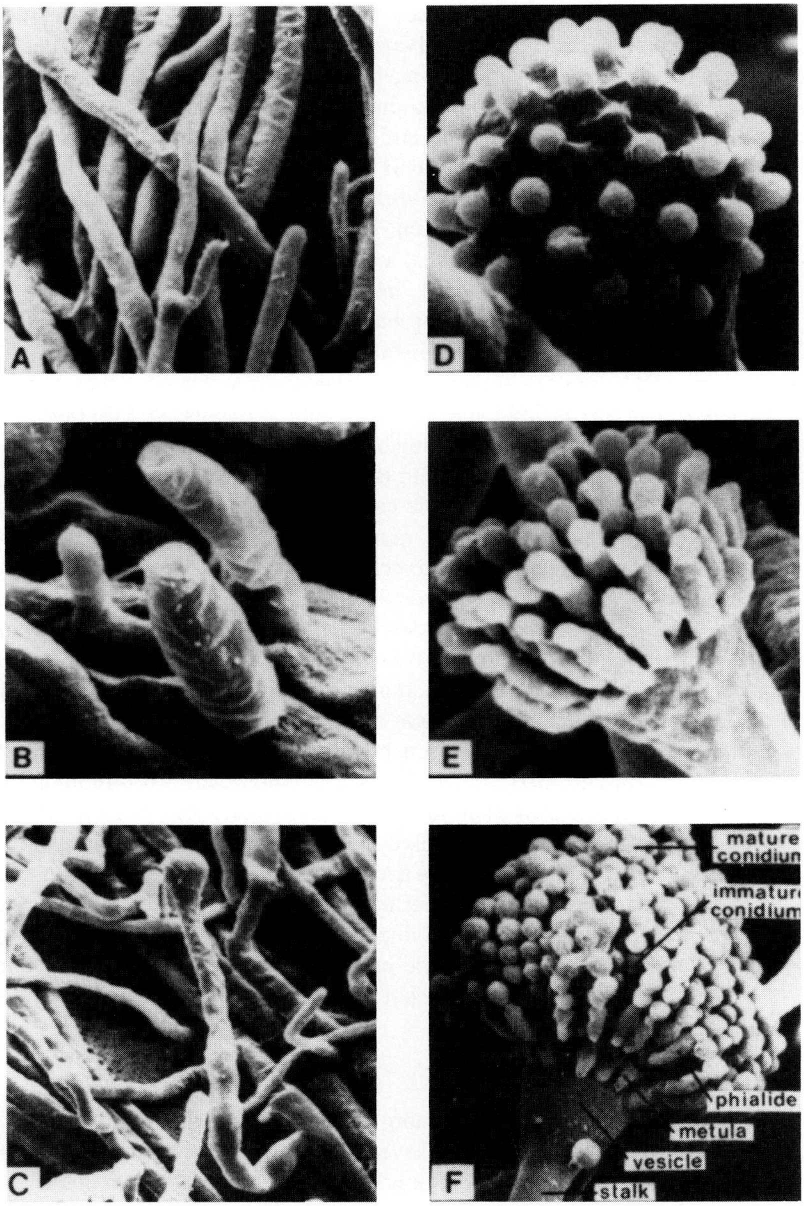


FIGURE 5. Morphological changes during *A. nidulans* conidiation. Conidiating cultures (FGSC4) were examined by scanning electron microscopy, and conidiophores at various stages of development were identified. (A) undifferentiated hyphae. (B) young aerial stalks. (C) conidiophore stalks with swollen tips. (D) metullae budding from the swollen tips. (E) phialide formation. (F) mature conidiophore. Reproduced with permission.[75]

replacement of medium does not inhibit asexual reproduction. Nutrient limitation actually inhibits development, although starvation of at least some cells may follow developmental induction since the conidiophore is lifted away from the source of nutrients.[57] Foot cells can be clearly distinguished by the presence of a darkly pigmented, two-layered cell wall. The foot cell branches and grows by apical extension, similar to vegetative cells, to produce an aerial hypha, called the conidiophore stalk. Cytoplasmic vesicles, presumably carrying cell wall enzymes and constituents, are localized at the apical tip of the elongating hyphae. At about 100 μm, apical growth ceases and the tip of the conidiophore stalk swells to about 10 μm in diameter. During this stage of development, vesicles become distributed around the periphery of the tip. The height of the conidiophore stalk appears to vary little in response to environmental conditions, indicating that the height is genetically determined.[4] As the tip of the conidiophore stalk grows, the cytoplasm becomes densely packed with nuclei, which then crowd into the swelling tip. Whether the foot cell is uninucleate or multinucleate is unclear, however conidiophores from heterokaryotic colonies rarely produce mixed conidia. Uniform pits form on the inner surface of the swollen tip wall and spherical buds that become the primary sterigmata or metulae are synchronously extruded. Mims et al.[56] reported that nuclei align below the initial buds and undergo oriented synchronous divisions. One daughter nucleus enters the bud and the other is retained in the tip. The metulae are then partitioned off by the formation of a septum at their base. The septum retains a pore that maintains cytoplasmic communication with the conidiophore stalk. Metulae grow to 5 to 7 μm long, and then, in turn, initiate the formation of secondary sterigmata or phialides. The phialide nucleus is derived from a mitotic division of the metula nucleus, after which a septum is formed with a central connecting pore. Often a second phialide develops from the metulae.

Conidia are formed by a cytoplasmic swelling from the neck of the phialide. The phialide nucleus moves to the neck region and divides mitotically. One daughter nucleus enters the developing conidium, where it arrests in $G_1$, while the other moves out of the neck into the phialide cell. At about 3 μm in diameter, the conidium forms a septum by invagination of the cell wall at the neck region. As the septum wall thickens, the central pore closes. The process of conidium formation is repeated by the phialide cells, resulting in a chain of conidia with the oldest conidia being at the distal end. Conidia mature as they move away from the phialide cell, and the cell walls thicken and become green-pigmented. Phialide cells are, in some respects, reminiscent of *Bacillus* spore mother cells, which provide components for the outer wall of the spore. Conidial laccase, which accumulates in maturing conidia, is transcribed in the phialide cells. This implies that phialide cells secrete enzymes needed for condial cell wall formation. Ultimately, spores become quiescent and enter dormancy, a process that is not well understood. At 37°C, wild-type cultures which are induced to synchronously conidiate produce aerial hyphae in about 4 h. The conidiophore stalk elongates and swells by 10 h, producing metulae and phialides between 12 to 15 h. Conidia are first observed at 16 to 18 h.

### 2. Utility of Developmental Mutants

Conidiation is a dispensible function in the laboratory and has greatly facilitated genetic studies. Aconidial mutants can be readily manipulated and maintained as ascospore cultures. Many genes other than those expressed during conidiogenesis can affect conidiophore formation. Auxotrophic strains, such as those with a defective ornithine-transcarbamylase gene (*argB*) conidiate poorly even when supplemented with arginine. The role of these genes in conidiation has largely been neglected. Clutterbuck,[58] in an extensive genetic analysis of conidiation, excluded auxotropic mutants from further consideration. With such mutants excluded, quantitative studies of conidiation mutants indicate that only 45 to 100 loci are involved in this developmental process. More recent estimates by DNA hybridization techniques indicate about 1050 mRNAs are unique to sporulating cultures and an additional 300 sequences are unique to conidia themselves.[59] This large discrepancy may reflect the fact that many genes developmentally expressed have a minor role or are dispensible. Indeed, as will be discussed later, there are several examples of gene clusters that are not required for normal laboratory growth and development.

The majority of the mutants identified by Clutterbuck were aconidial, that is, they did not form foot cells. Others, however, were postinductive mutants that formed abnormal

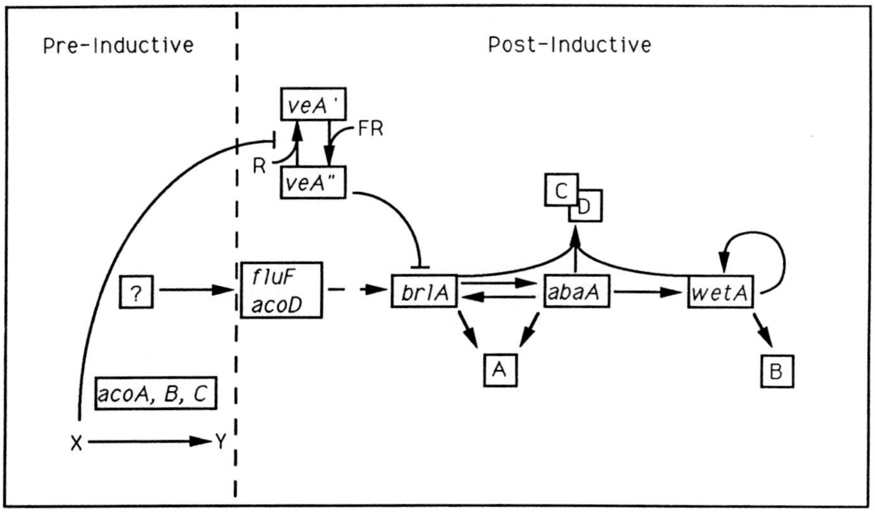


FIGURE 6. Genetic interactions regulating the central regulatory pathway underlying conidiation in *A. nidulans*. Redrawn with permission.[63]

conidiophores or affected conidia maturation. Clutterbuck divided the mutations into four categories based on function.[60] Preinductive mutations, i.e., loci that regulate the conversion from vegetative growth to conidiation, were designated ''strategic''. Loci that regulate the morphological development of the conidiophore or conidium were designated as ''tactical''. Loci that modified the structure or physiological characteristics of the conidiophore or conidium were designated as ''auxilliary''. The last class was designated ''support'' loci, and represented genes necessary for normal vegetative growth and conidiophore development. Although the consensus as to which functional class particular loci belong is incomplete, this classification scheme has been supported by molecular analysis and has led to models for genetic regulation of conidiogenesis.

Genetic analysis of conidiation-specific mutants implicated *brlA, abaA,* and *wetA* as the major postinductive pathway-specific regulators of asexual development (tactical loci).[61] Mutations in the bristle loci, such as *brlA1,* cause the conidiophore stalk to continue elongating up to 2 to 3 mm in length. BRLA appears to control the switch from polar to anisotropic growth. Other mutations that map to *brlA* may represent increased BRLA activity, such as *brlA42* mutant, which has multi-tiered and branching conidiophores.

The second important pathway regulator is *abaA*. Abacus mutations such as *abaA1* fail to differentiate normal metulae and are aconidial. The metula nucleus does not appear to arrest in G1 and continues to divide to give rise to reiterated metulae-like structures. Wet-white mutants, like *wetA6,* appear to undergo normal maturation but autolyze when they reach approximately position eight in the conidial chain. WET may be necessary for the structural integrity of the conidia or be required for spores to become quiescent. In other words, *wetA6* mutations may allow precocious germination.

All three genes have been cloned and characterized.[62,63] Transcription of these genes is developmentally regulated and expression of *brlA, abaA,* and *wetA* appears sequentially during asexual reproduction. *brlA* transcript (2.4 kb) begins to accumulate about the time the tip of the conidiophore stalk begins to swell (10 h), while *abaA* (3.0 kb) accumulates as phialides develop at about 15 h. *wetA* transcript (1.8 kb) accumulates during conidial maturation at about 20 h. These results match well with those predicted, based on the epistatic relationships from the phenotypes of double mutants. The genes define a linear-dependent pathway: *brlA* to *abaA* to *wetA* (Figure 6). A second independent pathway modifies conidiophore structure. At least two loci that belong to this modifier pathway have been identified.[64]

Mutations in *stuA* result in the formation of shortened conidiophores. These mutants also lack metulae and phialides, and conidia bud directly off the swollen tip of the stalk. The other member is medusa (*medA*). A large number of *medA* mutants have been observed, and vary in the severity of branching to produce secondary metulae. Medusa mutants eventually form phialides and conidia. *stuA* mutants are epistatic to *medA* mutants. Double mutants of *stuA* in *brlA1* or *abaA1* backgrounds form stunted bristle and stunted abacus phenotypes, respectively. Double medusa and bristle mutants are complex and relationships are not readily interpretable.[54] This modifier pathway also plays a role in the sexual reproductive cycle. *stuA* mutations block the formation of thick walled cells within which the cleistothecium normally develops. *medA* lesions allow Hulle cell formation, but block cleistothecium formation and meiosis.

*brlA* is a primary regulator of asexual development.[62] The deduced BRLA protein is rich in proline (10%) and serine (13%), as is characteristic of a number of eukaryotic transcription factors, and also contains two repeated "zinc fingers." These sites strongly indicate that BRLA is a nucleic acid-binding protein. The direct targets and the biochemical activities of BRLA remain to be elucidated. Adams et al.[62] demonstrated that ectopic expression of BRLA induced certain aspects of asexual development, supporting its proported function as a regulatory factor. A second copy of *brlA* was placed under the control of the *alcA* promoter. This strain was grown in liquid medium, an environment that normally suppresses conidiation. When transferred to liquid medium containing threonine, the *alcA*(p) was activated, and *brlA* expression was detected within 2 to 3 h. Apical growth ceased, hyphal cells became vacuolated, cell walls thickened, and hyphal tips were transformed into reduced conidiophores that budded off viable conidia. In addition, *brlA* expression led to the activation of *abaA* and *wetA*, a number of structural genes (*yA* and *ivoB*), and genes with unknown functions that are developmentally regulated. BRLA is probably a positive regulator, although its molecular mechanism is presently unknown.

The conceptual ABAA protein has a molecular weight of 90 kDa. It has a potential for a "leucine zipper" but lacks the adjacent basic region needed for DNA binding.[63] ABAA may form a dimer, but its mechanism of DNA binding, if it does so at all, is unlike the mechanism proposed in yeast for the *Fos-Jun* heterodimeric transcription factor.[65] Deletions of the *cis*-acting elements regulating *abaA* revealed multiple regulatory elements, some that repress expression in vegetative cells, and other regions that activate transcription.[66] BRLA has been proposed to displace a repressor from *abaA* promoter during development. However, direct evidence is unavailable. Threonine-induced expression of *alcA(p)::abaA* halted growth and stimulated the formation of septa and extensive vacuolization, but not conidium formation. *abaA* expression also led to the activation of *wetA* and, unexpectedly, the activation of *brlA*. *brlA* and *abaA* are reciprocal inducers, and *abaA* appears to be a feedback inducer of *brlA* expression. However, *brlA* must be expressed prior to *abaA* expression for conidiation to occur.

*wetA* has an open reading frame (ORF) sufficient to encode a 59-kDa protein, however, the function of WETA is unknown. Hyphal activation of *wetA* inhibits conidiation but has little effect on growth. *wetA* induction does not lead to *abaA* or *brlA* expression. *wetA* is at least partially autoregulated, since during normal conidiation, *wetA* mRNA accumulation requires *wetA* activity.[67]

Numerous other sporulation-specific genes, in addition to those discussed, have been cloned by differential screening of a cDNA library comprised of poly(A)$^+$ RNA from conidiating cultures.[61] These CAN clones have been divided into four categories based on their temporal expression during normal conidiation and on their expression in a series of *brlA*$^-$, *abaA*$^-$, and *wetA*$^-$ strains in which *alcA*(p)::*brlA*, *alcA*(p)::*abaA*, or *alcA*(p)::*wetA* was induced. Class A genes require *brlA* or *abaA* or both, but not *wetA*. Class B genes are activated by *wetA,* independent of *brlA* or *abaA*. Class C and D genes require the combined activation of of *abaA, brlA,* and *wetA*. Class C and D can be distinguished; Class D genes, but not class C genes, are expressed independently of *brlA*$^+$ activity during normal development.[63] These gene classes have been incorporated into a model depicting the central regulatory pathway underlying asexual reproduction (Figure 6).

One class A member, *CAN41,* has been studied in some detail. The ectopic expression of both *brlA* and *abaA* led to high levels of *CAN41* expression and growth cessation. Initially these two observations were thought to be related, but it has since been shown that ectopic expression of *CAN41* caused only limited growth inhibition. However, deletion experiments revealed a more interesting role for this gene. The nucleic acid sequence of *CAN41* predicts a slightly acidic small protein of 14 kDa. It is likely that the protein is membrane spanning. Over 50% of the amino acid residues are hydrophobic, and large blocks of between 14 to 20 residues are found at the $NH_4$ and COOH termini, with an additional block near the middle.[68,69] Strains with *CAN41* deleted by site-directed mutagenesis produce conidiophores that appear darker and are encased in a droplet of liquid.[70] Electron micrographs of the conidia clearly show that the conidial wall lacks the hydrophobic rodlet.[71] Additional experiments are being actively pursued to determine whether *CAN41* codes for a structural protein or regulates the formation of this layer.

The simple molecular model proposed above probably does not include all the factors involved in postinductive differentiation of the conidiophore. Additional regulatory networks involving *stuA* and *medA* are undoubtedly required for cellular diversity and structural complexity. *stuA* has been cloned and characterized. The *stuA* product is a 64-kDa protein that, like BRLA, is rich in proline (13%) and serine (11%).[54] The effects of forced expression in different genetic backgrounds is currently under investigation.

### 3. Developmentally Expressed Genes

In the course of selecting chromosomal recombinant DNA clones that contained genes that were expressed developmentally, Zimmerman et al.[72] noted that sporulation-specific genes were not randomly distributed in the genome. Detailed molecular analysis of one particular genomic clone, lambda anSpoC1, revealed that it encompassed several developmentally expressed transcripts.[73] Orr and Timberlake.[74] calculated that 80% of the isolated sporulation-specific genomic clones were clustered. One cluster, SpoC1, encompassed a region of 38 kbp. This region encodes at least 14 developmentally regulated transcripts and is bordered by a 1.1-kbp repeat sequence.[75] The regulatory mechanism may have a chromatin level component. Miller et al.[76] tested the effect of chromosomal position on expression of SpoC1 genes. With one exception, SpoC1 genes were expressed in hyphae when placed outside of the cluster. Furthermore, expression of *argB* was partially repressed in hyphae when placed within the SpoC1 cluster. It is not clear whether chromatin structure is completely responsible for the coordinated control of transcription, since several other genes found within the region are less developmentally regulated. Chromatin control remains an attractive area for further study.

Continued research on this gene cluster is likely to receive low priority, because deletion by site-directed mutagenesis of the entire region caused no readily detectable phenotypic alteration.[77] This is a significant loss and represents approximately 0.15% of genome DNA and 2% of the mRNA mass in conidia. There are several other well-documented examples of dispensable developmental genes, particularly in studies on yeast ascospore formation.[78] These again serve as a reminder that working with genes with unknown functions may not yield the kind of insight into the process under investigation as expected.

Few genes have a well-defined biochemical function during fungal development. An exception in *A. nidulans* is the product of the *yA* gene, which encodes the enzyme *p*-diphenol oxidase, or laccase (EC 1.10.32), that is needed to convert yellow pigment in the conidia cell walls to a dark green-colored pigment. The composition of these pigments is unknown. The *yA* gene has been cloned and is regulated at the level of mRNA accumulation. As discussed earlier, *yA* transcripts accumulate specifically in phialide cells.[79]

Conidial pigmentation requires the activities of other identified genes. Mutations in *wA* result in white spores and are epistatic to *yA* mutations. It has been suggested that *wA* and *yA* encode enzymes that sequentially catalyze pigment synthesis from a colorless precursor. The *wA* gene product has not been identified, but, because the gene has been recently cloned, can now be tested.[80]

### 4. Conidiation Induction Signals

The signals that induce conidiation in *A. nidulans* are only beginning to be understood. Induction appears to be controlled both endogenously and by environmental factors. Hyphae growing in submerged cultures will not normally produce conidophores or sexual structures, but instead will grow vegetatively until reaching a stationary phase. However, upon exposure to an air interface, the hyphal mat on a solid support will produce condia.[81] This phenomenon is used routinely in studies of conidiation, since it promotes synchronous development. The inductive signal has not been clearly identified. Exposure to air may cause osmotic shock; however, conidiation occurs even when the environment is saturated with water. Furthermore, osmotic shock in a liquid culture caused by changing the osmoticum does not induce sporulation.[82] The time required to produce conidiophore stalks is dependent on the time spent in submerged culture. If cultures are transferred at any time prior to 20 h, conidiation always begins at about 25 h. However, after 20 h of vegetative growth, the time following induction for development to occur remains constant at about 5 h. These results indicate that cultures become responsive to induction, or competent, after 20 h of vegetative growth.[81,83] Competence appears to be endogenously controlled, since cell density or replacement of medium has no effect on the time to conidiate.[84] Aquisition of competence is temperature-dependent.[85]

An interesting class of genes that regulate competence has been identified. Temperature-sensitive (ts) mutants were isolated based on the premise that preinductive, or competence, genes might be nonessential for development once a strain has become competent. Such mutants were temperature-insensitive after the preinductive phase and conidiated. These unlinked genes *acoA, acoB,* and *acoC* have been defined and partially characterized.[85-87] Furthermore, mutations in these genes block sexual as well as asexual reproduction and lead to the production of phenolic compounds, such as the antibiotic diorcinol. Champe et al.[88] isolated from these mutants a compound referred to as psi factor that stimulated sexual reproduction while inhibiting asexual reproduction in wild-type strains. These results may indicate that these genes are part of a metabolic pathway that produce compounds controlling reproductive development.

Another class of mutants that does not respond to conidiation-inductive signals was isolated following exposure to low levels of 5-azacytidine (5-AC). These mutants occur in high frequency; 20% of the colonies show a ''fluffy'' phenotype. This phenotype is characterized by large, cottonlike colonies that, unlike wild-type strains, overgrow adjacent colonies. These recessive mutations have been mapped to a single locus (*fluF*) on the right arm of chromosome VIII.[89] Although *A. nidulans* has undetectable levels of methylcytosine, an interpretation of these observations is that 5-AC leads to the demethylation of *fluF*. *fluF* may become fully methylated following germination, and only in this state does *fluF* respond to inductive signals.

Recently, red light has been found necessary for conidiation, and, like phytochrome-mediated responses in plants, conidiation is at least partially reversible by far-red light.[90] This phenomenon is only observed in strains carrying a wild-type copy of the velvet (*veA*) locus. Most laboratory strains of *A. nidulans* have a *veA* mutation (*veA1*), since they conidiate more abundantly and lack aerial hyphae.[91] A number of differences in asexual and sexual development between *veA*$^+$ and *veA1* strains have been observed, indicating that the mutation has pleiotropic effects. *veA* probably acts as a repressor, since mutations are recessive in diploids. *veA* appears to interact with *aco* genes as *aco* mutations are suppressed by *veA1*. *veA* represses conidiation prior to *brlA* expression. *brlA* is activated in a *veA*$^+$ strain only after exposure to light, and ectopic expression of *brlA* in a *veA*$^+$ strain leads to development in both the dark and light. How all of these pathways are interconnected remains to be determined. Certainly, many of the intricacies will be resolved when the genes are cloned and characterized. A simple model to explain competence involves the disappearance of a developmental repressor. This repressor may have been stored in the spore. *acoA, acoB*, and *acoC* could be required for its degradation to below a critical threshold level allowing red light inactivation of *veA*. Thus, inactivated *veA* would no longer repress the expression of *brlA* and other postinduction developmental genes (Figure 6).

### C. *MAGNAPORTHE GRISEA*

Many phytopathogenic fungi are of great economic significance and have been responsible for mass human migrations. However, the understanding of how fungi infect plants is limited when compared to other aspects of fungal and cellular development, as previously discussed. A major and valid explanation stems from the need to study a vast and diverse group of agronomically important fungi which means that few plant-pathogen interactions have been studied in great detail. Also, there are few suitable pathogens that are amenable to Mendelian genetics and molecular genetic manipulation. Furthermore, many phytopathogenic fungi are difficult to work with and grow poorly or not at all in culture. Many lack a sexual stage. Although it may not be possible to study the fine details of pathogenesis in just a few experimental systems, detailed investigations of a limited group should provide frameworks and hypotheses that can be tested in other important plant-pathogen interactions. This kind of reasoning has been most eloquently argued by Valent.[92]

The basidiomycete *Ustilago maydis*, the causative agent of corn smut, is one such model system currently being developed. The fungus has two distinct forms: a unicellular nonpathogenic haploid form that grows like yeast, and a filamentous dikaryotic form, whose growth is dependent on the plant and causes destructive tumors on the leaves and ears. Space does not permit detailed description of this system; however, the subject has been recently reviewed.[93] In brief, the two forms are controlled by the mating type loci **a** and **b**. There appear to be only two alleles of the **a** locus but at least 25 alleles of **b**. For mating to occur, strains must carry different alleles of **a** and **b**. It is not understood why there are so many **b** alleles. Four alleles of **b** have recently been cloned and sequenced and show similarity with DNA-binding proteins.[94] Comparisons of open reading frames reveals that each contains a variable domain in the N-terminal 110 amino acids (40% identity), and that the rest of the open reading frame is highly conserved (93% identity). This conserved region shows similarity with DNA-binding proteins.

Another phytopathogenic fungus that is beginning to receive considerable attention is *M. grisea*. This fungus causes rice blast, one of the most devastating diseases of rice. Annual world losses of rice due to blast exceed 20 million tons per year.[95] With the recent development of sexually compatible rice strains and the development of a DNA-mediated transformation system, *M. grisea* offers the experimental advantages of its close relative *A. nidulans*. *Magnaporthe grisea* not only presents opportunities to gain insight into the mechanisms controlling infection structure formation, but also host-species and host-cultivar specificity. The fungus is a pathogen of many other commercial cereals (wheat, barley) and wild grasses. Individual isolates of *M. grisea* have restricted host range and can only infect a few species of grasses. For example, isolates that infect weeping lovegrass (*Eragrostis carvula*) do not generally infect rice. Rice isolates can be further divided into races based on the particular cultivars of rice they parasitize. Natural populations appear to be highly variable, and new races appear with surprising frequency. The source of this variability has not been established, but the development of stable laboratory strains should now allow this question to be addressed.

### 1. Development of *Magnaporthe grisea* for Genetic Analysis

*Magnaporthe grisea,* unlike many fungi used for genetic studies, grows rapidly and sporulates abundantly on defined medium. The fungus grows as a haploid mycelium by hyphal tip extension and cell division. Between one and twenty asexually produced conidia are formed on the top of a branch-shaped conidiophore. Each of the three cells of the teardrop-shaped conidium contains a single haploid nucleus. All three nuclei have a common origin and thus are genetically identical, thereby permitting the selection of auxotrophic, antibiotic-resistant or morphological mutants.[96-98] For many fungi a sexual stage has never been observed. Indeed, only relatively recently has the teleomorphic or sexual stage of *M. grisea* been described.[99] The fungus was formerly known by the anamorph *Pyricularia oryzae*.[100,101] The presence of a sexual state is particularly valuable because the ability to perform genetic crosses has facilitated the generation of restriction fragment length polymorphism maps[92,102] and permits linkage analysis of pathogenicity genes and other loci. Fully fertile strains of *M. grisea* are hermaphroditic, but most rice isolates are female-sterile.[103] Highly fertile strains that are pathogenic on rice have been produced by mating rice isolates with isolates from finger millet (*Eleusine coracana*), goose grass (*E. indica*), or weeping lovegrass (*Eragrostis curvula*).[92] In addition, crosses with a rice isolate from French Guyana have also yielded isolates that are highly pathogenic on rice and highly fertile.[104] The two mating types *MATa* (*MAT1*) and *MATA* (*MAT2*) have been mapped to a single genetic locus.[105] The development of these fully fertile laboratory strains represents a significant advantage for genetic analysis compared to other fungal plant pathogens.

Mating occurs when strains of the opposite mating type are paired on oatmeal agar. Perithecia are produced on the surface or slightly embedded in the medium within 2 weeks. As the the perithecia mature, they darken, and after about another week, contain many hundreds of tapered, slightly curved asci. Each ascus contains eight ascospores arranged linearly that serves to identify the order of meiosis. Thus, it is possible to perform both random spore and tetrad analysis. In addition, genetic analysis is enhanced by the ability to produce heterokaryons and transient diploids.[96,97,106,104]

DNA transformation of *M. grisea* is essentially the same as performed for *A. nidulans*. Hyphal fragments are treated enzymically to release protoplasts that are collected by centrifugation. Exogenous DNA is forced into the protoplasts by $CaCl_2$ and polyethyleneglycol (PEG) precipitation. Transformants are then selected by plating on an appropriate medium. Plasmid and cosmid vectors carrying the *argB* gene from *A. nidulans* have been useful to transform $Arg^-$ *M. grisea* mutants to arginine prototrophy.[107] DNA integrates into the genome both homologously as well as randomly. These techniques have been used to clone the *LYS* locus. By complementing an $Arg^-$, $Lysine^-$requiring strain with a genomic library in the cosmid vector pArC, *argB* and DNA inserts were carried into *M. grisea*. The complementing cosmid was recovered by adding DNA from the *M. grisea* transformant to a lambda bacteriophage packaging extract and transducing *E. coli*. These cosmids were subsequently shown to recomplement the mutation. Southern hybridization and Mendelian genetic analysis demonstrated that recombination occurred at the *LYS* locus.[107,108] The availability of techniques for mutant selection and DNA-mediated transformation will allow the identification of genes that are essential for infection structure formation.

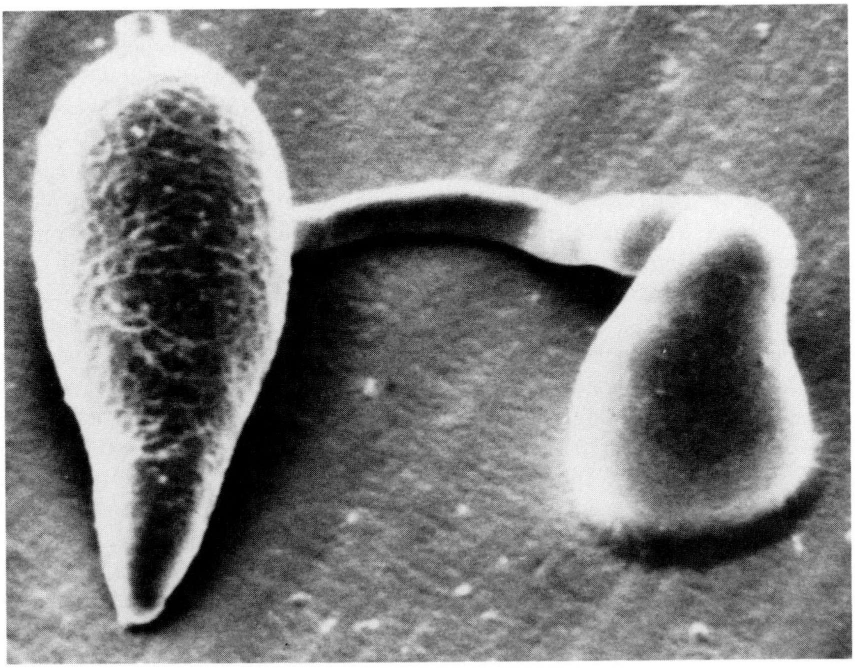

FIGURE 7. Scanning electron micrograph of mature appressorium (right) and conidium (left) of *M. grisea*. As the appressorium developed, the interconnecting germ tube lifted off the substratum. Magnification approximately × 37,000. Reproduced with permission of R. Howard and Springer-Verlag.

### 2. Infection Structure Development

Many phytopathogenic fungi including *M. grisea* have evolved the capacity to recognize potential hosts and, upon contact, produce specialized structures (appressoria) that allow them to attach to the plant surface and to penetrate the cell wall. This process has been intensively investigated for many fungi, including *M. grisea,* at both the biochemical and ultrastructural level *in planta,* as well as *in vitro.*[109,110]

Conidia are released explosively from the condiophore by osmotic rupture of a small cell beneath the conidium on exposure to free moisture. Under conditions of high humidity these propagules infect plants. Upon contact with a hard surface, the hydrated conidium releases a mucilage from the apical cell. This spore tip mucilage, "STM," attaches the spore to the leaf surface.[111] Each cell of the three-celled conidia is capable of giving rise to a germ tube, but germination from the middle cell is less frequent. Within a few hours the tip of the germ tube hooks to form the appressorium initial. Following this formation, the apex swells rapidly into the domed-shaped appressorium structure (Figure 7). Mitosis occurs at the proximal end of the germ tube. One daughter nucleus remains in the conidium, the other migrates down the germ tube to the appressorium, and a septum forms at the site of mitosis. Mature appressoria only contain a single nucleus. During appressoria formation, vesicles become concentrated along the substratum interface. An additional cross-linking layer is laid down over the entire appressorium, particularly at the substratum interface. As the appressorium continues to expand, this layer at the substratum interface slowly thins out. By about 8 h, and after the formation of the septa, a dense double layer of melanin forms between the plasma membrane and the preexisiting appressorium wall, except across the wall-less appressorium pore. At this time, the plasma membrane is in direct contact with the substratum. Before the penetration peg emerges at about 24 h, this 4.5 μm diameter pore becomes covered with wall material. One layer appears to be continuous with the appressorium cell wall; the other ends where the penetration peg emerges. A ring of dense material, 1- to 2-μm wide, has been noted around the pore rim and may represent an area of appressorium attachment to the plant surface. The cytoplasm of the penetration peg is composed primarily of microfilaments that lie parallel with the direction of penetration. Few vesicles and ribosomes are observed in these structures.

The function of appressorial melanin in infection has received considerable attention. Buff-colored mutants that lack melanin are defective in host penetration. *buf* mutations in *M. grisea* are defined as a single locus; therefore, it has been concluded that melanin is required for infection. Melanin appears to act as a barrier to solute efflux and supports a high osmotic potential. The function of melanin in the appressoria cell wall may be to provide support for osmotically generated hydrostatic pressure, which is essential for the pathogen to penetrate its host.[110] Consequently, mechanical force is postulated to be the primary means of penetrating the epidermal cell wall; however, extracellular hydrolytic enzymes may also be required. The absence of melanin may interfere with the production of cell wall-degrading enzymes.. Recently, cutinase-defective mutants of *M. grisea* have been made by site-directed mutagenesis. A normal copy was replaced by a disrupted copy of the gene. Pathogenicity tests of these cutinase-deficient mutants do not indicate that cutinase is required for infection as it is for other fungi, such as, *Fusarium solani pisi*.[112] The roles of other hydrolytic enzymes strongly implicated in pathogensis, such as pectinases and xylanases, remain to be elucidiated.

### 3. Gene Expression During Infection Structure Formation

As in other developmental processes, it is likely that a few of the genes expressed during infection encode products that are key regulators. However, a word of caution is necessary, since spores of *Colletotrichum lagenarium* germinate and form appressoria in the presence of protein synthesis inhibitors.[113] This may indicate that translation and possible gene expression are not required for some aspects of development. Few studies to determine gene expression during infection structure formation have been initiated.[114] Infection-specific cDNA clones have been isolated from a limited number of phytopathogenic fungi. One gene from *Uromyces appendiculatus* has been cloned and characterized.[115] *INF24* encodes a 16.4-kDa protein that is induced during infection structure formation on oil-collodion membranes; its function is unknown. Only in rare instances have roles for infection specific genes been deduced. For example, the genes encoding certain cell wall-degrading enzymes, such as polygalacturonase, have been cloned and are developmentally regulated.[116]

One of the most successful strategies for investigating the formation of infection structures is likely to involve obtaining developmental mutants. UV-induced mutagenesis has been used to isolate prototrophic mutants of *M. grisea* that are defective in spore attachment.[98] Mutants were screened for their inability to attach to a Teflon membrane. Several mutants formed abnormally shaped conidia and appressoria. This phenotype, termed $Smo^-$ for abnormal spore morphology, maps to a single genetic locus. In addition, this mutation affects ascus shape and permits conidia to differentiate appressoria in conditions not normally conducive. Despite these dramatic changes in morphology, $Smo^-$ mutants do not appear to have greatly altered pathogenicity. Attempts to clone *SMO* by functional complementation have not yet been successful. The locus has been mapped to a specific chromosomal region by linkage to *M. grisea* repeat elements (MGR). Efforts are proceeding to complement the mutation with chromosomal DNA fragments that bridge these MGR elements.[117]

### 4. Infection Structure Induction Signals

Phytopathogenic fungi have evolved mechanisms to ensure that germination and subsequent developmental processes are regulated by endogenous and exogenous controls that together optimize their chances of successfully infecting their hosts. Factors that regulate development have been most actively studied in rust fungi.[118,119] How these signals are perceived and how the information is relayed to induce gene expression in phytopathogenic fungi remains a mystery.

The touch, or thigmotrophic, response appears to be one of the major stimuli for appressorium formation. Many fungi can be "tricked" into development by allowing spores to germinate on a hard surface. Urediospores of *U. appendiculatus* require critical topographic features to develop, specifically a groove or ridge of 0.5 μm. Other fungi, including *M. grisea,* are less particular and develop on a variety of hard synthetic surfaces. The efficiency of appressorium development can be influenced by a number of compounds from cuticular extracts. Wax esters, aldehydes, and alcohols, having polar groups and low contact angles, promote appressoria formation in *M. grisea,* but alkanes, non-polar molecules having high contact angles, have no effect.[120] These effects may not be due to the chemical nature of the wax components, but to physical properties, such as improved wettability of the polar wax molecules, promoting better substratum contact. Components from the wax layer have been reported to directly influence gene expression in *F. solani pisi.* In conjunction with a soluble protein extract, cutin monomers selectively activated cutinase gene transcription in isolated nuclei.[121] Whether these monomers directly interact with DNA, or act indirectly by interacting with transcription factors or second messengers analogous to steroid hormone action in animals, has not been determined.[122] Elucidating these mechanisms is likely to provide significant insight into signal transduction pathways controlling fungal development and penetration. Scientific literature is filled with numerous reports of various plant extracts affecting fungal development, but rarely have the active molecules been identified. α-Tocopherol (vitamin E) isolated from *Silene abba* and *Pastinaea sativa* has been shown to influence mating in *Uromyces* species.[123] Macko et al.[124] demonstrated that 2-propenal can induce appressorium formation of *Puccinia graminis tritici* in the absence of thigmotrophic stimuli. How 2-propenal functions in conjunction with the stimulus has not been elucidated.

Light undoubtedly has an effect on phytopathogen development, including sporulation. Light is required for efficient asexual reproduction of *M. grisea* as well as *A. nidulans.* In at least some plant-pathogen interactions, fungal infection is influenced by light. Light generally has a negative affect on infection structure development, probably because during daylight, temperatures rise and evaporate the free moisture required for the germination of the spores of many fungal species. Uredospore germination of *P. graminis tritici* is photosensitive, and high light intensities can completely inhibit germination.[125] Blue and far-red light are the most inhibitory. Inihibition can be photoreversed by exposure to red light, indicating that phytochrome may be involved. Photoinhibition may be linked with the self-inhibitor, methyl-*cis*-ferulate, of uredospore germination. One hypothesis is that as spores hydrate the germination inhibitor is slowly inactivated to a threshold level below which germination can proceed. In the light the inactive inhibitor is photoconverted to another inhibitory compound. Although this model is consistent with the available data, it remains to be rigorously tested.

## II. FUTURE PROGRESS

The availability of classical genetic analysis coupled with sophisticated techniques of molecular biology and biochemistry have led to new and exciting insights into our understanding of developmental processes in fungi. Much is not only understood about what genes are developmentally important, but also how environmental signals are transduced to control developmental gene expression. Attempts to understand formation of infection structures in phytopathogenic fungi have been frustrated by a number of factors. Now, with the development of DNA-mediated transformation, suitable sexually compatable strains, and a redefining of research efforts, rapid achievements matching those in yeast and *A. nidulans* will be forthcoming. Important regulatory genes from *M. grisea* that control the formation of the appressorium and penetration peg will be identified. Strategies similar to those used successfully to isolate genes that form the central regulatory pathway of asexual reproduction in *A. nidulans* are now being employed in studies with phytopathogenic fungi. Although much remains to be understood about fungal development and signal transduction pathways in yeast and *A. nidulans,* the application of recent technological achievements to selected phytopathogenic fungi presents a unique and powerful opportunity to make major inroads into a developmental process that is required by many fungi to infect plants and identification of signal molecules that control differentiation and infection.

## ACKNOWLEDGMENTS

I am grateful to Nancy Doubrava, Jennifer Frick, and Becky Lee for assistance with the preparation of this manuscript. My appreciation is also extended to Drs. W. E. Timberlake, University of Georgia, and E. Zehr and other members of the Department of Plant Pathology and Physiology, Clemson University, for critical reading of this manuscript. Technical Contribution No. 3130 of the South Carolina Agricultural Experiment Station, Clemson University.

## REFERENCES

1. **Herskowitz, I.,** Life cycle of the budding yeast *Saccharomyces cerevisiae, Microbiol. Rev.,* 52, 536, 1988.
2. **Herskowitz, I.,** A regulatory hierarchy for cell specialization in yeast, *Nature (London),* 342, 749, 1989.
3. **Nasmyth, K. and Shore, D.,** Transcriptional regulation in the yeast life cycle, *Science,* 237, 1162, 1987.
4. **Timberlake, W. E.,** Molecular genetics of *Aspergillus* development, *Annu. Rev. Genet.,* 24, 36, 1990.
5. **Pringle, J. R. and Hartwell, L. H.,** The *Saccharomyces cerevisiae* cell cycle, in *The Molecular Biology of the Yeast Saccharomyces: Life Cycle and Inheritance,* Strathern, J. N., Jones, E. W., and Broach, J. R., Eds., Cold Spring Harbor Laboratory, Cold Spring Harbor, NY, 1981, 97.
6. **Duntze, W., McKay, V. L., and Manney, T. R.,** *Saccharomyces cerevisiae:* a diffusible sex factor, *Science,* 168, 1472, 1970.
7. **Betz, R., Crabb, J. W., Meyer, H. E., Wittig, R., and Duntze, W.,** Amino acid sequences of **a**-factor mating peptides from *Saccharomyces cerevisiae, J. Biol. Chem.,* 262, 546, 1987.
8. **Kurjan, J. and Herskowitz, I.,** Structure of a yeast pheromone gene (*MF*$\alpha$): a putative $\alpha$-factor precursor contains four tandem copies of mature $\alpha$-factor, *Cell,* 30, 933, 1982.
9. **Nakayama, N., Miyajima, A., and Arai, K.,** Nucleotide sequences of *STE2* and *STE3,* cell type-specific sterile genes from *Saccharomyces cerevisiae, EMBO J.,* 4, 2643, 1985.
10. **Dixon, R. A. F., Kobilka, B. K., Strader, D. J., Benovic, J. L., Dohlman, H. G., Frielle, T., Bolanowski, M. A., Bennet, C. D., Rands, E., Diehl, R. E., Mumford, R. A., Slater, E. E., Sigal, I. S., Caron, M. S., Lefkavitz, R. J., and Strader, C. D.,** Cloning the gene and cDNA for mammalian $\beta$-adrenergic receptor and homology with rhodopsin, *Nature (London),* 321, 75, 1986.
11. **Nathans, J. and Hogness, D. S.,** Isolation, sequence analysis, and intron-exon arrangement of the gene encoding bovine rhodopsin, *Cell,* 34, 807, 1983.
12. **Cole, G. M., Stone, D. E., and Reed, S. I.,** Stoichiometry of G protein subunits affects the *Saccharomyces cerevisiae* mating pheromone signal transduction pathway, *Mol. Cell. Biol.,* 10, 510, 1990.
13. **Whiteway, M., Hougan, L., and Thomas, D. Y.,** Overexpression of the *STE4* gene leads to mating response in haploid *Saccharomyces cerevisiae, Mol. Cell Biol.,* 10, 217, 1990.
14. **Stone, D. E. and Reed, S. I.,** G protein mutations that alter the pheromone response in *Saccharomyces cerevisiae, Mol. Cell. Biol.,* 10, 4439, 1990.
15. **Neer, E. J. and Clapham, D. E.,** Roles of G protein subunits in transmembrane signaling, *Nature (London),* 333, 129, 1988.
16. **Errede, B. and Ammerer, G.,** *STE12,* a protein involved in cell type-specific transcription and signal transduction in yeast, is part of protein DNA complexes, *Genes Dev.,* 3, 1349, 1989.

17. **Strathern, J., Hicks, J., and Herskowitz, I.,** Control of cell type in yeast by the mating type locus: The α1-α2 hypothesis, *J. Mol. Biol.,* 147, 357, 1981.
18. **Sprague, G. F., Jr., Jensen, R., and Herskowtiz, I.,** Control of yeast cell type by the mating type locus: Positive regulation of the α-specific *STE3* gene by the MATα1 product, *Cell,* 32, 409, 1983.
19. **Wilson, K. L. and Herskowitz, I.,** Negative regulation of *STE6* gene expression by the α2 product *Saccharomyces cerevisiae, Mol. Cell. Biol.,* 4, 2420, 1984.
20. **Goutte, C. and Johnson, A. D., a1** protein alters the DNA binding specificity of α2 repressor, *Cell,* 52, 875, 1988.
21. **Trueheart, J., Boeke, J. D., and Fink, G. R.,** Two genes required for cell fusion during yeast conjugation: Evidence for a pheromone-induced surface protein, *Mol. Cell. Biol.,* 7, 2316, 1987.
22. **Rose, M. D. and Fink, G. R.,** *KAR1,* a gene required for function of both intranuclear and extranuclear microtubules in yeast, *Cell,* 48, 1047, 1987.
23. **Bender, A. and Sprague, G. F., Jr.,** MATα1 protein, a yeast transcription activator, binds synergistically with a second protein to a set of cell type-specific genes, *Cell,* 50, 681, 1987.
24. **Keleher, C. A., Goutte, C., and Johnson, A. D.,** The yeast cell-type-specific repressor α2 acts cooperatively with a non-cell type-specific protein, *Cell,* 53, 927, 1988.
25. **Passmore, S., Elble, R., and Tye, B.-K.,** A protein involved in minichromosome maintenance in yeast binds a transcriptional enhancer conserved in eukaryotes, *Genes Dev.,* 3, 921, 1989.
26. **Jarvis, E. E., Hagen, D. C., and Sprague, G. F., Jr.,** Identification of a DNA segment that is necessary and sufficient for α specific gene control in *Saccharomyces cerevisiae:* Implications for regulation of α-specific and **a**-specific genes, *Mol. Cell. Biol.,* 8, 309, 1988.
27. **Mitchell, A. P. and Herskowitz, I.,** Activation of meiosis and sporulation by repression of the *RME1* product in yeast, *Nature (London),* 319, 738, 1986.
28. **Kassir, Y., Granot, D., and Simchen, G.,** *IME1,* a positive regulator gene of meiosis in *S. cerevisiae, Cell,* 52, 853, 1988.
29. **Smith, H. E. and Mitchell, A. P.,** A transcriptional cascade governs entry into meiosis in *Saccharomyces cerevisiae, Mol. Cell. Biol.,* 9, 2142, 1989.
30. **Mitchell, A. P., Driscoll, S. E., and Smith, H. E.,** Positive control of sporulation-specific genes by the *IME1* and *IME2* products in *Saccharomyces cerevisiae, Mol. Cell. Biol.,* 10, 2104, 1990.
31. **Strich, R., Slater, M. R., and Esposito, R. E.,** Identification of negative regulatory genes that govern the expression of early meiotic genes in yeast, *Proc. Natl. Acad. Sci. U.S.A.,* 86, 10018, 1989.
32. **Matsumoto, K., Uno, I., and Ishikawa, T.,** Initiation of meiosis in yeast mutants defective in adenylate cyclase and cyclic AMP-dependent protein kinase, *Cell,* 32, 417, 1983.
33. **Toda, T., Uno, I., Ishikawa, T., Powers, S., Kataoka, T., Broek, D., Cameron, S., Broach, J., Matsumoto, K., and Wigler, M.,** In yeast, RAS proteins are controlling elements of adenylate cyclase, *Cell,* 40, 27, 1985.
34. **Powers, S., O'Neill, K., and Wigler, M.,** Dominant yeast and mammalian RAS mutants that interfere with the CDC25-dependent activation of wild-type RAS in *Saccharomyces cerevisiae, Mol. Cell. Biol.,* 9, 390, 1989.
35. **Tanaka, K., Matsumoto, K., and Toh-e, A.,** *IRA1,* an inhibitory regulator of the RAS-cyclic AMP pathway in *Saccharomyces cerevisiae, Mol. Cell. Biol.,* 9, 757, 1989.
36. **Malone, R. E.,** Dual regulation of meiosis in yeast, *Cell,* 61, 375, 1990.
37. **Lindegren, C. C. and Lindegren, G.,** A new method for hybridizing yeast, *Proc. Natl. Acad. Sci. U.S.A.,* 29, 306, 1943.
38. **Klar, A. J. S., McIndoo, J., Hicks, J. B., and Strathern, J. N.,** Precise mapping of the homothallism genes, *HML* and *HMR,* in yeast *Saccharomyces cerevisiae, Genetics,* 96, 315, 1980.
39. **Takano, I., Kusumi, T., and Oshima, Y.,** An α mating type allele insensitive to the mutagenic action of the homothallic gene system in *Saccharomyces diastaticus, Mol. Gen. Genet.,* 126, 19, 1973.
40. **Strathern, J. N., Spatola, E., McGill, C., and Hicks, J. B.,** The structure and organization of transposable mating type cassettes in *Saccharomyces cerevisiae, Proc. Natl. Acad. Sci. U.S.A.,* 77, 2389, 1980.
41. **Nasmyth, K. A. and Tatchell, K.,** The structure of transposable yeast mating type loci, *Cell,* 19, 753, 1980.
42. **Kostriken, R., Strathern, J. N., Klar, A. J. S., Hicks, J., and Heffron, F.,** A site-specific endonuclease essential for mating-type switching in *Saccharomyces cerevisiae, Cell,* 35, 167, 1983.
43. **Malone, R. E. and Esposito, R. E.,** The *RAD52* gene is required for homothallic interconversion of mating types and spontaneous mitotic recombination in yeast, *Proc. Natl. Acad. Sci. U.S.A.,* 77, 503, 1980.
44. **Klar, A. J. S., Strathern, J. N., and Hicks, J. B.,** Developmental pathways in yeast, in *Microbial Development,* R. Losick and L. Shapiro, Eds., Cold Spring Harbor Laboratory, Cold Spring Harbor, NY, 1984, 151.
45. **Rine, J. and Herskowitz, I.,** Four genes responsible for a position effect on expression from *HML* and *HMR* in *Saccharomyces cerevisiae, Genetics,* 116, 9, 1987.
46. **Nasmyth, K.,** Molecular analysis of a cell lineage, *Nature (London),* 302, 670, 1983.

47. **Breeden, L. and Nasmyth, K.**, Cell cycle control of the yeast *HO* gene: cis- and trans-acting regulators, *Cell,* 48, 389, 1987.
48. **Sternberg, P. W., Stern, M. J., Clark, I., and Herskowitz, I.**, Activation of the yeast *HO* gene by release from multiple negative controls, *Cell,* 48, 567, 1987.
49. **Miller, A. M., MacKay, V. L., and Nasmyth, K. A.**, Identification and comparison of two sequence elements that confer cell-type specific transcription in yeast, *Nature (London),* 314, 738, 1985.
50. **Stillman, D. J., Bankier, A. T., Seddon, A., Groenhout, G., and Nasmyth, K. A.**, Characterization of a transcription factor involved in mother cell specific transcription of the yeast *HO* gene, *EMBO J.*, 7, 485, 1988.
51. **Nasmyth, K. A., Seddon, A., and Ammerer, G.**, Cell cycle regulation of *SWI5* is required for mother-cell-specific *HO* transcription in yeast, *Cell,* 49, 549, 1987.
52. **Nasmyth, K. A., Adolf, G., Lydall, D., and Seddon, A.**, The identification of a second cell cycle control on the *HO* promoter in yeast: cell cycle regulation of *SWI5* nuclear entry, *Cell,* 62, 631, 1990.
53. **Timberlake, W. E. and Marshall, M. A.**, Genetic engineering of filamentous fungi, *Science,* 244, 1313, 1989.
54. **Miller, B. L.**, The developmental genetics of asexual reproduction in *Aspergillus nidulans, Sem. Dev. Biol.,* 1, 207, 1990.
55. **Cole, G. T.**, Models of cell differentiation in conidial fungi, *Microbiol. Rev.*, 50, 95, 1986.
56. **Mims, C. W., Richardson, E. A., and Timberlake, W. E.**, Ultrastructural analysis of conidiophore development in the fungus *Aspergillus nidulans* using freeze-substitution, *Protoplasma,* 144, 132, 1988.
57. **Adams, T. H. and Timberlake, W. E.**, Developmental repression of growth and gene expression in *Aspergillus, Proc. Natl. Acad. Sci. U.S.A.*, 87, 5405, 1990.
58. **Clutterbuck, A. J.**, A mutational analysis of conidial development in *Aspergillus nidulans, Genetics,* 63, 317, 1969.
59. **Timberlake, W. E.**, Developmental gene regulation in *Aspergillus nidulans, Dev. Biol.*, 78, 497, 1980.
60. **Clutterbuck, A. J.**, The genetics of conidiation in *Aspergillus nidulans,* in *Genetics and Physiology of Aspergillus,* Smith, J. E. and Pateman, J. A., Eds., Academic Press, San Diego, 1977, 305.
61. **Boylan, M. T., Mirabito, P. M., Willett, C. E., Zimmerman, C. R., and Timberlake, W. E.**, Isolation and physical characterization of three essential conidiation genes from *Aspergillus nidulans, Mol. Cell. Biol.*, 7, 3113, 1987.
62. **Adams, T. H., Boylan, M. T., and Timberlake, W. E.**, *brlA* is necessary and sufficient to direct conidiophore development in *Aspergillus nidulans, Cell,* 54, 353, 1988.
63. **Mirabito, P. M., Adams, T. H., and Timberlake, W. E.**, Interactions of three sequentially expressed genes control temporal and spatial specificity in *Aspergillus* development, *Cell,* 57, 859, 1989.
64. **Martinelli, S. D.**, Phenotypes of double conidiation mutants of *Aspergillus nidulans, J. Gen. Microbiol.*, 114, 277, 1979.
65. **Abate, C., Patel, L., Rauscher, III, F. J., and Curran, T.**, Redox regulation of *Fos* and *Jun* DNA-binding activity *in vitro, Science,* 249, 1077, 1990.
66. **Adams, T. H. and Timberlake, W. E.**, Upstream elements repress premature expression of an *Aspergillus* developmental regulatory gene, *Mol. Cell. Biol.*, 10, 4912, 1990.
67. **Marshall, M. A. and Timberlake, W. E.**, unpublished data, 1990.
68. **Dean, R. A., Stringer, M., Sewell, T., and Timberlake, W. E.**, Targeting sporulation specific genes for disease control, *Phytopathology,* 80, 993, 1990.
69. **Dean, R. A. and Timberlake, W. E.**, unpublished data, 1990.
70. **Stringer, M. and Timberlake, W. E.**, unpublished data, 1990.
71. **Sewell, T. C., Mims, C. W., and Timberlake, W. E.**, unpublished data, 1990.
72. **Zimmerman, C. R., Orr, W. C., Leclerc, R. F., Barnard, E. C., and Timberlake, W. E.**, Molecular cloning and selection of genes regulated in *Aspergillus* development, *Cell,* 21, 709, 1980.
73. **Timberlake, W. E. and Barnard, E. C.**, Organization of a gene cluster expressed specifically in the asexual spores of *A. nidulans, Cell,* 26, 29, 1981.
74. **Orr, W. C. and Timberlake, W. E.**, Clustering of spore-specific genes in *Aspergillus nidulans, Proc. Natl. Acad. Sci. U.S.A.*, 79, 5676, 1982.
75. **Gwynne, D. I., Miller, B. L., Miller, K. Y., and Timberlake, W. E.**, Structure and regulated expression of the *SpoC1* gene cluster from *Aspergillus nidulans, J. Mol. Biol.*, 180, 91, 1984.
76. **Miller, B. L., Miller, K. Y., Roberti, K. A., and Timberlake, W. E.**, Position-dependent and -independent mechanisms regulate regulate cell-specific expression of the *SpoC1* gene cluster of *Aspergillus nidulans, Mol. Cell. Biol.*, 7, 427, 1987.
77. **Aramayo, R., Adams, T. H., and Timberlake, W. E.**, A large cluster of highly expressed genes indispensable for growth and development in *Aspergillus nidulans, Genetics,* 122, 65, 1989.
78. **Percival-Smith, A. and Segall, J.**, Characterization and mutational analysis of a cluster of three genes expressed preferentially during sporulation of *Saccharomyces cerevisiae, Mol. Cell. Biol.*, 6, 2443, 1986.

79. **O'Hara, E. B. and Timberlake, W. E.,** Molecular characterization of the *Aspergillus nidulans yA* locus, *Genetics,* 129, 249, 1989.
80. **Mayorga and Timberlake,** unpublished data, 1990.
81. **Axelrod, D. E., Gealt, M., and Pastushok, M.,** Gene control of developmental competence in *Aspergillus nidulans, Dev. Biol.,* 34, 9, 1973.
82. **Timberlake, W. E. and Hamer, J. E.,** Regulation of gene activity during conidiophore development in *Aspergillus nidulans, Gen. Eng.,* 8, 1, 1986.
83. **Champe, S. P., Kurtz, M. B., Yager, L. N., Butnick, N. J., and Axelrod, D. E.,** Spore formation in *Aspergillus nidulans:* competence and other developmental processes, in *The Fungal Spore: Morphogenetic Controls,* Turian, G. and Hohl, H. R., Eds., Academic Press, San Diego, 1981.
84. **Pastushok, M. and Axelrod, D. E.,** Effect of glucose, ammonium, and media maintenance on the time of conidiophore initiation by surface colonies of *Aspergillus nidulans, J. Gen. Microbiol.,* 94, 221, 1976.
85. **Yager, L. N., Kurtz, M. B., and Champe, S. P.,** Temperature-shift analysis of conidial development in *Aspergillus nidulans, Dev. Biol.,* 93, 92, 1982.
86. **Butnick, N. Z., Yager, L. N., Hermann, T. E., Kurtz, M. B., and Champe, S. P.,** Mutants of *Aspergillus nidulans* blocked at an early stage of sporulation secrete an unusual metabolite, *J. Bacteriol.,* 160, 533, 1984.
87. **Butnick, N. Z., Yager, L. N., Kurtz, M. B., and Champe, S. P.,** Genetic analysis of mutants of *Aspergillus nidulans* blocked at an early stage of sporulation, *J. Bacteriol.,* 160, 541, 1984.
88. **Champe, S. P., Rao, P., and Chang, A.,** An endogenous inducer of sexual development in *Aspergillus nidulans, J. Gen. Microbiol.,* 133, 1383, 1987.
89. **Tamame, M., Antequera, F., and Santos, E.,** Developmental characterization and chromosomal mapping of the 5-azacytidine-sensitive *fluF* locus of *Aspergillus nidulans, Mol. Cell. Biol.,* 8, 3043, 1988.
90. **Mooney, J. and Yager, L. N.,** personal communication, 1990.
91. **Käfer, E.,** Origins of translocations in *Aspergillus nidulans, Genetics,* 52, 217, 1965.
92. **Valent, B.,** Rice blast as a model system for plant pathology, *Phytopathology,* 80, 33, 1990.
93. **Banuett, F. and Herskowitz, I.,** *Ustilago maydis,* smut of maize, in *Genetics of Plant Pathogenic Fungi, Advances in Plant Pathology,* Vol. 6, Sidhu, G. S., Ed., Academic Press, San Diego, 1988, 427.
94. **Schulz, B., Banuett, F., Dahl, M., Schlesinger, R., Schäfer, W., Martin, T., Herskowitz, I., and Kahmann, R.,** The b alleles of *U. maydis,* whose combinations program pathogenic development, code for polypeptides containing a homeodomain-related motif, *Cell,* 60, 295, 1990.
95. **Thurston, H. D.,** *Tropical Plant Diseases,* The American Phytopathological Society, MN, 1984, 24.
96. **Crawford, M. S., Chumley, F. G., Weaver, C. G., and Valent, B.,** Characterization of the heterokaryotic and vegetative diploid phases of *Magnaporthe grisea, Genetics,* 114, 1111, 1986.
97. **Valent, B., Crawford, M. S., Weaver, C. G., and Chumley, F. G.,** Genetic studies of fertility and pathogenicity in *Magnaporthe grisea (Pyricularia oryzae), Iowa State J. Res.,* 60, 569, 1986.
98. **Hamer, J. E., Valent, B., and Chumley, F. G.,** Mutations at the *SMO* genetic locus affect the shape of diverse cell types in the rice blast fungus, *Genetics,* 122, 351, 1989.
99. **Yaegashi, H. and Udagawa, S.,** The taxonomical identity of the perfect state of *Pyricularia grisea* and its allies, *Can. J. Bot.,* 56, 180, 1978.
100. **Herbert, T. T.,** The perfect stage of *Pyricularia grisea, Phytopathology,* 61, 83, 1971.
101. **Yaegashi, H.,** On sexuality of blast fungus, *Pyricularia* spp., *Ann. Phytopathol. Soc. Jpn.,* 43, 432, 1977.
102. **Leong, S. A. and Holden, D. W.,** Molecular genetic approaches to the study of fungal pathogenesis, *Ann. Rev. Phytopathol.,* 27, 463, 1989.
103. **Itoi, S., Mishima, T., Araise, S., and Noza, M.,** Mating behavior of Japanese isolates of *Pyricularia oryzae, Phytopathology,* 73, 155, 1983.
104. **Ellingboe, A. H., Bai-Chai, Wu and Robertson, W.,** Inheritance of avirulence/virulence in a cross of two isolates of *Magnaporthe grisea* pathogenic to rice, *Phytopathology,* 80, 108, 1990.
105. **Leung, H. and Williams, P. H.,** Genetic analysis of electrophoretic enzyme variants, mating type, and hermaphroditism in *Pyricularia oryzae* Cavara, *Can. J. Genet. Cytol.,* 27, 697, 1985.
106. **Kolmer, J. A. and Ellingboe, A. H.,** Genetic relationships between fertility and pathogenicity and virulence to rice in *Magnaporthe grisea, Can. J. Bot.,* 66, 891, 1988.
107. **Parsons, K. A., Chumley, F. G., and Valent, B.,** Genetic transformation of the fungal pathogen responsible for rice blast disease, *Proc. Natl. Acad. Sci. U.S.A.,* 84, 4161, 1987.
108. **Parsons, K. A.,** The development of genetic transformation in the ascomycete *Magnaporthe grisea* and the cloning of the *LYS1+* gene. Ph.D. thesis, University of Colorado, Boulder, CO, 1988.
109. **Bourett, T. M. and Howard, R. J.,** *In vitro* development of penetration structures in the rice blast fungus *Magnaporthe grisea, Can. J. Bot.,* 68, 329, 1990.
110. **Howard, R. J. and Ferrari, M. A.,** Role of melanin in appressorium function, *Exp. Mycol.,* 13, 403, 1989.
111. **Hamer, J. E., Howard, R. J., Chumley, F. G., and Valent, B.,** A mechanism for surface attachment in spores of a plant pathogenic fungus, *Science,* 239, 288, 1988.

112. **Kolattukudy, P. E., Podila, G. K., and Mohan, R.,** Molecular basis of the early events in plant-fungus interaction, *Genome,* 31, 342, 1989.
113. **Suzuki, K., Furusawa, I., Ishida, N., and Yamamoto, M.,** Protein synthesis during germination and appresorium formation of *Colletotrichum lagenarium* spores. *J. Gen. Microbiol.,* 128, 1035, 1981.
114. **Michelmore, R. W. and Hulbert, S. H.,** Molecular markers for genetic analysis of phytopathogenic fungi, *Annu. Rev. Phytopathol.,* 25, 383, 1987.
115. **Bhairi, S. M., Staples, R. C., Freve, P., and Yoder, O. C.,** Characterization of an infection structure-specific gene from the rust fungus *Uromyces appendiculatus, Gene,* 81, 237, 1989.
116. **Scott-Craig, J. S., Panaccione, D. G., Cervone, F., and Walton, J. D.,** Endopolygalacturonase is not required for pathogenicity of *Cochliobolus carbonum* on maise, *Plant Cell,* 2, 1191, 1990.
117. **Hamer, J. E.,** personal communication, 1990.
118. **Hoch, H. C. and Staples, R. C.,** Structural and chemical changes among the rust fungi during appressorium development, *Annu. Rev. Phytopathol.,* 25, 231, 1987.
119. **Staples, R. C. and Hoch, H. C.,** Infection structures—form and function, *Exp. Mycol.,* 11, 163, 1987.
120. **Uchiyama, T. and Okuyama, K.,** Participation of *Oryza sativa* leaf wax in appressorium formulation by *Pyricularia oryzae, Phytochemistry,* 29, 91, 1990.
121. **Podila, G. K., Dickman, M. B., and Kolattukudy, P. E.,** Transcriptional activation of a cutinase gene in isolated fungal nuclei by plant cutin monomers, *Science,* 242, 922, 1988.
122. **Yamamoto, K. R.,** Steroid receptor regulated transcription of specific genes and gene networks, *Annu. Rev. Genet.,* 19, 209, 1985.
123. **Castle, A. J. and Day, A. W.,** Isolation and identification of a α-tocopherol as an inducer of the parasitic phase of *Ustilago violacae, Phytopathology,* 74, 1184, 1984.
124. **Macko, V., Renwick, J. A. A., and Rissler, J. F.,** Acrolein induces differentiation of infection structures in the wheat stem rust fungus, *Science,* 199, 442, 1978.
125. **Lucas, J. and Knights, I.,** Spores on leaves: endogenous and exogenous control of development, in *Fungal Infection of Plants,* 1st ed., Pegg, G. F. and Ayres, P. G., Eds., Cambridge University Press, Cambridge, 1987, chap. 3.

# ANALYSIS OF TRANSCRIPTION CONTROL SEQUENCES OF FUNGAL GENES

**Peter J. Punt and Cees A. M. J. J. van den Hondel**

## TABLE OF CONTENTS

# I. INTRODUCTION

Filamentous fungi, in particular the genetically and biochemically well-characterized species *Aspergillus nidulans* and *Neurospora crassa,* have a number of properties that make them very attractive for molecular biological studies of eukaryotic gene organization and regulation of gene expression. First, filamentous fungi (especially many plant pathogenic fungi) show distinct cellular differentiation of the vegetative mycelium and a complex life cycle,[1] which makes them clearly distinct from taxonomically related unicellular yeasts, such as, *Saccharomyces cerevisiae* and *Schizosaccharomyces pombe*. Second, fungi generally have enormous metabolic versatility and a wealth of genetic and biochemical data is available for many biosynthetic and catabolic pathways in various fungal species.[2-4] Finally, many species can be cultivated in simple growth media, which makes them easy to work with in the laboratory.

The development of genetic transformation techniques was a major breakthrough for molecular biological research in filamentous fungi. The first report about genetic transformation of a filamentous fungus dates from the late 1970s, when Case et al.[5] transformed a *N. crassa qa-2* mutant with a vector containing the cloned *qa-2* gene encoding dehydroquinase. Since then a large number of reports have been published describing genetic transformation of more than 50 different fungal species, using different auxotrophic and dominant selection markers.[6] The availability of transformation techniques made possible the development of various genetic manipulation techniques required for further molecular biological research on the regulation of gene expression.

In the last few years many fungal genes have been isolated and characterized based on these techniques. Extensive data have accumulated about the primary structure of these genes including 5′- and 3′-flanking sequences. In this chapter we will focus on an analysis of 5′-flanking sequences and their role in regulation of gene expression, in particular, transcription control. Obviously, 3′-flanking sequences can also be involved in the regulation of gene expression, e.g., by determining the site of transcription termination and polyadenylation. However, no experimental data concerning the role of 3′-flanking sequences of fungal genes in gene expression are yet available. The first step in identifying sequences involved in transcription control is generally the comparison of 5′-flanking sequences, with 5′-flanking sequences of genes of *S. cerevisiae* and higher eukaryotes.

For *S. cerevisiae* and higher eukaryotes, a considerable amount of data regarding the organization of expression signals already exists.[7,8] A generalized scheme of yeast transcription control regions is presented in Figure 1. Three types of sequences are indicated: (1) upstream activating/repressing sequences (UAS/URS); (2) TATA sequences; and (3) transcription initiation sequences.

1. UAS/URS — These are short DNA sequences of 10 to 30 nt located at various distances upstream of the mRNA start site. These sequences are involved in the regulation of gene expression, in most cases as (putative) target sites for transacting regulatory proteins.
2. TATA sequences — These are located 40 to 120 nt upstream of the mRNA start site. They are involved in determining the efficiency of transcription initiation in many yeast genes. Furthermore, in several genes the TATA sequence is also involved in determining the correct transcription initiation.
3. Transcription initiation sequences — Many different sequences can function as the initiation site. In the case of efficiently expressed yeast genes, initiation predominantly occurs at PyAAG sequences. In general, initiation sequences are not involved in determining transcription efficiency.

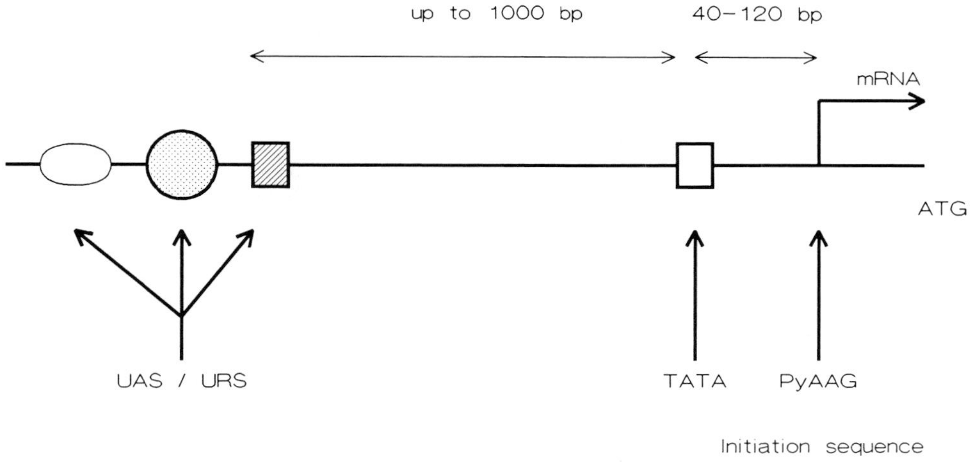


FIGURE 1. Scheme of yeast transcription control sequences.

A similar organization is also found for transcription control regions in higher eukaryotes, although, in this case, the TATA sequence is located at a more or less fixed position about 30 nt upstream of the mRNA start site. Furthermore, an additional, conserved sequence, the CAAT box, located at 70 to 90 nt upstream of the mRNA start site is observed in a number of genes. Many fungal genes lack either TATA-, CAAT-, or PyAAG-like sequences,[9,10] indicating that filamentous fungi differ in the organization of their transcription control sequences from *S. cerevisiae* (and related yeasts) and higher eukaryotes. A further indication that filamentous fungi and yeasts, although taxonomically related, differ in their transcription control sequences, is that successful use of yeast transcription control sequences in filamentous fungi has never been reported. Moreover, in only a few cases were fungal transcription control sequences functional in *S. cerevisiae*.[11] An overview about approaches for analysis of fungal transcription control sequences will be provided in this chapter. The results obtained using these approaches will be discussed.

# II. ANALYSIS OF TRANSCRIPTION CONTROL SEQUENCES

## A. SEQUENCE ANALYSIS

The most simple approach to analyze (cloned) transcription control regions is a comparison of the 5′-flanking DNA sequences of different genes. For example, comparison of the 5′-flanking sequences of coregulated genes may be used to identify regions of similar sequence. Such regions could constitute binding sites for either general or specific regulatory proteins. The usefulness of this approach was demonstrated by Gwynne et al.,[12] in their study of the genes involved in ethanol metabolism in *A. nidulans*. From genetic and biochemical data it was already known that the *alcR* gene product was involved in the regulation of expression of both *alcA* (encoding alcohol dehydrogenase) and *aldA* (encoding aldehyde dehydrogenase).[13] Comparison of the 5′-flanking sequences of both genes resulted in the identification of putative binding sites for the regulatory protein encoded by the *alcR* gene.[12] Also, in the study of the gene cluster involved in quinic acid metabolism in *N. crassa*[14,15] and *A. nidulans,*[16] sequence comparison of 5′-flanking sequences from coregulated genes led to the identification of putative regulatory sequences. Recently, a number of elements of similar sequence has also been identified in the nitrate gene cluster of *A. nidulans* at both sides of the intergenic region between the coregulated *niiA* and *niaD* genes, encoding nitrite and nitrate reductase, respectively.[17]

Sequence comparison of the 5′-flanking regions of genes encoding the same protein, in related fungal species, can also lead to the identification of putative transcription control sequences, as can be illustrated for the *A. niger* and *A. nidulans gpdA* and *oliC* genes, encoding glyceraldehyde-3-phosphate dehydrogenase and an ATPase subunit, respectively. The 5′-flanking sequences of both *gpdA* genes show an overall similarity of about 70%. However, a 50-nt region of much higher similarity (95%) was identified about 250 nt upstream of the transcription initiation site (Figure 2A).[18,19] In the 5′-flanking sequences of both *oliC* genes, which show about 50% overall similarity, a 30-nt region of about 85% similarity was identified about 60 nt upstream of the (major) transcription initiation site (Figure 2B).[20] Indications that these regions are functional transcription control sequences have been obtained by *in vivo* analysis (see Section C.2 below).

## B. PROTEIN-DNA-BINDING ANALYSIS

In general, regulation of gene expression at the transcriptional level is based on the action of regulatory proteins.[7,8,21-24] The most simple representation for this type of regulation is binding of a regulatory protein to sequences in the 5′-flanking regions of the gene of interest and subsequent interaction of the bound protein with the transcription initiation complex. Fungal DNA sequences which bind to regulatory proteins can be identified by similar *in vitro* analysis methods as has been described for the analysis of protein-DNA binding in *S. cerevisiae* and higher eukaryotes.[22,23] In most cases, the strategy in identifying the sequences involved in protein binding comprises the following steps: (1) detection of specific binding of (regulatory) protein(s) to particular transcription control sequences by so-called bandshift experiments. In these experiments, DNA fragments containing putative control sequences are incubated with (nuclear) protein preparations or (partially) purified regulatory protein; (2) identification of the specific sequences involved in DNA binding by detailed footprint analyses, such as DNAse I and methylation protection assays. In most cases, the latter techniques require the availability of purified regulatory protein.

Extensive biochemical and genetic studies about the regulation of expression of the genes of several metabolic and differentiation pathways in filamentous fungi have resulted in the isolation and characterization of a number of regulatory mutants.[2-4] Based on these mutants, several of the corresponding regulatory genes have been cloned and the products of these genes characterized (Table 1).[25] The availability of these genes permits isolation of sufficiently large amounts of the regulatory protein by overexpression of the cloned gene to identify DNA-binding sequences in the transcription control sequences of the (cloned) genes of the various pathways. As summarized in Table 1, in most cases, binding of the regulatory protein to transcription control sequences is suggested by the presence of DNA-binding motifs (leucine zipper, zinc finger, helix-loop-helix) in the regulatory proteins.[22] For some regulatory proteins, the role of these motifs in regulation of gene expression was tested by analysis of the activity of mutant proteins from which these DNA-binding motifs have been deleted. In all cases tested, these mutant proteins were no longer functional (Table 1, column 3, indicated with +).[29,34,47,49,53] In some cases, the functionality of the DNA-binding motif was indicated by sequence alterations in the DNA-binding motif of mutant alleles of the regulatory gene [Table 1, column 3, indicated with (+)].[27,30,31,41,42,49,50,51] Currently, only for a few regulatory proteins specific binding to transcription control regions of genes regulated by these proteins have been demonstrated by bandshift analysis (Table 1, column 4).[28-31,33,41,42,45,48] In four cases (all *N. crassa*), the DNA sequences involved in binding were further identified by footprint analysis (Table 1, column 4).[29,31,33,41,42] Surprisingly, the "leucine zipper" regulatory proteins encoded by *cpc-1* and *cys-3*, which are involved in regulation of two distinct metabolic pathways, bind to a similar target sequence (Table 1, ATGactCAT).[29,31]

Even if the regulatory DNA-binding protein is not available in a purified form or if its

A.

```
TCCAAATATCGTGCCTCTCCTGCTTTGCCCGGTGTATGAAACCGGAAAGG -235 A. nidulans gpdA
*************  ***********************************
TCCAAATATCGTGAGTCTCCTGCTTTGCCCGGTGTATGAAACCGGAAAGG -283 A. niger gpdA
```

B.

```
GGTGAAAAAAGGGCGAAAATTAAGCGGGAGA  -60     A. nidulans oliC
**  ******  ********* *********
GGGCAAAAAACCGCGAAAATTTAGCGGGAGA  -50/60  A. niger oliC
```

FIGURE 2. Elements of similar sequence in the upstream region of the (A) *gpdA* and (B) *oliC* genes of *A. nidulans* and *A. niger*. The distance from the transcription initiation site is given (in nt). In the case of the *A. niger oliC* gene, the exact position of the transcription initiation site is unknown.

**TABLE 1**
**Cloned Regulatory Genes of Filamentous Fungi**

| Regulatory gene | Regulatory function | Functional characterization | | | Ref. |
|---|---|---|---|---|---|
| | | DNA-binding motif[a] | Functional analysis[b] of DNA-binding motif | *In vitro* DNA binding[c] (target sequence) | |
| *Neurospora crassa* | | | | | |
| *cpc-1* | Amino acid catabolism | L | (+)+ | + (ATGACTCAT) | 26-29 |
| *cys-3* | Sulfur metabolism | L | (+) | + ($ATGN_{1\text{-}4}CAT,TTC^{T}/_{G}C^{T}/_{G}$) | 30,31 |
| *nit-2* | Nitrogen catabolism | Z | + | + (repeated TATCTA)* | 32-34 |
| *nit-4* | Nitrate assimilation | Z | NT | NT | 35,36 |
| *nmr-1* | Nitrogen catabolism | ? | NT | NT | 37-39 |
| *nuc-1* | Phosphorus metabolism | HLH | NT | NT | 40 |
| *qa-1F* | Quinate catabolism | Z | (+) | + ($GG^{A}/_{G}TAA^{A}/_{G}{}^{C}/_{T}{}^{A}/_{G}{}^{C}/_{T}TTATCC$) | 41,42 |
| *qa-1S* | Quinate catabolism | ? | NT | NT | 42 |
| *Aspergillus nidulans* | | | | | |
| *abaA* | Conidiospore development | L | NT | NT | 1,43 |
| *alcR* | Ethanol metabolism | Z | NT | + | 44,45 |
| *amdR* | Amide, ω-amino acid and lactam catabolism | Z | + | + | 46-48 |
| *areA* | Nitrogen catabolism | Z | (+)+ | NT | 49-51 |
| *brlA* | Conidiospore development | Z | + | NT | 43,52,53 |
| *creA* | Carbon catabolism | — | — | — | 54,55 |
| *facB* | Amide and acetate metabolism | — | — | — | 56 |
| *nirA* | Nitrate assimilation | Z | NT | NT | 45 |
| *prnA* | Proline catabolism | Z | NT | NT | 45,57 |
| *qutA* | Quinate catabolism | Z | NT | NT | 58 |
| *qutR* | Quinate catabolism | ? | NT | NT | 59 |
| *uay* | Purine catabolism | Z | NT | NT | 45 |
| *wetA* | Conidiospore development | — | — | — | 1,43 |

[a] Based on sequence comparison, various DNA binding motifs were indicated: Z, zinc finger (different types); L, leucine zipper (including basic region); HLH, helix-loop-helix motif; ?, no homology to either of the known motifs; —, no sequence data available.

[b] Functionality of the putative DNA-binding motif was tested by functional *(in vivo)* analysis of the products of mutant genes, which were constructed by deletion- and site-directed mutagenesis, indicated by +. In some cases, functionality of the DNA-binding motif was indicated by sequence analysis of various mutant alleles, indicated by (+); NT, not tested.

[c] *In vitro* DNA binding was analyzed by bandshift analysis. If a target sequence is indicated, footprint analysis was also carried out. *, in this case footprint analysis was carried out with only a part of the Nit2 protein. NT, not tested.

identity is unknown, protein-DNA-binding analysis may lead to identification of transcription control sequences. Frederick et al.[60] have used bandshift experiments to identify protein-binding sequences in the 5′-flanking region of the *N. crassa am* gene, encoding glutamate dehydrogenase. Indications that these binding sequences are also functional *in vivo* was obtained by *in vivo* analysis (see below, Section C.2).[60,73] Binding of a protein factor to sequences of the 5′-flanking region of the cutinase gene from *Fusarium solani (cutA)* was also observed.[61] Preliminary results from bandshift analyses showed that several sequence elements present at both sides of the intergenic region of the coregulated *A. nidulans niiA* and *niaD* genes are bound by protein(s).[17] With one of these elements, a copy of which is present at about 170 nt upstream of the transcription initiation site of each gene (Figure 3A), binding of one or more proteins of a nuclear extract from nitrate-induced cultures was observed, whereas no binding was observed with a nuclear extract from uninduced (or ammonium repressed) cultures (Figure 3B). This result suggests that this sequence is involved in the induction of the *niiA* and *niaD* genes by nitrate. A similar sequence element is also present at 225 nt upstream of the *crnA* gene, encoding nitrate permease (Figure 3A).[62] This gene, which is clustered with *niiA* and *niaD,* is also induced by nitrate, further indicating a role for the identified sequence element in nitrate induction.

It is important to note that bandshift, DNAse I protection, and related types of assays do not give information about the *in vivo* function of the identified sequences. Therefore, additional *in vivo* analysis (see Section C below) is necessary to obtain conclusive data about the sequences which are involved in regulation of gene expression at the transcriptional level (and the mechanisms by which regulatory proteins work).

## C. *IN VIVO* ANALYSIS

### 1. Titration Analysis

In general, introduction of multiple copies of a gene into a host strain leads to an increase in the amount of corresponding protein. However, if the expression of this gene is regulated through the action of regulatory protein(s) and the genes encoding these proteins are not concomitantly amplified, shortage of the regulatory protein(s) could result. This so-called titration of regulatory proteins could result in a nonlinear relation between the level of gene expression and the copy number. Titration of a positively acting regulatory protein would lead to a decrease in the amount of gene product per gene copy. In the case of a negatively acting regulatory protein, an increase in the amount of gene product per gene copy is to be expected. Evidence for titration of regulatory proteins was reported in expression studies of the *A. nidulans qutE,*[63,64] *alcA,*[65] and *amdS*[66] genes. In the case of the *qutE* gene (encoding dehydroquinase), transformants containing multiple copies of this gene showed wild type (wt) expression levels, suggesting (very tight) titration.[63] Expression of the interferon α2 gene driven by the transcription control region of the *alcA* gene and, thus, controlled by the *alcR* gene product, resulted in production of interferon α2 in *A. nidulans*. However, in multicopy transformants a decrease in *alcA* (and *aldA,* which is also controlled by the *alcR* gene product) expression, compared to untransformed *A. nidulans* was observed, indicating that the amount of regulatory protein (*alcR* gene product) was limiting through titration.[65]

Of special interest is the titration analysis of the complex regulation of the *A. nidulans amdS* gene, encoding acetamidase. In this case, the titration analysis has led to the identification and precise localization of transcription control sequences, which are involved in binding of regulatory proteins. Genetic data had already shown that several regulatory proteins are involved in the expression of the *amdS* gene. These proteins are also involved in the regulation of several other genes (Table 2). The presence of multiple copies of the *amdS* gene or *amdS* upstream sequences in an *A. nidulans* strain resulted in a change in growth properties of this strain, indicating titration of the regulatory proteins encoded by *amdR, facB, amdA,* and possibly *areA.*[56,67-69] Introduction into *A. nidulans* of defined regions

```
TCT-TGATTTG    -170/180  niaD
**   ***** *
TCGCTGATTCG    -168      niiA
***  ****  *
TCG-TGATCGG    -225      crnA
```

A

FIGURE 3. Protein-DNA-binding analysis of a sequence element in the upstream region of the *A. nidulans niaD* gene. (A) Elements of similar sequence in the upstream region of all three genes of the nitrate gene cluster. The distance from the transcription initiation site is given (in nt). In the case of the *niaD* gene the exact position of the transcription initiation site is unknown. (B) Bandshift analysis with a double-stranded oligonucleotide containing the *niaD* element and nuclear extracts from *A. nidulans* cultures grown in medium containing $NO_3^-$ [induced (I)], glutamate or proline [noninduced (NI)] or $NH_4^+$ [repressed (R)] as sole nitrogen source.

**TABLE 2**
**Regulatory Genes Involved in Expression of the *Aspergillus nidulans amdS* Gene**

| Regulatory gene | Regulation by | Other regulated genes/pathways |
|---|---|---|
| *amdA* | Acetate | *aciA* |
| *amdR* | ω-Amino acids | *gabA, gatA, lamA, lamB* |
| *creA/B/C* | Carbon catabolites (e.g., glucose) | Carbon catabolism (e.g., ethanol and acetate metabolism) |
| *facB* | Acetate | *acuD, acuE, facA* |
| *areA* | Nitrogen metabolites ($NH_4^+$, glutamine) | Nitrogen catabolism (e.g., purine and proline catabolism, nitrate assimilation) |

From references 48 and 66.

of the 5′-flanking sequences of the *amdS* gene led to the localization of the FacB, AmdA and AmdR target sequences.[68] Due to only weak titration effects, the AreA target sequence(s) were not precisely localized.[69]

Similar titration effects through *amdS* sequences are also observed in fungal species related to *A. nidulans*.[70,71] *Aspergillus niger* grows very poorly on agar plates containing acetamide as the sole nitrogen source. However, transformation of *A. niger* with a vector containing the *A. nidulans amdS* gene results in transformants which show strong growth on agar plates containing acetamide.[70] Transformants with multiple copies of the *amdS* gene can be selected by their strong growth on agar plates with acrylamide as the nitrogen source (B13, B38; Figure 4).[70,71] These multicopy transformants showed impaired growth on plates with ω-amino acids, such as γ-aminobutyric acid, as the sole carbon source (Figure 4), implying that the expression of genes involved in degradation of ω-amino acids is impaired.[72] This result indicates titration of a regulatory protein from *A. niger* with similar characteristics as AmdR of *A. nidulans*.

Strong support for titration of regulatory proteins was obtained from so-called antititration experiments. In these experiments, multiple copies of the gene encoding the regulatory protein were introduced into strains already containing multiple copies of the gene of interest.

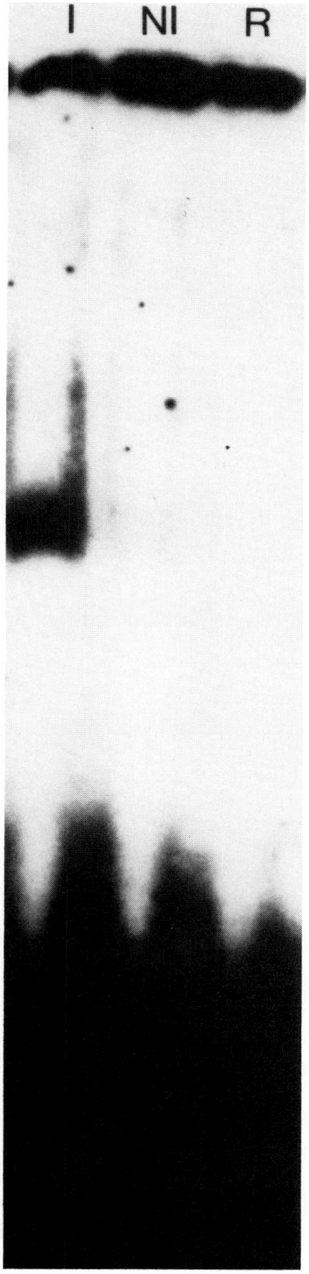


FIGURE 3B.

In all three cases mentioned in this section *(qutE, alcA, amdS)*, introduction of multiple copies of the relevant regulatory gene resulted in increased expression of the genes of interest.[46,56,64,65]

In conclusion, the results described in this section indicate that, with the aid of titration analysis, 5′ upstream sequences which are involved in binding of regulatory proteins can be localized.

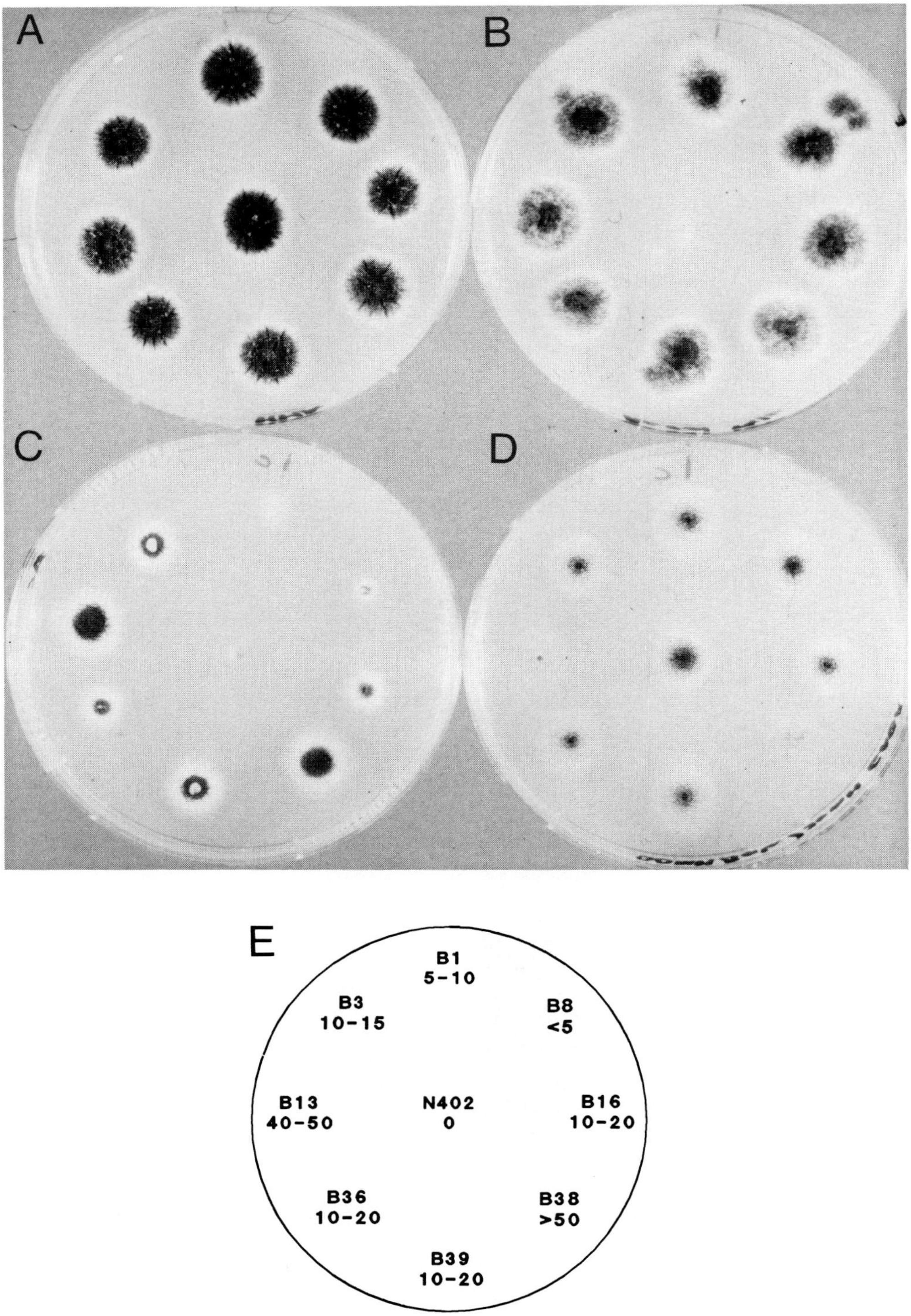


FIGURE 4. *Aspergillus niger* AmdS$^+$ transformants plated on (A) nonselective medium, (B) acetamide medium, (C) acrylamide medium, (D) γ-aminobutyric acid medium. (E) From all transformants (B1 to B39) the number of *amdS* gene copies was determined by Southern analysis. *Aspergillus niger* N402 is the recipient strain without any *amdS* gene copy.

## 2. Mutation Analysis

In this type of analysis, the effects of mutations in a specific upstream region on transcription efficiency/regulation and transcription initiation are studied. In all cases, the amount of gene product (either of the gene corresponding to the upstream region[60,73,74] or of a reporter gene[19,48,75-85]) is used as a measure for transcription efficiency. Transcription initiation is analyzed by nuclease S1 or primer extension analyses. From the data obtained by these analyses, transcription control sequences can be identified, which are involved in the efficiency, regulation, or initiation of transcription.

The first objective in many ''promoter'' analysis studies is to define the region involved in transcription control. For this purpose, unidirectional deletion of upstream sequences is carried out and the effect of the resulting deletion mutants on the level of gene expression is analyzed. The results of two examples of this type of analysis are presented in Figure 5. Deletion mutants of the upstream sequences of both the *A. nidulans gpdA* and *oliC* gene were fused to the *E. coli lacZ* gene. Subsequently, *A. nidulans* transformants containing a single copy of the fusion genes integrated at a specific site in the genome (*argB* locus) were generated and the amount of β-galactosidase activity was determined in mycelial extracts from these transformants. As can be seen in Figure 5, sequences as far upstream as 700 to 1000 nt from the transcription initiation site contribute to the level of gene expression.[19,86] In several other fungal genes, upstream regions of similar size, or even larger, are involved in transcription control.[73,76,83] However, only very small regions, even less than 50 bp upstream of the transcription initiation site, are capable of driving significant expression, as was concluded from deletion analysis of the transcription control region of the *A. nidulans trpC, argB, oliC, abaA,* and *gpdA* genes (see also Figure 5).[19,74,76,79,82]

### *a. Characterization of Sequences Involved in Transcription Initiation*

For *S. cerevisiae*, TATA-like elements, PyAAG sequences, and, in some cases, pyrimidine-rich regions in the vicinity of the transcription initiation site are involved in transcription initiation.[9,10]

Only a few studies with filamentous fungi have been reported in which these types of sequences were analyzed by deletion analysis.[19,76,79,82]

In the transcription control region of the *A. nidulans gpdA* and *oliC* genes, pyrimidine-rich regions, so-called *ct* boxes, were shown to be involved in determining the site of transcription initiation. Deletion of one of the *ct* boxes abolished transcription initiation from the site directly downstream of this box (compare d1 with d896 in Figure 6A and d0 with d10/d104 in Figure 6B).[19,79] Transcription initiation in that case occurred at sites downstream of other *ct* boxes. For transcription initiation of the *A. nidulans trpC* gene (encoding a trifunctional protein involved in tryptophan biosynthesis) and *abaA* gene, deletion analysis also indicated the involvement of a pyrimidine-rich region in transcription initiation.[76,82]

Deletion of the TATA box in the transcription control region of the *oliC* gene resulted in a change of the wt pattern of transcription initiation sites. Two of the initiation sites used in the wt upstream region were not used when the TATA box was deleted (compare d0 and d101 in Figure 6B).[79] In the *A. nidulans gpdA* and *abaA* genes, deletion of TATA-like sequences did not result in any change in transcription initiation sites.[19,82]

Based on the results available at present, we conclude that pyrimidine-rich regions are clearly involved in determining transcription initiation sites, whereas TATA-like sequences are not or only to a minor extent.

### *b. Characterization of Sequences Involved in Transcription Efficiency and Regulation of Gene Expression*

Unidirectional deletion experiments, to define the DNA region involved in transcription control, also provide information about the identification of specific sequences involved in

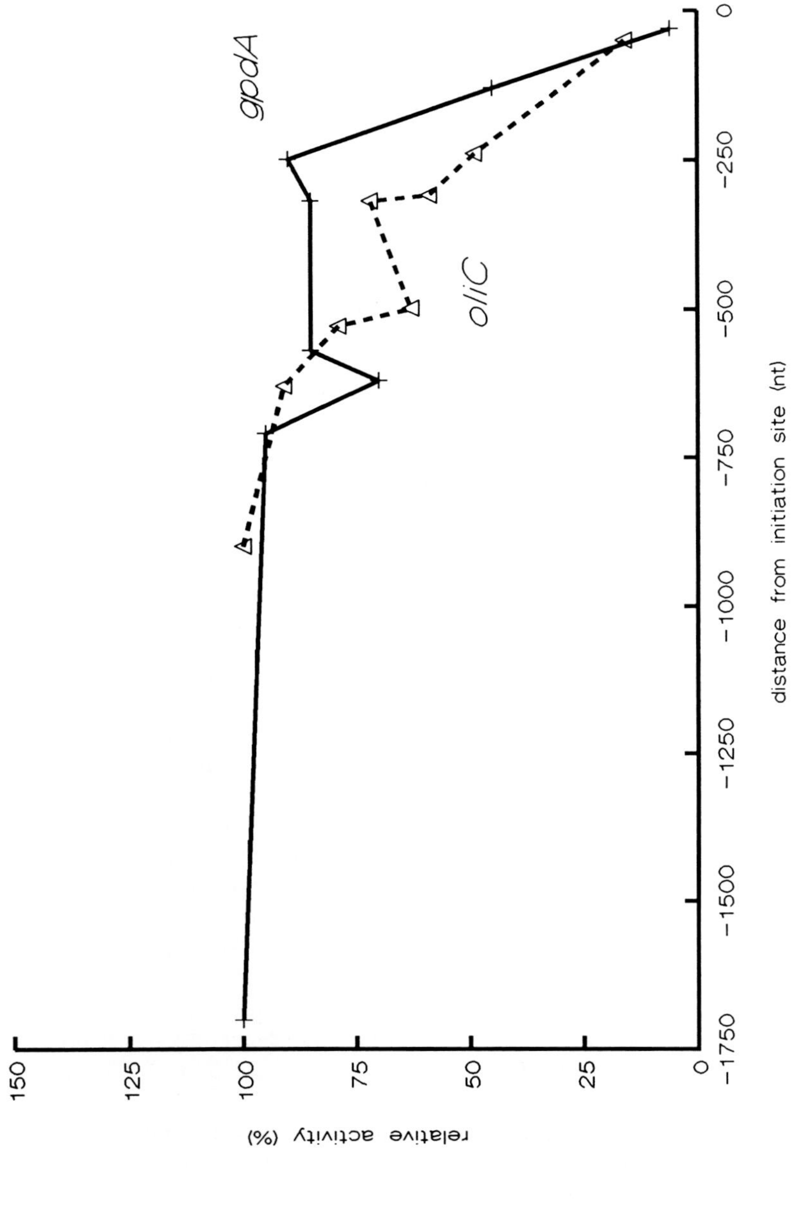


FIGURE 5. Relative β-galactosidase activity of *A. nidulans* transformants containing *gpdA* or *oliC* upstream sequences of various length fused to the *E. coli lacZ* gene. The distance of the 5′ endpoint of the various upstream sequences from the transcription initiation site is given (in nt).

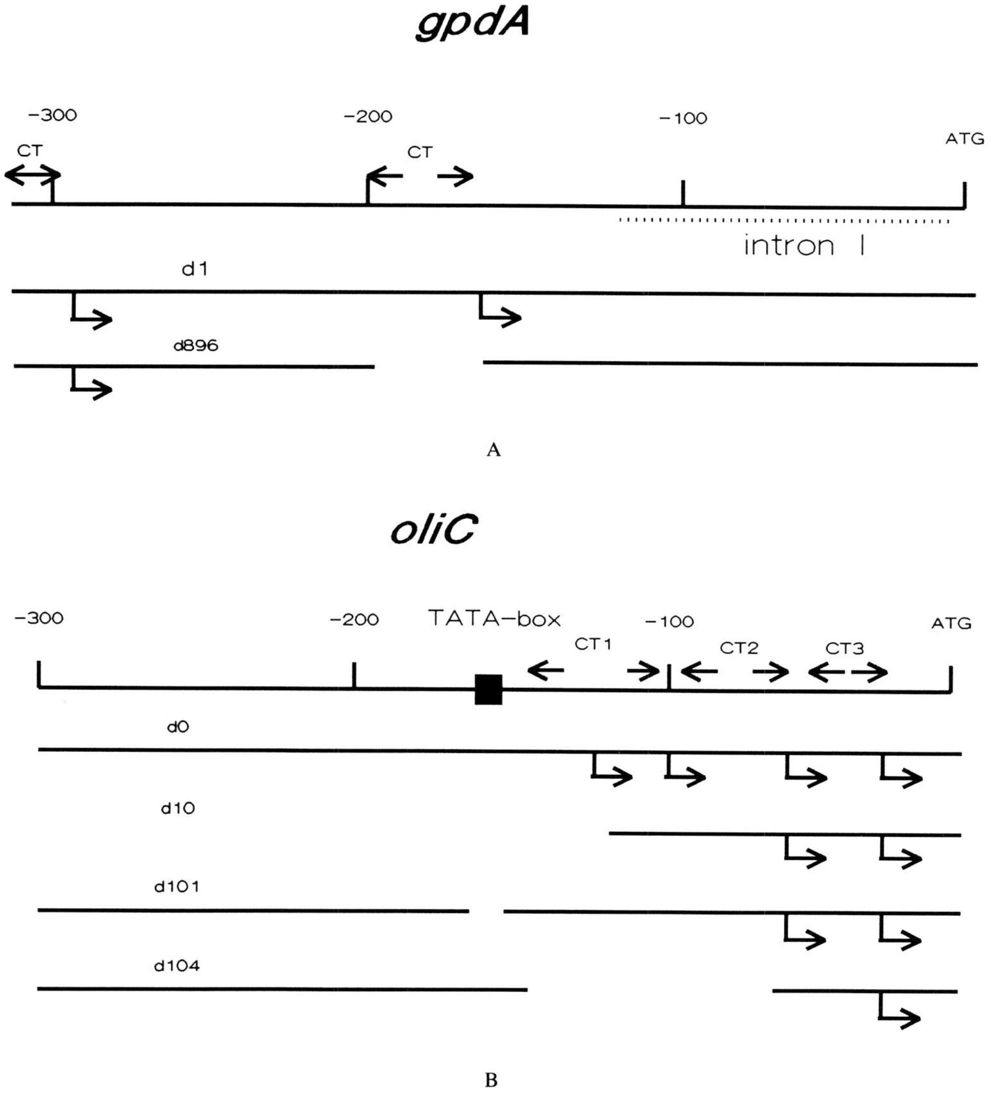


FIGURE 6. Transcription initiation analysis of total RNA from *A. nidulans* transformants containing mutant *gpdA* (A) or *oliC* (B) upstream sequences fused to the *E. coli lacZ* gene. Transcription initiation sites, indicated with ↳ , were determined by primer extension analysis. CT, pyrimidine-rich region (*ct* box). The distance from the ATG codon is given (in nt). In (A) the position of an intron in the untranslated region of the *gpdA* transcript is indicated.

transcription control. The regions involved in regulation of expression of the *A. nidulans argB* gene (encoding ornithine carbamoyl transferase)[74] and the nitrogen metabolite repression of expression of the *Penicillium chrysogenum pcbC* gene (encoding isopenicillin N synthetase)[83] were localized by this type of analysis.

An approach to localize such upstream elements more precisely is the analysis of mutant upstream sequences obtained by deletion of small specific sequences. Putative control sequences, indicated by either sequence comparison or titration analysis, can be precisely deleted from the 5′-flanking region of the gene of interest with the aid of *in vitro* mutagenesis protocols, including PCR methods.[87,88] Using this approach, functional elements in the flanking region of the *gpdA* and *oliC* gene of *A. nidulans* have been identified.[19,89] A mutant *A. nidulans gpdA* flanking region, which missed a conserved sequence present in the upstream region of the wt *gpdA* gene of *A. nidulans* and *A. niger* (*gpd* box, Figure 2), and which was fused to the *E. coli lacZ* gene, resulted in lower expression levels (50%) compared to the levels obtained with the wt flanking region. This result indicates that this box contains (part of) a functional element.[19] Also, in *A. niger,* a lower expression level (30%) was found with the *A. nidulans* mutant flanking region, supporting the idea that the *gpd* box contains a functional element.[90]

To determine whether an element identified by deletion analysis contains all sequences essential for transcription activation/regulation, it is necessary to demonstrate that this element per se is functional. This can be done by introduction of the element into a 5′-flanking region of another gene and subsequent analysis of the effects on gene expression. The activity of a putative repressing element of the *abaA* gene was verified by this approach after introduction of this element into the transcription control region of the *trpC* gene. A significant decrease of the expression of the *lacZ* gene regulated by the *trpC::abaA* control sequences was observed, compared to the expression of the *lacZ* gene driven by the *trpC* control sequences.[82] Similarly, introduction of the *gpd* box into the *A. nidulans amdS* flanking region fused to *lacZ* resulted in increased levels of expression (up to 30-fold).[91] In another study, sequence elements, indicated by titration analysis, which are present in the upstream regions of the *amdS* and *gatA* gene, were introduced into the 5′-flanking region of the *amdS* gene from which most transcription control sequences were deleted. Expression from the mutant transcription control region, when fused to the *E. coli lacZ* gene, was considerably decreased compared to the expression from wt *amdS* upstream region. Introduction of the *amdS* and *gatA* sequence elements (partially restored expression of the fusion gene.[48,80]

As already mentioned, in all cases described, the amount of gene product is used as a measure for transcription efficiency. It is important to note that only when transcripts derived from different (mutant) upstream regions from a particular gene are identical the amount of gene product faithfully represents the transcription efficiency. In several cases, transcription initiation analysis was carried out to analyze the transcripts derived from different mutant upstream regions. In general, deletion of a sequence from an upstream region, which resulted in a change in transcription efficiency, did not result in any changes in transcription initiation,[19,76,79,82] If both transcription efficiency and transcription initiation are changed as a consequence of a specific deletion, direct analysis of the efficiency of transcription initiation, without interference of differences in stability and translation efficiency of the different transcripts, has to be carried out to be able to draw reliable conclusions. To our knowledge this type of experiment has not yet been carried out for filamentous fungi.

The approaches for identification and characterization of transcription control sequences, described thus far, are mostly based on *in vitro* generated mutations and subsequent *in vivo* analysis of the effects of these mutations. However, from genetic research on the regulation of metabolic pathways, several regulation mutants were obtained.[3] Several of the corresponding mutations map in the 5′-flanking region of the relevant genes. From these results, it was concluded that altered regulation of expression of these genes was due to mutations in their transcription control sequences. Comparison of the upstream sequences of the *qa-2* gene of *N. crassa,*[92] the *amdS*[3,72,93,94] and *uapA* (encoding uric acid-xanthine permease)[3,95] genes of *A. nidulans*, isolated from wild-type and mutant strains, respectively, confirmed this conclusion. At the same time, this comparison also revealed the putative site of action of the different regulatory proteins involved in expression of these genes.[42,93,95]

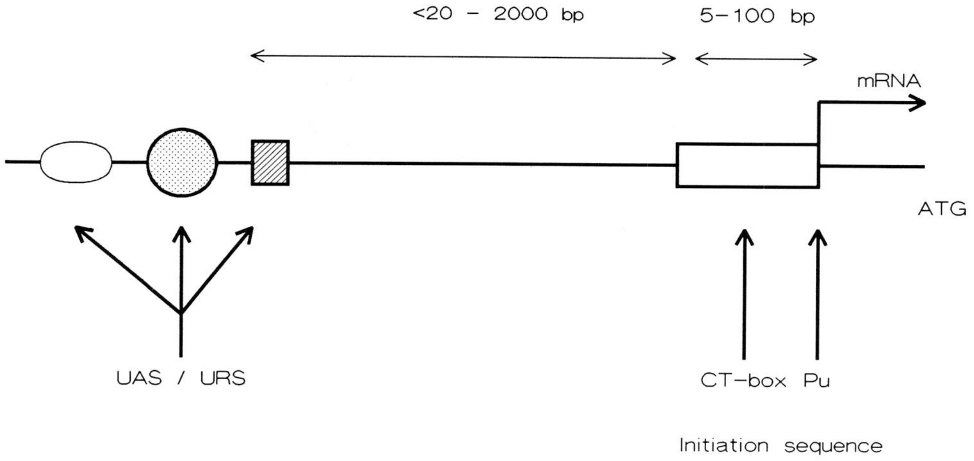


FIGURE 7. Scheme of fungal transcription control sequences. In most cases transcription initiation occurs at the first purine (Pu) base downstream of pyrimidine-rich regions.

## III. CONCLUSIONS AND FUTURE PROSPECTS

In this chapter we have described a number of approaches for the analysis of transcription control sequences. From the results reviewed, it is clear that, to date, only a limited amount of data concerning fungal transcription control sequences is available.

Until now, research has primarily been focused on the organization of these sequences. A few interesting points concerning the organization of fungal transcription control sequences emerge from these data. First, the region involved in transcription control can be large, extending more than 500 nt upstream of the major start site, as indicated by deletion analysis.[19,73,76,83,86] Protein-DNA-binding analysis also indicated putative transcription control sequences at far upstream positions.[31,33,41,60] However, at least some regulatory sequences, identified by *in vitro* or *in vivo* analysis, are localized much closer to the transcription initiation site.[12,15,17,41,42,72,74,76,82] Second, pyrimidine-rich sequences are clearly involved in determining transcription initiation sites.[19,72,76,82] Authentic transcription initiation was observed from transcription control sequences consisting of only a pyrimidine-rich sequence.[19,82] The involvement of a TATA-like element in this process is indicated in only one instance,[79] whereas in a few other cases no involvement was observed.[19,82] A scheme of the 5′-flanking sequences of a fungal gene is depicted in Figure 7. Clearly, the overall structure resembles the generalized structure of eukaryotic (including yeast) transcription control regions.[7,8]

Obviously, further research on transcription regulation will not only be aimed at the elucidation of the organization of transcription control sequences, but particularly at a better understanding of the mechanisms of gene expression. Of interest in this respect, are complex regulatory circuits as, for example, those governing carbon and nitrogen metabolism, and genetic regulation of highly complex phenomena, such as, differentiation and host-pathogen interactions by pathogenic fungi.

We feel that only an integrated approach consisting of both *in vitro* and *in vivo* analysis, as described in this chapter, may lead to a further understanding of the mechanisms of gene expression. An example that clearly illustrates the necessity to combine *in vitro* and *in vivo* approaches is taken from the work of Hynes and co-workers. Sequence comparison of the 5′-flanking sequences of the *gatA* gene[80] suggested as many as four AmdR target sites. However, only one of these sequences gave positive results in titration analysis.[96] On the other hand, functional transcription control sequences may also be missed by sequence comparison, as demonstrated for FacB target sites in the *amdS* upstream sequences. Titration analysis identified a FacB site in the upstream region of the *amdS* gene not identified by sequence comparison.[96]

During the last few years molecular biological research in filamentous fungi has extended into two very interesting fields, namely, biotechnology and plant pathology.

Filamentous fungi have been used for several decades in the fermentation industry for the production of proteins and primary and secondary metabolites. Based on the knowledge accumulated from molecular biological research on gene expression, the potential of fungi, especially black *Aspergillus* species and *A. oryzae,* to produce heterologous (mammalian) proteins is being investigated in several laboratories around the world.[97] Some of the results obtained from the analysis of transcription control sequences, as described in this chapter, have already been used for optimization of the production of heterologous proteins.[65]

In the area of plant pathology, molecular biological research will focus on unravelling the very complex fungus-host interactions. The first steps in this direction are being made.[61,98] We expect that in this area the approaches described in this chapter will be very useful.

## ACKNOWLEDGMENTS

We wish to acknowledge all fungal scientists who communicated their unpublished data to us. We especially wish to thank Jan Verdoes, Cas Kramer, Jim Kinghorn, and Geoff Turner for their contribution to this paper. Furthermore, Peter Pouwels is acknowledged for critical reading of the manuscript and stimulating discussions.

## REFERENCES

1. **Boylan, M. T., Mirabito, P. M., Willett, C. E., Zimmerman, C. R., and Timberlake, W. E.,** Isolation and physical characterization of three essential conidiation genes from *Aspergillus nidulans, Mol. Cell. Biol.,* 7, 3113, 1987.
2. **Cove, D. J.,** Genetic studies of nitrate assimilation in *Aspergillus nidulans, Biol. Rev.,* 54, 291, 1979.
3. **Arst, H. N. and Scazzochio, C.,** Formal genetics and molecular biology of the control of gene expression in *Aspergillus nidulans,* in *Gene Manipulations in Fungi,* Bennett, J. W. and Lasure, L. L., Eds., Academic Press, NY, 1985, 309.
4. **Marzluf, G. A. and Fu, Y.-H.,** Molecular analyses of the nitrogen and the sulfur regulatory circuits of *Neurospora crassa,* in *Genetics and Molecular Biology of Industrial Microorganisms,* Hershberger, C. L., Queener, S. W., and Hegeman, G., Eds., American Society for Microbiology, Washington, DC, 1989, 279.
5. **Case, M. E., Schweizer, M., Kushner, S. R., and Giles, N. H.,** Efficient transformation of *Neurospora crassa* by utilizing hybrid plasma DNA, *Proc. Natl. Acad. Sci. U.S.A.,* 76, 5259, 1979.
6. **Van den Hondel, C. A. M. J. J. and Punt, P. J.,** Gene-transfer systems and vector development for filamentous fungi, in BSPI Symposium, Vol. 18, *Applied Molecular Genetics of Fungi,* Peberdy, J. F., Ed., British Mycological Society, Cambridge University Press, U.K., in press.
7. **Struhl, K.,** Yeast promoters, in *Maximizing Gene Expression,* Biotechnol. Ser. 9, Reznikoff, W. and Gold, L., Eds., Open University Press, Milton Keynes, 1986, 35.
8. **Wasylik, B.,** Protein coding genes of higher eukaryotes: promoter elements and *trans*-acting factors, in *Maximizing Gene Expression,* Biotechnol. Ser. 9, Reznikoff, W. and Gold, L., Eds., Open University Press, Milton Keynes, 1986, 79.
9. **Ballance, D. J.,** Sequences important for gene expression in filamentous fungi, *Yeast,* 2, 229, 1986.
10. **Gurr, S. J., Unkles, S. E., and Kinghorn, J. R.,** The structure and organisation of nuclear genes of filamentous fungi, in *Gene Structure in Eukaryotic Microbes,* SGM Spec. Publ., Vol. 23, Kinghorn, J. R., Ed., IRL Press, Oxford, 1988, 93.
11. **Rambosek, J. and Leach, J.,** Recombinant DNA in filamentous fungi: Progress and prospects, *CRC Crit. Rev. Biotechnol.,* 6, 357, 1987.

12. **Gwynne, D. I., Buxton, F. P., Sibley, S., Davies, R. W., Lockington, R. A., Scazzochio, C., and Sealy-Lewis, H. M.,** Comparison of the *cis*-acting control regions of two coordinately controlled genes involved in ethanol utilization in *Aspergillus nidulans, Gene,* 51, 205, 1987.
13. **Lockington, R. A., Sealy-Lewis, H. M., Scazzochio, C., and Davies, R. W.,** Cloning and characterization of the ethanol utilization regulon in *Aspergillus nidulans, Gene,* 33, 137, 1985.
14. **Alton, N. K., Buxton, F., Patel, V., Giles, N. H., and Vapnek, D.,** 5′-Untranslated sequences of two structural genes in the *qa* gene cluster of *Neurospora crassa, Proc. Natl. Acad. Sci. U.S.A.,* 79, 1955, 1982.
15. **Hawkins, A. R., Lamb, H. K., Smith, M., Keyte, J. W., and Roberts, C. F.,** Molecular organisation of the quinic acid utilization *(QUT)* gene cluster of *Aspergillus nidulans, Mol. Gen. Genet.,* 214, 224, 1988.
16. **Baum, J. A. and Giles, N. R.,** DNAse I hypersensitive sites within the inducible (qa gene cluster of *Neurospora crassa, Proc. Natl. Acad. Sci. U.S.A.,* 83, 6533, 1986.
17. **Kinghorn, J. R.,** personal communications, 1990.
18. **Punt, P. J., Dingemanse, M. A., Jacobs-Meijsing, B. J. M., Pouwels, P. H., and Van den Hondel, C. A. M. J. J.,** Isolation and characterization of the glyceraldehyde-3-phosphate dehydrogenase gene of *Aspergillus nidulans, Gene,* 69, 49, 1988.
19. **Punt, P. J., Dingemanse, M. A., Kuyvenhoven, A., Soede, R. D. M., Pouwels, P. H., and Van den Hondel, C. A. M. J. J.,** Functional elements in the promoter region of the *Aspergillus nidulans gpdA* gene, encoding glyceraldehyde-3-phosphate dehydrogenase, *Gene,* 93, 101, 1990.
20. **Ward, M., Wilson, L. J., Carmona, C. L., and Turner, G.,** The *oliC3* gene of *Aspergillus nidulans;* isolation, sequence and use as a selectable marker for transformation, *Curr. Genet.,* 14, 37, 1988.
21. **Guarente, L.,** Regulatory proteins in yeast, *Annu. Rev. Genet.,* 21, 425, 1987.
22. **Johnson, P. F. and McKnight, S. L.,** Eukaryotic transcriptional regulatory proteins, *Annu. Rev. Biochem.,* 58, 799, 1989.
23. **Struhl, K.,** Molecular mechanisms of transcriptional regulation in yeast, *Annu. Rev. Biochem.,* 58, 1051, 1989.
24. **Polyanovsky, O. L. and Stepchenko, A. G.,** Eukaryotic transcription factors, *BioEssays,* 12, 205, 1990.
25. **Davis, M. A. and Hynes, M. J.,** Regulatory genes in *Aspergillus nidulans, Trends Genet.,* 5, 14, 1989.
26. **Paluh, J. L., Orbach, M. J., Legerton, T. L., and Yanofsky, C.,** The cross-pathway control gene of *Neurospora crassa, cpc-1* encodes a protein similar to GCN4 of yeast and the DNA-binding domain of the oncogene *v-jun*-encoded protein, *Proc. Natl. Acad. Sci. U.S.A.,* 85, 3728, 1988.
27. **Paluh, J. L., Plamann, M., Krüger, D., Barthelmess, I. B., Yanofsky, C., and Perkins, D. D.,** Determination of the inactivating alterations in two mutant alleles of the *Neurospora crassa* cross-pathway control gene *cpc-1, Genetics,* 124, 599, 1990.
28. **Ebbole, D. J., Paluh, J. L., Plamann, M., Sachs, M. S., and Yanofsky, C.,** *Cpc-1,* the general regulatory gene for genes of amino acid biosynthesis in *Neurospora crassa,* is differentially expressed during the asexual life cycle, *Mol. Cell Biol.,* 11, 928, 1991.
29. **Paluh, J. L. and Yanofsky, C.,** Characterization of *Neurospora cpc-1,* a *b*ZIP DNA binding protein that does not require aligned heptad leucines for dimerization, *Mol. Cell Biol.,* 11, 935, 1991.
30. **Fu, Y.-H., Paietta, J. V., Mannix, D. G., and Marzluf, G. A.,** *Cys-3,* the positive-acting sulfur regulatory gene of *Neurospora crassa,* encodes a protein with a putative leucine zipper DNA-binding element, *Mol. Cell. Biol.,* 9, 1120, 1989.
31. **Fu, Y.-H. and Marzluf, G. A.,** *Cys-3,* the positive-acting sulfur regulatory gene of *Neurospora crassa,* encodes a sequence-specific DNA-binding protein, *J. Biol. Chem.,* 265, 11942, 1990.
32. **Fu, Y.-H. and Marzluf, G. A.,** *Nit-2,* the major nitrogen regulatory gene of *Neurospora crassa,* encodes a protein with a putative zinc finger DNA-binding domain, *Mol. Cell. Biol.,* 10, 1056, 1990.
33. **Fu, Y.-H. and Marzluf, G. A.,** *Nit-2,* the major positive-acting nitrogen regulatory gene of *Neurospora crassa,* encodes a sequence-specific DNA-binding protein, *Proc. Natl. Acad. Sci. U.S.A.,* 87, 5331, 1990.
34. **Fu, Y.-H. and Marzluf, G. A.,** Site-directed mutagenesis of the "zinc-finger" DNA-binding domain of the nitrogen-regulatory protein NIT2 of *Neurospora, Mol. Microbiol.,* 4, 1847, 1990.
35. **Fu, Y.-H., Kneesi, J. Y., and Marzluf, G. A.,** Isolation of *nit-4,* the minor nitrogen regulatory gene which mediated nitrate induction in *Neurospora crassa, J. Bacteriol.,* 171, 4067, 1989.
36. **Yuan, G. F. and Marzluf, G. A.,** personal communication, 1990.
37. **Fu, Y.-H., Young, J. L., and Marzluf, G. A.,** Molecular cloning and characterization of a negative-acting nitrogen regulatory gene of *Neurospora crassa, Mol. Gen. Genet.,* 214, 74, 1988.
38. **Sorger, G. J., Brown, D., Farzannejad, M., Guerra, A., Jonathan, M., Knight, S., and Sharda, R.,** Isolation of a gene that down-regulates nitrate assimilation and influences another regulatory gene in the same system, *Mol. Cell. Biol.,* 9, 4113, 1989.
39. **Young, J. L., Jarai, G., Fu, Y.-H., and Marzluf, G. A.,** Nucleotide sequence and analysis of *NMR,* a negative-acting regulatory gene in the nitrogen circuit of *Neurospora crassa, Mol. Gen. Genet.,* 222, 120, 1990.

40. **Kang, S. and Metzenberg, R. L.,** Molecular analysis of *nuc-1*$^{+}$, a gene controlling phosphorus acquisition in Neurospora crassa, *Mol. Cell. Biol.*, 10, 5839, 1990.
41. **Baum, J. A., Geever, R., and Giles, N. H.,** Expression of qa-1F activator protein: identification of upstream binding sites in the *qa* gene cluster and localization of the DNA-binding domain, *Mol. Cell. Biol.*, 7, 1256, 1987.
42. **Geever, R. F., Huiet, L., Baum, J. A., Tyler, B. M., Patel, V. B., Rutledge, B. J., Case, M. E., and Giles, N. H.,** DNA sequence, organization and regulation of the *qa* gene cluster of *Neurospora crassa, J. Mol. Biol.*, 207, 15, 1989.
43. **Mirabito, P. M., Adams, T. H., and Timberlake, W. E.,** Interactions of three sequentially expressed genes control temporal and spatial specificity in *Aspergillus* development, *Cell*, 57, 859, 1989.
44. **Felenbok, B., Sequeval, D., Mathieu, M., Sibley, S., Gwynne, D. I., and Davies, R. W.,** The ethanol regulon in *Aspergillus nidulans:* characterization and sequence of the positive regulatory gene *alcR, Gene*, 73, 385, 1988.
45. **Scazzochio, C. and Felenbok, B.,** personal communication, 1990.
46. **Andrianopoulos, A. and Hynes, M. J.,** Cloning and analysis of the positively acting regulatory gene *amdR* from *Aspergillus nidulans, Mol. Cell. Biol.*, 8, 3532, 1988.
47. **Andrianopoulos, A. and Hynes, M. J.,** Sequence and functional analysis of the positively acting regulatory gene *amdR* from *Aspergillus nidulans, Mol. Cell. Biol.*, 10, 3194, 1990.
48. **Hynes, M. J., Andrianopoulos, A., Davis, M. A., Van Heeswijck, R., Katz, M. E., Littlejohn, T. G., Richardson, I. B., and Saleeba, J. A.,** Multiple circuits regulating the *amdS* gene of *Aspergillus nidulans,* in *Proceedings of the EMBO-Alko Workshop on Molecular Biology of Filamentous Fungi,* Nevalainen, H. and Penttilä, M., Eds., Foundation for Biotechnical and Industrial Fermentation Research, Helsinki, 1989, 63.
49. **Kudla, B., Caddick, M. X., Langdon, T., Martinez-Rossi, N. M., Bennett, C. F., Sibley, S., Davies, R. W., and Arst, H. N.,** The regulatory gene *areA* mediating nitrogen metabolite repression in *Aspergillus nidulans*. Mutations affecting specificity of gene activation alter a loop residue of a putative zinc finger, *EMBO J.*, 9, 1355, 1990.
50. **Scazzochio, C.,** Of moulds and men, or two fingers are not better than one, *Trends Genet.*, 6, 311, 1990.
51. **Caddick, M. X. and Arst, H. N.,** Nitrogen regulation in *Aspergillus:* are two fingers better than one?, *Gene,* 95, 123, 1990.
52. **Adams, T. H., Boylan, M. T., and Timberlake, W. E.,** *BrlA* is necessary and sufficient to direct conidiophore development in *Aspergillus nidulans, Cell,* 54, 353, 1988.
53. **Adams, T. H., Deising, H., and Timberlake, W. E.,** *BrlA* requires both zinc fingers to induce development, *Mol. Cell. Biol.*, 10, 1815, 1990.
54. **Dowzer, C. A. E. and Kelly, J. M.,** Cloning of the *creA* gene from *Aspergillus nidulans:* a gene involved in carbon catabolite repression, *Curr. Genet.*, 15, 457, 1989.
55. **Felenbok, B., Sophianopoulou, V., Mathieu, M., Sequeval, D., Kulmberg, P., Diallinas, G., and Scazzochio, C.,** Regulation of genes involved in the utilization of carbon sources in *Aspergillus nidulans,* in *Proceedings of the EMBO-Alko Workshop on Molecular Biology of Filamentous Fungi,* Nevalainen, H. and Penttilä, M., Eds., Foundation for Biotechnical and Industrial Fermentation Research, Helsinki, 1989, 73.
56. **Katz, M. E. and Hynes, M. J.,** Isolation and analysis of the acetate regulatory gene, *facB,* from *Aspergillus nidulans, Mol. Cell. Biol.*, 9, 5696, 1989.
57. **Hull, E. P., Green, P. M., Arst, H. N., and Scazzocchio, C.,** Cloning and physical characterization of the L-proline catabolism gene cluster of *Aspergillus nidulans, Mol. Microbiol.*, 3, 553, 1989.
58. **Beri, R. K., Whittington, H., Roberts, C. F., and Hawkins, A. R.,** Isolation and characterization of the positively acting regulatory gene *qutA* from *Aspergillus nidulans, Nucleic Acids Res.*, 15, 7991, 1987.
59. **Lamb, H. K., Hawkins, A. R., Smith, M., Harvey, I. J., Brown, J., Turner, G., and Roberts, C. F.,** Spatial and biological characterization of the complete quinic acid utilisation gene cluster in *Aspergillus nidulans, Mol. Gen. Genet.*, 223, 17, 1990.
60. **Frederick, G. D. and Kinsey, J. A.,** Nucleotide sequence and nuclear protein binding of the two regulatory sequences upstream of the *am* (GDH) gene in *Neurospora, Mol. Gen. Genet.*, 221, 148, 1990.
61. **Podila, G. K., Dickman, M. B., Rogers, L. M., and Kolattuduky, P. E.,** Regulation of expression of fungal genes by plant signals, in *Proceedings of the EMBO-Alko Workshop on Molecular Biology of Filamentous Fungi,* Nevalainen, H. and Penttilä, M., Eds., Foundation for Biotechnical and Industrial Fermentation Research, Helsinki, 1989, 129.
62. **Unkles, S. E., Hawker, K. L., Grieve, C., Campbell, E. I., and Kinghorn, J. R.,** *CrnA* encodes a eukaryotic nitrate transporter, *Proc. Natl. Acad. Sci., U.S.A.*, 88, 204, 1991.
63. **Da Silva, A. J. F., Whittington, H., Clements, J., Roberts, C., and Hawkins, A. R.,** Sequence analysis and transformation by the catabolic 3-dehydroquinase *(qutE)* gene from *Aspergillus nidulans, Biochem. J.*, 240, 481, 1986.

64. **Beri, R. K., Grant, S., Roberts, C. F., Smith, M., and Hawkins, A. R.,** Selective overexpression of the *qutE* gene encoding catabolic 3-dehydroquinase in multicopy transformants of *Aspergillus nidulans, Biochem. J.,* 265, 337, 1990.
65. **Gwynne, D. I., Buxton, F. P., Gleeson, M. A., and Davies, R. W.,** (1987c). Genetically engineered secretion of foreign proteins from *Aspergillus species* in *Protein Purification: Micro to Macro.* Burgess, R., Ed., Alan R. Liss, NY, 355-365.
66. **Hynes, M. J. and Davis, M. A.,** The *amdS* gene of *Aspergillus nidulans:* control by multiple regulatory signals, *BioEssays,* 5, 123, 1986.
67. **Kelly, J. M. and Hynes, M. J.,** Multiple copies of the *amdS* gene of *Aspergillus nidulans* cause titration of *trans*-acting regulatory proteins, *Curr. Genet.,* 12, 21, 1987.
68. **Hynes, M. J., Corrick, C. M., Kelly, J. M., and Littlejohn, T. G.,** Identification of the sites of action for regulatory genes controlling the *amdS* gene of *Aspergillus nidulans, Mol. Cell. Biol.,* 8, 2589, 1988.
69. **Hynes, M. J. and Andrianopoulos, A.,** Transformation studies of gene regulation in filamentous fungi, in *Genetics and Molecular Biology of Industrial Microorganisms,* Hershberger, C. L., Queener, S. W., and Hegeman, G., Eds., American Society for Microbiology, Washington, DC, 1989, 304.
70. **Kelly, J. M. and Hynes, M. J.,** Transformation of *Aspergillus niger* by the *amdS* gene of *Aspergillus nidulans, EMBO J.,* 4, 475, 1985.
71. **Verdoes, J. and Van den Hondel, C. A. M. J. J.,** unpublished results, 1990.
72. **Verdoes, J.,** personal communication, 1990.
73. **Frederick, G. D. and Kinsey, J. A.,** Distant upstream regulatory sequences control the level of expression of the *am* (GDH) locus of *Neurospora crassa, Curr. Genet.,* 18, 53, 1990.
74. **Goc, A. and Weglenski, P.,** Regulatory region of the *Aspergillus nidulans argB* gene, *Curr. Genet.,* 14, 425, 1988.
75. **Soliday, C. L., Dickman, M. B., and Kolattukudy, P. E.,** Structure of the cutinase gene and detection of promoter activity in the 5′-flanking region by fungal transformation, *J. Bacteriol.,* 171, 1942, 1989.
76. **Hamer, J. E. and Timberlake, W. E.,** Functional organization of the *Aspergillus nidulans trpC* promoter, *Mol. Cell. Biol.,* 7, 2352, 1987.
77. **Van Gorcom, R. F. M., Punt, P. J., Pouwels, P. H., and Van den Hondel, C. A. M. J. J.,** A system for the analysis of expression signals in *Aspergillus, Gene,* 48, 211, 1986.
78. **Davis, M. A., Cobbett, C. S., and Hynes, M. J.,** An *amdS-lacZ* fusion for studying gene regulation in *Aspergillus, Gene,* 63, 199, 1988.
79. **Turner, G., Brown, J., Kerry-Williams, S., Bailey, A. M., Ward, M., Punt, P. J., and Van den Hondel, C. A. M. J. J.,** Analysis of the *oliC* promoter of *Aspergillus nidulans,* in *Proceedings of the EMBO-Alko Workshop on Molecular Biology of Filamentous Fungi,* Nevalainen, H. and Penttilä, M., Eds., Foundation for Biotechnical and Industrial Fermentation Research, Helskini, 1989, 101.
80. **Richardson, I. B., Hurley, S. K., and Hynes, M. J.,** Cloning and molecular characterisation of the *amdR* controlled *gatA* gene of *Aspergillus nidulans, Mol. Gen. Genet.,* 217, 118, 1989.
81. **Gomez-Pardo, E. and Penalva, M. A.,** The upstream region of the *IPNS* gene determines expression during secondary metabolism in *Aspergillus nidulans, Gene,* 89, 109, 1990.
82. **Adams, T. H. and Timberlake, W. E.,** Upstream elements repress premature expression of an *Aspergillus* developmental regulatory gene, *Mol. Cell. Biol.,* 10, 4912, 1990.
83. **Kolar, M., Holzmann, K., Weber, G., Leitner, E., and Schwab, H.,** Molecular characterization and functional analysis in *Aspergillus nidulans* of the 5′-region of the *Penicillium chrysogenum* isopenicillin N synthetase gene, *J. Biotechnol.,* 17, 67, 1991.
84. **Greaves, P. A. and Kinghorn, J. R.,** personal communication, 1990.
85. **Sachs, M. S. and Ebbole, D.,** The use of *lacZ* gene fusions in *Neurospora crassa, Fungal Genet. Newsl.,* 37, 39, 1990.
86. **Punt, P. J.,** unpublished results, 1990.
87. **Kunkel, T. A.,** Rapid and efficient site-specific mutagenesis without phenotypic selection, *Proc. Natl. Acad. Sci. U.S.A.,* 82, 488, 1985.
88. **Innes, M. A., Gelfand, D. H., Sninsky, J. J., and White, T. J., Eds.,** *PCR Protocols. A Guide to Methods and Applications,* Academic Press, San Diego, 1990.
89. **Turner, G.,** personal communication, 1990.
90. **Van Gorcom, R. F. M.,** personal communication, 1990.
91. **Punt, P. J. and Kramer, I.,** unpublished results, 1990.
92. **Tyler, B. M., Geever, R. F., Case, M. E., and Giles, N. H.,** *Cis*-acting and *trans*-acting regulatory mutations define two types of promoters controlled by the *qa-1F* gene of *Neurospora, Cell,* 36, 493, 1984.
93. **Corrick, C. M., Twomey, A. P., and Hynes, M. J.,** The nucleotide sequence of the *amdS* gene of *Aspergillus nidulans* and the molecular characterization of 5′ mutations, *Gene,* 53, 63, 1987.
94. **Katz, M. E., Saleeba, J. A., Sapats, S. I., and Hynes, M. J.,** A mutation affecting *amdS* expression in *Aspergillus nidulans* contains a triplication of *cis*-acting regulatory sequence, *Mol. Gen. Genet.,* 220, 373, 1990.

95. **Diallinas, G. and Scazzochio, C.,** A gene coding for uric acid-xanthine permease of *Aspergillus nidulans:* inactivational cloning, characterization, and sequence of a *cis*-acting mutation, *Genetics,* 122, 341, 1989.
96. **Hynes, M. J.,** personal communication, 1990.
97. **Van den Hondel, C. A. M. J. J., Punt, P. J., and Van Gorcom, R. F. M.,** Heterologous gene expression in filamentous fungi, in *More Gene Manipulation in Fungi,* Bennett, J. W. and Lasure, L. L., Eds., Academic Press, San Diego, 1991, 396.
98. **De Wit, P. J. G. M. and Oliver, R. P.,** The interaction between *Cladosporium fulvum* (syn. *Fulvia fulva*) and tomato: A model system in molecular plant pathology, in *Proceedings of the EMBO-Alko Workshop on Molecular Biology of Filamentous Fungi,* Nevalainen, H. and Penttilä, M., Eds., Foundation for Biotechnical and Industrial Fermentation Research, Helsinki, 1989, 227.

# GENE EXPRESSION IN SUSCEPTIBILITY AND RESISTANCE OF FUNGAL PLANT DISEASES

**Hachiro Oku**

## TABLE OF CONTENTS

# I. INTRODUCTION

As stated by Kuć[1] "preventing a general resistance or immune response apparently is a rare occurrence in nature, and this may account for the fact that susceptibility is the exception rather than the rule." This seems to be common to the mechanism of susceptibility and resistance in all plant diseases.

The defense reactions of plants against microbial attack is thought to be a rejection reaction such that living beings do not accept cells or tissues from other species or even from other individuals of the same species. This type of rejection reaction might be very important not only to maintain the purity of the species but also to avoid invasion by other deleterious organisms.

Resistance mechanisms in plants are classified into two types. One is static resistance or so-called preformed resistance that is dependent upon the characteristics of normal, uninfected plants, such as, hardness and thickness of cuticle,[2] antimicrobial compounds contained in varieties of plant species,[3] inhibitors of cell wall-degrading enzymes, which are produced by parasitic fungi,[4] and so on. The other is dynamic resistance or the so-called active defense reaction, which is expressed after microbial invasion, such as papillae formation,[5-7] phytoalexin production, production of infection inhibitor,[8,9] lignification of parenchyma walls,[10] and accumulation of hydroxyproline-rich glycoprotein (HRGP).[11,12] In order to establish infection on plants, pathogenic fungi must overcome all of the resistance mechanisms.

The static resistance factors were considered to be governed by horizontal resistance genes that are already operating before attack by pathogens. The dynamic defense reactions result from the activation of vertical resistance genes stimulated by microbial attack, and which are not expressed in healthy plant cells or tissues.

This chapter summarizes, from a biological, biochemical, and genetic prospective, up-to-date concepts of defense mechanisms used by higher plants when attacked by microorganisms, and how pathogenic fungi (compatible pathogen) overcome these defense mechanisms in order to establish infection on their host plants.

# II. DYNAMIC RESISTANCE OF PLANTS AGAINST MICROORGANISMS

Plants defend themselves from microbial attack by means of a series of events. Expression of the defense reaction by host cells is triggered by some metabolites of invading microorganisms or abnormal metabolites that result from host-parasite interactions.

The earliest morphological response might be the formation of papilla inside the epidermal wall of the attacked cell. The role of papilla as the barrier of infection has been the matter of much discussion.[13-18] However, conclusions cannot be drawn as to the importance of this structure, since there is no guarantee that some antimicrobial compounds might not be deposited in the papilla. Furthermore, there is evidence that fluorescent antibiotic substances accumulate in this structure.[16-18]

The other morphological change in resistant host cells is the elicitation of the hypersensitive reaction in both invaded and neighboring cells. The first record of hypersensitive cell death may have been by Ward (1902)[19] who noted that brome grass immune to brown rust was characterized by a rapid necrosis of the invaded and few surrounding cells. Since then, many plant pathologists have focused their attention on the hypersensitive reaction and on obligate and nonobligate parasitic diseases, since the key to solving the mechanism of resistance might be harbored in these phenomena. Notwithstanding these efforts, the exact mechanism of the role of hypersensitive reaction in resistance has not yet been fully elucidated.

In the chemical defense reaction, phytoalexin biosynthesis has been proved in many host-parasite combinations and a number of phytoalexins have been characterized, including isoflavonoids, sesquiterpenoids, polyacetylenes, and stilbenoids. The substances which induce the phytoalexin biosynthesis were initially termed elicitors,[20] but the term is now more widely used to include, for example, substances which induce other defense reactions. Many exogenous (fungal origin, etc.) elicitors were isolated from spore germination fluid,[8,21] fungal culture filtrates, cell wall preparations of pathogenic fungi, and chemically characterized peptides, hepta-β-glucoside, chitosan, glucan, polysaccharides containing glucose, mannose, and arabinose, eicosapentanoic acid, and arachidonic acid. Endogenous elicitors which are produced as the result of host response to pathogens have also been reported.

Phytoalexins and their elicitors were reviewed by Darvill and Albersheim.[22] Elicitors of biological origin are sometimes called biotic elicitors. Many metal ions such as $Cu^{+}$ and $Hg^{2+}$, synthetic organic compounds, and UV light etc., also induce phytoalexin biosynthesis and are called abiotic elicitors. However, this classification includes some contradictions. Ascochitine, a toxin isolated from culture filtrate of *Ascochyta fabae,* induces pisatin biosynthesis in pea endocarp,[23] but can also be chemically synthesized, as can many other compounds of natural origin. Thus, using this terminology, the chemically synthesized ascochitine would be called an abiotic elicitor, while the fungal product would be considered a biotic elicitor.

The importance of so-called biotic elicitors is to elucidate the mechanism of expression of defense reactions. Other than phytoalexins, the treatment of pear leaves with spore germination fluid of a nonpathogenic strain of *Alternaria alternata,*[8] or treatment of pea leaves with the elicitor isolated from *Mycosphaerella pinodes* induced substances which do not have an antibiotic activity but inhibit the infection by pathogenic fungi. These substances are called "infection-inhibiting factor"[8] or "infection inhibitor".[9] Therefore, in pea leaves, at least, two processes of chemical defense are possible. One is the production of infection inhibitor at 1 to 2 h, and then pisatin production at 6 to 9 h after treatment with elicitor from *M. pinodes.*[9]

Rapid lignification of the parenchymatous cell wall of radish infected by downy mildew fungus serves a barrier for invading hyphae. The lignin synthesized as a result of the parasitic attack is composed mainly of guaiacylpropane units and apparently differs from syringyl lignin present in healthy tissues, and was called diseased lignin.[10] In this case the lignification-inducing factor (elicitor) was isolated and identified as a glycopeptide which is bound to the cell wall in an inactive form in healthy tissue. Fungal or other stimuli served to activate the release of this factor from the cell wall.[24] Thus, the factor is an endogenous elicitor. Keen and Yoshikawa,[25] and Yoshikawa et al.[26] demonstrated that the elicitor from *Phytophthora megasperma* f. sp. *glycinea* was solubilized and liberated from fungal cell wall by the action of 1,3-endoglucanase, which is present in host tissue.

The phenomena analogous to acquired immunity in animals have been reported in many plant virus diseases. Loebenstein and Van Praagh[27] extracted an "interferon-like substance" from *Datura stranonium* leaves infected by tobacco mosiac virus (TMV) or tobacco necrosis virus (TNV). The experimental data suggested that the agent was a protein whose molecular weight was less than 30,000. It is now widely known that new proteins which are not found in healthy plant appear in many plants infected not only by viruses but also by bacterial and fungal plant pathogens.

The relationship between induced resistance and these new proteins has been the matter of extensive discussion. Many scientists have reported a positive correlation between the amount of new proteins and the degree of induced resistance.

These proteins called b-proteins or pathogenesis-related proteins (PR-proteins), were characterized by Van Loon[28]: low molecular weight, acid-soluble, low isoelectric point, resistant to proteolytic digestion, secreted into intercellular fluid, and no apparent enzymic activity. Recently, however, four PR-proteins were identified as β-1,3-glucanases[29] and four others were identified chitinases.[30]

Ye et al.[31] found that stem injected with sporangiospores of *Peronospora tabacina* or inoculation with TMV induced systemic protection in tobacco against diseases caused by the same fungus and TMV. The protected leaves accumulated PR-proteins. This suggested the possibility that PR-proteins, which have β-1,3-glucanase activity, degraded the mycelial wall of *P. tabacina* distributed in the intercellular spaces of the tobacco leaves. Tomato leaves infected by *Phytophthora infestans* or *Fulva fulva* also accumulate PR-proteins.[32] The proteins were also induced by treatment with UV light, by injecting leaves with indoleacetic acid (IAA), and also by treatment with elicitor from *Phytophthora megasperma* f. sp. *glycinea*. These leaves showed increased resistance to *P. infestans*. Based on these studies, PR-proteins are believed to be produced as the result of metabolic disturbance caused by various pathogens and the other stimuli, such as, chemical and physical treatments.

Several scientists[33,34] believe that there are receptors in plant host cells that recognize fungal elicitors. We[35] have recently found that protein kinase seems to play an important role in the elicitation of the defense reaction in pea plant. Verapamil, a $Ca^{2+}$ channel blocker, and K-252a, a strong inhibitor of protein kinase, inhibited pisatin accumulation in pea epicotyl which had been treated with elicitor isolated from *M. pinodes*. Since $LaCl_3$ and EGTA did not inhibit pisatin accumulation, and furthermore, since verapamil inhibited pisatin accumulation even if it was applied 6 h after the elicitor treatment, at a time when the pisatin biosynthetic pathway had already been activated, it can hardly be considered that $Ca^{2+}$ plays a role as the second messenger for signaling of pisatin biosynthesis in pea tissue induced by elicitor. On the other hand, K-252a inhibited pisatin accumulation only when it was applied to pea epicotyl before elicitor treatment. This fact suggests that signal transduction occurs very rapidly after elicitor treatment, and that protein kinase may play a key role in signal transduction for pisatin biosynthesis because K-252a markedly inhibits *in vitro* phosphorylation of pea plasma membrane proteins.

## III. SUPPRESSION OF THE DEFENSE REACTION OF HOST PLANTS BY COMPATIBLE PATHOGENS

### A. BIOLOGICAL EVIDENCE FOR SUPPRESSION OF THE DEFENSE REACTION

Notwithstanding the resistance mechanisms of host plants, plant pathogens can establish infection on their host plants. The host cells may be killed by deleterious toxins or enzymes, which then allows the pathogen to live on the dead cells in a saprophytic manner. However, this is not the case in obligate parasites.

Table 1 shows the result of double inoculation experiments done in our laboratory using a barley cultivar, H.E.S.4, the compatible and incompatible races of *Erysiphe graminis* f. sp., and the other genus of powdery mildew fungi, *Sphaerotheca*.[36-38] *Erysiphe graminis* f. sp. *hordei* race Hh 4 is compatible to the barley cultivar, H.E.S.4, incompatible or less compatible to the barley cultivar, Russian 74, incompatible to wheat, and nonpathogenic to melon. *Erysiphe graminis hordei* race Hr 74 is the pathogen on barley cv. Russian 74 and *E. graminis tritici* race t2 is a pathogen on wheat cv. Norin No. 4. *Sphaerotheca fuliginea* is a melon fungus and nonpathogenic to barley and wheat.

The powdery mildew fungi, *E. graminis* and *S. fuliginea,* are surface parasites. The barley cultivar, H.E.S.4, was inoculated with a fungus, and 48 h later, the fungal material was wiped off with a wet cotton ball, and reinoculated with a challenger fungus. Forty eight hours after challenger inoculation, the infection frequency and the length of the secondary hyphae were measured under a microscope.

As shown in Table 1 (Experiment 3), the preliminary inoculation with incompatible or less compatible race, t2 or Hr 74, reduced the infection frequency, and the length of the

**TABLE 1**
**Induction of Susceptibility and Resistance in Barley Cultivar, H.E.S.4 by Compatible and Incompatible Races of *Erysiphe graminis***

| | Inducer | | | Affinity indices of challenger | |
|---|---|---|---|---|---|
| **Experiment** | **Race** | **Inducer infectivity (ESH frequency[a] %)** | **Challenger race** | **ESH frequency (%)** | **SH length (μm)[b]** |
| 1 | None | | Hh4 | 34.2 | 264 |
| | Hh4 | 23.4 | Hh4 | 34.7 | 260 |
| | Hr74 | 12.8 | Hh4 | 17.5[d] | 201[d] |
| 2 | None | | Hr74 | 13.1 | 87 |
| | Hh4 | 21.3 | Hr74 | 20.1[d] | 130[d] |
| | Hr74 | 13.0 | Hr74 | 9.0 | 85 |
| 3 | None | | Hh4 | 12.2 | 129 |
| | Hr74 | | Hh4 | 6.7[d] | 105[d] |
| | t2 | | Hh4 | 4.6[d] | 102[d] |
| 4 | None | | t2 | 2.0 | 112 |
| | Hh4 | 22.4 | t2 | 14.1[d] | 173[d] |
| | Hr74 | 8.6 | t2 | 4.1 | 104 |
| | t2 | 2.7 | t2 | 1.0 | 92 |
| 5 | None | | S.f.[c] | 0 | — |
| | Hh4 | 33.6 | S.f. | 48.6[d] | 315[d] |
| | Hr74 | 25.3 | S.f | 8.2[d] | 165[d] |
| | t2 | 3.6 | S.f. | 1.7[d] | 140[d] |

*Note:* Challenger race was inoculated 48 h after the inoculation with inducer race and the affinity indices were determined after 48 h of additional incubation.

[a] ESH frequency, frequency of spores elongating the secondary hyphae.
[b] SH length, length of the secondary hyphae.
[c] S.f., *Sphaerotheca fuliginea*.
[d] Significantly higher or lower than the respective control ($p = 0.05$).

From refs. 36-38.

secondary hyphae of the compatible challenger, race Hh 4. Preliminary inoculation with the incompatible race induced resistance against the originally compatible race.

Table 1 (Experiment 4) also shows the preliminary inoculation with the compatible race induced susceptibility against the originally incompatible fungus, a wheat pathogen, t2. Experiment 5 (Table 1) shows that the preliminary inoculation with *E. graminis* induced susceptibility against the melon fungus, *S. fuliginea,* a nonpathogen of barley. It is noteworthy that the infection frequency of *S. fuliginea* as the challenger correlates to the degree of compatibility of the inducer fungus. Thus, the most virulent inducer, Hh 4, increases the infection frequency to melon fungus by 50%. The wheat fungus, t2, also induced susceptibility to the melon fungus. Melon leaves inoculated with melon fungus became susceptible to barley fungus. In the melon leaf cells predisposed by *S. fuliginea,* the barley powdery mildew fungus, *E. graminis,* formed characteristic globe-shaped haustorium, continued to grow, and finally formed abundant conidia on melon leaves. The conidia of the barley fungus thus produced on melon leaves were only pathogenic to barley but not to melon.

These results suggest that the initial contact of invading microorganism conditions the host cell as incompatible or compatible. By contacting the incompatible fungi, host cells exhibit a series of defense reactions, and, as a result, the originally compatible pathogen cannot infect the host cells. The induction of susceptibility in plant cells to the incompatible or nonpathogen by inoculation with compatible pathogen indicates that the compatible pathogen has some mechanisms that suppress the defense response of its own host.

**TABLE 2**
**Suppressors of Defense Reaction Found in Plant Pathogenic Fungi**

| Pathogenic fungus | Composition | Host plant | References |
|---|---|---|---|
| *Phytophthora infestans* | Glucan, Phosphoglucan | Potato | 45 |
| *Phytophthora infestans* | ? | Tomato | 46 |
| *Mycosphaerella pinodes* | Peptide or Glycopeptide | Pea | 47 |
| *Mycosphaerella melonis* | Peptide ? | Cucumber | 48 |
| *Mycosphaerella ligulicola* | Peptide ? | Chrysanthemum | 48 |
| *Ascochyta rabiei* | Glycoprotein | Chick pea | 50 |
| *Phytophthora megasperma* f.sp. *glycinea* | Mannan-glycoprotein | Soybean | 51 |

Similar results were obtained by Tsuchiya and Hirata.[39] Neighboring cells of the *E. graminis*-infected cells of barley became susceptible to many nonpathogenic powdery mildew fungi. Kunoh et al.[40,41] carried out more detailed experiments to determine the timing for induced susceptibility (accessibility) and resistance by inoculating *E. graminis* as a compatible pathogen and *E. pisi* as a nonpathogen on the same coleoptile cells of barley using a micromanipulator. Inoculation with *E. pisi* alone never infects barley coleoptile cell, but penetration efficiency increased about 30% if the coleoptile cell was inoculated with *E. graminis* 60 min or more prior to the inoculation with *E. pisi*. When *E. pisi* was inoculated 60 min earlier than *E. graminis* on the same coleoptile cell, the penetration efficiency of *E. graminis* was reduced from 75 to 28.6%.

To distinguish the general concept of induced susceptibility which operates at the tissue or whole plant level, Ouchi et al.[36] proposed the term ''induced accessibility'' because the susceptibility caused by the compatible pathogen occurs at the cellular level. The use of this terminology was accepted by Kunoh et al.[41]

In potato late blight disease, the inoculation with a compatible race of *Phytophthora infestans* suppressed phytoalexin accumulation and the hypersensitive response.[42] The active mechanism which suppressed these defense responses was considered to lead to susceptibility.[1]

## B. MECHANISM OF SUPPRESSION OF THE DEFENSE REACTION

### 1. Suppressors of the Defense Reaction Produced by Pathogenic Fungi

Several plant pathogenic fungi have been shown to produce substances which suppress the defense response of host cells (Table 2). These substances are called suppressors. A water-soluble fraction from mycelia of a compatible race of *P. infestans* suppressed the hypersensitive reaction and phytoalexin production in potato discs; the fraction isolated from the incompatible race was less active.[43] The component of the water-soluble fraction was composed of 17 to 24 $\beta(1\rightarrow3)$, $\beta(1\rightarrow6)$-linked glucose units.[44] Doke et al.[45] considered that the glucan plays a role as the determinant of specificity in the potato-*P. infestans* interactions. Storti et al.[46] also reported that substances released by germinating sporangia of *Phytophthora infestans,* but not by the mycelium, suppressed the hypersensitive reaction and phytoalexin production in tomato.

The pea pathogen, *Mycosphaerella pinodes,* secretes elicitor and suppressor for the biosynthesis of pisatin, a pea phytoalexin, into the spore germination fluid.[47,48] The elicitor was found in a high-molecular weight fraction of the fluid and identified as a polysaccharide (MW ca. 70,000). Two kinds of suppressors, found in a low-molecular weight fraction, were named F2 and F5. Both suppressors were ninhydrin-positive and F5 was more active than F2. F5 contains serine, aspartic acid, and glycine at a ratio of 2:1:1. Since treatment with proteolytic enzymes reduced the activity of F5, an oligopeptide seemed to be an essential part of the F5 molecule. Only in the presence of F5, could the conidia of eight species of nonpathogens to pea establish infection on pea leaves. Among these, *Alternaria alternata* 15B, an avirulent isolate of pear pathogen, and *Stemphylium sarcinaeforme,* a red clover pathogen, colonized on pea and formed conidia on the F5-treated leaves 2 weeks later. Furthermore, *Alternaria* 15B could establish infection on five species of leguminous plant to which *M. pinodes* was pathogenic in the presence of F5, but not on other plant species. In other words, the host specificity of the F5 producer, *M. pinodes,* coincided completely with the specificity of the biological activity of F5. Thus, it was found that F5 might be a determinant for pathogenicity of *M. pinodes,* by suppressing the general defense reaction.

F5 did not cause any visible injury to pea leaf, stem, and isolated protoplast (prepared from pea leaf), therefore, it can hardly be a "host-specific toxin." The kinetic analysis indicated that the mechanism of action of suppressor was not competitive to the elicitor,[49] similar to the suppressor isolated from *Ascochyta rabiei,* a chick pea pathogen.[50,52]

A mannan glycoprotein isolated from culture fluid of *Phytophthora megasperma* f. sp. *glycinea* suppresses glyceollin accumulation in soybean induced by elicitor from the same fungus.[51] This inhibition was reported to be race-specific.

Our recent results[49] indicate that the suppressor from *M. pinodes* inhibits the plasma membrane ATPase by blocking the formation of the phosphorylated intermediate of the enzyme. Since the suppressor inhibits the decline of pH of the droplet of elicitor solution or water placed on the pea leaf, the suppressor may function by blocking the proton pump ATPase in pea plasma membrane. This inhibitory activity of the suppressor on pea leaves is temporary since it disappears within 2 to 3 h after treatment.

### 2. Host-Specific Toxins as Suppressors of the Defense Reaction

*Alternaria alternata* produces various kinds of host-specific toxins, and the specificity of the toxins to plant species determines the host range of the fungus.[53] The early event caused by the host-specific toxin has been proved to be the suppression of the defense reaction of the host plant.[8,54] For example, the spore germination fluid of *A. alternata* produces a nonspecific polysaccharide elicitor whose molecular weight is approximately 40,000. This elicitor induces resistance in pear leaves to the invasion by a virulent *A. alternata.* However, the activity of the elicitor was completely suppressed when it was treated with AK-toxin (host-specific toxin produced by *A. alternata,* Japanese pear pathotype). Thus, an important role of AK-toxin in the early state of pathogenesis is to suppress the defense reaction in host plant.[54]

### 3. Suppression of the Defense Reaction Against Microorganisms by Host Components

Some plants contain factors that suppress the defense reaction against plant pathogenic fungi. The extract from rust-infected soybean leaves induced haustoria formation of cowpea rust fungus in soybean leaves.[55] Intercellular fluid from *Cladosporium fulvum*-infected tomato suppresses the hypersensitive reaction and callose deposition in tomato leaf induced by nonspecific elicitors from *C. fulvum.* Since the same suppressing activity is detected in some, but not in all, intercellular fluids of uninoculated tomato, the suppressor seems to be of host origin.[56]

Healthy pea leaves contain substances that suppress the pisatin accumulation induced by a fungal elicitor. These substances are called endogenous suppressors.[57,58] The molecular weight of the endogenous suppressors are ca. 15,000 and 5,000. The latter was identified as glycopeptide. The treatment of pea leaves with these substances induces susceptibility to avirulent fungi of pea, *Mycosphaerella ligulicola* and *M. melonis.* These substances do not suppress the defense reaction of other leguminous plants, such as, kidney bean or cowpea. One of these substances ($M_r$ = 5000 Da) inhibited the plasma membrane ATPase activity similar to the suppressor from *M. pinodes.*

Barley leaves contain a substance that enhances the infection frequency of incompatible powdery mildew fungi on barley.[59] The substance was isolated, purified, and characterized as a glycopeptide of $M_r$ 3000 to 3500.[60] The glycopeptide is composed of glycine, asparagine, glutamic acid, serine, and threonine in the peptide moiety, and acetylglucosamine, mannose, fucose, galactose, neuramic acid, and xylose in the sugar moiety. The infection-enhancing factor was also isolated from barley leaves infected with powdery mildew and proved to be identical with that from healthy leaves. A larger amount of the factor diffused from barley tissue when inoculated with a compatible race of powdery mildew fungus compared with those inoculated with incompatible races. The infection of wheat by *Erysiphe graminis* f. sp. *tritici* or pea leaves by *E. pisi* was not enhanced by the factor isolated from barley leaves. Thus, the factor seems to be responsible for the basic compatibility as suggested by Bushnell[61] and Heath[62,63] between barley and *E. graminis* at species-species level. The factor also inhibits plasma membrane ATPase of bean leaves.

### C. DETOXIFICATION OF PHYTOALEXINS BY PATHOGENIC FUNGI

In some plant-parasite interactions, the detoxicification mechanism plays an important role in the pathogenicity of pathogenic fungi. Cruickshank[64] reported his experimental results using purified pisatin: 5 fungal strains out of 50 tested were tolerant to pisatin, and all 5 were pea pathogens. Only 1 of 45 sensitive strains was a pea pathogen. Uehara[65] found that a pea pathogen, *Ascochyta pisi,* metabolized pisatin to a less toxic product. According to Nonaka,[66] pea pathogens *Fusarium oxysporum* f. sp. *pisi* and *A. pisi* degrade pisatin but nonpathogens of pea do not. He found positive correlations between pathogenicity to pea and the pisatin-degrading ability of fungi.

The first step of pisatin degradation by many pea pathogens was shown to be the removal of 3-*O*-methyl group to yield 6a-hydroxymaakiain. VanEtten[67,68] proved that the demethylated product was less toxic to many plant pathogenic fungi than pisatin. This was confirmed by Nonaka.[69] The importance of phytoalexin detoxification ability in pathogenicity was reviewed recently by VanEtten.[70]

## IV. EXPRESSION OF GENES INVOLVED IN THE DEFENSE REACTION

There are many reports that infection of a plant by pathogenic fungi, especially in incompatible fungi, induces various kinds of isoenzymes not normally present in healthy plants, and that these newly formed isoenzymes have been shown to be the result of net protein synthesis. This fact suggests that the infection causes activation of genes encoding these proteins.

Cultured plant cells are suitable experimental material to study the response of plant cells in some plant diseases because the signal molecules distribute simultaneously to each cell. Isolated bean cells cultured in the dark were treated with elicitor prepared from *Colletotrichum lindemuthianum,* and the amount of mature mRNA accumulated in the cells was determined by Northern blot-hybridization analysis with cDNAs of phenylalanine ammonia-lyase (PAL), chalcone synthase (CHS), and chalcone isomerase (CHI), which are key enzymes for isoflavonoid phytoalexin biosynthesis. Results showed that all mRNAs (PAL, CHS, and CHI) accumulated very rapidly, reaching a maximum level 3 h after treatment with elicitor, and then decreasing to the original level.[71-75]

The mRNA encoding cinnamyl-alcohol dehydrogenase, an enzyme specific to the synthesis of lignin monomers, accumulates in cultured bean cells very rapidly following treatment with elicitor from *C. lindemuthianum.* It reaches maximum levels at 1.5 to 2.0 h, and then decreases to the original level after 4 h.[75,76] Similarly, the transcription of genes encoding chitinase, which is responsible for the defense reaction by degrading the cell wall of pathogenic fungi, is activated very rapidly following treatment with fungal elicitor.[75,77]

The rapid activation of these gene expressions in plant cells after elicitor treatment or fungal attack indicates that in plant cells very rapid steps may be involved in the signal transduction system from the recognition of microorganisms to the transcriptional activation of these genes.[75]

Expression patterns of several defense-related genes involved in phenylpropanoid biosynthesis in nonhost resistance were studied by *in situ* RNA hybridization.[78] All mRNAs tested accumulated transiently and locally around the infection site of the primary leaf of parsley inoculated with *Phytophthora megasperma* f. sp. *glycinea*, to which parsley is nonhost-resistant.

Enzymes responsible for phenylpropanoid biosynthesis which are induced by different kinds of elicitors and those in unelicited bean cells were compared by *in vitro* translation using an mRNA-dependent rabbit reticulocyte lysate system and the polysomal mRNA isolated from elicitor-treated and untreated cells.[79] Chromatofocusing analysis revealed that culture filtrate and cell wall elicitors prepared from *C. lindemuthianum* induced similar patterns of PAL and CHS isoforms which were, however, different from those observed in unelicited cells.

Hydroxyproline-rich glycoproteins (HRGPs) are usually found in low amounts in the cell wall of higher plants. However larger accumulations were found in melon seedlings infected by *Colletotrichum lagenarium*.[11] The same phenomena have been found in many host-parasite combinations[12] and it is believed that HRGPs play a role in resistance of plants to pathogens by acting as a structural barrier and as an agglutinin.[80,81] A remarkable increase in HRGPs occurs not only during infection but also following wounding and elicitor treatment.[82,83] Recently, several mRNAs that hybridize to a genomic clone of HRGPs have been reported.[84,85] Transcriptional activation of genes encoding HRGPs begins 2 h after elicitor treatment. The activation of these genes is slower than compared with the PAL or CHS genes, which suggests the involvement of a secondary signal substance which originates indirectly from host cells.[75,85]

Rumeau et al.[86] revealed by an *in vitro* translation experiment with RNA from melon plants that several alterations of patterns were observed during infection. In particular, they found that the cytosine-rich RNA separated from poly(A)-RNA codes for proline-rich peptides. Two peptides of molecular weight 54,500 and 56,000 were synthesized by the *in vitro* translation system, and they suggest that these peptides are precursors of the peptide moiety of HRGP because the molecular weights are similar to that of melon HGRP (55,000) and the time course of their appearance *in vitro* coincides with the *in vivo* accumulation of HGRP.

Recently, the pattern of activation, structure, and genomic organization of a gene *(prpl)* encoding a pathogenesis-related protein (PR1) in potato cultivar carrying resistance gene *R1* elicited by infection of *Phytophthora infestans* or culture filtrate-elicitor were clarified.[87] According to the run-off transcription assay with isolated nuclei from potato leaves, the gene, *prp1*, is activated very rapidly after the elicitor treatment, reaches a maximum level of expression 1 h later, and then decreases. The pattern is very similar to that of 4-coumarate-CoA lygase. The coding sequence of the *prp1* and the deduced amino acid sequence are strikingly similar to that of a 26-kDa heat shock protein from soybean. An *in situ* RNA hybridization assay revealed that the *prp1* transcript accumulated around the infection site.

As described above, genes encoding enzymes concerned with phenylpropanoid biosynthesis are activated very rapidly after the elicitor treatment, but are restored to the original level within several hours. The mechanism of this rapid decline of the accumulation of the gene transcripts is not clear, but Dixon[88] suggested the possibility that *trans*-cinnamic acid, the product of PAL, may play a role as a regulation signal in gene expression of the phenylpropanoid biosynthetic pathway, because *trans*-cinnamic acid inhibits the transcription of PAL and CHS genes in cultured bean cells.

## V. SUPPRESSION OF GENE EXPRESSION INVOLVED IN THE DEFENSE REACTION

In spite of many studies on gene expression for host defense reactions, few data are available on the suppression of the gene activation by signal molecules produced by pathogenic fungi, in order to allow their establishment on their host(s). In our laboratory,[89] the suppressor prepared from spore germination fluid of *M. pinodes,* a pea pathogen, delays the expression of PAL and CHS genes encoding key enzymes for pisatin biosynthetic pathway. Pea epicotyl grown in the dark has minimal amounts of PAL and CHS. However, treatment of epicotyl with elicitor prepared from *M. pinodes* induces the accumulation of PAL and CHS mRNA within 1 h. Under the concomitant presence of suppressor with elicitor, initiation of the transcriptional activation of these genes is delayed 3 h (Figure 1). A 3-h delay of the activation of these genes results in a 6 to 9 h delay of pisatin accumulation in pea epicotyl. Thus, the suppressor of the pathogenic fungus suppresses the expression of genes responsible for the defense reaction.[89]

Cuypers et al.[90] conducted *in situ* hybridization experiments of PAL mRNA with $^{32}$P-labeled PAL antisense RNA as a probe at the infection site of potato leaves inoculated with compatible or incompatible races of the late blight fungus. They found that the remarkable accumulation of PAL mRNA in leaves inoculated with the compatible race delayed 3 h compared with that inoculated with incompatible race. The coincidence of the 3-h delay in this experiment supports the concept that the suppressor produced by the pathogenic fungus plays a key role as the determinant of pathogenicity not only at the species-species level but also at the cultivar-race level.

## VI. IMPORTANCE OF GENES FOR PHYTOALEXIN DEGRADATION IN PATHOGENICITY

The importance of phytoalexin detoxification in pathogenicity was proved in a pea pathogen-pea system. Pisatin, a pea phytoalexin, is detoxified by the demethylation ability of pea pathogens, *Ascochyta pisi* or *Nectria heamatococca.*

Recently, a 3.2-kb DNA fragment was isolated from *N. heamatococca*[91] which confers on *Aspergillus nidulans* the ability to demethylate pisatin. The transformation of this fragment (gene encoding pisatin demethylase) into an avirulent strain of *N. heamatococca* or the maize pathogen, *Cochliobolus heterostropus,* rendered these fungi virulent on pea.[92] These results indicate that, in addition to the importance of phytoalexin degrading ability in the pathogenicity of some pathogenic fungi, pisatin is the key compound responsible for resistance of pea plant.

## VII. STRUCTURE AND REGULATION OF GENES ENCODING PHYTOALEXIN SYNTHETIC ENZYMES

Phenylalanine ammonia-lyase in cultured bean cells is encoded by a small gene family. The nucleotide sequences of two PAL genes, *gPAL2* and *gPAL3*, were determined and compared.[93] Both genes are interrupted with an intron, and showed 59% sequence similarity in exon I and 74% in exon II. There are extensive divergence in the introns 5′- and 3′-flanking regions. *gPAL2* is activated by elicitor but *gPAL3* is not, whereas both are activated by wounding of the hypocotyl. The transcription starting sites of both genes are 99 bp and 35 bp upstream from initiation codon ATG. The 5′-flanking region of both genes contain TATA and CAAT boxes at 30 and 74 bp upstream from the transcription starting site, respectively. In addition, both genes contain sequences resembling SV40 sequences enhancer core.

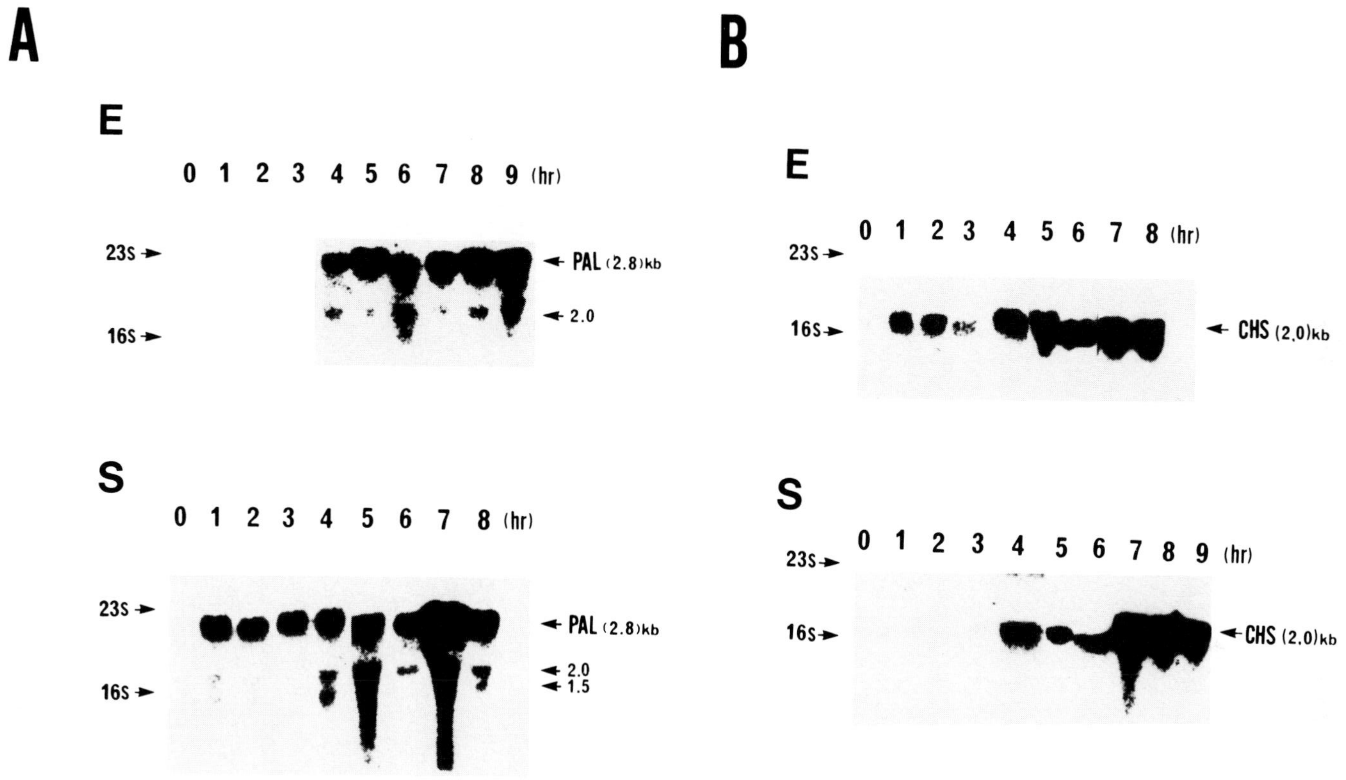


FIGURE 1. Induction of PAL mRNA (A) or CHS mRNA in pea epicotyl tissues treated with elicitor or elicitor + suppressor. Total RNA (10 μg) was isolated from epicotyl tissues treated with elicitor (E), or elicitor + suppressor (E + S). Samples were blot hybridized with $^{32}P$-labeled bean PAL or CHS cDNA fragment.[89] Numbers on the top indicate time (h) of incubation after the treatment. Position of the putative pea mRNA is indicated with an arrow. Modified from Yamada, T., et al., *Mol. Plant-Microbe Interact.*, 2, 259, 1989.

PAL is also encoded by a small family of at least four genes in parsley, and the transcription of three identified PAL genes is activated by treatment with fungal elicitor or UV light in cultured cell.[94] These stress-responsive genes show over 90% similarity at the nucleotide level. The promoter regions of parsley PAL-1 gene activated by UV or fungal elicitor involve two specific nucleotide sequences. The consensus motifs are conserved in promoter regions of the other genes involved in phenylpropanoid metabolism of other plant species.[93,95,96]

In addition to these sequence similarities, the fact that the other genes encoding enzymes in phytoalexin biosynthesis are coordinately regulated by elicitor[71-75,89] and suppressor[89] suggests that these motifs are elements responsible for interacting with *trans*-acting factors induced in plant cells by different stresses.[94]

## VIII. CONCLUSION

In this chapter, the ways in which plants defend themselves are described from a biological, biochemical, and genetic perspective. How plant pathogens overcome the defense reactions of host plants in order to establish infection is also discussed.

A series of defense reactions are expressed by recognizing some signal molecules produced by pathogenic fungi or as a result of host-parasite interactions. These defense reactions in plants are thought to be one of the general properties of living beings, which have been acquired during the long evolutional process to avoid invasions of deleterious organisms and to maintain the purity of the species. The defense reactions seem to have evolved from the expression of genes encoding enzymes necessary for those reactions. As a general concept, the expression of these genes should be sufficient to support defense reactions which would prevent infection by invading organisms, that is, the state of resistance is the rule for living beings, including plants.

However, during the coevolutionary process, parasitic organisms, including plant pathogenic fungi, have acquired strategies to overcome the defense reactions exerted by their host or hosts such that they are able to parasitize and derive food materials from them. Some kill the host cell by producing toxins and live saprophytically on the dead cells. However, this is not the case in obligate parasites. Others acquired the ability to inactivate the defense substances produced by hosts, such as detoxification of phytoalexins. Some plant pathogens produce substances to suppress the host defense reactions at the level of gene expression. When these strategies of fungi are sufficiently effective to overcome host defense reactions, the plant becomes susceptible. Where such fungal strategies are ineffective, the host plant exhibits resistance.

That the elicitor directly activates a variety of genes for defense reactions simultaneously is unlikely. Rather, the so-called vertical resistance gene recognizes the elicitor signals either directly or indirectly, and a second message activates the regulatory regions of the genes for defense reaction.

## REFERENCES

1. **Kuć, J.,** Phytoalexins, *Annu. Rev. Phytopathol.,* 10, 207, 1972.
2. **Hawkins, L. A. and Harvey, D. B.,** Physiological study of the parasitism of *Pythium debaryanum* Hesse on potato tuber, *J. Agric. Res.,* 18, 275, 1919.
3. **Ingham, J. L.,** Disease resistance in higher plants. The concept of pre-infectional and post-infectional resistance, *Phytopathol. Z.,* 78, 314, 1973.
4. **Albersheim, P. and Anderson, A.,** Proteins from plant cell walls inhibit polygalacturonases secreted by plant pathogens, *Proc. Natl. Acad. Sci. U.S.A.,* 68, 1815, 1971.

5. **Aist, J. R.,** Papillae and related wound plugs of plant cells, *Annu. Rev. Phytopathol.,* 14, 145, 1976.
6. **Aist, J. R.,** Mechanically induced wall appositions of plant cells can prevent penetration by a parasitic fungus, *Science,* 197, 568, 1977.
7. **Aist, J. R., Kunoh, H., and Israel, H. W.,** Challenge appressoria of *Erysiphe graminis* fail to breach preformed papillae of a compatible barley cultivar, *Phytopathology,* 69, 1245, 1979.
8. **Hayami, N., Otani, H., Nishimura, S., and Kohmoto, K.,** Induced resistance in pear leaves by spore germination fluid of nonpathogens to *Alternaria alternata* Japanese pear pathotype and suppression of the induction by AK-toxin, *J. Fac. Agric. Tottori Univ.,* 17, 9, 1982.
9. **Yamamoto, Y., Oku, H., Shiraishi, T., Ouchi, S., and Koshizawa, K.,** Non-specific induction of pisatin and local resistance in pea leaves by elicitors from *Mycosphaerella pinodes, M. molonis* and *M. ligulicola* and the effect of suppressor from *M. pinodes, J. Phytopathol.,* 117, 136, 1986.
10. **Asada, Y., Oguchi, T., and Matsumoto, I.,** Induction of lignification in response to fungal infection, in *Recognition and Specificity in Plant Host-Parasite Interaction,* Daly, J. M. and Uritani, I., Eds., Japan Scientific Press, Tokyo, University Park Press, Baltimore, MD, 1979, 99.
11. **Esquerré-Tugayé, M. T., Lafitte, C., Mazau, D., Toppan, A., and Touzé, A.,** Cell surfaces in plant-microorganism interactions. II. Evidence for the accumulation of hydroxyproline-rich glycoproteins in the cell wall of diseased plants as a defense mechanism, *Plant Physiol.,* 64, 320, 1979.
12. **Mazau, D. and Esquerré-Tugayé, M. T.,** Hydroxyproline-rich glycoprotein accumulation in the cell walls of plants infected by various pathogens, *Physiol. Mol. Plant Pathol.,* 29, 147, 1986.
13. **Aist, A. and Israel, H. W.,** Cytological aspects of host responses to primary penetration by fungi, in *Biochemistry and Cytology of Plant Parasite Interaction,* Tomiyama, K., Daily, J. M., Uritani, I., Oku, H., and Ouchi, S., Eds., Kodansha, Tokyo, Elsevier, Amsterdam, Oxford, New York, 1976, 26.
14. **Skou, J. P., Jørgensen, J. H., and Lilholt, U.,** Comparative studies on callose formation in powdery mildew compatible and incompatible barley, *Phytopathol. Z.,* 109, 147, 1984.
15. **Koga, H., Mayama, S., and Shishiyama, J.,** Correlation between the deposition of fluorescent compounds in papillae and resistance in barley against *Erysiphe graminis hordei, Can. J. Bot.,* 58, 536, 1980.
16. **Mayama, S. and Shishiyama, J.,** Localized accumulation of fluorescent and U.V.-absorbing compounds at penetration sites in barley leaves infected with *Erysiphe graminis hordei, Physiol. Plant Pathol.,* 13, 347, 1978.
17. **Sahashi, N. and Shishiyama, J.,** Increased papilla formation, a major factor of induced resistance in the barley-*Erysiphe graminis* f. sp. *hordei* system, *Can. J. Bot.,* 64, 2178, 1986.
18. **Sahashi, N. and Shishiyama, J.,** Defense reaction of barley cultivars to non-pathogenic fungi, *Ann. Phytopathol. Soc. Jpn.,* 53, 242, 1987.
19. **Ward, H. M.,** On the relations between host and parasite in the bromes and their brown rust, *Puccinia dispersa* (Erikss.), *Ann. Bot.,* 16, 233, 1902.
20. **Keen, N. T.,** Specific elicitors of plant phytoalexin production: determinants of race specificity in pathogens, *Science,* 187, 74, 1975.
21. **Shiraishi, T., Oku, H., Yamashita, M., and Ouchi, S.,** Elicitor and suppressor of pisatin induction in spore germination fluid of pea pathogen, *Mycosphaerella pinodes, Ann. Phytopathol. Soc. Jpn.,* 44, 659, 1978.
22. **Darvill, A. G. and Albersheim, P.,** Phytoalexins and their elicitors—defense against microbial infection in plants, *Annu. Rev. Plant Physiol.,* 35, 243, 1984.
23. **Oku, H., Nakanishi, T., Shiraishi, T., and Ouchi, S.,** Phytoalexin induction by some agricultural fungicides and phytotoxic metabolites of pathogenic fungi, *Sci. Rep. Fac. Agric. Okayama Univ.,* 42, 17, 1973.
24. **Asada, Y.,** Induced lignification and elicitors, *Abstr. 5th ICPP,* 1988, 216.
25. **Keen, N. T. and Yoshikawa, M.,** β-1,3-Endoglucanase from soybean releases elicitor-active carbohydrates from fungus cell wall, *Plant Physiol.,* 71, 460, 1983.
26. **Yoshikawa, M., Matama, M., and Masago, H.,** Release of a soluble phytoalexin elicitor from mycelial walls of *Phytophthora megasperma* var. *sojae* by soybean tissues, *Plant Physiol.,* 67, 1032, 1981.
27. **Loebenstein, G. and Van Praagh, T.,** Extraction of a virus interfering agent induced by localized and systemic infection, in *Host-Parasite Relations in Plant Pathology,* Kiraly, Z. and Ubrizsy, G., Eds., Res. Inst. Plant Protection, Budapest, 1964, 53.
28. **Van Loon, L. C.,** Pathogenesis-related proteins, *Plant Mol. Biol.,* 4, 111, 1985.
29. **Kaufman, S., Legrand, M., Geoffroy, P., and Fritig, B.,** Biological function of pathogenesis-related proteins: Four PR-proteins of tobacco have 1,3-glucanase activity, *EMBO J.,* 6, 3209, 1987.
30. **Legrand, M., Kaufman, S., Geoffroy, P., and Fritig, B.,** Biological function of pathogenesis-related proteins: four tobacco pathogenesis-related proteins are chitinases, *Proc. Natl. Acad. Sci. U.S.A.,* 34, 6750, 1987.
31. **Ye, X. S., Pan, S. Q., and Kuć, J.,** Pathogenesis-related proteins and systemic resistance to blue mould and tobacco mosaic virus induced by tobacco mosaic virus, *Peronospora tabacina* and aspirin, *Physiol. Mol. Plant Pathol.,* 35, 161, 1989.

32. **Christ, U. and Mosinger, E.,** Pathogenesis-related proteins of tomato: I. Induction by *Phytophthora infestans* and other biotic and abiotic inducers and correlation with resistance, *Physiol. Mol. Plant Pathol.,* 35, 53, 1989.
33. **Yoshikawa, M., Keen, N. T., and Wang, M. C.,** A receptor on soybean membranes for a fungal elicitor of phytoalexin accumulation, *Plant Physiol.,* 73, 497, 1983.
34. **Ebel, J. and Grisebach, H.,** Defense strategies of soybean against the fungus *Phytophthora megasperma* f. sp. *glycinea:* a molecular analysis, *Trends Biochem. Sci.,* 13, 23, 1988.
35. **Shiraishi, T., Hori, N., Yamada, T., and Oku, H.,** Suppression of pisatin accumulation by an inhibitor of protein kinase, *Ann. Phytopathol. Soc. Jpn.,* 56, 261, 1990.
36. **Ouchi, S., Oku, H., Hibino, C., and Akiyama, I.,** Induction of accessibility and resistance in leaves of barley by some races of *Erysiphe graminis, Phytopathol. Z.,* 79, 24, 1974.
37. **Ouchi, S., Oku, H., Hibino, C., and Akiyama, I.,** Induction of accessibility to a non-pathogen by preliminary inoculation with a pathogen, *Phytopathol. Z.,* 79, 142, 1974.
38. **Oku, H. and Ouchi, S.,** Host plant accessibility to pathogens, *Rev. Plant Prot. Res.,* 9, 58, 1976.
39. **Tsuchiya, K. and Hirata, K.,** Growth of various powdery mildew fungi on the barley leaves infected preliminarily with the barley powdery mildew fungus, *Ann. Phytopathol. Soc. Jpn.,* 396, 1973.
40. **Kunoh, H., Hayashimoto, A., Harumi, M., and Ishizaki, H.,** Induced susceptibility and enhanced resistance at the cellular level in barley coleoptile. I. The significance of timing of fungal invasion, *Physiol. Plant Pathol.,* 27, 43, 1985.
41. **Kunoh, H., Katsuragawa, N., Yamaoka, N., and Hayashimoto, A.,** Induced accessibility and enhanced inaccessibility at the cellular level in barley coleoptiles. III. Timing and localization of enhanced inaccessibility in a single coleoptile cell and its transfer to an adjacent cell, *Physiol. Mol. Plant. Pathol.,* 33, 181, 1988.
42. **Varns, J. L. and Kuć, J.,** Suppression of rishitin and phytuberin accumulation and hypersensitive response in potato by compatible races of *Phytophthora infestans, Phytopathology,* 61, 178, 1971.
43. **Garas, N. A., Doke, N., and Kuć, J.,** Suppression of hypersensitive reaction in potato tubers by mycelial components from *Phytophthora infestans, Physiol. Plant Pathol.,* 15, 117, 1979.
44. **Kuć, J., Henfling, J., Garas, N., and Doke, N.,** Control of terpenoid metabolism in potato-*Phytophthora infestans* interaction, *J. Food Prot.,* 42, 508, 1979.
45. **Doke, N., Garas, N. A., and Kuć, J.,** Effect of host hypersensitivity suppressors released during the germination of *Phytophthora infestans* cystospores, *Phytopathology,* 70, 35, 1980.
46. **Storti, E., Pelucchini, D., Tegli, S., and Scala, A.,** A potential defense mechanism of tomato against the late blight disease is suppressed by germinating sporangia-derived substances from *Phytophthora infestans, J. Phytopathol.,* 121, 275, 1988.
47. **Oku, H., Shiraishi, T., Ouchi, S., Ishiura, M., and Matsueda, R.,** A new determinant of pathogenicity in plant disease, *Naturwissenschaften,* 67, 310, 1980.
48. **Oku, H., Shiraishi, T., and Ouchi, S.,** Role of specific suppressors in pathogenesis of *Mycosphaerella* species, in *Molecular Determinants of Plant Diseases,* Nishimura, S., Vance, C. P., and Doke, N., Eds., Japan Scientific Press, Tokyo, Springer-Verlag, Berlin, 1987, 145.
49. **Shiraishi, T., Yamada, T., Oku, H., and Yoshioka, H.,** Suppressor production as a key factor for fungal pathogenesis, in *Molecular Strategies of Pathogen and Host Plants,* Patil, S., Ouchi, S., Mills, D., and Vance, C., Eds., Springer-Verlag, New York, 1991, 151.
50. **Kessmann, H. and Barz, W.,** Elicitation and suppression of phytoalexin and isoflavone accumulation in cotyledons of *Cicer arietinum* L. as caused by wounding and by polymeric components from the fungus *Ascochyta rabiei, J. Phytopathol.,* 117, 321, 1986.
51. **Ziegler, E. and Pontzen, R.,** Specific inhibition of glucan-elicited glyceollin accumulation in soybeans by an extracellular mannan-glycoprotein of *Phytophthora megasperma* f. sp. *glycinea, Physiol. Plant Pathol.,* 20, 321, 1982.
52. **Barz, W.,** personal communication, 1988.
53. **Nishimura, S., Kohmoto, K., Otani, H., Ramachandran, P., and Tamura, F.,** Pathological and epidemiological aspects of *Alternaria alternata* infection depending on a host-specific toxin, in *Plant Infection,* Asada, Y., Bushnell, W. R., Ouchi, S., and Vance, C. P., Eds., Japan Scientific Press, Tokyo, Springer-Verlag, Berlin, 1982, 199.
54. **Otani, H., Kohmoto, K., Kodama, M., and Nishimura, S.,** Suppression of resistance by toxins, *Abstr. 5th ICPP,* 1988, 220.
55. **Heath, M. C.,** Effect of infection by compatible species or injection of tissue extracts on the susceptibility of nonhost plants to rust fungi, *Phytopathology,* 70, 356, 1980.
56. **Peever, T. L. and Higgins, V. J.,** Suppression of the activity of non-specific elicitor from *Cladosporium fulvum* by intercellular fluids from tomato leaves, *Physiol. Mol. Plant. Pathol.,* 34, 471, 1989.
57. **Hori, N., Shiraishi, T., Yamada, T., and Oku, H.,** The role of endogenous suppressors in phytoalexin production by pea plant, *Abstr. 5th ICPP,* 1988, 231.

58. **Shiraishi, T., Nasu, K., Yamada, T., Ichinose, Y., and Oku, H.,** Suppression of defense reaction and accessibility induction in pea by substances from healthy pea leaves, in *Molecular Strategies of Pathogen and Host Plants,* Patil, S., Ouchi, S., Mills, D., and Vance, C., Eds., Springer-Verlag, New York, 1991, 252.
59. **Shiraishi, T., Miyazaki, T., Yamada, T., and Oku, H.,** Infection enhancing factor for *Erysiphe graminis* prepared from healthy barley seedlings, *Ann. Phytopathol. Soc. Jpn.,* 55, 357, 1989.
60. **Oku, H., Shiraishi, T., Miyazaki, T., Yamada, T., and Ichinose, Y.,** Infection enhancing factor in barley, a substance possibly responsible for basic compatibility with *Erysiphe graminis,* in *Molecular Strategies of Pathogens and Host Plants,* Patil, S., Ouchi, S., Mills, D., and Vance, C., Eds., Springer-Verlag, New York, 1991, 253..
61. **Bushnell, W. R.,** The nature of basic compatibility between pistil-pollen and host-parasite interaction, in *Recognition and Specificity in Host-Parasite Interaction,* Daily, J. M. and Uritani, I., Eds., Japan Scientific Press, Tokyo, University Park Press, Baltimore, MD., 1979, 211.
62. **Heath, M. C.,** A general concept of host-parasite specificity, *Phytopathology,* 71, 1121, 1981.
63. **Heath, M. C.,** Evolution of plant resistance and susceptibility to fungal invaders, *Can. J. Plant Pathol.,* 9, 389, 1987.
64. **Cruickshank, I. A. M.,** Studies on phytoalexins. IV. The antimicrobial spectrum of pisatin, *Aust. J. Biol. Sci.,* 15, 147, 1962.
65. **Uehara, K.,** Relationship between host specificity of pathogen and phytoalexin, *Ann. Phytopathol. Soc. Jpn.,* 29, 103, 1964.
66. **Nonaka, F.,** Inactivation of pisatin by pathogenic fungi, *Agric. Bull. Saga Univ.,* 24, 109, 1967.
67. **VanEtten, H. D., Pueppke, S. G., and Kelsey, T. C.,** 3,6a-Dihydroxy-8,9-methylenedioxypterocarpan as a metabolite of pisatin produced by *Fusarium solani* f. sp. *pisi, Phytochemistry,* 14, 1103, 1975.
68. **VanEtten, H. D. and Pueppke, S. G.,** Isoflavonoid phytoalexins, in *Biochemical Aspects of Plant Parasitic Relationships,* Friend, J. and Threifall, D., Eds, *Ann. Proc. Phytochem. Soc.,* 239, 13, 1976.
69. **Nonaka, F.,** Studies on the mechanism of plant disease resistance, especially on the phytoalexins, *Ann. Phytopathol. Soc. Jpn.,* 47, 284, 1981.
70. **VanEtten, H. D., Matthews, D. E., and Matthews, P. S.,** Phytoalexin detoxification: Importance for pathogenicity and practical implications, *Annu. Rev. Phytopathol.,* 27, 143, 1989.
71. **Hahlbrock, K. and Scheel, D.,** Physiology and molecular biology of phenylpropanoid metabolism, *Annu. Rev. Plant Physiol.,* 40, 347, 1989.
72. **Edwards, K., Cramer, C. L., Bolwell, G. P., Dixon, R. A., Schuch, W., and Lamb, C. J.,** Rapid transient induction of phenylalanine ammonia-lyase mRNA in elicitor-treated bean cells, *Proc. Natl. Acad. Sci. U.S.A.,* 82, 6731, 1985.
73. **Mehdy, M. C. and Lamb, C. J.,** Chalcone isomerase cDNA cloning and mRNA induction by fungal elicitor, wounding and infection, *EMBO J.,* 6, 1527, 1987.
74. **Ryder, T. B., Cramer, C. J., Bell, J. N., Robbins, M. P., Dixon, R. A., and Lamb, C. J.,** Elicitor rapidly induces chalcone synthase mRNA in *Phaseolus vulgaris* cells at the onset of the phytoalexin defense response, *Proc. Natl. Acad. Sci. U.S.A.,* 81, 5724, 1984.
75. **Lamb, C. J., Lawson, M. A., Dron, M., and Dixon, R. A.,** Signals and transduction mechanisms for activation of plant defenses against microbial attack, *Cell,* 56, 215, 1989.
76. **Walter, M. H., Grima-Pettenati, J., Grand, C., Boudet, A. M., and Lamb, C. J.,** Cinnamyl-alcohol dehydrogenase, a molecular marker specific for lignin synthesis: cDNA cloning and mRNA induction by fungal elicitor, *Proc. Natl. Acad. Sci. U.S.A.,* 85, 5546, 1988.
77. **Hedrick, S. A., Bell, J. N., Boller, T., and Lamb, C. J.,** Chitinase cDNA cloning and mRNA induction by fungal elicitor, wounding and infection, *Plant Physiol.,* 86, 182, 1988.
78. **Schmelzer, E., Kruger-Lebus, S., and Hahlbrock, K.,** Temporal and spatial patterns of gene expression around sites of attempted fungal infection in parsley leaves, *Plant Cell,* 1, 993, 1989.
79. **Hamadan, A. M. S. and Dixon, R. A.,** Differential patterns of protein synthesis in bean cells exposed to elicitor fractions from *Colletotrichum lindemuthianum, Physiol. Mol. Plant. Pathol.,* 31, 105, 1987.
80. **Lamport, D. T. A.,** Structure, biosynthesis and significance of cell wall glycoproteins, in *Recent Advances in Phytochemistry,* Loews, F. and Rumeckles, V. C., Eds., Plenum Press, NY, 1979, 79.
81. **Leach, J. E., Cantrell, M. A., and Sequeira, L.,** Hydroxyproline-rich bacterial agglutinin from potato: extraction, purification and characterization, *Plant Physiol.,* 70, 1353, 1982.
82. **Esquerré-Tugayé, M. T., Mazau, D., Toppan, A., and Roby, D.,** Elicitation via l'éthylène, de la synthèse de glycoprotéins pariétales associées à la défense des plantes, *Ann. de Phytopathol.,* 12, 403, 1980.
83. **Roby, D., Toppan, A., and Esquerré-Tugayé, M. T.,** Cell surfaces in plant microorganism interactions. V. Elicitor of fungal and of plant origin trigger the synthesis of ethylene and of cell wall hydroxyproline-rich glycoproteins in plants, *Plant Physiol.,* 77, 700, 1985.
84. **Chen, J. and Varner, J. E.,** An extracellular matrix protein in plants: characterization of a genomic clone for carrot extensin, *EMJO J.,* 4, 2145, 1985.

85. **Showalter, A. M., Bell, J. N., Cramer, C. L., Bailey, J. A., Varmer, J. E., and Lamb, C. J.,** Accumulation of hydroxyproline-rich glycoprotein mRNA in response to fungal elicitor and infection, *Proc. Natl. Acad. Sci. U.S.A.*, 82, 6551, 1985.
86. **Rumeau, D., Mazau, D., and Esquerré-Tugayé, M. T.,** Cytocine-rich RNAs from infected melon plants and their *in vitro* translation Products, *Physiol. Mol. Plant. Pathol.*, 31, 305, 1987.
87. **Taylor, J. L., Fritzmeier, K., Hauser, I., Kombrink, E., Rohwer, F., Schroder, M., Strittmatter, G., and Hahlbrock, K.,** Structural analysis and activation by fungal infection of a gene encoding a pathogenesis-related protein in potato, *Mol. Plant-Microbe Interact.*, 3, 72, 1990.
88. **Dixon, R. A.,** The phytoalexin response: elicitation, signalling and the control of host gene expression, *Biol. Rev.*, 61, 239, 1986.
89. **Yamada, T., Hashimoto, H., Shiraishi, T., and Oku, H.,** Suppression of pisatin, phenylalanine ammonia-lyase mRNA, and chalcone synthase mRNA accumulation by a putative pathogenicity factor from the fungus *Mycosphaerella pinodes, Mol. Plant-Microbe Interact.*, 2, 256, 1989.
90. **Cuypers, B., Schmelzer, E., and Hahlbrock, K.,** *In situ* localization of rapidly accumulated phenylalanine ammonia-lyase mRNA around penetration sites of *Phytophthora infestans* in potato leaves, *Mol. Plant-Microbe Interact.*, 1, 157, 1988.
91. **Weltring, K-M., Turgeon, B. G., Yorder, O. C., and VanEtten, H. D.,** Isolation of a phytoalexin-detoxifying gene from the plant pathogenic fungus *Nectria haematococca* by detecting its expression in *Aspergillus nidulans, Gene,* 68, 335, 1988.
92. **Yorder, O. C.,** Altered virulence in recombinant fungal pathogens, *Abstr. 5th ICPP,* 1988, 226.
93. **Cramer, C. L., Edwards, K., Dron, M., Liang, X., Dildine, S. L., Bolwell, G. P., Dixon, R. A., Lamb, C. J., and Schuch, W.,** Phenylalanine ammonia-lyase gene organization and structure, *Plant Mol. Biol.*, 12, 367, 1989.
94. **Lois, R., Dietrich, A., Hahlbrock, K., and Schulz, W.,** A phenylalanine ammonia-lyase gene from parsley: structure, regulation and identification of elicitor and light responsive *cis*-acting elements, *EMBO J.*, 8, 1641, 1989.
95. **Dron, M., Clouse, S. D., Dixon, R. A., Lawton, M. A., and Lamb, C. J.,** Glutathione and fungal elicitor regulation of a plant defense gene promoter in electroporated protoplasts, *Proc. Natl. Acad. Sci. U.S.A.*, 85, 6738, 1988.
96. **Sommer, H. and Seadler, H.,** Structure of the chalcone synthase gene of *Artirrhinum majus. Mol. Gen. Genet.*, 202, 429, 1986.

# PLANT-FUNGAL COMMUNICATIONS THAT TRIGGER GENES FOR BREAKDOWN AND REINFORCEMENT OF HOST DEFENSIVE BARRIERS

**P. E. Kolattukudy**

## TABLE OF CONTENTS

# I. INTRODUCTION

Fungal interactions with plants that result in plant diseases cause one of the most serious problems in the production of food and fiber. Other types of fungi (Mycorrihizae) that establish mutually beneficial symbiotic relationships with plants enhance plant nutrition. Characteristics of these interactions clearly suggest that the fungal spores and plants begin to communicate with each other as soon as they get close to each other. Thus, plant signals may trigger germination of fungal spores.[1-3] Such signals could involve specific chemicals that trigger the expression of the genes involved in germination or the plant components may meet a more general need for nutrients for germination. Both the physical features of the plant surface and chemicals from the plant have been suggested to promote germination of the fungal spores and the early differentiation processes, such as the formation of structures that are needed for penetration into the host. However, the molecular basis of such interactions is very poorly understood. Plant signals also trigger fungal gene expression that allows the fungus to penetrate through the preexisting defensive barriers of the host. Fungal attack triggers expression of defense genes in plants. In recent years, much information has become available about the nature of such genes and how they might be regulated. Such aspects are reviewed elsewhere in this book. In the present chapter the discussion is confined to how the plant signals trigger fungal gene expression that assists the fungus to penetrate through the protective barriers of plants, and fungal signals that trigger host gene expression that results in reinforcement of plant defensive barriers to prevent further ingress of the fungus into the host.

# II. PLANT SIGNALS THAT CAUSE APPRESSORIUM FORMATION

In fungi that penetrate into the host through stomata, sensing of the physical structure of stomata was reported to cause formation of appressoria. The rust fungi form appressorium when the tip of the fungal germ tube contacts the stomata or a replica of the stomata.[4-7] The bending of the tip that occurs when the germ tube encounters the stomatal aperture is somehow transmitted, probably through the cytoskeletal structure, to the gene expression machinery involved in differentiation into appressoria. On the other hand, in a fungus that directly penetrates through the cuticle,[8] a chemical signal from the cuticular wax of the plant was recently found to cause appressorium formation. Thus, germinating spores of *Colletotrichum gloeosporioides*, a pathogen of avocado, are induced to form appressoria by the cuticular wax of avocado fruit (Figure 1) but not by cuticular wax of a variety of other plants.[9] Furthermore, avocado wax was not found to be capable of inducing appressorium formation in *Colletotrichum* species that attack other species of hosts. This specificity suggests that a host-specific chemical signal induces appressorium formation that is essential for pathogenesis. Chromatographic fractionation of the cuticular wax of avocado fruits showed that the fatty alcohol fraction of the wax induced appressorium formation. When synthetic fatty alcohols were tested, the very long-chain alcohols ($\geq C_{24}$) were found to cause appressorium formation whereas shorter ones could not. Chemical examination of the avocado fatty alcohols showed that $C_{24}$ to $C_{32}$ alcohols were major components. Mixing experiments suggest that some plant surface components can inhibit appressorium formation. Thus it appears that an appropriate balance between inducers and inhibitors in the surface components of the host may determine whether the host surface will allow the induction of appressorium formation. The host signal-induced fungal proteins associated with appressorium formation have been detected by two-dimensional gel analysis and genes that code for such proteins are being identified using subtractive libraries.[9a]

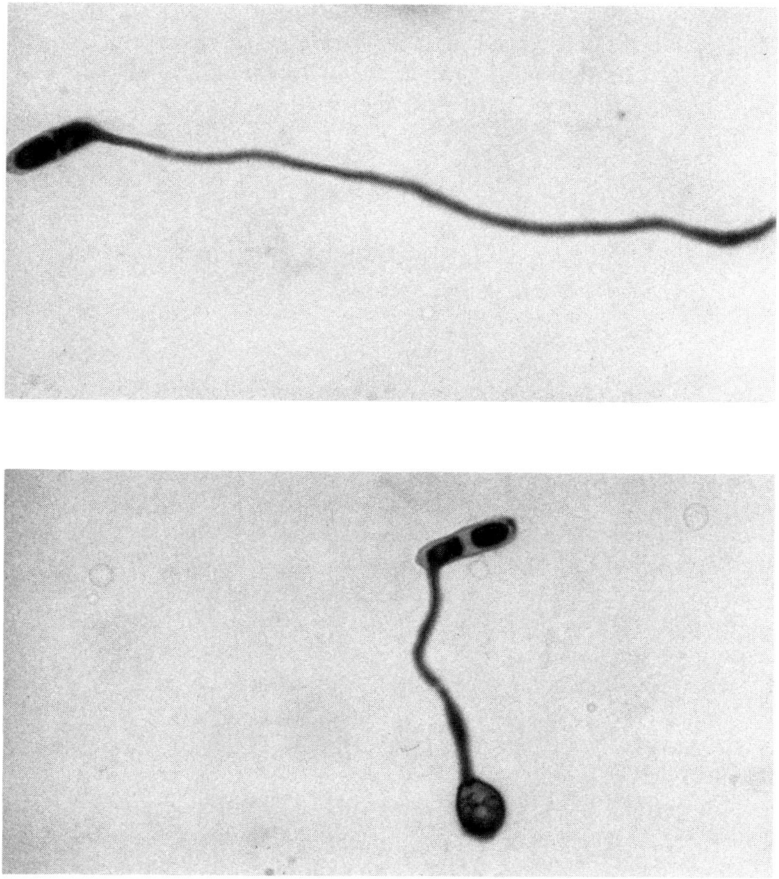

FIGURE 1. *Colletotrichum gloeosporioides* spores germinated in the absence (top) and presence (bottom) of cuticular wax from avocado fruits.

# III. PLANT SIGNAL INDUCTION OF A FUNGAL GENE NECESSARY FOR PENETRATION THROUGH THE HOST CUTICLE

## A. ROLE OF CUTINASE IN FUNGAL INFECTION

We have provided evidence that certain fungi use an extracellular cutinase to gain access into the host cuticle.[10,11] It was demonstrated immunocytochemically that a fungal spore penetrating into its host secretes cutinase.[12] Immunochemical localization of cutinase secretion by germinating fungal spores provided evidence that the enzyme secretion was localized at the point of fungal penetration. In fungi (for example, *Fusarium solani pisi*), in which a germ tube penetrates directly into the host without the formation of the appressorium, cutinase secretion occurs from the tip of this germ tube. On the other hand, in fungi (for example, *C. gloeosporioides*), in which the end of the germ tube differentiates into appressorium that generates a penetrating infection peg, cutinase secretion occurs at the tip of this infection peg and not at the tip of the original germ tube.[13] Selective inhibition of cutinase using specific antibodies or chemical inhibitors was found to prevent infection of several plant hosts by their fungal pathogens (Table 1).[14,15] Cutinase-deficient mutants of directly penetrating pathogens were found to be incapable of infecting their hosts without the addition

**TABLE 1**
**Host-Pathogen Pairs in which Specific Inhibition of Cutinase Has Been Found to Protect the Host**

| Fungus | Host |
|---|---|
| *Fusarium solani pisi* | Pea[a] |
| *Venturia inequalis* | Apple[b] |
| *Colletotrichum graminicola* | Corn[b] |
| *Colletotrichum gloeosporioides* | Papaya[c] |
| *Colletotrichum capsici* | Pepper[d] |

[a] Ref. 14.
[b] R. Chacko and P. E. Kolattukudy, unpublished, 1983.
[c] Ref. 15.
[d] W. F. Ettinger and P. E. Kolattukudy, unpublished, 1985.

of exogenous cutinase.[16,17] *Mycosphaerella*, a pathogen that requires a wound for infection of papaya fruits, could infect only when exogenous cutinase was included in the inoculum.[15] When this fungus was transformed with the cutinase gene from *F. solani pisi*, the transformants that could respond to the plant signal by induction of cutinase gene could also infect intact papaya fruits without requiring a wound.[18] Thus, these as well as other lines of evidence show fungal pathogens need cutinase to infect intact organs directly through the cuticle.

## B. CUTINASE INDUCTION

The cutinase produced in the extracellular fluid of *F. solani pisi* cultures grown on cutin as the sole source of carbon was purified.[19] Induction of cutinase under such conditions was presumably caused by the low amounts of cutin hydrolyzate that would be generated by the small amount of cutinase, one of the many extracellular hydrolytic enzymes generated in small quantities upon depletion of readily utilizable carbon sources. In support of such a hypothesis, it was demonstrated that in glucose-grown *F. solani pisi* cutinase could be detected at extremely low levels upon glucose depletion and addition of low levels of cutin hydrolyzate induced high levels of cutinase synthesis upon glucose depletion; glucose strongly repressed cutinase synthesis.[20]

When a fungal spore lands on a plant surface the contact with cutin might trigger the expression of the cutinase gene if the spore has the ability to generate monomers, which would be the real inducers of cutinase. It has been demonstrated that spores of highly virulent strains of pathogens that are capable of penetration into an intact host have cutinase on the surface.[21] When such spores carrying cutinase come in contact with insoluble cutin, cutinase is induced; cutinase transcripts are detectable within 15 min after contact with cutin.[22] That this induction was due to the monomers generated by the cutinase of the spore was shown by the observation that cutin hydrolyzate and isolated individual cutin monomers induced cutinase. The most effective inducers were the unique monomers of cutin, dihydroxy $C_{16}$ acid and trihydroxy $C_{18}$ acid. Examination of a number of isolates of *F. solani pisi* showed that those with higher amounts of cutinase on the spore could be readily induced to generate cutinase[23] and these strains were highly pathogenic on the intact host.[24] On the other hand, those with little cutinase on the spore showed poor inducibility and could not infect an intact host.[21] Thus, the unique cutin monomer is the signal that the fungus uses to trigger the expression of the cutinase gene and this process is highly significant to pathogenesis.

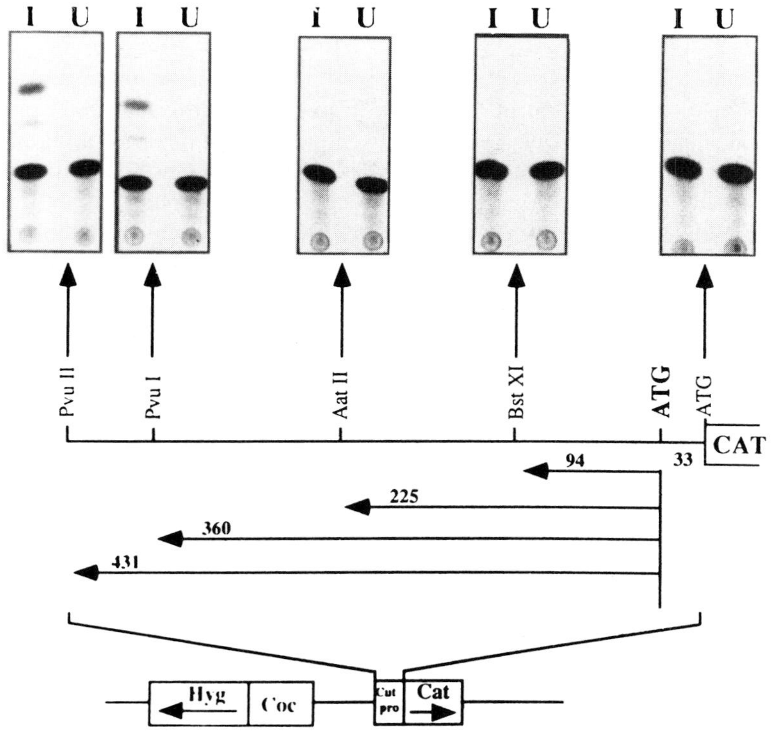


FIGURE 2. Promoter analysis of the cutinase gene from *Fusarium solani pisi*. CAT activity of *Fusarium* transformants containing the indicated segments of the cutinase promoter were assayed after induction by the addition of cutin hydrolyzate. I, induced; U, uninduced.

## C. CUTINASE GENE PROMOTER INDUCIBLE BY CUTIN MONOMER

To determine how the unique cutin monomers of a plant trigger the expression of the fungal cutinase gene, the cutinase promoter was identified. The 5′-flanking region of the cutinase gene from *F. solani pisi* was tested for promoter activity. When a flanking region of cutinase gene was inserted at the 5′-end of a promoter-less hygromycin-resistance gene and this plasmid was used to transform *F. solani pisi*, hygromycin-resistant transformants were obtained.[25] A 360-bp region of the 5′-flanking segment of cutinase gene was sufficient to generate hygromycin-resistant transformants. Since these transformants could not be used to quantitatively measure promoter activity and inducibility by the cutin monomers, a different approach was recently used. A constititutive promoter was used to drive the hygromycin resistance gene while in the same plasmid the putative cutinase promoter segments were fused at the 5′-end of the coding regions for chloramphenicol acetyltransferase (CAT). When this plasmid was used to transform *F. solani pisi* by electroporation hygromycin-resistant transformants were obtained.[26] In these transformants, CAT activity was inducible by cutin monomers and this induction was repressed by glucose. The 5′-flanking region of the cutinase gene was progressively shortened using convenient restriction sites and the transformants obtained with such constructs were tested for CAT induction by cutin monomers and for glucose repression. The results showed that the inducible promoter activity resides in the −225 to −360 region of the cutinase gene (Figure 2).[26] This segment contained consensus sequences for an SP1 site and a cAMP responsive element that might be involved in catabolite repression. Such consensus sequences were also found in the 5′-flanking regions of the cutinase gene in *C. gloeosporioides* and *C. capsici*,[27] suggesting that such elements may be a general feature of the cutinase gene that is involved in the regulation of the gene.

## D. CUTINASE TRANSCRIPTION ACTIVATION IN ISOLATED FUNGAL NUCLEI

Nuclear run-on experiments with nuclei from *F. solani pisi*, that was already induced to produce cutinase by treatment of the fungus with cutin hydrolyzate for various periods of time, showed that cutin hydrolyzate stimulated cutinase gene expression at a transcriptional level (Figure 3A). The labeled cutinase transcripts produced in these nuclear run-on experiments had already been initiated at the time of isolation of the nuclei as indicated by the finding that novobiocin, an inhibitor of initiation of transcription, did not inhibit the labeling of cutinase (Figure 3B).

To help elucidate the mechanism by which the plant signal triggers the expression of the cutinase gene, transcription in isolated fungal nuclei was examined. Nuclei isolated from glucose-grown *F. solani pisi* (not induced to produce cutinase), produced cutinase transcripts when they were incubated with cutin monomers and a soluble factor from the fungal extract, whereas neither the monomer alone nor the supernatant alone stimulated cutinase gene transcription (Figure 3C).[28] To test whether the stimulation of cutinase gene transcription in isolated nuclei by the cutin monomer and the protein factor is the result of some generalized effect or a specific activation of the cutinase gene, the effect of cutin monomer and the protein factor on transcription of the actin gene was examined. While the actin gene was transcribed in fungal nuclei, labeling of actin transcripts was found to be unchanged by the addition of cutin monomer and the protein factor. These results strongly suggest that the observed stimulation of cutinase gene transcription was a selective stimulation. Novobiocin severely inhibited this transcription (Figure 3C), strongly suggesting that the cutin monomer and protein factor stimulated initiation of transcription of the cutinase gene[28] unlike the results obtained with the run-on experiments with nuclei from already induced fungal cultures indicated above (Figure 3B). The *in vitro* transcript generated by the isolated nuclei was identical in size to the cutinase mRNA generated *in vivo,* showing that the transcription induced in the isolated nuclei by the cutin monomer and protein factor was properly initiated and terminated. If the activation of cutinase gene expression observed in the isolated nuclei is relevant to the mechanism by which the fungal spore senses its contact with the plant surface, the isolated nuclear preparation should show a requirement for the structural elements of the unique monomers of cutin. To test this possibility, several analogs of cutin monomer were tested. When the mid-chain hydroxyl group of 10,16-dihydroxyhexadecanoic acid was removed, the transcription-stimulating effect decreased dramatically; the ω-hydroxy acid and the α-hydroxy fatty acid were ineffective (Figure 4). The ω-hydroxyl group was found to be essential for the transcription-activation effect, since ricinoleic acid, which does have a mid-chain hydroxyl group but lacks an ω-hydroxyl, could not substitute for the cutin monomer. 9,10,18-Trihydroxyoctadecanoic acid, another unique cutin monomer, also stimulated cutinase transcription. 9,10,16-Trihydroxyhexadecanoic acid, a molecule very similar to the trihydroxy-$C_{18}$ cutin monomer, showed some stimulation of cutinase gene transcription. Thus, cutinase gene transcription activation requires the unique structural elements found in the cutin monomers that have not been found heretofore elsewhere in nature. The isolated nuclei provide a valid and convenient system to elucidate the molecular mechanism involved in activation of cutinase gene in the spore by the plant cutin components.

## E. *TRANS*-ACTING FACTORS INVOLVED IN THE REGULATION OF EXPRESSION OF CUTINASE GENE

The nature of the soluble factor required for activation of transcription of the cutinase gene by the dihydroxy fatty acid of cutin should provide insight into the mechanism of regulation of cutinase gene expression. Since boiling or protease treatment of the factor destroyed the transcription-activating effect of the factor, it was concluded that the factor was a protein. Fractionation of transcription-stimulating protein suggested that it aggregated

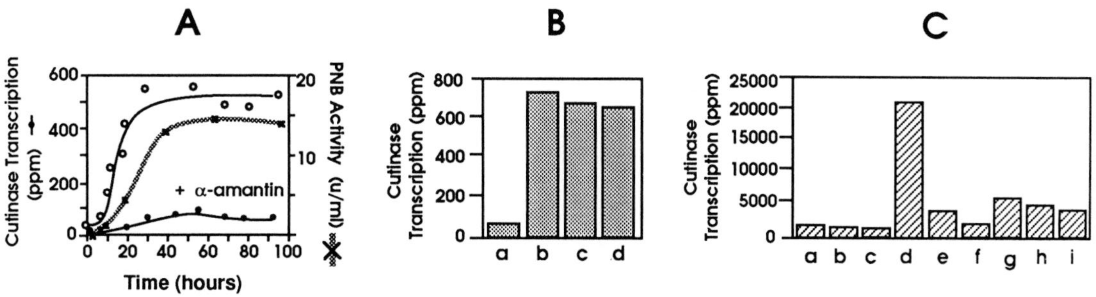


FIGURE 3. (A) Nuclear run-on experiments showing cutinase transcription rates in nuclei from *F. solani pisi* induced with cutin monomers for the indicated periods. α-Amanitin inhibition of transcription is indicated. Cutinase activity of the extracellular fluid of the culture at the time of isolation of nuclei is shown by the hydrolysis of *p*-nitrophenyl butyrate (PNB). (B) Effect of novobiocin on the formation of labeled cutinase gene transcript in nuclear run-on experiments with isolated nuclei from induced cultures of *F. solani pisi*. Lane a, nuclei from uninduced culture; lanes b,c, and d, nuclei from induced cultures with 0.1, 1, and 10 m*M* novobiocin. (C) Activation of cutinase gene transcription in isolated nuclei from glucose-grown *F. solani pisi*. Transcription reaction mixtures contained $^{32}$P-labeled UTP and other components needed for transcription, as well as: a, nuclei only; b, nuclei + 10,16-dihydroxyhexadecanoic acid (monomer) only; c, nuclei + cell extract supernatant only; d, nuclei + monomer + cell extract supernatant; e, as in d except that cell extract supernatant was boiled; f, as in d except that cell extract was pretreated with immobilized protease; g, h, and i, as in lane d with 0.1, 1, and 10 m*M* novobiocin, respectively.

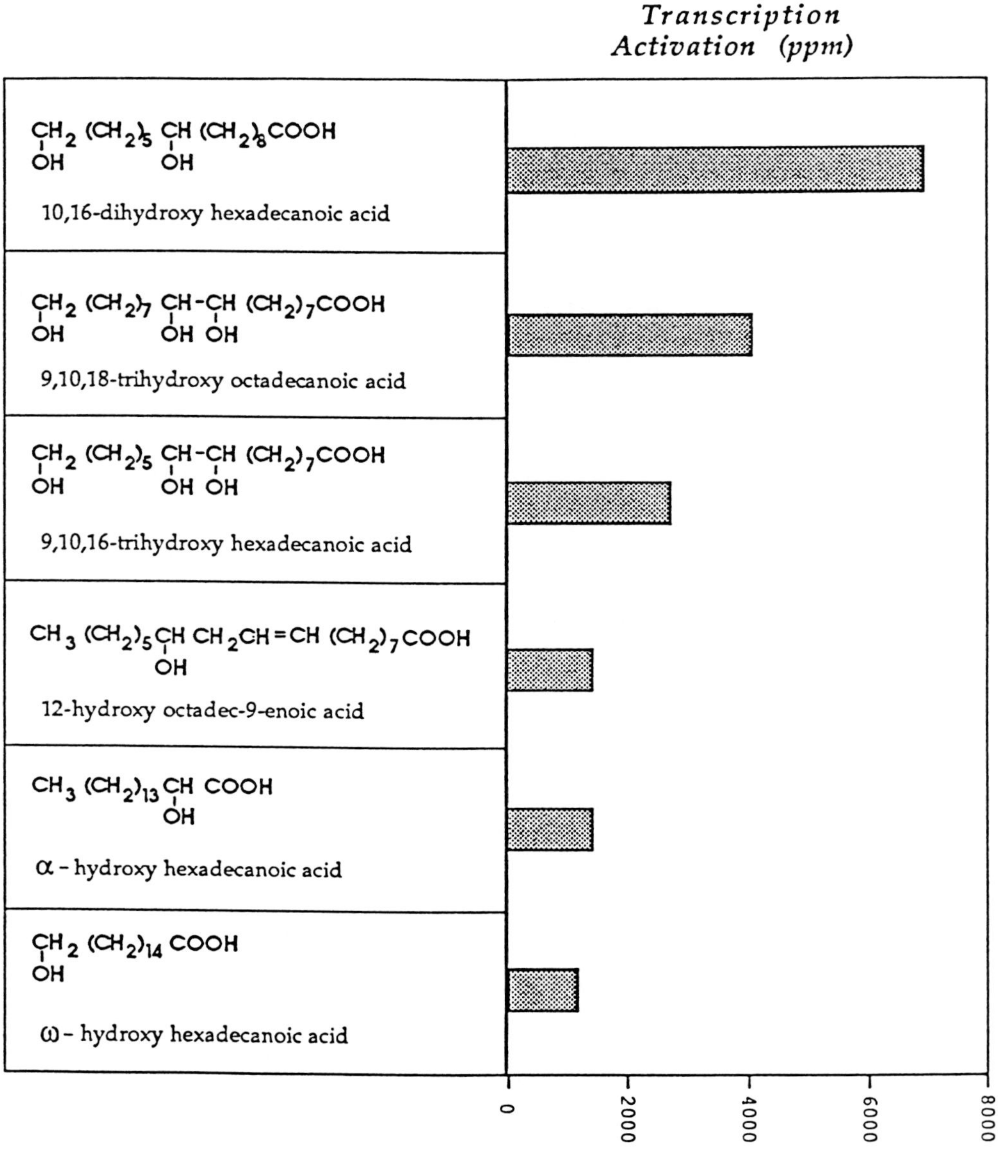


FIGURE 4. Activation of cutinase gene transcription in isolated nuclei of *F. solani pisi* by cutin monomers and their analogs in the presence of protein factor.

at low ionic strength. Gel filtration at high ionic strength showed that transcription-activating factor was a ~100-kDa protein (Figure 5). Among the many possibilities that can be postulated for the mechanism by which the protein factor might stimulate the transcription of the cutinase gene is its direct binding to a regulatory element in the gene. To test for such a possibility, each fraction from the gel filtration of the factor was tested for binding to the 5′-flanking region of the cutinase gene by a nitrocellulose binding assay. Although this assay is critically dependent on the experimental conditions, the protein fraction that stimulated cutinase gene transcription in isolated nuclei showed binding to the $^{32}P$-labeled 360-bp segment of the 5′-flanking region of the cutinase gene. The binding activity was associated with the protein fractions that showed transcription-stimulating activity. A similar size fragment from the 3′-flanking region or other DNA fragments from other sources did not show any binding (Figure 5). These results strongly suggested that the protein factor might be a DNA-binding transcription factor. Certain characteristics of the transcription-activating factor, such as, aggregation in the absence of high ionic strength, suggested that the protein might be of nuclear origin. In fact, nuclear extract from *F. solani pisi* substituted for the protein factor in the cutinase transcription assays with isolated nuclei[28a] (J. Kämper and P. E. Kolattukudy, unpublished). Electrophoresis of the 360 bp segment of the 5′-flanking region of cutinase gene after incubation with either this nuclear extract or the protein factor from the supernantant showed the same gel retardation band.[28b] This result strongly suggests that the protein factor is a nuclear protein that binds to the upstream region of the cutinase gene.

If the observed binding of the nuclear protein is, in fact, involved in the activation of transcription of cutinase gene, it might bind at or near the promoter regions identified to be necessary for the inducible expression by the dihydroxy acid of cutin. When the various deletion fragments from the 5′-flanking region of cutinase gene were subjected to electrophoresis after incubating with nuclear extracts, it was found that the 360-bp fragment showed a retardation band whereas deletion of the 135-bp 5′-segment of this DNA abolished gel retardation (Figure 6). Thus, the nuclear protein binds with the same 135-bp segment as that found to be required for inducible expression of the gene in the transformants.

How the dihydroxy acid of the plant functions with the fungal nuclear protein to activate transcription of the cutinase gene remains unclear. Time course of activation of transcription of the cutinase gene by the protein factor and dihydroxy $C_{16}$ acid in the isolated nuclei showed a lag of 30 min and near maximal stimulation was observed in about 1 h.[28] Presumably the biochemical events that occur in this early period are necessary to activate transcription. Therefore, an understanding of these events is the key to understanding the mechanism of activation of the cutinase gene by the cutin monomer. Preincubation of protein factor and the monomer before the addition of nuclei and $^{32}P$-labeled UTP did not affect the lag indicating that the protein factor did not convert the monomer into a biologically functional transcription activator. Preincubation of both the monomer and the protein factor, but neither alone, with the nuclei before the addition of $^{32}P$-labeled UTP eliminated the lag period. Thus it appears that the biochemical reaction for cutinase transcription activation requires the monomer and the protein factor together.

## F. EVIDENCE THAT PHOSPHORYLATION IS INVOLVED IN THE ACTIVATION OF CUTINASE GENE TRANSCRIPTION BY CUTIN MONOMERS

Incubation of nuclei with protein factor and monomer in the presence of $^{32}P$-labeled ATP showed phosphorylation of a ~50-kDa protein, whereas neither protein factor nor monomer alone was adequate to achieve maximum phosphorylation of the protein by nuclei (Figure 7, left). That phosphorylation is required for transcription was indicated by the observation that when the isolated nuclei were preincubated with the protein factor and cutin monomer together with protein kinase inhibitor H-7 before the addition of transcription components including $^{32}P$-labeled UTP, cutinase gene transcription was severely inhibited. When the inhibitor was added after the preincubation of nuclei with monomer and protein factor very little inhibition of transcription occurred suggesting that the inhibitor prevented the reaction essential for activation of transcription that would normally occur during the preincubation. Preincubation of the nuclei with phosphotyrosine antibodies, but not control IgG, also inhibited cutinase gene transcription (Figure 7, right). These results suggest that the cutin monomer is involved in causing phosphorylation of a protein, possibly at tyrosine, and that the phosphyorylation is required for activation of the transcription of the cutinase gene. One of the key reactions that occur upon preincubation of nuclei with monomer and

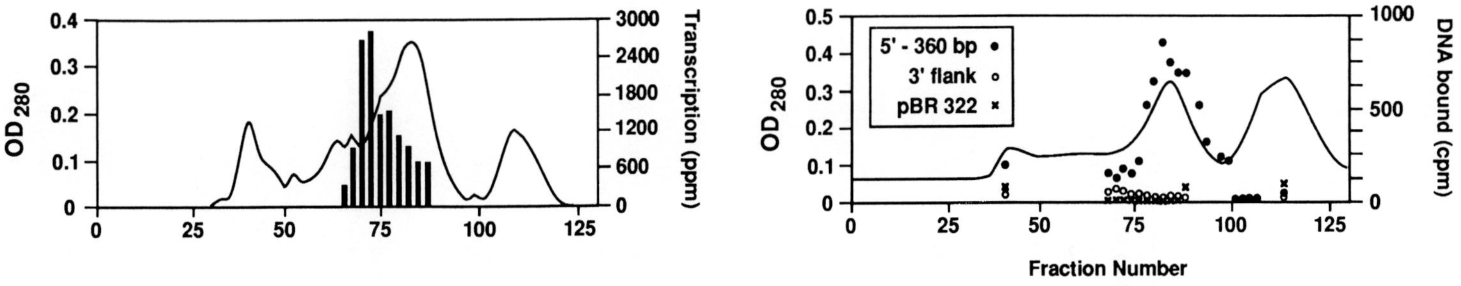


FIGURE 5. Sepharose 6B chromatography of soluble protein from *F. solani pisi* under high salt (0.1 *M* KC1) conditions showing (left) transcription activation as indicated by solid bars (assayed with isolated *F. solani pisi* nuclei and 10,16-dihydroxyhexadecanoic acid) and (right) binding of labeled DNA fragments from the 5′-flanking (●), 3′-flanking (○) regions of the cutinase gene, or from a HAE II restriction of pBR322 DNA (x), as measured by a filter binding assay.

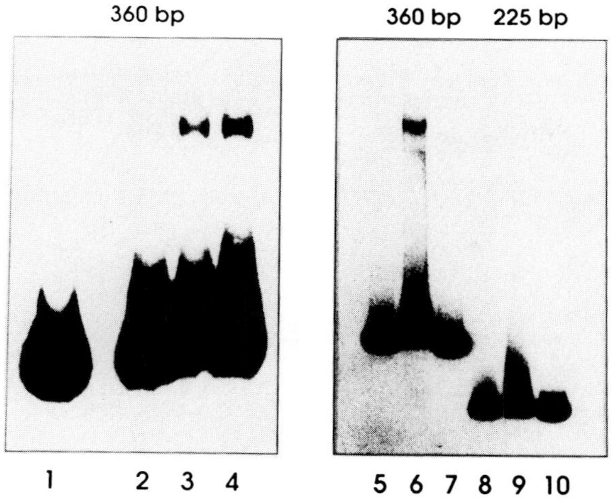


FIGURE 6. Gel retardation of 5′-flanking region of cutinase gene by interaction with nuclear proteins from *F. solani pisi* and prevention of this interaction by phosphatase treatment of the nuclear protein. $^{32}$P-labeled 360-bp fragment (lanes 1-7) or 225-bp fragment (lanes 8-10) of the cutinase gene was incubated for 30 min with (lanes 2, 6, 7, 9, and 10) or without (lanes 1, 5, and 8) nuclear protein isolated from *F. solani pisi* and the mixture subjected to electrophoresis. In lane 2, 100-fold excess of unlabeled 360-bp was added; in lanes 7 and 10 the nuclear protein was treated for 30 min with immobilized alkaline phosphatase and the phosphatase removed by centrifugation prior to incubation with labeled DNA.

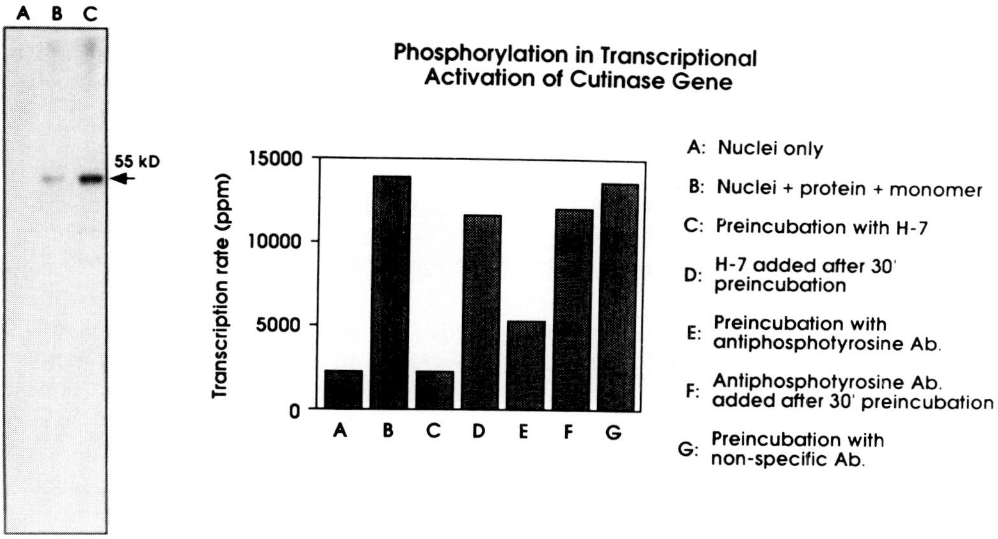


FIGURE 7. (Left) Phosphorylation of proteins by: A, isolated nuclei from *F. solani pisi;* B, nuclei in the presence of cutin monomer; C, nuclei in the presence of monomer and soluble protein extract from *F. solani pisi*. After incubation of all assay components with [$\gamma^{32}$P]ATP, the proteins were analyzed by SDS-PAGE and autoradiography. (Right) Effect of the phosphorylation state of the protein factor on transcription of the cutinase gene by isolated nuclei. All assays except A contained nuclei, cutin monomer, and soluble protein extract; specific assay variations are described at the right.

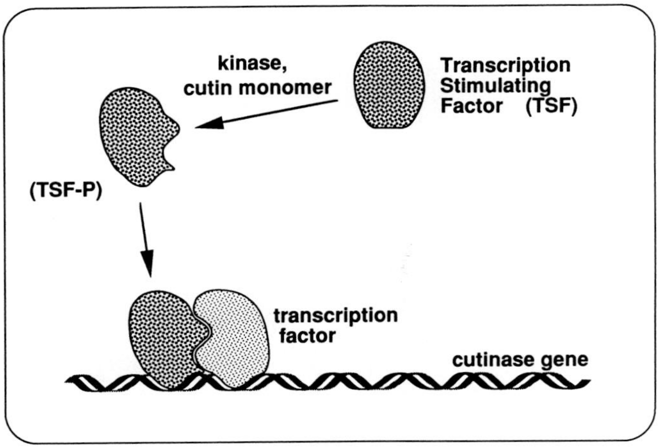


FIGURE 8. Proposed model for the transcriptional activation of cutinase gene by cutin monomer.

protein factor that eliminates the lag in transcription activation is probably this essential phosphorylation event.

If the phosphorylation is on a transcription factor, it is possible that binding of the factor to the inducible promoter regions of the cutinase gene would require phosphorylation. In fact, treatment of the nuclear protein fraction with immobilized phosphatase prevented binding the 5′-flanking segment of cutinase gene as detected by gel retardation (Figure 6). Thus, all of the evidence so far obtained strongly suggests that the hydroxy fatty acid monomer of cutin causes phosphorylation of a transcription factor that activates transcription of cutinase gene by binding to the −225 to −360 region of the cutinase gene only in the phosphorylated form (Figure 8). Since SDS gel electrophoresis of the phosphorylated form showed one half the size of the native transcription-activating protein, it seems likely that phosphorylated transcription factor functions as a dimer, like many other transcription factors.[29,30]

# IV. FUNGAL SIGNAL INDUCTION OF A PLANT GENE INVOLVED IN REINFORCEMENT OF THE HOST CELL WALL

## A. NATURE OF SUBERIN AND THE ROLE OF PEROXIDASE IN SUBERIZATION

As the invading fungus breaks down the protective barriers of plants some of the products released in this process, including signals derived from the cell walls of the pathogens, trigger many defense reactions in the host. Among these defensive processes is reinforcement of the host cell walls to make them resistant to the extracellular degradative enzymes secreted by the pathogen. It has been known for a long time that pathogen attack causes deposition of phenolic materials on the cell wall.[31-33] The polymeric material containing the phenolics deposited on the wall has been variously designated as lignin, wound-lignin, suberin, etc., and this identification has been based mainly on staining characteristics of the deposits. Such staining has been directed at the phenolics and little information is available on the possible association of aliphatic components characteristic of suberin. In the absence of reliable information about the chemical composition, it has not been possible to accurately designate the wall deposits caused by fungal infections. It was suggested that, in many cases, fungal infection might be triggering suberization,[10] a universal response to wounding in many plant organs.[34]

**TABLE 2**
**Suberin in the Petiole of Susceptible and Resistant Tomato Line Inoculated with *Verticillium albo-atrum***

| Source | Octadecene-1,18-diol (arbitrary area units) |
|---|---|
| Susceptible control | 9,500 |
| Susceptible inoculated | 9,700 |
| Resistant control | 7,300 |
| Resistant inoculated | 44,000 |

Even though the precise structures present in suberized wall are not known, a model representing the general picture of the chemical nature of suberized walls has emerged from a series of chemical and biochemical studies.[10,35,36] According to this model, a phenolic matrix is deposited on the cell walls and aliphatic components, often consisting of ω-hydroxy fatty acids, dicarboxylic acids, and very long-chain acids and alcohols, are attached to the phenolic matrix.[36] Biochemical studies suggested that the phenolic materials are polymerized and cross-linked by a highly anionic peroxidase.[37] Using immunocytochemical methods, it was found that this peroxidase is localized exclusively in the walls of suberized cells and that this enzyme appears only during suberization.[38] When suberization is either inhibited or promoted by mineral deficiency[39,40] or hormones,[41] these changes in suberization were associated with appropriate changes in the level of anionic peroxidase, as expected from its involvement in suberization. The cDNA for this peroxidase was cloned and sequenced using mRNA from suberizing potato tuber.[42] With this cDNA as the probe, the induction of suberization was found to be associated with the induction of peroxidase gene transcription.

## B. SUBERIZATION AS A DEFENSE AGAINST FUNGAL INVASION

A clear example of suberization as a defense mechanism against fungal attack is the case of defense of tomato against *Verticillium albo-atrum*. In two near-isogenic tomato lines that differ in their ability to defend against *V. albo-atrum*, the resistant one was found to deposit a vascular coating material that seemed to limit the spread of the fungus whereas the susceptible one was unable to defend in this manner.[43] Chemical analysis of the depolymerization products of the polymeric materials by combined capillary GLC-MS showed that the aliphatic components characteristic of suberin appeared in the resistant but not in the susceptible tomato line (Table 2).[44] This strong chemical evidence for suberization as a defense against *V. albo-atrum* was strongly supported by the finding that upon inoculation of tomato petioles with *V. albo-atrum* spores through the transpirational stream, the transcription of highly anionic peroxidase gene appeared in the resistant line in a manner consistent with the time course of appearance of the cytochemically detectable vascular coating; in the susceptible line, very little peroxidase transcript was found in the same time period (Figure 9). Introduction of an elicitor fraction from *V. albo-atrum* or abscisic acid, an agent known to induce suberization in tomato and potato tissue cultures,[45] also caused production of aliphtic components of suberin and induction of anionic peroxidase transcripts in the resistant, but not in the susceptible, tomato line. Thus the cytochemical, chemical, and molecular biological lines of evidence suggested that the susceptible tomato line is unable to respond to fungal attack or abscisic acid by depositing a defensive suberin barrier.

Cell cultures from these two near-isogenic tomato lines also responded differently to the fungal signal. When cell cultures were treated with a *V. albo-atrum* wall elicitor preparation, the anionic peroidase gene expression was triggered in the resistant line but the susceptible line failed to respond (Figure 10).[46] In the cells from the resistant tomato line

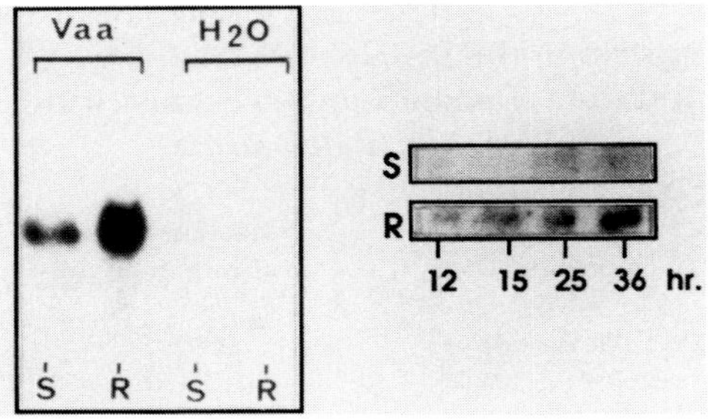


FIGURE 9. Peroxidase RNA appearance (Northern blot) in the petioles of resistant (R) and a near-isogenic susceptible (S) tomato lines due to infection with *V. albo-atrum*. (Left) comparison of petioles that received spores of *V. albo-atrum* (Vaa) or water ($H_2O$) incubated for 72 h; (Right) time course of induction.

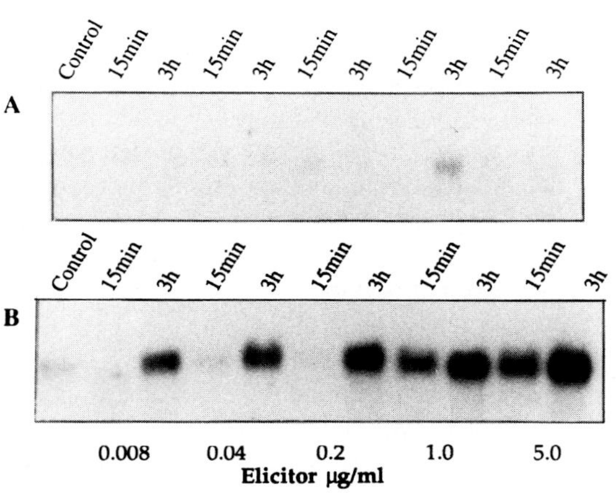


FIGURE 10. Northern blots of RNA from cell cultures from susceptible (A) and a near-isogenic resistant (B) tomato line treated with the indicated concentrations of cell wall elicitor prepared from *V. albo-atrum* for the indicated periods.

very low levels (nanogram per milliliter) of the elicitor preparation triggered the peroxidase gene expression whereas cells from the susceptible line required several orders of magnitude higher concentration before the gene expression could be triggered. Within 15 min after the application of the elicitor, peroxidase transcripts could be readily detected in the cells from the resistant tomato line. Most probably this highly anionic peroxidase, that is generated and secreted into the cell wall only in response to physical injury or signals generated by fungal attack, catalyzes the formation of cross-links in the preexisting cell wall components involving the phenolics. As more phenolics are deposited, further cross-links are generated, ultimately completing the formation of a suberized defensive barrier that is virtually impermeable to most pathogenic fungi. Thus the cross-linking by peroxidase probably plays a role from the early period of attack or injury until the barrier formation is completed.

## C. GENE FOR ANIONIC PEROXIDASE

Since regulation of expression of the gene for anionic peroxidase is important for both wound healing and resistance to fungal attack, we cloned and sequenced the gene encoding this enzyme from tomato using the highly homologous potato peroxidase cDNA as the probe.[45] Two copies of the gene in tandem were found. Each had three exons and the two copies were highly (84%) homologous.

### D. ANIONIC PEROXIDASE PROMOTER

Since peroxidase gene expression is an early event triggered by signals generated by fungal attack and physical injury, the 5′-flanking region of this gene might have the structural elements that allow the gene to respond to the signal(s). To test this possibility, peroxidase gene including ≈1.8-kb upstream region was introduced into tobacco, that does not seem to be able to generate any peroxidase transcripts which hybridize with the potato (or tomato) anionic peroxidase cDNA. When these transgenic tobacco plants were wounded, the tomato anionic peroxidase transcripts appeared.[47] Both leaves and stems responded similarly to wounding by expressing the peroxidase transcripts. Unwounded leaves and stems of these transgenic tobacco plants did not contain detectable amounts of the peroxidase transcripts. The expression of this peroxidase was limited to the immediate vicinity of the wound. The time course of expression of the peroxidase transcript was identical to that observed for the production of the transcripts in wound-healing potato and tomato. Thus, this peroxidase gene contained wound-inducible promoter.

Transgenic tobacco plants containing a 767-bp 5′-segment of the tomato anionic peroxidase gene fused to the β-glucuronidase (GUS) gene were also used to study the wound-inducible promoter of the peroxidase gene. These transgenic plants expressed GUS activity upon wounding.[47a] Time-course of expression of activity showed a biphasic response, the latter phase being similar to the expression of peroxidase observed during suberization. The early phase that started within 15 min after wounding reached a maximum in 12 h. This early response probably reflects peroxidase production for cross-linking of the preexisting phenolics and the phenolics deposited in the wall soon after wounding. When the transgenic tobacco plants were infected by fungi, GUS activity appeared at the site of infection.

### E. TRANSGENIC TOBACCO PLANTS THAT CONSTITUTIVELY EXPRESS HIGH LEVELS OF ANIONIC PEROXIDASE

To study the effect of expression of high levels of the suberization-associated, highly anionic peroxidase, a CMV-35S promoter was fused to the tomato peroxidase gene and introduced into tobacco plants. The transgenic tobacco plants thus generated expressed the tomato peroxidase gene at widely differing levels, generating plants that showed peroxidase activity ranging from 4- to 400-fold of that found in the control untransformed tobacco plants (Table 3).[48] These plants showed normal growth characteristics. The plants that expressed the highest levels of peroxidase grew and matured more slowly; however, the ultimate size of the plant was normal. In no case did wilting occur during the flowering period or at any other stage in any of the plants that expressed the gene at modest to extremely high levels. It is clear that high expression of the peroxidase does not cause physiological changes that results in wilting reported to be caused by overexpression of ''lignin-synthesizing'' peroxidase from tobacco in tobacco.[49] Since the suberization-associated peroxidase is a distinctly different enzyme with limited homology to other peroxidases including the ''lignin-synthesizing'' peroxidase, the observed differences in the behavior of the plants that overexpress the two peroxidases are not surprising, unless the term peroxidase is loosely and erroneously taken to mean the same enzyme.

### F. ANTISENSE APPROACH TO SUPPRESS PEROXIDASE EXPRESSION

If the expression of the highly anionic peroxidase is involved in resistance to *V. albo-atrum*, as strongly suggested by the results obtained so far, suppression of expression of this gene might render susceptibility to the fungal attack. One approach to test this possibility

**TABLE 3**
**Peroxidase Activity Assays of *Nicotiana tabaccum* (35 S)$_2$/TAP-1[a] Expressors**

| Transgenic plants | Specific activity[b] |
|---|---|
| B | 407 |
| 7 | 240 |
| P | 154 |
| A | 146 |
| L | 127 |
| 5 | 110 |
| 100 | 109 |
| 16 | 99.2 |
| D | 65.3 |
| **Nontransformed controls** | |
| cv. NC2326 | 1.02 |
| cv. Coker | 2.94 |

[a] *N. tobaccum* transformants generated with a double CMV-35S promoter.
[b] S.A., $OD_{470}$/min/mg total protein.

is to produce antisense RNA by introducing antisense DNA in a transcriptionally competent manner with a promoter. Since there are two highly homologous anionic peroxidase genes in tomato, the 5′-end of both genes were fused to CMV-35S promoter and introduced into tomato plants. Transgenic tomato plants thus generated showed production of the antisense transcripts as revealed by Northern blot analysis. When ABA or spores of *V. albo-atrum* were introduced into the petioles of the transgenic plants, they failed to generate the peroxidase message whereas the control plant produced substantial amounts of the peroxidase message (Figure 11). This evidence strongly suggests that the antisense approach is effective in suppressing the level of mRNA for the peroxidase gene even when the expression of the gene is triggered by agents known to strongly induce the expression of this gene.[50] The degree of susceptibility of these transgenic tomato plants to *V. albo-atrum* has yet to be assessed.

## V. CONCLUSION

Since fungal interaction with plants can have major beneficial or detrimental consequences, a clear understanding of the process could lead to new ways of modifying these interactions. Such an approach might help to better utilize mycorrhizal fungi to enhance plant nutrition and other beneficial effects. Interventions at some crucial stages in the interaction between pathogenic fungi with their hosts could be an effective way to protect plants. Significant beginnings are being made in understanding these processes, especially in the case of the pathogenic interactions. Only two aspects of these interactions that involve breaking of the preformed defensive barriers and reinforcement of the host cell walls triggered by the fungal signal are reviewed in this chapter. The regulation of a fungal pathogenesis gene (cutinase) by the plant signal appears to be analogous to the regulation of other genes in animals by lipid-like regulators of gene expression such as steroid hormones retinoic acid, thyroxine, etc.[51,52] Recent progress in other aspects of fungal interaction with plants, especially those involving the host responses, are described in other chapters. However, much

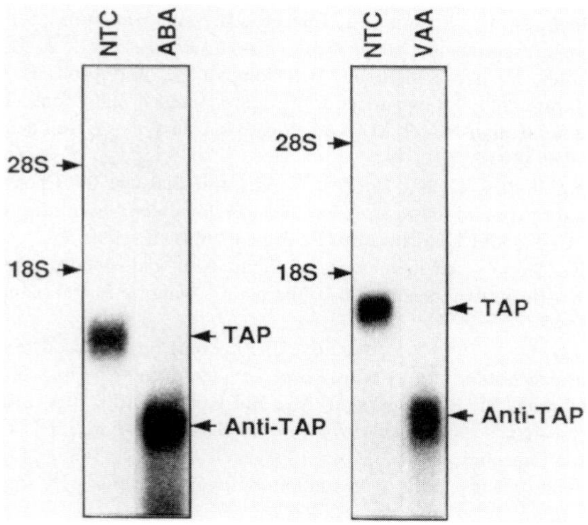


FIGURE 11. Northern analysis of expression of tomato anionic peroxidase (TAP) in abscisic acid-treated (ABA) and *V. albo-atrum* inoculated (Vaa) petioles of transgenic tomato plants producing the antisense TAP transcript. NTC, nontransformed control.

remains unclear. For example, little is known about the molecular mechanisms by which the host triggers the programmed differentiation of the fungal spore that ultimately results in pathogenesis. Molecular aspects of mycorrhizal interactions with plants is another area that needs elucidation. With the recent availability of the modern molecular tools it is possible to elucidate the molecular basis of the many facets of fungal interaction with plants. Thus this area holds much potential for progress in both understanding the process and in devising new methods to modify the process for human benefit.

## ACKNOWLEDGMENTS

This work was supported in part by grant DCB-8819008 from the National Science Foundation. The assistance of Linda M. Rogers in the preparation of this manuscript is deeply appreciated.

## REFERENCES

1. **Staples, R. C. and Macko, V.,** Formation of infection structures as a recognition response in fungi, *Exp. Mycol.,* 4, 2, 1980.
2. **Macko, V.,** Inhibitors and stimulants of spore germination and infection structure formation in fungi, in *The Fungal Spore,* Turian, G. and Hohl, H. R., Eds., Academic Press, San Diego, 1981, 565.
3. **French, R. C.,** The bioregulatory action of flavor compounds on fungal spores and other propagules, *Annu. Rev. Phytopathol.,* 23, 173, 1985.
4. **Wynn, W. K.,** Appressorium formation over stomates by the bean rust fungus: response to a surface contact stimulus, *Phytopathology,* 66, 136, 1976.
5. **Wynn, W. K. and Staples, R. C.,** Tropisms of fungi in host recognition, in *Plant Disease Control: Resistance and Susceptibility,* Staples, R. C. and Toenniessen, G. A., Eds., John Wiley & Sons, NY, 1981, 45.

6. **Staples, R. C., Hoch, H. C., Epstein, L., Laccetti, L., and Hassouna, S.,** Recognition of host morphology by rust fungi: responses and mechanisms, *Can. J. Plant Pathol.*, 7, 314, 1985.
7. **Hoch, H. C., Staples, R. C., Whitehead, B., Comeau, J., and Wolf, E. D.,** Signaling for growth orientation and cell differentiation by surface topography, *Science*, 235, 1659, 1987.
8. **Binyamini, A. and Schiffman-Nadel, M.,** Latent infections in avocado fruit due to *Colletotrichum gloeosporioides, Phytopathology*, 62, 592, 1972.
9. **Kolattukudy, P. E., Podila, G. K., Rogers, L. M., and Prusky, D.,** Chemical signals from avocado surface wax trigger formation of appressoria in *Colletotrichum gloeosporioides*, manuscript in preparation.
9a. **Pyee, J., Hwang, C.-S., and Kolattukudy, P. E.,** unpublished, 1990.
10. **Kolattukudy, P. E.,** Biopolyester membranes of plants: cutin and suberin, *Science*, 208, 990, 1980.
11. **Kolattukudy, P. E.,** Enzymatic penetration of the plant cuticle by fungal pathogens, *Annu. Rev. Phytopathol.*, 23, 223, 1985.
12. **Shaykh, M., Soliday, C. L., and Kolattukudy, P. E.,** Proof for the production of cutinase by *Fusarium solani f. pisi* during penetration into its host, *Pisum sativum, Plant Physiol.*, 60, 170, 1977.
13. **Podila, G. K., Dickman, M. B., Rogers, L. M., and Kolattukudy, P. E.,** Regulation of expression of fungal genes by plant signals, in *Molecular Biology of Filamentous Fungi*, H. Nevalainen and M. Penttila, Eds., Foundation for Biotechnical and Industrial Research, Helsinki, 1989, 217.
14. **Maiti, I. B. and Kolattukudy, P. E.,** Prevention of fungal infection of plants by specific inhibition of cutinase, *Science*, 205, 507, 1979.
15. **Dickman, M. B., Patil, S. S., and Kolattukudy, P. E.,** Purification, characterization and role in infection of an extracellular cutinolytic enzyme from *Colletotrichum gloeosporioides* Penz. on *Carica papaya* L., *Physiol. Plant Pathol.*, 20, 333, 1982.
16. **Dickman, M. B. and Patil, S. S.,** Cutinase deficient mutants of *Colletotrichum gloeosporioides* are nonpathogenic to papaya fruit, *Physiol. Mol. Plant. Pathol.*, 28, 235, 1986.
17. **Dantzig, A. H., Zuckerman, S. H., and Andonov-Roland, M. M.,** Isolation of a *Fusarium solani* mutant reduced in cutinase activity and virulence, *J. Bacteriol.*, 168, 911, 1986.
18. **Dickman, M. B., Podila, G. K., and Kolattukudy, P. E.,** Insertion of cutinase gene into a wound pathogen enables it to infect intact host, *Nature (London)*, 342, 446, 1989.
19. **Purdy, R. E. and Kolattukudy, P. E.,** Hydrolysis of plant cuticle by plant pathogens. Purification, amino acid composition, and molecular weight of two isoenzymes of cutinase and a nonspecific esterase from *Fusarium solani* f. *pisi, Biochemistry*, 14, 2824, 1975.
20. **Lin, T. S. and Kolattukudy, P. E.,** Induction of a biopolyester hydrolase (cutinase) by low levels of cutin monomers in *Fusarium solani* f. sp. *pisi, J. Bacteriol.*, 133, 942, 1978.
21. **Koller, W., Allen, C. R., and Kolattukudy, P. E.,** Role of cutinase and cell wall degrading systems in infection of *Pisum sativum* by *Fusarium solani* f. sp. *pisi, Physiol. Plant Pathol.*, 20, 47, 1982.
22. **Woloshuk, C. P. and Kolattukudy, P. E.,** Mechanism by which contact with plant cuticle triggers cutinase gene expression in the spores of *Fusarium solani* f. sp. *pisi, Proc. Natl. Acad. Sci. U.S.A.*, 83, 1704, 1986.
23. **Kolattukudy, P. E.,** Lipid-derived defensive polymers and waxes and their role in plant-microbe interaction, in *The Biochemistry of Plants*, Vol. 9, *Lipids: Structure and Function*, Stumpf, P. K., Ed., Academic Press, San Diego, 1987, 291.
24. **Kolattukudy, P. E., Soliday, C. L., Woloshuk, C. P., and Crawford, M.,** Molecular biology of the early events in the fungal penetration into plants, in *Molecular Genetics of Filamentous Fungi*, Alan R. Liss, NY, 1985, 421.
25. **Soliday, C. L., Dickman, M. B., and Kolattukudy, P. E.,** Structure of the cutinase gene and detection of promoter activity in the 5′-flanking region by fungal transformation, *J. Bacteriol.*, 171, 1942, 1989.
26. **Bajar, A., Podila, G. K., and Kolattukudy, P. E.,** Identification of a fungal cutinase promoter that is inducible by a plant signal via a phosphorylated transacting factor, *Proc. Natl. Acad. Sci. U.S.A.*, 88, in press, 1991.
27. **Ettinger, W. F., Thukral, S. K., and Kolattukudy, P. E.,** Structure of cutinase gene, cDNA, and the derived amino acid sequence from phytopathogenic fungi, *Biochemistry*, 26, 7883, 1987.
28. **Podila, G. K., Dickman, M. B., and Kolattukudy, P. E.,** Transcriptional activation of a cutinase gene in isolated fungal nuclei by plant cutin monomers, *Science*, 242, 922, 1988.
28a. **Kämper, J. and Kolattukudy, P. E.,** unpublished, 1990.
28b. **Podila, G. K., Rogers, L. M., and Kolattukudy, P. E.,** unpublished, 1990.
29. **Fawell, S. E., Lees, J. A., White, R., and Parker, M. G.,** Characterization and colocalization of steroid binding and dimerization activities in the mouse estrogen receptor, *Cell*, 60, 953, 1990.
30. **Glass, C. K., Lipkin, S. M., Devary, O. V., and Rosenfeld, M. G.,** Positive and negative regulation of gene transcription by a retinoic acid-thyroid hormone receptor heterodimer, *Cell*, 59, 697, 1989.
31. **Kahl, G.,** Metabolism in plant storage tissues, *Bot. Rev.*, 40, 263, 1974.
32. **Friend, J.,** Lignification in infected tissue, in *Biochemical Aspects of Plant-Parasite Relationship*, Friend, J. and Threlfall, D. R., Eds., Academic Press, San Diego, 1976, 195.

33. **Uritani, I. and Oba, K.,** The tissue slice system as a model for studies of host-parasite relationships, in *Biochemistry of Wounded Plant Tissues,* Kahl, G., Ed., De Gruyter, Berlin, 1978, 287.
34. **Dean, B. B. and Kolattukudy, P. E.,** Synthesis of suberin during wound-healing in jade leaves, tomato fruit, and bean pods, *Plant Physiol.,* 58, 411, 1976.
35. **Kolattukudy, P. E.,** Lipid polymers and associated phenols: their chemistry biosynthesis, and role in pathogenesis, *Recent Adv. Phytochem.,* 11, 185, 1977.
36. **Kolattukudy, P. E. and Espelie, K. E.,** Chemistry, biochemistry and function of suberin and associated waxes, in *Natural Products of Woody Plants, Chemicals Extraneous to the Lignocellulosic Cell Wall,* Rowe, J., Ed., Springer-Verlag, Berlin, 1989, 304.
37. **Kolattukudy, P. E.,** Structure, biosynthesis, and biodegradation of cutin and suberin, *Annu. Rev. Plant Physiol.,* 32, 539, 1981.
38. **Espelie, K. E., Franceschi, V. R., and Kolattukudy, P. E.,** Immunocytochemical localization and time-course of appearance of an anionic peroxidase associated with suberization in wound-healing of potato tuber tissue, *Plant Physiol.,* 81, 487, 1986.
39. **Pozuelo, J. M., Espelie, K. E., and Kolattukudy, P. E.,** Magnesium deficiency results in increased suberization in endodermis and hypodermis of corn roots, *Plant Physiol.,* 74, 256, 1984.
40. **Sijmons, P. C., Kolattukudy, P. E., and Bienfait, H. F.,** Iron deficiency decreases suberization in bean roots through a decrease in suberin-specific peroxidase activity, *Plant Physiol.,* 78, 115, 1985.
41. **Cottle, W. and Kolattukudy, P. E.,** Abscisic acid stimulation of suberization: induction of enzymes and deposition of polymeric components and associated waxes in tissue cultures of potato tuber, *Plant Physiol.,* 70, 775, 1982.
42. **Roberts, E., Kutchan, T., and Kolattukudy, P. E.,** Cloning and sequencing of cDNA for a highly anionic peroxidase from potato and the induction of its mRNA in suberizing potato tuber and tomato fruits, *Plant Mol. Biol.,* 11, 15, 1988.
43. **Robb, J., Powell, D. A., and Street, P. F. S.,** Vascular coating: a barrier to colonization by the pathogen in *Verticillium* wilt of tomato, *Can. J. Bot.,* 67, 600, 1989.
44. **Robb, J., Lee, S. W., Mohan, R., and Kolattukudy, P. E.,** Chemical characterization of stress induced vascular coating in tomato, *Plant Physiol.,* in press.
45. **Roberts, E. and Kolattukudy, P. E.,** Molecular cloning, nucleotide sequence, and abscisic acid induction of a suberization-associated highly anionic peroxidase, *Mol. Gen. Genet.,* 217, 223, 1989.
46. **Mohan, R. and Kolattukudy, P. E.,** Differential activation of expression of a suberization-associated anionic peroxidase gene in near-isogenic resistant and susceptible tomato lines by elicitors of *Verticillium albo-atrum, Plant Physiol.,* 921, 276, 1990.
47. **Bajar, A. and Kolattukudy, P. E.,** Wound-inducible expression of suberization-associated anionic peroxidase gene from tomato in transgenic tobacco plants, manuscript submitted.
47a. **Mohan, R. and Kolattukudy, P. E.,** manuscript in preparation.
48. **Sherf, B. A., Bajar, A., and Kolattukudy, P. E.,** Constitutive expression of suberization-associated anionic peroxidase gene from tomato in transgenic tobacco, manuscript submitted.
49. **Lagrimini, L. M., Bradford, S., and Rothstein, S.,** Peroxidase-induced wilting in transgenic tobacco plants, *Plant Cell,* 2, 7, 1990.
50. **Sherf, B. A., Bajar, A., and Kolattukody, P. E.,** manuscript in preparation.
51. **Beato, M.,** Gene regulation by steroid hormones, *Cell,* 56, 335, 1989.
52. **Evans, R. M.,** The steroid and thyroid hormone receptor superfamily, *Science,* 240, 889, 1988.

# NONHOST RESISTANCE IN PLANT-FUNGAL INTERACTIONS

**Lee A. Hadwiger**

## TABLE OF CONTENTS

# I. INTRODUCTION

Through evolution only a relatively small group of microorganisms have found their pathological niche on a given plant species. Fortunately, the same niche is usually unavailable for different microbe groups on the same plant species. Thus, the plant species is said to express "nonhost" resistance[1] to groups of microorganisms which are unsuccessful. This mismatch is often manifested as an intense host response (hypersensitive response) based on the activation of a selection of host genes.[2,3] Nonhost resistance has recently attained new importance with the advent of genetic engineering, because it is now possible to transfer genes from a species which actively resists a given plant pathogen to a species that is infected by that pathogen.[4] Theoretically, there is a larger number of potential disease resistance genes existing in all of the plant species on earth simply waiting to be transferred as needed. What is missing is the definition of how these genes are controlled and which genes or combination of genes are needed to code for that portion of the intense host response which actually suppresses the pathogen and preserves the viability of host cells[5] adjacent to the point of pathogen challenge.

# II. RELATION OF NONHOST RESISTANCE TO RACE-SPECIFIC AND INDUCED RESISTANCE

Two other aspects of disease resistance, race-specific resistance and induced resistance, relate to nonhost resistance. To contend with pathogens which find a biological niche, disease resistance can evolve in plant species through mutation and reassortment of traits via interspecific crossing and subsequent selection; "single Mendelian traits" sometimes referred to as R-genes often provide the plant with race-specific resistance[6] to races evolving within a given microbe species. Resistance is often attributable to a match-up of a single dominant trait in the host with a single dominant trait in the pathogen (gene-for-gene interaction).[7]

In addition, plants which have no genetically identifiable traits for disease resistance can often generate a response by the phenomenon called "induced resistance."[8] Induced resistance occurs when a plant, prechallenged by a nonvirulent microbe, is subsequently inoculated with a virulent pathogen. As a result, the virulent pathogen is often resisted. Another profound aspect of induced resistance is the observation that a challenge by a compatible virulent pathogen can also induce a resistance response which retards infection of the same pathogen subsequently inoculated on adjacent leaves.[9] Induced resistance can also occur following the application of a compound capable of eliciting a disease resistance response.[10] Often disease resistance does not function properly if metabolic inhibitors, heat shock, etc., are utilized to interfere with the development of a resistance response.[11-17] Such observations clearly demonstrate that a plant has the biochemical capability to resist nearly any pathogen if its response is generated both quickly and completely toward the challenging organism.

# III. NONHOST RESISTANCE

A definition of nonhost resistance in terms of classical genetics, as that presented by McIntosh[6] for the R-genes of wheat, is not presently possible, because it is seldom possible to make interspecies crosses and follow the chromosomal inheritance of resistance traits within a progeny. However, it is possible to follow genes whose mRNA and protein products accumulate in abundance as nonhost resistance is being expressed. This chapter will primarily review the evidence acquired comparing the nonhost resistance reaction of pea to *Fusarium solani* f. sp. *phaseoli* with the pea's susceptible reaction to *F. solani* f. sp. *pisi*. It will also examine the potential for utilizing the cloned response genes of the nonhost resistance response and their regulators in engineering other plants to respond quickly and intensely to plant pathogenic strains normally infecting these species.

Nonhost resistance may, in the simplest terms, be viewed as an intense cell incompatibility between the plant cell and a foreign cell because the plant tissue is probably not preprogrammed by design to contend with every different cell it encounters. Features of the host tissue such as cuticle or hairlike structures may nonspecifically preclude encounters involving molecular exchange. Intense incompatibility may also be by-passed as a result of inadequate traits. For example, many saprophytes lack the pathogenicity factors necessary for penetration.[18-21] Pea endocarp tissue without the protection of a cuticle layer responds intensely to the presence of foreign pollen, saprophytes, other plant tissue, bacteria, fungi, and even mouse tumor cells.[15] The pea cell and the challenging cells both experience change. The interaction of pea with *F. solani* f. sp. *phaseoli* (a pathogen of bean and incompatible on pea) can be negated if the fungal cell is physically separated from the pea tissue (cuticle-free) by a dialysis or filter membrane, even though a continuous aqueous phase is available to transport an array of soluble small and large molecules.[22,23] Normally, when this incompatible fungus is in direct contact with the tissue, it is immediately inhibited, but when separated by the filter barrier, grows prolifically. Further, the pea tissue on the other side of the barrier, does not respond to generate resistance or a hypersensitive reaction.

## IV. COMPONENTS OF THE HOST-PARASITE INTERACTION

Not surprisingly, the microbes categorized as plant pathogens have evolutionarily acquired enzymes to digest various plant components such as the cuticle, cell wall, middle lamella, phytoalexins, etc, and even cytoplasmic components.[21,23-26] Some saprophytes and nonbiotrophic parasites can penetrate and grow well in plant leaf tissue that has been killed by autoclaving[18] and thus is relatively barrier-free. Plant pathogens which have the tools to remove external plant barriers must be resisted by an active plant response. Finally if the plant response is subjugated or diminished by toxins,[27] or artificially reduced by heat shock or metabolic inhibitors,[18] many plant pathogens previously excluded can grow prolifically through this tissue. This brings us to the questions: what is the nature of the host responses, how are they signaled and how do they suppress the parasite?

The physical contact between plant and fungal tissue enables hydrolytic enzymes, such as β-glucanase[28,29] and chitinase,[29,30] to contact their substrate polymers, β-1,3 glucan and chitin in the fungus, fungal pectolytic enzymes to contact plant pectins,[21] and fungal nuclease[26] and protease to contact plant nucleic acid and protein substrates thus releasing an array of fragments which can serve as signals.[31] The hydrolytic action may also release signal compounds from the plant tissue itself.[32] Further, the chemical composition of both the exposed surfaces change as the development of the host-parasite interaction progresses.[33-36] As indicated above, the β-glucanases and chitinases of plants,[32] have been convincingly shown to possess the potential to act at the plant-pathogen interface. In peas, the combined β-glucanase/chitinase families of enzymes[37] or the interaction per se has been shown capable of releasing the chitosan-like signal from *F. solani* f. sp. *phaseoli,* which, in turn, can trigger induced resistance and/or directly inhibit fungal growth.[38]

## V. BIOCHEMICAL EXPRESSION OF DISEASE RESISTANCE — HOW IS THE PATHOGEN SUPPRESSED?

In the pea-*Fusarium* system of nonhost resistance, pea endocarp tissue reacts to completely resist the growth of the bean pathogen *Fusarium solani* f. sp. *phaseoli* within 4 to 6 h after inoculation. This laboratory assayable interaction has been used to develop an overview of the gene products which accumulate as the pea tissue is generating a resistance response. There are more than 20 major protein products that selectively accumulate in temporal correlation with the cytologically observable suppression of the bean pathogen.[3]

As indicated above, if one prevents either the accumulation of these protein products or their mRNAs with specific inhibitors[15,22] the tissue becomes susceptible. Further, a heat shock treatment of 40°C for 1 h is sufficient to initiate the synthesis of the classical heat shock proteins. This consequently prevents the pathogen from triggering the pattern of proteins observed in the resistance response, and resistance is broken.[12] If the fungal challenge is delayed for 9 h following heat shock to allow the tissue to recover, the fungus can then reinduce the protein pattern and thus regenerate disease resistance. If the tissue is challenged by the bean pathogen 4 h prior to heat shock then disease resistance response proteins (and eventually some heat shock proteins) appear and resistance is maintained (even though there is no significant production of phytoalexin). Thus, some portion of this protein synthesis response appears to be required for disease resistance.

Functional roles have been found for some of these disease resistance response proteins. Some are necessary for secondary pathways[39] leading to phenolic products such as lignin and phytoalexins. Phytoalexins do not appear to have a determining role in this nonhost resistance reaction,[15] however, the phytoalexin, pisatin, appears to manipulate virulence expressed by an authentic pea pathogen because the strains of this pathogen that can best reduce pisatin toxicity cause larger lesions.[40] This virulence potential when transferred to a pathogen of maize can enhance its virulence on pea, normally a nonhost.[41]

Some of the pea responses overlap into the classification of pathogenesis related (PR) proteins, e.g., chitinases and β-1,3-glucanases, mentioned above. PR-proteins from other systems and their functions will be discussed elsewhere. We now direct our attention to the disease resistance-response proteins which are currently without defined functions.[42] These proteins, which may be important in a resistance response because they are abundant, accumulate as resistance is expressed or have a *potential* function. The 17-kDa protein, (product of the pea disease resistance response gene 49), accumulates very abundantly in fungal-challenged peas.[43] There are also homologs of this protein in potatoes,[44] soybean,[45] parsley,[46] birch,[47] alder,[47] and bean.[48] The *in situ* localization of this protein product in pea tissue is primarily in the heterochromatin of the nucleus especially in cells near the point of challenge, but also in regions 11 to 18 cell layers away.[49] There are also cysteine-rich proteins produced by a family of plant genes which are processed to mature proteins of ~5 kDa.[50] Some of these cysteine-rich proteins, termed thionins, may possess antifungal properties.[51]

Thus, an arsenal of proteins accumulate in a nonhost resistance response and are responsible for increases in phytoalexins, lignification, saponins, hydrolytic enzymes, toxic proteins, and proteins of unknown function. Some of these latter proteins may accumulate incidently or simply for the well-being of the host cell. For example, the 17-kDa protein mentioned above may increase in the heterochromatin of cells adjacent to fungal challenge to stabilize the nuclear structure and thus the viability of these cells. The nuclei of these cells retain their viability as can be shown by fluoresce with Hoesht 33258 stain, while the nuclei in cells directly in contact with the fungal spore do not. In addition, one of the early observations in infected pea tissue is that the tissue destined to be resistant to the bean pathogen maintains cell viability in cells adjacent to those directly challenged, whereas in pea tissue destined to succumb to *F. solani* f. sp. *pisi,* more cells adjacent to the fungus experience loss of viability.[5] Part of this effect may occur because the pea pathogen releases more viability-decreasing or toxic material which can suppress the entire host response. However, if the host has an inherently quick response or the benefit of a head start, e.g., a preinduced resistance response, it may quickly negate the detrimental effects of the pathogen. This action could reduce the loss of cell viability, by simply checking fungal growth and its influence on adjacent tissue. This concept is diagramed in Figure 1. Two enzyme families in plants, β-glucanase and chitinase, can directly affect fungal growth. Adequate levels of the two purified enzymes can digest the glucan and chitin wall polymers of the

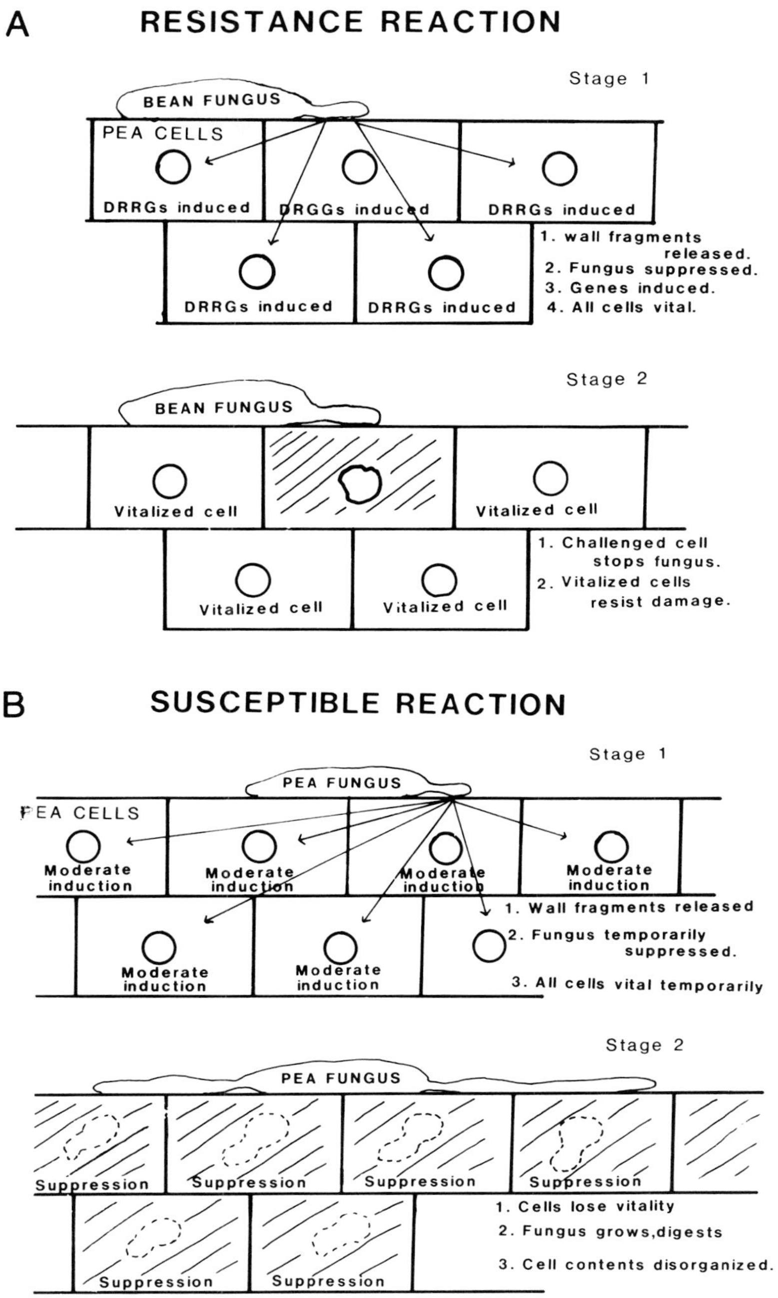


FIGURE 1. The proposed stages of response within the pea endocarp tissue resistant to *Fusarium solani* f. sp. *phaseoli* (A) or susceptible to *F. solani* f. sp. *pisi* (B). In the resistant reaction (A) the surface cells are viable and genes are induced to accumulate the disease resistance response proteins which we propose assist in maintaining the cytologically observed viability of most cells, except those in direct contact with the fungus, through stage 2. In the susceptibility reaction (B) response genes in surface cells are induced to moderate levels and fungal growth is only temporarily checked. In stage 2 a broader radius of cells adjacent to the fungal spore lose viability. Also the accumulations of response proteins diminish, fungal growth continues, and the cellular organization of host cells deteriorates.

expanding hyphal tips and cause lysis of the fungal cell.[52] This extensive wall destruction has been shown to occur at later stages in some infections, but did not distinguish resistant from susceptible interactions.[30] We have demonstrated that these two pure pea enzymes assist the release of chitosan, a component of *Fusarium* cell walls, in pea-*F. solani* interactions.[38] Chitosan can also inhibit the germination and/or the growth of *F. solani* at approximately 7 ppm.[10] Thus, the resisted pathogen may initially be stopped by chitosan-like polymers released in the interaction. The fungal polymers simultaneously trigger host responses which directly (or indirectly) improve cell viability. The longer host viability is maintained in nearby cells, the longer the pathogen growth is suppressed. Callose, lignin, saponins, and phytoalexins, which accumulate during this extended period, may moderate virulence expression. Thus, the functional aspects of disease resistance may be carried out by a rather large component of ''slave genes.'' The most compelling evidence supporting this is that the enhanced levels of specific disease-resistance response proteins are maintained as disease resistance is being expressed and, as previously stated, interference with the accumulation of these proteins promotes susceptibility. In addition, some of these same proteins are enhanced when the incompatible reaction is controlled by various single-dominant genes for disease resistance (master genes) which functionally link nonhost resistance to race-specific resistance.[53] These ''master genes,'' must in some way potentiate the hyperaccumulation of these ''slave'' proteins. The coregulation of the slave genes must involve *trans*-acting factors since, in at least one case (peas), a master gene and its slave genes reside on different chromosomes (see Figure 2). In peas, the single Mendelian trait for disease resistance to *Fusarium oxysporum* f. sp. *pisi,* race 1, has been mapped to chromosome IV,[54] whereas gene 49, which is preferentially enhanced in temporal association with the expression of resistance in the seedling, has been tentatively mapped to chromosome VI.[54a]

The resolution of the protein functions along with the control expressed by *trans*-acting factors may enable the genetic manipulation of disease resistance in plants which do not possess the appropriate master genes. That is, it may be possible to potentiate through genetic engineering the expression of the proper slave gene(s) at the temporal period they are needed following inoculation. Some of the mechanics for manipulating these resistance response genes are available and strategies have been devised.[4] The genes differentially expressed can be sorted out from those which are not, by plus-minus hybridization techniques.[55] Each gene differentially expressed is likely to have a useful regulatory sequence in its promoter region[42,56] which, when properly signaled, may function appropriately. There are reliable procedures for purifying elicitor components from pathogenic organisms capable of triggering induced resistance.[37,56-58] Thus, a slave gene from a plant such as peas which readily responds to create resistance against many potato pathogens, should possess 5′ or 3′ sequences which would respond quickly to elicitors from potato pathogens. A potato plant containing a functional slave gene constructed in line with such a sequence should now respond to the presence of the potato pathogen's elicitors in a manner likely to enhance disease resistance. Unfortunately, if resistance cannot be manipulated by the transfer of one or two genes, substantially greater technical problems exist. That is, the transfer and regulation of *multiple* genes requires multiple antibiotic selection genes, and often there is a low potential for uniform regulation of the new genes in the transgenic plant.[59]

## VI. MODE OF REGULATION OF NONHOST DISEASE RESISTANCE RESPONSE GENES IN PEAS

An understanding of the regulatory features of the slave genes is pertinent since, without having cloned any of the master genes, these constitute the only plant disease resistance genes available for study. The slave genes given the most attention are those involved as enzymes in secondary pathways toward phytoalexins, lignins, tannins, etc. The role of these

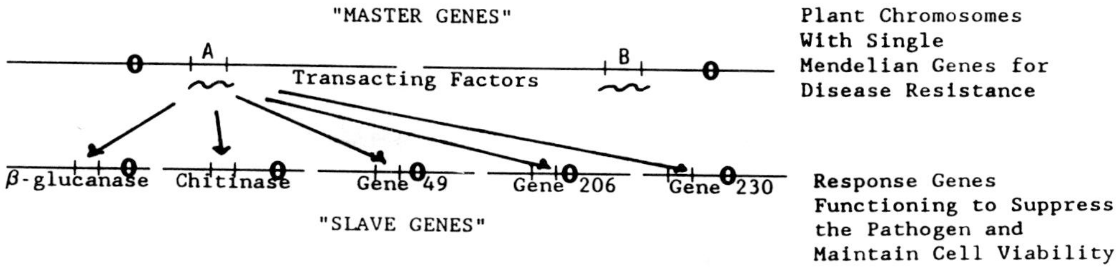


FIGURE 2. A diagrammatic representation of the proposed relationship between single dominant genes for resistance (master genes) and disease resistance-response genes (slave genes). Classic chromosomal mapping has designated the location of many single dominant genes on specific chromosomes. Preliminary RFLP mapping combined with other map site markers have indicated that the response genes can be on separate chromosomes. Since we have observed an enhanced expression of some slave genes only when the appropriate master gene is present, some sort of transacting factor must be required for this long-distance control.

will be discussed elsewhere, however their 5′- and 3′-sequence regions may be useful in the manipulated regulation of the other slave genes which are currently without defined functions and vice versa.

The disease resistance-response genes (DRRG) 49, 176, 206, 39, and 230, selected from the total slave genes in peas, on the basis of their expression showing the best correlation with resistance expression,[2,53] possess features which are useful in understanding the regulation of the nonhost response.[42] For example, the 5′-region of pea gene 49 contains an SV40 core enhancer, consensus topoisomerase II cleavage sequences, an AP-1 binding site, potential Z-DNA-forming regions, and sequences in common with some of the sequences within the promoter of the PR2 gene of parsley necessary for elicitor-mediated gene expression.[42,60] Also the region 3′ to pea gene 49 contains another topoisomerase consensus site, an AP-1 site, and, again, sequences in common with those of the parsley PR2 gene, whose structural gene shares 40% homology with that of the pea gene 49. The topoisomerase II sites are of interest because of the recent realization that these sites are often associated with DNA regions at which the chromatin strands attach to the chromosomal scaffolding.[61] If clusters of the gene 49 topoisomerase II sites do attach to pea scaffolding, the remaining <2-kb segment containing the structural gene would be suspended as a very short chromosomal loop, and be very sensitive to torsional stresses on the DNA within this loop.[61-63] Indeed, the *in vivo* activity of gene 49 is enhanced by compounds that alter DNA conformation, e.g., cross-linked psoralen compounds, chromomycin $A_3$, actinomycin D, and chitosan (a natural elicitor).[4,64] These compounds elicit enhanced accumulation of mRNA and proteins in precisely the same proportions as do the living spores of *Fusarium solani*.[65] Interestingly, the presence of scaffold attachments on each side of the structural gene may also stabilize the expression of genes especially when included in a transformation vector.[66]

The presence of the AP-1 sites 5′ and 3′ of the pea structural gene 49 provide a potential for additional control. The AP-1 consensus site TGA$^{C}_{G}$TCA is a binding domain for leucine zipper-forming dimers of the oncoproteins produced by the *fos* and *jun* genes of mammalian systems.[67] Although proteins totally homologous with fos and jun have not been reported in plants, proteins with amino acid sequence homologies in the DNA-binding basic region and part of the leucine zipper motif have been found.[68]

We have observed (unpublished data) that the product of gene 49 is selectively enhanced in the pea cultivar Vantage in response to a challenge by *Fusarium oxysporum* f. sp. *pisi* race 1. Disease resistance to race 1 is controlled by a single dominant gene on chromosome IV of the pea genome.[54] Although no defined role for the gene 49 protein has been documented, this system demonstrates a potential for a single Mendelian gene (master gene) to exert some control over gene 49 (slave gene) located on a separate chromosome (Figure 2).

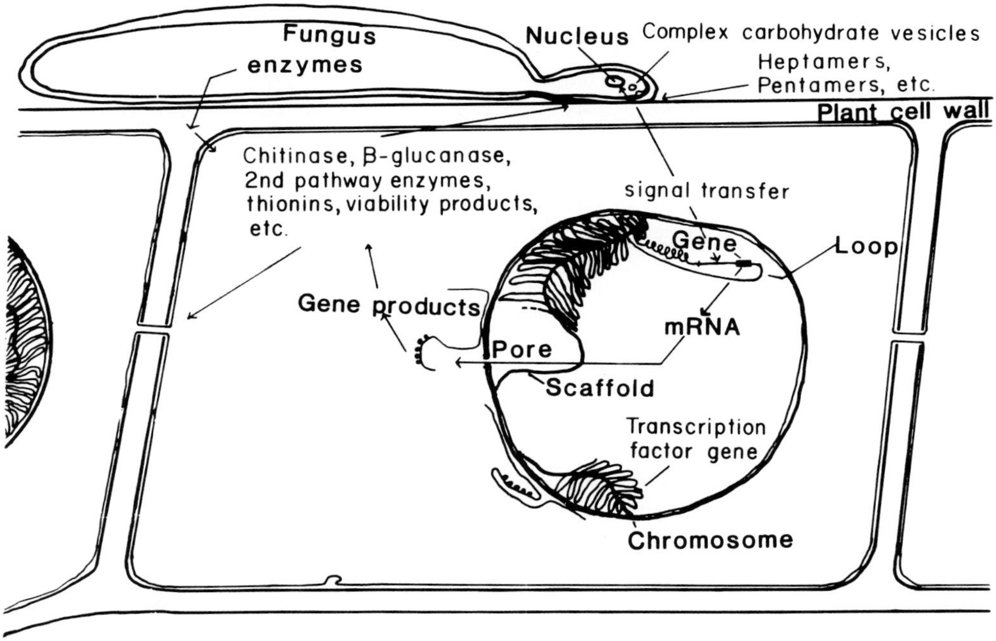


FIGURE 3. A proposed interaction of some of the components of the pea endocarp-*Fusarium solani* f. sp. *phaseoli* nonhost resistant reaction. Chitosan oligomers, primarily heptamers and pentamers, are synthesized from *N*-acetylglucosamine probably by way of chitin, deacetylated,[38,69] and released early in the interaction. The heptamer, but not the pentamer, is capable of inhibiting the fungus and inducing a host resistance response.[38] The chitosan oligomer reaches the plant nucleus with 30 min.[70,71] Chitosan has been proposed to associate with the cellular DNA and has the potential to alter DNA conformation.[72] Chitosan elicits the accumulation of selected response proteins including chitinase, β-glucanase,[73] enzymes of secondary plant metabolism,[39] cysteine-rich (thionin-like) proteins,[50] and other proteins without designated functions.[65] Accumulations of chitinase and β-glucanase have the potential to increase further the release of fungal wall oligomers. The proposed[74] activation of the response gene is related to conformational changes[75] on chromosome loops with unique features (e.g., small loops, specific DNA sequences) and/or by the action of induced transcription factor genes which need not reside on the same chromosome.

# VII. NONHOST DISEASE RESISTANCE IN PEAS — AN OVERVIEW

Our long-term observations on nonhost resistance in the pea-*Fusarium solani* system have been summarized as a model presented in Figure 3. This bean pathogen germinates visibly within 3 h on pea endocarp tissue, however within 30 min chitosan heptamers are released[38] and chitosan appears within the nuclei of the surface plant cells.[70] At this point in time definite physical changes occur in the nuclei of these host cells.[71] Since chitosan oligomers, heptamers and larger, but not the pentamers, can enhance the same protein increases as do the intact fungal spores, chitosan appears to be the signal involved in response induction.[38] The intense DNA affinity[10] and the DNA conformation-changing potential[72] of chitosan suggest a possibility for a direct action[74] on the plant DNA. However, other workers have proposed its effect is on the cell membrane.[76]

The response genes appear to be subject to multiple controls which may include, local DNA stresses,[64] a single dominant gene's effect from the same or other chromosomes,[53,54] elicitors from fungal[38] and bacterial cells,[77] transcription factors, and multiple synthetic DNA-specific compounds. The protein products accumulating from these genes may be incidental or vital to the resistance of the tissue. Since one of the more obvious differences between the resistance and susceptible reactions is the maintenance of cell viability,[5] this may be enhanced in the resistance reaction by some of these proteins.

The chitosan heptamer released in the pea-*F. solani* f. sp. *phaseoli* interaction is singly able to suppress both the germination and growth of *Fusarium* macroconidia probably by inhibiting RNA and/or DNA synthesis.[78] Growth can resume, however, if chitosan concentrations are reduced to below 10 μg/ml.[38] Thus, the initial suppression of the pathogen must be and is maintained in the resistance reaction but not in the susceptible reaction. Phytoalexin accumulations[15] or lignifications,[74] etc., in pea tissue do not develop soon enough to contribute to the initial pathogen suppression occurring within 6 h. Chitinase and β-glucanase activities accumulate equally in resistant and susceptible reactions within 24 h[73] and thus do not appear to determine the reaction type. Although the production of pectolytic enzymes by the pea pathogen, *F. solani* f. sp. *pisi* is enhanced after 20 to 30 h, which contributes to the deterioration of the host tissue, this increase is not observed within 30 h following challenges with the bean pathogen.[79] The presence of these and other hydrolytic enzymes as well as secondary metabolite accumulations[40,80] appear to contribute to the ultimate virulence and symptom expression of the interaction.

Not knowing which of the proteins, whose synthesis is enhanced, actually contributes to disease resistance remains a problem. However, the worth of these individual gene products may be testable by transferring each of the response genes to another species. For example, peas readily "nonhost-resist" many of the plant pathogens normally virulent on potatoes. Thus, if the proper gene or combination of pea genes is transferred to potato via genetic engineering, the transgenic plants can be tested for improved resistance to potato pathogens.[4] Preliminary research in this direction has confirmed that the promoter of pea gene 49 does indeed enhance marker gene expression in potato tissue when challenged by a plant pathogen.[81] On the plus side, "nonhost resistance" in nature is extremely stable to virulence changes in plant pathogenic fungi. On the downside, the transfer of useful resistance traits to transformable economic plant species still appears to be a formidable task. The mechanism of nonhost resistance may differ in each host-pathogen system and the transfer of these genes to other plants may provide an answer to this question.

## REFERENCES


1. **Heath, M. C.,** Non-host resistance, in *Plant Disease Control; Resistance and Susceptibility,* Staples, R. C. and Toenniessen, G. H., Eds., John Wiley and Sons, NY, 1981, 201.
2. **Fristensky, B. W., Riggleman, R. C., Wagoner, W., and Hadwiger, L. A.,** Gene expression in susceptible and disease resistant interactions of peas induced with *Fusarium solani* pathogens and chitosan, *Physiol. Plant Pathol.,* 27, 15, 1986.
3. **Hadwiger, L. A. and Wagoner, W.,** Electrophoretic patterns of pea and *Fusarium solani* proteins synthesized *in vitro* which characterize the compatible and incompatible interactions, *Physiol. Plant Pathol.,* 23, 153, 1983.
4. **Hadwiger, L. A., Chiang, C. C., Pettinger, A., and Chang, M. M.,** Strategy to improve disease resistance by transferring "non-host" disease resistance genes from peas to potatos, in *Biotechnology and Food Safety,* Bills, D. D. and Kung, S.-D., Eds., Butterworth-Heinemann, Boston, 1990, 241.
5. **Kendra, D. F. and Hadwiger, L. A.,** Cell death and membrane leakage not associated with the induction of disease resistance in peas by chitosan or *Fusarium solani* f. sp. *phaseoli, Phytopathology,* 77, 100, 1987.
6. **McIntosh, R. A.,** A catalog of gene symbols for wheat, *Proc. 6th Int. Wheat Genet. Symp. Kyoto Jpn.,* p. 1197, 1983.
7. **Flor, H. H.,** Current status of the gene-for-gene concept, *Annu. Rev. Phytopathol.,* 9, 275, 1971.
8. **Kuć, J. and Preisig, C.,** Fungal regulation of disease resistance mechanisms in plants, *Mycologia,* 76, 767, 1984.
9. **Kuć, J. and Richmond, S.,** Aspects of the protection of cucumber against *Colletotrichum lagenarium* by *Colletotrichum lagenarium, Phytopathology,* 67, 1290, 1977.
10. **Hadwiger, L. A. and Beckman, J. M.,** Chitosan as a component of pea-*F. solani* interactions, *Plant Physiol.,* 66, 2305, 1980.

11. **Chamberlain, D. W. and Gerdemann, J. W.,** Heat-induced susceptibility of soybeans to *Phytophthora megasperma* var. *sojae, Phytophthora cactorum,* and *Helminthosporium sativum, Phytopathology,* 5670, 1966.
12. **Hadwiger, L. A. and Wagoner, W.,** Effect of heat shock on the mRNA-directed disease resistance response of peas, *Plant Physiol.,* 72, 553, 1983.
13. **Heath, M. C.,** Effects of heat shock, actinomycin D, cycloheximide, and blasticidin S on nonhost interactions with rust fungi, *Physiol. Plant Pathol.,* 15, 211, 1979.
14. **Tani, T., Yamamoto, H., Kadota, G., and Naito, N.,** Development of rust fungi in oat leaves treated with blasticidin S, a protein synthesis inhibitor, *Tech. Bull. Fac. Agr. Kagawa Univ.,* 27, 95, 1976.
15. **Teasdale, J., Daniels, D., Davis, W. C., Eddy, Jr., R., and Hadwiger, L. A.,** Physiological and cytological similarities between disease resistance and cellular incompatibility responses, *Plant Physiol.,* 54, 690, 1974.
16. **Vance, C. P. and Sherwood, R. T.,** Cycloheximide treatments implicate papilla formation in resistance of reed canarygrass to fungi, *Phytopathology,* 66, 498, 1976.
17. **Yarwood, C. E.,** Heat-induced susceptibility of beans to some viruses and fungi, *Phytopathology,* 46, 523, 1956.
18. **Fernandez, M. R. and Heath, M. C.,** Interactions of the nonhost French bean plant *(Phaseolus vulgaris)* with parasitic and saprophytic fungi. I. Fungal development on and in killed, untreated, heat-treated, or blasticidin S treated leaves, *Can. J. Bot.,* 67, 661, 1989.
19. **Kolattukudy, P. E. and Crawford, M. S.,** The role of polymer degrading enzymes in fungal pathogenesis, in *Molecular Determinants of Plant Diseases,* S. Nishimura et al., Eds., Japanese Scientific Society, Tokyo, Springer-Verlag, Berlin, 1987, 75.
20. **Bell, A. A. and Wheeler, M. H.,** Biosynthesis and function of fungal melanin, *Ann. Rev. Phytopathol.,* 24, 411, 1986.
21. **Collmer, A. and Keen, N. T.,** The role of polymer degrading enzymes in fungal pathogenesis, in *Molecular Determinants of Plant Diseases,* S. Nishimura et al., Eds., Japanese Scientific Society, Tokyo, Springer-Verlag, Berlin, 1981, 75.
22. **Hadwiger, L. A. and Loschke, D. C.,** Molecular communication in host-parasite interactions, hexosamine polymers (chitosan) as regulator compounds in race-specific and other interactions, *Phytopathology,* 71, 756, 1981.
23. **Nichols, E. J., Beckman, J. M., and Hadwiger, L. A.,** Glycosidic enzyme activity in pea tissue and pea-*Fusarium solani* interactions, *Plant Physiol.,* 66, 199, 1980.
24. **Mackintosh, S. F., Matthews, D. E., and Van Etten, H. D.,** Two additional genes for pisatin demethylation and their relationship to pathogenicity of *Nectria haematococca* on pea, *Mol. Plant-Microbe Interact.,* 2, 354, 1989.
25. **Barna, B., Ibenthal, W. D., and Heitefuss, R.,** Extra-cellular RNAse activity in healthy and rust infected wheat leaves, *Physiol. Mol. Plant Pathol.,* 35, 151, 1989.
26. **Gerhold, D., Pettinger, A., and Hadwiger, L. A.,** A nuclease from *Fusarium solani* is host-induced, *J. Cell Biochem.,* 15A, 56, 1991.
27. **VanAlfen, N. K.,** Reassessment of plant wilt toxins, *Annu. Rev. Phytopathol.,* 27, 533, 1989.
28. **Benhamou, N., Grenier, J., Asselin, A., and Legrand, M.,** Immunogold localization of β-glucanases in two plants infected by vascular wilt fungi, *Plant Cell,* 1, 1209, 1989.
29. **Mauch, F. and Staehelin, L. A.,** Functional implications of the subcellular localization of ethylene-induced chitinase and β-1,3-glucanase in bean leaves, *Plant Cell,* 1, 447, 1989.
30. **Benhamou, N., Joosten, M. H. A. J., and DeWitt, P. J. G. M.,** Subcellular location of chitinase and of its potential substrate in tomato root tissues infected by *Fusarium oxysporum* f. sp. *radicis-lycopersici, Plant Physiol.,* 92, 1108, 1990.
31. **Ryan, C. A.,** Oligosaccharide signalling in plants, *Annu. Rev. Cell Biol.,* 3, 295, 1987.
32. **Doares, S. H., Bucheli, P., Albersheim, P., and Darvill, A. G.,** Host parasite interactions. XXXIV. A heat labile activity secreted by a fungal phytopathogen releases fragments of plant cell walls that kill plant cells, *Mol. Plant-Microbe Interact.,* 2, 346, 1989.
33. **Kapooria, R. G. and Mendgen, K.,** Infection structures and their surface changes during differentiation in *Uromyces fabae, Phytopathol. Z.,* 113, 317, 1985.
34. **Wycoff, K. L., Jellison, J., and Ayers, A. R.,** Monoclonal antibodies to glycoprotein antigens of a fungal plant pathogen, *Phytophthora megasperma* f. sp. *glycinea, Plant Physiol.,* 85, 5088, 1987.
35. **Hadwiger, L. A. and Line, R. F.,** Hexosamine accumulations are associated with the terminated growth of *Puccina striiformis* on wheat isolines, *Physiol. Plant Pathol.,* 19, 249, 1981.
36. **Delmer, D. P.,** Cellulose biosynthesis, *Annu. Rev. Plant Physiol.,* 38, 259, 1987.
37. **Mauch, F., Hadwiger, L. A., and Boller, T.,** Antifungal hydrolases in pea tissue, *Plant Physiol.,* 87, 325, 1988.
38. **Kendra, D. F., Christian, D., and Hadwiger, L. A.,** Chitosan from *Fusarium solani* pea interactions, chitinase/β-glucanase digestion of sporelings and from fungal wall chitin actively inhibit growth and enhance disease resistance, *Physiol. Mol. Plant. Pathol.,* 35, 215, 1989.

39. **Loschke, D. C., Hadwiger, L. A., Schroder, J., and Hahlbrock, K.,** Effects of pathogenic and non-pathogenic races of *Fusarium solani* and of light in peas: Rate of synthesis and activity of phenylalanine ammonia-lyase, *Plant Physiol.*, 68, 680, 1981.
40. **Van Etten, H. D., Matthews, D. E., and Matthews, P. S.,** Phytoalexin detoxification: importance for pathogenicity and practical implications, *Annu. Rev. Phytopathol.*, 27, 143, 1989.
41. **Schafer, W., Straney, D., Ciuffetti, L., Van Etten, H. D., and Yoder, O. C.,** One enzyme makes a fungal pathogen, but not a saprophyte, virulent on a new host plant, *Science,* 246, 247, 1989.
42. **Fristensky, B., Horovitz, D., and Hadwiger, L. A.,** cDNA sequences for pea disease resistance response genes, *Plant Mol. Biol.*, 11, 713, 1988.
43. **Chiang, C. C. and Hadwiger, L. A.,** Cloning and characterization of a disease resistance response gene in pea inducible by *Fusarium solani, Mol. Plant-Microbe Interact.*, 3, 78, 1990.
44. **Matton, D. P. and Brisson, N.,** Cloning, expression, and sequence conservation of pathogenesis-related gene transcripts of potato, *Mol. Plant-Microbe Interact.*, 2, 325, 1989.
45. **LeGuay, J. J., Piecoup, M., Puckett, J., and Jouanneau, J. P.,** Common responses of cultured soybean cells to 2,4-D starvation and fungal elicitor treatment, *Plant Cell Rep.*, 7, 19, 1988.
46. **Somssich, I. E., Schmelzer, E., Kawalleck, P., and Hahlbrock, K.,** Gene structure and *in situ* transcript localization of pathogenesis-related protein I in parsley, *Mol. Gen. Genet.*, 213, 93, 1988.
47. **Breiteneder, H., Pettenburger, K., Bito, A., Valenta, R., Kraft, D., Rumpold, H., Scheiner, O., and Breitenbach, M.,** The gene coding for the major birch pollen allergen *Betv* I is homologous to a pea disease resistance response gene, *EMBO J.*, 8, 1935, 1989.
48. **Walter, M. H., Liu, J.-W., Grand, C., Lamb, C. J., and Hess, D.,** Bean pathogenesis-related (PR) proteins deduced from elicitor-induced transcripts are members of a ubiquitous new class of conserved PR proteins including pollen allergens, *Mol. Gen. Genet.*, 222, 353, 1990.
49. **Allaire, B. S., Chiang, C. C., and Hadwiger, L. A.,** Product of the disease resistance response gene 49 in *Pisum sativum* inoculated with *Fusarium solani* is localized in nuclei and xylem elements, unpublished observations.
50. **Chang, C. C. and Hadwiger, L. A.,** The *Fusarium solani*-induced expression of a pea gene family encoding high cysteine content proteins, *Plant Mol. Biol.*, 4, 324, 1991.
51. **Bohlmann, H., Clausen, S., Behnke, S., Giese, H., Hiller, C., Reimann-Phillip, U., Schrader, G., Barkholt, V., and Apel, K.,** Leaf specific thionins of barley-a novel class of cell wall proteins toxic to plant-pathogenic fungi and possibly involved in the defense mechanism of plants, *EMBO J.*, 7, 1449, 1988.
52. **Mauch, F., Mauch-Mani, B., and Boller, T.,** Antifungal hydrolases in pea tissue, II. Inhibition of fungal growth by combinations of chitinase and β-1,3-glucanase, *Plant Physiol.*, 88, 936, 1988.
53. **Daniels, C. H., Fristensky, B., Wagoner, W. W., and Hadwiger, L. A.,** Pea genes associated with non-host disease resistance to *Fusarium* are also active in race-specific disease resistance to *Pseudomonas, Plant Mol. Biol.*, 8, 309, 1986.
54. **Charchar, M. and Kraft, J. M.,** Response of near-isogenic pea cultivars to infection by *Fusarium oxysporum* f. sp. *pisi* races 1 and 5, *Can. J. Plant Sci.*, 69, 1335, 1989.
54a. **Weedon, N.,** personal communication, 1990.
55. **Riggleman, R. C., Fristensky, B. W., and Hadwiger, L. A.,** The disease resistance response in pea is associated with increased transcription of specific mRNAs, *Plant Mol. Biol.*, 4, 81, 1985.
56. **Lois, R., Dietrich, A., Hahlbrock, K., and Schulz, W.,** A phenylalanine ammonia-lyase gene from parsley: structure, regulation and identification of elicitor and light responsive *cis*-acting elements, *EMBO J.*, 8, 1641, 1989.
57. **Keen, N. T., Tamaki, S., Kobayashi, D., Gerhold, D., Stayton, M., Shen, H., Gold, S., Lorang, J., Thordal-Christensen, H., Dahlbeck, D., and Staskawicz, B.,** Bacteria expressing avirulence gene D produce a specific elicitor of the soybean hypersensitive reaction, *Mol. Plant-Microbe Interact.*, 3, 112, 1990.
58. **DeWitt, P. J. G. M., Hofman, A. E., Velthuis, G. C. M., and Kuć, J. A.,** Isolation and characterization of an elicitor of Necrosis isolated from inter-cellular fluids of compatible interactions of *Cladosporium falvum* (Syn. *Fulbia fulva)* and tomato, *Plant Physiol.*, 77, 642, 1985.
59. **Fang, R. X., Nagy, F., Sivasubramaniam, S., and Chua, N. H.,** Multiple *cis* regulatory elements for maximal expression of cauliflower mosaic virus 35S promoter in transgenic plants, *Plant Cell,* 1, 141, 1989.
60. **Van deLocht, U., Meier, I., Hahlbrock, K., and Somssich, I. E.,** A 125 bp promoter fragment is sufficient for strong elicitor-mediated gene activation in parsley, *EMBO J.*, 9, 2945, 1990.
61. **Newport, J. W. and Forbes, D. J.,** The nucleus: structure function and dynamics, *Annu. Rev. Biochem.*, 56, 535, 1987.
62. **Hadwiger, L. A.,** Possible role of nuclear structure in disease resistance of plants, *Phytopathology,* 78, 1009, 1988.
63. **Gross, D. S. and Garrad, W. T.,** Poising chromatin for transcription, *Trends Biochem. Sci.*, 12, 293, 1987.

64. **Hadwiger, L. A., Chiang, C. C., and Pettinger, A.,** Topoisomerase activity and the expression of disease resistance response genes (DRRG) of peas, *J. Cell Biochem.,* 14B, 131, 1990.
65. **Loschke, D. C., Hadwiger, L. A., and Wagoner, W.,** Comparison of mRNA populations coding for phenylalanine ammonia-lyase and other peptides from pea tissue treated with biotic and abiotic phytoalexin inducers, *Physiol. Plant Pathol.,* 23, 163, 1983.
66. **Stief, A., Winter, D. M., Stratling, W. H., and Sippel, A. E.,** A nuclear DNA attachment element mediates elevated and position-independent gene activity, *Nature (London),* 341, 343, 1989.
67. **Druker, B. J., Mamon, H. J., and Roberts, T. M.,** Oncogenes, growth factors and signal transduction, *New Engl. J. Med.,* 321, 1983, 1989.
68. **Singh, K., Dennis, E. S., Ellis, J. G., Llewellyn, J., Tokuhisa, T. G., Wahleither, J. A., and Peacock, W. J.,** OCSBF-1, a maize OCS enhancer binding factor: isolation and expression during development, *Plant Cell,* 2, 891, 1990.
69. **Davis, L. L. and Bartnicki-Garcia, S.,** Chitosan synthesis by the tandem action of chitin synthetase and chitin deacetylase from *Mucor rouxii, Biochem. J.,* 23, 1065, 1984.
70. **Hadwiger, L. A., Beckman, J. M., and Adams, M. J.,** Localization of fungal components in the pea-*Fusarium* interaction detected immuno-chemically with anti-chitosan and anti-fungal cell wall antisera, *Plant Physiol.,* 67, 170, 1981.
71. **Hadwiger, L. A. and Adams, M.,** Nuclear changes associated with the host-parasite interaction between *Fusarium solani* and peas, *Physiol. Plant Pathol.,* 12, 63, 1978.
72. **Kendra, D. F., Fristensky, B., Daniels, C. H., and Hadwiger, L. A.,** Disease resistance response genes in plants: expression and proposed mechanism of induction, in *Molecular Strategies for Crop Protection,* Arntzen, C. J. and Ryan, C., Eds., Alan R. Liss, NY, 1987, 13.
73. **Mauch, F., Hadwiger, L. A., and Boller, T.,** Ethylene: Symptom not signal for the induction of chitinase and β-1,3-glucanase in pea pods by pathogens and elicitors, *Plant Physiol.,* 76, 607, 1984.
74. **Hadwiger, L. A., Chiang, C., Victory, S., and Horovitz, D.,** Molecular biology of chitosan in plant/pathogen interaction and its application in agriculture, in *Chitin and Chitosan: Sources Chemistry, Physical Properties and Applications,* SkjakBraek, G., Anthonsen, T., and Sanford, P., Eds., Elsevier Applied Sci., London, 1989, 119.
75. **Hadwiger, L. A.,** Possible role of nuclear structure in disease resistance of plants, *Phytopathology,* 78, 1009, 1988.
76. **Kohle, H., Jeblick, W., Poten, F., Blaschek, W., and Kauss, H.,** Chitosan elicited callose synthesis in soybean cells as a $Ca^{2+}$-dependent process, *Plant Physiol.,* 77, 544, 1985.
77. **Daniels, C. H., Cody, Y. S., and Hadwiger, L. A.,** Host-responses in peas to challenge by wall components of *Pseudomonas syringae* pv. *pisi* races 1, 2, and 3, *Phytopathology,* 78, 1451, 1988.
78. **Hadwiger, L. A., Kendra, D. F., Fristensky, B. W., and Wagoner, W.,** Chitosan both activates genes in plants and inhibits RNA synthesis in fungi, in *Chitin in Nature and Technology,* Muzzarelli, R., Jeuniaux, C., and Gooday, G. W., Eds., Plenum Press, NY, 1986, 209.
79. **Buell, R. C.,** The role of pectin enzymes in the *Fusarium solani*-pea endocarp system. M.S. Thesis, Washington State University, Pullman, WA, 1988, 29.
80. **Christian, D. A. and Hadwiger, L. A.,** Pea saponins in the pea-*Fusarium solani* interaction, *Exp. Mycol.,* 13, 419, 1989.
81. **Chang, M. M., Chiang, C. C., and Hadwiger, L. A.,** Transformation of a disease resistance response gene from pea to potato, *in vitro, Cell Dev. Biol.,* 26, 69A, 1990.

# A MODEL SYSTEM FOR THE STUDY OF PLANT-FUNGAL INTERACTIONS: TOMATO LEAF MOLD CAUSED BY *CLADOSPORIUM FULVUM*

**Richard P. Oliver**

## TABLE OF CONTENTS

# I. INTRODUCTION

Fungal pathogens are responsible for the lion's share of damage to crop yields and have, therefore, been the subject of intense study for the past 100 years. A cursory glance at the phytopathological literature reveals that a huge range of fungi are under active study, the choice of which appears to depend on the extent of damage caused rather than any scientific considerations of the amenability of the organisms to study. It is likely that this diversity of subject matter has limited the speed with which fundamental knowledge of the genetic and biochemical details of the pathology has accumulated, although it must be appreciated that workable and satisfactory strategies for limiting disease based on resistant cultivars, fungicides, and cultural practices have been developed for most pathogens. The development of fundamental fungal genetics has undoubtedly been aided by the choice of model system organisms. It is clear that the enormous diversity of the fungal plant pathogens will mean that no single organism can ever be regarded as exemplary of the entire group. Nevertheless, the advantages of pooling of effort between laboratories into a limited number of paradigmatic fungi are such that it may well be essential if significant progress in the understanding of genetic and molecular mechanisms is to be made. It is the purpose of this chapter to demonstrate that *Cladosporium fulvum* has the attributes to be a model system for the study of plant-fungal interactions at the molecular level.

# II. NOMENCLATURE

*Cladosporium fulvum* was first described by M.C. Cooke in 1883 on samples of tomato leaves sent from South Carolina.[1] It is assumed that the pathogen originated in South America where its host had been cultivated for over 300 years. It is likely that it followed the tomato as it obtained world-wide cultivation first in Italy in 1544 and then in diverse locations.

It was placed in the *Cladosporium* genus, a heterogeneous collection of saprophytic and phytopathogenic fungi. More recently, two novel names for the fungus have been suggested. Ciferri[2] created the genus *Fulvia* and called *C. fulvum, Fulvia fulva.* He put forward a number of reasons to isolate *C. fulvum* from the *Cladosporia,* many of which are hard to understand. For example, the genus is called *Fulvia* because Ciferri described the fungus as fulvous or yellow. All samples of the fungus in my laboratory are olive green except when grown on high sucrose. Nevertheless *Fulvia fulva* is the accepted name for this pathogen. More recently, von Arx[3] suggested shifting *C. fulvum* into the *Mycovellosiella* genus.

As *C. fulvum* is the name used by almost all the plant pathologists, and mycologists who have studied it as well as plant breeders and tomato growers, I shall continue to use it. It is clear that the nomenclature of Deuteromycete fungi is based almost entirely on cultural and morphological characteristics with the result that name changes are frequent. Hopefully the application of molecular techniques to fungal taxonomy will provide a sound basis for a reliable nomenclature and throw light on the phylogeny and evolution of this diverse group.

## A. EARLY STUDIES OF THE *CLADOSPORIUM FULVUM*-TOMATO INTERACTION

Leaf mold is found occasionally on tomatoes grown outdoors in subtropical regions but became a serious pathogen when intensive greenhouse cultivation became widespread. By 1908 it had been observed in the United States, Canada, England, Australia, France, Germany, Italy, Denmark, and The Netherlands.[4]

Resistance was noted as early as 1908 by Lind in Denmark[5] and in 1914 by Nortin in North America.[6] These observations made little impact on commercial growing of tomatoes

**TABLE 1**
**Interaction of *Cladosporium fulvum* Races with Tomato Cultivers**

| Race | Tomato cultivar[a] | | | | | | | Presumed Fungal Genotype[b] |
|---|---|---|---|---|---|---|---|---|
| | Cf0 | Cf2 | Cf4 | Cf2.4 | Cf5 | CF9 | CF11 | |
| 0 | S | R | R | R | R | R | R | A2, A4, A5, A9, A11 |
| 2 | S | S | R | R | R | R | R | a2, A4, A5, A9, A11 |
| 4 | S | R | S | R | R | R | R | A2, a4, A5, A9, A11 |
| 2,4 | S | S | S | S | R | R | R | a2, a4, A5, A9, A11 |
| 5 | S | R | R | R | S | R | R | A2, A4, a5, A9, A11 |
| 2,4,5 | S | S | S | S | S | R | R | a2, a4, a5, A91, A11 |
| 2,4,9,11 | S | S | S | S | R | S | S | a2, a4, A5, a9, a11 |
| 2,4,5,9,11 | S | S | S | S | S | S | S | a2, a4, a5, a9, a11 |

[a] Indicates specific *C. fulvum* resistance genes (*Cf* genes) carried by cultivar. S, susceptible; R, resistant.
[b] A2 indicates avirulence to Cf2, a2 virulence to Cf2, etc.

until Langford[7] began breeding resistance into commercial varieties. The results of his efforts, a variety called Stirling Castle, was resistant to some, but not all, isolates of *C. fulvum*. This established the notion of variety-specific races of the pathogen. Langford differentiated total susceptibility, two levels of partial resistance, and immunity in various race-variety combinations. Further breeding of tomato by Bailey[8] and Williams[9] involving hybridization with the resistant species *Lycopersicon pimpinellifolium* resulted in the commercially successful variety "Vetomold" as well as the identification of a total of seven physiological races of the fungus. Bailey proposed that the newly identified races arose via spontaneous mutation during the course of his work rather than by selection by the resistant plants, a controversy which is still ongoing.

Further breeding involved other *Lycopersicon* species including *L. hirsutum, L. peruvianum*[10] and *L. glandulosum*.[11] With monotonous regularity, the introduction of novel resultant varieties was followed by the isolation of novel virulent races of the fungus. To date, some 25 resistant genes have been recognized and mapped with varying accuracy to the tomato genotype. Races virulent on all tomato cultivars except those carrying the Cf6 resistance gene have been isolated.[12]

## B. GENETIC STUDIES

The gene-for-gene hypothesis states that dominant plant genes for resistance are matched by corresponding dominant fungal genes for avirulence. All available evidence suggests that *C. fulvum*-tomato interaction fits this model. For example, in his pioneering studies, Day induced mutants in an avirulent strain that were virulent.[13] Races of *C. fulvum* are named according to the resistance genes of tomato they can overcome. Thus Race 0 can only infect plants carrying no resistance genes. These are known as Cf0 plants. Cf2 plants, carrying resistance gene 2, are resistant to all races except race 2 and the races virulent to more than one resistance gene such as Race 2,4. Day mutated a race 0 isolate to behave like a race 2 isolate. The strain was genetically marked by virtue of its red pigmentation so it is unlikely that this result was caused by a cross-contamination. It is unfortunate that these strains are now dead, as we have been unable to repeat this observation on a sample of over 2000 survivors of mutagenesis when screening for loss of three separate avirulence genes.

The interaction of *C. fulvum* races with cultivars of tomato carrying resistance genes giving immunity is illustrated in Table 1. The race nomenclature refers to the presumed recessive virulence alleles. As the concept of recessive alleles in a haploid organism is somewhat bizarre, the list of presumed dominant avirulence alleles carried by each race is included.

## C. PHYSIOLOGICAL STUDIES

The first systematic investigation of the microscopic features of the infection were undertaken by T.E.T. Bond in Cambridge in 1938.[14] He established that only a subset of the *Lycopersicon* genus was susceptible. Other plants, both within and outside the Solanaceae were found to be resistant. Only *L. esculentum, L esculentum* var. *cerasiforme, L. chmielski* and *Solanum penelli* have been shown to be susceptible. The taxonomic status of *S. penelli* is controversial,[15] and its susceptibility to *C. fulvum* supports the view that it should be placed in the *Lycopersicon* genus.

On the microscopic level, Bond[14] and Langford[7] established that the route of infection was via the stomata. Conidia alighting on the abaxial surface germinate to produce hyphae which extend randomly. Penetration of the stomata involves no special morphological structures. Subsequent progress of the infection depended on whether the plant is susceptible, immune, or intermediate. Resistance is expressed soon after penetration with the formation of microscopic necrotic lesions typical of a hypersensitive response. In compatible interaction, the penetrating hyphae produce extensive ramified mycelium. The mycelium remains intercellular with no obvious structures connecting to the mesophyll cells. Interestingly, the diameter of the intercellular mycelium (4 to 5 $\mu$m) is significantly greater than the primary hyphae on the leaf surface (2 to 3 $\mu$m). Intermediate resistance is expressed as a limitation of the rate and extent of mycelial colonization of the apoplastic space. In fully susceptible combinations, hyphae re-emerge through the stomata and the formation of conidia occurs. The whole process takes a minimum of 10 to 11 days. The requirement for stomatal penetration and conidia explain the optimal conditions for infection. High humidity is required for germination while subsequent lower and variable humidity (85 to 100%) promotes periodic stomatal opening.

The observations of these workers clearly established that *C. fulvum* is a biotrophic pathogen. Bond[14] stated that "*C. fulvum* only appears to flourish so long as the tissues are alive." The combination of biotrophy throughout the pathogen life cycle and axenic cultivability is rare and a major factor in its adoption as a model system. Given the biotrophic mode of nutrition it was not expected that extracellular hydrolytic enzymes would be parts of *C. fulvum*'s armory. Nonetheless, a number of enzymes have been detected during *in vivo* growth.[14a] These include several proteases (probably at least three), cellulase, and pectinase. The role of these enzymes is unclear, particularly since there is wide race-to-race variation in activities. For example race 0, unlike all the other races, excretes virtually no protease when grown on Czapeks Dox supplemented with skimmed milk, yet is fully pathogenic.[14b] Although there is quantitative variation in between the races, there is no qualitative variation indicating that protease plays no part in cultivar specificity.

Mannitol dehydrogenase has been detected in intercellular fluids of infected plants, where its accumulation during infection is correlated with a depletion of apoplastic sucrose and glucose.[16] It seems that *C. fulvum* can satisfy its carbon requirements simply by converting mono- and disaccharides into a form unavailable to the plant. This seems a very simple strategy and raises the question of why so few fungi can infect tomato leaves given that sufficient carbohydrate seems to be available.

## D. MECHANISM OF RESISTANCE

The modern era of *C. fulvum* research can be said to have started with the studies of Higgins[18,19] and de Wit.[17] Microscopic observations confirmed and extended the conclusions of Langford and Bond. The resistance response was associated with callose deposition. Electrolyte leakage from plant cells undergoing the resistance response was first noted by van Dijkman and Kaars Sijpesteijn.[20] Originally this was ascribed to high-molecular weight compounds produced in culture in a race-specific manner.[21] Dow and Callow[22] were unable to observe the race specificity but went on to purify a glycoprotein from culture filtrates that was a nonspecific elicitor of the resistance response. These observations were confirmed by Lazarovitz and Higgins[23,24] and by de Wit and Roseboom.[25] These nonspecific elicitors also induced lipoxygenase activity and lipid peroxidation.[26]

Resistance is also associated with the induction of phytoalexins,[27] which were subsequently shown to include the polyacetylene falcarindiol. It is not clear whether sufficient falcarindiol is produced to account for the cessation of growth.

Proteins are also accumulated during the resistance response. This work was aided by the development of a simple technique to isolate intercellular fluid.[28] These pathogenesis-related (PR) proteins are frequently excreted into the apoplast and include the so-called P14,[29] chitinase, and β-1,3-glucanase[30] (see also chapter by Oku, this volume). These proteins are of plant origin and in combination with the phytoalexin and callose deposition provide a plausible basis for the resistance expression.

Gel electrophoresis of apoplast material revealed another protein which is most probably of fungal origin.[31] Unlike the PR problems, it is only found in compatible interactions of *C. fulvum* and tomato. Tomato infected with the blight pathogen, *P. infestans,* lack the protein. It is suggested that this protein may be important in establishing basic compatibility. Testing this hypothesis will require a technique such as gene replacement.

The original purpose of isolating intercellular fluids was to establish whether they contained race-specific elicitors — the putative products of the avirulence genes. This was tested by injecting the fluids into uninfected leaves of the various cultivars.[28,32] A highly consistent and reproducible pattern of responses is induced in the leaves. The infection of Cf4 plants with race 4 is a compatible reaction. As the genotype of race 4 is A2a4A5A9, injection of intercellular fluid from the race 4-Cf4 interaction into uninfected Cf2, Cf5, or Cf9 plants results in a hypersensitive reaction visualised as a chlorosis for Cf2 and Cf5 and necrosis for Cf9. Injection into Cf0 or Cf4 plants gives no reaction above that induced by wounding. This was clearly a landmark observation demonstrating for the first time the presence of race- and cultivar-specific communication molecules.The suggestion that these elicitors were of fungal origin was strengthened by using plants that carry two resistance genes (Cf2, Cf4) and fungal races that lack two or more avirulence genes — race 2,4, race 2,4,5, and race 2,4,5,9.[33]

The rapidity and extent of the reaction of Cf9 plants indicated that purification of the putative avirulence gene 9 product was the priority.[34] The product proved to be a highly stable peptide which has been purified and completely sequenced. The peptide is produced by all races except those virulent on Cf9 plants. It has six cysteine residues as disulfide bonds. The purified peptide elicits a strong necrotic reaction on Cf9 plants but surprisingly a synthetic peptide failed to elicit any reaction. This may be due to incorrect folding and disulfide bridge formation. The primary sequence allowed the construction of an oligonucleotide probe. This was used to isolate a cDNA clone.[34a] Hybridization of the cDNA to DNA of the different races reveals bands consistent with a single gene in all races except race 2.4.5.9 which lacks all hybridization. Thus the absence of avirulence is accounted for by a null allele for the avirulence gene. The cause of this deletion is under active study.The further analyses of this gene, including its transfer into races virulent on Cf9 plants, are awaited with interest.

## E. METHODS FOR THE STUDY OF *CLADOSPORIUM FULVUM*

*Cladosporium fulvum* is easily grown on common culture media including fully defined minimal media. Nutrient requirements for maximal growth are low, perhaps reflecting the low growth rate. Optimum sporulation is obtained on a modified Czapeks Dox medium containing 0.5g $l^{-1}$ $NaNO_3$ and 3 g $l^{-1}$ sucrose.[34b] It grows well in liquid on solid media (2% agar) with optimum growth rates at about 25°C. Alternative carbon sources include mannitol, fructose, and glucose. Alternative nitrogen sources include $NH_3$, glutamine, and hypoxanthine. It can also be grown in potato dextrose broth (PDB) or $V_8$ juice agar (diluted × 4).

*Cladosporium fulvum* sporulates profusely after 8 to 14 days. The spores can be collected by washing the surface of the plate with distilled water. Spores can be stored for at least 5 years by freezing in 10% w/v glycerol at −80°C.

A small-scale pathogenicity test has been developed which permits the testing of many hundreds of isolates in growth rooms or in the greenhouse.[34c] Surface-sterilized tomato seeds are placed in a bottle containing 7 ml of modified Hoagland's solution[35] solidified with 0.7% agar. The bottles are incubated in a growth room for 14 days at 23°C on a 12-h light-dark cycle. Illumination by mercury vapor lamps gives a light intensity of 300 $\mu$Mol·$m^{-2}$·$s^{-1}$. After 14 days, 25 $\mu$l of conidia at a concentration of $5 \times 10^5$ ml are pipetted into the underside of the cotyledons. Surface tension holds the droplets in place. After an additional 14 days, symptoms become apparent. In the case of compatible interactions, sporulation is observed. In the incompatible interaction, some chlorosis is common but sporulation is absent. In doubtful cases, clearing and staining will reveal whether intramatrical mycelium is present. Alternatively the full-sized plant test can be performed as described by de Wit.[16]

Molecular and genetic techniques for *C. fulvum* have also been developed. High-quality DNA and RNA can be prepared using slight modifications of standard techniques.[36] Protoplasts can be prepared from mycelium and used for transformation or fusion experiments.[37] Conidia ($10^8$) are germinated in 50 ml potato dextrose broth in 250-ml Erlenmeyer flasks for 24 h. Mycelium is collected onto sterile cellulose nitrate membranes (3.0 $\mu$m). Mycelium (250 mg wet wt) is resuspended in 10 ml of 7.5 mg $ml^{-1}$ Novozyme 234 in 50 m*M* sodium maleate, pH 5.8, 1.0 *M* $MgSO_4$. The enzyme solution is previously sterilized and clarified by centrifugation at 17,000 rpm for 1 h. Protoplasting is allowed to proceed for about 4 h at 28°C. Samples may be removed at intervals for microscopic examination. Protoplasting has been found to be variable depending on the age of mycelium, the age of conidia, and the batch of Novozyme and remains somewhat empirical. When satisfactory protoplasting has been achieved, the suspension is filtered through sterile scintered glass and the filtrate centrifuged at 10,000 rpm for 15 min. The pelleted mycelial debris is discarded. To reduce the density of the medium, 2 vol of 1.0 *M* NaCl, 50 m*M* sodium maleate, pH 5.8 (NS buffer) is added. This permits the pelleting of the protoplasts after 15 min, by centrifuging at 5000 rpm for 15 min, which are subsequently resuspended in 1 ml NS buffer.

## F. GENETIC ANALYSIS

The isolation of auxotrophic mutants by conventional screening techniques proved very unrewarding perhaps because *C. fulvum* is so adept at scavenging trace amounts of nutrients. Positive selection of mutants, resistant to fluoroacetate, selenate, various fungicides, as well as chlorate has been fruitful.

The absence of any sexual stage in the life cycle of *C. fulvum* precluded the use of traditional genetic methods. The isolation of mutants permitted the testing of various strategies for parasexual genetic analysis. Despite the earlier claim of success, we were unable to induce the formation of heterokaryons following hyphal anastomosis.[38] In contrast, protoplast fusion of strains carrying mutations in the nitrate-assimilating pathways has been very successful.[39] Protoplasts from each "parental" mutant are prepared and mixed. The protoplasts are pelleted in a microfuge and resuspended in 1 ml of 30% polyethyleneglycol (PEG), 0.01 *M* $CaCl_2$ which had been prewarmed to 30°C. The suspension is mixed thoroughly and incubated for 15 min at 30°C. The PEG is removed by diluting the suspension with 6 ml of NS buffer and centrifuging for 10 min at 3500 rpm. The supernatant is discarded and the procedure repeated two more times. The protoplasts are finally resuspended in 2 ml MS buffer (1.0 *M* $MgSO_4$, 0.05 *M* sodium maleate, pH 5.8). Serial dilutions were made in MS buffer and 100-$\mu$l aliquots plated onto selective and control media.

A rapid physical method to determine ploidy was also required. This was developed using *Schizosaccharomyces pombe* and is based on video imaging of DAPI-stained nuclei.[40] Using this technique it was shown that transient diploids of *C. fulvum* are obtained which frequently spontaneously haploidize. The haploid products are nearly all recombinant, permitting rapid linkage analysis. In the absence of morphological markers, a number of restriction fragment length polymorphism (RFLP) loci have been studied. More recently, the use of Random Amplified Polymorphic DNA, a novel PCR technique, has shown great promise. The extension of current studies to larger numbers of progeny and markers will permit the creation of an extensive linkage map and open up the possibility of cloning linked genes by chromosome walking.

Linkage analysis by progeny segregation has been complemented by analysis of electrophoretically separated chromosomes.[40a] Pelleted protoplasts are resuspended in 1% LMP Agarose (Sigma) in plug-support buffer (PSM)[41] containing 1.2 *M* sorbitol, 0.02 *M* EDTA, 0.014 *M* β-mercaptoethanol, which had been previously cooled to 45°C, to give a final concentration of $1 \times 10^7$ protoplasts per milliliter. The suspension is pipetted into plastic plug molds which had previously been washed in 0.25 *M* HCl for 1 h at 70°C and rinsed with sterile distilled water. The plugs are allowed to solidify for 30 min at 4°C. They are then transferred to lysis buffer (NDS)[42] containing 0.5 *M* EDTA, 0.1 *M* Tris, 1% (w/v) lauroylsarcosine (pH 9.5), which had been preheated to 50°C and were incubated at this temperature with gentle shaking in the presence of 1 mg $ml^{-1}$ proteinase K (sigma) for 48 h. After this, plugs were washed in 0.05 *M* EDTA and stored in NDS buffer at 4°C until use. Electrophoresis was performed in an LKB Pharmacia "Pulsaphor" system using a hexagonal array of electrodes giving DNA separation analogous to CHEF separations as described by Chu et al.[43] Gels are 1% agarose (Sigma) in modified 1 × TBE buffer (containing 50 m*M* Tris, 45 m*M* boric acid and 100 μ*M* EDTA). Electrophoresis was performed in modified 1 × TBE at 12°C at 45 V for 14 days. These techniques have revealed nine chromosomal bands ranging in size from about 2 to 5 Mbp. The largest two bands contain more than one molecule indicating a minimum karyotype of 11 chromosomes, consistent with microscopic observation of mitosis. Hybridization of cloned probes to blots of these gels permits unambiguous assignment of genes to bands and considerably speeds up genetic analysis.

The pulsed-field gel electrophoresis has revealed a chromosome-length polymorphism dividing the available isolates into two groups. Races 2.5.9.11 and 2.4.5.9.11 possess 4.8 and 2.5 Mbp bands and lack 5 and 2 Mbp bands, which all the other races have. It appears that these races have undergone a translocation. The possibility that this translocation is linked to the loss of the avirulence gene 9 is being actively pursued.

DNA-mediated transformation of *C. fulvum* is performed in an adaptation of the technique described by Oliver et al.[44] Protoplasts prepared as described are resuspended in 1 ml MTC buffer (1.0 *M* $MgSO_4$ ,0.05 *M* Tris, 0.05 *M* $CaCl_2$, pH 7.5) to give a concentration of $10^6$ protoplasts per milliliter. Plasmid DNA (4 μg) was added and maintained at room temperature for 10 min. One milliliter of 20% w/v PEG 4000 in 0.05 *M* Tris-$CaCl_2$, pH 7.5 is then added, mixed, and incubated for a further 15 min at room temperature. Aliquots 0.3 ml were then resuspended in 20 ml CM (0.5 × minimal media supplemented with yeast extract, 1 g $l^{-1}$, 12 g $l^{-1}$ potato dextrose broth, 1 g $l^{-1}$ mycological peptone, and 1 g $l^{-1}$ acid casein hydrolyzate) molten agar at 45°C and plates poured. After an 18-h incubation at 23°C plates were overlaid with CM media containing 200 mg ml hygromycin B or other antibiotic as appropriate.The modified procedure for transformation improved the recovery of transformants to about 100 per microgram, although results are quite variable The selectable markers used have been hygromycin B[45] and phleomycin.[46]

The transformation system was used to develop a reporter gene system for the fungus based on β-glucuronidase.[47] Expression of GUS under the control of fungal promoters is easily detected. This system has the advantage that endogenous GUS activities seem undetectable. GUS-expressing fungi have been used to quantify fungal invasion of plant tissue.The vector system forms the basis of a strategy designed to isolate plant-induced promoters.

## G. RETROTRANSPOSONS IN THE GENOME OF *CLADOSORIUM FULVUM*

One of our current strategies for the isolation of fungal genes involved in pathogenesis is the analysis of protein of fungal origin present in intracellular fluids. Antibodies raised by P. de Wit to compatible intracellular fluid substances were used to screen a lambda gt11 library of *C. fulvum* constructed using *Eco*R1*-digested genomic DNA. The first positive clone analyzed turned out to be part of a multicopy sequence with homology to retrotransposon reverse transcriptase sequences.[48] Subsequent cloning has revealed the presence of about 30 copies of a long-terminal repeat (LTR) retrotransposon with strong similarities to the gypsy group of retrotransposons which are found in insect, yeast, and plant species. Sequencing of the LTRs has revealed primer-binding sites, target site duplications, and terminal inverted duplications. The presence of identical LTRs is evidence of recent transposition. Reverse transcriptase activity can also be assayed and copurifies with a viruslike particle similar in morphology to the yeast Ty particles. The particles contain RNA complementary to the cloned reverse transcriptase sequence. Furthermore an HIV RT antibody recognizes epitopes in the fungal mycelium.

Southern hybridization analysis of the different races reveals polymorphism when probed with LTR sequences, although the majority of bands are shared. The polymorphism of these bands will provide a means to reconstruct the evolution of the races, as well being of great value in constructing genetic maps.

The possible role of the retrotransposon in generating variation or disrupting avirulence genes is being investigated. The strong relationships of the elements sequence and structure to retrotransposons in organisms as diverse as *Drosophila,* yeast, lily, and pine is strongly suggestive of horizontal gene transfer. It will be interesting to determine if related elements are present in tomato.

## H. FUTURE PROSPECTS

The physiological, biochemical, and genetic bases for the study of *C. fulvum* are now in place. A number of strategies for the further elucidation of the mechanisms of pathogenicity are available. These can be divided into directed and blind methods. Directed methods include those aimed at an analysis of mRNAs and proteins present in the interaction between plant and pathogen. Blind methods include the isolation of mutants by integrative transformation and the search for plant-induced promoters.The discovery that the A9 peptide's gene is absent in races not expressing the phenotype opens up the possibility of using genomic subtraction techniques to isolate other avirulence genes. Analysis of the distribution of repetitive sequence elements such as the recently discovered retrotransposon, will allow an analysis of the evolution of the current range of races. Finally the cloned genes, such as the A9 peptide, can be analyzed using the GUS system. We can look forward to a rewarding period of discovery and increased understanding.

# REFERENCES


1. **Cooke, M. C.,** New American Fungi, *Grevillea,* XII, 32, 1883.
2. **Ciferri, R.,** A few critical Italian fungi, *Atti Ser.,* 5, X, 245, 1952.
3. **von Arx, J. A.,** *Mycosphaerella* and its anamorphs, *Proc. K. Ned. Akad. Wet.,* 86, 47, 1983.
4. **Makemson, W. K.,** The Leaf mould of tomatoes caused by *Cladosporium fulvum, Mich. Acad. Sci. Annu. Rep.,* 20, 309, 1918.
5. **Lind, J.,** En tomatsort der ikke angribes af sygdom tomatbladenes flojlsplet, *Gartner Tidende,* xxv, 201, 1909.
6. **Norton, J. B. S.,** Resistance to *Cladosporium fulvum* in tomato varieties, *Phytopathology,* 4, 398, 1914.
7. **Langford, A. N.,** The parasitism of *Cladosporium fulvum* and the genetics of resistance to it, *Can. J. Res.,* 15, 35, 1937.

8. **Bailey, D. L.,** Studies on the racial trends and constancy in *Cladosporium fulvum* Cooke, *Can. J. Res.,* 28, 535, 1950.
9. **Williams, P. H.,** Tomato leaf mould *Cladosporium fulvum, Rep. Exp. Res. Sta. Cheshunt,* 28, 30, 1943.
10. **Kerr, E. A.,** Breeding for resistance to leaf mould in greenhouse tomatoes, *Ont. Hort. Exp. Sta. Rep.,* 1955.
11. **Kerr, E. A. and Bailey, D. L.,** Resistance to *Cladosporium fulvum* Cke obtained from wild species of tomato, *Can. J. Bot.,* 42, 1541, 1964.
12. **Lindhout, P., Korta, W., Cislik, M., Vos., I., and Gerlagh, T.,** Further identification of races of *Cladosporium fulvum (Fulvia fulva)* on tomato originating from the Netherlands, France and Poland, Neth. *Plant Pathol.,* 95, 143, 1981.
13. **Day, P. R.,** Mutations to virulence in *Cladosporium fulvum, Nature (London),* 179, 1141, 1957.
14. **Bond, T. E. T.,** Infection experiments with *Cladosporium fulvum* Cooke and related species, *Ann. Appl. Biol.,* 25, 277, 1938.
14a. **Kenyon, L.,** unpublished results, 1990.
14b. **Roberts, I. N., and Oliver, R. P.,** unpublished, 1990.
15. **Rick, C. M.,** Hybridisation between *Lycopersicon esculentum* and *Solanum penelli.* Phylogenetic and cytogenetic significance, *Proc. Natl. Acad. Sci. U.S.A.,* 46, 78, 1960.
16. **Joosten, M. H. A. J., Hendrickx, L. J. M., and de Wit, P. J. G. M.,** Carbohydrate composition of apoplastic fluids isolated from tomato leaves inoculated with virulent or avirulent races of *Cladosporium fulvum* syn *Fulvia fulva, Neth. J. Plant Pathol.,* in press, 1990.
17. **de Wit, P. J. G. M.,** A light and scanning-electron microscope study of infection of tomato plants by virulent and avirulent races of *Cladosporium fulvum, Neth. J. Plant Pathol.,* 83, 109, 1977.
18. **Lazarovitz, G. and Higgins, V. J.,** Histological comparison of *Cladosporium fulvum* race 1 on immune resistant and susceptible varieties, *Can. J. Bot.,* 54, 224, 1976.
19. **Lazarovitz, G. and Higgins, V. J.,** Ultrastructure of susceptible, resistant and immune reactions of tomato to races of *Cladosporium fulvum, Can. J. Bot.,* 54, 235, 1976.
20. **van Dijkman, A. and Sijpesteijn, K. A.,** Leakage of pre-absorbed $^{32}P$ from tomato leaf disks infiltrated with high molecular weight products of incompatible races of *Cladosporium fulvum, Physiol. Plant Pathol.,* 3, 57, 1973.
21. **van Dijkman, A. and Sijpesteijn, K. A.,** A biochemical mechanism for the gene-for-gene resistance of tomato to *Cladosporium fulvum, Neth. J. Plant Pathol.,* 77, 14, 1971.
22. **Dow, J. M. and Callow, J. A.,** Leakage of electrolytes from isolated leaf mesophyll cells of tomato induced by glycopeptides from culture filtrates of *Fulvia fulva* Cooke Ciferri syn *Cladosporium fulvum, Physiol. Plant Pathol.,* 15, 27, 1979.
23. **Lazarovitz, G. and Higgins, V. J.,** Biological activity and specificity of a toxin produced by *Cladosporium fulvum, Physiol. Biochem.,* 69, 1057, 1979.
24. **Lazarovitz, G. and Higgins, V. J.,** Purification and partial characterisation of a glycoprotein toxin produced by *Cladosporium fulvum, Physiol. Biochem.,* 69, 1063, 1979.
25. **de Wit, P. J. G. M. and Roseboom, P. H. M.,** Isolation partial characterisation and specificity of glycoprotein elicitors from culture filtrates mycelium and cell walls of *Cladosporium fulvum* syn *Fulvia fulva, Physiol. Plant Pathol.,* 16, 391, 1980.
26. **Peever, T. L. and Higgins, V. J.,** Electrolyte leakage lipoxygenase and lipid peroxidation induced tomato leaf tissue by specific and non-specific elicitors from *Cladosporium fulvum, Plant Physiol.,* 90, 867, 1989.
27. **de Wit, P. J. G. M. and Flach, W.,** Differential accumulation of phytoalexins in tomato leaves but not in fruits after inoculation with virulent and avirulent races of *Cladosporium fulvum, Physiol. Plant Pathol.,* 15, 257, 1979.
28. **de Wit, P. J. G. M. and Spikman, G.,** Evidence for the occurrence of race and cultivar-specific elicitors of necrosis in the intercellular fluids of compatible interactions of *Cladosporium fulvum* and tomato, *Physiol. Plant Pathol.,* 21, 1, 1982.
29. **de Wit, P. J. G. M. and van der Meer, F. E.,** Accumulation of the pathogenesis-related tomato leaf protein P14 as an early indicator of incompatibility in the interaction between *Cladosporium fulvum* syn *Fulvia fulva* and tomato, *Physiol. Mol. Plant Pathol.,* 28, 203, 1986.
30. **Joosten, M. H. A. J. and de Wit, P. J. G. M.,** Identification of several pathogenesis-related proteins in tomato leaves inoculated with *Cladosporium fulvum* syn *Fulvia fulva* as 13β-glucanases and chitinases, *Plant Physiol.,* 89, 945, 1989.
31. **Joosten, M. H. A. J. and de Wit, P. J. G. M.,** Isolation, purification and preliminary characterisation of a protein specific for compatible *Cladosporium fulvum* syn *Fulvia fulva*-tomato interactions, *Physiol. Mol. Plant Pathol.,* 33, 142, 1988.
32. **de Wit, P. J. G. M., Buurlage, M. B., and Hammond, K. E.,** The occurrence of host-pathogen and interaction-specific proteins in the apoplast of *Cladosporium fulvum* syn *Fulvia fulva* infected tomato leaves, *Physiol. Mol. Plant Pathol.,* 29, 159, 1986.

33. **de Wit, P. J. G. M.,, Hofman, J. E., and Aarts, J. M. M. J. G.,** Origin of specific elicitors of chlorosis and necrosis occurring in intercellular fluids of compatible interactions of *Cladosporium fulvum* syn *Fulvia fulva, Physiol. Plant Pathol.,* p. 24, 1984.
34. **Scholtens-Toma, I. M. J. and de Wit, P. J. G. M.,** Purification and primary structure of a necrosis-inducing peptide from the apoplastic fluids of tomato with *Cladosporium fulvum* syn *Fulvia fulva, Physiol. Mol. Plant Pathol.,* 33, 59, 1988.
34a. **Van Kan, J., van den Ackerveken, G., and de Wit, P. J. G. M.,** 1990.
34b. **Talbot, N. J.,** unpublished, 1990.
34c. **Harling, R., Kenyon, L., and Turner, J. G.,** unpublished, 1990.
35. **Mitchell, R. E.,** Isolation and structure of a chlorosis-inducing toxin of *Pseudomonas phaseolicola, Phytochemistry,* 15, 1941, 1976.
36. **Raeder, U. and Broda, P.,** Rapid preparation of DNA from filamentous fungi, *Lett. Appl. Microbiol.,* 1, 17, 1985.
37. **Harling, R., Kenyon, L., Lewis, B. G., Oliver, R. P., Turner, J. G., and Coddington, A.,** Conditions for efficient isolation and regeneration of protoplasts from *Fulvia fulva, J. Phytopathol.,* 122, 143, 1988.
38. **Barr, R. and Tomes, M. L.,** Variation in the tomato leaf mold organism, *Cladosporium fulvum, Am. J. Bot.,* 48, 512, 1953.
39. **Talbot, N. J., Coddington, A., Roberts, I. N., and Oliver, R. P.,** Diploid construction by protoplast fusion in *Fulvia fulva* syn. *Cladosporium fulvum* genetic analysis of an imperfect fungal plant pathogen, *Curr. Genet.,* 14, 567, 1988.
40. **Talbot, N. J., Rawlins, D., and Coddington, A.,** A rapid method for ploidy determination in fungal cells, *Curr. Genet.,* 14, 51, 1988.
40a. **Talbot, N. J. and Oliver, R. P.,** unpublished, 1990.
41. **Southern, E. M., Anand, R., Brown, W. R. A., and Fletcher, D. S.,** A model for the separation of large DNA molecules by crossed field gel electrophoresis, *Nucl. Acids Res.,* 15, 7865, 1987.
42. **Orbach, M. J., Vollrath, D., Davis, R. W., and Yanofsky, C.,** An electrophoretic karyotype for Neurospora crassa, *Mol. Cell. Biol.,* 8, 1469, 1988.
43. **Chu, G., Vollrath, D., and Davis, R. W.,** Separation of large DNA molecules by contour-clamped homogeneous electric fields, *Science,* 234, 1582, 1986.
44. **Oliver, R. P., Roberts, I. N., Harling, R., Kenyon, L., Punt, P. J., Dingemanse, M. A., and van den Hondel, C. A. M. J. J.,** Transformation of *Fulvia fulva,* a fungal pathogen of tomato to hygromycin B resistance, *Curr. Genet.,* 12, 231, 1987.
45. **Punt, P. J., Oliver, R. P., Dingemanse, M. A., Pouwels, P. H., and van den Hondel, C. A. M. J. J.,** Transformation of *Aspergillus* based on the hygromycin B resistance marker from *Escherichia coli, Gene,* 56, 117, 1987.
46. **Mattern, I. E., Punt, P. J., and van den Hondel, C. A. M. J. J.,** A vector for *Aspergillus* transformation conferring phleomycin resistance, *Fungal Genet. Newsl.,* 36, 25, 1989.
47. **Roberts, I. N., Oliver, R. P., Punt, P. J., and van den Hondel, C. A. M. J. J.,** Expression of the *Escherichia coli*β-glucuronidase gene in industrial and phytopathogenic filamentous fungi, *Curr. Genet.,* 15, 177, 1989.
48. **McHale, M. T., Roberts, I. N., Talbot, N. J., and Oliver, R. P.,** Expression of reverse transcriptase genes in *Fulvia fulva. Mol. Plant-Microbe Interact.,* 1989, 4, 2.

# *Section II: Bacterial-Plant Interactions (Pathogenic)*

# OPINES IN *AGROBACTERIUM* BIOLOGY

**Yves Dessaux, Annik Petit, and Jacques Tempé**

## TABLE OF CONTENTS

## FOREWORD

There is little doubt that the studies on the phytopathogenic bacteria *Agrobacterium,* have been among the most rewarding in the plant-microbe interaction field. The discovery of this remarkable example of interkingdom genetic transfer has provided scientists with a powerful tool to study and eventually modify the plant genome. These studies also allowed pathologists to understand the molecular basis of an interaction between a plant and a pathogen. This understanding results, in part, from the studies of low-molecular weight molecules termed opines, first discovered in crown gall tumors. According to a theory known as the opine concept, they are produced by the host plant in response to the action of bacteria and contribute to the growth and the dissemination of the pathogen. The opine concept, which first dealt with the *Agrobacterium tumefaciens*-plant interaction, has been extended to the *A. rhizogenes*-plant system. The discovery of opine-like compounds in nodules incited by several *Rhizobium* strains indicates that this concept may also apply to other plant-microbe interactions (see the chapter by Murphy and Saint in this volume).

In this chapter, we present an update of the information on the biological roles of opines in the *Agrobacterium*-plant interaction, and discuss this information in view of the opine concept. In addition, data on the chemistry and analysis of opines is provided.

## I. INTRODUCTION

Bacteria of the *Agrobacterium* genus[1,2] are Gram-negative, soil microbes, which incite crown gall[3] (*A. tumefaciens*) and hairy root[4] (*A. rhizogenes*) diseases on a variety of dicotyledonous plants.[5-7] In nature or in the field, the infection process begins at a wound site[8-10] and eventually leads to formation of typical overgrowths (Figures 1 and 2). Since *Agrobacterium* lives in the soil and since roots are often wounded during growth, tumors are generally localized on the underground system of the plant. They are commonly observed at the junction of the stem and root (crown), on the lower part of the stem, at grafting points, and, when plants are uprooted, on the roots.[11,12] The disease affects several economically important woody species such as stone-fruit trees, poplar, grapevine, etc.[11-14] Although, in nature, strains may exhibit limited host range,[15] this is generally not the case in experimental inoculations. Some strains isolated from grapevine tumors do exhibit a limited host range in the laboratory.[16-20] In grapevine, and probably in a few other species, bacteria are capable of invading the vascular system of the plant.[21,22]

Crown gall cells are cancerous: grafted on healthy plants they multiply to form a tumor.[7,23] Furthermore, whereas uninfected tissues generally need exogenous growth hormones to grow *in vitro,* crown gall tissues and hairy roots can be indefinitely maintained in axenic cultures on hormone-free media.[7,23-26] To account for this stable phenotype, the existence of a "tumor-inducing principle" (TIP), transferred from the bacteria to the plant cell, was postulated.[23,27]

The nature of the TIP was elucidated only after the emergence and improvement of molecular biology techniques. Analysis of the DNA content of various *Agrobacterium* strains showed that all pathogenic isolates contained a large plasmid which was absent from avirulent strains.[28-31] Virulence was lost upon plasmid curing and recovered when a similar extrachromosomal element was introduced by conjugation or transformation into avirulent bacteria.[31-33] These plasmids were called Ti (tumor-inducing) in *A. tumefaciens* and Ri (root-inducing) in *A. rhizogenes.* It was shown, by DNA hybridization techniques, that plasmid DNA sequences, called T-DNA, are present in the genome of the host plant[34-40] where they are expressed and maintained.[41-43]

The mechanism by which T-DNA is integrated into the plant chromosomal DNA is not fully understood. It involves, in part, functions encoded on these plasmids by the nontransferred *vir* (virulence) genes.[44-48] In response to a favorable environment,[49] and upon induction[50]

FIGURE 1. A crown gall tumor formed on *Datura stramonium* following inoculation with a wild-type *A. tumefaciens* strain. (Photograph by Daniel Meur, ISV-CNRS, Gif/Yvette.)

by plant metabolites such as acetosyringone,[51,52] the products of the *vir* genes promote excision and transfer of the T-DNA sequences (for reviews, see References 53 to 58).

T-DNA genes responsible for plant cell proliferation fall in two groups: (1) those encoding synthesis of plant growth hormones (auxin,[59-61] cytokinin,[62-65] reviews[53,54,57,66,67]) and (2) those modulating the response of plant cells to hormones.[68-70] Overproduction of cytokinin and auxin in crown gall cells is responsible for tumor growth.[66,71,72] On the other hand, hairy root proliferation results from a more subtle mechanism, involving (1) bacterial factors, as T-DNA genes (encoding auxin production and/or conferring the high sensitivity to auxin) and, (2) plant factors, e.g., the endogenous level of auxin in plant cells.[68,69,73-79]

## II. OPINES AND OPINE CONCEPT

### A. GENESIS OF THE CONCEPT

Long before the discovery of the pathogenic plasmids and T-DNA, crown gall cells were shown to produce unusual amino acid derivatives that were not present in normal tissues of the same plant.[80-82] These compounds were purified and identified as lysopine,[83] octopine,[84] octopinic acid,[85] and nopaline[86] (Figure 3), hence they were given the name "opine"[87,88] (see review[89]).

FIGURE 2. Hairy root induced on *Kalanchoe daigremontiana* following inoculation with an *A. rhizogenes* strain. (Photograph by the authors and Daniel Meur, ISV-CNRS, Gif/Yvette.)

The significance of the synthesis of these molecules remained unclear for several years. Three observations contributed to the understanding of their biological role: (1) opines can be degraded by strains of *Agrobacterium*;[90] (2) the nature of the opines synthesized in tumors depends on the causative bacteria, not on the plant;[91,92] and (3) the nature of opines synthesized (in crown gall tumors) and degraded (by the bacteria) is strain-specific.[92] In other words, only opines whose synthesis were induced in tumors could be degraded by the inciting Agrobacterium (see reviews[93,94]). The correlation between opine degradation by agrobacteria and opine synthesis in plant cells is strict: one could predict the tumor opine type from knowing the degradative capabilities of the bacteria. These observations led to the hypothesis that a fraction of the bacterial genome was transferred from the bacteria to the plant cells.[92] Genetic support for this hypothesis was provided after the discovery of Ti plasmids, when it was shown that genes involved in both opine degradation and opine synthesis were Ti plasmid-borne.[95,96] As mentioned above, physical demonstration that Ti plasmid sequences, including genes involved in opine synthesis, were present and expressed in crown gall cells resulted from hydridization experiments and also from transposon and deletion mutagenesis and sequencing experiments.[43,45-47,97-105] As early as 1970, the correlation between opine synthesis in tumors and opine degradation by the bacteria was taken as an indication of gene transfer. Only years later was this correlation understood as a relevant feature of the host-pathogen interaction. It was then proposed that opine production by crown gall cells provided the pathogen with selective growth substrates favoring its propagation.[106,107]

The opine concept[106] and the genetic colonization theory[107] combined the previous results with earlier data on *in planta* transfer of virulence,[108,109] and findings that Ti plasmids were transmissible plasmids[110] whose conjugal transfer is induced by some of the opines.[111] These theories propose that opines play a fundamental role in the *Agrobacterium*-plant interaction. Synthesized at the expense of plant metabolic pools, they support multiplication of the virulent bacteria and favor the dissemination of the pathogenic plasmid by conjugal transfer.[112,113] Thus, opines are defined as chemical mediators of parasitism. The opine concept makes no assumptions on the mechanism by which the metabolism of plant cells is redirected toward opine production; on the other hand, the genetic colonization theory adds the precision that this is achieved by gene transfer from the bacterium to the plant cell.

### B. VALIDITY AND EXTENSION OF THE OPINE CONCEPT

In the beginning, the opine concept was based on properties of the octopine- and nopaline-type Ti plasmids. At the time, a third type of plasmid was known: the so-called "null-type" plasmid which incited formation of tumors containing no known opines.[114] If the opine concept was valid, these tumors had to synthesize at least one opine. Using plant extracts as differential growth substrates, it was shown that extracts from null-type tumors, but not of corresponding normal tissues, did contain substances that were specific catabolic substrates for the bacteria that had incited the tumors.[115] It was, therefore, assumed that the null-type tumor contained an unidentified opine. Four new opines, the mannityl opines (also called opines of the agropine family, Figure 3) were finally characterized in these tumors.[116,117] Similarely, opines of the succinamopine family (Figure 3) were detected in an other kind of null-type tumor.[118,119]

A major step in the extension of the opine concept came from the work on the *A. rhizogenes*-plant interaction. Similarities between hairy root and crown gall[31,33,38-40,120-122] suggested that opines could be involved in the *A. rhizogenes*-plant interaction. Indeed, opines were detected in hairy root tissue cultures[26] providing the first evidence that hairy root, like crown gall, is a natural instance of gene transfer. These opines belong to two classes: those of the agropine family,[26,116,117,123] and those of the cucumopine and mikimopine family[124-126] (see Table 3, Figure 3). Here also, opine synthesis and opine degradation are plasmid-borne functions.[103-105,123,127] These findings brought strong experimental support to the view that opines are essential features in the *Agrobacterium*-plant interaction.

Attempts to extend the theory outside the genus *Agrobacterium* met with limited success and preliminary investigations on other systems have so far been discouraging except for a few *Rhizobium* sp.-legume interactions. Studying the *Rhizobium*-legume symbiosis was a reasonable choice since *Rhizobium* and *Agrobacterium* belong to the family Rhizobiaceae and since large plasmids are often involved in the nodule formation (see several other chapters in this volume and reviews[128,129]). Also, strain-specific substances had been previously observed in *Lotus corniculatus*[130] nodules. Further investigations on this system yielded indications that these substances could be opines.[131] Similar studies undertaken on the *R. meliloti-Medicago sativa* symbiosis led to indentification of a compound which had properties of opines.[132] However, the biological significance and importance of these compounds are not fully understood. For more information, the reader should refer to the chapter by Murphy and Saint in this volume.

## III. CURRENT STATUS OF RESEARCH ON OPINES

About twenty opines have now been identified (Figure 3) in various *Agrobacterium*-plant interactions. Their presence in tumors constitutes the basis of the Ti and Ri plasmid

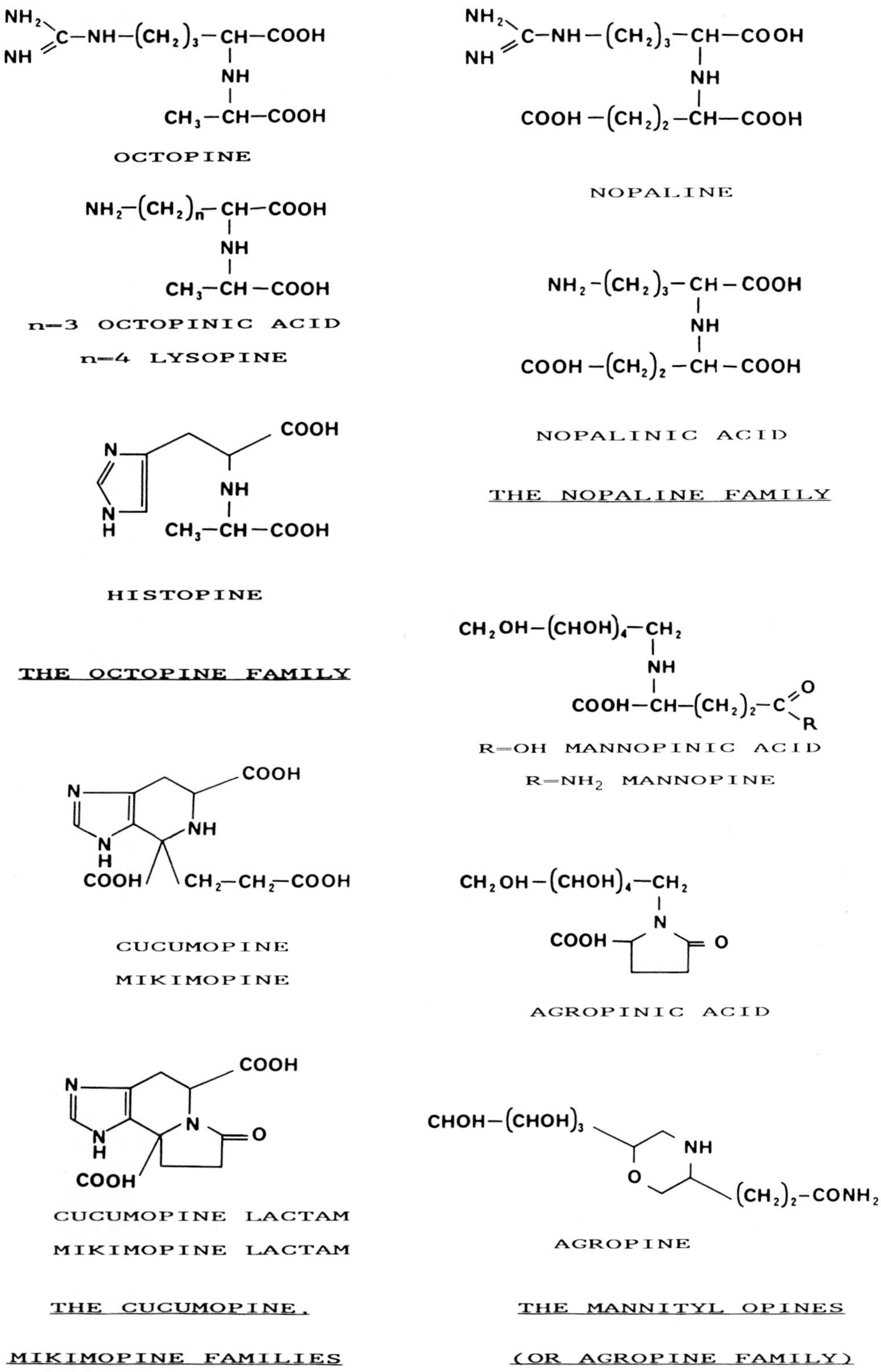


FIGURE 3. Structural formulas of opines.

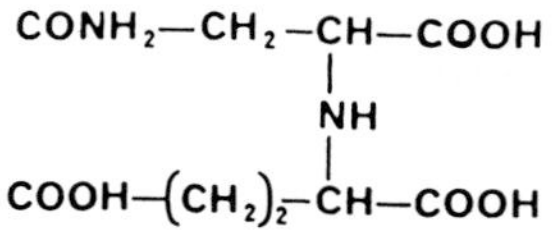

SUCCINAMOPINE

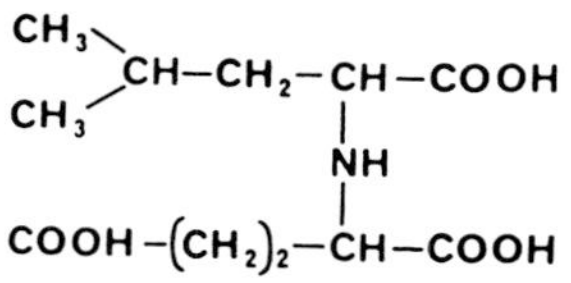

LEUCINOPINE

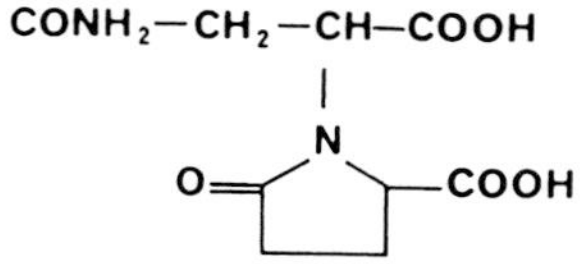

SUCCINAMOPINE LACTAM

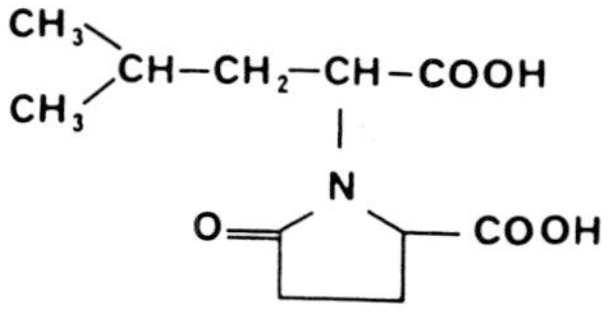

LEUCINOPINE LACTAM

THE SUCCINAMOPINE, LEUCINOPINE FAMILIES

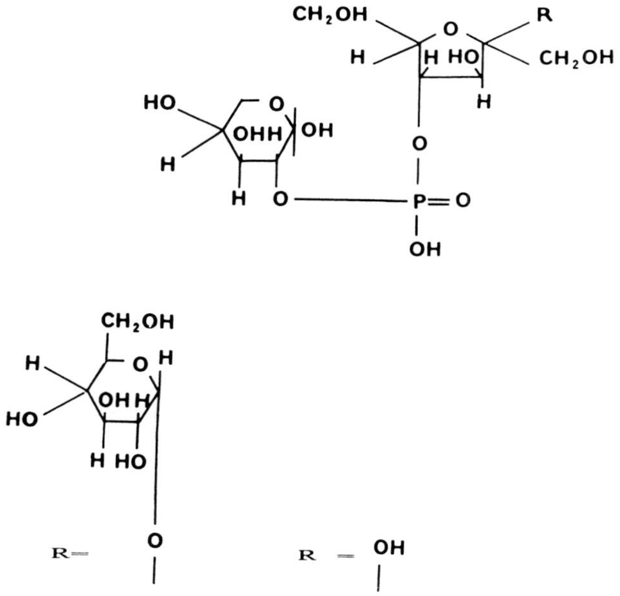

AGROCINOPINE A

AGROCINOPINE B

THE AGROCINOPINE FAMILY

FIGURE 3 (continued) Structural formulas of opines.

**TABLE 1**
**Classification of Opines**

| Characteristic chemical bond | Substitution on the imine group | Type of α-Ketoacid | Opine class | Opines |
|---|---|---|---|---|
| Phosphodiesters | n.a.[a] | n.a. | Agrocinopines | Agrocinopines A + B probably agrocinopine C + D |
| Imines | Sugar (mannose) | n.a. | Mannityl opines | Agropine, mannopine, mannopinic and, agropinic acids |
| | α-Ketoacids | Pyruvic acid | Octopine | Octopine, octopinic acid, lysopine, histopine |
| | | α-Ketoglutarate | Nopaline | Nopaline, nopalinic acid |
| | | | Succinamopine | Succinamopine, succinamopine lactam, succinopine |
| | | | Cucumopine | Cucumopine, cucumopine lactam, mikimopine, mikimopine lactam |

[a] n.a., Not applicable.

classifications. New *Agrobacterium* strains are constantly being isolated, among which some do not belong to previously described types; therefore, the data are in constant evolution. The current situation is presented below.

## A. OPINE STRUCTURES AND CLASSIFICATION

Opine structures and their classification are presented in Table 1 and Figure 3. According to their structure, they fall in two classes: sugar-phosphodiesters (agrocinopines[133,134-136]) and imines (all the others). In the most important class, in terms of members, the imines are all primary amine derivatives, and can be divided into subclasses according to the nature of the substitution on the NH group of the amine residue. The biosynthetic pathway leading to the imines likely involves one common first step, the establishment of a chemical bond between the nitrogen atom of a primary imine and the carbon atom of an aldehyde or keto group. A Schiff base is probably formed first which then undergoes one of three possible reactions: reduction (octopine, nopaline, succinamopine families), Amadori rearrangement (mannityl opines), or the Pictet-Spengler reaction (cucumopine family). The keto or aldehyde component could be either a sugar (mannityl opines[115-117,137]), an α-ketoacid, such as pyruvate (lysopine,[83] octopine,[84] octopinic acid,[85] histopine,[138] etc.), or α-ketoglutarate (nopaline,[86] succinamopine,[118,119] cucumopine, cucumopine lactam,[124,125] mikimopine, and mikimopine lactam[126]). The new opine vitopine[139] and a parent compound,[140] which have not yet been entirely characterized, appear to be derivatives of putrescine and α-ketoglutarate.[140]

## B. OPINE ANALYSIS AND DETECTION

Opines are small, water-soluble, generally stable, ionizable molecules, carrying characteristic functional groups (guanido, amino, glycol, acid, etc.). As a consequence, they are easy to extract from biological material. Routine separation of opines from other compounds is readily achieved by high-voltage paper electrophoresis (HVPE), at 50 to 100 V/cm,[116,123,133] essentially at pH 1.9 or 9.2. In principle, all of those belonging to the most important group, the imines, could be detected with a reagent specific for the imino group. However, such a reagent is not very sensitive and for practical purposes it is more convenient to base their detection on the presence in the molecule of other functional groups. Methods for visualization of opine spots, therefore, vary according to the structure of the opine. Appendix A summarizes extraction, separation, and detection procedures.

**TABLE 2**
**Classification of Ti Plasmids**

| Opine-type | Characteristic opine markers | Opine inducing conjugal transfer of the plasmid |
|---|---|---|
| Octopine | Octopine, octopinic acid, lysopine, histopine, agropine, mannopine, agropinic and mannopinic acids | Octopine, lysopine, octopinic acid |
| Nopaline | Nopaline, nopalinic acid, agrocinopines A + B | Agrocinopines A + B |
| Agropine | Agropine, mannopine, agropinic and mannopinic acids, agrocinopines C + D, leucinopine, leucinopine lactam, L,L-succinamopine | Agronopines C + D |
| Succinamopine | D,L-Succinamopine, succinamopine lactam, succinopine, nopaline[b] | n.d.[a] |
| Lippia | Agrocinopines[b] | n.d. |
| Grapevine I | Octopine,[c] cucumopine | n.d. |
| Grapevine II | Nopaline | n.d. |
| Grapevine III | Vitopine, unknown parent compound | n.d. |

[a] n.d., not determined.
[b] Probably degraded by the bacteria; these opines have not been detected in tumors.
[c] Presence of other opines of the octopine class (octopinic acid, lysopine, histopine) has not been investigated.

## C. OPINE-BASED CLASSIFICATION OF TI AND RI PLASMIDS

The Ti plasmids are classified according to their opine type, i.e., the nature of opines present in the tumors and degraded by the bacteria (Table 2). All opines synthesized in tumors are degraded by the bacteria, including leucinopine and leucinopine lactam[141-143] (Figure 3), but with the exception of *N*-carboxyethylmethionine[144] (not shown). As a consequence, this latter compound should not be regarded as an opine but rather an opine analog.[144] The reverse proposal is not true. Several *Agrobacterium* strains utilize opines whose synthesis is not induced in the tumor.[145,146] This phenomenon generally involves catabolic functions encoded by plasmids other than the pathogenic ones.[114,147-149] Thus, some avirulent strains can also degrade opines.[145-148]

Most of the Ti plasmids exhibit a wide host range, at least in experimental inoculations: in the *Agrobacterium* chromosomal background, they incite crown galls on numerous dicots.[5,12,16] However, some strains isolated from grapevine,[17-20,150] as well as from *Lippia* sp.,[15] exhibit a limited (or narrow) host range. The mechanism accounting for the host range limitation is not fully understood;[18,151,152] it may involve genes located in both the *vir* region and in the T-DNA.[17,20,153-156]

There are four classes of Ri plasmids (Table 3). Two of them appear to be closely related since they are both characterized by mannityl opines.[123] The agropine type, which is characterized by the four opines of the family present the unusual features that the opine catabolic genes for mannopine, mannopinic, and agropinic acids are not located on the Ri plasmid but on another large plasmid.[123] In the two other cases, Ri plasmids determine synthesis and utilization of epimeric opines, cucumopine and mikimopine, which, although their biosynthesis has not been formally established, probably both derive from histidine and α-ketoglutarate.[124-126] Cucumopine-type Ri plasmids do not encode synthesis or degradation of mikimopine, and vice versa.[157,158] Whether these plasmids and their opine degradation and synthesis functions are related is not known. A similar observation could be made regarding the occurrence of succinamopine as L,L or D,L forms, in tumors induced by two different types of Ti plasmids.[118,119,159]

**TABLE 3**
**Classification of Ri Plasmids**

| Opine type | Characteristic opine markers | Opines inducing conjugal transfer of the plasmid |
|---|---|---|
| Agropine | Agropine, mannopine, agropinic acid, mannopinic acid, agrocinopine A[a] | n.d.[b] |
| Mannopine | Mannopine, agropinic acid, mannopinic acid, agrocinopine C[a] | n.d. |
| Cucumopine | Cucumopine, cucumopine lactam | n.d. |
| Mikimopine | Mikimopine, mikimopine lactam, mannopine[a] | n.d. |

[a] Probably degraded by the bacteria; these opines have not been detected in roots induced by these strains.
[b] n.d., not determined.

# IV. THE BIOLOGICAL ROLE OF OPINES

Tumor or hairy root formation may be seen as the result of a plant-microbe interaction involving signal exchanges between the two partners. These signals include the opines, synthesized by the plant and degraded by the bacteria. This description[160] of the *Agrobacterium*-plant interaction points out the biological role of opines: through opine utilization (*opc*) and plasmid transfer (*tra*), opine synthesis (*ops*) serves to provide the pathogen with an environment favoring its growth and dissemination. Although several plasmid-borne regions (such as the *vir* region, the T-DNA, or the origin of replication) may be "vital" for the plasmids, the opine concept predicts that the opine-related functions (opine synthesis, opine degradation, and conjugal transfer) are essential since they probably provide the selective pressure responsible for the evolution and maintenance of Ti and Ri plasmids.

## A. OPINE SYNTHESIS IN PLANT: HOW BACTERIA DIVERT ENERGY AND NUTRIENTS FROM THE HOST

As indicated above, opines are synthesized at the expense of the pool of metabolites of the transformed cells, and result from the expression of *ops* (see Appendix B for designations of opine-related functions) genes of the T-DNA. Among all *ops* genes, only a few are well known: essentially these are *nos* (nopaline synthase),[45,161] *ocs* (octopine synthase),[45,162] and the *mas*/*ags* (mannopine synthase/agropine synthase)[100-102] cluster of genes, involved in synthesis of mannityl opines. The *mas*/*ags* group of genes is found on octopine-type Ti plasmids[100-102] and agropine-type Ri plasmids[105,163,164] whereas the mannopine-type Ri plasmid only carries the two *mas* genes.[165] All these genes have been sequenced and analyzed.[105,160,161,164,166-169] So far, as reported for all other known T-DNA genes,[170] all *ops* genes exhibit an open reading frame (ORF) devoid of introns and flanked by eukaryotic regulatory sequences.[164-167,169]

*ocs* and *nos* genes, their respective polyA signals, and promoters, have been widely used in the construction of chimeric genes. *ocs* and *nos* genes are suitable marker genes for plant transformation (as reviewed[171-173]). Transcription of *ops* genes has been studied,[174,175] and enhancer sequences in *ocs* and *mas* promoters of both octopine-type Ti and agropine-type Ri plasmids have been characterized.[176-183] Interestingly, transcription of the *mas* genes proceeds from a bidirectional promoter region[103,105,184,185] which is, on octopine-type Ti plasmids, inducible by wounding and auxin,[186] a property that may be useful in constructing new plant gene vectors.

The locus encoding agrocinopine A synthase (*acs*) has been localized on the nopaline-type T-DNA and on agropine-type Ri plasmids.[98,187] The gene encoding cucumopine synthesis has been mapped close to the right T-DNA border of the corresponding Ri plasmid[104] and on the $T_B$-DNA of the octopine/cucumopine grapevine Ti plasmids.[158]

At the protein level, only the octopine synthase and nopaline synthase have been purified.[188-192] The octopine synthase recognizes arginine and lysine, histidine, and ornithine as alternative substrates. Therefore the activity of the enzyme leads to synthesis of, respectively, octopine and lysopine, histopine, and octopinic acid[188,189] (as reviewed[89,192]). In these cases, the opines are synthesized in one step. On the contrary, mannopine and mannopinic acid synthesis is achieved in two steps, from glucose and, respectively, glutamine and glutamate, via deoxyfructosyl intermediates.[101,102] The activity of the product of the *ags* gene is responsible for the cyclization of mannopine to agropine.[101,102,193] Interestingly, agropinic acid results from spontaneous rearrangements of mannopine and agropine.[116,117] Finally, the biochemical pathway leading to biosynthesis of agrocinopine A is not known but probably involves formation of the phosphodiester bond between a molecule of arabinose and the fructose residue of sucrose[134]. Although not formally demonstrated, agrocinopine B is probably a degradation product of agrocinopine A.[134]

*ops* genes are generally localized next to the right T-DNA border on Ti and Ri plasmids.[162,163,166-168] Since T-DNA transfer starts at the right border (reviews[53,54,56-58]), these *ops* genes are, therefore, the first T-DNA genes to be transferred to plant cells. This feature supports the assumption that opine synthesis functions are essential since their localization ensures the highest probability "to reach their target", the plant cell.

## B. OPINE DEGRADATION: HOW BACTERIA UTILIZE THE ENERGY AND METABOLITES FROM THE PLANT

The interest in opine degradation is multifaceted. For the pathologist these functions are essential attributes of the pathogen. They also constitute a means of identification and classification for *Agrobacterium* strains and plasmids. Also, as we shall see later (Section IV.C) agrocinopine degradation by strains harboring nopaline Ti plasmids is associated with sensitivity to agrocin 84, the antibiotic responsible for biological control of crown gall. For the microbiologist, opine degradation also presents some quite interesting features: (1) opines are unusual growth substrates, (2) their degradation is plasmid-encoded, and (3) in some cases controlled by complex regulatory systems.

Early studies have shown that octopine catabolism *(occ)* is inducible by octopine, octopinic acid, and lysopine[87,96,112,194,195] as is nopaline catabolism *(noc)* by nopaline.[194] Octopine, octopinic acid, and lysopine are degraded by membrane-associated enzymes, respectively, to arginine and pyruvate, ornithine and pyruvate, and lysine and pyruvate.[87,90,92,96,194,196,197] Another catabolic function, arginine degradation *(arc)*, coregulated with octopine and and nopaline degradation,[112,198] is associated with octopine and nopaline Ti plasmids.[198] Analysis of this function showed that arginine is degraded by an arginase to ornithine,[199,200] which is, in turn, converted directly into proline[201] by an unusual enzytic activity: ornithine cyclodeaminase (OCDase).[200,202,203] From an evolutionary point of view, this pathway is quite interesting since *Agrobacterium* and *Rhizobium* are the only Gram-negative bacteria to degrade arginine via the arginase pathway (see review[204]), and among the few that have OCDase[204]. Early genetic experiments indicated that genes responsible for octopine degradation and arginine catabolism are organized as an operon.[112,198] Molecular genetics and sequencing data confirmed this result and showed related, but not identical, organization of catabolic functions on the octopine-type and on the nopaline-type Ti plasmid.[203,205] From the right to the left, genes on both catabolic regions, are: (1) the regulatory gene, probably an activator, separately transcribed and probably constitutively expressed, (2) genes for opine transport and, finally, (3) the degradative enzymes (opine oxidase, arginine catabolic function).[206-208] However, transcription of genes involved in transport, and degradation differ whether they are on nopaline- or octopine-type plasmids.[205-208] In addition, there is no gene for arginase on octopine-type Ti plasmids.[205] Sequence homology searches have revealed that the nopaline- and octopine-type OCDase genes are closely related,[203] as are the NocD regulatory protein and the activator protein LysR (lysine biosynthesis).[208,209] Strikingly, transcription of the *lys*R gene in *Escherichia coli* proceeds from a divergent promoter,[210] as for the regulatory gene of *Agrobacterium*.[208] The homology searches have also demonstrated that genes encoding the transport systems for the two opines resemble other genes encoding various components of basic amino acid transport systems, such as *his*P (histidine permease) and *his*J/*arg*T (histidine and arginine periplasmic binding proteins).[207,208] Interestingly, octopine (and nopaline), lysopine, and histopine are, respectively, arginine, lysine, and histidine derivatives.

Mannityl-opine degradation is a property of several pathogenic plasmids, such as the octopine- and agropine-type Ti plasmids,[115-117] and mannopine- and agropine-type Ri plasmids.[123] However, only the catabolic functions of the octopine-type Ti plasmid have been investigated. Regulatory systems for these functions are probably complex:[211] mannopinic and agropinic acid catabolic functions show cross-inducibility, whereas agropine and mannopine induce the utilization of all of the four mannityl opines.[212] Mannopinic acid degradation probably yields mannose and glutamate, mannose being isomerized to fructose.[212] There is no other data on degradation of these opines, except for mannopine which is converted *in vivo* and *in vitro* to agropine by a plasmid-encoded enzyme termed catabolic mannopine cyclase.[213] The role of this enzyme in the degradation of mannopine is not known since the activity seems to be associated with the ability to degrade agropine.[213] The catabolic agropine cyclase shows unusual stability and has, therefore, been used to prepare agropine from mannopine.[213] On the octopine-type Ti plasmid, the region encoding the degradation of the four mannityl opines spans a 43 kb region representing ca. 25% of the plasmid length.[214] Genes involved in degradation of each mannityl opine are clustered, thus defining five overlapping regions, two of those being involved in mannopine utilization.[212,214,215] One set of genes responsible for mannopine degradation recognizes mannopinic acid as an non inducing substrate.[212] Interestingly, the catabolic mannopine cyclase activity[213] is very similar to the activity responsible for agropine synthesis *(ags)*,[101,102] *in planta*. Both genes were sequenced and found to be closely related.[216]

## C. OPINE AND SENSITIVITY TO AGROCINE K84: FURTHER ELABORATION OF THE OPINE CONCEPT

Degradation of agrocinopines A and B have also been investigated. The catabolic pathway and the related enzymic activities remain unknown. However, the transport system for agrocinopines A and B presents an interesting feature: it is responsible for the uptake of a toxic metabolite[133,217-219] called agrocin K84. Produced by the nonpathogenic *Agrobacterium* strain K84 successfully used in biocontrol of crown gall,[220-223] agrocin K84 is an adenosine nucleoside analog that contains 3′-deoxyarabinose in place of 2′-deoxyribose.[224-225] The antibiotic probably acts as a dideoxynucleotide by blocking DNA chain elongation.[226] As a consequence, *Agrobacterium* strains that have the ability to take up agrocinopine are often sensitive to the toxic molecule. These include nopaline- and agropine-type strains, several *A. rhizogenes* strains, and even some avirulent *Agrobacterium* isolates.[31,222,225,227,228]

In the nopaline-strain C58, genes responsible for uptake and catabolism of agrocinopines A and B (and, therefore, sensitivity to agrocin 84) are localized between coordinates 127 and 137 in the 8 o'clock region of the Ti plasmid.[45,133,221,229] They are moderately expressed in the absence of agrocinopine.[229] Presence of agrocinopine further induces the catabolic genes, thus rendering the strain more sensitive to agrocin K84.[133,218,229] Strain K84 harbors three plasmids. The first one is a large cryptic plasmid. The second plasmid, pAtK84b, is 190 kb in size, and specifies utilization of agrocinopines A and B and nopaline.[114,230] Homology studies revealed that this plasmid is related to the nopaline-type Ti plasmid, though it lacks T-DNA and virulence functions.[147,148] The third plasmid, the 48-kb plasmid pAgK84 encodes production and immunity to agrocin 84.[221,231,232] These functions map in the same 20 kb region.[233]

These properties probably confer upon strain K84 the ability to act as a superparasite, or an "ecological pirate",[234,235] killing the pathogen through agrocin production to colonize its ecological niche: the nopaline-producing tumor. Since some of the nopaline-type *Agrobacterium* strains are partially constitutive for agrocinopine degradation, this selective advantage may also extend to opine-free environments such as the rhizosphere of uninfected plants. This idea is supported by experimental results showing that strain K84 efficiently colonizes plant root systems[114,236] and survives in a rhizosphere where the pathogenic population declines.[236]

### D. OPINES AS INDUCERS OF PLASMID CONJUGATION: HOW OPINES FAVOR DISSEMINATION OF THE PATHOGENIC DETERMINANTS

Looking at Ti and Ri plasmids as catabolic plasmids, it makes sense that their conjugal transfer can be induced by the catabolic substrate(s). This property ensures that, under conditions where it is advantageous for the bacteria to carry the plasmid, it will readily be disseminated by conjugation to plasmidless recipients. Although inducibility of transfer has not been demonstrated for all Ti/Ri plasmid types, this conjugal behavior fits well with the opine concept since plasmid transfer to nonpathogenic *Agrobacterium* will, *in fine,* increase the proportion of pathogens in the bacterial population.

First detected *in planta,* the conjugal transfer of the Ti plasmid[108,109] occurs spontaneously *in vitro,*[110] but transfer frequencies can be increased by growing the donor strain in a medium containing certain opines.[110,111,113,233] Well-studied on octopine Ti plasmids, the regulation of conjugal transfer was shown to be linked to that of octopine catabolism.[111,194,195] Three classes of regulatory mutants for octopine degradation or conjugation were obtained: some were constitutive for both functions, whereas members of the two other classes were constitutive either for octopine degradation or conjugal transfer.[113,237] From these results, it was inferred that a common gene was regulating both processes.[113,195,237,238] Transposon mutagenesis allowed mapping of *tra* genes between coordinates 29 and 39 of the octopine-type Ti plasmids.[47] However, all the analyzed mutants were still able to transfer pTi at a low frequency, in a noninducible way, suggesting that other *tra* genes may map elsewhere on the octopine-type Ti plasmid.[47]

Transfer of nopaline-type Ti plasmids is inducible by agrocinopine A.[218,233] Isolates harboring a Ti plasmid, partly constitutive for transfer, have been described.[233] The 190-kb plasmid pAtK84b from strain K84, involved in nopaline and agrocinopine degradation and related to nopaline-type Ti plasmids, is self-transferable; transfer of this plasmid is inducible by both nopaline and agrocinopine.[233]

Several studies reported interesting results on the C58 *tra* functions. To summarize, three *tra* regions were identified on the C58 Ti plasmid. One was located to the right of the nopaline catabolic (*noc*) region (ca. between coordinates 12 to 18), the second and third ones lay close to the agrocinopine catabolic (*acc*) locus, in the 7 o'clock region of this plasmid.[45,239,240] Interestingly, a *tra* constitutive mutant was isolated that was also supersensitive to agrocin K84.[218,229] This suggests that both functions share some common regulatory system,[218] an hypothesis consistent with the observation that agrocinopines A and B induce both Ti plasmid transfer and agrocinopine utilization.[218,233,234]

No data on inducibility of *tra* functions on other Ti and Ri plasmids have been reported, except for the agropine-type Ti plasmid, on which *tra* functions are inducible by agrocinopine C and D.[233]

Recently, transfer of an octopine-type Ti plasmid was shown to promote chromosome mobilization.[241] The biological significance of this observation is not established. Since essential genes for pathogenicity are located on the chromosome of the bacterium (reviews[53,55,56]) it may be relevant that such genes can be mobilized concomitantly with the Ti plasmid. The tumor may thus be a place where both chromosomal and plasmid determinants of the disease could be transferred between *Agrobacterium* cells. Indirectly, this emphasizes the ecological role of opines in the dissemination of the pathogen.

### E. OTHER OPINE RELATED FUNCTIONS

Since opines are specific markers of the transformed cells, it was proposed that *Agrobacterium* may interpret presence of these molecules as indicative of a successful transformation event, thus leading to repression of the expression of the *vir* genes.[160,242] On the contrary, opines were recently shown to enhance expression of virulence genes in *Agrobacterium*.[160,242] This observation could be related to an earlier result demonstrating that tumors incited by octopine strains were larger when the bacterium was inoculated in a medium containing octopine.[243]

The biological significance of this observation is not understood, however the phenomenon may find some application in facilitating transformation of recalcitrant plant species.

## V. EVOLUTION OF THE OPINE CONCEPT

Since the opine concept or the genetic colonization theory were first formulated,[106,107] the discovery of over a dozen opines and characterization of several new *Agrobacterium* Ti/Ri plasmids (with respect to their opine type) have validated these theories, and left little doubt that opines are essential to the biology of *Agrobacterium* and its pathogenic plasmids.

Recent observations, however, have shown that bacteria or fungi outside the genus *Agrobacterium* can catabolize opine.[244-246] These non-agrobacterial isolates utilize one, two, or, occasionally, three opines as growth substrates. The ecological importance of this observation is exemplified by the detection, in crown gall tumors, of microorganisms belonging to the genus *Rhizobium* and *Pseudomonas*.[244,246] As a consequence, opine molecules may not be as specific for *Agrobacterium* as they were thought to be. Consistent with this, competition experiments performed with opine-utilizing *Agrobacterium* and *Pseudomonas* strains showed that, in tumor extract, nopaline was degraded faster by the *Pseudomonas* strain.[247] These bacteria could, therefore, constitute a threat for pathogenic *Agrobacterium* strains should those saprophytes deplete the availability of opines to a level where they no longer constitute a selective advantage for the pathogen. However, in nature, competition between the pathogenic *Agrobacterium* and the competitor strain might still be favorable to the inciting agrobacteria since, as mentioned earlier, the competitor strain generally utilizes only a limited number of opines.[244,246]

These findings, however, do not invalidate the opine concept. The tumor, which is supposed to result from the action of a few bacterial cells on a plant cell, indeed remains a place where these few inciting *Agrobacterium* do multiply to an important population, and where Ti plasmid transfer occurs. In an environment where the availability of carbon and nitrogen is limited, opines appear to be valuable sources of carbon and, for some, nitrogen. Thus, it is not surprising that other soil microbes have evolved opine catabolic systems. In fact, the presence of opines in nature probably provides the selective pressure which accounts for occurrence of opine catabolic functions outside the genus *Agrobacterium*. Consistent with this, octopine-degrading microorganisms have been isolated from sea mollusks.[248] Strikingly, octopine is produced by mollusks as a result of the anaerobic metabolism of their muscles.[249] Similar comments could be made regarding the existence of opine-degrading isolates of avirulent *Agrobacterium*. These include some isolates harboring the small plasmid of *A. rhizogenes,* and strains such as K84.[114,123,233]

One weak joint in the opine concept resides in the way the concept has been built. The role of opines has been deduced from their generalized occurrence in crown gall tumors and hairy root overgrowths and the success of the theory comes from the experimental support that the predictions always received. In other words, and with the exception of *in planta* transfer of Ti plasmids,[108,109,110] there is no direct evidence (as, for example, ecological studies) showing that presence of opines in tumors does confer a selective advantage to the inciting strain.

## VI. FUTURE DIRECTIONS OF RESEARCH ON OPINES

The above discussion emphasizes the role of opines in the plant-*Agrobacterium* interaction: they provide the bacteria with a suitable ecological niche. This favorable environment results from the expression, in the plant and in the bacteria, of several classes of genes: *ops, opc,* and *tra* genes, and in a way which is not yet entirely understood, *vir* genes. To investigate these functions should, therefore, provide fundamental information on the *Agrobacterium*-plant interaction, leading eventually to a global description of the system.

First, studies on opine-related functions should allow scientists to directly demonstrate the validity of the opine concept. With the availibility of mutants affected in *ops, opc* and *tra* functions, it should be possible to perform ecological studies, and establish whether opine-related functions and opine-induced plasmid transfer confer a selective advantage to the wild-type pathogen. Attempts to apply the opine concept to create a trophic relationship between a nonpathogenic bacterium able to catabolize opines and opine-synthesizing transgenic plants should also provide information on the validity of the concept. Such experiments, which could eventually lead to creation of artificial symbiosis, are in progress in several laboratories. Large-scale application of such systems could eventually more dramatically influence the ecology than *Agrobacterium* and its plasmids have done so far.

Also, studies on opine-related functions should provide a convenient way to address the exciting topic of the evolution of pathogenic plasmids. This proposal is exemplified by results concerning the octopine and nopaline catabolic functions and their homology to other catabolic functions, such as those involved in arginine, lysine, and histidine metabolism.[207,208] Similar comments could also be made for the duplication of the mannopine cyclase gene.[216]

Regarding the question of the origin of the plasmids, we currently believe that they all derive from at least one common ancestor which provided the bacteria with the ability to transfer genes to plant cells. This assumption is supported by experimental results obtained with the broad-host range plasmid pRSF1010.[250] In an appropriate *Agrobacterium* background, where *vir* genes are present, the *ori*T and *mob* genes of the small nonconjugative plasmid RSF1010 promote transfer of plasmid DNA to plant cells. Indeed, there are striking homologies of functions between the two systems: (1) the *ori*T region of RSF1010 may act as a T-DNA border; (2) *mob* genes on RSF1010 encode endonuclease(s) that nick DNA at the *ori*T locus, whereas *vir*D genes encode an endocuclease that nicks Ti plasmid at the T-DNA border.[53-58,250]

However, one of the mysteries in the evolution of the *Agrobacterium* plasmid remains the acquisition of the couple functions *ops/opc*. Homology studies indicate that it is unlikely that both functions were simultaneously acquired. Similarly, acquisition of only ancestral *opc* functions does not make too much sense in terms of selective pressure. Therefore, it could be proposed that the *ops* genes arose first. Here also, these events had to be favorable to the survival of the bacteria to be selected during the course of the evolution. In other words, the ancestral opine, produced by plant cells, had to be readily degradable by the bacteria. If true, the ancestral *opc* genes were not really acquired but were always resident in the bacteria, possibly as chromosomal genes. A few experimental results support these assumptions. First, nonpathogenic agrobacteriae (or rhizobiae) isolates may utilize some of the known opines *following a mutation event*.[247,251] This is specially true for octopine, which still can serve as a nitrogen source to mutants of the Ti plasmid-free strain C58C1.[251] Similarly, mannopine-utilizing mutants of the same strain have been obtained.[247]

The origin of *ops* genes remains unclear since they have to be expressible in the plant cell and not in the bacteria. The same is true for the other T-DNA genes, including genes involved in the control of the cell division. A previous hypothesis proposed that these genes were "captured" later by the "protopathogenic plasmids," as a way to amplify the *ops* functions.[235] *Agrobacterium* strains harboring such protoplasmids may still be present in nature, but their detection is not easy since they should produce no visible symptoms. However, study of the octopine-type Ti plasmid supports this idea. On this plasmid, there are two T-DNAs.[36,42,166] The right one, called TR-DNA, carries only genes involved in mannityl opine synthesis, and no functions for cell proliferation.[101,102,166] In the appropriate *Agrobacterium* background, this T-DNA is capable of independant transfer and integration in the plant genome. Transformed cells harboring the only TR-DNA should produce opines, but do not develop as a tumor.[252] Obviously, several couples of *ops/opc* genes have been acquired, leading to nopaline/octopine-, agropine-, or succinamopine-type plasmids. Consistent with our model, they all harbor very related *vir* regions (see review[253]). Although this whole scheme is almost entirely speculative, it provides some directions for future studies on opine-related functions.

Finally, a recurring question remains. How "extensible" is the opine concept? In this respect, additional work on systems related to the *Agrobacterium*-plant model is required. Interactions involving other *Rhizobium, Bradyrhizobium,* or *Frankia* strains may be suitable to further studies.

## ACKNOWLEDGMENTS

The authors thank Jean Brevet and Geneviève Hansen for their help in gathering references, Stephen Farrand and Patrice Dion for helpful discussions and communication of unpublished material, Lise Jouanin for the Kalanchoe plants, Claudine Deforeit for typing part of this manuscript, and Daniel Meur for excellent photographic artwork.

## APPENDIX A: EXTRACTION, SEPARATION, AND DETECTION OF OPINES

### EXTRACTION OF OPINES (MODIFIED[123])

Samples (10 to 300 mg fresh weight) of recently harvested plant tissues are placed in a microtube. Distilled water (3 ml/g of sample) is added, and the tube is heated at 100°C for 10 min.

Soften tissues are crushed, briefly vortexed, and separated from the liquid phase by centrifugation, for 5 min. at 13,000 *g* at room temperature. The supernatant is collected and rotary-evaporated at 40°C under vacuum, to yield a dried plant extract. This extract is resuspended in distilled water (0.5 μl/mg of sample) and kept frozen for further studies if necessary.

### SEPARATION OF OPINES[116,123,133]

Plant extract (1 to 6 μl) are spotted on high-quality chromatography paper (e.g., Whatman 3MM), and submitted to high-voltage paper electrophoresis, at 50 to 100 V/cm.

Essentially, two buffer systems are used: (1) 1.1 *M* acetic acid/0.7 *M* formic acid, pH 1.9; (2) 0.1 *M* ammonium bicarbonate/ammonium hydroxide, pH 9.2. Following electrophoresis, papers are dried in a stream of hot air.

### DETECTION OF OPINES

Detection techniques vary according to the nature of the characteristic chemical group of the investigated opine(s). Thus, methods (all derived from ref. 254) are presented below as a function of the opine(s) to be analyzed.

### Detection of Octopine and Nopaline

Detection of these compounds is achieved using phenanthrenequinone,[255] a reagent specific for monosubstituted guanidines.[255] Thus, arginine also reacts. Paper is first examined under UV light (312 nm) to visualize all fluorescent compounds already present in the plant extract. The paper is then carefully dipped in the phenanthrenequinone reagent. After partial drying of the paper, guanidines are visualized under UV light as fluorescent spots.

Preparation of the reagent:

- Solution I: 0.1% phenanthrenequinone in 90% ethanol. (the solution is stable several months at 4°C, in the dark)
- Solution II: 6 *N* NaOH (in water)
- Staining solution: dilute 10 ml of solution I with 90 ml of 95% ethanol. Add 25 ml solution B. Mix thoroughly. This solution could be kept 2 h (possibly carcinogenic, corrosive).

### Detection of Lysopine, Octopinic Acid, and Nopalinic Acid

Detection of these compounds is achieved using ninhydrin, a reagent specific for primary amines (review[256]). Thus, amino acids and free amino sugars also react. The electrophoregram is dipped carefully in the ninhydrin reagent, and placed in a stream of warm air. Since plant extracts contain various amino acids, a standard of opine should be run along with the samples. Lysopine and octopinic acid give purple/pink spots on a white background. To keep the background clear, it is not recommended to bring the paper to high temperature.

Preparation of the reagent: 0.25% ninhydrin in 95% ethanol (can be kept several weeks at 4°C in the dark).

### Detection of Cucumopine, Cucumopine Lactam, Mikimopine, Mikimopine Lactam, and Histopine

Detection of these compounds is achieved using the sulfanilic acid/sodium nitrite (Pauly's) reagent.[257] This reagent detects the imidazole moiety of the molecule but is not highly specific. Thus compounds such as histidine, tyrosine, and indole also react. A standard of opine should be run along with the samples. After extensive drying, the electrophoregram is lightly sprayed with cool sulfanylic acid/sodium nitrite reagent, allowed to dry, and resprayed with the sodium carbonate solution until the spots appear. Histidine, histopine, and tyrosine generally give pink color; cucumopine, cucumopine lactam, and mikimopine give colors ranging from bright orange to intense pink.

Preparation of the reagents:

- Solution I: 0.5% sulfanilic acid in 1 *N* HCl (keep at 4°C in the dark)
- Solution II: 5% sodium nitrite in water (keep at 4°C in the dark)
- Solution III: 15% sodium carbonate in water
- Staining solution: Mix equal volumes of cold solutions I and II under a fume hood, and let stand for 5 min before spraying.

### Detection of Mannityl Opines

Detection of these compounds is achieved using alkaline silver nitrate, a reagent that detects α-diols[258]. Thus, sugars, aminosides, uronic acids, and polyols react. A standard of opines should be run along with the samples. Dried paper is first dipped into the silver nitrate reagent (solution I), allowed to dry in a stream of cold air, and dipped again into the sodium hydroxide solution (II). After a few seconds at room temperature, α-diols appear as dark brown spots on a light brown background. To increase the sensitivity, electrophoregrams can be redipped in the sodium hydroxide solution, a few minutes later and passed a few times through a stream of steam. Finally, electrophoregrams should be fixed by dipping into diluted photofixative (solution III) followed by extensive washing (generally overnight) in water. These two latter steps reduce the background color, and allow long-term conservation of the documents.

Preparation of the reagents:

- Solution I: 0.4% silver nitrate in 99:1 acetone:water mixture (*highly flammable and corrosive;* should be kept in a safe cool place, in the dark; stable several weeks). When preparing the solution, first dissolve $AgNO_3$ in water and then add acetone.
- Solution II: 2% NaOH in 90% ethanol (can be kept at least 2 weeks; efficient even if the solution turns turbid)
- Solution III: water-diluted (1:4) photofixative solution (e.g., Kodak AL-4)

**Detection of Agrocinopines**

Detection of agrocinopines can be achieved as described above for the mannityl opines, though with a low sensitivity. An alternative technique uses the molybdate reagent.[259] After drying, the paper is dipped in the reagent, heated 15 min at 60 to 65°C, and exposed to UV light for 30 min. Phosphate-containing molecules appear as yellow to blue spots on a light blue background. Nucleosides also react.

Preparation of the reagent:

- Solution I: 20% sodium molybdate in water
- Solution II: 1 *N* HCl
- Solution III: 72% perchloric acid (very *caustic,* may deflagrate when in contact with oxidizable substances)
- Staining solution: to 140 ml of 75% acetone, add 5 ml of solution I, 10 ml solution II, and 5 ml solution III. Prepare fresh each time.

**Detection of Other Opines**

Other opines which do not react with previouly described reagents could be detected using a modification of the silver nitrate staining (see above: Detection of Mannityl Opines). The procedure to follow is exactly the same, except that reagent II now contains 20 to 40 mg/l mannitol.[118] Consequently, a brown background appears on electrophoregrams, on which silver-chelating substances produce a white spot.

Opines such as those of the succinamopine group can also be detected using the bromocresol green reagent as a pH indicator.[118] Well-dryed electrophoregrams are rapidly dipped in the reagent and allowed to lay flat on another sheet of paper. Organic acids are seen as yellow spots on a blue-green background. This technique is not very sensitive and does not work if the pH 9.2 buffer is used for running the HVPE.

Preparation of the reagents:

- Solution I: 0.05% bromocresol green in 95% ethanol. Add diluted NaOH until color turns blue.
- Solution II: 75% acetone in water
- Staining reagent: dilute 1 vol solution I with 3 vol solution II *(highly flammable).*

## APPENDIX B: DESIGNATION OF OPINE RELATED FUNCTIONS AND GENES

*opc*: opine catabolism
*ops*: opine synthesis
*tra*: conjugative plasmid transfer

Within the *ops* genes and functions, the following designations are used:

*acs*: agrocinopine synthesis
*ags*: agropine synthesis
*cus*: cucumopine synthesis
*mas*: mannopine/mannopinic acid synthesis
*nos*: nopaline synthesis
*ocs*: octopine synthesis

Within the *opc* genes and functions, the following designations are used:

*acc*: agrocinopine catabolism
*arc*: arginine catabolism
*noc*: nopaline catabolism
*nox*: nopaline oxidase
*occ*: octopine catabolism
*oox*: octopine oxidase

Within the *opc* genes and functions, the following designations are proposed:

*aac*: agropinic acid catabolism
*agc*: agropine catabolism
*amc*: agropine/mannopine catabolism
*cuc*: cucumopine degradation
*mpc*: mannopine catabolism
*mac*: mannopinic acid catabolism

## REFERENCES

1. **Kersters, K. and De Ley, J.,** *Agrobacterium,* in *Bergey's manual of systematic bacteriology,* Vol. 1, Krieg, N. R. and Holt, J. G., Eds., Williams & Wilkins, Baltimore, MD, 1984, 244.
2. **Elkan, G. H.,** The taxonomy of the Rhizobiaceae, in *International Review of Cytology* Suppl. 13: *Biology of the Rhizobiaceae,* Giles, K. L. and Atherly, A. G., Eds., Academic Press, San Diego, 1981, 1.
3. **Smith, E. and Townsend, C.,** A plant tumor of bacterial origin, *Science,* 24, 671, 1907.
4. **Riker, A., Banfield, W., Wright, W., Keitt, G., and Sagen, H.,** Studies on infectious hairy root of nursery apple trees, *J. Agric. Res.,* 41, 887, 1930.
5. **De Cleene, M. and De Ley, J.,** The host range of crown gall, *Bot. Rev.,* 42, 389, 1976.
6. **De Cleene, M. and De Ley, J.,** The host range of infectious hairy root, *Bot. Rev.,* 47, 147, 1981.
7. **Braun, A. C.,** A history of the crown gall problem, in *Molecular Biology of Plant Tumors,* Kahl, G. and Schell, J. S., Eds., Academic Press, San Diego, 1982, 155.
8. **Lippincott, B. B. and Lippincott, J. A.,** Bacterial attachment to a specific wound site as an essential stage in tumor initiation by *Agrobacterium tumefaciens, J. Bacteriol.,* 97, 620, 1969.
9. **Lippincott, B. B., Whatley, M. H., and Lippincott, J. A.,** Tumors induction by Agrobacterium involves attachment of the bacterium to a site on the host plant cell wall, *Plant Physiol.,* 59, 388, 1977.
10. **Matthysse, A. G.,** Initial interaction of *Agrobacterium tumefaciens* with plant host cells, *CRC Crit. Rev. in Microbiol.,* 13, 281, 1986.
11. **Moore, L. W. and Cooksey, D. A.,** Biology of *Agrobacterium tumefaciens*:plant interaction, in *International Review of Cytology,* Suppl. 13: *Biology of the Rhizobiaceae,* Giles, K. L. and Atherly, A. G., Eds., Academic Press, San Diego, 1981, 15.
12. **El-Fiki, F. and Giles, K. L.,** *Agrobacterium tumefaciens* in agriculture and research, in *International Review of Cytology,* Suppl. 13: *Biology of the Rhizobiaceae,* Giles, K. L. and Atherly, A. G., Eds., Academic Press, San Diego, 1981, 47.

13. **Dochinger, L. S.,** *Agrobacterium* galls of hybrid poplar trees in Iowa, *Phytopathology,* 59, 1024, 1969.
14. **Panagopoulos, C. G. and Psallidas, P. G.,** Characteristics of greek isolates of *Agrobacterium tumefaciens* (E. F. Smith & Townsend) Conn., *J. Appl. Bacteriol.,* 36, 233, 1973.
15. **Unger, L., Ziegler, S. F., Huffman, G. A., Knauf, V. C., Peet, R., Moore, L. W., Gordon, M. P., and Nester, E. W.,** New class of limited host-range *Agrobacterium* mega-tumor-inducing plasmids lacking homology to the transferred DNA of a wide host-range, tumor-inducing plasmid, *J. Bacteriol.,* 164, 723, 1985.
16. **Loper, J. E. and Kado, C. I.,** Host-range conferred by the virulence-specifying plasmid of *Agrobacterium tumefaciens, J. Bacteriol.,* 139, 591, 1979.
17. **Knauf, V. C., Yanofsky, M. F., Montoya, A. L., and Nester, E. W.,** Physical and functional map of an *Agrobacterium tumefaciens* tumor-inducing plasmid that confers a narrow host-range, *J. Bacteriol.,* 160, 319, 1984.
18. **Thomashow, M. F., Panagopoulos, C. G., Gordon, M. P., and Nester, E. W.,** Host range of *Agrobacterium tumefaciens* is determined by the Ti plasmid, *Nature (London),* 283, 794, 1980.
19. **Paulus, F., Huss, B., Bonnard, G., Ridé, M., Szegedi, E., Tempé, J., Petit, A., and Otten, L.,** Molecular systematics of biotype III Ti plasmids of *Agrobacterium tumefaciens. Mol. Plant-Microbe Interact,* 2, 64, 1989.
20. **Petit, A., Martin, L. A., Larribe, M., Brevet, J., and Tempé, J.,** Diversity of T-DNA functions encoding plant cell proliferation in *Agrobacterium* pathogenic plasmids, in *Plant gene transfer, Proc. UCLA Symp.,* Lamb, C. J. and Beachy, R. N., Eds., Wiley-Liss, New York, 1990, 13.
21. **Lehoczky, J.,** Spread of *Agrobacterium tumefaciens* in the vessels of grapevine, after natural infection. *Phytopathol. Z.,* 63, 239, 1968.
22. **Tarbah, F. and Goodman, R. N.,** Systemic spread of *Agrobacterium tumefaciens* biovar III in the vascular system of grapes, *Phytopathology,* 77, 915, 1987.
23. **White, P. R. and Braun, A. C.,** A cancerous neoplasm of plants. Autonomous bacteria-free crown gall tissue. *Cancer Res.,* 2, 97-617, 1942.
24. **Braun, A. C .,** The activation of two growth substance systems accompanying the conversion of normal tumor cells in crown gall, *Cancer Res.,* 16, 53, 1956.
25. **Braun, A. C.,** A physiological basis for autonomous growth of the crown gall tumor cells, *Proc. Natl. Acad. Sci. U.S.A.,* 44, 344, 1958.
26. **Tepfer, D. A. and Tempé, J .,** Production d'agropine par des racines formées sous l'action d'*Agrobacterium rhizogenes,* souche A4, *C.R. Acad. Sci.,* 292, 153, 1981.
27. **Braun, A.C., and White, P.R .,** Bacteriological sterility of tissues derived from secondary crown gall tumors. *Phytopathology,* 33, 85, 1943.
28. **Zaenen, I., Van Larebeke, N., Teuchy, H., Van Montagu, M., and Schell, J.,** Supercoiled circular DNA in crown gall inducing *Agrobacterium* strains, *J. Mol. Biol.,* 86, 109, 1974.
29. **Van Larebeke, N., Engler, G., Holsters, M., Van den Elsacker, S., Zaenen, I., Schilperoort, R. A., and Schell, J.,** Large plasmid in *Agrobacterium tumefaciens* essential for crown gall inducing ability. *Nature (London),* 252, 169, 1974.
30. **Watson, B., Currier, T. C., Gordon, M. P., Chilton, M. D., and Nester, E. W.,** Plasmid required for virulence of *Agrobacterium tumefaciens, J. Bacteriol.,* 123, 255, 1975.
31. **Moore, L. W., Warren, G., and Strobel, G.,** Involvement of a plasmid in the hairy root disease of plants caused by *Agrobacterium rhizogenes, Plasmid,* 2, 617, 1979.
32. **Van Larebeke, N., Genetello, C., Schell, J., Schilperoort, R. A., Hermans, A. K., Hernalsteens, J.-P., and Van Montagu, M.,** Acquisition of tumour-inducing ability by non-oncogenic agrobacteria as a result of plasmid transfer, *Nature, (London),* 255, 742, 1975.
33. **White, F. F. and Nester, E. W.,** Hairy root: plasmid encodes virulence traits in *Agrobacterium rhizogenes, J . Bacteriol.,* 141, 134, 1980.
34. **Chilton, M. D., Drummond, M. H., Merlo, D. J., Sciaky, D., Montoya, A. L., Gordon, M. P., and Nester, E. W.,** Stable incorporation of plasmid DNA into higher plant cells: the molecular basis of crown gall tumorigenesis, *Cell,* 11, 263, 1977.
35. **De Beuckeleer, M., De Block, M., De Greve, H., Depicker, A., De Vos, R., De Vos, G., De Wilde, M., Dhaese, P., Dobbelaere, M. R., Engler, G., Genetello, C., Hernalsteens, J.-P., Holsters, M., Jacobs, A., Schell, J., Seurinck, J., Silva, B., Van Haute, E., Van Montagu, M., Van Vliet, F., Villaroel, R., and Zaenen, I.,** The use of Ti plasmid as a vector for the introduction of foreign DNA into plants, in *Proc. 4th Int. Conf. Plant Pathogenic Bacteria,* Ridé, M., ed., INRA, Angers, 1978, 115.
36. **Thomashow, M. F., Nutter, R., Montoya, A. L., Gordon, M. P., and Nester, E. W.,** Integration and organization of Ti plasmid sequences in crown gall tumors, *Cell,* 19, 729, 1980.
37. **Willmitzer, L., De Beuckeleer, M., Lemmers, M., Van Montagu, M., and Schell, J.,** The Ti plasmid-derived T-DNA is present in the nucleus, and absent from plastids of plant crown gall cells, *Nature, (London),* 287, 359, 1980.

38. **Chilton, M. D., Tepfer, D. A., Petit, A., Casse-Delbart, F., and Tempé, J.,** *Agrobacterium rhizogenes* inserts T-DNA into the genomes of the host plant root cells, *Nature (London),* 295, 432, 1982.
39. **White, F. F., Ghidossi, G., Gordon, M. P., and Nester, E. W.,** Tumor induction by *Agrobacterium rhizogenes* involves the transfer of plasmid DNA to the plant genome, *Proc. Natl. Acad. Sci. U.S.A.,* 79, 3139, 1982.
40. **Spano, L., Pomponi, M., Costantino, P., Van Slogteren, G. M. S., and Tempé, J.,** Identification of T-DNA in the root-inducing plasmid of the agropine-type *Agrobacterium rhizogenes* 1855, *Plant Mol. Biol.,* 1, 291, 1982.
41. **Drummond, M. H., Gordon, M. P., Nester, E. W., and Chilton, M. D.,** Foreign DNA of bacterial origin is transcribed in crown gall tumors. *Nature (London),* 269, 535, 1977.
42. **Gurley, W. B., Kemp, J. D., Albert, M. J., Sutton, D. W., and Callis, J.,** Transcription of Ti plasmid-derived sequence in three octopine-type crown gall tumor lines. *Proc. Natl. Acad. Sci. U.S.A.,* 77, 2828, 1979.
43. **Willmitzer, L., Sanchez-Serrano, J., Buschfield, E., and Schell, J.,** DNA from *Agrobacterium rhizogenes* is transferred to and expressed in axenic hairy root plant tissues, *Mol. Gen. Genet.,* 186, 16, 1982.
44. **Garfinkel, D. J., and Nester, E. W.,** *Agrobacterium tumefaciens* mutants affected in crown gall tumorigenesis and octopine catabolism, *J. Bacteriol.,* 144, 732, 1980.
45. **Holsters, M., Silva, B., Van Vliet, F., Genetello, C., De Block, M., Dhaese, P., Depicker, A., Inzé, D., Engler, G., Villaroel, R., Van Montagu, M., and Schell, J.,** The functionnal organization of the nopaline *A. tumefaciens* plasmid pTiC58, *Plasmid,* 3, 212, 1980.
46. **Ooms, G., Klapwijck, P. M., Poulis, J. A., and Schilperoort, R. A.,** Characterization of Tn904 insertions in octopine Ti plasmid mutants of *Agrobacterium tumefaciens, J. Bacteriol.,* 144, 82, 1980.
47. **De Greve, H., Decraemer, H., Seurinck, J., Lemmers, M., Van Montagu, M., and Schell, J.,** The functional organization of the octopine *Agrobacterium tumefaciens* plasmid pTiB6S3, *Plasmid,* 6, 235, 1981.
48. **Iver, V. N., Klee, H. J., and Nester, E. W.,** Units of genetic expression in the virulence region of a plant-tumor inducing plasmid of *Agrobacterium tumefaciens, Mol. Gen. Genet.,* 188, 418, 1982.
49. **Alt-Mörbe, J., Kühlmann, H., and Schröder, J.,** Differences in induction of Ti plasmid virulence genes virG and virD and continued control of virD expression by four external factors, *Mol. Plant Microbe Interact.,* 2, 301, 1989.
50. **Bolton, G. W., Nester, E. W., and Gordon, M. P.,** Plant phenolic compounds induce expression of the *Agrobacterium tumefaciens* loci needed for virulence, *Science,* 232, 983, 1986.
51. **Stachel, S. E., Nester, E. W., and Zambryski, P. C.,** A plant cell factor induces *A. tumefaciens vir* gene expression, *Proc. Natl. Acad. Sci. U.S.A.,* 83, 379, 1986.
52. **Stachel, S. E., Messens, E., Van Montagu, M., and Zambryski, P.,** Identification of the signal molecules produced by wounded plant cells which activate the T-DNA transfer process in *Agrobacterium tumefaciens, Nature (London),* 318, 624, 1986.
53. **Melchers, L. S., and Hooykaas, P. J. J.,** Virulence of *Agrobacterium,* in *Oxford Surveys of Plant Molecular and Cell Biology,* Miffin, B. J., Ed., Oxford University Press, Oxford, 1987, 167.
54. **Koukolíková-Nicola, Z., Albright, L., and Hohn, B.,** The mechanism of T-DNA transfer from *Agrobacterium tumefaciens* to the plant cell, in *Plant Gene Research: Plant DNA Infectious Agents,* Hohn, T. and Schell, J. S., Eds, 1987, 109.
55. **Binns, A. N. and Thomashow, M. F.,** Cell Biology of Agrobacterium infection and transformation of plants, *Annu. Rev. Microbiol.,* 42, 575, 1988.
56. **Zambryski, P.,** Basic processes underlying *Agrobacterium*-mediated DNA transfer to plant cells, *Annu. Rev. Genet.,* 22, 1, 1988.
57. **Zambryski, P., Tempé, J., and Schell, J.,** Transfer and function of T-DNA genes from *Agrobacterium* Ti and Ri plasmids in plants, *Cell,* 56, 193, 1989
58. **Zambryski, P.,** Chapter in this volume.
59. **Inzé, D., Follin, A., Van Lijsbettens, M., Simoens, C., Genetello, C., Van Montagu, M., and Schell, J.,** Genetic analysis of the individual T-DNA genes of *Agrobacterium tumefaciens;* further evidence that two genes are involved in indole-3-acetic acid production, *Mol. Gen. Genet.,* 194, 265, 1984.
60. **Schröder, G., Waffenschmidt, S., Weilter, E. W., and Schröder, J.,** The T-region of Ti plasmids codes for an enzyme synthesizing indole-3-acetic acid, *Eur. J. Biochem.,* 138, 387, 1984.
61. **Thomashow, L. S., Reeves, S., and Thomashow, M. F.,** Crown gall oncogenesis: evidence that a T-DNA gene from *Agrobacterium* Ti plasmid pTiA6 encodes an enzyme that catalyses synthesis on indoleacetic acid. *Proc. Natl. Acad. Sci. U.S.A.,* 81, 5071, 1984.
62. **Akiyoshi, D. E., Klee, H., Amasino, R. M., Nester, E. W., and Gordon, M. P.,** T-DNA of *Agrobacterium tumefaciens* encodes an enzyme of cytokinine biosynthesis, *Proc. Natl. Acad. Sci. U.S.A.,* 81, 5994, 1984.
63. **Barry, G. F., Rogers, D. A., Fraley, R. T., and Brandt, L.,** Identification of a cloned cytokinin biosynthetic gene, *Proc. Natl. Acad. Sci. U.S.A.,* 81, 4776, 1984.

64. **Buchanan, I., Marner, F. J., Schröder, G., Waffenschmidt, S., and Schröder, J.,** Tumor genes in plants: T-DNA encoded cytokinin biosynthesis, *EMBO J.*, 4,853, 1985.
65. **Beaty, J. S., Powell, G. K., Lica, L., Regier, D. A., Mc Donald, E. M. S., Hommes, N. G., and Morris, R. O.,** Tzs, a nopaline Ti plasmid gene from *Agrobacterium tumefaciens* associated with *trans*-zeatin biosynthesis, *Mol. Gen. Genet.*, 203, 274, 1986.
66. **Gheysen, G., Dhaese, P., Van Montagu, M., and Schell, J.,** DNA flux across genetic barriers: the crown gall phenomenon, in *Plant Gene Research: Genetic Flux in Plants,* Hohn, B. and Dennis, E. S., Eds., Springer-Verlag, NY, 1985, 11.
67. **Morris, M. O.,** Genes specifying auxin and cytokinin biosynthesis in pathogens, *Annu. Rev. Plant Physiol.*, 37, 509, 1986.
68. **Shen, W. H., Petit, A., Guern, J., and Tempé, J.,** Hairy root are more sensitive to auxin than normal roots, *Proc. Natl. Acad. Sci. U.S.A.*, 85, 3417, 1988.
69. **Tinland, B., Huss, B., Paulus, F., Bonnard, G., and Otten, L.,** *Agrobacterium tumefaciens* 6b genes are strain-specific and affect the activity of auxin as well as cytokinin genes, *Mol. Gen. Genet.*, 219, 217, 1989.
70. **Shen, W. H., Davioud, E., David, C., Barbier-Brygoo, H., Tempé, J., and Guern, J.,** High sensitivity to auxin is a common feature of hairy root, *Plant Physiol.*, 94, in press, 1990.
71. **Amasino, R. M. and Miller, C. O.,** Hormonal control of tobacco crown gall tumor morphology, *Plant Physiol.*, 69, 389, 1982.
72. **Akiyoshi, D. E., Morris, R. O., Hinz, R., Mischke, B. S., Kosuge, T., Garfinkel, D. J., Gordon, M. P., and Nester, E. W.,** Cytokinin/auxin balance in crown gall tumors is regulated by specific loci in the T-DNA., *Proc. Natl. Acad. Sci. U.S.A.*, 80, 407, 1983.
73. **Vilaine, F. and Casse-Delbart, F.,** Independant induction of transformed roots by the TL and TR regions of the Ri plasmid of agropine-type *Agrobacterium rhizogenes.*, *Mol. Gen. Genet.*, 206, 17, 1987.
74. **Vilaine, F., Charbonnier, C., and Casse-Delbart, F.,** Further insight concerning the TL region of the Ri plasmid of *Agrobacterium rhizogenes* strain A4: transfer of a 1.9 kb fragment is sufficient to induce transformed roots on tobacco leaf fragments, *Mol. Gen. Genet.*, 210, 411, 1987.
75. **Cardarelli, M., Spano, L., Mariotti, D., Mauro, M. L., Van Sluys, M. A., and Costantino, P.,** The role of auxin in hairy root induction, *Mol. Gen. Genet.*, 208, 457, 1987.
76. **Cardarelli, M., Mariotti, D., Pomponi, M., Spano, L., Capone, I., and Costantino, P.,** *Agrobacterium rhizogenes* genes capable of inducing hairy root phenotype, *Mol. Gen. Genet.*, 209, 475, 1987.
77. **Spena, A., Schmülling, T., Koncz, C., and Schell, J.,** Independent and synergistic activity of the rolA, B, and C loci in stimulating abnormal growth in plants, *EMBO J.*, 6, 3891, 1987.
78. **Capone, I., Spano, L., Cardarelli, M., Bellincampi, D., Petit, A., and Costantino, P.,** Induction and growth properties of carrot discs with different complements of *Agrobacterium rhizogenes* T-DNA, *Plant Mol. Biol.*, 13, 43, 1989.
79. **Maurel, C., Barbier-Brygoo, H., Brevet, J., Spena, A., Tempé, J., and Guern, J.,** *Agrobacterium rhizogenes* T-DNA genes and sensitivity of plant protoplasts to auxin, in *Proc. 5th Int. Symp. on the Mol. Genet. Plant-Microbe Interact.*, Interlaken, 1990, in press.
80. **Lioret, C.,** Sur la mise en évidence d'un acide aminé non identifié particulier aux tissus de crown gall, *Bull. Soc. Fr. Physiol. Veg.*, 2, 76, 1956.
81. **Morel, G.,** Métabolisme de l'arginine par les tissus de crown gall de topinambour, *Bull. Soc. Fr. Physiol. Veg.*, 2, 75, 1956.
82. **Lioret, C.,** Les acides aminés libres des tissus de crown gall. Mise en évidence d'un acide aminé particulier ces tissus., *C.R. Acad. Sci.*, 244, 2171, 1957.
83. **Biemann, K., Lioret, C., Asselineau, J., Lederer, E., and Polonski, J.,** Sur la structure chimique de la lysopine, nouvel acide aminé isolé des tissus de crown gall, *Bull. Soc. Chim. Fr.*, 17, 979, 1960.
84. **Ménagé, A. and Morel, G.,** Sur la présence d'octopine dans les tissus de crown gall, *C.R. Acad. Sci.*, 259, 4795, 1964.
85. **Ménagé, A. and Morel, G.,** Sur la présence d'un acide aminé nouveau dans le tissu de crown-gall, *C.R. Soc. Biol.*, 159, 561, 1965.
86. **Goldmann, A., Thomas, D. W., and Morel, G.,** Sur la structure de la nopaline, métabolite anormal de certaines tumeurs de crown gall, *C.R. Acad. Sci. Ser. D*, 268, 852, 1969.
87. **Petit, A.,** Recherches sur la signification de la présence d'opines dans les tumeurs provoquées par *Agrobacterium radiobacter tumefaciens,* Ph.D. Thesis, University of Paris VII, Paris, 1977.
88. **Tempé, J. and Schell, J.,** Is crown gall a natural instance of gene transfer?, In *Translation of Natural and Synthetic Polynucleotides,* Legocki, A. B., Ed., University of Agriculture, Poznan, 1977, 415.
89. **Tempé, J. and Goldmann, A.,** Occurrence and biosynthesis of opines, in *Molecular Biology of Plant Tumors,* Kahl, G., and Schell, J. S., Eds., Academic Press, San Diego, 1982, 427.
90. **Lejeune, B. and Jubier, M.-F.,** Etude de la dégradation de la lysopine par *Agrobacterium tumefaciens, C.R. Acad. Sci. Ser. D,* 264, 1803, 1968.

91. **Goldmann, A., Tempé, J., and Morel, G.,** Quelques particularités de diverses souches d'*Agrobacterium tumefaciens, C.R. Soc. Biol.*, 162, 630, 968.
92. **Petit, A., Delhaye, S., Tempé, J., and Morel, G.,** Recherches sur les guanidines des tissus de crown gall. Mise en évidence d'une relation spécifique entre les souches d'*Agrobacterium tumefaciens* et les tumeurs qu'elles induisent, *Physiol. Veg.*, 8, 205, 1970.
93. **Tempé, J. and Petit, A.,** La piste des opines, in *Molecular Genetics of the Bacteria Plant Interaction,* Pühler, A., Ed., Springer-Verlag, Berlin, 1983, 14.
94. **Petit, A. and Tempé, J.,** The function of T-DNA in nature, in *Molecular Form and Function of the Plant Genome,* Van Vloten-Doting, L., Groot, G. P. S., and Hall, T. C., Eds., Plenum Press, New York, 1985, 625.
95. **Bomhoff, G. H., Klapwijck, P. M., Kester, H. C., Schilperoort, R. A., Hernalsteens, J.-P., and Schell, J.,** Octopine and nopaline synthesis and breakdown genetically controlled by a plasmid of *Agrobacterium tumefaciens, Mol. Gen. Genet.*, 145, 177, 1976.
96. **Montoya, A., Chilton, M. D., Gordon, M. P., Sciaky, D., and Nester, E. W.,** Octopine and nopaline metabolism in *Agrobacterium tumefaciens* and crown gall tumor cells: role of plasmid genes, *J. Bacteriol.*, 129, 101, 1977.
97. **Garfinkel, D. J., Simpson, B., Ream, L. W., White, F. F., Gordon, M. P., and Nester, E. W.,** Genetic analysis of crown gall: fine structure map of the T-DNA by site directed mutagenesis, *Cell,* 27, 143, 1981.
98. **Joos, H., Inzé, D., Caplan, A., Sormann, M. Van Montagu, M., and Schell, J.,** Genetic analysis of T-DNA transcripts in nopaline crown galls, *Cell,* 32, 1057, 1983.
99. **Ooms, G., Hooykass, P. J. J., Van Veen, J. M., Van Belen, P., Regensburg-Tuink, T. J. G., and Schilperoort, R. A.,** Octopine Ti plasmid deletion mutants of *Agrobacterium tumefaciens* with emphasis on the right side of the T-region, *Plasmid,* 7, 15, 1982.
100. **Velten, J., Willmitzer, L., Leemans, J., Ellis, J., Deblaere, R., Van Montagu, M., and Schell, J.,** TR genes involved in agropine production, in *Molecular Genetics of the Bacteria-Plant Interaction,* Pühler, A., Ed., Springer-Verlag, Berlin, 1983, 303.
101. **Salomon, F., Deblaere, R., Leemans, J., Hernalsteens, J.-P., Van Montagu, M., and Schell, J.,** Genetic identification of functions of TR-DNA transcripts in octopine crown galls, *EMBO J.*, 3, 141, 1984.
102. **Ellis, J. G., Ryder, M. H., and Tate, M. E.,** *Agrobacterium tumefaciens* TR-DNA encodes a pathway for agropine biosynthesis, *Mol. Gen. Genet.*, 195, 66, 1984.
103. **De Paolis, A., Mauro, M. L., Pomponi, M., Cardarelli, M., Spano, L., and Costantino, P.,** Localization of agropine-synthesizing functions in the TR region of the root-inducing plasmid of *Agrobacterium rhizogenes* 1855, *Plasmid,* 13, 1, 1985.
104. **Brevet, J., Borowski, D., and Tempé, J.,** Identification of the region encoding opine synthesis and of a region involved in hairy root induction on the T-DNA of cucumber-type Ri plasmid, *Mol. Plant-Microbe Interact.*, 1, 75, 1988.
105. **Bouchez, D. and Tourneur, J.,** Organization and nucleotide sequence of the agropine synthesis region on the T-DNA of the Ri plasmid from *Agrobacterium rhizogenes, Plasmid,* 1990, in press.
106. **Tempé, J., Guyon, P., Tepfer, D., and Petit, A.,** The role of opines in the ecology of the Ti plasmids of *Agrobacterium,* in *Plasmids of Medical, Environmental, and Commercial Importance,* Timmis, K. N., and Pühler, A., Eds., Elsevier/North Holland Biomedical Press, Amsterdam, 1979, 353.
107. **Schell, J., Van Montagu, M., De Beuckeleer, M., De Block, M., Depicker, A., De Wilde, M., Engler, G., Genetello, C., Hernalsteens, J.-P., Holsters, M., Seurinck, J., Silva, B., Van Vliet, F., and Villaroel, R.,** Interaction and DNA transfer between *Agrobacterium tumefaciens,* the Ti plasmid and the host, *Proc. Roy. Soc. London, B,* 204, 251, 1979.
108. **Kerr, A.,** Transfer of virulence between isolates of *Agrobacterium, Nature (London),* 223, 1175, 1969.
109. **Kerr, A.,** Acquisition of virulence by non-pathogenic isolates of *Agrobacterium, Physiol. Plant Pathol.* 1, 241, 1971.
110. **Kerr, A., Manigault, P., and Tempé, J.,** Transfer of virulence *in vivo* and *in vitro* in *Agrobacterium, Nature (London),* 65, 560, 1977.
111. **Petit, A., Tempé, J., Kerr, A., Holsters, M., Van Montagu, M., and Schell, J.,** Substrate induction of conjugative activity of *Agrobacterium tumefaciens* Ti plasmids, *Nature (London),* 271, 5645, 1978.
112. **Petit, A., Dessaux, Y., and Tempé, J.,** The biological significance of opines. I. A study of opine catabolism by *Agrobacterium tumefaciens,* in *Proc. 4th Int. Conf. Plant Pathol. Bacterial,* Ridé, M., Ed., INRA-Angers, 1978, 43.
113. **Tempé, J., Estrade, C., and Petit, A.,** The biological significance of opines. II. The conjugative activity of the Ti plasmids of *Agrobacterium tumefaciens,* in *Proc. 4th Int. Conf. on Plant Pathol. Bacterial.*, Ridé, M., Ed., INRA-Angers, 1978, 153.
114. **Sciaky, D., Montoya, A.L., and Chilton, M.D.,** Fingerprints of *Agrobacterium* Ti plasmids, *Plasmid,* 1, 238, 1978.

115. **Guyon, P., Chilton, M. D., Petit, A., and Tempé, J.,** Agropine in "null-type" crown gall tumours: evidence for the generality of the opine concept. *Proc. Natl. Acad. Sci. U.S.,A.,* 77, 2693, 1980.
116. **Firmin, J. L. and Fenwick, R. G.,** Agropine–a major new plasmid-determined metabolite in crown gall tumours, *Nature (London),* 276, 842, 1977.
117. **Tate, M. E., Ellis, J. G., Kerr, A., Tempé, J., Murray, K. E., and Shaw, K. J.,** Agropine: a revised structure. *Carbohydr. Res.,* 104, 105, 1982.
118. **Chilton, W.S., Tempé, J., Matzke, M., and Chilton, M. D.,** Succinamopine: a new crown gall opine, *J. Bacteriol.,* 157, 357, 1984.
119. **Chilton, W. S., Rinehart, K. L., Jr., and Chilton, M. D.,** Structure and stereochemistry of succinamopine, *Biochemistry,* 23, 3290, 1984.
120. **Birot, A.-M., Bouchez, D., Casse-Delbart, F., Durand-Tardiff, M., Jouanin, L., Pautot, V., Robaglia, C., Tepfer, D., Tepfer, M., Tourneur, J., and Vilaine, F.,** Studies and uses of the Ri plasmids of *Agrobacterium rhizogenes. Plant Physiol. Biochem.,* 25, 323, 1987.
121. **Tempé, J. and Casse-Delbart, F.,** Plant gene vectors and genetic transformation: *Agrobacterium* Ri plasmids, in *Cell Culture and Somatic Cell Genetics of Plant,* Vol. 6, Schell, J. and Vasil, I. K., Eds., Academic Press, San Diego, 1989, 25.
122. **Tepfer, D.,** Ri T-DNA from *Agrobacterium rhizogenes:* a source of genes having applications in rhizosphere biology and plant development, ecology and evolution, in *Plant-microbe interactions: molecular and genetic perspectives,* Vol. 3, Kosuge, T., and Nester, E. W., Eds., McGraw-Hill, NY, 1989, 294.
123. **Petit, A., David, C., Dahl, G. A., Ellis, J. G., Guyon, P., Casse-Delbart, F., and Tempé, J.,** Further extension of the opine concept: plasmids in *Agrobacterium rhizogenes* cooperate for opine degradation, *Mol. Gen. Genet.,* 190, 204, 1983.
124. **Davioud, E., Quirion, J.-C., Tate, M. E., Tempé, J., and Husson, H.-P.,** Structure of cucumopine, a new crown gall and hairy root opine, *Heterocycles,* 27, 2423, 1988.
125. **Davioud, E., Petit, A., Tate, M. E., Ryder, M. H., and Tempé, J.,** Cucumopine — a new T-DNA-encoded opine in hairy root and crown gall, *Phytochemistry,* 27, 2429, 1989.
126. **Isogai, A., Fukuchi, N., Hayashi, M., Kamada, H., Harada, H., and Suzuki, A.,** Structure of a new opine, mikimopine, in hairy root induced by *Agrobacterium rhizogenes, Agric. Biol. Chem.,* 52, 3225, 1988.
127. **Brevet, J. and Clérot, D.,** personnal communication, 1990.
128. **Long, S. R.,** Genetics of Rhizobium nodulation, in *Plant Microbe Interactions: Molecular and Genetic Perspectives,* Vol. 1, Kosuge, T., and Neter, E. W., Eds., Macmillan, NY, 1984, 265.
129. **Long, S. R.,** Rhizobium-legume nodulation: life together in the underground, *Cell,* 56, 203.
130. **Greenwood, R.M. and Bathurst, N.O.,** Effect of *Rhizobium* strains and host on the aminoacids of *Lotus* root nodules, *N. Z. J. Sci.,* 11, 280, 1968.
131. **Scott, D. B., Wilson, R., Shaw, G. J., Petit, A., and Tempé, J.,** Biosynthesis and degradation of nodule-specific *Rhizobium loti* compounds in *Lotus* nodules, *J. Bacteriol.,* 169, 278, 1987.
132. **Tempé, J., Petit, A., and Bannerot, H.,** Présence de composés semblables à des opines dans les nodosités de luzerne, *C. R. Acad. Sci. Ser. C,* 95, 413, 1982.
133. **Ellis, J. G. and Murphy, P. J.,** Four new opines from crown gall tumors–their detection and properties, *Mol. Gen. Genet.,* 186, 275, 1981.
134. **Ryder, M., Tate, M. E., and Jones, G. P.,** Agrocinopine A, a tumor-inducing plasmid-coded enzyme product, is a phosphodiester of sucrose and arabinose, *J. Biol. Chem.,* 259, 9704, 84.
135. **Franzkowiak, M. and Thiem, J.,** Synthesen von agrocinopin A und B, *Justus Liebigs Ann. Chem.,* 1987, 1065.
136. **Lindberg, M. and Norberg, T.,** Synthesis of sucrose 4′-(L-arabinose-2-yl phosphate)(agrocinopine A) using an arabinose 2-H-phosphate intermediate, *J. Carbohydr. Chem.,* 7, 749, 1988.
137. **Tempé, J., Guyon, P., Petit, A., Ellis, J. G., Tate, M. E., and Kerr, A.,** Préparation et propriétés de nouveaux substrats cataboliques pour deux types de plasmides oncognes d'*Agrobacterium tumefaciens, C. R. Acad. Sci. Ser. D,* 290, 1173, 1980.
138. **Kemp, J. D.,** A new amino acid derivative present in crown gall tumor tissue, *Biochem. Biophys. Res. Commun.,* 74, 862, 1977.
139. **Szegedi, E., Czakó, M., Otten, L., and Koncz, C.,** Opines in crown gall tumors induced by biotype 3 isolates of *Agrobacterium tumefaciens, Physiol. Mol. Plant Pathol.,* 32, 237, 1988.
140. **Petit, A., Dessaux, Y., and Tempé, J.,** unpublished results, 1990.
141. **Chang, C. C. and Cheng, C. M.,** Evidence for the presence of *N*-(1,3-dicarboxypropyl)-L-amino acids in crown gall tumors induced by *Agrobacterium tumefaciens* strains 181 and EU6 *FEBS Lett.,* 162, 432, 1983.
142. **Chang, C. C., Cheng, C. M., Adams, B. R., and Trost, B. M.,** Leucinopine, a characteristic compound of some crown gall tumors, *Proc. Natl. Acad. Sci. U.S.A.,* 80, 3573, 1983.
143. **Chilton, W. S., Hood, E., and Chilton, M. D.,** Absolute stereochemistry of leucinopine, a crown gall opine, *Phytochemistry,* 24, 221, 1985.

144. **Firmin, J. L., Stewart, I. M., and Wilson, K. E.,** $N^2$-(1-carboxyethyl)methionine, a "pseudo-opine" in octopine-type crown gall tumours, *Biochem. J.*, 232, 431, 1985.
145. **Sayad, A., Farrand, S. K., Dion, P., and Nautiyal, C. S.,** Relationship between genes of mannopine catabolism from virulent and avirulent agrobacteria, presented at 5th Int. Symp. Mol. Gene. of Plant-microbe Interact., Interlaken, September, 9 to 14, 1990.
146. **Petit, A., Martin, L., Desaux, Y.,** unpublished results, 1990.
147. **Merlo, D. J. and Nester, E. W.,** Plasmids in avirulent strains of *Agrobacterium, J. Bacteriol.*, 129, 76, 1977.
148. **Clare, B. G., Kerr, A., and Jones, D. A.,** Characteristics of the nopaline catabolic plasmid in *Agrobacterium* strains K84 and K1026 used for biological control of crown gall disease, *Plasmid,* 23, 126, 1990.
149. **Wabiko, H., Kagaya, M., and Sano, H.,** Various nopaline catabolism genes located outside the Ti plasmids in *Agrobacterium tumefaciens* strains, *J. Gen. Microbiol.*, 136, 97, 1990.
150. **Ma, D., Yanofski, M. F., Gordon, M. P., and Nester, E. W.,** Characterization of *Agrobacterium tumefaciens* strains isolated from grapevine tumors in China, *Appl. Environ. Microbiol.*, 53, 1338, 1987.
151. **Thomashow, M. F., Knauf, V. C., and Nester, E. W.,** Relationship between the limited and wide host range octopine-type Ti plasmids of *Agrobacterium tumefaciens, J. Bacteriol.*, 146, 484, 1981.
152. **Knauf, V. C., Panagopoulos, C. G., and Nester, E. W.,** Comparison of Ti plasmids from three different biotypes of *Agrobacterium tumefaciens* isolated from grapevines, *J. Bacteriol.*, 153, 1535, 1983.
153. **Yanofsky, M., Lowe, B., Montoya, A. Rubin, R., Krul, W., Gordon, M., and Nester, E.,** Molecular and genetic analysis of factors controlling host range in *Agrobacterium tumefaciens, Mol. Gen. Genet.*, 201, 237, 1985.
154. **Yanovsky, M., Montoya, A., Knauf, V., Lowe, B., Gordon, M., and Nester, E.,** Limited-host-range plasmid of *Agrobacterium tumefaciens:* molecular and genetic analyses of transferred DNA, *J. Bacteriol.*, 163, 341, 1985.
155. **Leroux, B., Yanofsky, M. F., Winans, S. C., Ward, J. E., Ziegler, S. F., and Nester, E. W.,** Characterization of the virA locus of *Agrobacterium tumefaciens:* a transcriptional regulator and host range determinant, *EMBO J.*, 6, 849, 1987.
156. **Petit, A. and Martin, L. A.,** personal communication, 1989.
157. **Kamada, H., Hayashi, M., Tsuji, Y., Ohtomo, Y., and Harada, H.,** Characterization of a new isolate of *Agrobacterium rhizogenes* exerting an ability to produce a new opine, mikimopine, in plant cells, in preparation.
158. **Petit, A.,** personal communication, 1990.
159. **Chilton, W. S., Hood, E., Rinehart Jr., K. L., and Chilton, M.-D.,** L,L-Succinamopine: an epimeric crown gall opine, *Phytochemistry,* 24, 2945, 1985.
160. **Gelvin, S. B.,** chapter in this volume.
161. **Depicker, A., Stachel, S., Dhaese, P., Zambriski, P., and Goodman, H. M.,** Nopaline synthase: transcript mapping and DNA sequence, *J. Mol. Appl. Genet.* 1, 61, 1982.
162. **DeGreve, H., Dhaese, P., Seurinck, J., Lemmers, M., Van Montagu, M., and Schell, J.,** Nucleotide sequence and transcript map of *Agrobacterium tumefaciens* Ti-plasmid encoded octopine synthase gene, *J. Mol. Appl. Genet.*, 1, 499, 1982.
163. **Jouanin, L.,** Restriction map of an agropine-type Ri plasmid and its homology with Ti plasmids, *Plasmid,* 12, 91, 1984.
164. **Huffmann, G. A., White, F. F., Gordon, M. P., and Nester, E. W.,** Hairy root inducing plasmid: physical map and homology to tumor-inducing plasmid, *J. Bacteriol.*, 157, 69, 1984.
165. **Hansen, G., Larribe, M., Vaubert, D., Tempé, J., Chilton, M.-D., and Brevet, J.,** *Agrobacterium rhizogenes* pRi8196 T-DNA: mapping and sequence of functions involved in mannopine synthesis and hairy root differentiation, in preparation.
166. **Barker, R. F., Idler, K. B., Thompson, D. V., and Kemp, J. D.,** Nucleotide sequence of the T-DNA region from the *Agrobacterium tumefaciens* octopine type plasmid pTi15955 *Plant Mol. Biol.*, 2, 5, 1983.
167. **Gielen, J., De Beuckeleer, M., Seurinck, J., Deboeck, F., De Greve, H., Lemmers, M., Van Montagu, M., and Schell, J.,** The complete nucleotide sequence of the TL-DNA of the *Agrobacterium tumefaciens* plasmid pTiAch5, *EMBO J.*, 3, 835.
168. **Bevan, M., Barnes, W. M., and Chilton, M.-D.,** Structure and transcription of the nopaline synthase gene region of T-DNA, *Nucleic Acids Res.*, 11, 369, 1983.
169. **Monneuse, M.-O. and Rouzé, P.,** Sequence comparisons between *Agrobacterium tumefaciens* T-DNA-encoded octopine and nopaline dehydrogenases and other nucleotide-requiring enzymes: structural and evolutionary implications, *J. Mol. Evol.*, 25, 46, 1987.
170. **Slightom, J. L., Durand-Tardiff, M., Jouanin, L., and Tepfer, D.,** Nucleotide sequence analysis of TL DNA of *Agrobacterium rhizogenes* agropine-type plasmid. Identification of open reading frames, *J. Biol. Chem.*, 261, 108, 1986.
171. **Klee, H., Horsch, R., and Rogers, S.,** *Agrobacterium*-mediated plant transformation and its further applications to plant biology, *Annu. Rev. Plant Physiol.*, 38, 467, 1987.

172. **Draper, J., Scott, R., Armitage, P., and Walden, R. (Eds.),** *Plant Genetic Transformation and Gene Expression, A Laboratory Manual,* Blackwell Scientific, Oxford, 1988.
173. **Rogers, S. G., Klee, H. J., Horsch, R. B., and Fraley, R. T.,** Improved vectors for plant transformation: expression cassette vectors and new selectable markers, *Methods Enzymol.,* 153, 253, 1987.
174. **An, G., Ebert, P. R., Yi, B. Y., Choi, C. H.,** Both TATA box and upstream regions are required for the nopaline synthase promoter activity in transformed tobacco cells, *Mol. Gen. Genet.,* 203, 245, 1986.
175. **An, G., Costa, M., Mitra, A., Ha, S. B., and Marton, L.,** Organ-specific and developmental regulation of the nopaline synthase promoter in trangenic tobacco plants, *Plant Physiol.,* 88, 547, 1988.
176. **Ebert, P. R., Ha, S. B., and An, G.,** Identification of an essential upstream element in the nopaline synthase promoter by stable and transient assays, *Proc. Natl. Acad. Sci. U.S.A.,* 84, 5745, 1987.
177. **Ellis, J. G., Llewellyn, D. J., Walker, J. C., Dennis, E. S., and Peacocok, W. J.,** The ocs element: a 16 base pair palindrome essential for activity of the octopine synthase enhancer, *EMBO J.,* 6, 3203, 1987.
178. **Bruce, W. B., Bandyopadhyay, R., and Gurley, W. B.,** An enhancer-like element present in the promoter of a T-DNA gene from the Ti plasmid of *Agrobacterium tumefaciens, Proc. Acad. Sci. U.S.A.* , 85, 4310, 1988.
179. **Leisner, S. M., and Gelvin, S. B.,** Structure of the octopine synthase upstream activator sequence, *Proc. Natl. Aad. Sci. U.S.A.,* 85, 2553, 1989.
180. **Bandyopadhyay, R. S., Bruce, W. B., and Gurley, W. B.,** Regulatory elements within the agropine synthase promoter of T-DNA, *J. Biol. Chem.,* 264, 19399, 1989.
181. **Fromm, H., Katagiri, F., and Chua, N. H.,** An octopine synthase enhancer element directs tissue-specific expression and binds ASF-1, a factor from tobacco nuclear extracts, *Plant Cell,* 1, 977, 1989.
182. **Mitra, A. and An, G.,** Three distinct regulatory elements comprise the upstream promoter region of the nopaline synthase gene, *Mol. Gen. Genet.,* 215, 294, 1989.
183. **Bouchez, D., Tokuhisa, J. G., Llewellyn, D. J., Dennis, E. S., and Ellis, J. G.,** The ocs-element is a component of the promoters of several T-DNA and plant viral genes, *EMBO J.,* 8, 4197, 1989.
184. **Velten, J., Velten, L., Hain, R., and Schell, J.,** Isolation of a dual plant promoter fragment from the Ti plasmid of *Agrobacterium tumefaciens, EMBO J.,* 3, 2723, 1984.
185. **DiRita, V. J., and Gelvin, S. B.,** Deletion analysis of the mannopine synthase gene promoter in sunflower crown gall tumors and *Agrobacterium tumefaciens, Mol. Gen. Genet.,* 207, 233, 1987.
186. **Langridge, W. H. R., Fitzgerald, K. J., Koncz, C., Schell, J., and Szalay, A. A.,** Dual promoter of *Agrobacterium tumefaciens* mannopine synthase genes is regulated by plant growth hormones, *Proc. Natl. Acad. Sci. U.S.A.,* 86, 3219, 1989.
187. **Bouchez, D.,** L'ADN transféré du plasmide Ri d'*Agrobacterium rhizogenes:* aspects structuraux et fonctionnels de quelques fonctions portées par le TR-DNA du plasmide à agropine pRiA4, Université de Paris-sud, Centre d'Orsay, Orsay, 1990.
188. **Goldmann, A.,** Octopine and nopaline dehydrogenase in crown gall tumors, *Plant Sci. Lett.,* 10, 49, 1977.
189. **Otten, L. A. B. M., Vreugdenhil, D., and Schilperoort, R. A.,** Properties of D(+)-lysopine dehydrogenase from crown gall tumor tissue, *Biochim. Biophys. Acta,* 485, 268, 1977.
190. **Sutton, D., Kemp, J. D., and Hack, E.,** Characterization of the enzyme responsible for nopaline and ornaline dehydrogenase in sunflower crown gall tissues, *Plant Physiol.,* 62, 363, 1978.
191. **Kemp, J. D., Sutton, D. W., and Hack, E.,** Purification and characterization of the crown gall specific enzyme nopaline synthase, *Biochemistry,* 18, 3755, 1979.
192. **Kemp, J. D.,** Enzymes in octopine and nopaline metabolism, in *Molecular Biology of Plant Tumors,* Kahl, G. and Schell, J. S., Eds., Academic Press, San Diego, 1982, 461.
193. **Hong, S. B., Dessaux, Y., Guyon, P., Tempé, J., and Farrand, S. K.,** unpublished data, 1990.
194. **Klapwijk, P. M., Hooykaas, P. J. J., Kesters, H. C. M., Schilperoort, R. A., and Rorsch, A.,** Isolation and characterization of *Agrobacterium tumefaciens* mutants affected in the utilization of octopine, octopinic acid and lysopine, *J. Gen. Microbiol.,* 96,155,1976.
195. **Petit, A., and Tempé, J.,** Isolation of *Agrobacterium* Ti-plasmid regulatory mutants, *Mol. Gen. Genet.,* 167, 147, 1978.
196. **Jubier, M.-F.,** Degradation of lysopine by an inducible membrane-bound oxidase in *Agrobacterium tumefaciens, FEBS Lett.,* 28, 129, 1972.
197. **Bomhoff, G.,** Studies on crown gall–a plant tumor. Investigations on protein composition and on the use of guanidine compounds as a marker for transformed cells, Ph.D. Thesis, University of Leyden, The Netherlands.
198. **Ellis, J. G., Kerr, A., Tempé, J., and Petit, A.,** Arginine catabolism, a new function of both octopine and nopaline Ti plasmids of *Agrobacterium, Mol. Gen. Genet.,* 145, 177, 1979.
199. **Wu, L. and Unger, L.,** Utilization of octopine by *Agrobacterium tumefaciens,* Book of abstracts of the EMBO Workshop on Plant Tumor Research, Noodwijkerhout, N. L., 1978
200. **Dessaux, Y., Petit, A., Tempé, J., Demarez, M., Legrain, C., and Wiame, J.-M.,** Arginine catabolism in *Agrobacterium* strains: role of the Ti plasmid, *J. Bacteriol.,* 166, 44, 1986.

201. **Farrand, S. K. and Dessaux, Y.,** Proline biosynthesis encoded by the noc and occ loci of *Agrobacterium* Ti plasmid, *J. Bacteriol,* 167, 732, 1986.
202. **Sans, N., Schröder, G., and Schröder, J.,** The noc-region of Ti plasmid C58 codes for arginase and ornithine cyclodeaminase, *Eur. J. Biochem.,* 173, 123, 1987.
203. **Schindler, U., Sans, N., and Schröder, J.,** Ornithine cyclodeaminase from octopine Ti plasmid Ach5: identification, DNA sequence, enzyme properties, and comparision with gene and enzyme from nopaline Ti plasmid C58, *J. Bacteriol.,* 171, 847, 1989.
204. **Stalon, V.,** Evolution of arginine metabolism, in *Evolution of Prokaryotes,* Schleifer, K. H. and Stackebrandt, E., Eds., Academic Press, San Diego, 1985, 277.
205. **Schröder, J., von Lintig, J., and Zanker, H.,** Ti plasmid encoded catabolism of octopine and nopaline, presented at the 5th Int. Symp. Mol. Gene. Plant-Microbe Interact., Interlaken, September 9 to 14, 1990, P38.
206. **von Lintig, J., Zanker, H., Alt-Mörbe, J., and Schröder, J.,** Promoters in the noc-region of *Agrobacterium* nopaline Ti plasmid C58, presented at the 5th Int. Symp. Mol. Genet. Plant-Microbe Interact., Interlaken, September 9 to 14, 1990, P40.
207. **Zanker, H., von Lintig, J. and Schröder, J.,** Octopine catabolism (occ) in *Agrobacterium tumefaciens,* presented at the 5th Int. Symp. Mol. Genet. Plant-Microbe Interact., Interlaken, September 9 to 14, 1990, P43.
208. **White, D. W. R., Pritchard, M., and Marincs, F.,** Genetic and sequence analysis of *Agrobacterium* opine catabolism genes, presented at the 5th Int. Symp. Mol. Genet. Plant-Microbe Interact., Interlaken, September 9 to 14, 1990, P39.
209. **Stragier, P., Richaud, F., Borne, F., and Patte, J.-C.,** Regulation of diaminopimelate-decarboxylase synthesis in *Escherichia coli.* I. Identification of a lysR gene encoding an activator of the lysA gene, *J. Mol. Biol.,* 168, 307, 1983.
210. **Stragier, P. and Patte, J.-C.,** Regulation of diaminopimelate-decarboxylase synthesis in *Escherichia coli.* III. Nucleotide sequence and regulation of the lysR gene, *J. Mol. Biol.,* 168, 333, 1983.
211. **Chilton, W. S. and Chilton, M.-D.,** Mannityl opine analogs allow isolation of catabolic pathway regulatory mutants, *J. Bacteriol.,* 158, 650, 1984.
212. **Dessaux, Y., Guyon, P., Petit, A., Tempé, J., Demarez, M., Legrain, C., Tate, M. E., and Farrand, S. K.,** Opine utilization by *Agrobacterium* spp.: octopine-type Ti plasmids encode two pathways for mannopinic acid degradation, *J. Bacteriol.,* 170, 2939, 1988.
213. **Dessaux, Y., Guyon, P., Farrand, S. K., Petit, A., and Tempé, J.,** *Agrobacterium* Ti and Ri plasmids specify enzymic lactonization of mannopine to agropine, *J. Gen. Microbiol.,* 132, 2549, 1986.
214. **Dessaux, Y., Tempé, J., and Farrand, S. K.,** Genetic analysis of mannityl opine catabolism in octopine-type *Agrobacterium tumefaciens* strain 15955, *Mol. Gen. Genet.,* 208, 301, 1987.
215. **Farrand, S. K., Tempé, J., and Dessaux, Y.,** Localization and characterization of the region encoding catabolism of mannopinic acid from the octopine-type Ti plasmid pTi15955, *Mol. Plant-Microbe Interact.,* 3, 259, 1990.
216. **Hong, S.-B., Guyon, P., Tempé, J., Dessaux, Y., and Farrand, S. K.,** unpublished results, 1990.
217. **Murphy, P. M. and Roberts, W. P.,** A basis for agrocine 84 sensitivity in *Agrobacterium radiobacter, J. Gen. Microbiol.,* 114, 207, 1979.
218. **Ellis, J. G., Murphy, P. M., and Kerr, A.,** Isolation and properties of transfer regulatory mutants of the nopaline-type Ti plasmid pTiC58, *Mol. Gen. Genet.,* 186, 275, 1982.
219. **Cooksey, D. A.,** A spontaneous insertion in the agrocin sensitivity region of the Ti plasmid of *Agrobacterium tumefaciens* C58, *Plasmid,* 16, 222, 1986.
220. **Kerr, A. and Htay, K.,** Agrocins and the biological control of crown gall through bacteriocin production, *Physiol. Plant Pathol.,* 4, 37,1974.
221. **Ellis, J. G., Kerr, A., Van Montagu, M., and Schell, J.,** *Agrobacterium:* genetic studies on agrocin 84 production and the biological control of crown gall, *Physiol. Plant Pathol.,* 15, 311,1979.
222. **New, P. B. and Kerr, A.,** Biological control of crown gall: field measurements and glasshouse experiments, *J. Appl. Bacteriol.,* 35, 279, 1972.
223. **Farrand, S. K.,** *Agrobacterium radiobacter* strain K84: a model biocontrol system, in *New Directions in Biological Control: Alternatives for Suppressing Agricultural Pests and Diseases,* Baker, R. R. and Dunn, P. E., Eds., Alan R. Liss, NY, 1990, 679.
224. **Roberts, W. P., Tate, M. E., and Kerr, A.,** Agrocin K84 is a 6-N-phosporamidate of an adenine nucleotide analogue, *Nature (London),* 265, 379, 1977.
225. **Tate, M. E., Murphy, P. J., Roberts, W. P., and Kerr, A.,** Adenine N6-substituent of agrocine 84 determines its bacteriocin-like specificity, *Nature (London),* 280, 697, 1979.
226. **Kerr, A. and Tate, M. E.,** Agrocins and the biological control of crown gall, *Microbiol. Sci.,* 1, 1, 1984.
227. **Roberts, W. P. and Kerr, A.,** Crown gall induction: serological reactions, isoenzyme patterns, ans sensitivity to mitomycin C and to bacteriocin, of pathogenic and nonpathogenic strains of *Agrobacterium radiobacter, Physiol. Plant Pathol.,* 4, 81, 1974.

228. **Hayman, G. T. and Farrand, S. K.,** *Agrobacterium* plasmids encode structurally and functionally different loci for catabolism of agrocinopine-type opines, *Mol. Gen. Genet.,* 223, 465, 1990.
229. **Hayman, G. T. and Farrand, S. K.,** Characterization and mapping of the agrocinopine-agrocin K84 locus on the nopaline Ti plasmid C58. *J. Bacteriol.,* 170, 1759, 1988.
230. **Ellis, J. G., Kerr, A., Petit, A., and Tempé, J.,** Conjugal transfer of nopaline and agropine Ti-plasmids — the role of agrocinopines, *Mol. Gen. Genet.,* 186, 269, 1982.
231. **Ellis, J. G. and Kerr, A.,** Transfer of agrocin 84 production from strain 84 to pathogenic recipients: a comment on the previous paper, in *Soil-borne Plant Pathogens,* Schippers, B. and Gams, W., Eds., Academic Press, San Diego, 1979, 579.
232. **Slota, J. E. and Farrand, S. K.,** Genetic isolation and physical characterization of pAgK84, the plasmid responsible for agrocin 84 production, *Plasmid,* 8, 175, 1982.
233. **Ryder, M. H., Slota, J. E., Scarim, A., and Farrand, S. K.,** Genetic analysis of agrocin 84 production and immunity in *Agrobacterium* spp., *J. Bacteriol.,* 169, 4184, 1990.
234. **Hayman, G. T.,** Genetic mapping and characterization of *acc,* the agrocinopine-agrocin 84 region, on pTi C58, the nopaline-type Ti plasmid of *Agrobacterium tumefaciens* strain C58, Ph.D. Thesis, Loyola University, Chicago, 1989.
235. **Tempé, J., Petit, A., and Farrand, S. K.,** Induction of cell proliferation by *Agrobacterium tumefaciens* and *Agrobacterium rhizogenes:* a parasite's point of view, in, *Plant Gene Research: Genes Involved in Microbe-Plant Interactions,* Verma, D. P. S. and Hohn, T., Eds., Springer-Verlag, Berlin, 1984, 271.
236. **Kerr, A.,** Biological control of crown gall: seed inoculation, *J. Appl. Bacteriol.,* 35, 493, 1972.
237. **Klapwijk, P. M., Scheulderman, T., and Schilperoort, R. A.,** Coordinated regulation of octopine degradation and conjugative transfer of Ti plasmids in *Agrobacterium tumefaciens:* evidence for a common regulatory gene and separate operons, *J. Bacteriol.,* 136, 75, 1978.
238. **Klapwijk, P. M. and Schilperoort, R. A.,** Negative control of octopine degradation and transfer genes of octopine Ti plasmids in *Agrobacterium tumefaciens, J. Bacteriol.,* 139, 424, 1979.
239. **Beck von Bodman, S., McCutchan, J. E., and Farrand, S. K.,** Characterization of conjugal transfer functions of *Agrobacterium tumefaciens* Ti plasmid pTiC58, *J. Bacteriol.,* 171, 5281, 1989.
240. **Beck von Bodman, S. and Farrand, S. K.,** Primary negative and secondary positive regulation of pTiC58-specific conjugal transfer in *Agrobacterium tumefaciens,* presented at the 5th Int. Symp. Mol. Genet. Plant-Microbe Interact., Interlaken, September 9 to 14th, 1990, P21.
241. **Dessaux, Y., Petit, A., Ellis, J. G., Legrain, C., Demarez, M., Wiame, J.-M., Popoff, M., and Tempé, J.,** Ti plasmid-controlled chromosome transfer in *Agrobacterium tumefaciens, J. Bacteriol.,* 171, 63, 1989.
242. **Veluthambi, K., Krishnan, M., Gould, J. H., Smith, R. H., and Gelvin, S. B.,** Opines stimulate induction of the *vir* genes of the *Agrobacterium tumefaciens* Ti plasmid, *J. Bacteriol.,* 171, 3696, 1989.
243. **Swain, L. W.,** Crown-gall enhancement by octopine, *Bull. Torrey Bot. Club,* 99, 31, 1972.
244. **Tremblay, G., Gagliardo, R., Chilton, W. S., and Dion, P.,** Diversity among opine-utilizing bacteria: identification of coryneform isolates, *Appl. Environ. Microbiol.,* 53, 1519, 1987.
245. **Beauchamp, C. J., Chilton, W. S., Dion, P., and Antoun, H.,** Fungal catabolism of crown gall opines, *Appl. Environ. Microbiol.,* 56, 150, 1990.
246. **Nautiyal, C. S. and Dion, P.,** Characterization of the opine-utilizing microflora associated with samples of soil and plants, *Appl. Environ. Microbiol.,* 56, 2576, 1990.
247. **Dion, P.,** personnal communication, 1990.
248. **Dion, P.,** Utilization of octopine by marine bacteria isolated from mollusks, *Can. J. Microbiol.,* 32, 959, 1986.
249. **Thoai, N. V. and Robin, Y.,** Métabolisme des dérivés guanidylés. VIII–Biosynthèse de l'octopine et répartition de l'enzyme opérant chez les invertébrés, *Biochim. Biophys. Acta,* 35, 446, 1959.
250. **Buchanan-Wollaston, V., Passiatore, J. E., and Cannon, F.,** The *mob* and *ori*T mobilization functions of a bacterial plasmid promote its transfer to plants, *Nature (London),* 328, 172, 1987.
251. **Petit, A., Dessaux, Y., and Tempé, J.,** unpublished observations, 1982.
252. **Savka, M. A. and Farrand, S. K.,** personnal communication, 1988.
253. **Hille, J., Hoekema, A., Hooykaas, P. and Schilperoort, R.,** Gene organization of the Ti-plasmid, in *Plant Gene Research: Genes Involved in Microbe-Plant Interactions,* Verma, D. P. S. and Hohn, T., Eds., Springer-Verlag, Berlin, 1984, 287.
254. **Zweig, C. and Sherma, J.,** *Handbook of Chromatography,* CRC Press, Cleveland, Ohio, 1972, 107.
255. **Yamada, S. and Itano, H. A.,** Phenanthrene quinone as an analytical reagent for arginine and other monosubstituted guanidines, *Biochim. Biophys. Acta,* 133, 538, 1966.
256. **Breton, J.,** *Etudes Chimiques et Biologiques sur la Ninhydrine, Réactif Des Aminoacides,* Imprimerie E. Drouillard, Bordeaux, 1958.
257. **Dawson, R. M. C., Elliot, D. C., Elliot, W. H., and Jones, K. M.,** *Data for biochemical research,* 2nd ed, Clarendon Press, Oxford, 1968, 530.
258. **Trevelyan, W. E., Procter, D. P., and Harrison, J. P.,** Detection of sugars on paper chromatograms, *Nature (London),* 166, 444, 1950.
259. **Harrap, F. E. G.,** The detection of phosphate esters on paper chromatograms, *Analyst,* 85, 452, 1960.

# CHEMICAL SIGNALING BETWEEN *AGROBACTERIUM* AND ITS PLANT HOST

**Stanton B. Gelvin**

## TABLE OF CONTENTS

# I. INTRODUCTION

Phytopathogenic and symbiotic microbes can interact in complex ways with their hosts through chemicals released by either the plant or the microbe. In some instances, substances detected by the microbes can serve as substrates for extracellular enzymes secreted by the pathogen. These substrates can induce genes encoding the degradative enzymes. Examples of this kind of interaction include the induction of the cutinase gene by cutin monomers in the fungus *Fusarium solani* sp. *pisi*,[1] and the induction of genes encoding pectolytic enzymes in several *Erwinia* species[2] (see also chapter by Kolattukudy, this volume). In phytointeractive bacteria such as various *Rhizobium* species, compounds such as flavones and flavanones secreted by the plant can induce certain nodulation (*nod*) genes in the bacteria.[3-6] The induced bacteria subsequently synthesize and secrete complex and unusual molecules that can induce root hair curling and nodule formation on the host plant.[7] Excellent reviews of plant-microbe interactions can be found in Halverson and Stacey[8] and in Peters and Verma.[9]

*Agrobacterium tumefaciens* is a phytopathogen that causes crown gall disease on a wide variety of dicotyledonous plants. It can also induce tumors on certain monocots and gymnosperms. The infection cycle of *Agrobacterium* is complex, involving a number of chemical signals emitted by both the pathogen and its host. Infection commences when the bacteria are attracted to a number of phenolic compounds secreted by wounded plant cells. These chemicals synthesized by the wounded cells induce genes on the Ti (tumor-inducing) plasmid harbored by all virulent strains of *Agrobacterium.* The bacteria bind to the plant cells and transfer a portion of this Ti plasmid, called the T-(transferred) DNA, to the plant cell. The T-DNA covalently integrates into plant nuclear DNA, where genes encoded by the T-DNA direct the synthesis of the phytohormones auxin and cytokinin, as well as novel low-molecular weight metabolites called opines. Opines are secreted from the tumor cells and can be used as carbon and sometimes nitrogen sources by the inciting strain of *Agrobacterium.* Certain of these opines induce the conjugal transfer of the Ti plasmid between *Agrobacterium* cells (see chapter by Dessaux et al. in this volume). In addition, some opines can also interact in complex ways with the *vir* (virulence) genes located on the Ti plasmid that regulate T-DNA processing and transfer. Certain sugars, both neutral and pectic, affect gene activity in *Agrobacterium.*

Some of the chemical signaling that occurs between *Agrobacterium* and the wounded host, as well as the genetic events associated with them, are indicated in Figure 1. This chapter does not attempt to detail all the molecular and genetic events associated with T-DNA processing and transfer, or does it attempt to characterize in detail the biochemistry and genetic regulation of opine synthesis by the plant and catabolism by the bacterium. These subjects are covered in other chapters in this volume, as well as in many excellent recent reviews.[10-14] Rather, it will focus on the nature of the numerous chemical signals exchanged between the plant cell and the infecting *Agrobacterium* cell, and the biochemical and genetic responses to these signals.

# II. CHEMOTAXIS

As a first step in the bacterial-plant interaction, bacteria may respond chemotactically to chemicals released into the rhizosphere by plant cells. Many strains of *Agrobacterium* are highly motile[15] and are attracted to a number of sugars and amino acids.[15,16] Some of the chemoattractant sugars, such as sucrose, glucose, and fructose, are prevalent in wounded plant tissues. In addition, *Agrobacterium* is attracted toward root tips excised from pea, maize, bean, and cotton plants, and from cells isolated from pea root tips.[16] Hawes et al.[16] isolated a number of mutant *Agrobacterium* strains, generated by Tn*5* mutagenesis, that did not respond chemotactically to root exudates. Among these mutants were strains that were

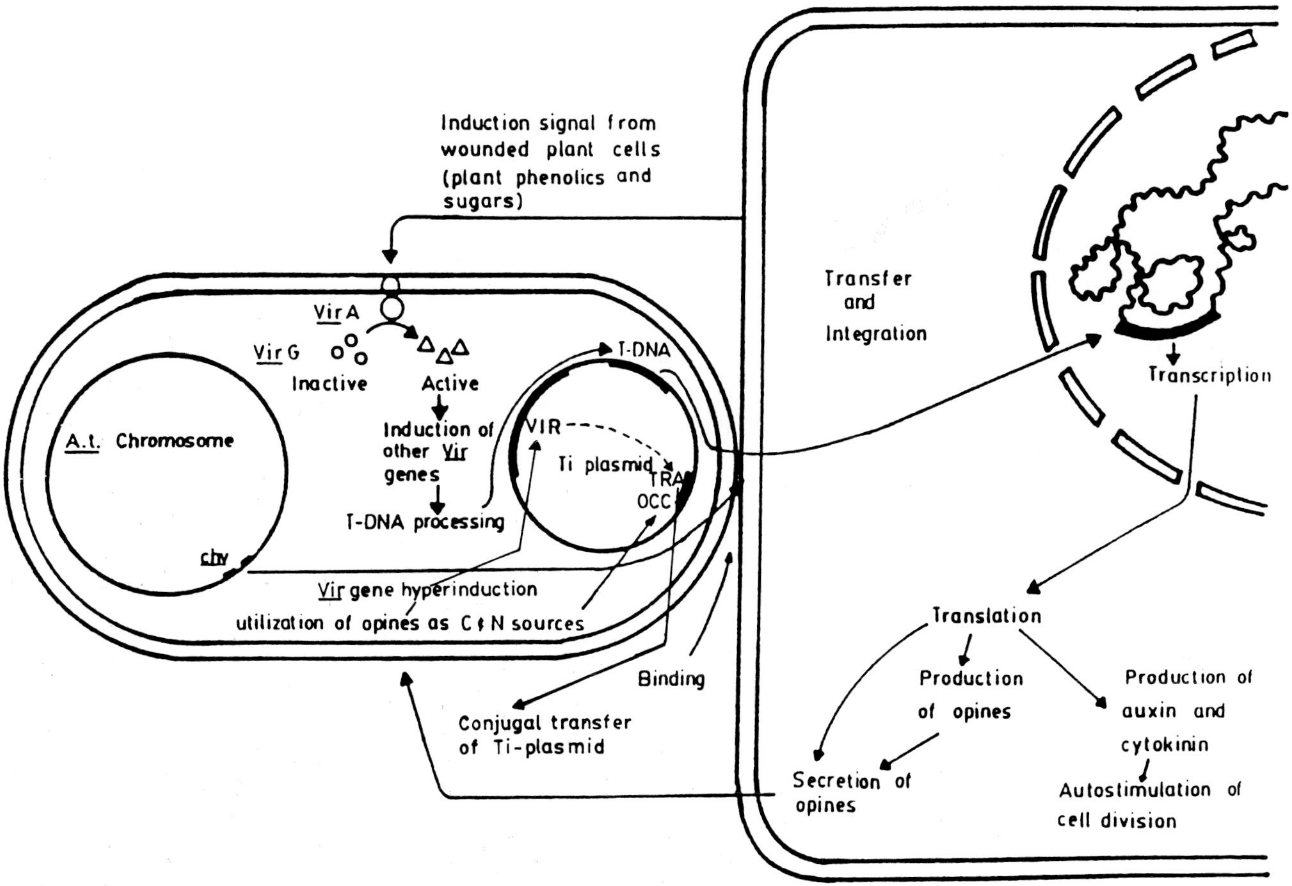


FIGURE 1. Schematic representation of the signals involved in mediating *Agrobacterium* infection processes. For the purposes of this chapter, sugars, phenolic compounds, the T-DNA, and opines are considered chemical signals that influence various phases of tumorigenesis.

nonmotile, mutants that were slow-migrating, and mutants that exhibited normal motility but were not attracted to root exudates. In addition, one mutant was attracted to excised pea root tips but not to root cap cells derived from these tissues.

The relevance of chemotaxis to tumorigenesis by *Agrobacterium tumefaciens* was demonstrated by Hawes and Smith.[17] Mutant *Agrobacterium* strains deficient in motility and chemotaxis were fully virulent when inoculated directly onto wounded pea seedlings. When wounded plants were grown in either sand or soil that had been previously inoculated with the mutant *Agrobacterium* strains, however, different results were obtained. The mutant bacteria were almost as virulent as the wild-type controls on plants grown in inoculated sand. However, the mutant bacteria were avirulent on plants grown in soil. These results suggested that motility and chemotaxis were important in situations where the soil was more compact, but less so in situations where bacterial motility was less restricted.

A controversial observation suggests that *Agrobacterium* may be positively chemotactic to phenolic compounds that induce the *vir* genes of the *Agrobacterium* Ti plasmid (see below). Data from Shaw's laboratory indicated that highly motile strains of *A. tumefaciens* C58C[1] harboring either an octopine-type or a nopaline-type Ti plasmid were attracted to very low concentrations ($10^{-7}$ *M*) of acetosyringone and related compounds.[18-20] These authors showed that such chemotaxis depends upon the genes *virA* and *virG,* and that phenolic compounds that are the most effective *vir* gene inducers (such as acetosyringone, sinapinic acid, and syringic acid) were better chemoattractants than were weaker *vir* gene inducers (such as catechol and vanillyl alcohol). These authors postulated that *Agrobacterium* cells respond positively to very low concentrations of these wound-induced phenolics by swimming toward the source of these compounds. When the bacteria detect higher concentrations ($10^{-5}$ *M*) of these compounds, the *vir* genes are induced, eventually resulting in the excision of the T-DNA from the Ti plasmid and its subsequent transfer to the wounded plant.

These observations were refuted by several laboratories, however. Parke et al.[21] claimed that acetosyringone was a chemoattractant only at high concentrations ($10^{-2}$ *M*), and that this chemotaxis was independent of the presence of a Ti plasmid in the bacterium. Hawes et al.[16] were not able to demonstrate chemotaxis of *Agrobacterium* to acetosyringone over a broad range of concentrations ($10^{-2}$ to $10^{-12}$ *M*). Ashby et al.[19] suggested that the conflicting results between Shaw's laboratory and the reports of Parke et al.[21] and Hawes et al.[16] may have resulted from differences in bacterial strains that the groups used.

## III. *VIR* GENE INDUCERS

The excision of the T-DNA from the Ti plasmid and its subsequent transfer from *Agrobacterium* to plant cells is mediated by a group of genes, called *vir* (virulence) genes, found on the Ti plasmid.[22-25] Except for the genes *virA* and, to a limited extent, *virG,* the *vir* genes are transcriptionally inactive when *Agrobacterium* grows vegetatively in either rich broth or in minimal medium. All *vir* genes, including *virA* and *virG,* are induced when *Agrobacterium* is incubated in a suitable medium in the presence of extracts from wounded plant cells. This induction is easily monitored when *vir* genes are linked as transcriptional fusions to readily assayable reporter genes, such as *lacZ*[26] or a gene encoding bacterial luciferase (*lux*).[27]

Stachel et al.[28] first described such induction of the *vir* genes by incubating *Agrobacterium* harboring the *lacZ* fusion transposon Tn*3*-HoHo1 with NT1 tobacco cells or *Nicotiana tabacum* root cultures. The inducing factor(s) derived from these plants was small (it could pass through a dialysis membrane), stable to heat and cold treatments, and partially hydrophobic (as determined by its elution behavior from a Sep-Pak C-18 column). Although the compound itself was stable to extremes of pH, optimal induction conditions required incubation of the bacteria at a pH less than 6.0. If the plant cells were first incubated with a translation inhibitor, cycloheximide, the subsequent induction of *Agrobacterium vir* genes was markedly inhibited. This result suggested that the induction of the *vir* genes was not merely a response of *Agrobacterium* to necrotic plant cells, but required active plant cellular metabolism. In addition to inducing the *vir* genes, incubation of *Agrobacterium* cells with tobacco cells, or the phenolic *vir* gene inducer acetosyringone (see below), resulted in the accumulation in the bacterium of circular T-DNA molecules (T-circles) that could be rescued and amplified by *in vitro* transfer to *Escherichia coli*.[29-31]

Stachel et al.[32] identified two major *vir* gene inducers in tobacco exudates as the phenolic compounds acetosyringone (3,5-methoxy-4-hydroxyacetophenone; AS) and hydroxyaceto-syringone (OH-AS). They noted that such exudates contained 0.5 to 1.0 $\mu M$ AS, a concentration sufficient to induce the *vir* genes of *Agrobacterium*. Optimal induction of the *vir* genes, however, required higher concentrations of the inducing compound. A preliminary examination of related chemical compounds indicated that two methoxy groups at the ring carbon atoms 3 and 5 rendered the compound more potent in *vir* gene induction than compounds containing only one methoxy substituent. The finding that sinapinic acid, a lignin precursor, was an excellent *vir* gene inducer, and that wounded leaf sections served as better inducers than intact tobacco leaves, suggested that many potent *vir* gene inducers were derived from defense-related phenolic compounds synthesized by wounded plant cells to ward off microbial attack or to strengthen plant cell walls. It thus appeared that *Agrobacterium* had adapted itself to recognize such plant defense compounds as both inducers of the genes necessary to infect plant cells and, as mentioned above, chemoattractants.

Subsequent to the initial findings of Stachel et al.,[28,32] a number of other plant-derived phenolic compounds were identified as *vir* gene inducers. Bolton et al.[33] showed that a mixture of commonly found plant phenolics, including catechol, gallic acid, pyrogallic acid, *p*-hydroxbenzoic acid, protocatechuic acid, β-resorcylic acid, and vanillin, stimulated *vir* gene activity. Zerback et al.[34] found that certain flavonol glycosides from *Petunia* pollen could induce *vir* genes. These compounds were only active at extremely high concentrations (5 m*M*), however, and may not play a role as inducers in nature. Spencer and Towers[35] and Melchers et al.[36] investigated a large number of chemical compounds related to AS, and were able to establish a number of chemical "rules" governing the phenolic compounds that served as active *vir* gene inducers. In general, the most potent inducers contained a guaiacyl or syrinyl substitution on a benzene ring, and an electron withdrawing group, such as a carbonyl or a carbon-carbon double-bond substituent, para to the hydroxyl group on the ring. A number of compounds more commonly found in plants than AS were potent *vir* gene inducers. These compounds included methyl sinapate, methyl ferulate and 5-hydroxy-methyl ferulate, methyl syringate, syringaldehyde, and coniferyl alcohol. Two compounds in particular (2′,4′,4-trihydroxy-3-methoxychalcone and 2′,4′,4-trihydroxy-3,5-dimethoxy-chalcone) were 10 to 100-fold more potent inducers than was AS.

Although many *Agrobacterium* strains are virulent on a wide variety of dicotyledonous (and some monocotyledonous) plants, most monocots, many Gymnosperms, and even some dicots are relatively recalcitrant to infection. Soon after the discovery of phenolic *vir* gene inducers in extracts of wounded dicots, the question was raised as to whether or not the lack of infectivity of *Agrobacterium* on many monocots resulted from the absence of such inducing compounds in the extracts or exudates of these plants. Although an initial report suggested that seedling exudates of many monocots did not contain *vir* gene inducers,[37] subsequent reports indicated that extracts of certain monocots did. Usami et al.[38] found that extracts from certain tissues of wheat and oats contained substances that could induce *vir* genes. That these substances were perhaps different from those previously characterized from dicots was indicated by their relatively high molecular weight and hydrophilic nature. The authors suggested that the inducers may be a phenolic molecule conjugated to a hydrophilic molecule. Messens et al.[39] identified a potent *vir* gene inducer in *Triticum monococcum* suspension cultures as ethyl ferulate.

The findings of *vir* gene inducers in extracts of monocotyledonous plants refractory to *Agrobacterium* infection suggested that this block to infection may occur at a step different from or subsequent to *vir* gene induction. That this may in part be true was suggested by an observation of Sahi et al.[40] These authors discovered that homogenates of maize seedlings contained a substance that both inhibited the growth of *Agrobacterium* and inhibited the induction of *vir* genes by AS. This substance was identified as 2,4-dihydroxy-7-methoxy-2*H*-1,4-benzoxazin-3(4*H*)-one (DIMBOA). DIMBOA is a potent antimicrobial compound produced by many cultivars of maize, and is frequently present at concentrations at least tenfold greater than that necessary to inhibit *vir* gene induction by AS. In addition to its effects on *Agrobacterium* growth and *vir* gene induction, the authors found that low concentrations (0.5 m*M*) of DIMBOA could inhibit crown gall tumorigenesis on *Kalanchoe,* a dicotyledonous plant normally susceptible to *Agrobacterium* infection.

Different strains of *A. tumefaciens* display a broad range of virulence characteristics on many gymnosperms. Morris and Morris[41] found that some strains, such as B3/73, were highly virulent on *Pseudotsuga menziesii* (Douglas fir), whereas other strains, such as MFM83.4, were only weakly tumorigenic. *vir* genes were induced by acetosyringone in both strains, but only the strongly tumorigenic strain responded to extracts from Douglas fir. The *vir* gene-inducing compound found in these extracts was identified as the phenylpropanoid glucoside coniferin. Because *Agrobacterium* strains that contained high β-glucosidase activity were the most virulent on Douglas fir and displayed the highest degree of *vir* gene induction by extracts from these plants, the authors suggested that the virulence of *Agrobacterium* on this host was, in part, determined by the ability of the bacterium to convert an inert *vir* gene-inducing precursor to an active molecule.

There have been a few reports of increasing the virulence of *Agrobacterium* by the application of plant extracts or known *vir* gene inducers at the time of infection. Bouckaert-Urban and Vendrig[42] reported enhanced virulence of *A. tumefaciens* B6 on *Helianthus giganteus* by the application of extracts from 6-day-old cotyledons, and Schafer et al.[43] were able to induce crown gall tumors on the monocot *Dioscorea bulbifera* only when the inciting *Agrobacterium* strain C58 was pretreated with wound extracts from potatoes. Similarly, Owens and Smigocki[44] increased the number of tumors formed on soybean cotyledons by the weakly virulent *A. tumefaciens* strain A348 by treatment of the bacteria with syringaldehyde. Veluthambi et al.[45] reported an increased transformation frequency of cotton when the wounded apical meristems were incubated with acetosyringone or acetosyringone plus nopaline (see below) at the time of infection. Such reports of increasing the virulence of *Agrobacterium* by incubation with *vir* gene inducers may not provide a general solution to the problem of limited infectivity of *Agrobacterium* on many recalcitrant plants.

## IV. T-DNA TRANSFER AND EXPRESSION

Although not usually regarded as such, the T-DNA region of the Ti (tumor-inducing) or Ri (root-inducing) plasmid can be considered a chemical signal that travels from *Agrobacterium* to plant cells. The T-DNA (10 to 20 kbp), is the region of these large (200 to 800 kbp) plasmids destined to be transferred to plant cells and represents approximately 5 to 10% of the plasmid DNA. Some Ti plasmids contain only one T-DNA region, whereas other Ti and Ri plasmids contain two such regions that can be transferred to plants independently.[46-59] DNA sequence analyses of the T-region of several Ti and Ri plasmids and the junctions of T-DNA and plant DNA in crown gall and hairy root tumors suggested that specific DNA sequences delimited the T-region. These sequences, termed T-DNA border repeat sequences (or T-DNA borders), have been found in all Ti and Ri plasmids investigated to date. These T-DNA borders are made up of 25-bp DNA sequences that are highly conserved, and the borders are in a directly repeated orientation.[60-75]

The importance of these border repeat sequences in tumorigenesis was shown by mutational analysis. The alteration of some nucleotides within the border repeat sequence resulted in an "inactive" (i.e., unable to be processed or to transfer the T-DNA) border.[74,75] Deletion of both borders from the T-DNA region of the Ti plasmid resulted in an avirulent bacterial strain.[67,72] In addition, the inversion of one of the T-DNA borders, resulting in a T-DNA delimited by inverted repeats, also rendered the *Agrobacterium* strain harboring such a plasmid weakly virulent.[62,64,72] Experiments that removed only one T-DNA border suggested that right borders play an essential role in T-DNA transfer. The deletion of the right border rendered the resulting *Agrobacterium* strain virtually avirulent, especially on tobacco plants,[61-64,66,69,72] whereas deletion of just the left border had very little effect upon virulence.[67,68,76] In addition, Ti plasmid derivatives or "binary" plasmids harboring a replicon and a marker selectable in plant tissues only required a right T-DNA border for transfer of the marker to plant cells.[66,67,69,70,72] These experiments suggested a polarity of these borders, indicating that certain molecular events of T-DNA processing initiated at the right border and terminated at the left border.[52,62,64,67,68,70] Ream and co-workers[64,77] and others[72] identified a DNA sequence, called "overdrive," that was present near right borders investigated but not near left borders. This sequence was important for efficient T-DNA transfer because deletion of the overdrive sequences resulted in an *Agrobacterium* strain that was only weakly virulent. Restoration of the natural or a synthetic overdrive sequence resulted in enhanced virulence. Overdrive, as a T-DNA transfer enhancer, had properties similar to known eukaryotic transcriptional enhancers. It could function in both orientations and in a position-independent manner.[77,78]

The processing of the T-DNA from the Ti plasmid and its transfer to the plant cell is a complex process which depends upon genes encoded both by the Ti plasmid and the *Agrobacterium* chromosome. Garfinkel and Nester[22] originally defined, by Tn*5* transposon mutagenesis, a region of the Ti plasmid that was essential for virulence. This region, termed the *vir* (virulence) region, is composed of 6 to 10 genetic complementation groups, depending upon the Ti plasmid investigated.[23-25,79] More highly refined mutagenesis studies, using the *lacZ* transcriptional fusion transposon Tn*3*-HoHo1,[26] mapped both the extent and the direction of transcription of each of these complementation groups in the octopine-type Ti plasmid pTiA6.[80,81] Similar experiments, using bacterial luciferase (*lux*) as a reporter of gene activity, identified *vir* operons in the nopaline-type Ti plasmid pTiC58.[27] The gene *virA* was active to a high level, and *virG* to a low level, in *Agrobacterium* cells grown in rich meda.[27,80,81] The genes *virB, virC, virD, virE, virH* (formerly called *pinF*[82]), and *virF*[83] found on octopine-type Ti plasmids, and similar genes harbored by nopaline-type Ti plasmids (although these plasmids contain a *tzs* gene[84] and lack a *virF* and a *virH* gene), are relatively inactive in bacteria grown in broth.[27,80,81] Upon incubation of *Agrobacterium* cells in a medium of low pH (5.0 to 5.8) containing phenolic compounds such as acetosyringone, these genes are induced to high levels.[27,80,81] In addition, *virG* and *virA* can also be induced severalfold.[85] Because of their common mode of regulation by acetosyringone and related molecules, these genes have been termed the *vir* regulon. A map of the *vir* region of pTiA6 is shown in Figure 2.

Genetic analysis of the various *vir* genes suggested that *virA* and *virG* comprised a two-component system necessary for the expression of the *vir* regulon. The VirA protein was postulated to be a sensor of the phenolic-inducing molecules, and the VirG protein a transcriptional activator of the *vir* genes.[81,86-95] Similar two-component environment sensing-transcription activating systems had previously been described in other bacteria. Such systems regulate the response of bacteria to environmental cues, resulting in the activation of genes involved in, among other cellular processes, phosphate and nitrogen utilization, osmotic regulation, chemotaxis, and sporulation.[96-99]

Subsequent molecular analyses indicated that the VirA protein was localized to the inner

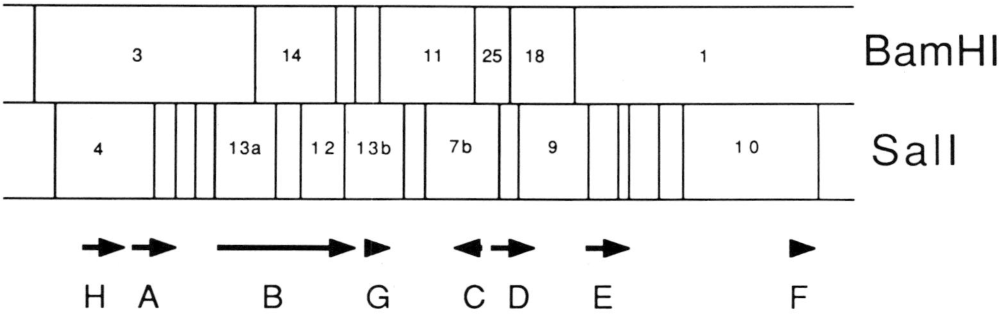


FIGURE 2. Map of the *vir* region of a typical wide host range octopine-type Ti plasmid. The direction of transcription of each of the *vir* operons is indicated by the arrows. This map is based upon data from Stachel and Nester[80], Stachel and Zambryski[81], Kanemoto et al.[82], and Melchers et al.[83]

(periplasmic) membrane of *Agrobacterium* cells.[87] DNA sequence analyses suggested that VirA contained an amino terminal region localized within the cell cytoplasm, two membrane spanning regions delimiting a periplasmic domain, and a carboxy terminal region within the cytoplasm.[89] Protein engineering experiments indicated that the periplasmic domain was responsive to pH and temperature, and that the second membrane-spanning domain was responsive to acetosyringone. Following induction by acetosyringone, a specific histidine residue within the carboxy-terminal region of the protein was autophosphorylated.[92,100,101] This phosphate group could subsequently be transferred to an aspartic acid residue in the amino terminal region of the VirG protein.[95,102] It has been suggested that such phosphorylation of VirG may activate the protein to serve as a transcriptional activator. VirG protein can bind to *vir* boxes (DNA sequence motifs upstream of all the *vir* genes[103,104] and necessary for induced expression by acetosyringone) *in vitro*.[93,94] However, such *in vitro* binding is independent of phosphorylation of the VirG protein.

Although the *vir* regulon is controlled by the genes *virA* and *virG,* the processing of the T-DNA is carried out by enzymes encoded by the *virD* operon.[105-112] Both of the first two open reading frames (orf), and only these, are essential for T-DNA processing.[107] The endonuclease activity encoded by *virD* cleaves the T-DNA between nucleotides 3 and 4 of the border repeat sequences.[75,106] Such cleavage can result either in the single-stranded nicking of the ''bottom'' T-DNA strand,[75,78,106,113] or in the double-stranded cleavage of the border repeat sequence.[107,109,114] VirD ORF1 has topoisomerase activity,[115] and, following cleavage, the VirD ORF2 polypeptide covalently associates with the 5′-ends of the processed T-DNA molecules.[116-120] The VirD2 protein contains within both its amino- and carboxy-terminal regions amino acid sequences homologous to motifs that, in animal and fungal cells, target proteins to the nucleus.[105,107] Indeed, Herrella-Estrella et al.[120a] have recently shown that a translational fusion of the amino terminal VirD2 region, containing this putative nuclear-targeting domain, fused to *E. coli* β-galactosidase directs β-galactosidase activity to the nucleus of transgenic tobacco plants.

Following induction of the *Agrobacterium vir* genes by acetosyringone or by incubation of the bacteria with plant cells, several different forms of ''processed'' T-DNA accumulate within the bacteria. Early reports suggested that circular T-DNA molecules (T-circles), that could be recovered by plasmid rescue in *E. coli,* played a prominent role in T-DNA transfer.[29] It is now generally accepted, however, that such molecules are rare and may result from recombination events between the similar T-DNA borders.[121] However, two other forms of the T-DNA can accumulate to significant extents in induced *Agrobacterium* cells. These are full-length double-stranded and full-length single-stranded linear T-DNA molecules.

The predominant form of processed T-DNA found by most investigators is the full-length linear single-stranded T-DNA molecule, termed the T-strand.[45,78,106-123] These molecules most likely result from the single-stranded nicking by the VirD endonuclease of the "bottom" T-DNA strand between nucleotides 3 and 4 of the border repeat sequences, followed by the replicative displacement of the bottom T-DNA strand. In Ti plasmids that contain multiple T-DNA regions, all possible combinations of borders could serve to delineate the T-strand molecules.[108,123] Upon phenol extraction of DNA from induced *Agrobacterium* cells, the T-strands tended to accumulate at the phenol-aqueous interphase, suggesting that they were complexed with protein. VirD2 protein is tightly (probably covalently) associated with the 5′-ends of T-strands,[116-118,120] and it was demonstrated that this VirD2 protein "cap" can protect the 5′-end of T-strands against exonucleolytic degradation.[119] As mentioned above, the VirD2 protein may also direct the T-strand to the nucleus of the plant cell. In addition, T-strands are noncovalently coated by a single-stranded DNA-binding protein encoded by the *virE2* gene.[124] This protein cooperatively binds to any single-stranded DNA molecule,[125-129] and may protect the T-strand from endonucleolytic cleavage during its passage from *Agrobacterium* to the plant cell nucleus. This complex of T-strands capped at their 5′- ends by VirD2 protein and coated by the VirE2 protein has been termed the T-complex,[130] and it has been postulated that this is the form of the T-DNA that is transferred to the plant cell. These authors have suggested that the transfer of the single-stranded T-DNA molecule from *Agrobacterium* to plant cells is an adaptation of the mechanism by which *E. coli* conjugally transfers F plasmids. However, such a "conjugation" now uses an eukaryotic cell as the recipient.[10] Although this model has been supported by a large amount of circumstantial evidence, there is no direct evidence that T-strands actually are the transferred T-DNA intermediate.

The presence of other forms of processed T-DNA in acetosyringone-induced *Agrobacterium* cells has been noted by a number of laboratories. That double-stranded T-DNA molecules may accumulate was first suggested by experiments conducted by Veluthambi et al.,[109,123] in which they found the relatively efficient double-stranded cleavage of the T-DNA borders following incubation of *Agrobacterium* with tobacco protoplasts or acetosyringone. Such double-stranded cleavage was also directed by the VirD endonuclease[107,109] and, in some Ti plasmids, was reported to be the major processing activity.[114] Durrenberger et al.[119] demonstrated that full-length double-stranded T-DNA molecules could accumulate in acetosyringone-induced *Agrobacterium* cells, and that these molecules also were capped at the 5′- end by VirD2 protein. The relative abundance of single- and double-stranded T-DNA molecules in such induced bacteria depended upon a number of experimental parameters, such as the method of DNA extraction from the cells and the processes of electrophoresis and DNA blotting used.

Experiments in our laboratory have confirmed the presence of full-length double-stranded T-DNA molecules from octopine-type Ti plasmids in acetosyringone-induced *Agrobacterium* cells.[123a] Figure 3 shows an analysis of DNA extracted from such cells. Both single- and double-stranded T-DNA molecules could be detected in approximately equal molar ratios (Figure 3C). These molecules were derived from the T-DNA flanked by borders A and C as well as by borders A and B. In an attempt to determine which form of the T-DNA (single- or double-stranded) was the true transferred intermediate, these authors followed the kinetics of disappearance of the various forms of processed T-DNA from preinduced *Agrobacterium* cells following transfer of the bacteria to medium lacking acetosyringone and either containing or lacking tobacco protoplasts. Figure 3B shows that, in the absence of protoplasts, all forms of processed T-DNA began to disappear at approximately the same rate after 4 h. The bacteria incubated with protoplasts were divided into two groups, those that bound tightly to plant cells and those that remained unbound. Figure 4 shows that the disappearance of all forms of processed T-DNA from *Agrobacterium* that were not tightly associated with plant cells followed closely the same kinetics as the disappearance of the T-DNA from

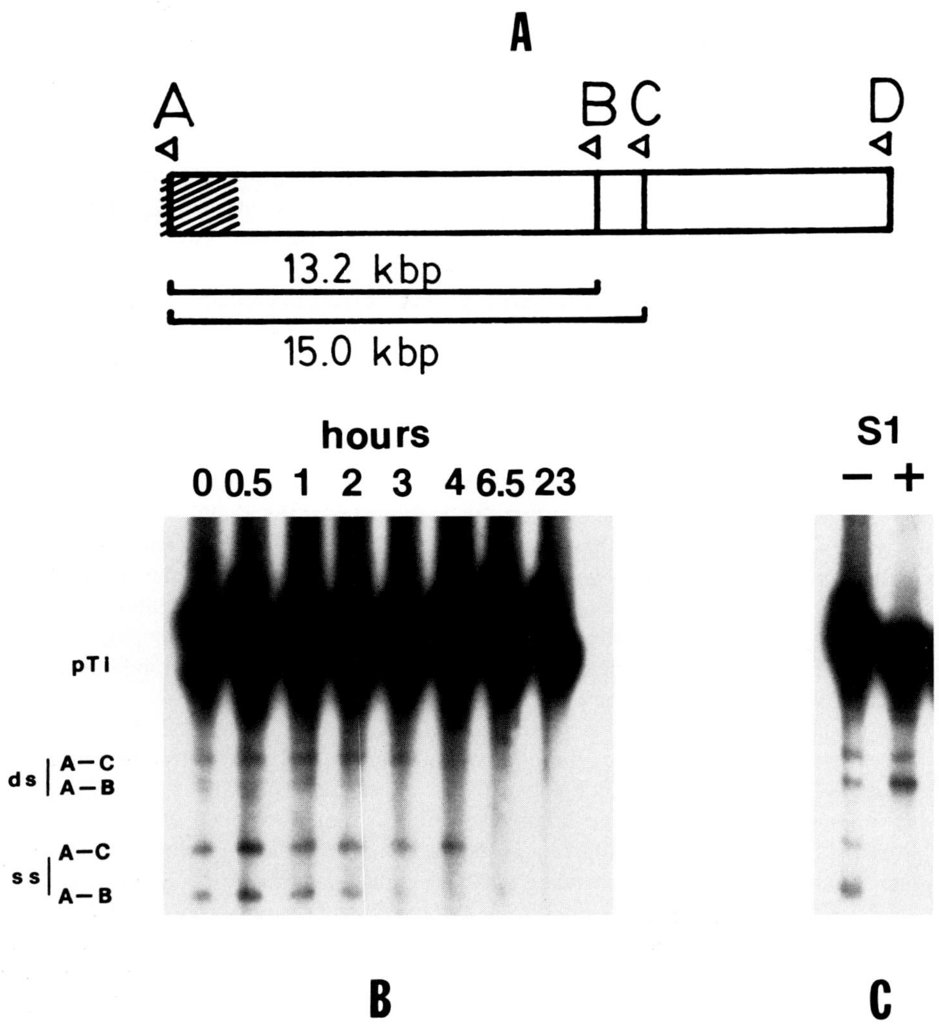


FIGURE 3. Characterization and decay kinetics of processed T-DNA molecules in *A. tumefaciens* harboring a mutation in the gene *pinF*. (A) Schematic map of the T-DNA region of a typical wide-host range octopine-type Ti plasmid. The triangles represent the four T-DNA border repeat sequences (A-D), and the shaded area represents the hybridization probe (*Hin*dIII fragment 18C) used for these analyses. The brackets beneath the map indicate the sizes of the expected processed T-DNA molecules. (B) DNA blot of total cellular DNA extracted from *Agrobacterium* cells. The cells were preinduced overnight with acetosyringone and octopine, then washed and resuspended in K3 tissue culture medium in the absence of the inducers. Hybridization signals from the unprocessed Ti plasmid (pTi) and the processed T-DNA (single-(ss) and double-stranded (ds) T-DNA molecules from borders A to B and from borders A to C) indicate that the decay of all forms of the processed T-DNA is approximately equal, and is not significant until approximately 4 h following removal of the inducer. (C) Following incubation with S1 nuclease (+), the two fastest migrating processed T-DNA molecules disappear, indicating that they are single-stranded.

bacteria incubated in the absence of plant cells. The disappearance of T-DNA from the bound bacteria was quite different, however. Single-stranded T-DNA molecules began to disappear rapidly (within the first half hour of incubation with plant cells), whereas the double-stranded T-DNA forms remained relatively stable during the first 2 h. These data suggest that, in bacteria in which the T-DNA transfer apparatus has been preinduced, the transfer of T-DNA to plant cells can occur rapidly. In addition, these data suggest that the

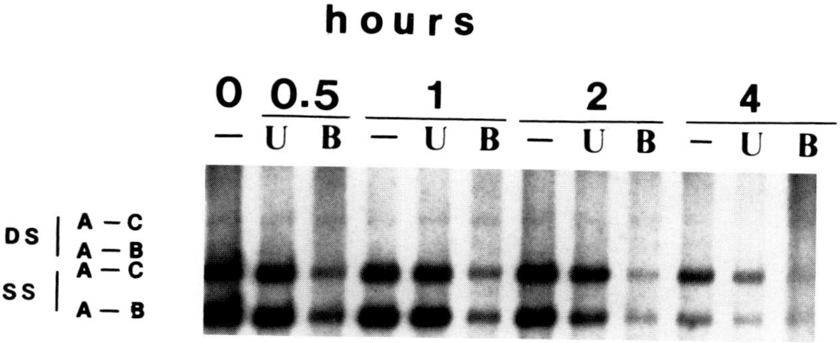


FIGURE 4. Decay kinetics of processed T-DNA molecules in *A. tumefaciens* harboring a mutation in the gene *pinF* following cocultivation with tobacco protoplasts. Following preinduction of the bacteria, DNA was extracted from bacteria incubated for various times in the absence of protoplasts (-), or from bacteria either bound to plant cells (B) or from unbound bacteria (U). In this particular experiment, only low amounts of double-stranded processed T-DNA intermediates were detected using as a hybridization probe *Hin*dIII fragment 18C. Nevertheless, the double-stranded (ds) molecules were more stable than were the single-stranded molecules. Loss of single-stranded (ss) molecules from the bound bacteria was rapid, and could be detected within 0.5 h after the bacteria were incubated with the protoplasts.

single-stranded form of the T-DNA is lost most quickly, and may, therefore, be the true transferred intermediate. Although other explanations for these data are possible (for example, the induction by plant cells of an S1-like nuclease activity in *Agrobacterium*), the relative stability of T-strands in bacteria incubated with, but not attached to plant cells suggests that this alternative explanation of the data is not likely.

Although the *vir* genes may play a major role in T-DNA processing and transfer from the bacteria, other genes on the *Agrobacterium* chromosome may also play important roles in these processes. A number of *Agrobacterium* chromosomal genes encode functions important in determining the surface characteristics of the bacterium. The gene *chvB* encodes a very large protein responsible for the biosynthesis of a β-1,2-cyclic glucan that is secreted from the bacterial cell.[131,132] Both the secretion and modification of this glucan are mediated by the product of the *chvA* locus.[133-135] Mutagenesis of either of these genes resulted in *Agrobacterium* strains that were avirulent or highly attenuated in virulence, depending upon the host plant species that was inoculated.[136-141] These bacteria did not bind well to plant cells,[136,137] and the effect of mutations in these genes upon virulence may have resulted from this altered binding behavior. Mutations in the gene *pscA* also resulted in bacterial strains that were severely impaired in virulence.[142,143] This gene encodes functions involved in the synthesis and secretion by the bacterial cell of succinoglycan molecules.[144] In addition, genes involved in cellulose biosynthesis by the bacteria (*cel* genes) may play a more subtle role in binding of the bacteria to plant cells. Although *Agrobacterium* strains mutant in these genes showed normal virulence characteristics when the bacteria were inoculated directly into wound sites, these mutant strains demonstrated decreased virulence when the wound was washed subsequent to inoculation.[145] Matthysse and co-workers[146] suggested that the cellulose fibrils encoded by the products of the *cel* genes may entrap bacteria and plant cells in a loose association. The latter may be important in natural environments.

Several other genes encoded by the *Agrobacterium* chromosome are also involved in virulence. The gene *chvE* encodes a glucose/galactose-binding protein.[147] Mutation of this gene resulted in a severely attenuated induction of the *vir* genes by acetosyringone, paralleled by a decrease in virulence upon certain plant species.[148,149] This gene will be discussed in greater detail below. In addition to *chvE,* we have identified two other *Agrobacterium* mutants involved in *virB* induction.[149a] Although the effects of mutations in these strains were not

**TABLE 1**
**Induction of *virB*::*lacZ* in At44, At44#9, and At44#37[a]**

| Strain | Miller units | Percentage of control |
|---|---|---|
| At44 | 445 +/- 81 | 100 |
| At44#9 | 9 +/- 4 | 2 |
| At44#37 | 49 +/- 9 | 11 |

[a] *A. tumefaciens* strains were induced for 12 h in AB medium containing 0.5% sucrose and 100 m*M* acetosyringone. These experiments represent replicates of three independent inductions.

as severe as were mutations in *chvE* (Table 1), they nevertheless resulted in decreases in virulence (Figure 5).

In addition to the chromosomal genes described above, a gene termed *ros* (for rough surface) has been identified in *A. tumefaciens* strain C58. The mutation in *ros* had pleiotropic effects, including altered bacterial growth characteristics, colony morphology, and *vir* gene induction.[150] Unlike the situation in wild-type *Agrobacterium* chromosomal backgrounds, in which the genes *virC* and *virD* were not expressed in the absence of acetosyringone, *Agrobacterium* strains harboring a mutant *ros* gene expressed the *virC* and *virD* genes constitutively. This effect was only upon these two *vir* genes, suggesting alternate mechanisms of regulation of the various *vir* genes.[151] These genes could be further induced by acetosyringone.

Following the transfer of the T-DNA from *Agrobacterium* to plant cells, the T-DNA covalently integrates into plant nuclear DNA.[48,50,152-154] Genes encoded by the T-DNA, although expressed weakly in the bacteria,[155,156] are transcribed to a much greater extent in the plant cell. The expression of the T-DNA in the plant is controlled by transcriptional regulatory elements similar to those associated with other eukaryotic genes. The transcription of the T-DNA is mediated by RNA polymerase II,[157] and typical TATA and CAAT boxes precede most of these genes.[158] In addition, many T-DNA genes contain upstream transcriptional activating elements similar to UAS sequences preceding many yeast promoters.[159-168]

T-DNA genes code for enzymes that mediate two fundamental crown gall processes: tumorigenesis and opine biosynthesis and secretion. Two genes encode a novel pathway for auxin biosynthesis. These genes were first defined by transposon mutagenesis,[169,170] and encode the enzymes tryptophan monooxygenase (*tms1*, or *iaaM*) and indoleacetamide hydrolase (*tms2*, or *iaaH*).[171-176] The gene *tmr* or (*ipt*) encodes an isopentenyl transferase responsible for cytokinin biosynthesis.[177-179] Additional T-DNA genes (gene 5 and 6b) may encode functions that mediate the plants cell's sensitivity to cytokinin or auxin, respectively.[180-183] The Ri plasmid of *A.rhizogenes* contains the T-DNA genes *rolA, rolB,* and *rolC* that encode functions that sensitize plant cells to endogenous concentrations of auxin.[184-190]

## V. OPINES

Crown gall tumor cells proliferate rapidly, both on the plant and in tissue culture. Although these tumors may severely debilitate the plant, the tumorous state per se probably does not benefit *Agrobacterium*. What is of benefit to the inciting strain of *Agrobacterium*

VIRULENCE OF MUTANTS AS TESTED BY TOBACCO LEAF DISK INOCULATIONS

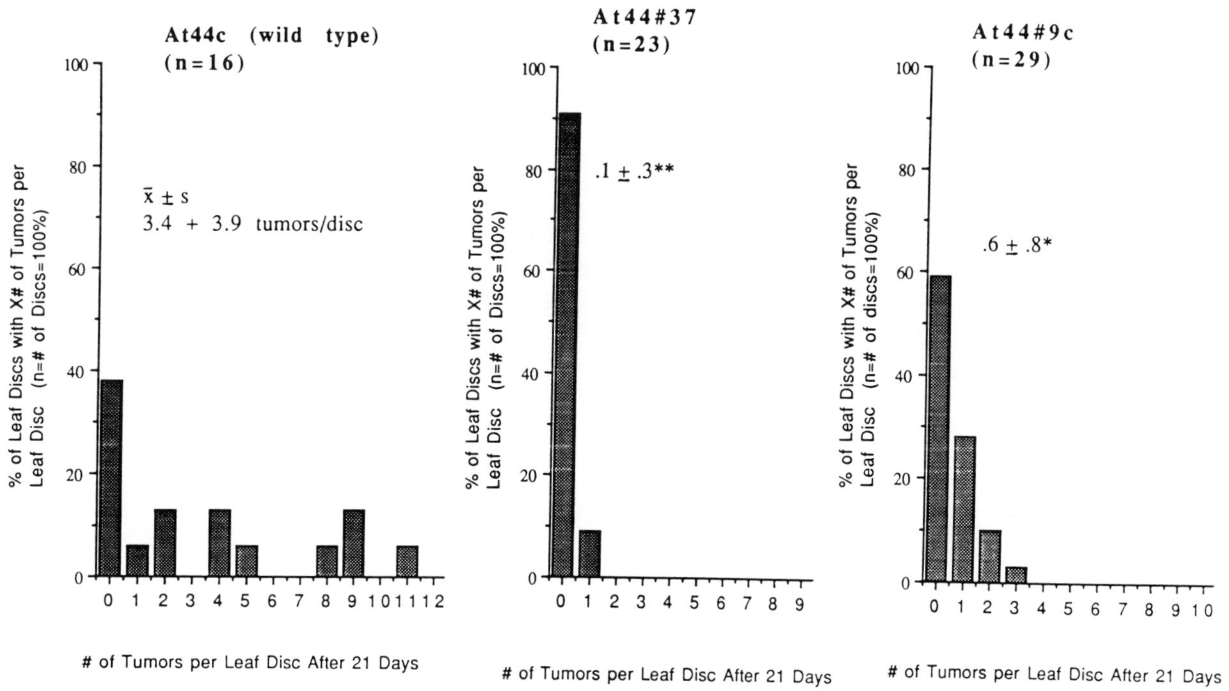


*: MEAN SIGNIFICANTLY DIFFERENT FROM WILD TYPE AT 95% LEVEL OF PROBABILITY
**: MEAN SIGNIFICANTLY DIFFERENT FROM WILD TYPE AT 99% LEVEL OF PROBABILITY

FIGURE 5. Virulence of *A. tumefaciens* At44c (harboring a wild-type C58 chromosome) and *A. tumefaciens* At44-37c and At44-9c (harboring Tn*5* insertions in the chromosome) upon tobacco leaf disks. The bacteria were grown to a Klett of 100; tumors were scored 21 days after inoculation.

most likely is the large quantity of opines synthesized by the tumors. Opines are low-molecular weight compounds that are, in general, formed from amino acids, α-ketoacids, and sugars. Very high concentrations (up to 5 to 10 m*M*) of opines can accumulate in crown gall tumors.[191-193] The production of these opines is mediated by opine biosynthetic genes encoded by the T-DNA.[194-201] In addition, the T-DNA from certain Ti plasmids harbors a gene (*ons*) that specifies octopine or nopaline secretion.[202] Although there have been a few reports of the production of opines by nontransformed plants under certain conditions,[203] it is generally considered that opine biosynthesis by plants is a crown gall (or hairy root) tumor-specific process. Plant cells cannot utilize opines. *Agrobacterium* cells, although harboring a Ti plasmid that contains opine biosynthetic genes, do not produce opines. However, the Ti plasmid harbored by the strain of *Agrobacterium* that incites a particular type of crown gall tumor generally carries opine catabolic genes on a non-T-DNA region of the plasmid.[22,204] These genes enable particular *Agrobacterium* strains to utilize opines as the sole source of carbon and, in some instances, nitrogen.[194,195,205-208] Furthermore, only those opines that are directed to be produced in the plant by a given Ti plasmid generally are catabolized by *Agrobacterium* strains harboring that Ti plasmid. Although some species of *Pseudomonas* and some Gram-positive soil bacteria can also catabolize opines,[193] the directed infection of plants for the presumed purpose of the synthesis of opines that will be utilized by the inciting *Agrobacterium* strain has been termed an example of "genetic colonization." The "opine concept" implies that *Agrobacterium* "genetically engineers" plant cells to produce compounds, the opines, that (to a first approximation) only it can utilize. *Agrobacterium* thereby is able to establish an ecological niche for itself, with the plant now supplying the nutrients directed to the establishment of this colonization.

A detailed description of the biosynthesis and catabolism of opines can be found in Chapter 7 by Dessaux et al. in this volume. For the purposes of this chapter, the opines can be considered a signal from the tumor to the *Agrobacterium* cell. Opines can serve at least three functions for the bacterium. First, they can be used as a carbon and, in some cases, nitrogen source for the inciting strain of *Agrobacterium*. Opine catabolism and transport genes are transcriptionally silent in the absence of the appropriate opine inducer, but are activated in the presence of the opine.[155] Genetic data formerly suggested that the genes responsible for octopine or nopaline catabolism and transport were controlled both positively by the opine and negatively by a repressor.[205,209-213] Some of these authors also suggested that the regulation of conjugal Ti plasmid transfer by certain opines (see below) was under the control of this same repressor system.[205] More recent data suggest, however, that octopine and nopaline catabolism and transport genes are regulated only positively by a protein homologous to the *lysR* family of DNA-binding proteins.[213a] In an *Agrobacterium* strain that harbors no complete copy of any gene from the octopine catabolism region of the Ti plasmid, the promoter of one of the opine catabolic genes, that directs the expression of a *lacZ* fusion, was silent. This promoter was induced by octopine.[213b] Because octopine transport is inducible by octopine, the question arises as to how molecules of octopine initially enter the bacterial cell in order to induce the transport and catabolism genes. Although a possible explanation may be that the opine transport gene promoter is slightly "leaky," resulting in a low level of production of the octopine transport machinery, a very likely explanation for this paradox has been proposed by this author's laboratory. Using highly radioactive $^{14}C$-labeled octopine or $^{14}C$-labeled nopaline, Krishnan et al.[213a] found that both of these compounds could be transported into *A. tumefaciens* strains with the C58 chromosomal background even in the absence of a Ti plasmid (Figure 6). This transport required active cellular metabolism in that it was inhibited by incubation of the bacteria with sodium azide. The transport of radioactive octopine or nopaline could be competed by either the heterologous opine or by arginine (Figure 7). These data suggest that the Ti plasmid-independent transport of octopine and nopaline into *Agrobacterium* is mediated by a chromosomally encoded

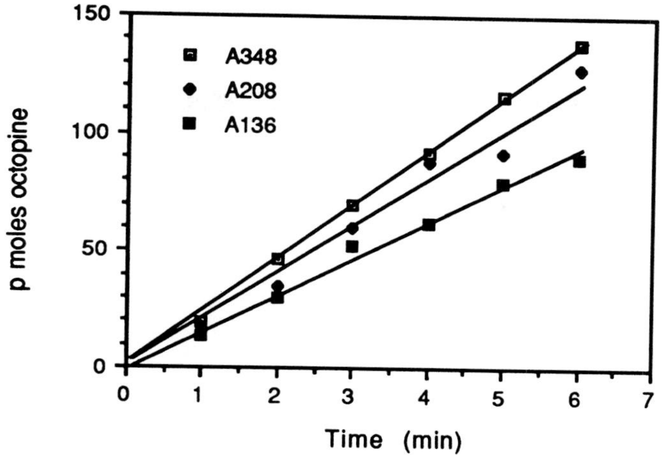


A

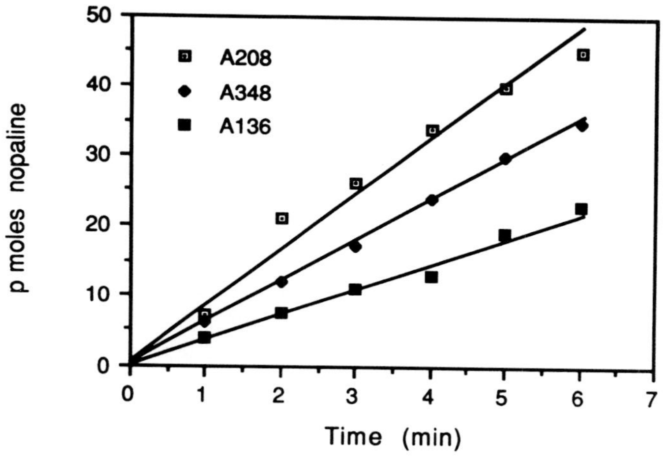


B

FIGURE 6. Transport of octopine (A) or nopaline (B) by *A. tumefaciens* A136 (lacking a Ti plasmid), A348 (harboring the octopine-type Ti plasmid pTiA6), or A208 (harboring the nopaline-type Ti plasmid pTiT37). The bacteria were grown to a Klett of 100, resuspended at a concentration of 2 x $10^8$ cells/ml in K3 medium containing 5 m*M* MES-NaOH[45] and incubated with 50,000 cpm/ml octopine (270 mCi/mmol) plus 1 m*M* octopine (A), or 50,000 cpm/ml (270 mCi/mmol) nopaline plus 1 m*M* nopaline (B). Samples were collected by filtration, washed, and counted.

arginine transport system. Although this system is much less efficient in transporting octopine or nopaline than are the Ti plasmid-encoded opine transport systems, it is sufficient to permit the initial transport of the opines into the cell to induce the opine catabolic and transport genes.

The Ti plasmid is a conjugal plasmid. Transfer of the entire Ti plasmid between *Agrobacterium* cells is normally repressed. However, in the presence of certain opines, termed conjugal opines, conjugation of the Ti plasmid is induced. Not all opines are conjugal opines. For example, although octopine-type Ti plasmids encode functions for the catabolism of

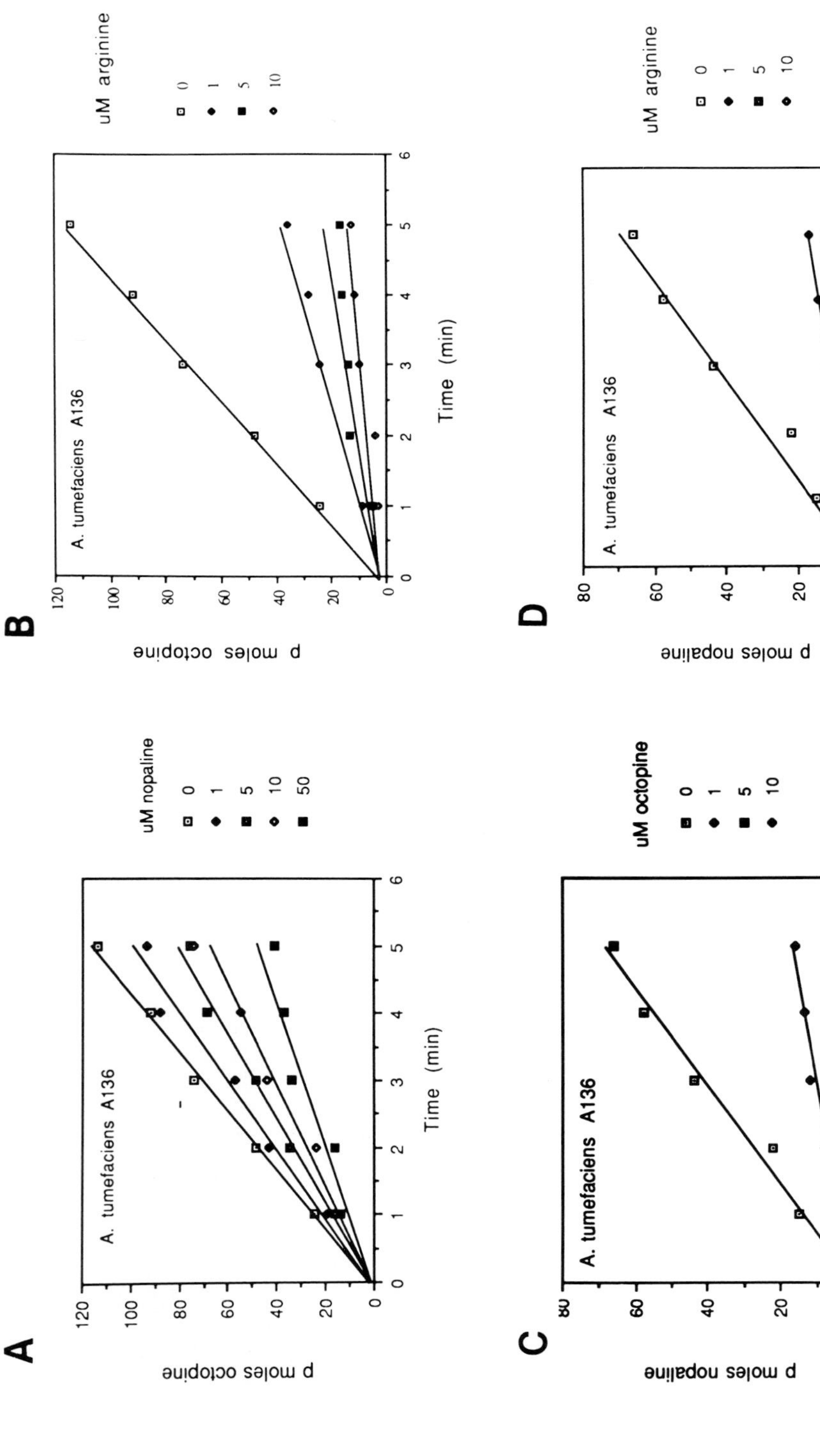


FIGURE 7. Opine transport by *A. tumefaciens* A136 in the presence of heterologous opines or arginine. (A) Octopine transport in the presence of nopaline; (B) octopine transport in the presence of arginine; (C) nopaline transport in the presence of octopine; (D) nopaline transport in the presence of arginine.

both octopine and the mannityl opines (mannopine, mannopinic acid, agropine, and agropinic acid), only octopine serves as a conjugal opine.[205,209,211,214,215] Similarly, nopaline-type Ti plasmids use agrocinopines A and B, but not nopaline, as the conjugal opine.[216] Mannopine/agropine-type Ti plasmids utilize agrocinopines C and D, but not the mannityl opines, as the conjugal opine.[216] The conjugal transfer of a copy of the Ti plasmid from a virulent *Agrobacterium* cell to a cell lacking such a plasmid makes sense in light of the opine hypothesis. Because the opine catabolic genes reside on the Ti plasmid, these *Agrobacterium* cells lacking a Ti plasmid cannot partake in the benefits of "genetic colonization;" i.e., they cannot utilize the opines produced by the crown gall tumors. The conjugal receipt of such a plasmid from a virulent donor subsequently allows the recipient to utilize these opines. In addition, the spread of a copy of the Ti plasmid to these bacteria disperses the genetic determinant of tumorigenesis and opine production to other *Agrobacterium* cells, allowing the tumorigenic cycle to continue.

Although Tempé et al.[217] once suggested that the transfer of the T-DNA from *Agrobacterium* to plant cells may utilize the same transfer system as is used in the conjugal transfer of the Ti plasmid between *Agrobacterium* cells, we now know that this is most likely not true. *Agrobacterium* strains in which the conjugal transfer genes were mutated are virulent, and strains lacking T-DNA could still transfer the Ti plasmid conjugally between *Agrobacterium* cells.[212,213] However, a possible link between T-DNA transfer to plant cells and conjugal Ti plasmid transfer between bacterial cells was recently suggested by Gelvin and Habeck[218] and Steck and Kado.[219] Using octopine- and nopaline-type *Agrobacterium* strains, these authors showed that mutations in certain *vir* genes (necessary for T-DNA transfer) could affect the efficiency of conjugal Ti plasmid transfer. Gelvin and Habeck[218] showed that some mutations in the genes *virA, virB, virC,* and *virG* could depress the frequency of octopine-induced conjugal transfer of the Ti plasmid by up to four orders of magnitude. This effect was not absolute, however, in that it was manifested only under limiting times of octopine induction of the conjugal transfer apparatus (18 to 24 h). When the bacteria were induced by octopine for longer times (48 to 72 h), the inhibitory effect of mutations in these *vir* genes diminished. These authors suggested that, although *vir* gene functions were not essential for conjugal Ti plasmid transfer, these functions may aid the conjugation process under limiting conditions. These authors therefore established a genetic link between the T-DNA transfer process, mediated by the *vir* genes, and the opine-induced conjugal plasmid transfer process.

A third possible function for certain opines as a signal for mediating tumorigenesis was recently described by Veluthambi et al.[45] We originally posed the question, would *Agrobacterium* cells induce their *vir* genes when in the presence of a crown gall tumor? The bacteria could sense such a tumor by the presence of crown gall tumor-specific compounds, opines, synthesized by the plant cells. Our original hypothesis suggested that, in the presence of opines, the *vir* genes may not be induced by the plant phenolic compounds. To our surprise, the induction of the *vir* genes by acetosyringone, as determined by β-galactosidase activity of *vir::lacZ* fusions, was enhanced by certain opines. The concentrations of octopine necessary for this potentiation of *vir* gene induction were high (1 to 10 m*M*), but within the range of concentration of octopine and octopinic acid found in mature crown gall tumors.[192,193] Although the strains examined (derivatives of A348, an octopine-type *Agrobacterium* strain) could utilize octopine and the mannityl opines, but not nopaline, nopaline had a greater synergistic effect upon *vir* gene induction by acetosyringone than did octopine. Two other nonmetabolizable opines, leucinopine and succinamopine, also were better potentiators of *vir* gene induction than was octopine. Mannopine, however, did not hyperinduce the *vir* genes in the presence of acetosyringone. Apparently, amino di- and tri-acid opines (but not arginine, a constituent of octopine and nopaline), were effective compounds in potentiating the induction of the *vir* genes by acetosyringone.

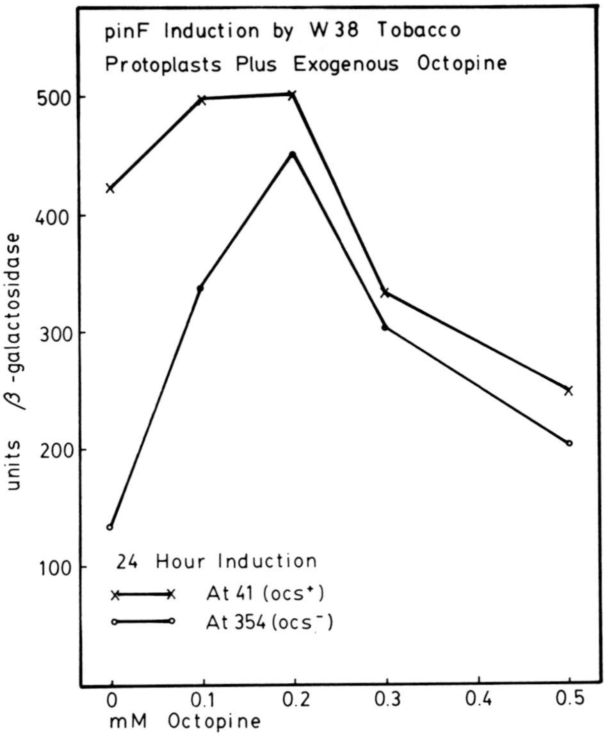


FIGURE 8. Effect of exogenous octopine upon the induction of the *pinF* gene in *Agrobacterium* strains that harbor a wild-type (At41) or mutant (At335) octopine synthase gene. Although the maximal hyperinduction by octopine of the *pinF* gene in strain At41 is approximately 15%, octopine at 200 μ*M* hyperinduces the *pinF* gene in strain At335 by 300%. Concentrations of octopine in excess of 200 μ*M* repress the induction of the *pinF* locus. These higher concentrations of octopine match those found in mature crown gall tumors.

The hyperinduction of the *vir* genes by opines and acetosyringone required no Ti plasmid-encoded genes other than *virA* and *virG*. It was possible, however, that chromosomal genes could also mediate this process. The potentiation of *vir* gene induction by opines resulted in the enhanced accumulation of T-strands within the induced *Agrobacterium* cell. The double-stranded cleavage of the T-DNA border repeat sequences was also increased. In addition, the application of both opines and acetosyringone, but not opines or acetosyringone alone, to the cut plant apical meristem resulted in an increased transformation efficiency of cotton, a plant relatively recalcitrant to *Agrobacterium* transformation and regeneration.[45]

Although the hyperinduction of the *vir* genes by opines increased with opine concentration (up to at least 10 to 20 m*M*) when the bacteria were induced in a test tube by acetosyringone, the induction of *Agrobacterium* cells by tobacco protoplasts resulted in a different pattern of potentiation by octopine. Using an *Agrobacterium* strain harboring a *lacZ* transcriptional fusion in the gene *pinF* (also known as *virH*,[82] a gene inducible by acetosyringone but not necessary for transformation of tobacco[80]), Krishnan and Gelvin showed that relatively low concentrations of octopine (100 to 200 μ*M*) could have a stimulatory effect upon induction. Higher concentrations of octopine, however, repressed *vir* gene activity (Figure 8). In addition, the *pinF* gene of *Agrobacterium* strains harboring a Tn*5* insertion in the octopine synthase (*ocs*) gene was induced two- to three-fold less by

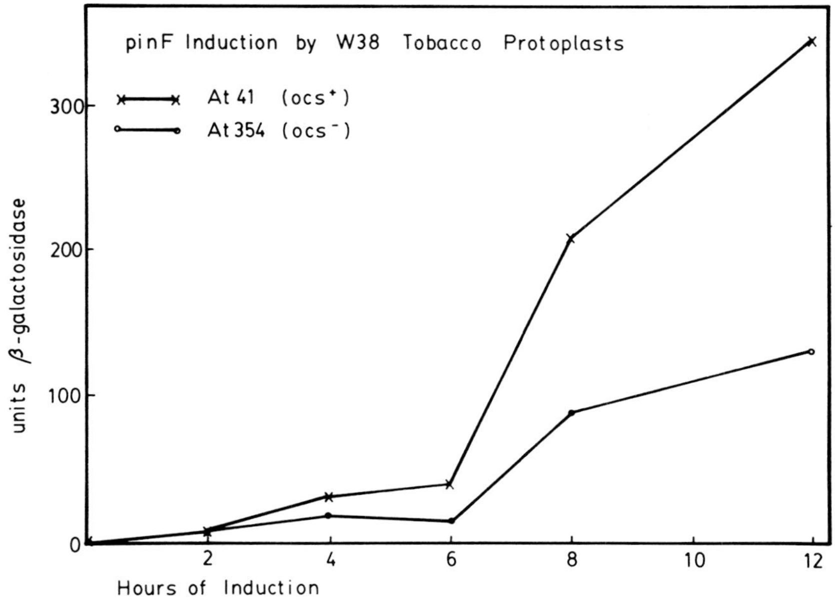


FIGURE 9. Induction by wild-type tobacco protoplasts of the *pinF* gene in *Agrobacterium* strains that harbor a wild-type (At41) or mutant (At335) octopine synthase gene. Although the kinetics of induction of the *pinF* gene are similar in the two bacterial strains, $ocs^+$ bacteria induce the *pinF* gene to a threefold greater extent than do $ocs^-$ bacteria.

protoplasts than was the *pinF* gene in a strain harboring a wild-type *ocs* gene (Figure 9). These two strains behaved identically, however, when in the presence of tobacco protoplasts derived from transgenic plants that synthesized octopine (Figure 10). Data from Dion's laboratory[193] indicated that in the extracellular fluid of mature crown gall tumors, octopine concentrations reached approximately 200 $\mu M$. These results suggested that, in the presence of plant cells, *Agrobacterium* could respond to physiological concentrations of octopine.

We have hypothesized that the concentration of opines, among other chemicals in the wound environment, may influence the physiological state of the *Agrobacterium* cell. During the initial stages of infection, when the concentration of phenolic *vir* gene-inducing molecules is high, low concentrations of opine synthesized by the nascent tumor may indicate to the bacterium that infection has been successful. The bacteria may be "encouraged" to continue the infection process. As the wound heals and the concentration of phenolic molecules decreases, the limitation of inducing molecules may be partially compensated by increased opine concentrations. We previously showed that, in the presence of limiting acetosyringone concentrations, T-strand production was increased by the addition of certain opines.[45] Finally, the concentration of phenolic-inducing molecules may decrease to such an extent, and the concentration of opines produced by a mature crown gall tumor increase to such a degree, that *vir* gene activity is no longer induced to a great extent. The "physiology" of the bacterial cell alters from an "inductive" to a "vegetative" state, whereby *Agrobacterium* now concentrates its efforts upon metabolizing opines. Such a model would predict that *Agrobacterium* cells harboring T-DNA that is unable to direct the production of certain opines may be less virulent than strains that could induce opine-producing tumors. That this may be true is suggested by the data presented in Table 2. *Agrobacterium* strains harboring a wild-type *ocs* gene are approximately 40% more virulent, as measured by a potato disk assay, than a near-isogenic strain harboring a mutation in the *ocs* gene.

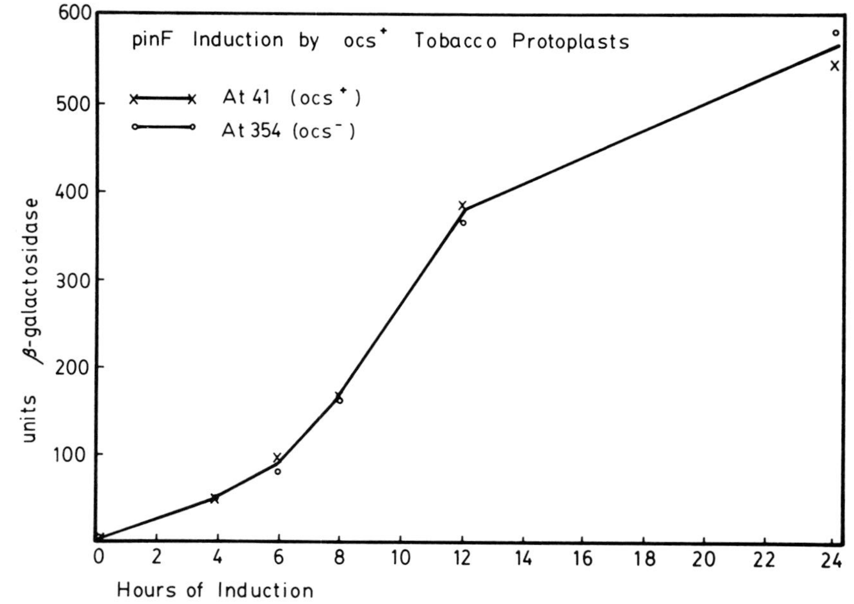


FIGURE 10. Induction by transgenic tobacco protoplasts of the *pinF* gene in *Agrobacterium* strains that harbor a wild-type (At41) or mutant (At335) octopine synthase gene. Note that when the protoplasts synthesize octopine, there is no difference in the induction of the *pinF* gene between the two strains.

**TABLE 2**
**Virulence of *Agrobacterium tumefaciens* At41 (*ocs*$^+$) and At355 (*ocs*$^-$) on Potato Tuber Discs**[a]

| | | Average number of tumors/disk | |
|---|---|---|---|
| **Experiment** | **Bacteria/ml** | **At41** | **At335** |
| 1 | $10^6$ | 49.8 (n=6) | 40.2 (n=6) |
| 2 | $10^7$ | 14.2 (n=47) | 8.4 (n=48) |
| 3 | $10^6$ | 11.2 (n=49) | 7.4 (n=52) |

[a] Bacteria (50 μl) were inoculated onto each disk, and the disks scored 18 days after infection. *n*, Number of disks scored.

# VI. NEUTRAL SUGARS

Although the importance of such phenolic molecules as acetosyringone in *vir* gene induction has been well established, recent data indicate that certain neutral sugars may also play an important role in virulence and in *vir* gene induction. Garfinkel and Nester[22] originally reported mutations in the chromosome of *A. tumefaciens* strains A1008, A1059, and A1068 that reduced the host range. In a more recent report, the nature of these mutations was investigated.[148] None of these strains formed tumors on *Kalanchoe* leaves, pea hypocotyls, tomato stems, or *N. glauca* leaf disks, although they maintained weak virulence on sunflower stems, zinnia stems, and *N. glauca* stems. The mutant strains had normal plant cell binding activities in several assays. The authors therefore investigated the extent of *vir* gene induction in each of these strains and found greatly reduced induction of a *virB*::*lacZ* fusion, especially under conditions of limiting acetosyringone concentrations (10 to 50 μ*M*). This new virulence locus, called *chvE,* was characterized both genetically and by DNA sequence analysis. The sequence analysis indicated that the ChvE protein was homologous to two *E. coli* periplasmic proteins, a ribose-binding protein and a galactose binding protein. Indeed, the ChvE protein was identical in sequence to a glucose/galactose-binding protein identified in *Agrobacterium* by Cornish et al.[147] These data suggested that certain sugars may play a role in *vir* gene induction and virulence.

Additional support for the idea that certain sugars play a role in *vir* gene induction came from two laboratories. Shimoda et al.[220] showed that, under limiting acetosyringone induction conditions, such neutral sugars as L-arabinose, D-xylose, D-lyxose, D-glucose, D-mannose, D-idose, D-galactose, D-talose, 2-deoxy-D-glucose, and the acidic sugar D-galacturonic acid, could greatly enhance the induction of several *vir* genes. This response to sugars was eliminated by deleting the periplasmic domain of the VirA protein.

Cangelosi et al.[149] also showed the importance of certain sugars for the acetosyringone induction of *vir* genes. D-Galactose, D-glucose, L-arabinose, D-fucose, D-xylose, cellobiose, 2-deoxy-D-glucose, and 6-deoxy-D-glucose could stimulate *vir* gene induction in the presence of limiting concentrations of acetosyrigone. D-Galacturonic acid and D-glucuronic acid were even better coinducers than were the neutral sugars.[221] This stimulation was dependent upon an intact copy of the *chvE* gene. In addition, a deletion of the periplasmic domain of the VirA protein also resulted in loss of stimulation by these sugars.[149] Certain sugars, such as L-arabinose and D-fucose, could even induce a *virE::lacZ* fusion in the absence of acetosyringone following extended incubation in the presence of the sugar. This induction was likewise dependent upon *chvE* and *virA* function. The mutation of the gene *chvE,* as well as eliminating the coinduction by sugars and acetosyringone of the *vir* genes, also eliminated chemotaxis toward many of these ''inducing'' sugars. Mutations that deleted the periplasmic domain of the VirA protein reduced the host range of *Agrobacterium* in the same way as did mutation of the gene *chvE.* These results suggest that the periplasmic domain of the VirA protein may interact with the ChvE ribose binding protein in a way as to allow coinduction of the *vir* regulon by both certain phenolic molecules as well as certain sugars. The response to sugars is dependent upon the host plant, however. The virulence of *Agrobacterium* on certain plants requires a functional *chvE* gene, whereas the tumorigenic response of other plants is independent of such gene function.

## VII. ACIDIC SUGARS

Acidic polysaccharides, such as polygalacturonic acid, make up a large part of the pectic portion of the plant cell wall. Our laboratory has recently discovered that the uronic acid fraction of extracts from carrot root shavings can alter gene expression in *Agrobacterium* cells exposed to such extracts.[222] Protein labeling experiments indicated that at least 10 proteins were induced, and the synthesis of one repressed. Transposon MudI-1681 mutagenesis identified at least ten loci that were induced by these extracts, and one strain, At156, which was inducible to a particularly high level, was chosen for further study.

The mutant strain At156 did not show any detectable difference in either growth rate on minimal or complete medium, or any reduction in virulence, compared to the parental wild-type strain. When wild-type *A. tumefaciens* A136 was incubated with pea root cap cells using the assay of Hawes and Pueppke,[223] the bacteria bound to the plant cell surface but did not aggregate to any great extent within 16 h at the concentration of bacteria tested. The mutant strain At156, however, formed long aggregates, or ropes, under similar incubation conditions. Neither the mutant nor the wild-type strain formed ropes in the absence of plant cells. These results suggest that the gene disrupted by MudI-1681 in strain At156 may regulate the surface properties of *Agrobacterium.* DNA sequence analysis of the locus disrupted by the transposon indicated no extensive sequence homology with any genes or proteins in several data bases, suggesting that this locus (called *picA*, for *p*lant *i*nducible *c*hromosomal) is a novel gene not yet described in the literature.[224]

Chemical analysis of the inducing fraction of the carrot root extract indicated that it contained greater than 80% uronic acid sugars, with a small amount of xylose and arabinose.[225] These data suggest that the inducing substance is derived from the pectic portion of the plant cell wall. Interestingly, polygalacturonic acid (PGA) also induces the *picA* locus. The most effective inducers are PGA molecules with a degree of polymerization of 6 to 16 sugar residues. Using these data, and assuming that the inducing compound is a galacturonic acid-like molecule with a chain length of approximately 10, we calculated that the inducer is effective at concentrations of 1 to 10 $\mu M$. This low effective concentration, coupled with the fact that *A. tumefaciens* A136 cannot metabolize polygalacturonic acid,[226] suggests that the inducing substance is a signal molecule and not merely a metabolite. It is interesting to note that a molecule from the plant cell surface appears to induce an *Agrobacterium* chromosomal gene that may regulate the properties of the bacterial surface that, in turn, interacts with the plant cell. Experiments are in progress to identify the function of the *picA* locus.

## VIII. CONCLUSION

During the processs of tumorigenesis, *Agrobacterium* both receives from the plant and transmits to the plant numerous chemical signals. Sugars and defense-related phenolic compounds secreted by wounded plant cells serve both as chemoattractants for the bacterium, and as inducers of the *vir* genes. The activation of the *vir* genes results in the processing of the T-DNA from the Ti plasmid and its transfer to the plant. Transcription of the integrated T-DNA, and the subsequent translation of T-DNA encoded products, leads to the production of phytohormones responsible for the proliferation of tumors, and the production and secretion of opines. The bacteria transport these opines via at least two mechanisms, a Ti plasmid-independent pathway and a Ti plasmid-encoded pathway that is inducible by the opine. The opines are used as energy sources by the bacteria, and some opines can induce the conjugal transfer of the Ti plasmid between *Agrobacterium* cells. The recipient cells can thereby utilize the opines, and participate in further rounds of infection. Certain opines can influence *vir* gene induction in complex ways. The activity of some *vir* genes may play a role in the conjugal transfer of the Ti plasmid. In addition, certain sugars play important roles in *vir* gene induction and other processes encoded by the *Agrobacterium* chromosome.

Thus, *Agrobacterium* is able to receive and interpret numerous chemical signals in the rhizosphere. The bacterium monitors these chemicals and most likely uses this information to direct bacterial gene expression in order to adapt its physiology to the environment.

## ACKNOWLEDGMENTS

The author wishes to thank John Gray, Juan Wang, and Drs. Todd Steck and M. Krishnan for use of unpublished data; Drs. Susan Karcher, Walt Ream, and Todd Steck for critical reading of the manuscript; Dr. Susan Karcher for help with the figures; and Ms. Wilma Foust for help in the preparation of the manuscript. Work in the author's laboratory has been funded by the U.S. Department of Agriculture, the National Science Foundation, the Midwest Plant Biotechnology Consortium, Dow/Elanco, and the Dupont Co.

# REFERENCES


1. **Lin, T. S. and Kolattukudy, P. E.,** Induction of a biopolyester hydrolase (cutinase) by low levels of cutin monomers in *Fusarium solani* f. sp. *pisi, J. Bacteriol.,* 133, 942, 1978.
2. **Tsuyuma, S. and Chatterjee, A. K.,** Pectin lyase production in *Erwinia chrysanthemi* and other soft-rot *Erwinia* species, *Physiol. Plant Pathol.,* 24, 291, 1984.
3. **Downie, J. A. and Johnston, A. W. B.,** Nodulation of legumes by *Rhizobium:* The recognized root?, *Cell,* 47, 153, 1986.
4. **Peters, N. K., Frost, J. W., and Long, S. R.,** A plant flavone, lutiolin, induces expression of *Rhizobium meliloti* nodulation genes, *Science,* 233, 977, 1986.
5. **Shearman, C. A., Rossen, L., Johnston, A. W. B., and Downie, J. A.,** The *Rhizobium leguminosarum* nodulation gene *nodF* encodes a polypeptide similar to acyl-carrier protein and is regulated by *nodD* plus a factor in pea root exudate, *EMBO J.,* 5, 647, 1986.
6. **Rossen, L., Davis, E. O., and Johnston, A. W. B.,** Plant-induced expression of *Rhizobium* genes involved in host specificity and early stages of nodulation, *Trends Biochem. Sci.,* 12, 430, 1987.
7. **Lerouge, P., Roche, P., Faucher, C., Maillet, F., Truchet, G., Prome, J. C., and Denarie, J.,** Symbiotic host-specificity of *Rhizobium meliloti* is determined by a sulphated and acylated glucosamine oligosaccharide signal, *Nature (London),* 344, 781, 1990.
8. **Halverson, L. J. and Stacey, G.,** Signal exchange in plant-microbe interaction, *Microbiol. Rev.,* 50, 193, 1986.
9. **Peters, N. K. and Verma, D. P. S.,** Phenolic compounds as regulators of gene expression in plant-microbe interactions, *Mol. Plant-Microbe Interact.,* 3, 4, 1990.
10. **Stachel, S. E. and Zambryski, P. C.,** *Agrobacterium tumefaciens* and the susceptible plant cell: A novel adaptation of extracellular recognition and DNA conjugation, *Cell,* 47, 155, 1986.
11. **Binns, A. N. and Thomashow, M. F.,** Cell biology of *Agrobacterium* infection and transformation of plants, *Annu. Rev. Microbiol.,* 42, 575, 1988.
12. **Zambryski, P.,** Basic processes underlying *Agrobacterium*-mediated DNA transfer to plant cells, *Annu. Rev. Genet.,* 22, 1, 1988.
13. **Ream, W.,** *Agrobacterium tumefaciens* and interkingdom genetic exchange, *Annu. Rev. Phytopathol.,* 27, 583, 1989.
14. **Zambryski, P., Tempé, J., and Schell, J.,** Transfer and function of T-DNA genes from *Agrobacterium* Ti and Ri plasmids in plants, *Cell,* 56, 193, 1989.
15. **Loake, G. J., Ashby, A. M., and Shaw, C. H.,** Attraction of *Agrobacterium tumefaciens* C58C1 towards sugars involves a highly sensitive chemotaxis system, *J. Gen. Microbiol.,* 134, 1427, 1988.
16. **Hawes, M. C., Smith, L. Y., and Howarth, A. J.,** *Agrobacterium tumefaciens* mutants deficient in chemotaxis to root exudates, *Mol. Plant-Microbe Interact.,* 1, 182, 1988.
17. **Hawes, M. C. and Smith, L. Y.,** Requirement for chemotaxis in pathogenicity of *Agrobacterium tumefaciens* on roots of soil-grown pea plants, *J. Bacteriol.,* 171, 5668, 1989.
18. **Ashby, A. M., Watson, M. D., and Shaw, C. H.,** A Ti-plasmid determined function is responsible for chemotaxis of *Agrobacterium tumefaciens* towards the plant wound product acetosyringone, *FEMS Microbiol. Lett.,* 41, 189, 1987.
19. **Ashby, A. M., Watson, M. D., Loake, G. J., and Shaw, C. H.,** Ti plasmid-specified chemotaxis of *Agrobacterium tumefaciens* C58C1 toward *vir*-inducing phenolic compounds and soluble factors from monocotyledonous and dicotyledonous plants, *J. Bacteriol.,* 170, 4181, 1988.
20. **Shaw, C. H., Ashby, A. M., Brown, A., Royal, C., Loake, G. J., and Shaw, C. H.,** *virA* and *virG* are the Ti-plasmid functions required for chemotaxis of *Agrobacterium tumefaciens* towards acetosyringone, *Mol. Microbiol.,* 2, 413, 1988.
21. **Parke, D., Ornston, L. N., and Nester, E. W.,** Chemotaxis to plant phenolic inducers of virulence genes is constitutively expressed in the absence of the Ti plasmid in *Agrobacterium tumefaciens, J. Bacteriol.,* 169, 5336, 1987.
22. **Garfinkel, D. J. and Nester, E. W.,** *Agrobacterium tumefaciens* mutants affected in crown gall tumorigenesis and octopine catabolism, *J. Bacteriol.,* 144, 732, 1980.
23. **Klee, H. J., White, F. F., Iyer, V. N., Gordon, M. P., and Nester, E. W.,** Mutational analysis of the virulence region of an *Agrobacterium tumefaciens* Ti plasmid, *J. Bacteriol.,* 153, 878, 1983.
24. **Hooykaas, P. J. J., Hofker, M., den Dulk-Ras, H., and Schilperoort, R. A.,** A comparison of virulence determinants in an octopine Ti plasmid, a nopaline Ti plasmid, and an Ri plasmid by complementation analysis of *Agrobacterium tumefaciens* mutants, *Plasmid,* 11, 195, 1984.
25. **Lundquist, R. C., Close, T. J., and Kado, C. I.,** Genetic complementation of *Agrobacterium tumefaciens* Ti plasmid mutants in the virulence region, *Mol. Gen. Genet.,* 193, 1, 1984.
26. **Stachel, S. E., An, G., Flores, C., and Nester, E. W.,** A Tn3 *lacZ* transposon for the random generation of β-galactosidase gene fusions: Application to the analysis of gene expression in *Agrobacterium, EMBO J.,* 4, 891, 1985.

27. **Rogowsky, P. M., Close, T. J., Chimera, J. A., Shaw, J. J., and Kado, C. I.,** Regulation of the *vir* genes of *Agrobacterium tumefaciens* plasmid pTiC58, *J. Bacteriol.*, 169, 5101, 1987.
28. **Stachel, S. E., Nester, E. W., and Zambryski, P. C.,** A plant cell factor induces *Agrobacterium tumefaciens vir* gene expression, *Proc. Natl. Acad. Sci. U.S.A.*, 83, 379, 1986.
29. **Koukolikova-Nicola, Z., Shillito, R. D., Hohn, B., Wang, K., Van Montagu, M., and Zambryski, P.,** Involvement of circular intermediates in the transfer of T-DNA from *Agrobacterium tumefaciens* to plant cells, *Nature (London)*, 313, 191, 1985.
30. **Alt-Moerbe, J., Rak, B., and Schroder, J.,** A 3.6-kbp segment from the *vir* region of Ti plasmids contains genes responsible for border sequence-directed production of T region circles in *E. coli*, *EMBO J.*, 5, 1129, 1986.
31. **Machida, Y., Usami, S., Yamamoto, A., Niwa, Y., and Takebe, I.,** Plant-inducible recombination between the 25 bp border sequences of T-DNA in *Agrobacterium tumefaciens*, *Mol. Gen. Genet.*, 204, 374, 1986.
32. **Stachel, S. E., Messens, E., Van Montagu, M., and Zambryski, P.,** Identification of the signal molecules produced by wounded plant cells that activate T-DNA transfer in *Agrobacterium tumefaciens*, *Nature (London)*, 318, 624, 1985.
33. **Bolton, G. W., Nester, E. W., and Gordon, M. P.,** Plant phenolic compounds induce expression of the *Agrobacterium tumefaciens* loci needed for virulence, *Science*, 232, 983, 1986.
34. **Zerback, R., Dressler, K., and Hess, D.,** Flavonoid compounds from pollen and stigma of *Petunia hybrida*: Inducers of the *vir* region of the *Agrobacterium tumefaciens* Ti plasmid, *Plant Sci.*, 62, 83, 1989.
35. **Spencer, P. A. and Towers, G. H. N.,** Specificity of signal compounds detected by *Agrobacterium tumefaciens*, *Phytochemistry*, 27, 2781, 1988.
36. **Melchers, L. S., Regensburg-Tuink, A. J. G., Schilperoort, R. A., and Hooykaas, P. J. J.,** Specificity of signal molecules in the activation of *Agrobacterium* virulence gene expression, *Mol. Microbiol.*, 3, 969, 1989.
37. **Usami, S., Morikawa, S., Takebe, I., and Machida, Y.,** Absence in monocotyledonous plants of the *diffusible* plant factors inducing T-DNA circularization and *vir* gene expression in *Agrobacterium*, *Mol. Gen. Genet.*, 209, 221, 1987.
38. **Usami, S., Okamoto, S., Takebe, I., and Machida, Y.,** Factor inducing *Agrobacterium tumefaciens vir* gene expression is present in monocotyledonous plants, *Proc. Natl. Acad. Sci. U.S.A.*, 85, 3748, 1988.
39. **Messens, E., Dekeyser, R., and Stachel, S. E.,** A nontransformable *Triticum monococcum* monocotyledonous culture produces the potent *Agrobacterium vir*-inducing compound ethyl ferulate, *Proc. Natl. Acad. Sci. U.S.A.*, 87, 4368, 1990.
40. **Sahi, S. V., Chilton, M.-D., and Chilton, W. S.,** Corn metabolites affect growth and virulence of *Agrobacterium tumefaciens*, *Proc. Natl. Acad. Sci. U.S.A.*, 87, 3879, 1990.
41. **Morris, J. W. and Morris, R. O.,** Identification of an *Agrobacterium tumefaciens* virulence gene inducer from the pinaceous gymnosperm *Pseudotsuga menziesii*, *Proc. Natl. Acad. Sci. U.S.A.*, 87, 3614, 1990.
42. **Bouckaert-Urban, A.-M. and Vendrig, J. C.,** Enhancement of crown-gall tumor initiation by extract fractions from tumor forming cotyledons of *Helianthus giganteus* L., *Z. Pflanzenphysiol.*, 105, 395, 1982.
43. **Schafer, W., Gorz, A., and Kahl, G.,** T-DNA integration and expression in a monocot crop plant after induction of *Agrobacterium*, *Nature (London)*, 327, 529, 1987.
44. **Owens, L. D. and Smigocki, A. C.,** Transformation of soybean cells using mixed strains of *Agrobacterium tumefaciens* and phenolic compounds, *Plant Physiol.*, 88, 570, 1988.
45. **Veluthambi, K., Krishnan, M., Gould, J. H., Smith, R. H., and Gelvin, S. B.,** Opines stimulate induction of the *vir* genes of the *Agrobacterium tumefaciens* Ti plasmid, *J. Bacteriol.*, 171, 3696, 1989.
46. **Lemmers, M., DeBeuckeleer, M., Holsters, M., Zambryski, P., Depicker, A., Hernalsteens, J. P., Van Montagu, M., and Schell, J.,** Internal organization, boundaries and integration of Ti-plasmid DNA in nopaline crown gall tumors, *J. Mol. Biol.*, 144, 353, 1980.
47. **Merlo, D. J., Nutter, R. C., Montoya, A. L., Garfinkel, D. J., Drummond, M. H., Chilton, M.-D., Gordon, M. P., and Nester, E. W.,** The boundaries and copy numbers of Ti plasmid T-DNA vary in crown gall tumors, *Mol. Gen. Genet.*, 177, 637, 1980.
48. **Thomashow, M. F., Nutter, R., Montoya, A. L., Gordon, M. P., and Nester, E. W.,** Integration and organization of Ti plasmid sequences in crown gall tumors, *Cell*, 19, 729, 1980.
49. **Zambryski, P., Holsters, M., Kruger, K., Depicker, A., Schell, J., Van Montagu, M., and Goodman, H. M.,** Tumor DNA structure in plant cells transformed by *A. tumefaciens*, *Science*, 209, 1385, 1980.
50. **DeBeuckeleer, M., Lemmers, M., DeVos, G., Willmitzer, L., Van Montagu, M., and Schell, J.,** Further insight on the transferred-DNA of octopine crown gall, *Mol. Gen. Genet.*, 183, 283, 1981.
51. **Ooms, G., Bakker, K. A., Molendijk, L., Wullems, G. J., Gordon, M. P., Nester, E. W., and Schilperoort, R. A.,** T-DNA organization in homogeneous and heterogeneous octopine-type crown gall tissues of *Nicotiana tabacum*, *Cell*, 30, 589, 1982.
52. **Zambryski, P., Depicker, A., Druger, K., and Goodman, H. M.,** Tumor induction by *Agrobacterium tumefaciens*: Analysis of the boundaries of T-DNA, *J. Mol. Appl. Genet.*, 1, 361, 1982.

53. **Byrne, M. C., Koplow, J., David, C., Tempe, J., and Chilton, M.-D.,** Structure of T-DNA in roots transformed by *Agrobacterium rhizogenes, J. Mol. Appl. Genet.*, 2, 201, 1983.
54. **Hepburn, A. G., Clarke, L. E., Blundy, K. S., and White, J.,** Nopaline Ti-plasmid pTiT37 T-DNA insertions into a flax genome, *J. Mol. Appl. Genet.*, 2, 211, 1983.
55. **Holsters, M., Villarroel, R., Gielen, J., Seurinck, J., DeGreve, H., Van Montagu, M., and Schell, J.,** An analysis of the boundaries of the octopine TL-DNA in tumors induced by *Agrobacterium tumefaciens, Mol. Gen. Genet.*, 190, 35, 1983.
56. **Knauf, V., Yanofsky, M., Montoya, A., and Nester, E.,** Physical and functional map of an *Agrobacterium tumefaciens* tumor inducing plasmid that confers a narrow host range, *J. Bacteriol.*, 160, 564, 1984.
57. **Taylor, B. H., Amasino, R. M., White, F. F., Nester, E. W., and Gordon, M. P.,** T-DNA analysis of plants regenerated from hairy root tumors, *Mol. Gen. Genet.*, 201, 554, 1985.
58. **Yanofsky, M., Montoya, A., Knauf, V., Lowe, B., Gordon, M., and Nester, E.,** Limited-host-range plasmid of *Agrobacterium tumefaciens:* Molecular and genetic analyses of transferred DNA, *J. Bacteriol.*, 163, 341, 1985.
59. **Vilaine, F. and Casse-Delbart, F.,** Independent induction of transformed roots by the TL and TR regions of the Ri plasmid of agropine type *Agrobacterium rhizogenes, Mol. Gen. Genet.*, 206, 17, 1987.
60. **Yadav, N. S., Van der Leyden, J., Bennett, D. R., Barnes, W. M., and Chilton, M.-D.,** Short direct repeats flank the T-DNA on a nopaline Ti plasmid, *Proc. Natl. Acad. Sci. U.S.A.*, 79, 6322, 1982.
61. **Shaw, C. H., Watson, M. D., Carter, G. H., and Shaw, C. H.,** The right hand copy of the nopaline Ti-plasmid 25 bp repeat is required for tumour formation, *Nucleic Acids Res.*, 12, 6031, 1984.
62. **Wang, K., Herrera-Estrella, L., Van Montagu, M., and Zambryski, P.,** Right 25 bp terminus sequence of the nopaline T-DNA is essential for and determines direction of DNA transfer from *Agrobacterium* to the plant genome, *Cell*, 38, 455, 1984.
63. **Hepburn, A. G. and White, J.,** The effect of right terminal repeat deletion on the oncogenicity of the T-region of pTiT37, *Plant Mol. Biol.*, 5, 3, 1985.
64. **Peralta, E. G. and Ream, L. W.,** T-DNA border sequences required for crown gall tumorigenesis, *Proc. Natl. Acad. Sci. U.S.A.*, 82, 5112, 1985.
65. **Gardner, R. C. and Knauf, V. C.,** Transfer of *Agrobacterium* DNA to plants requires a T-DNA border but not the *virE* locus, *Science*, 231, 725, 1986.
66. **Horsch, R. B. and Klee, H. J.,** Rapid assay of foreign gene expression in leaf discs transformed by *Agrobacterium tumefaciens:* Role of T-DNA borders in the transfer process, *Proc. Natl. Acad. Sci. U.S.A.*, 83, 4428, 1986.
67. **Jen, G. C. and Chilton, M.-D.,** Activity of T-DNA borders in plant cell transformation by mini-T plasmids, *J. Bacteriol.*, 166, 491, 1986.
68. **Jen, G. C. and Chilton, M.-D.,** The right border region of pTiT37 is intrinsically more active than the left border region in promoting T-DNA transformation, *Proc. Natl. Acad. Sci. U.S.A.*, 83, 3895, 1986.
69. **Caplan, A. B., Van Montagu, M., and Schell, J.,** Genetic analysis of integration mediated by single T-DNA borders, *J. Bacteriol.*, 161, 655, 1985.
70. **Rubin, R. A.,** Genetic studies on the role of octopine T-DNA border regions in crown gall tumor formation, *Mol. Gen. Genet.*, 202, 312, 1986.
71. **Slightom, J. L., Durand-Tardif, M., Jouanin, L., and Tepfer, D.,** Nucleotide sequence analysis of TL-DNA of *Agrobacterium rhizogenes* agropine type plasmid, *J. Biol. Chem.*, 261, 108, 1986.
72. **Van Haaren, M. J. J., Pronk, J. T., Schilperoort, R. A., and Hooykaas, P. J. J.,** Functional analysis of the *Agrobacterium tumefaciens* octopine Ti-plasmid left and right T-region border fragments, *Plant Mol. Biol.*, 8, 95, 1987.
73. **Van Haaren, M. J. J., Sedee, N. J. A., Krul, M., Schilperoort, R. A., and Hooykaas, P. J. J.,** Function of heterologous and pseudo border repeats in T region transfer via the octopine virulence system of *Agrobacterium tumefaciens, Plant Mol. Biol.*, 11, 773, 1988.
74. **Van Haaren, M. J. J., Sedee, N. J. A., de Boer, H. A., Schilperoort, R. A., and Hooykaas, P. J. J.,** Mutational analysis of the conserved domains of a T-region border repeat of *Agrobacterium tumefaciens, Plant Mol. Biol.*, 13, 523, 1989.
75. **Wang, K., Stachel, S. E., Timmerman, B., Van Montagu, M., and Zambryski, P. C.,** Site-specific nick in the T-DNA border sequence as a result of *Agrobacterium vir* gene expression, *Science*, 235, 587, 1987.
76. **Joos, H., Inze, D., Caplan, A., Sormann, M., Van Montagu, M., and Schell, J.,** Genetic analysis of T-DNA transcripts in nopaline crown galls, *Cell*, 32, 1057, 1983.
77. **Peralta, E. G., Helmiss, R., and Ream, W.,** Overdrive, a T-DNA transmission enhancer on the *A. tumefaciens* tumour-inducing plasmid, *EMBO J.*, 5, 1137, 1986.
78. **Culianez-Macia, F. A. and Hepburn, A. G.,** Right-border sequences enable the left border of an *Agrobacterium tumefaciens* Ti-plasmid to produce single-stranded DNA, *Plant Mol. Biol.*, 389, 1988.
79. **Iyer, V. N., Klee, H. J., and Nester, E. W.,** Units of genetic expression in the virulence region of a plant tumor-inducing plasmid of *Agrobacterium tumefaciens, Mol. Gen. Genet.*, 188, 418, 1982.

80. **Stachel, S. E. and Nester, E. W.,** The genetic and transcriptional organization of the *vir* region of the A6 Ti plasmid of *Agrobacterium tumefaciens, EMBO J.,* 5, 1445, 1986.
81. **Stachel, S. E. and Zambryski, P. C.,** *virA* and *virG* control the plant-induced activation of the T-DNA transfer process of *A. tumefaciens, Cell,* 46, 325, 1986.
82. **Kanemoto, R. H., Powell, A. T., Akiyoshi, D. E., Regier, D. A., Kerstetter, R. A., Nester, E. W., Hawes, M. C., and Gordon, M. P.,** Nucleotide sequence and analysis of the plant-inducible locus *pinF* from *Agrobacterium tumefaciens, J. Bacteriol.,* 171, 2506, 1989.
83. **Melchers, L. S., Maroney, M. J., den Dulk-Ras, A., Thompson, D. V., van Vuuren, H. A. J., Schilperoort, R. A., and Hooykaas, P. J. J.,** Octopine and nopaline strains of *Agrobacterium tumefaciens* differ in virulence; molecular characterization of the *virF* locus, *Plant Mol. Biol.,* 14, 249, 1990.
84. **John, M. C. and Amasino, R. M.,** Expression of an *Agrobacterium* Ti plasmid gene involved in cytokinin biosynthesis is regulated by virulence loci and induced by plant phenolic compounds, *J. Bacteriol.,* 170, 790, 1988.
85. **Winans, S. C., Kerstetter, R. A., and Nester, E. W.,** Transcriptional regulation of the *virA* and *virG* genes of *Agrobacterium tumefaciens, J. Bacteriol.,* 170, 4047, 1988.
86. **Melchers, L. S., Thompson, D. V., Idler, K. B., Schilperoort, R. A., and Hooykaas, P. J. J.,** Nucleotide sequence of the virulence gene *virG* of the *Agrobacterium tumefaciens* octopine Ti plasmid: significant homology between *virG* and the regulatory genes *ompR, phoB,* and *dye* of *E. coli, Nucleic Acids Res.,* 14, 9933, 1986.
87. **Leroux, B., Yanofsky, M. F., Winans, S. C., Ward, J. E., Ziegler, S. F., and Nester, E. W.,** Characterization of the *virA* locus of *Agrobacterium tumefaciens:* A transcriptional regulator and host range determinant, *EMBO J.,* 6, 849, 1987.
88. **Powell, B. S., Powell, G. K., Morris, R. O., Rogowsky, P. M., and Kado, C. I.,** Nucleotide sequence of the *virG* locus of the *Agrobacterium tumefaciens* plasmid pTiC58, *Mol. Microbiol.,* 1, 309, 1987.
89. **Melchers, L. S., Regensburg-Tuink, T. J. G., Bourret, R. B., Sedee, N. J. A., Schilperoort, R. A., and Hooykaas, P. J. J.,** Membrane topology and functional analysis of the sensory protein VirA of *Agrobacterium tumefaciens, EMBO J.,* 8, 1919, 1989.
90. **Morel, P., Powell, B. S., Rogowsky, P. M., and Kado, C. I.,** Characterization of the *virA* virulence gene of the nopaline plasmid, pTiC58 of *Agrobacterium tumefaciens, Mol. Microbiol.,* 3, 1237, 1989.
91. **Powell, B. S., Rogowsky, P. M., and Kado, C. I.,** *virG* of *Agrobacterium tumefaciens* plasmid pTiC58 encodes a DNA-binding protein, *Mol. Microbiol.,* 3, 411, 1989.
92. **Huang, Y., Morel, P., Powell, B., Kado, C. I.,** VirA, a coregulator of Ti-specified virulence genes, is phosphorylated *in vitro, J. Bacteriol.,* 172, 1142, 1990.
93. **Jin, S., Roitsch, T., Christie, P. J., and Nester, E. W.,** The regulatory VirG protein specifically binds to a *cis*-acting regulatory sequence involved in transcriptional activation of *Agrobacterium tumefaciens* virulence genes, *J. Bacteriol.,* 172, 531, 1990.
94. **Pazour, G. J. and Das, A.,** *virG,* an *Agrobacterium tumefaciens* transcriptional activator, initiates translation at a UUG codon and is a sequence-specific DNA-binding protein, *J. Bacteriol.,* 172, 1241, 1990.
95. **Roitsch, T., Wang, H., Jin, S., and Nester, E. W.,** Mutational analysis of the VirG protein, a transcriptional activator of *Agrobacterium tumefaciens* virulence genes, *J. Bacteriol.,* 172, 6054, 1990.
96. **Nixon, B. T., Ronson, C. W., and Ausubel, F. M.,** Two-component regulatory systems responsive to environmental stimuli share strongly conserved domains with the nitrogen assimilation regulatory genes *ntrB* and *ntrC, Proc. Natl. Acad. Sci. U.S.A.,* 83, 7850, 1986.
97. **Winans, S. C., Ebert, P. R., Stachel, S. E., Gordon, M. P., and Nester, E. W.,** A gene essential for *Agrobacterium* virulence is homologous to a family of positive regulatory loci, *Proc. Natl. Acad. Sci. U.S.A.,* 83, 8278, 1986.
98. **Ronson, C. W., Nixon, B. T., and Ausubel, F. M.,** Conserved domains in bacterial regulatory proteins that respond to environmental stimuli, *Cell,* 49, 579, 1987.
99. **Albright, L. M., Huala, E., and Ausubel, F. M.,** Prokaryotic signal transduction mediated by sensor and regulatory protein pairs, *Annu. Rev. Genet.,* 23, 311, 1989.
100. **Jin, S., Roitsch, T., Ankenbauer, R. G., Gordon, M. P., and Nester, E. W.,** The VirA protein of *Agrobacterium tumefaciens* is autophosphorylated and is essential for *vir* gene regulation, *J. Bacteriol.,* 172, 525, 1990.
101. **Morel, P., Powell, B. S., and Kado, C. I.,** Mise en evidence de trois domaines fonctionnels responsables d'une activite kinase chez VirA, une proteine sensorielle transmembranaire codee par le plasmide Ti d' *Agrobacterium tumefaciens, C.R. Acad. Sci. Ser. III,* 310, 21, 1990.
102. **Jin, S., Prusti, R. K., Roitsch, T., Ankenbauer, R. G., and Nester, E. W.,** Phosphorylation of the VirG protein of *Agrobacterium tumefaciens* by the autophosphorylated VirA protein: Essential role in the biological activity of VirG, *J. Bacteriol.,* 172, 4945, 1990.
103. **Winans, S. C., Jin, S., Komari, T., Johnson, K. M., and Nester, E. W.,** The role of virulence regulatory loci in determining *Agrobacterium* host range, in *Plant Molecular Biology,* von Wettstein, D. and Chua, N.-H., Eds., Plenum Press, NY, 1987, 573.

104. **Steck, T. R., Morel, P., and Kado, C. I.,** *Vir* box sequences in *Agrobacterium tumefaciens* pTiC58 and A6, *Nucleic Acids Res.*, 16, 8736, 1988.
105. **Yanofsky, M. F., Porter, S. G., Young, C., Albright, L. M., Gordon, M. P., and Nester, E. W.,** The *virD* operon of *Agrobacterium tumefaciens* encodes a site-specific endonuclease, *Cell*, 47, 471, 1986.
106. **Albright, L. M., Yanofsky, M. F., Leroux, B., Ma, D., and Nester, E. W.,** Processing of the T-DNA of *Agrobacterium tumefaciens* generates border nicks and linear, single-stranded T-DNA, *J. Bacteriol.*, 169, 1046, 1987.
107. **Jayaswal, R. K., Veluthambi, K., Gelvin, S. B., and Slightom, J. L.,** Double-stranded cleavage of T-DNA and generation of single-stranded T-DNA molecules in *Escherichia coli* by a *virD*-encoded border-specific endonuclease from *Agrobacterium tumefaciens*, *J. Bacteriol.*, 169, 5035, 1987.
108. **Stachel, S. E., Timmerman, B., and Zambryski, P.,** Activation of *Agrobacterium tumefaciens vir* gene expression generates multiple single-stranded T-strand molecules from the pTiA6 T-region: Requirement for 5′ *virD* gene products, *EMBO J.*, 6, 857, 1987.
109. **Veluthambi, K., Jayaswal, R. K., and Gelvin, S. B.,** Virulence genes A, G, and D mediate the double stranded border cleavage of T-DNA from the *Agrobacterium* Ti plasmid, *Proc. Natl. Acad. Sci. U.S.A.*, 84, 1881, 1987.
110. **Yamamoto, A., Iwahashi, M., Yanofsky, M. F., Nester, E. W., Takebe, I., and Machida, Y.,** The promoter proximal region in the *virD* locus of *Agrobacterium tumefaciens* is necessary for the plant-inducible circularization of T-DNA, *Mol. Gen. Genet.*, 206, 174, 1987.
111. **De Vos, G. and Zambryski, P.,** Expression of *Agrobacterium* nopaline-specific VirD1, VirD2, and VirC1 proteins and their requirement for T-strand production in *E. coli*, *Mol. Plant-Microbe Interact.*, 2, 43, 1989.
112. **Wang, K., Herrera-Estrella, A., and Van Montagu, M.,** Overexpression of *virD1* and *virD2* genes in *Agrobacterium tumefaciens* enhances T-complex formation and plant transformation, *J. Bacteriol.*, 172, 4432, 1990.
113. **Culianez-Macia, F. A. and Hepburn, A. G.,** The kinetics of T-strand production in a nopaline-type helper strain of *Agrobacterium tumefaciens*, *Mol. Plant-Microbe Interact.*, 1, 207, 1988.
114. **Steck, T. R., Close, T. J., and Kado, C. I.,** High levels of double-stranded transferred DNA (T-DNA) processing from an intact nopaline Ti plasmid, *Proc. Natl. Acad. Sci. U.S.A.*, 86, 2133, 1989.
115. **Ghai, J. and Das, A.,** The *virD* operon of *Agrobacterium tumefaciens* Ti plasmid encodes a DNA-relaxing enzyme, *Proc. Natl. Acad. Sci. U.S.A.*, 86, 3109, 1989.
116. **Herrera-Estrella, A., Chen, Z.-M., Van Montagu, M., and Wang, K.,** VirD proteins of *Agrobacterium tumefaciens* are required for the formation of a covalent DNA-protein complex at the 5′ terminus of T-strand molecules, *EMBO J.*, 7, 4055, 1988.
117. **Ward, E. R. and Barnes, W. M.,** VirD2 protein of *Agrobacterium tumefaciens* very tightly linked to the 5′ end of T-strand DNA, *Science*, 242, 927, 1988.
118. **Young, C. and Nester, E. W.,** Association of the VirD2 protein with the 5′ end of T strands in *Agrobacterium tumefaciens*, *J. Bacteriol.*, 170, 3367, 1988.
119. **Durrenberger, F., Crameri, A., Hohn, B., and Koukolikova-Nicola, Z.,** Covalently bound VirD2 protein of *Agrobacterium tumefaciens* protects the T-DNA from exonucleolytic degradation, *Proc. Natl. Acad. Sci. U.S.A.*, 86, 9154, 1989.
120. **Howard, E. A., Winsor, B. A., De Vos, G., and Zambryski, P.,** Activation of the T-DNA transfer process in *Agrobacterium* results in the generation of a T-strand-protein complex: Tight association of VirD2 with the 5′ ends of T-strands, *Proc. Natl. Acad. Sci. U.S.A.*, 86, 4017, 1989.
120a. **Herrera-Estrella, A., Van Montagu, M., and Wang, K.,** A bacterial peptide acting as a plant nuclear targeting signal: The amino-terminal portion of *Agrobacterium* VirD2 protein directs a β-galactosidase fusion protein into tobacco nuclei, *Proc. Natl. Acad. Sci. U.S.A.*, 87, 9534, 1990.
121. **Timmerman, B., Van Montagu, M., and Zambryski, P.,** *vir*-induced recombination in *Agrobacterium*: Physical characterization of precise and imprecise T-circle formation, *J. Mol. Biol.*, 203, 373, 1988.
122. **Stachel, S. E., Timmerman, B., and Zambryski, P.,** Generation of single-stranded T-DNA molecules during the initial stages of T-DNA transfer from *Agrobacterium tumefaciens* to plant cells, *Nature (London)*, 322, 706, 1986.
123. **Veluthambi, K., Ream, W., and Gelvin, S. B.,** Virulence genes, borders, and overdrive generate single stranded T-DNA molecules from the A6 Ti plasmid of *Agrobacterium tumefaciens*, *J. Bacteriol.*, 170, 1523, 1988.
123a. **Steck, T. and Gelvin, S.,** unpublished, 1991.
124. **Christie, P. J., Ward, J. E., Winans, S. C., and Nester, E. W.,** The *Agrobacterium tumefaciens virE2* gene product is a single-stranded DNA-binding protein that associates with T-DNA, *J. Bacteriol.*, 170, 2659, 1988.
125. **Gietl, C., Koukolikova-Nicola, Z., and Hohn, B.,** Mobilization of T-DNA from *Agrobacterium* to plant cells involves a protein that binds single-stranded DNA, *Proc. Natl. Acad. Sci. U.S.A.*, 84, 9006, 1987.

126. **Citovsky, V., De Vos, G., and Zambryski, P.**, Single-stranded DNA binding protein encoded by the *virE* locus of *Agrobacterium tumefaciens, Science,* 240, 501, 1988.
127. **Citovsky, V., Wong, M. L., and Zambryski, P.**, Cooperative interaction of *Agrobacterium* VirE2 protein with single-stranded DNA: Implications for the T-DNA transfer process, *Proc. Natl. Acad. Sci. U.S.A.,* 86, 1193, 1989.
128. **Das, A.**, *Agrobacterium tumefaciens virE* operon encodes a single-stranded DNA-binding protein, *Proc. Natl. Acad. Sci. U.S.A.,* 85, 2909, 1988.
129. **Sen, P., Pazour, G. J., Anderson, D., and Das, A.**, Cooperative binding of *Agrobacterium tumefaciens* VirE2 protein to single-stranded DNA, *J. Bacteriol.,* 171, 2573, 1989.
130. **Howard, E. and Citovsky, V.**, The emerging structure of the *Agrobacterium* T-DNA transfer complex, *BioEssays,* 12, 103, 1990.
131. **Zorreguieta, A. and Ugalde, R. A.**, Formation in *Rhizobium* and *Agrobacterium* spp. of a 235-kilodalton protein intermediate in β-D(1,2)glucan synthesis, *J. Bacteriol.,* 167, 947, 1986.
132. **Zorreguieta, A., Geremia, R. A., Cavaignac, S., Cangelosi, G. A., Nester, E. W., and Ugalde, R. A.**, Identification of the product of an *Agrobacterium tumefaciens* chromosomal virulence gene, *Mol. Plant-Microbe Interact.,* 1, 121, 1988.
133. **Cangelosi, G. A., Martinetti, G., Leigh, J. A., Lee, C. C., Theines, C., and Nester, E. W.**, Role of *Agrobacterium tumefaciens* ChvA protein in export of β-1,2-glucan, *J. Bacteriol.,* 171, 1609, 1989.
134. **Inon de Iannino, N. and Ugalde, R. A.**, Biochemical characterization of avirulent *Agrobacterium tumefaciens chvA* mutants: Synthesis and excretion of β-(1,2)glucan, *J. Bacteriol.,* 171, 2842, 1989.
135. **O'Connell, K. P. and Handelsman, J.**, *chvA* locus may be involved in export of neutral cyclic β-1,2 linked D-glucan from *Agrobacterium tumefaciens, Mol. Plant-Microbe Interact.,* 2, 11, 1989.
136. **Douglas, C. J., Halperin, W., and Nester, E. W.**, *Agrobacterium tumefaciens* mutants affected in attachment to plant cells, *J. Bacteriol.,* 152, 1265, 1982.
137. **Douglas, C. J., Staneloni, R. J., Rubin, R. A., and Nester, E. W.**, Identification and genetic analysis of an *Agrobacterium tumefaciens* chromosomal virulence region, *J. Bacteriol.,* 161, 850, 1985.
138. **Puvanesarajah, V., Schell, F. M., Stacey, G., Douglas, C. J., and Nester, E. W.**, Role for 2-linked β-D-glucan in the virulence of *Agrobacterium tumefaciens, J. Bacteriol.,* 164, 102, 1985.
139. **Cangelosi, G. A., Hung, L., Puvanesarajah, V., Stacey, G., Ozga, D. A., Leigh, J. A., and Nester, E. W.**, Common loci for *Agrobacterium tumefaciens* and *Rhizobium meliloti* exopolysaccharide synthesis and their roles in plant interactions, *J. Bacteriol.,* 169, 2086, 1987.
140. **Cangelosi, G. A., Martinetti, G., and Nester, E. W.**, Osmosensitivity phenotypes of *Agrobacterium tumefaciens* mutants that lack periplasmic β-1,2-glucan, *J. Bacteriol.,* 172, 2172, 1990.
141. **Hawes, M. C. and Pueppke, S. G.**, Variation in binding and virulence of *Agrobacterium tumefaciens* chromosomal virulence *(chv)* mutant bacteria on different plant species, *Plant Physiol.,* 91, 113, 1989.
142. **Marks, J. R., Lynch, T. J., Karlinsey, J. E., and Thomashow, M. F.**, *Agrobacterium tumefaciens* virulence locus *pscA* is related to the *Rhizobium meliloti exoC* locus, *J. Bacteriol.,* 169, 5835, 1987.
143. **Thomashow, M. F., Karlinsey, J. E., Marks, J. R., and Hurlbert, R. E.**, Identification of a new virulence locus in *Agrobacterium tumefaciens* that affects polysaccharide composition and plant cell attachment, *J. Bacteriol.,* 169, 3209, 1987.
144. **Uttaro, A. D., Cangelosi, G. A., Geremia, R. A., Nester, E. W., and Ugalde, R. A.**, Biochemical characterization of avirulent *exoC* mutants of *Agrobacterium tumefaciens, J. Bacteriol.,* 172, 1640, 1990.
145. **Matthysse, A. G.**, Role of bacterial cellulose fibrils in *Agrobacterium tumefaciens* infection, *J. Bacteriol.,* 154, 906, 1983.
146. **Deasey, M. C. and Matthysse, A. G.**, Interactions of wild-type and a cellulose-minus mutant of *Agrobacterium tumefaciens* with tobacco mesophyll and tobacco tissue culture cells, *Phytopathology,* 74, 991, 1984.
147. **Cornish, A., Greenwood, J. A., and Jones, C. W.**, Binding-protein-dependent sugar transport by *Agrobacterium radiobacter* and *A. tumefaciens* grown in continuous culture, *J. Gen. Microbiol.,* 135, 3001, 1989.
148. **Huang, M.-L., Cangelosi, G. A., Halperin, W., and Nester, E. W.**, A chromosomal *Agrobacterium tumefaciens* gene required for effective plant signal transduction, *J. Bacteriol.,* 172, 1814, 1990.
149. **Cangelosi, G. A., Ankenbauer, R. G., and Nester, E. W.**, Sugars induce the *Agrobacterium* virulence genes through a periplasmic binding protein and a transmembrane signal protein, *Proc. Natl. Acad. Sci. U.S.A.,* 87, 6708, 1990.

149a. **Gray, J., Wang, J., and Gelvin, S.**, unpublished, 1991.

150. **Close, T. J., Tait, R. C., and Kado, C. I.**, Regulation of Ti plasmid virulence genes by a chromosomal locus of *Agrobacterium tumefaciens, J. Bacteriol.,* 164, 774, 1985.
151. **Close, T. J., Rogowsky, P. M., Kado, C. I., Winans, S. C., Yanofsky, M. F., and Nester, E. W.**, Dual control of *Agrobacterium tumefaciens* Ti plasmid virulence genes, *J. Bacteriol.,* 169, 5113, 1987.
152. **Chilton, M.-D., Drummond, M. H., Merlo, D. J., Sciaky, D., Montoya, A. L., Gordon, M. P., and Nester, E. W.**, Stable incorporation of plasmid DNA into higher plant cells: The molecular basis of crown gall tumorigenesis, *Cell,* 11, 263, 1977.

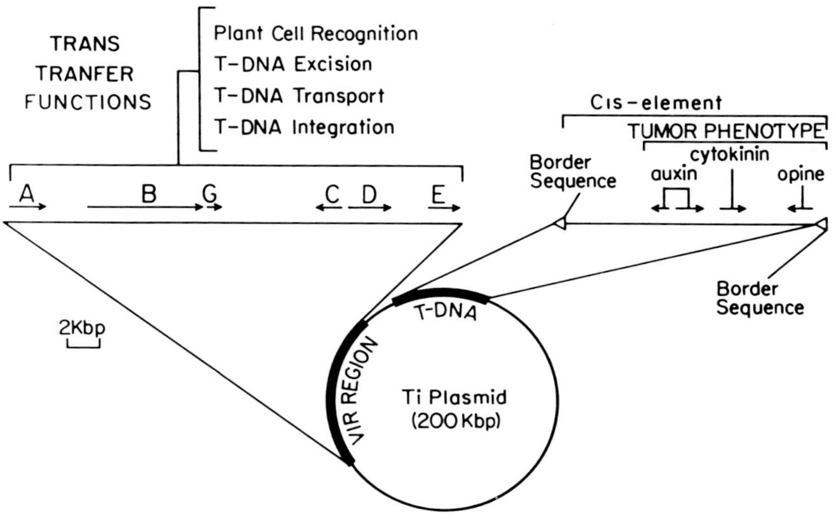


FIGURE 1. *Agrobacterium tumefaciens* Ti plasmid and the genetic components of virulence. The transcriptional and functional organization of the virulence (*vir*) region is taken from Stachel and Nester.[4] The T-DNA region is that of nopaline Ti plasmid, which contains a single T-DNA segment; the polar 25-bp T-DNA border repeats are indicated by the open arrows at the end of the element. From Zambryski, P., in *Mobile DNA*, Berg, D. and Howe, M., Eds., American Society for Microbiology, Washington, DC, 1989, 309. With permission.

positive regulation by the *virA/virG* gene products in response to signal molecules, and (2) negative regulation of two *vir* loci by a chromosomally encoded product, *ros*. The *ros* gene product regulates the divergent promoter region upstream of *virC* and *virD*.[27-29] *ros* mutants constitutively express *virC* and *virD* at 10 to 100-fold higher levels than levels in uninduced wild-type cells; the expression of other *vir* loci is unaffected. *virC* and *virD* are still positively regulated by *virA* and *virG* so that in the presence of phenolics, their expression is further increased. The chromosomal *ros* locus, then, appears to encode a negative regulator of two *vir* loci.

The major impact on *vir* expression is that of the positive regulatory system consisting of VirA and VirG proteins and plant signal molecules.[4,30] The expression of the *vir* loci requires VirA and VirG since mutations in *virA* and *virG* block transcriptional activation at other *vir* loci.[30] *virA* and *virG* are consititutively expressed in free-living bacteria, although in the presence of plant inducer molecules, the transcription of *virG* increases in a *virG*-dependent manner. In contrast, the transcription of *virA* is unaffected or slightly induced in the presence of inducer.[8,30] The *virA/virG* system is a member of a large family of two component positive regulatory systems.[31-33] As illustrated in Figure 2, the sensory component, typically membrane bound, responds to an environmental signal by phosphorylating the regulatory component. The regulatory component then functions as a transcriptional activator to regulate the gene(s) necessary for the cell's response to the stimulus.

## A. SIGNAL MOLECULES AFFECTING THE REGULATION OF *VIR* GENE EXPRESSION

*vir* genes, with the exception of *virA* and *virG,* are not expresssed in free-living bacteria but are transcriptionally activated in the presence of host plant exudates.[30,34,35] Three types of inducers and enhancers of *vir* gene expression from plant exudates have been identified: phenolics, opines, and sugars. The best characterized and most effective inducers are phenolics. Stachel et al.[34] found that monocyclic plant phenolic compounds induce expression

**TABLE 1**
***vir*-Specific Protein Products**[a]

| Locus | Size (kb) | General role | Protein products: Size (kDa)[b] | Protein products: Amount | Protein products: Location | Function | References |
|---|---|---|---|---|---|---|---|
| *virA* | 2.0 | Absolutely essential for virulence, regulatory system (virA, virG) | 92 | + | M | AS sensor; sugar sensor; protein kinase | 4, 30, 41, 78, 100, 105, 106 |
| *virG* | 1.0 | | 30 | ++ | C(?) | *vir* transcriptional regulator | 4, 30, 31, 80, 81, 86-89, |
| *virC* | 2.0 | Attenuated virulence, increased host range (virC, virE) | 26, 23 | + | ? | Enhancement of T-strand production | 4, 40, 133, 203 |
| *virE* | 2.0 | | 7, 60.5 | +++ | C/M(?) | VirE2–single-stranded DNA-binding protein; unfolding the T-strand | 4, 144-150, 202, 204, 206 |
| *virD* | 4.5 | Absolutely essential for virulence, T-DNA transfer machinery (virD, virB) | 16, 47, 21, 75 | + | C(?) | VirD1–topoisomerase; helicase (?); VirD2 -T-DNA border endonuclease; nuclear localization; helicase (?); primase (?); integrase (?) | 4, 123, 127, 129-132, 135, 139, 205 |
| *virB* | 9.5 | | 26, 12, 12, 87, 23, 32, 6, 26, 32, 41, 38 | +/+++ | M—VirB1, B4, B5<br>IM—VirB10<br>C/IM—VirB11 | T-DNA transfer structure (?); membrane channel (?); VirB11-ATPase; protein kinase | 4, 150, 157-163, 167, 168 |

[a] Abbreviations: (AS), acetosyringone; (C); cytoplasm; (IM), inner membrane; (M), membrane.
[b] Protein sizes are listed in order of arrangement relative to their respective promoters.

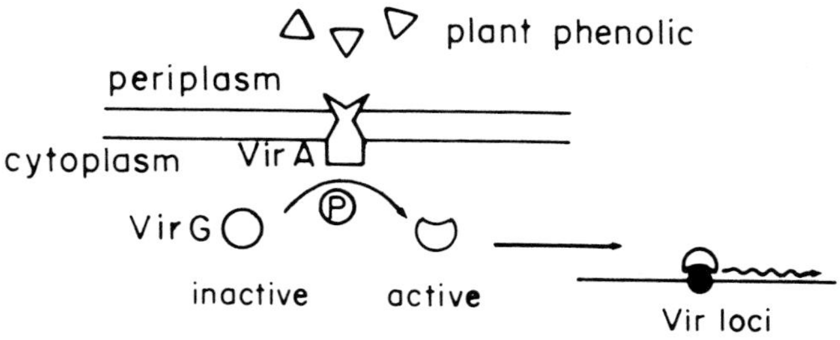


FIGURE 2. Diagram of the VirA/VirG regulatory system.

of several *vir* genes; the most prevalent inducer found in metabolically active wounded tobacco cells was acetosyringone (AS). The range of *vir*-inducing plant phenolics is not limited to acetophenones but includes chalcones and cinnamic acid derivatives.[34,36-38] The molecules responsible for induction are not detected or are detected at low levels in uninjured plants but the amounts are significantly increased in wounded plants.[34] These inducers are related in structure to precursors of flavonoids or lignin, involved in plant defense or cell wall biosynthesis, respectively.

Lignin/flavonoid precursors are ubiquitous in plants, yet different *Agrobacterium* strains show strong host specificity among dicots. Isolates of *Agrobacterium* which utilize a wide range of dicotyledonous plants as hosts (wide host range, WHR) probably respond to compounds common among these susceptible plants. Both WHR and LHR (limited host range) strains are generally sensitive to AS; however, LHR isolates are less sensitive and, consequently, less inducible by AS than are WHR isolates.[37] The lack or low levels of the inducer molecules necessary for infection in different host plants may inhibit the virulence of a particular *Agrobacterium* strain; for example, *Arabidopsis* may have a low level of *vir*-inducing phenolics since its transformation by *Agrobacterium* can be increased by the addition of AS.[39] Host range of WHR and LHR isolates may be affected by the differences in the structure of the VirA protein. On some host plants, VirA encoded by a LHR Ti plasmid could not complement mutations in *virA* encoded by a WHR Ti plasmid, suggesting that LHR VirA may not recognize the same inducer molecules as the WHR VirA.[40] Further, VirA from a LHR strain that was avirulent on tobacco did not induce *vir* expression in response to either AS or cocultivation with tobacco cells.[41] Although VirA from WHR and LHR strains are structurally related overall, the proteins are quite divergent in their amino termini, the region of the protein involved in signal recognition.[41] Recent data also shows that in addition to containing *vir*-inducing compounds, various plants produce compounds that are inhibitory to *vir* induction.[37] Thus, differences in host specificity and transformation efficiency could result from the competition of *vir* inducers and inhibitors for receptors in *Agrobacterium*.[37]

Monocot plants, such as wheat and oats, also produce *vir*-inducing compounds but remain refractory to transformation by *Agrobacterium*.[42] Messens et al.[43] found inducer compounds in exudates of both *Agrobacterium*-susceptible and- nonsusceptible monocotyledenous plants and subsequently isolated and chemically identified the active *vir*-inducing compound from a wheat suspension culture line resistant to infection. Low concentrations of this inducer, ethyl ferulate, were more effective for *vir* induction than were low concentrations of AS, and the wheat culture produced ethyl ferulate in quantities that should cause induction of *vir* genes. Thus, the resistance of some monocot species is not due to the lack of *vir* inducers, but, instead, may result from the different physiological and growth characteristics of monocots and dicots. For example, the wound response of monocots and dicots differs; in monocots, cells around the wound differentiate into lignified or sclerified cells without apparent cell division.[44] Since cell division and DNA synthesis are probably involved in the incorporation of T-DNA into the plant genome, the absence of cell division may prevent successful T-DNA transfer.

Although it is not a general occurrence, monocots can be transformed by *Agrobacterium*. *Agrobacterium* stably transforms *Asparagus officinalis,* a monocot that is easily grown in culture.[45, 46] Also wound exudates from a dicotyledonous host plant allowed *Agrobacterium*-mediated transformation of yam, a monocotyledonous crop plant.[47] Raineri et al.[48] pre-incubated a supervirulent (VirG-overproducing) strain of *Agrobacterium* with AS to initate the induction process and to allow efficient *Agrobacterium*-mediated DNA transfer to rice culture embryos. This rice cultivar had good tissue culture potential as did *Asparagus*. The observation that at least two monocots that are efficiently transformed by *Agrobacterium* are amenable to tissue culture suggests that the specificity of *Agrobacterium* transformation is closely connected with the physiology and growth of the plant.

*Agrobacterium* also can infect gymnosperms. Generally, strains pathogenic on herbaceous plants are considerably less pathogenic on conifers so that transformation of conifiers by typical *Agrobacterium* strains has been very inefficient.[49,50] Recent data, however, show that specific *Agrobacterium* strains transform conifers at high frequency.[51] Morris and Morris[51] characterized the major native phenolic *vir* inducer for these specific *Agrobacterium* strains from Douglas fir. The inducer, coniferin, is a phenylpropanoid glucoside; two other inducers have been detected at lower concentrations and are also glucosides. Morris and Morris[51] hypothesize that the functional inducer is actually coniferyl alcohol and that the strains effective on conifers produce a glucosidase that converts coniferin to coniferyl alcohol. In fact, the strains most tumorigenic on Douglas fir contained high levels of glucosidase activity. AS and coniferyl alcohol induced *vir* expression at equal levels in all strains tested but coniferin induced *vir* expression effectively only in strains with active glucosidases.[51] These data suggest *Agrobacterium* can convert a normal plant metabolite, a glucoside, to a signal that initiates *vir* induction, implying that processing enzymes such a β-glucosidase may play a vital role in the pathogenic potential of the bacterium.

Opines stimulate AS induction of *vir* genes but opines alone have no major effect on induction.[52] Opine catabolism is not required for stimulation, since opines that could not be metabolized by the bacterium also stimulated AS-mediated induction. Veluthambi et al.[52] found that the rapid decrease in production of T-DNA transfer intermediates (see below) following removal of AS from the incubation media can be circumvented by addition of certain opines in the incubation media. Low concentrations ($<0.2$ m*M*) of opines stimulate the expression of a *vir* gene (*pinF* = *virH*) while high ($>0.2$ m*M*) concentrations of opines repress expression.[53] Measurements of opines within and excreted by mature crown gall tumors exceed the opine concentrations that appear to repress *vir* expression.[54,55] Previous research has shown that conditions which promote *vir* induction are poor for bacterial growth and conditions which support vegetative bacterial growth are unsuitable for *vir* induction.[56,57] Krishnan and Gelvin propose that, in the plant-wound environment containing low opine concentrations, *Agrobacterium* remains in a state capable of *vir* induction and resultant T-DNA transfer while high opine concentrations following tumor formation restore *Agrobacterium* to vegetative growth and allow the opine to be used as a carbon and nitrogen source.[53] Opines produced in initial stages of transformation also may signal the presence of a susceptible host to other *agrobacteria*.

Recent results suggest some sugars directly enhance the plant phenolic-mediated induction of *vir* gene expression.[58,59] Since nonmetabolized sugars also enhanced *vir* induction, this enhancement does not result from improvement of bacterial growth. Monosaccharides, such as glucose and galactose, significantly enhance *vir* expression only when AS is limited or absent.[58,59] The decreased enhancement by sugars when high levels of AS are present suggests *vir* expression is a saturable process.

The synergistic induction by sugars and phenolics occurs by a regulatory pathway that includes *virA* and *chvE* and may be distinct from the *virA*/*virG* regulatory pathway involved in AS-mediated induction.[58] Huang et al. found that ChvE, a chromosomally encoded protein, is required for wild-type levels of *vir* induction by AS.[9] ChvE also is required for *Agrobacterium* chemotaxis toward sugars and *vir* induction in response to sugars.[58] Sequence data indicates *chvE* is homologous to the periplasmic ribose-binding protein of *E. coli* and galactose/glucose-binding proteins of *E. coli* and *Agrobacterium radiobacter*.[58] The Ti plasmid-encoded component of sugar induction, VirA, consists of a cytoplasmic domain and two transmembrane segments separated by a periplasmic domain (see also below). This VirA structure resembles the methyl-accepting signal protein (Trg) of *E. coli* that functions in both chemotaxis and sugar uptake.[60, 61] The periplasmic region of VirA contains a 15-amino acid region that is strongly homologous to the region in the Trg protein which interacts with either galactose-binding protein or ribose-binding protein and which is implicated in the chemotactic response to sugar. Recent data, however, indicate VirA is not required for *Agrobacterium* chemotaxis toward sugars.[58] Thus, the Trg homologous amino acid domain may perform a different, as yet unknown, sugar-signal response function in VirA. *Agrobacterium* mutants lacking *chvE* or the periplasmic domain of VirA (which is not required for AS-mediated induction) do not exhibit enhanced *vir* expression by sugars and have limited host range for infection, suggesting *Agrobacterium* infection may require sugar-enhanced AS induction for efficient transformation in some plants.[58,59] Interestingly, in the absence of sugars, *vir* induction in mutants lacking the periplasmic domain of VirA was higher than induction of *vir* genes in wild-type strains. Although not required for signaling, this periplasmic domain of VirA may regulate the activity of the protein in response to signal molecules.

Almost all sugars effective in *vir* induction are presursors of the major component of higher plant cell wall polysaccharides.[62] The process of wounding can release active sugars from the cell walls; for example, wound exudates from some plants have been found to contain glucose, one such active sugar.[59] *vir* induction is increased when substances extracted from wheat or tobacco seedlings are combined with either partially purified monocot factors or low AS concentrations; circumstantial evidence suggests these inducers are sugars.[59] *vir* genes may be sufficiently activated to produce T-DNA transfer by combining weak phenolic inducers, such as syringaldehyde and acetovanillone, with specific sugars; Shimoda et al. have found that weak inducers plus monosaccharides can produce levels of *vir* expression that correspond to levels produced by AS.[59] In addition to other breakdown/degradation products of cell walls such as phenolics, sugars may signal the presence of susceptible wounded plants to the bacterium and, by synergistic action with the phenolic inducers, determine the level of *vir* induction. Plants which allow tumor formation by *chvE* mutant *Agrobacterium* strains may have either high levels of or very active phenolic inducers to overcome the lack of sugar induction; alternatively, the plants may be susceptible to *Agrobacterium* following low levels of *vir* activation.

In addition to their role in *vir* induction, some inducer compounds appear to act as chemotatic agents and recruit *Agrobacterium* to wounded plant tissue.[63-66] The bacterium is attracted to susceptible wounded plant cells by following a concentration gradient of the inducer compounds. Among different *Agrobacterium* strains, the basis of chemotaxis varies. For example, a highly motile strain is attracted to phenolic compounds (AS) by a Ti plasmid-derived function which requires both *virA* and *virG*.[63, 65] Some strains lacking a Ti plasmid also are attracted toward phenolic compounds,[64] indicating a chromosomally encoded function for chemotaxis in these strains. Ti plasmid-independent chemotaxis has been observed using root exudates,[67] or plant cell homogenates.[65] In addition to phenolic compounds, chemoattractants in plant extracts also may be sugar molecules.[68] Thus, Ti-independent chemotaxis toward plant extracts, in part, may be a function of ChvE activity. The presence of both Ti plasmid-dependent and -independent chemotaxis may play a role in initiation and maintenance of *Agrobacterium* pathogenicity. Via Ti plasmid-dependent chemotaxis, strains capable of T-DNA transfer are attracted to the phenolic compounds released by a wounded plant; the chemotactic response directs the bacterium to a susceptible host cell, resulting in activation of *vir* genes and subsequent T-DNA transfer. Strains displaying Ti plasmid-independent chemotactic behavior move toward the wounded region also, and either utilize the opines produced by the tumor or participate in conjugative transfer of the Ti plasmid. *Agrobacterium* strains that are chemotactic toward AS do not need to carry oncogenic plasmids; such strains only need nononcogenic plasmids carrying loci of opine catabolism and opine-inducible conjugal transfer.[69,70] The conservation of chemotaxis toward inducer molecules in both strains with and without Ti plasmids ensures survival of both the bacterium and the Ti plasmid.

Although *Agrobacterium* is chemoattracted to inducer molecules and uses inducer molecules as signals for transcriptional activation of *vir* loci, at high concentrations most of the *vir*-inducing phenolics are bacteriostatic and, therefore, possibly harmful to the bacterium.[37] A plant-inducible locus in the virulence region, termed *virH* or *pinF,* may be involved in detoxification of harmful substances found in the environment of the wounded plant.[5] Although the *virH* locus generally is not essential for virulence, *virH* mutants show attenuated pathogenicity on specific dicots. In the presence of decreased numbers of bacteria, this attenuated virulence is more pronounced. Like other *vir* loci, *virH* is induced by plant cells and plant phenolics in a *virA*/*virG*-dependent manner. Two of the four open reading frames (ORFs) in *virH* share sequence homology with each other and with cytochrome P-450 enzymes,[71] in which a heme moiety is associated with the carboxy region of the protein. Both *virH* ORFs contain the conserved heme-binding cysteine and the invariant phenylalanine and glycine residues associated with this heme-binding site. The *virH* locus, then, appears to encode two inducible P-450 enzymes. P-450 enzymes catalyze the NADH-dependent addition of oxygen to different substrates, such as aromatic hydrocarbon, steroids, and camphor. The *virH*-encoded enzymes may be involved in metabolism of plant compounds during the infection process; however, VirH does not appear to be involved in AS catabolism. *virH* enzymes may enable *Agrobacterium* to survive in the presence of bacteriocidal or bacteriostatic plant compounds, such as high concentrations of phenolics released by wounded plants. For example, extracts of the plant fungal pathogen *Nectria haematococca* contain P-450 activity which is thought to be involved in demethylation of pisatin, the major pea phytoalexin, to a nontoxic substance.[72] Isolates of the fungus lacking demethylase activity have attenuated pathogenicity while isolates with the inducible demethylase are pathogenic.[73,74] *virH* mutants show similar attenuated virulence, possibly due to an inability to detoxify the protective compounds released by the wounded plant.

In summary, the range of substances that affect *vir* gene expression indicates the *vir* induction process is complex. Potentially new positive and/or negative signal molecules will be discovered as research in this area continues. Also, it is clear that this complexity reflects a sensitivity that makes biological sense. *Agrobacterium* infects a wide array of plant species. It is expected that the exact nature as well as relative amounts of specific wound products will vary among different plants. Thus, *Agrobacterium* has evolved to recognize and take advantage of different plant compounds. The discovery that *Agrobacterium* can recognize and utilize plant compounds to its advantage has lead to the concept of plant-bacterial communication. For example, it was subsequently shown that *Rhizobium* also utilizes plant compounds to initiate root nodule formation.[75-77] Besides plant molecules initiating *vir* induction per se, there are also plant defense molecules that *Agrobacterium* somehow has evolved to either ignore or metabolize. This latter idea has only recently come to our attention with the finding that *virH* might be involved in detoxifying plant substances. No doubt further investigation of the chemical environment during the initial stages of the *Agrobacterium*-plant interaction will yield additional features that are important for the interaction of *Agrobacterium* as well as other bacteria with plant cells.

## B. MOLECULAR MECHANISM OF *VIR* GENE INDUCTION

The above discussion describes plant specific products that affect *vir* gene expression. In addition, there are more subtle environmental factors, such as pH, temperature, and inorganic phosphate levels, that affect *vir* gene expression via the VirA/VirG system.[8,35,78,79] These latter effectors have been studied in relation to the actual transcription of *vir*-specific genes. As already mentioned, *vir* transcription requires the VirA and VirG products which are homologous to other two-component regulatory systems that sense and respond to changes in the bacterial environment. VirA is the sensor that recognizes the plant signal(s), and phosphorylates the regulatory component, VirG. VirG then acts as a transcriptional activator of the *vir* genes.

The *virG* product, a 27-kDa protein, functions as the regulator component of the VirA/VirG system.[30] The sequence of the *virG* gene product shares amino acid homology with regulator counterparts such as *ompR* and *ntrC* as well as many other regulator proteins.[31,33,80,81] *virG* in octopine strains of *Agrobacterium* is transcribed from two tandem promoters, 50 bp apart,[30] that respond to three stimuli: (1) plant phenolic compounds, (2) inorganic phosphate starvation, and (3) acidic media.[82] The shorter transcript made from the proximal promoter (P2) is produced in free-living bacteria; the longer transcript made by the distal promoter (P1) is produced during the induction process.[30]

The distal promoter (P1) is preceded by three copies of the *vir* box, a conserved 12-bp sequence (TNCAATTGAAAPy) found in the promoter regions of inducible *vir* genes.[82-85] The *vir* box is a *cis*-acting regulatory sequence that serves as the VirG protein-binding site.[86] Deletion studies show that removal of the *vir* box abolishes induction of *vir* genes.[86,87] Both the presence and relative position of the *vir* box appears important in transcriptional activation. All highly inducible *vir* genes (*virB, virC, virD, virE, and virH*) have at least one box at the same position relative to the inducible transcript start, but one *vir* box is sufficient for activation.[86] The low levels of induction of *virA* and *virG* transcription may be due to improper positioning of the *vir* box. VirG protein, produced in *E.coli,* binds preferentially to the *vir* box and mutational analysis shows that the DNA-binding activity of VirG is in the carboxy-terminal region of the protein.[88] The specificity of binding *in vitro* does not differ between phosphorylated and unphosphorylated forms of VirG.[86,87]

The albeit incomplete dyad symmetry of the *vir* box suggests VirG binds as a dimer.[89] Each *vir* box is situated in nearly opposite phase relative to the DNA helix in the −35 and −10 region of most *vir* promoters so that binding of VirG to the *vir* box sequence should not interfere with DNA-RNA polymerase interactions.[89] Aoyama and Oka [90] propose that VirG and RNA polymerase could bind at the same time to a *vir* promoter without steric hindrance; following the initial binding of VirG, additional VirG molecules may bind cooperatively at other *vir* boxes, stimulating the binding of RNA polymerase on the other side of the helix and leading to activation of weak promoters. Data show that in promoters with multiple *vir* boxes, the binding affinity of VirG for the *vir* box varies; of the three *vir* boxes preceding a *virG* promoter (P1), VirG binds *vir* box I and III but not II.[82, 87]

The best studied *vir* promoter is that preceding *virG* itself; transcription of *virG* is initiated from two promoters as mentioned above. Transcripts from the distal promoter P1 of *virG* are produced in response to AS induction and/or phosphate starvation. The *vir* boxes I and III, the sequences bound by VirG protein, are required for AS induction.[82] A sequence overlapping *vir* box III is required for phosphate starvation induction;[82] this sequence is similar to and in the same relative position from the *virG* promoter as the *pho* box found in *E.coli* promoters inducible by phosphate starvation.[91] The proximal *virG* promoter (P2) responds to acidic media. This promoter (P2) lacks similarity to *E.coli* vegetative promoters but shows strong similarity to heat shock-inducible promoters.[82,92] The heat shock promoters actually respond to a wide variety of environmental stresses, a characteristic shared by the *virG* P2 promoter.[82,93] Both promoters (P1 and P2) have a role in the induction process since

```
-200          vir box I            vir box II                      vir box III
AAGCGT[TTCACTTGTAAC][AACGATTGAGAA]TTTTTGTCA[TAAAATTGAAAT]ACTT
                                         [          "pho box"          ]
                                          CTGTCA  TAAAACTGTCAT

-125  Promoter 1                                                  Promoter 2
GGTTCGCATTTTTGTCATCCGCGGTCAGCCGCAATTCTGACGAACTGCCCATTTAGCTGGA
                          [  heat shock pr  ]          [heat shock pr]
                            -35 homology                -10 homology
                          TNTCNCCCTTGAA    13-15 bp    CCCCATTTA

-64          +1
GATGATTG....TTG
             MET
```

FIGURE 3. Sequence of the *virG* promoter region (beginning at −200 bp from the translational start). The positions of the two transcriptional start sites[30] termed Promoter 1 (P1) and Promoter 2 (P2) are underlined and the three proposed *vir* box sequences upstream of Promoter 1 are bracketed in the *virG* sequence. Consensus sequences for the *E. coli pho* box and heat shock promoter and the proposed regions of homology between the *virG* promoter and these sequences are shown below the *virG* sequence. From Winans, S. C., *J. Bacteriol.*, 172, 2433, 1990. With permission.

mutations in either promoter attenuate *vir* expression. Figure 3 shows the two *virG* promoters and the regions involved in response to the three environmental stimuli examined.

The regulation of *virG* by three different systems (VirA/VirG, phosphate starvation, heat shock/environmental stress) may allow the facultatively pathogenic *Agrobacterium* to ensure its survival under a variety of environmental conditions.[82] *Agrobacterium* may respond to environmental stresses by initiating the infection process, thus increasing the amount of VirG in the cell and, subsequently, enhancing the speed of T-DNA transfer when a favorable environment is encountered. For example, a wounded plant is located in an environment low in inorganic phosphate. Since transcripts of *virG* are initiated from P1 by either AS induction or phosphate starvation, the lack of inorganic phosphate may have already resulted in production of *virG* transcripts in *Agrobacterium* in this environment prior to release of plant signal molecules, thereby increasing *virG* RNA levels in the cell and leading to faster production of VirG protein. The crown gall tumor that results after infection might then provide inorganic phosphate to allow the survival of the bacteria in this environment where inorganic phosphate normally is scarce.[82] Further, *virG* P2, a heat shocklike promoter, may activate transcription in response to toxic compounds released by wounded plants that are not recognized by VirA. A chromosomal virulence gene, *chvD,* is also required for induction of *virG* in response to environmental stress, indicating other chromosomal loci may function in the *vir* induction process.[8]

The effects of increasing the VirG pool can be observed in some *Agrobacterium* strains which express *virG* at levels higher than those seen in wild-type strains. The elevated *virG* activity causes increased expression at other *vir* loci, resulting in a supervirulent phenotype.[94] The tumors induced by supervirulent strains are larger than tumors produced by strains with the typical WHR Ti plasmid;[94-96] supervirulent strains also induce tumor formation on normally recalcitrant hosts. The transfer of either the entire *vir* region or the specific supervirulent *virG* locus into strains carrying a wild-type Ti plasmid can extend the host range and increase the tumorigenic potential of the recipient strain.[94,96]

Insertional and deoxynucleotide-directed mutagenesis and protein sequencing show that VirG translation is initiated at a UUG codon in both nopaline and octopine strains.[87,97] Translation at the UUG codon is inefficient in *E. coli* and presumably in *Agrobacterium.*[98] *Agrobacterium* may use this mechanism to control the pool of VirG. Since *virG* can be transcriptionally activated without AS-mediated induction, this translational control may help prevent promiscuous activation of *vir* loci caused by transient or minor changes in the environment unrelated to compounds released by wounded plants. Thus, the control of VirG production by use of an inefficient initiator codon may ensure that higher amounts of VirG only are produced when sufficient mRNA has accumulated. In other words, VirG is produced only when *virG* transcription is strongly activated, as in the presence of plant signal compounds.

The transcriptional activation of other *vir* loci by VirG is increased by VirA and AS, indicating that modification of VirG by VirA is necessary for full activity.[87,99,100] As in other characterized two-component systems, VirG, the regulator component, is phosphorylated by VirA, the sensor component.[99,100] Phosphorylation is required for *in vivo* activation of *vir* gene expression.[100] The phosphate appears to be directly transferred from VirA to the amino-terminal region of VirG, a region conserved among regulator components.[99] Both biochemical and genetic methods indicate VirG is phosphorylated at the aspartate-52 residue.[99] When this aspartate residue was changed to an asparagine residue, which cannot be phosphorylated, VirG became biologically nonfunctional. Other regulator homologs such as CheY and NtrC, which share both sequence homology and phosphoreceptor function, contain this highly conserved aspartate residue.[101,102] A recent mutational analysis of VirG has shown that an additional conserved aspartic acid residue (Asp-8) in the amino terminal domain is important for VirA-mediated phosphorylation.[88] Unlike other characterized regulator components, phosphorylated VirG (VirG-P) is very stable.[99] The stability of VirG-P in *Agrobacterium* systems may be necessary for continued transcriptional activation of *vir* loci since the overall process of bacterial gene expression probably slows down in the poor growth conditions necessary for maximum *vir* induction.

The actual function of the phosphorylation of VirG has not been determined. After VirG binds to the promoter, presumably through the carboxy-terminal region of the protein, phosphorylation at the amino-terminal region may allow interactions with other components such as RNA polymerase. Although VirG can bind *vir* promoters preferentially without phosphorylation, modification may increase the DNA binding affinity without changing specificity as seen in *E. coli* with OmpR and NtrC.[103,104] Mutations in the amino-terminal regions of VirG which affected phosphorylation did not alter DNA binding activity of the carboxy-terminal region, indicating VirG has two independent functional domains.[88] VirG-P also may act as a phosphoryl transferase to a component of the transcription machinery to allow specific activation of *vir* genes. Phosphorylatable yet biologically inactive mutated VirG proteins have been identified, suggesting that other regions of VirG are required for activity.[88]

The *virA* product, a 92-kDa protein, functions as the sensor component of the VirA/VirG two-component regulatory system. Like some other sensor proteins, VirA is membrane-bound; the protein has two proposed transmembrane domains resulting in the localization of approximately 275 amino-terminal amino acids in the periplasmic space and the remainder of the protein, 554 amino acids, in the cytoplasm;[41,78,105] however, recent data suggest VirA inserts in the membrane only at the proposed second transmembrane domain.[105] The proposed second transmembrane segment is involved in inducer response, regions in the periplasmic domain are involved in responses to acidic media and to sugar enhancement, and the autophosphorylase/phosphotransfer function is located in the cytoplasmic domain.[58,78,100] VirA proteins from strains that infect different host plants are most divergent in the amino-terminal portion of the protein, presumably due to recognition of and/or interaction with different signal compounds.[41] The amino acid sequence of the VirA protein shows homology in the carboxy terminus to several bacterial sensor proteins which have autophosphorylase activity: EnvZ, PhoR, NtrB, and CpxA.[32,33] Like these sensor proteins, VirA, produced *in vitro,* functions as an autophosphorylase and as a phosphotransferase to VirG substrate.[99,100,106] The autophosphorylase activity is correlated with the biological activity of VirA in pathogenicity. Data strongly suggest the histidine-474 residue is the site of phosphorylation in

VirA; when histidine-474 was changed to glutamine, an amino acid structurally similar to histidine but unable to be phosphorylated, VirA lost autophosphorylase activity and could not induce *vir* gene expression or tumor formation.[100] The high-energy bond of phosphohistidine acts mainly as an intermediate in phosphate group transfer in both prokaryotes and eukaryotes,[107] and should allow efficient transfer of the phosphate from VirA to VirG. It is not yet known if plant molecules such as AS actually signal VirA autophosphorylase activity or transfer of the phosphate from preexisting VirA-P to VirG or another intermediate protein.

In summary, the VirA/VirG regulatory system in *Agrobacterium* shares both sequence and functional homology with other bacterial two-component positive regulatory systems. While using a common mechanism to sense changes in the environment and transmit the information to its transcriptional apparatus, *Agrobacterium* may have made unique modifications in this signal response system since the bacterium responds to compounds produced by wounded and susceptible plant cells. The observation that VirG-P is more stable than other phosphorylated regulator components suggests the molecular mechanism may be somewhat different in *Agrobacterium* than in other bacteria. The actual molecular mechanism of VirG activation by protein phosphorylation as well as interactions of VirG and potentially VirA with other proteins involved in initiation of *vir* transcription, however, has yet to be determined; the inactivation or potential negative regulation of the VirA/VirG regulatory system following infection of plants also has to be addressed and may uncover control mechanisms not found in other prokaryotic two-component systems. Although the molecular mechanisms are not completely understood, this signal response pathway of VirA and VirG is the first step in the production of gene products that will ultimately result in the transfer of a DNA segment from *Agrobacterium* to the plant nuclear genome.

## IV. GENERATION OF A TRANSFERABLE T-DNA COPY

### A. *CIS*-ACTING T-DNA TRANSFER SEQUENCES

Once *vir* genes are expressed, molecular reactions occur within *Agrobacterium* to mediate the production of a T-DNA copy capable of genetically transforming plant cells. The T-DNA is essentially the *cis* element that is acted on in *trans* by the *vir* products. T-DNA elements will be considered for the two most commonly studied types of Ti plasmids, nopaline and octopine. Transformation by *Agrobacterium* carrying either nopaline- or octopine-type Ti plasmids results in tumors that produce small sugar amino acid derivatives, nopaline or octopine, respectively. The T-DNA in the nopaline Ti plasmid is a contiguous stretch of about 22 kbp. The T-DNA in the octopine Ti plasmid is composed of three adjacent T-DNAs, a left T-DNA (TL) element of 13 kbp, a central T-DNA (TC) element of 1.5 kbp, and a right T-DNA (TR) element of 7.8 kbp. The borders of these T-DNAs are defined as those sequences which delimit the transferred segments. Studies which compared the sequences at the ends of the T-DNA in the Ti plasmid with T-DNA integrated in the plant DNA showed that, in all cases, the homology ends within or close to conserved 25-bp sequences which flank each type of T-DNA as direct repeats.[108-115]

Genetic studies using deletion mutants have shown that the right border is absolutely required for *Agrobacterium* pathogenicity, whereas deletions of the left border have little effect.[116-119] When restriction fragments overlapping the right border or clones carrying a synthetic 25-bp repeat were reintroduced at the right border deletion site, pathogenicity was restored.[118, 119] However, when either the right border region or the synthesized 25-bp repeat was reintroduced in reverse orientation to that found in the native T-DNA, tumorigenecity was greatly attentuated.[118,119] These results suggest that transfer of the T-DNA is polar from right to left, determined by the orientation of the 25 bp T-DNA border repeats. Thus, with T-DNA borders in their native orientation, leftward transfer initiating at the right border would result in transfer of the adjacent tumor-inducing genes. If the right border was reversed, the tumor-inducing genes would be the last DNA of the 200 kbp Ti plasmid to be transferred, resulting in low transformation efficiencies. In support of this model, mini plasmids, containing only the T-DNA region on small replicons in *trans* to the *vir* region, are pathogenic regardless of the right border orientation.[120] Such small plasmids containing the right border in a reverse orientation are also expected to transfer away from the T-DNA region; however, due to their small size, transfer could proceed around the entire plasmid to the tumor-inducing genes without adverse effect on transformation efficiencies.

In addition to an absolute requirement for the 25-bp repeat , sequences adjacent to T-DNA borders can either enhance or suppress T-DNA transfer. On the octopine Ti plasmid, an enhancer, designated *overdrive,* consists of a 24-bp sequence located to the right and within 60 bp of the right copies of the native 25-bp repeats of the TL and TR elements.[121] The *overdrive* sequence enhances T-DNA transfer even when it is placed up to 6 kbp from the borders in either orientation and either up- or downstream.[122] No sequences with obvious homology to *overdrive* have been identified near the nopaline T-DNA (although sequence data does not extend as far from the right border of the nopaline Ti plasmid as the octopine *overdrive* is observed to act, i.e., 6 kbp).[123] Sequence context has also been demonstrated to influence transfer of T-DNA from the nopaline Ti plasmid.[123] Clones containing synthesized 25-bp repeats from the right or left border are equally active in promoting T-DNA transfer (30% of wild type). Restriction fragments containing several kbp, however, of the native right or left border regions are either very (95% of wild type) or less (10 to 40% of wild type) active, respectively. These results suggest that the native right border contains sequences in addition to the 25-bp direct repeat which promote T-DNA transfer, whereas the native left border contains sequences which may have a negative effect on T-DNA transfer. The latter finding makes sense in that it may provide a mechanism for inhibition of nonproductive DNA transfer leftward from the T-DNA element.

In both the nopaline and the octopine Ti plasmids, the T-DNA itself (i.e., that DNA which is internal to the 25-bp repeats) has no effect on the efficiency of transfer. Indeed, nononcogenic Ti plasmids, with most of the internal sequences of the T-DNA replaced by DNA of interest, are widely used as vectors for genetic transformation of plants.[112,124,125]

## B. PRODUCTION OF T-STRANDS

The ability to induce *vir* gene expression with a purified factor, AS (i.e., in the absence of plant cells) facilitated the identification of novel T-DNA homologous molecules in *Agrobacterium* that could act as intermediates in the transfer process. The most abundant free T-DNA homologous molecule produced in *vir*-induced cells is a linear single-stranded (ss) copy of the T-DNA region, designated the T-strand.[126,127] The T-strand is produced at about 0.5 to 1 copy per induced *Agrobacterium,* and is homologous to the bottom strand of the T-DNA.[126] The polarity of the T-strand suggests it is generated in a 5′ to 3′ direction, initiating at the right T-DNA border and terminating at the left border; this property exactly fits with the genetic data summarized above, that T-DNA transfer is polar from right to left. *vir* induction also results in the detection of nicks at the T-DNA borders on the Ti plasmid.[126] These nicks have been mapped to between the 3rd and 4th base pair of the bottom strand of the 25-bp T-DNA border repeats.[128,129] Border nicks are presumed to represent initiation and termination sites of T-strand synthesis.

A single T-strand is produced from nopaline-type Ti plasmids carrying only two T-DNA border repeats. A more complex array of T-strands is produced from the four border octopine-type Ti plasmid. The four borders bracketing the three T-DNAs TL,TC, and TR, produce six distinct T-strands.[127,130] These linear ssDNAs correspond exactly to the molecules expected to arise if all border-to-border combinations are used for T-strand production. Thus, left and right borders can be used to both initiate and terminate T-strand synthesis. Also, borders can be skipped to produce composite molecules (TR-TC-TL, TR-TC, and TC-TL

T-strands). Further, each of the four borders are nicked independently of each other; thus, the T-strand made in a particular *Agrobacterium* cell is a function of which borders are cleaved on its Ti plasmid. T-strand synthesis likely must initiate at a border nick and proceeds left to right until it encounters a second nick; skipped borders of composite T-strands would be uncleaved on the parent Ti plasmid.

Genetic experiments have shown that only VirA,VirG, VirD1, and the amino-terminal half of VirD2 are required for either nicking or T-strand formation in *vir*-induced *Agrobacterium*.[127,131] As VirA and VirG are both required for overall *vir* gene expression, it is likely that VirD1 and VirD2 are the only *vir* proteins specifically required for T-strand generation. This conclusion has been further strengthened by the observation that only VirD1 and VirD2 are required in an *E. coli* heterologous system synthesizing the T-strand.[132,133] For example, in the heterologous system, *E. coli* cells contain a plasmid carrying a T-DNA region (i.e., DNA flanked by two borders) and a plasmid present in *trans* which contains the *virD1* and *virD2* genes under the control of an inducible promoter.[133] While VirD1 and VirD2 are the only absolute requirements for T-strand production, the heterologous system was useful to point out that the VirC1 product can enhance T-strand production when VirD1 and VirD2 are limiting (for example, when their production is under the control of a weak promoter).[133] It is interesting that so few *vir* (or *Agrobacterium*) products are necessary for T-strand production; this implies that other enzymes of DNA repair and metabolism which are expected to be required (such as DNA polymerase) must be constitititutively expressed products that are present, and compatible with T-strand synthesis, in either *E. coli* or *Agrobacterium*.

To better understand the molecular reactions involved in the generation of T-strands requires a reconstruction of T-strand production *in vitro*. Experiments to date have focused on analyzing VirD1 and VirD2. Since nicking is a prerequisite to T-strand production, it has been assumed that VirD1 and VirD2 are required for T-DNA border-specific nicking. Indeed, we have shown that purified VirD2 alone can act as a site-specific endonuclease of a substrate plasmid containing a 25-bp repeat.[134] *In vitro* experiments with a protein extract enriched for VirD1 suggest that VirD1 is a topoisomerase.[135] The topoisomerase activity of VirD1 is not specific for the 25-bp repeat, i.e., any supercoiled plasmid can act as a substrate for VirD1 topoisomerase. In agreement with these observations, the site-specific endonuclease activity of VirD2 is enhanced by either commercial wheat germ topoisomerase I or VirD1.[134] Taken together, these observations have led to the model shown in Figure 4 for the generation of T-strands. Future experiments should identify additional enzymic requirements for T-strand production.

## C. POSSIBLE STRUCTURE OF THE T-DNA TRANSFER INTERMEDIATE, THE T-COMPLEX

It is unlikely that the T-strand is transferred to the plant cell as a naked DNA molecule. The ss nature of the T-strand would render it particularly susceptible to endo- and exonucleolytic attack. The T-strand also needs a mechanism to target to and subsequently to traverse the bacterial and plant cell wall/cell membranes, and finally to localize to the plant cell nucleus and integrate into a plant chromosome. Some of these functions are provided by proteins bound to the T-strand to form a DNA-protein complex, designated the T-complex. Experimental evidence that the T-strand is indeed associated with protein within the *Agrobacterium* cell was provided by the observation that T-strands present in DNA prepared from *vir*-induced *Agrobacterium* cell extracts in the absence of pronase partition to the phenol phase. These studies also demonstrated that the VirD2 protein is bound tightly at or near the 5′ end of T-strands.[136-139] *In vitro* experiments using purified VirD2 confirmed that the VirD2 remains attached during phenol partitioning, indicative of a tight or possibly covalent linkage to the T-strand.[134] While both VirD1 and VirD2 are involved in border specific nicking leading to T-strand generation, there is no experimental evidence for VirD1 remaining stably attached to the T-strand product.

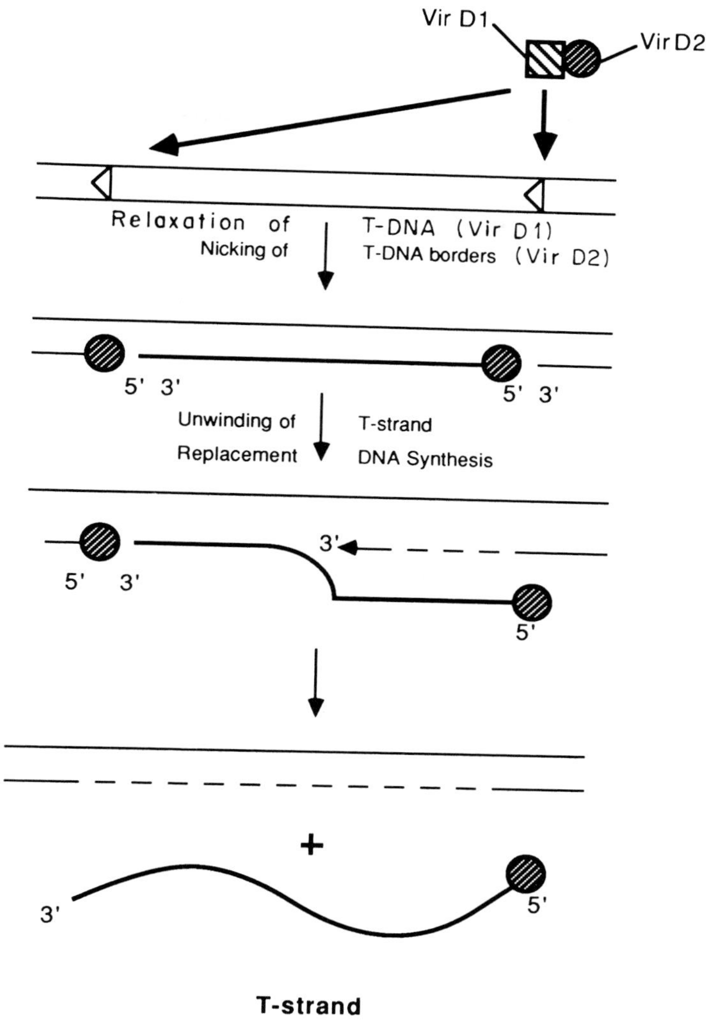


FIGURE 4. Replacement strand synthesis model for generation of T-strand from Ti plasmid. Modified from Howard, E. and Citovsky, V., *BioEssays*, 12, 103, 1990. With permission.

The attachment of VirD2 to the T-strand could provide several useful functions. For example, VirD2 may act as a helicase to facilitate unwinding of the T-strand from the T-DNA region on the Ti plasmid. Or, the VirD2 protein may act as a pilot protein to target the T-strand through various membranes during its transit from *Agrobacterium* to the plant cell. In fact, recent data indicate that the VirD2 protein may be involved in transferring the T-strand into the plant cell nucleus.[140] It had been previously noted that the VirD2 protein contains several stretches of amino acid sequences that have high homology to amino acid sequences known to act as nuclear localization signals in animal cells;[141,142] these sequences are usually basic containing predominantly arginine and lysine residues.[143] When constructs carrying *virD2* sequences fused in frame to a reporter gene are used to transfect plant cells, the reporter gene product can be detected in the plant cell nucleus.[140] The VirD2 nuclear localization signals are located at the carboxy terminus of the protein. These results provide a function for the carboxy terminus of the protein; as noted above, only the amino-terminus is required for site-specific nicking at the T-DNA borders,[127,131] while both amino- and carboxy-termini are required for tumorigenicity.[142] It will be interesting to determine whether VirD2 contains additional signal sequences to target, for example, to the bacterial or plant cell membranes.

Since the T-strand is a linear ssDNA, high levels of a (*vir*) specific ssDNA-binding protein (SSB) would be expected to be produced during the T-DNA transfer process. Such a protein would (1) protect the T-strand from cellular nucleases, and (2) potentially shape it into a conformation that allows transport through channels in the bacterial (see below) and plant cell membranes, as well as plant nuclear membrane pores. Such an SSB, was detected in *vir*-induced *Agrobacterium* and identified as the product of the *virE2* locus of the Ti plasmid.[144-147] The VirE2 protein has been found to preferentially bind ssDNA without sequence specificity as evidenced by gel mobility shift assays,[144-148] nitrocellulose filter binding,[148] and electron microscopy.[148,149] Lack of sequence specificity in VirE2 binding to ssDNA is consistent with observations that any DNA located between 25-bp borders on the Ti plasmid can function as the T-DNA and be transferred to plants.

Binding of VirE2 to ssDNA is a rapid process leading to almost instantaneous formation of stable protein-DNA complexes.[148] The strong binding affinity of VirE2 to ssDNA is reflected by the high concentrations of salt (1 to 1.5 *M* NaCl) required to inhibit formation of VirE2-ssDNA complexes.[148] The 69-kDa nopaline-specific VirE2 monomer occludes 28 to 30 nucleotides upon its binding to ssDNA;[148] its homolog, the 60.5-kDa octopine-specific VirE2 protein has a similar binding site of 34 to 36 nucleotides.[144] The actual binding of VirE2 to ssDNA is a complex process. Using gel mobility shift, electron microscopy, and binding kinetics studies, nopaline- and octopine-specific VirE2 proteins were shown to bind ssDNA in a highly cooperative manner.[148,149] Strong and cooperative binding of VirE2 to ssDNA implies that the T-strand molecules generated in *vir*-induced *Agrobacterium* are fully coated with this protein. Once the VirE2-T-strand complex is formed, it potentially will remain stable during the process of T-DNA transfer. That the T-strand molecules are fully coated with VirE2 also is inferred from its relative abundance in *vir*-induced *Agrobacterium.* For example, to completely coat a 22-kbp T-strand (produced from a wild-type nopaline-type Ti plasmid) would require about 700 molecules of VirE2 (assuming a 28 to 30 nucleotide binding site per protein monomer); the amount of *vir*E2 product observed in AS-induced *Agrobacterium* is consistent with this estimate.[148,150]

What are the properties of VirE2-ssSDNA complexes? Treatment of VirE2-ssDNA complexes with ssDNA-specific nucleases demonstrated that most of the protein-bound nucleic acid is not degraded by these enzymes.[148] Nucleolytic protection of ssDNA by VirE2 also was observed in DNase I footprinting experiments.[149] It is likely that cooperative binding of VirE2 to newly formed T-strands in *vir*-induced *Agrobacterium* also will form complexes resistant to cellular nucleases. Thus, VirE2-T-strand complexes likely are sufficiently protected to deliver T-strands across bacterial and plant cell membranes and cytoplasm into the nucleus.

Potentially, protection of transported T-strand molecules is necessary but not sufficient for T-DNA transfer to occur. By analogy with translocation of proteins through organelle membranes (reviewed in References 151 and 152), the T-strand molecules may need to be unfolded before transport through putative channels connecting *Agrobacterium* with its recipient plant cell. An ''unfoldase'' activity was postulated to be required for import of proteins into cell organelles [153] and for cell-to-cell movement of plant virus nucleic acids.[154] We suggest that VirE2 also has an unfoldase function. Free ssDNA molecules exist as irregular collapsed structures. Binding of VirE2 to ssDNA was shown to unfold these molecules to form extended (by about 50% relative to VirE2-free DNA molecules) and exceptionally thin protein-ssDNA complexes, measuring less than 2 nm in diameter.[148] Unfolding of ssDNA molecules to produce regularly shaped protein-ssDNA complexes also has been demonstrated for other SSBs (reviewed in Reference 155). Thus, the VirE2 protein may not only act to protect T-strands from cellular nucleases during the T-DNA transfer, but also may shape these T-strand molecules into an unfolded transferable form.

# V. EXPORT OF THE T-DNA

Newly formed T-complexes must be exported from *Agrobacterium* into recipient plant cells. Presumably, export of T-complexes occurs through specialized channels in the bacterial membrane. The likely candidates to produce such channel structures are the products of the *virB* locus, most of which are transmembrane or membrane-associated proteins. The following section summarizes our current knowledge about the *virB* operon and its protein products.

## A. IDENTIFICATION AND CHARACTERIZATION OF THE *VIRB* OPERON

The *virB* operon of pTiA6 was first identified as a complete transcriptional unit of approximately 9.5 kbp by transposon mutagenesis.[4] The operon lies between *virA* and *virG* loci on both the octopine Ti plasmid pTiA6 and the nopaline plasmid pTiC58 (see Figure 1).[4] Transposon mutagenesis studies indicated that distal sequences of the operon were absolutely required for virulence on plants and that *virB* was not required for production of T-strands.[4,127] A further study showed that three of the *virB* polypeptides fractionate with the bacterial envelope.[150] These latter data led to the hypothesis that *virB* gene products form a membrane pore or channel that could mediate export of T-strands from *Agrobacterium* to plant cells, and also further supported analogies with bacterial conjugation (see also below).[156]

The complete nucleotide sequence of the *virB* operon of the octopine Ti plasmids pTiA6 and pTi15955 and the nopaline plasmid pTiC58 has been determined.[157-160] After correcting for earlier sequencing errors,[161,162] the data indicate that the octopine and nopaline *virB* regions encode similar polypeptides. Both the pTiA6 and pTiC58 operons encode eleven open reading frames (ORF) for proteins of the following approximate molecular weights; VirB1, 26 kDa; VirB2, 12 kDa; VirB3, 12 kDa; VirB4, 87 kDa; VirB5, 23 kDa; VirB6, 32 kDa; VirB7, 6 kDa; VirB8, 26 kDa; VirB9, 32kDa; VirB10, 41 kDa; VirB11, 38 kDa (see also Table 1). The pTi15955 *virB* operon is identical except that the authors do not identify the small ORF VirB7 (6kDa).[158,162] The VirB ORFs will be referred to here using the most recent numbering system of Kuldau et al.[159] and Shirasu et al.[160] and the corrected version of the pTiA6 sequence of Ward et al.[161] Comparison of the predicted amino acid sequences of the *virB* ORFs from the different Ti plasmids shows that the ORFs are highly homologous suggesting their functions are both essential and conserved. Currently, there is no sequence analysis of the *virB* operon from an *Agrobacterium rhizogenes*-type plasmid. Analysis of a more distantly related *Agrobacterium* might reveal specific conserved domains within the VirB ORFs that are essential for function.

Analysis of the predicted amino acid sequences of the *virB* ORFs indicates some common structural features of the polypeptides. All the *virB* ORFs, except VirB4 and VirB11, have amino acid sequences of sufficient hydrophobicity to permit interaction with a membrane lipid bilayer. However, VirB4 and VirB11 are both also associated with the bacterial membrane. For example, VirB11 has an entirely hydrophilic hydropathy profile yet fractionates with the inner membrane of *Agrobacterium* cells.[163] Predictions for VirB4 vary from entirely hydrophilic to somewhat hydrophobic depending on the algorithm used; however, VirB4 fractionates with the *Agrobacterium* cell envelope.[150] VirB2 is the most hydrophobic ORF. It contains a contiguous section of approximately 70 hydrophobic amino acids representing 60% of the protein. VirB1, VirB5, VirB7, and VirB9 all have amino terminal sequences characteristic of bacterial membrane targeting signals.[157-160] Such sequences target proteins to the inner membrane, periplasm or outer membrane and contain the following four features;

(1) an amino terminal region containing 1 to 5 amino acids with a net average positive charge of +1.7, (2) a hydrophobic core region, and (3) a downstream polar region of 6 to 7 amino acids just preceding the (4) signal peptidase consensus cleavage site.[164,165] VirB7 has a signal cleavage site like those recognized by lipoprotein-specific signal peptidase II, which cleaves between a distal glycerol-modified cysteine and a proximal glycine.[157,159] Interestingly, Shirasu et al.[160] report limited amino acid sequence homology of VirB7 to *E. coli* proteins required for colicin export centering around the putative signal peptidase II cleavage site. While Thompson et al.[158] suggest VirB2 and VirB4 have signal sequences, neither of these putative signal sequences contains the characteristic hydrophobic core region, and the entire predicted VirB4 signal sequence is longer (38 amino acids) than the most "extreme" (36 amino acids) signal sequence reported by von Heijne.[165] The presence of a mononucleotide binding consensus site was noted for VirB4 and VirB11.[158,163,166]

Transposon mutagenesis was used to determine the 5′ and 3′ limits of the *virB* operon.[4] However, due to the polar nature of transposon insertions, the functional requirement of only the most distal portion (corresponding to VirB11) could be assessed for virulence. Indeed, in a more recent study using antibodies to VirB4, VirB9, VirB10, and VirB11, upstream transposon insertions were found to abolish downstream gene expression.[167] Complementation studies showed the requirement for *virB9, virB10,* and *virB11;* virulence was restored when wild-type levels of all three proteins were produced.[167] Except for VirB4, VirB9, VirB10, and VirB11, the expression of the other *virB* polypeptides remains to be determined. Also, further complementation studies will test the functional requirement for *virB* gene products upstream of *virB9.*

## B. LOCATION AND FUNCTION OF *VIRB* POLYPEPTIDES

While the precise role of the *virB* proteins in T-DNA transfer remains unknown, several studies have provided much useful information about their cellular location and characteristics. The initial studies of Engstrom et al.[150] showed that three of the 5′ *virB* polypeptides, corresponding in sizes to VirB1, VirB4, and VirB5, fractionate with the *Agrobacterium* cell envelope. Immunoblots of *Agrobacterium* cell fractions of inner membrane, outer membrane, periplasm, and cytosol probed with VirB10-specific antisera show that VirB10 is located in the inner membrane.[168] Translational fusions of VirB10 to *E. coli* alkaline phosphatase expressed in *Agrobacterium* cells demonstrated that VirB10 contains a periplasmic domain as well as a membrane-spanning region. Finally, high-molecular weight forms of VirB10 were observed after treatment of nonlysed *Agrobacterium* cells with cross-linking agents. These high molecular weight complexes contained VirB10 and possibly other proteins. These data suggest that VirB10 may be part of a membrane complex.

Cell fractionation studies have shown that the majority of VirB11 is associated with the inner membrane; a small portion is also found in the cytosol.[163] An alkaline phosphatase fusion to the carboxy terminus of VirB11 showed no activity indicating that this portion of the protein is not localized to periplasm.[163] Based on the VirB11 consensus nucleotide binding site, Christie and co-workers examined the enzymic properties of this protein. They found that VirB11 binds ATP, has an ATPase activity, and is autophosphorylated *in vitro.* Thus, VirB11 may regulate its own activity or the activity of other T-DNA transport proteins through phosphorylation.[163] Potentially, VirB11 located in the inner membrane is part of the "motor" which provides the energy to drive the T-strand through *Agrobacterium* membranes.

## C. COMPARISON TO OTHER DNA/PROTEIN IMPORT/EXPORT SYSTEMS

One characteristic common to T-DNA transfer as well as other cellular import/export processes is that they are mediated by specialized composite protein structures in cell membranes. Thus, we may obtain insight into T-DNA transfer by considering other types of membrane complexes involved in protein export and DNA import. For example, since T-strands are transferred as a DNA-protein complex, protein export machinery may facilitate their transport. One of the best studied protein export systems is the pullulanase operon of *Klebsiella*. PulC is an inner membrane protein that is required for the export of the secreted protein pullulanase.[169] Interestingly, the ATP-binding regions of VirB11 and PulC are highly homologous.[170]

Some *vir* functions may be similar to those required for the competence of bacteria to take in exogenous, ssDNA. In fact, *comG* of *Bacillus subtilis* is specifically required for the binding of exogenous DNA to the cell surface,[171] and the *comG* ORF1 gene has 14% amino acid sequence identity to *virB* ORF11.[166] Both proteins are predicted to be hydrophilic, and both have an ATP-binding domain. ComG ORF1 is a regulator of ComE expression, and is thought to be absolutely required for competence in this organism.[171] The *comG* operon has general similarities to the *virB* operon in that it encodes several membrane-spanning or membrane-associated proteins.

A further interesting correlation exists between the mechanism for encapsidation and export of filamentous bacteriophages (such as f1, fd, M13, and Ike) and a possible mechanism for T-complex export out of *Agrobacterium*. Inside the cell, the circular ssDNA of filamentous phages is coated with the gene V ssDNA-binding protein. The virus ssDNA-gene V protein structure is reminiscent of the *Agrobacterium* T-strand-VirE2 protein complex. The viral ssDNA genome is assembled into mature virions at the bacterial membrane.[172] During the assembly, a hairpin loop packaging signal on the viral ssDNA interacts with three membrane-associated proteins, gene I protein, $C^{VII}$ and $C^{IX}$, products of the genes VII and IX.[173] As the phage particles are extruded through the cell membrane, the gene V protein is replaced by viral structural proteins. The proteins $C^{VII}$ and $C^{IX}$ are attached to the virion at the emerging tip, followed by the major coat protein, B, of gene VIII, and proteins encoded by genes III and VI at the lagging tip. The gene I and gene IV proteins are not packaged with the viral DNA but are required for virus export. The gene I protein is implicated in the formation of the additional zones of adhesion from which mature virions are released, and the gene IV protein is thought to interact with gene I protein to form a channel for export of the assembled virus.[174,175] Likewise, one or more of the VirB proteins of *Agrobacterium* may form such a channel at zones of bacterium-plant attachment that mediates the export of the T-complex.

# VI. MECHANISM OF T-DNA TRANSFER TO PLANTS

## A. COMPARISON OF T-DNA TRANSFER TO BACTERIAL CONJUGATION

The polarity of function of the T-DNA borders coupled with the discovery of the (linear ssDNA) T-strand intermediate lead to the proposal that T-DNA transfer may be analogous to bacterial conjugation. There are eight characteristics of conjugation, exemplified by the well-studied F plasmid conjugation system.[176-178] Three steps are involved in the processing of F DNA: (1) site-specific nicking at the origin of transfer, (2) 5′ to 3′ directional transfer of ss F DNA from donor to the recipient, and (3) replacement strand synthesis in donor and recipient cells. Five other steps characterize the intercellular interactions between donor and recipient cells: (1) the synthesis (by donor) and recognition (by recipient) of the mating signal; (2) the elaboration of extended pilus structures; (3) formation of a stable mating pair presumably by association of membrane components between donor and recipient cells; (4) formation of a specialized channel for DNA transfer, and (5) exclusion of sibling mating. A detailed comparison between all these steps and T-DNA transfer has recently been compiled,[1] and, except for the elaboration of the pilus structure, most characteristics of F-mediated conjugation are shared by the T-DNA transfer process. This high degree of functional similarity leads to the expectation that the *tra* and *vir* protein products that mediate these two conjugal systems also might be homologous. However, the entire sequence of the F-transfer region has been determined, and none of the 38 predicted Tra proteins are homologous to any known proteins, including those encoded by the *vir* region.[179]

The well-studied F conjugation system provides the conceptual framework for understanding the steps in conjugative DNA transfer. However, other conjugation systems have recently provided more direct evidence for parallels with T-DNA transfer. Conjugation by the incompatibility group P (IncP) plasmids shows particular relation to T-DNA transfer. For example, the potential IncP transformation gene *kilB* and *virB11* share 28% amino acid sequence identity.[180] More importantly, the origin of transfer (oriT) from pRSF1010, an IncQ plasmid that is mobilizable by IncP plasmids, can direct DNA transfer to plant cells from *Agrobacterium.*[181] This hybrid transfer system also requires an intact *vir* region and the region of pRSF1010 encoding plasmid mobilization proteins. The oriT of pRSF1010 and its cognate mobilization proteins presumably generate a conjugative DNA transfer intermediate that is then transferred to plant cells using *vir*-specific functions. Recent results suggest that the (conjugative) transfer of pRSF1010 to plants directly involves interaction with the T-DNA transfer machinery, specifically with several *virB* proteins. Ward and colleagues[182] have shown that VirB9, VirB10, and VirB11 are essential for pRSF1010 transfer to plant cells. Furthermore, the presence of pRSF1010 in *Agrobacterium* can suppress the transfer of a T-DNA element from a coresident Ti plasmid, and this suppression can be overcome when a third plasmid that overproduces VirB9, VirB10, and VirB11, is introduced into the cell. These observations imply there is competition for a limited number of sites (presumably channels) for transfer of pRSF1010 or T-DNA-specific transfer intermediates, and that such sites are composed of VirB proteins.

There are parallels in the nicking reaction required for generation of donor DNA in plasmid and T-DNA transfer. First, the critical 10 bp overlapping the nick sites of the T-DNA borders and the oriT of the IncP plasmid RK2 show high DNA sequence identity.[183] Furthermore, in another IncP conjugative plasmid, RP4, the product of the *traI* gene is a site-specific endonuclease which becomes covalently bound to the 5′ side of the nick site after cleavage.[184] The analogous protein in the T-strand system may be the VirD2 product.[134] A prerequisite for nicking in the RP4 conjugative system is binding of the oriT by the product of the *traJ* gene.[185,186] The *traJ* product is thought to confer site specificity to the nicking reaction, although it possesses no intrinsic endonuclease activity. The proposed model for nicking at oriT in RP4 is that the oriT-TraJ complex is the actual substrate for nicking by the product of the *traI* gene.[187] The observation that the TraJ binding site and the TraI-specific nick site are on the same side of the DNA double helix is suggestive of TraJ-TraI protein-protein interaction.[186] A third protein, TraH, has also been implicated in the generation of a conjugative transfer molecule.[187] Stable oriT-TraI-TraJ complex formation requires the *traH* product. TraI-TraJ-TraH complexes also are observed in the absence of oriT, again indicative of protein-protein interactions.[187] In *Agrobacterium* T-strand transfer, the VirD1 product may be partially analogous to TraJ; since VirD1 is reported to be a topoisomerase, it may act on the substrate prior to binding by VirD2. Site-specific nicking, however, is observed *in vitro* without VirD1,[134] so, unlike TraJ, VirD1 does not confer (or at least is not solely responsible for conferring) site specificity to the nicking reaction. A protein in the *Agrobacterium* T-DNA transfer system analogous to the *traH* product of the RP4 plasmid may be the product of the *virC1* gene which enhances T-strand formation when VirD1 and VirD2 are present in limiting amounts.[133] Finally, in both conjugative and T-DNA transfer the 3′-end of the nick site remains accessible for replacement strand synthesis.

Other enzymic functions also must accompany the generation and transfer of the T-strand, e.g., an unwindase (helicase) to displace the T-strand from the Ti plasmid during its synthesis, and potentially a primase activity inside the plant nucleus to convert the T-strand to a double-stranded molecule prior to or coincident with integration. The TraI/TraY endonuclease of the F plasmid nicks at the oriT of F. The TraI component of this endonuclease has been shown to be also a helicase.[188,189] Possibly VirD1 or VirD2 provide this activity in T-strand formation. In the IncI-1 plasmid Collb-P9, the transferred strand is escorted into the recipient cell by multiple copies of two polypeptides specified by the *sog* gene.[190] The *sog* gene maps to the *tra* region and encodes a DNA primase that initiates DNA synthesis on the transferred strand following its transfer to the recipient cell. The *virD2* product which remains attached to the T-strand during transfer to the plant cell (see above) could potentially fulfill this additional role in the T-DNA transfer process. Other plant host or *vir* products (potentially *virB* specific) could also participate in these reactions.

Additional evidence that T-DNA transfer may be a conjugative process derives from experiments to test whether *vir* gene products can influence Ti plasmid transfer during *Agrobacterium* conjugation. Indeed, Steck and Kado[191] showed that mutants in *virA, virG, virB,* and *virE* resulted in a decrease of conjugation frequency up to 20-fold, and Gelvin and Habeck[192] showed that other mutants in *virA, virG, virB,* and *virC,* and could decrease transfer efficiency by 100- to 10,000-fold. In the latter work, the three mutants that gave the most severe effect mapped to *virB* ORFs 4 and11. ORF11 has been shown to have ATPase activity and is thought to be closely associated with the cytoplasmic side of the inner membrane; it has been proposed to function during transfer of the T-strand protein complex through specialized membrane channels in *vir*-induced *Agrobacterium.*[163] The function of VirB4 is not yet known; however, VirB4 localizes to the *Agrobacterium* cell envelope fraction and is predicted to play a role in forming a membrane channel for T-strand transport.[150] In contrast to evidence for the involvement of *vir* in Ti plasmid conjugation, a directed search for mutants that affect Ti plasmid conjugation in *Agrobacterium* did not lead to the detection of *vir* mutants;[193] and Ti plasmid conjugation-defective mutants are still fully virulent.[70,194] Evidently, *vir* genes can influence bacterial conjugation, but conjugation specific genes do not affect virulence.

While *Agrobacterium*-plant DNA transfer remains the only known example of a natural prokaryote-eukaryote conjugation system, additional support for conjugal-type transkingdom genetic exchange derives from experiments demonstrating that *E. coli* can transfer genes to yeast *in vitro. Saccharomyces cerevisiae* was a recipient for plasmids from two different incompatibility groups, a broad host range IncP plasmid, and the limited-host range F plasmid;[195] *S. pombeii* was also a recipient for the same IncP broad-host range plasmid.[196] Thus, there is strong evidence to suggest that conjugative DNA transfer may be even more widespread than expected. To date, the *Agrobacterium* T-DNA transfer system is the most unusual and intriguing given that its recipient is a higher eukaryotic (plant) cell. This fact is likely responsible for the additional complexity of the T-DNA transfer process, i.e., the major difference between bacterium-yeast and bacterium-plant DNA transfer is that bacterial conjugative functions are sufficient for plasmid transfer to yeast, but not to plants. Potentially *vir* genes have a special role in bypassing the host plant defense system, as well as their general role in mediating T-DNA transfer. Future studies of *Agrobacterium*-plant DNA transfer may provide novel insights into conjugative mechanisms in other systems as well.

## B. THE T-COMPLEX — A UNIQUE VECTOR FOR BACTERIUM-PLANT DNA TRANSFER

The interaction of *Agrobacterium* with plant cells is the only known natural example of interkingdom DNA transfer. Thus, general features of prokaryotic conjugative transfer of DNA are not likely to be sufficient to explain the complexity of the T-DNA transfer process. We have proposed that T-DNA transfer essentially combines two different mechanisms of DNA transfer, namely, bacterial conjugation and viral infection.[197,198] The early steps within *Agrobacterium* to generate a transferable T-strand molecule, and to export this molecule out of the bacterial cell, may resemble conjugation. However, the later steps of the transfer

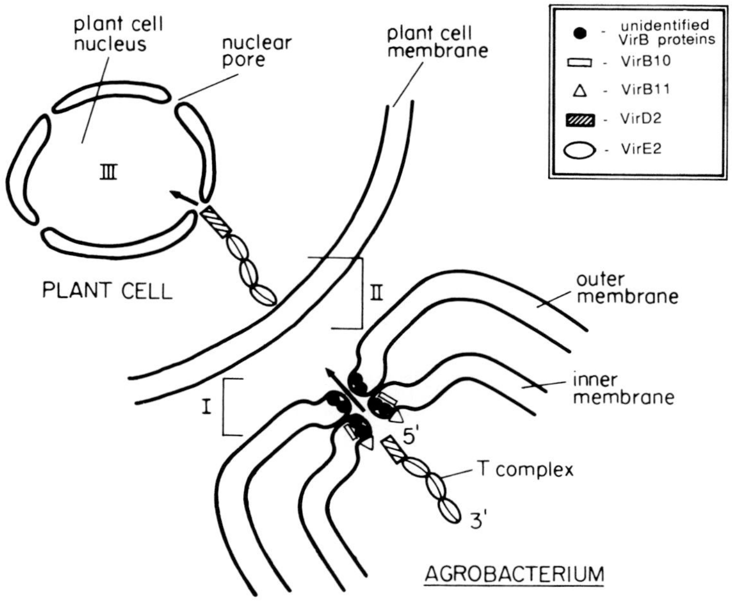


FIGURE 5. Transfer pathway of T-DNA from *Agrobacterium* into the plant cell. (I) association of *vir*-induced *Agrobacterium* with the recipient plant cell; (II) transport of the T-complex (an unfolded T-strand molecule with VirD2 attached to the 5′-end and cooperatively coated with VirE2 protein) across bacterial inner and outer membranes and cell wall (not drawn) through a specialized channel composed of VirB proteins, and its translocation through plant cell wall (not drawn) and cell membrane into the cytoplasm; (III) targeting of the T-complex inside the nucleus and integration of the T-strand into the plant genome.

process, penetration of the plant cell and nuclear membranes and integration into the plant nuclear genome, are classical features of virus infection. The T-complex (Figure 5) may, in fact, resemble a viral (or subviral) particle capable of plant cell transformation. By this analogy, VirE2 potentially functions as a coat protein of the T-complex particle, protecting it from external nucleases and possibly facilitating its movement through bacterial and plant cell membranes and cell walls. Since VirE2 is one of the largest prokaryotic SSBs identified to date, it may have different domains that function at different steps of the T-DNA transfer process, for example to facilitate transport of the T-complex through bacterial and plant cell membranes (Figure 5).

Several observations suggest that the T-complex is exported and is analogous to a viral particle. Coinoculation of *virE* insertional mutants with *Agrobacterium* carrying a wild-type *vir* region, resulted in extracellular complementation to wild-type virulence.[147, 199] These results can be explained if VirE2 is exported (as a free protein or in complex with the T-strand) from *Agrobacterium* into the extracellular space or directly into the plant cell. In addition, preliminary results from our laboratory suggest that the T-strand may be exported from AS-induced *Agrobacterium* even in the absence of the recipient plant cell . If *Agrobacterium* cells that are actively producing T-strands (i.e., 12 to 24 h after continuous growth in the presence of AS) are washed free of AS and then grown for a short period (2 to 4 h) in the absence of AS, T-strands are no longer detected in the bacterial cells.[200] Similar results have been obtained by S. Gelvin.[201] Concomitant with the disappearance of the T-strands, we observe that *vir* proteins presumed to be part of the T-complex (i.e., VirE2 and VirD2) also disappear following removal of AS.[200] These results imply that the T-strand with its associated proteins is rapidly exported from *Agrobacterium*. The alternative explanation, that the T-complex is rapidly degraded in the absence of continued *vir* expression, is unlikely, since it would require an efficient mechanism to specifically degrade VirE2-T-strand complexes even though such complexes have been observed to be very stable *in vitro*.[148,149]

Figure 5 presents a model for the sequence of events leading to the transport of the T-complex from *Agrobacterium* to plant cells. Following formation of the T-complex (an unfolded T-strand molecule with VirD2 attached to the 5′-end and cooperatively coated with VirE2 protein), it is transported across the bacterial inner and outer membranes, presumably through a specialized channel. This channel is likely composed of proteins specified by the *virB* locus. Since VirB11 has been localized to the cytoplasmic side of the inner bacterial membrane,[163] it may be the first protein of the channel to interact with the VirD2 protein on the leading end of the T-strand; furthermore, since VirB11 has ATPase activity,[163] it may provide the energy to move the T-complex out of the bacterial cell. Since VirB10 is hydrophobic and has a periplasmic domain,[168] it likely spans the cytoplasmic and periplasmic sides of the inner membrane part of the channel. Other VirB proteins, whose function and precise location have not yet been identified, may form the channel proper.

T-complexes exported from *Agrobacterium* are then transferred into the recipient plant cell. Presumably chromosomal virulence loci are involved in providing the initial contact with the plant cell surface. The actual transport may occur through preexisting plant membrane channels, or VirB (or other bacterial) proteins may directly function in transport of T-complexes across the plant cell membrane. Potentially the *Agrobacterium* channel for T-complex export may promote direct connection with the plant cell membrane. Once inside the recipient plant cell, nuclear localization domain(s) of the VirD2 protein target the T-complex specifically into the plant cell nucleus. Finally, the T-strand of the T-complex is integrated into the plant genome. Possibly, integration is facilitated by VirD2 and/or VirE2 components of the T-complex.

## VII. CONCLUSIONS AND PERSPECTIVES

It is noteworthy that the study of this relatively exotic system, the interaction of *Agrobacterium* with host plants, has provided basic information fundamental to other areas of biology. The discovery of plant signal molecules that induce the expression of the *vir* genes has lead to a fruitful area of investigation to identify similar communications between other soil bacteria and their plant hosts. The recent identification of nuclear localization signals in the VirD2 protein establishes that nuclear transport mechanisms are similar between plant and animal cells. The study of ssDNA transfer between donor and recipient cells is an important area of experimentation. The major unsolved problems in the *Agrobacterium*-plant cell interaction are (1) how is the T-strand-protein complex exported from the bacterium into the plant cell, and (2) how does the T-strand integrate into the plant cell genome. It is expected that research into these areas will provide information relevant both to the T-DNA transfer process and to fundamental mechanisms of DNA import and processing in plant cells.

## ACKNOWLEDGMENTS

We enthusiastically thank our colleagues for sending preprints and for sharing recent unpublished results. Our laboratory is currently funded by grants from NSF (DMB-8915613), DOE (DE-FG03-88ER13882), and USDA (90-37262-5291).

# REFERENCES

1. **Zambryski, P.,** *Agrobacterium* -plant cell DNA transfer, in *Mobile DNA,* Berg, D. and Howe, M., Eds., American Society of Microbiology, Washington, DC, 1989, 309.
2. **Zambryski, P.,** Basic processes underlying *Agrobacterium*-mediated DNA transfer to plant cells, *Annu. Rev. Genet.,* 22, 1, 1988.
3. **Zambryski, P., Tempe, J., and Schell, J.,** Transfer and function of T-DNA genes from *Agrobacterium* Ti and Ri plasmids in plant, *Cell,* 56, 193, 1989.
4. **Stachel, S. E. and Nester, E. W.,** The genetic and transcriptional organization of the *vir* region of the A6 Ti plasmid of *Agrobacterium, EMBO J.,* 5, 1445, 1986.
5. **Kanemoto, R. H., Powell, A. T., Akiyoshi, D. E., Regier, D. A., Kerstetter, R. A., Nester, E. W., Hawes, M. C., and Gordon, M. P.,** Nucleotide sequence and analysis of the plant-inducible locus *pinF* from *Agrobacterium tumefaciens, J. Bacteriol.,* 171, 2506, 1989.
6. **Douglas, C. J., Staneloni, R. J., Rubin, R. A., and Nester, E. W.,** Identification and genetic analysis of an *Agrobacterium tumefaciens* chromosomal virulence region, *J. Bacteriol.,* 161, 850, 1985.
7. **Zorreguieta, A., Geremia, R. A., Cavaignac, S., Cangelosi, G. A., Nester, E. W., and Ugalde, R. A.,** Identification of the product of an *Agrobacterium tumefaciens* chromosomal virulence gene, *Mol. Plant-Microbe Interact.,* 1, 121, 1988.
8. **Winans, S. C., Kerstetter, R. A., and Nester, E. W.,** Transcriptional regulation of the *virA* and *virG* genes of *Agrobacterium tumefaciens, J. Bacteriol.,* 170, 4047, 1988.
9. **Huang, M. L., Cangelosi, G. A., Halperin, W., and Nester, E. W.,** A chromosomal *Agrobacterium tumefaciens* gene required for effective plant signal transduction, *J. Bacteriol.,* 172, 1814, 1990.
10. **Binns, A. and Thomashow, M.,** Cell biology of *Agrobacterium* infection and transformation of plants, *Annu. Rev. Microbiol.,* 42, 575, 1988.
11. **Matthysse, A. G., Homes, K. V., and Gurlitz, R. G.,** Elaboration of cellulose fibrils by *Agrobacterium tumefaciens* during attachment to carrot cells, *J. Bacteriol.,* 145, 583, 1981.
12. **Douglas, C., Halperin, W., and Nester, E. W.,** *Agrobacterium tumefaciens* mutants affected in attachment to plant cells, *J. Bacteriol.,* 152, 1265, 1982.
13. **Pueppke, S. G. and Hawes, M. C.,** Understanding the binding of bacteria to plant surfaces, *Trends Biotechnol.,* 3, 310, 1985.
14. **Matthysse, A. G.,** Characterization of nonattaching mutants of *Agrobacterium tumefaciens, J. Bacteriol.,* 169, 313, 1987.
15. **Cangelosi, G. A., Hung, L., Puvanesarajah, V., Stacey, G., Ozga, D. A., Leigh, J. A., and Nester, E. W.,** Common loci for *Agrobacterium tumefaciens* and *Rhizobium meliloti* exopolysaccharide synthesis and their roles in plant interactions, *J. Bacteriol.,* 169, 2086, 1987.
16. **Thomashow, M. F., Karlinsey, J. E., Marks, J. R., and Hurlbert, R. E.,** Identification of a new virulence locus in *Agrobacterium tumefaciens* that affects polysaccharide composition and plant cell attachment, *J. Bacteriol.,* 169, 3209, 1987.
17. **Neff, N. T. and Binns, A. N.,** *Agrobacterium tumefaciens* interaction with suspension-cultured tomato cells, *Plant Physiol.,* 77, 35, 1985.
18. **Neff, N. T., Binns, A. N., and Brandt, C.,** Inhibitory effects of a pectin-enriched tomato cell wall fraction on *Agrobacterium tumefaciens* binding and tumor formation, *Plant Physiol.,* 83, 525, 1987.
19. **Kerr, A.,** Crown gall of stone fruit. I. Isolation of *Agrobacterium tumefaciens* and related species, *Aust. J. Biol.,* 22, 111, 1969.
20. **Lippincott, B. B. and Lippincott, J. A.,** Bacterial attachment to a specific wound site as an essential stage in tumor initiation by *Agrobacterium tumefaciens, J. Bacteriol.,* 97, 620, 1969.
21. **Glogowski, W. and Galsky, A. G.,** *Agrobacterium tumefaciens* site attachment as a necessary prerequisite for crown gall tumor formation on potato discs, *Plant Physiol.,* 61, 1031, 1978.
22. **Tanimoto, E., Douglas, C., and Halperin, W.,** Factors affecting crown gall tumorigenesis in tuber slices of Jerusalem artichoke *(Helianthus tuberosus L.), Plant Physiol.,* 63, 989, 1979.
23. **Gutlitz, R. H. G., Lamb, P. W., and Matthysse, A. G.,** Involvement of carrot cell surface proteins in attachment of *Agrobacterium tumefaciens, Plant Physiol.,* 83, 564, 1987.
24. **Pueppke, S. G. and Benny, U. K.,** Induction of tumors on *Solanum tuberosum* by *Agrobacterium tumefaciens:* quantitative analysis, inhibition by carbohydrates, and virulence of selected strains, *Physiol. Plant Pathol.,* 18, 169, 1981.
25. **Rao, S. S., Lippincott, B. B., and Lippincott, J. A.,** *Agrobacterium* adherence involves the pectic portion of the host cell wall and is sensitive to the degree of pectin methylation, *Plant Physiol.,* 56, 374, 1982.
26. **Robertson, J. L., Holliday, T., and Matthysse, A. G.,** Mapping of *Agrobacterium tumefaciens* chromosomal genes affecting cellulose synthesis and bacterial attachment to host cells, *J. Bacteriol.,* 170, 1408, 1988.
27. **Close, T. J., Tait, R. C., and Kado, C. I.,** Regulation of Ti plasmid virulence genes by a chromosomal locus of *Agrobacterium tumefaciens, J. Bacteriol.,* 164, 774, 1985.

28. **Close, T. J., Rogowsky, P. M., Kado, C. I., Winans, S. C., Yanofsky, M. F., and Nester, E. W.,** Dual control of *Agrobacterium tumefaciens* Ti plasmid virulence genes, *J. Bacteriol.*, 169, 5113, 1987.
29. **Tait, R. C. and Kado, C. I.,** Regulation of the *virC* and *virD* promoters of pTiC58 by the *ros* chromosomal mutation of *Agrobacterium tumefaciens, Mol. Microbiol.*, 2, 385, 1988.
30. **Stachel, S. E. and Zambryski, P. C.,** *virA* and *virG* control the plant-induced activation of T-DNA transfer process of *A. tumefaciens, Cell,* 46, 325, 1986.
31. **Winans, S. C., Ebert, P. R., Stachel, S. E., Gordon, M. P., and Nester, E. W.,** A gene essential for *Agrobacterium* virulence is homologous to a family of positive regulatory loci, *Proc. Natl. Acad. Sci. U.S.A.*, 83, 8278, 1986.
32. **Nixon, B. T., Ronson, C. W., and Ausubel, F. M.,** Two-component regulatory systems responsive to environmental stimuli share strongly conserved domains with the nitrogen assimilation regulatory genes *ntrB* and *ntrC, Nucleic Acids Res.*, 14, 8073, 1986.
33. **Ronson, C. W., Nixon, B. T., and Ausubel, F. M.,** Conserved domains in bacterial regulatory proteins that respond to environmental stimuli, *Cell,* 49, 579, 1987.
34. **Stachel, S. E., Messens, E., Van Montagu, M., and Zambryski, P.,** Identification of the signal molecules produced by wounded plant cells that activate T-DNA transfer in *Agrobacterium tumefaciens, Nature (London),* 318, 624, 1985.
35. **Stachel, S. E., Nester, E. W., and Zambryski, P.,** A plant cell factor induces *Agrobacterium tumefaciens vir* gene expression, *Proc. Natl. Acad. Sci. U.S.A.*, 83, 379, 1986.
36. **Spencer, P. A. and Towers, G. H. N.,** Specificity of signal compounds detected by *Agrobacterium tumefaciens, Phytochemistry,* 27, 2781, 1988.
37. **Spencer, P. A. and Towers, G. H. N.,** Virulence-inducing phenolic compounds detected by *Agrobacterium tumefaciens,* in *Plant Cell Wall Polymers: Biogenesis and Biodegradation,* Lewis, N. G. and Paice, M. G., Eds., American Chemical Society, Washington, DC, 1989, 383.
38. **Bolton, G. W., Nester, E. W., and Gordon, M. P.,** Plant phenolic compounds induce expression of the *Agrobacterium tumefaciens* loci needed for virulence, *Science,* 50, 983, 1986.
39. **Scheikholeslam, S. N. and Weeks, D. P.,** Acetosyringone promotes high efficiency transformation of *Arabidopsis thaliana* explants by *Agrobacterium tumefaciens, Plant Mol. Biol.*, 8, 291, 1987.
40. **Yanofsky, M., Lowe, B., Montoya, A., Rubin, R., Krul, W., Gordon, M., and Nester, E.,** Molecular and genetic analysis of factors controlling host range in *Agrobacterium tumefaciens, Mol. Gen. Genet.*, 201, 237, 1985.
41. **Leroux, B., Yanofsky, M. F., Winans, S. C., Ward, J. E., Ziegler, S. F., and Nester, E. W.,** Characterization of the *virA* locus of *Agrobacterium tumefaciens:* a transcriptional regulator and host range determinant, *EMBO J.*, 6, 849, 1987.
42. **Usami, S., Okamoto, S., Takebe, I., and Machida, Y.,** Factor inducing *Agrobacterium tumefaciens vir* gene expression is present in monocotyledonous plants, *Proc. Natl. Acad. Sci. U.S.A.*, 85, 3748, 1988.
43. **Messens, E., Dekeyser, R., and Stachel, S. E.,** A nontransformable *Triticum monococcum* monocotyledonous culture produces the potent *Agrobacterium vir*-inducing compound ethyl ferulate, *Proc. Natl. Acad. Sci. U.S.A.*, 87, 4386, 1990.
44. **Kahl, G.,** Molecular biology of wound healing: the conditioning phenomenon, in *Molecular Biology of Plant Tumors,* Kahl, G. and Schell, J., Eds., Academic, San Diego, 1982, 211.
45. **Bytebier, B., Deboeck, F., DeGreve, H., Van Montagu, M., and Hernalsteens, J.-P.,** T-DNA organization in tumor cultures and transgenic plants of the monocotyledon *Asparagus officinalis, Proc. Natl. Acad. Sci. U.S.A.*, 84, 5345, 1987.
46. **Hernalsteens, J. P., Thia-Toong, L., Schell, J., and Van Montagu, M.,** An *Agrobacterium* transformed cell culture from the monocot *Asparagus officinalis, EMBO J.*, 13, 3039, 1984.
47. **Schafer, W., Gorz, A., and Gunter, K.,** T-DNA integration and expression in an monocot crop plant after induction of *Agrobacterium, Nature (London),* 327, 529, 1987.
48. **Raineri, D. M., Bottino, P., Gordon, M. P., and Nester, E. W.,** *Agrobacterium*-mediated transformation of rice *(Oryza sativa L), BioTechnology,* 8, 33, 1990.
49. **Dandekar, A. M., Gupta, P. K., Durzan, D. J., and Knauf, V.,** Transformation and foreign gene expression in micropropagated Douglas fir *(Pseudotsuga menziesii), BioTechnology,* 5, 587, 1987.
50. **Sederoff, R., Stomp, A. M., Chilton, W. S., and Moore, L. W.,** Gene transfer into loblolly pine by *Agrobacterium tumefaciens, BioTechnology,* 4, 647, 1986.
51. **Morris, J. W. and Morris, R. O.,** Identification of an *Agrobacterium tumefaciens* virulence gene inducer from the pinaceous gymnosperm *Pseudotsuga menziesii, Proc. Natl. Acad. Sci. U.S.A.*, 87, 3614, 1990.
52. **Veluthambi, K., Krishnan, M., Gould, J. H., Smith, R. H., and Gelvin, S. B.,** Opines stimulate induction of the *vir* genes of the *Agrobacterium tumefaciens* Ti plasmid, *J. Bacteriol.*, 171, 3696, 1989.
53. **Krishnan, M. and Gelvin, S. B.,** Octopine influences the induction of *Agrobacterium tumefaciens vir* genes by tobacco protoplasts, submitted, 1990.
54. **Scott, I. M., Firmin, J. L., Butcher, D. N., Searle, L. M., Sogeke, A. K., Eagles, J., March, J. F., Self, R., and Fenwick, G. R.,** Analysis of a range of crown gall and normal plant tissues for Ti plasmid-determined compounds, *Mol. Gen. Genet.*, 176, 57, 1979.

55. **Saint-Pierre, B. and Dion, P.,** Growth of nopaline-utilizing *Agrobacterium* and *Pseudomonas* in extracts of crown-gall tumors, *Can. J. Microbiol.*, 34, 793, 1988.
56. **Vernade, D., Herrera-Estrella, A., Wang, K., and Van Montagu, M.,** Glycine betaine allows enhanced induction of the *Agrobacterium tumefaciens vir* genes by acetosyringone at low pH, *J. Bacteriol.*, 170, 1030, 1988.
57. **Culianez-Macia, F. A. and Hepburn, A. G.,** The kinetics of T-strand production in a nopaline-type helper strain of *Agrobacterium tumefaciens, Mol. Plant-Microbe Interact.*, 1, 207, 1988.
58. **Cangelosi, G. A., Ankenbauer, R. G., and Nester, E. W.,** Sugars induce the *Agrobacterium* virulence genes through a periplasmic binding protein and a transmembrane signal protein, *Proc. Natl. Acad. Sci. U.S.A.*, 87, 6708, 1990.
59. **Shimoda, N., Toyoda-Yamamoto, A., Nagamine, J., Usami, S., Katayama, M., Sakagami, Y., and Machida, Y.,** Control of expression of *Agrobacterium vir* genes by synergistic actions of phenolic signal molecules and monosaccharides, *Proc. Natl. Acad. Sci. U.S.A.*, 87, 6684, 1990.
60. **Bollinger, J., Park, C., Harayama, S., and Hazelbauer, G. L.,** Structure of the Trg protein: Homologies with and differences from other sensory transducers of *Escherichia coli, Proc. Natl. Acad. Sci. U.S.A.*, 81, 3287, 1984.
61. **Park, C. and Hazelbauer, G. L.,** Mutations specifically affecting ligand interaction of the Trg chemosensory transducer, *J. Bacteriol.*, 167, 101, 1986.
62. **McNeil, M., Darvill, A. G., Fry, S. C., and Albersheim, P.,** Structure and function of the primary cell walls of plants, *Annu. Rev. Biochem.*, 53, 625, 1984.
63. **Ashby, A. M., Watson, M. D., and Shaw, C. H.,** A Ti-plasmid determined function is responsible for chemotaxis of *Agrobacterium tumefaciens* towards the plant wound product acetosyringone, *FEMS Microbiol. Lett.*, 41, 189, 1987.
64. **Parke, D., Ornston, L. N., and Nester, E. W.,** Chemotaxis to plant phenolic inducers of virulence genes is constitutively expressed in the absence of the Ti plasmid in *Agrobacterium tumefaciens* [published erratum appears in *J. Bacteriol.* 170, 2002, 1988], *J. Bacteriol.*, 169, 5336, 1987.
65. **Ashby, A. M., Watson, M. D., Loake, G. J., and Shaw, C. H.,** Ti plasmid-specified chemotaxis of *Agrobacterium tumefaciens* C58C1 toward *vir*-inducing phenolic compounds and soluble factors from monocotyledonous and dicotyledonous plants, *J. Bacteriol.*, 170, 4181, 1988.
66. **Shaw, C. H., Ashby, A. M., Brown, A., Royal, C., Loake, G. J., and Shaw, C. H.,** *virA* and *virG* are the Ti-plasmid functions required for chemotaxis of *Agrobacterium tumefaciens* towards acetosyringone, *Mol. Microbiol.*, 2, 413, 1988.
67. **Hawes, M. C., Smith, L. Y., and Howarth, A. J.,** *Agrobacterium tumefaciens* mutants deficient in chemotaxis to root exudates, *Mol. Plant-Microbe Interact.*, 1, 82, 1988.
68. **Loake, G. J., Ashby, A. M., and Shaw, C. H.,** Attraction of *Agrobacterium tumefaciens* C58C towards sugars involves a highly sensitive chemotaxis system, *J. Gen. Microbiol.*, 134, 1427, 1988.
69. **Kerr, A. and Ellis, J. G.,** Conjugation and transfer of Ti plasmids in *Agrobacterium tumefaciens*, in *Molecular Biology of Plant Tumors*, Kahl, G. and Schell, J., Eds., Academic, San Diego, 1982, 321.
70. **Ellis, J., Kerr, A., and Tempe, J.,** Conjugal transfer of nopaline and agropine Ti-plasmids–The role of agrocinopines, *Mol. Gen. Genet.*, 186, 269, 1982.
71. **Nebert, D. W. and Gonzalez, F. J.,** P450 genes: structure, evolution and regulation, *Annu. Rev. Biochem.*, 56, 945, 1987.
72. **Matthews, D. E. and Van Etten, H. D.,** Detoxification of the phytoalexin pisatin by a fungal cytochrome P-450, *Arch. Biochem. Biophys.*, 224, 494, 1983.
73. **Kistler, H. C. and Van Etten, H. D.,** Three non-allelic genes for pisatin demethylation in the fungus *Nectria haematococca, J. Gen. Microbiol.*, 130, 2595, 1984.
74. **Kistler, H. C. and Van Etten, H. D.,** Regulation of pisatin demethylation in *Nectria haematococca* and its influence on pisatin tolerance and virulence, *J. Gen. Microbiol.*, 130, 2605, 1984.
75. **Peters, K. N., Frost, J. W., and Long, S. R.,** A plant flavone, luteolin, induces expression of *Rhizobium meliloti* nodulation genes, *Science*, 233, 977, 1986.
76. **Redmond, J. W., Bartley, M., Djordjevic, M. A., Innes, R. W., Kuempel, P. L., and Rolfe, B. G.,** Flavones induce expression of nodulation genes in *Rhizobium, Nature (London)*, 323, 632, 1986.
77. **Peters, N. K. and Long, S. R.,** Alfalfa root exudates and compounds which promote or inhibit induction of *Rhizobium meliloti* nodulation genes, *Plant Physiol.*, 88, 396, 1988.
78. **Melchers, L. S., Regensburg, T. T., Bourret, R. B., Sedee, N. J., Schilperoort, R. A., and Hooykaas, P. J.,** Membrane topology and functional analysis of the sensory protein VirA of *Agrobacterium tumefaciens, EMBO J.*, 8, 1919, 1989.
79. **Alt-Moerbe, J., Neddermann, P., von Lintig, J., Weiler, E. W., and Schroder, J.,** Temperature-sensitive step in Ti plasmid *vir*-region induction and correlation with cytokinin secretion by Agrobacteria, *Mol. Gen. Genet.*, 213, 1, 1988.

80. **Melchers, L. S., Thompson, D. V., Idler, K. B., Schilperoort, R. A., and Hooykaas, P. J.,** Nucleotide sequence of the virulence gene *virG* of the *Agrobacterium tumefaciens* octopine Ti plasmid: significant homology between *virG* and the regulatory genes *ompR, phoB* and *dye* of *E. coli, Nucleic Acids Res.*, 14, 9933, 1986.
81. **Powell, B. S., Powell, G. K., Morris, R. O., Rogowsky, P. M., and Kado, C. I.,** Nucleotide sequence of the *virG* locus of the *Agrobacterium tumefaciens* plasmid pTiC58, *Mol. Microbiol.*, 1, 309, 1987.
82. **Winans, S. C.,** Transcriptional induction of an *Agrobacterium* regulatory gene at tandem promoters by plant-released phenolic compounds, phosphate starvation, and acidic growth media, *J. Bacteriol.*, 172, 2433, 1990.
83. **Das, A., Stachel, S., Ebert, P., Allenza, P., Montoya, A., and Nester, E.,** Promoters of Agrobacterium tumefaciens Ti-plasmid virulence genes, *Nucleic Acids Res.*, 14, 1355, 1986.
84. **Steck, T. R., Morel, P., and Kado, C. I.,** *Vir* box sequences in *Agrobacterium tumefaciens* pTiC58 and A6, *Nucleic Acids Res.*, 16, 8736, 1988.
85. **Winans, S. C., Jin, S., Komari, T., Johnson, K. M., and Nester, E. W.,** The role of virulence regulatory loci in determining *Agrobacterium* host range, *Plant Molecular Biology,* von Wettstein, D., and Chua, N.-H., Eds., Plenum Press, NY, 1987, 573.
86. **Jin, S. G., Roitsch, T., Christie, P. J., and Nester, E. W.,** The regulatory VirG protein specifically binds to a cis-acting regulatory sequence involved in transcriptional activation of *Agrobacterium tumefaciens* virulence genes, *J. Bacteriol.*, 172, 531, 1990.
87. **Pazour, G. J. and Das, A.,** *virG.*, an *Agrobacterium tumefaciens* transcriptional activator, initiates translation at a UUG codon and is a sequence-specific DNA-binding protein, *J. Bacteriol.*, 172, 1241, 1990.
88. **Roitsch, T., Wang, H., Jin, S. G., and Nester, E. W.,** Mutational analysis of the *virG* protein, a transcriptional activator of *Agrobacterium tumefaciens* virulence genes, *J. Bacteriol.*, 172, 6054, 1990.
89. **Aoyama, T., Takanami, M., and Oka, A.,** Signal structure for transcriptional activation in the upstream regions of virulence genes on the hairy-root-inducing plasmid A4, *Nucleic Acids Res.*, 17, 8711, 1989.
90. **Aoyama, T. and Oka, A.,** A common mechanism of transcriptional activation by the three positive regulators, VirG, PhoB, and OmpR., *FEBS Lett.*, 263, 1, 1990.
91. **Makino, K., Shinagawa, H., Amemura, M., Kimura, S., Nakata, A., and Ishihama, A.,** Regulation of the phosphate regulon of *Escherichia coli.* Activation of *pstS* transcription by PhoB protein *in vitro, J. Mol. Biol.*, 203, 85, 1988.
92. **Cowing, D. E., Bardwell, J. C. A., Craig, E. A., Woolford, C., Hendrix, R. W., and Gross, C. A.,** Consensus sequence for *Escherichia coli* heat shock promoters, *Proc. Natl. Acad. Sci. U.S.A.*, 82, 2909, 1985.
93. **VanBogelen, R. A., Kelley, P. M., and Neidhardt, F. C.,** Differential induction of heat shock, SOS, and oxidation stress regulons and accumulation of nucleotides in *Escherichia coli, J. Bacteriol.,* 169, 26, 1987.
94. **Jin, S. G., Komari, T., Gordon, M. P., and Nester, E. W.,** Genes responsible for the supervirulence phenotype of *Agrobacterium tumefaciens A281, J. Bacteriol.*, 169, 4417, 1987.
95. **Guyon, P., Chilton, M. D., Petit, A., and Tempe, J.,** Agropine in "null" type crown gall tumors: evidence for the generality of the opine concept, *Proc. Natl. Acad. Sci. U.S.A.*, 80, 4803, 1980.
96. **Hood, E. E., Chilton, W. S., Chilton, M. D., and Fraley, R. T.,** T-DNA and opine synthetic loci in tumors incited by *Agrobacterium tumefaciens* A281 on soybean and alfalfa plants, *J. Bacteriol.*, 168, 1283, 1986.
97. **Aoyama, T., Hirayama, T., Tamamoto, S., and Oka, A.,** Putative start codon TTG for the regulatory protein VirG of the hairy-root-inducing plasmid pRiA4, *Gene,* 78, 173, 1989.
98. **Reddy, P., Peterkofsky, A., and McKenney, K.,** Translational efficiency of the *E. coli* adenylate cyclase gene: mutating the UUG initiation codon to GUG or AUG results in increased gene expression, *Proc. Natl. Acad. Sci. U.S.A.*, 82, 5656, 1985.
99. **Jin, S. G., Prusti, R. K., Roitsch, T., Ankenbauer, R. G., and Nester, E. W.,** Phosphorylation of the VirG protein of *Agrobacterium tumefaciens* by the autophosphorylated VirA protein: essential role in biological activity of VirG, *J. Bacteriol.*, 172, 4945, 1990.
100. **Jin, S., Roitsch, T., Ankenbauer, R. G., Gordon, M. P., and Nester, E. W.,** The VirA protein of *Agrobacterium tumefaciens* is autophosphorylated and is essential for *vir* gene regulation, *J. Bacteriol.*, 172, 525, 1990.
101. **Sanders, D., Gillece-Castro, B., Stock, A., Burlingame, A., and Koshland Jr., D.,** Identification of the site of phosphorylation of the chemotaxis response regulator protein, CheY, *J. Biol. Chem.*, 264, 21770, 1989.
102. **Weiss, V. and Magasanik, B.,** Phosphorylation of nitrogen regulator 1 (NR1) of *Escherichia coli, Proc. Natl. Acad. Sci. U.S.A.*, 85, 8919, 1988.
103. **Kato, M., Aiba, H., Tate, S. I., Nishimura, Y., and Mizuno, T.,** Location of phosphorylation site and DNA-binding site of a positive regulator, OmpR, involved in activation of the osmo-regulatory genes of *Escherichia coli, FEBS Lett.*, 249, 168, 1989.

104. **Ninfa, A. J., and Magasanik, B.,** Covalent modification of the *glnG* product, NRI, by the *glnL* product, NRII, regulates the transcription of the *gln ALG* operon in *Escherichia coli, Proc. Natl. Acad. Sci. U.S.A.*, 83, 5905, 1986.
105. **Winans, S. C., Kerstetter, R. A., Ward, J. E., and Nester, E. W.,** A protein required for transcriptional regulation of *Agrobacterium* virulence genes spans the cytoplasmic membrane, *J. Bacteriol.*, 171, 1616, 1989.
106. **Huang, Y., Morel, P., Powell, B., and Kado, C. I.,** VirA, a coregulator of Ti-specified virulence genes, is phosphorylated *in vitro, J. Bacteriol.*, 172, 1142, 1990.
107. **Postma, P. and Lengeler, J.,** Phosphoenolpyruvate: carbohydrate phosphotransfer system of bacteria, *Microbiol. Rev.*, 49, 232, 1985.
108. **Simpson, R. B., O'Hara, P. J., Kwok, W., Montoya, A. L., Lichenstein, C., Gordon, M. P., and Nestor, E. W.,** DNA from the A6S/2 crown gall tumor contains scrambled Ti plasmid sequences near its junctions with the plant DNA, *Cell*, 29, 1005, 1982.
109. **Yadav, N. S., Vanderleydan, J., Bennett, D. R., Barnes, W. M., and Chilton, M.-D.,** Short direct repeats flank the T-DNA on a nopaline Ti plasmid, *Proc. Natl. Acad. Sci. U.S.A.*, 79, 6322, 1982.
110. **Zambryski, P., Depicker, A., Kruger, K., and Goodman, H.,** Tumor induction by *Agrobacterium tumefaciens:* analysis of the boundaries of T-DNA, *J. Mol. Appl. Genet.*, 1, 361, 1982.
111. **Holsters, M., Villarroel, R., Gielen, J., Seurinck, J., De Greve, H., Van Montagu, M., and Schell, J.,** An analysis of the boundaries of the octopine TL-DNA in tumors induced by *Agrobacterium tumefaciens, Mol. Gen. Genet.*, 190, 35, 1983.
112. **Zambryski, P., Joos, H., Genetello, C., Leemans, J., Van Montagu, M., and Schell, J.,** Ti plasmid vector for the introduction of DNA into plant cells without alteration of their normal regeneration capacity, *EMBO J.*, 2, 2143, 1983.
113. **Kwok, W. W., Nester, E. W., and Gordon, M. P.,** Unusual plasmid DNA organization in an octopine crown gall tumor, *Nucleic Acids Res.*, 13, 459, 1985.
114. **Slightom, J. L., Jouanin, L., Leach, F., Drong, R. F., and Tepfer, D.,** Isolation and identification of TL-T-DNA/plant junctions in convolvulus arvenis transformed by *Agrobacterium rhizogenes* strain A4, *EMBO J.*, 4, 3069, 1985.
115. **Gheyson, G., Van Montagu, M., and Zambryski, P.,** Integration of *Agrobacterium tumefaciens* T-DNA involves rearrangements of target plant DNA sequences, *Proc. Natl. Acad. Sci. U.S.A.*, 86, 4017, 1987.
116. **Joos, H., Inze, D., Caplan, A., Sormann, M., Van Montagu, M., and Schell, J.,** Genetic analysis of T-DNA transcripts in nopaline crown galls, *Cell*, 32, 1057, 1983.
117. **Shaw, C. H., Watson, M. D., Carter, G. H., and Shaw, C. H.,** The right hand copy of the nopaline Ti plasmid 25 bp repeat is required for tumor formation, *Nucleic Acids Res.*, 12, 6031, 1984.
118. **Wang, K., Herrera-Estrella, L., Van Montagu, M., and Zambryski, P.,** Right 25 bp terminus sequences of the nopaline T-DNA is essential for and determines direction of DNA transfer from *Agrobacterium* to the plant genome, *Cell*, 38, 35, 1984.
119. **Peralta, E. G. and Ream, L. W.,** T-DNA border sequences required for crown gall tumorgenesis, *Proc. Natl. Acad. Sci. U.S.A.*, 82, 5112, 1985.
120. **Jen, G. and Chilton, M. D.,** The right border region of pTiT37 T-DNA is intrinsically more active than the left border region in promoting T-DNA formation, *Proc. Natl. Acad. Sci. U.S.A.*, 83, 3895, 1986.
121. **Peralta, E. G., Hellmiss, R., and Ream, W.,** Overdrive, a T-DNA transmission enhancer on the *A. tumefaciens* tumor-inducing plasmid, *EMBO J.*, 5, 1137, 1986.
122. **Van Haaren, M. J. J., Sedell, N. J. A., Schilperoort, R. A., and Hooykaas, P. J. J.,** Overdrive is a T-region transfer enhancer which stimulates T-strand production in *A. tumefaciens, Nucleic Acids Res.*, 15, 8983, 1987.
123. **Wang, K., Genetello, C., Van Montagu, M., and Zambryski, P.,** Sequence context of the T-DNA border repeat element determines its relative activity during T-DNA transfer to plant cells, *Mol. Gen. Genet.*, 210, 338, 1987.
124. **Lichenstein, C. P. and Fuller, S. L.,** Vectors for the genetic engineering of plants, in *Genetic Engineering*, Vol. 6, Academic Press, San Diego, 1987, 103.
125. **Rogers, S. G. and Klee, H.,** Pathways to plant genetic manipulation employing *Agrobacterium*, in *Plant DNA Infectious Agents*, Hohn, T. and Schell, J., Eds., Springer-Verlag, NY, 1987, 179.
126. **Stachel, S. E., Timmerman, B., and Zambryski, P.,** Generation of single-stranded T-DNA molecules during the initial stages of T-DNA transfer from *Agrobacterium tumefaciens* to plant cells, *Nature (London)*, 322, 706, 1986.
127. **Stachel, S. E., Timmerman, B., and Zambryski, P.,** Activation of *Agrobacterium tumefaciens vir* gene expression generates multiple single-stranded T-strand molecules from the pTiA6 T-region: requirement for 5′*virD* gene products, *EMBO J.*, 6, 857, 1987.
128. **Wang, K., Stachel, S. E., Timmerman, B., Van Montagu, M., and Zambryski, P.,** Site-specific nick occurs within the 25 bp transfer promoting border sequence following induction of *vir* gene expression in *Agrobacterium tumefaciens, Science*, 235, 587, 1987.

129. **Albright, L. M., Yanofsky, M. F., Leroux, B., Ma, D. Q., and Nester, E. W.,** Processing of the T-DNA of *Agrobacterium tumefaciens* generates border nicks and linear, single-stranded T-DNA, *J. Bacteriol.*, 169, 1046, 1987.
130. **Veluthambi, K., Ream, W., and Gelvin, S. B.,** Virulence genes, border and overdrive generate single-stranded T-DNA molecules from the A6 Ti plasmid of *Agrobacterium tumefaciens, J. Bacteriol.*, 170, 1523, 1988.
131. **Yanofsky, M. F., Porter, S. G., Young, C., Albright, L. M., Gordon, M. P., and Nester, E. W.,** The *virD* operon of *Agrobacterium tumefaciens* encodes a site-specific endonuclease, *Cell,* 47, 471, 1986.
132. **Jayaswal, R. K., Veluthambi, K., Gelvin, S. B., and Slightom, J. L.,** Double-stranded cleavage of T-DNA and generation of single-stranded T-DNA molecules in *Escherichia coli* by a *virD*-encoded border-specific endonuclease from *Agrobacterium tumefaciens, J. Bacteriol.*, 169, 5035, 1987.
133. **De Vos, G. and Zambryski, P.,** Expression of *Agrobacterium* nopaline specific VirD1, VirD2 and VirC1 proteins and their requirement for T-strand production in *E. coli, Mol. Plant-Microbe Interact.*, 2, 42, 1989.
134. **Howard, E. A., Warnick, D., and Zambryski, R.,** unpublished results, 1991.
135. **Ghai, J. and Das, A.,** The *virD* operon of *Agrobacterium tumefaciens* Ti plasmid encodes a DNA-relaxing enzyme, *Proc. Natl. Acad. Sci. U.S.A.*, 86, 3109, 1989.
136. **Herrera-Estrella, A., Chen, Z., Van Montagu, M., and Wang, K.,** VirD proteins of *Agrobacterium tumefaciens* are required for the formation of a covalent DNA protein complex at the 5′ terminus of T-strand molecules, *EMBO J.*, 7, 4055, 1988.
137. **Ward, E. and Barnes, W.,** VirD2 protein of *Agrobacterium tumefaciens* very tightly linked to the 5′ end of T-strand, *Science,* 242, 927, 1988.
138. **Young, C. and Nester, E. W.,** Association of the VirD2 protein with the 5′ end of T strands in *Agrobacterium tumefaciens, J. Bacteriol.*, 170, 3367, 1988.
139. **Howard, E. A., Winsor, B. A., De Vos, G., and Zambryski, P.,** Activation of the T-DNA transfer process in *Agrobacterium* results in the generation of a T-strand-protein complex: tight association of VirD2 with the 5′ ends of T-strands, *Proc. Natl. Acad. Sci. U.S.A.*, 86, 4017, 1989.
140. **Howard, E. A., Zupan, J., and Zambryski, P.,** The VirD2 protein of *Agrobacterium tumefaciens* contains a C-terminal bipartite nuclear localization signal: implications for nuclear uptake of DNA in plant cells, submitted, 1991.
141. **Stachel, S. E. and Zambryski, P.,** Generic trans-kingdom sex, *Nature (London),* 340, 190, 1989.
142. **Wang, K., Herrera-Estrella, A., and Van Montagu, M.,** Overexpression of *virD1* and *virD2* genes in *Agrobacterium tumefaciens* enhances T-complex formation and plant transformation, *J. Bacteriol.*, 172, 4432, 1990.
143. **Chelsky, D., Ralph, R., and Jonak, G.,** Sequence requirements for synthetic peptide-mediated translocation to the nucleus, *Mol. Cell Biol.*, 9, 2487, 1989.
144. **Gietl, C., Koukolikova, N. Z., and Hohn, B.,** Mobilization of T-DNA from *Agrobacterium* to plant cells involves a protein that binds single-stranded DNA, *Proc. Natl. Acad. Sci. U.S.A.*, 84, 9006, 1987.
145. **Citovsky, V., De Vos, G., and Zambryski, P.,** Single-stranded DNA binding protein encoded by the *virE* locus of *Agrobacterium tumefaciens, Science,* 240, 501, 1988.
146. **Das, A.,** *Agrobacterium tumefaciens virE* operon encodes a single-stranded DNA-binding protein, *Proc. Natl. Acad. Sci. U.S.A.*, 85, 2909, 1988.
147. **Christie, P. J., Ward, J. E., Winans, S. C., and Nester, E. W.,** The *Agrobacterium tumefaciens virE2* gene product is a single-stranded-DNA-binding protein that associates with T-DNA, *J. Bacteriol.*, 170, 2659, 1988.
148. **Citovsky, V., Wong, M. L., and Zambryski, P.,** Cooperative interaction of *Agrobacterium* VirE2 protein with single-stranded DNA: implications for the T-DNA transfer process, *Proc. Natl. Acad. Sci. U.S.A.*, 86, 1193, 1989.
149. **Sen, P., Pazour, G. J., Anderson, D., and Das, A.,** Cooperative binding of *Agrobacterium tumefaciens* VirE2 protein to single-stranded DNA, *J. Bacteriol.*, 171, 2573, 1989.
150. **Engstrom, P., Zambryski, P., Van Montagu, M., and Stachel, S.,** Characterization of *Agrobacterium tumefaciens* virulence proteins induced by the plant factor acetosyringone, *J. Mol. Biol.*, 197, 635, 1987.
151. **Eilers, M. and Schatz, G.,** Protein unfolding and the energetics of protein translocation across biological membranes, *Cell,* 52, 481, 1988.
152. **Neupert, W., Hartl, F.-U., Craig, E., and Pfanner, N.,** How do polypeptides cross the mitochondrial membranes?, *Cell,* 63, 447, 1990.
153. **Rothman, J. E. and Kornberg, R. D.,** An unfolding story of protein translocation, *Nature (London),* 322, 209, 1986.
154. **Citovsky, V., Knorr, D., Schuster, G. and Zambryski, P.,** The P30 movement protein of tobacco mosaic virus is a single-strand nucleic acid binding protein, *Cell,* 60, 637, 1990.
155. **Chase, J. W. and Williams, K. R.,** Single-stranded DNA binding proteins required for DNA replication, *Annu. Rev. Biochem.*, 55, 103, 1986.

156. **Stachel, S. E. and Zambryski, P. C.,** *Agrobacterium tumefaciens* and the susceptible plant cell: a novel adaptation of extracellular recognition and DNA conjugation, *Cell,* 47, 155, 1986.
157. **Ward, J. E., Akiyoshi, D. E., Regier, D., Datta, A., Gordon, M. P., and Nester, E. W.,** Characterization of the *virB* operon from an *Agrobacterium tumefaciens* Ti plasmid, *J. Biol. Chem.,* 263, 5804, 1988.
158. **Thompson, D. V., Melchers, L. S., Idler, K. B., Schilperoort, R. A., and Hooykaas, P. J.,** Analysis of the complete nucleotide sequence of the *Agrobacterium tumefaciens virB* operon, *Nucleic Acids Res.,* 16, 4621, 1988.
159. **Kuldau, G. A., De Vos, G., Owen, J., McCaffrey, G., and Zambryski, P.,** The *virB* operon of *Agrobacterium tumefaciens* pTiC58 encodes 11 open reading frames, *Mol. Gen. Genet.,* 221, 256, 1990.
160. **Shirasu, K., Morel, P., and Kado, C.,** Characterization of the *virB* operon of an *Agrobacterium tumefaciens* Ti-plasmid nucleotide sequence and protein analysis, *Mol. Microbiol.,* 4, 1153, 1990.
161. **Ward, J. E., Akiyoshi, D. E., Regier, D., Datta, A., Gordon, M. P., and Nester, E. W.,** Correction: characterization of the *virB* operon from *Agrobacterium tumefaciens* Ti plasmid, *J. Biol. Chem.,* 265, 4768, 1990.
162. **Hooykaas, P.,** personal communication, 1990.
163. **Christie, P. J., Ward, J. J., Gordon, M. P., and Nester, E. W.,** A gene required for transfer of T-DNA to plants encodes an ATPase with autophosphorylating activity, *Proc. Natl. Acad. Sci. U.S.A.,* 86, 9677, 1989.
164. **von Heijne, G.,** Patterns of amino acids near signal sequence cleavage sites, *Eur. J. Biochem.,* 133, 17, 1983.
165. **von Heijne, G.,** Signal sequences: the limits of variation, *J. Mol. Biol.,* 184, 99, 1985.
166. **Kuldau, G. and Zambryski, P.,** unpublished observations, 1990.
167. **Ward, J. J., Dale, E. M., Christie, P. J., Nester, E. W., and Binns, A. N.,** Complementation analysis of *Agrobacterium tumefaciens* Ti plasmid *virB* genes by use of a *vir* promoter expression vector: *virB9, virB10,* and *virB11* are essential virulence genes, *J. Bacteriol.,* 172, 5187, 1990.
168. **Ward, J. J., Dale, E. M., Nester, E. W., and Binns, A. N.,** Identification of a *virB10* protein aggregate in the inner membrane of *Agrobacterium tumefaciens, J. Bacteriol.,* 172, 5200, 1990.
169. **d'Enfert, C., Reyss, I., Wandersman, C., and Pugsley, A.,** Protein secretion by gram-negative bacteria, *J. Biol. Chem.,* 264, 17462, 1989.
170. **Thorstenson, Y.,** unpublished observations, 1990.
171. **Albano, M., Breitling, R., and Dubnau, D. A.,** Nucleotide sequence and genetic organization of the *Bacillus subtilis comG* operon, *J. Bacteriol.,* 171, 5386, 1989.
172. **Webster, R. E. and Lopez, J.,** Structure and assembly of the class I filamentous bacteriophage, in *Virus Structure and Assembly,* Casjens, S., Ed., Jones and Bartlett, Boston/Portola Valley, 1985, 235.
173. **Russel, M. and Model, P.** Genetic analysis of the filamentous bacteriophage packaging signal and of the proteins that interact with it, *J. Virol.,* 63, 3284, 1989.
174. **Model, P. and Russel, M.,** Filamentous bacteriophage, in *The Bacteriophages,* Calendar, R., Ed., Plenum Press, NY, 1988, 375.
175. **Brissette, J. L. and Russel, M.,** Secretion and membrane integration of filamentous phage-encoded morphogenic protein, *J. Mol. Biol.,* 211, 565, 1990.
176. **Willetts, N. and Wilkins, B.,** Processing of plasmid DNA during bacterial conjugation, *Microbiol. Rev.,* 48, 24, 1984.
177. **Ippen-Ihler, K. A. and Minkley, E. G. J.,** The conjugation system of F, the fertility factor of *Escherichia coli, Annu. Rev. Genet.,* 20, 593, 1986.
178. **Willetts, N. and Skurray, R.,** Structure and function of the F factor and mechanism of conjugation, in *Escherichia coli and Salmonella typhimurium: Cellular and Molecular Biology,* Neidhardt, F., Ingraham, J., Low, K., Magasanik, B., and Umbarger, H., Eds., American Society of Microbiology, Washington, DC, 1987, 1110.
179. **Frost, L. and Usher, K.,** personal communication, 1990.
180. **Thomas, C. M.,** personal communication, 1990.
181. **Buchanan-Wollaston, V., Passiatore, J., and Cannon, F.,** The mobilization functions (mob and ori-T) of a bacterial plasmid promote its transfer to plants, *Nature (London),* 328, 172, 1987.
182. **Ward, J., Dale, E., and Binns, A.,** Activity of *Agrobacterium* T-DNA transfer machinery is coordinately regulated by *virB9,10,11* gene products, *Proc. Natl. Acad. Sci. U.S.A.,* in press, 1991.
183. **Guiney, D.,** The transfer origin of the IncP plasmid RK2: Identification of functional domains and essential sequences, *Fallen Leaf Lake Conf. Abstr.,* 1990, 7.
184. **Pansegrau, W., Ziegelin, G., and Lanka, E.,** Covalent association of the *traI* gene product of plasmid RP4 with the 5′ terminal nucleotide at the relaxation nick site, *J. Biol. Chem.,* 265, 10637, 1990.
185. **Furste, J. P., Pansegrau, W., Ziegelin, G., Kroger, M., and Lanka, E.,** Conjugative transfer of promiscuous IncP plasmids. Interaction of plasmid-encoded products with the transfer origin, *Proc. Natl. Acad. Sci. U.S.A.,* 86, 1771, 1989.

186. **Zeiglin, G., Furste, J. P., and Lanka, E.**, TraJ protein of plasmid RP4 binds to a 19-base pair invert sequence repetition with the transfer origin, *J. Biol. Chem.*, 264, 11989, 1989.
187. **Pansegrau, W., Balzar, D., Kruft, V., Lurz, R., and Lanka, E.**, *In vitro* assembly of relaxosomes at the transfer origin of plasmid RP4, *Proc. Natl. Acad. Sci. U.S.A.*, 87, 6555, 1990.
188. **Traxler, B. A. and Minkley, E. G. J.**, Revised genetic map of the distal end of the F transfer operon: implications for DNA helicase I, nicking at ori T and conjugal DNA transport, *J. Bacteriol.*, 169, 3251, 1987.
189. **Traxler, B. A. and Minkley, E. G. J.**, Evidence that DNA helicase I and oriT site-specific nicking are both functions of the F TraI protein, *J. Mol. Biol.*, 204, 205, 1988.
190. **Rees, C. E. D. and Wilkins, B. M.**, Transfer of tra proteins into the recipient cell during conjugation mediated by plasmid Collb-P9, *J. Bacteriol*, 171, 3152, 1989.
191. **Steck, T. R. and Kado, C. I.**, Virulence genes promote conjugative transfer of the Ti plasmid between *Agrobacterium* strains, *J. Bacteriol.*, 172, 2191, 1990.
192. **Gelvin, S. B. and Habeck, L. L.**, *vir* genes influence conjugal transfer of the Ti plasmid of *Agrobacterium tumefaciens, J. Bacteriol.*, 172, 1600, 1990.
193. **Beck von Bodman, S., McCutchan, J. E., and Ferrand, S. K.**, Characterization of conjugal transfer functions of *Agrobacterium tumefaciens* Ti plasmid pTiC58, *J. Bacteriol.*, 171, 5281, 1989.
194. **Klapwijk, P., Scheulderman, T., and Schilperoort, R.**, Coordinated regulation of octopine degradation and conjugative transfer of Ti plasmids in *Agrobacterium tumefaciens:* Evidence for a common regulatory gene and separate operons, *J. Bacteriol.*, 136, 775, 1978.
195. **Heineman, J. and Sprague, G.**, Bacterial conjugative plasmids mobilize DNA transfer between bacteria and yeast, *Nature (London)*, 340, 205, 1989.
196. **Sikorski, R., Michaud, M., Levin, H., Boeke, J., and Hieter, P.**, Trans-kingdom promiscuity, *Nature (London)*, 345, 581, 1990.
197. **Citovsky, V., Howard, E., Winsor, B., and Zambryski, P.**, Proteins that mediate DNA transfer by *Agrobacterium tumefaciens* to plant cells: Summary and perspectives, in *Molecular Biology of Plant-Pathogen Interactions*, Staskawicz, B., Ahlquist, P., and Yoder, O., Eds., Alan R. Liss, NY, 1989, 3.
198. **Howard, E. and Citovsky, V.**, The emerging structure of the *Agrobacterium* T-DNA transfer complex, *BioEssays*, 12, 103, 1990.
199. **Otten, L., DeGreve, H., Leemans, L., Hain, R., Hooykass, P. and Schell, J.**, Restoration of virulence of *vir* region mutants of *A. tumefaciens* strain B6S3 by coinfection with normal and mutant *Agrobacterium* strains, *Mol. Gen. Genet.*, 195, 159, 1984.
200. **Citovsky, V., Howard, E., Franklin, A., and Zambryski, P.**, unpublished results, 1989.
201. **Steck, T. and Gelvin, S.**, personal communication, 1990.
202. **Hirooka, T. and Kado, C. I.**, Location of the right boundary of the virulence region on *Agrobacterium tumefaciens* plasmid pTiC58 and a host-specifying gene next to the boundary, *Proc. Natl. Acad. Sci. U.S.A.*, 83, 7850, 1986.
203. **Yanofsky, M. F. and Nester, E. W.**, Molecular characterization of a host-range-determining locus from *Agrobacterium tumefaciens, J. Bacteriol.*, 168, 237, 1986.
204. **Hirooka, T., Rogowsky, P. M., and Kado, C. I.**, Characterization of the *virE* locus of *Agrobacterium tumefaciens* plasmid pTiC58, *J. Bacteriol.*, 169, 1529, 1987.
205. **Porter, S. G., Yanofsky, M. F., and Nester, E. W.**, Molecular characterization of the *virD* operon from *Agrobacterium tumefaciens, Nucleic Acids Res.*, 15, 7503, 1987.
206. **Winans, S. C., Allenza, P., Stachel, S. E., McBride, K. E., and Nester, E. W.**, Characterization of the *virE* operon of the *Agrobacterium* Ti plasmid pTiA6, *Nucleic Acids Res.*, 15, 825, 1987.

# CHROMOSOMAL- AND TI PLASMID-MEDIATED REGULATION OF *AGROBACTERIUM* VIRULENCE GENES: MOLECULAR INTERCOMMUNICATION

**Clarence I. Kado**

## TABLE OF CONTENTS

# I. INTRODUCTION

Like many different types of bacteria that reside in specific ecological niches, those naturally associated with plants can respond quickly to various environmental stimuli. These adaptive responses produce transient changes in gene expression leading to diverse phenotypic changes. The response is the result of the detection of one or more signals generated in the environment that is sensed and transduced in the bacterium culminating in the initiation of the expression of genes whose products catalyze the phenotypic change(s) to best fit the particular environment. Likewise, chemical signals from the bacteria are detected by the defense system of the plant. The molecular mechanisms responsible for triggering this adaptive response in bacteria is often mediated by protein phosphorylation of the sensory protein, upon detection of an external environmental stimuli, and the receiver (or regulator) protein, which upon phosphorylation promotes transcription of specific genes. With the sensor/regulator (or "two-component") regulatory system, the phosphorylation is often catalyzed by a histidine kinase.[1-4] Signal transduction takes place by the transfer of the phosphate moiety from ATP to histidine residues in the kinase portion of the sensory protein followed by transphosphorylation from phosphohistidine to aspartic acid residues in the receiver portion of the regulator protein, and then from the phosphoaspartate side chain to water. Due to the high degree of similarities between two-component regulatory proteins among different bacteria,[2,3,5,6] there is a growing ensemble of these protein families involved in bacterial adaptive responses. Among these is the two-component regulatory system involved with the expression of the virulence phenotype of *Agrobacterium tumefaciens* and *A. rhizogenes*. The two-component regulatory system that controls the expression of virulence of *Agrobacterium* is comprised of *virA* and *virG* genes located within a 28.6-kp region on a large 200-kb extrachromosomal element called the Ti plasmid. The products of the *virA* and *virG* genes are the sensor and regulator pair, whereby VirG protein positively activates the transcription of four *vir* operons.

Superimposed on the two-component regulation in *A. tumefaciens* is a chromosomal gene *ros* that also regulates virulence expression.[7] The *ros* gene has recently been cloned and characterized.[9] Its product is a repressor that controls the transcription of operons *virC* and *virD* of the virulence regulon on the Ti plasmid. Another gene called *chvE* plays a role in promoting the expression of the virulence genes in response to external inducers identified as simple sugars which play a synergistic role in this induction phenomenon.[10,11]

# II. SIGNAL TRANSDUCTION AND TRANSCRIPTIONAL ACTIVATION BY TI PLASMID *virA/virG* GENES

It has been long known that wounding of the host was a prerequisite for successful infection by *A. tumefaciens* and *A. rhizogenes*. One of the effects of wounding is the production of precursor molecules of the lignin biosynthetic pathway.[12] Precursors such as sinapinic acid and its oxidation product acetosyringone have turned out to be powerful inducers of the virulence *(vir)* genes in *A. tumefaciens*. Other compounds such as phenylpropanoid glucoside from Douglas fir,[13] ethyl ferulate from wheat,[14] ethyl caffeate from wine,[15] and opines from crown galls[22] have also been reported as *vir* inducers or stimulators.

The molecular mechanism of *vir* gene induction involves the products of *virA* and *virG*, whereby VirA protein is associated with the bacterial membrane and periplasmic space and can sense the presence of these inducers. This hypothesis is based on sequence homologies of *virA/virG* to members of the two-component family, where the product of *virA* would logically belong to the sensor protein subgroup and the product of *virG* would be a member of the receiver/regulator subgroup (Table 1). The mechanism of specific interaction with the inducer has not been elucidated for VirA protein. The amino-terminal half of VirA is

**TABLE 1**
**Membrane-Associated Two-Component Gene Regulatory Systems**

| Bacterium | Sensory protein | Regulator protein | Regulated function | References |
|---|---|---|---|---|
| *Agrobacterium tumefaciens* | VirA | VirG | Virulence | 16-20,24,29, 30,33,37,39 |
| *Rhizobium leguminosarum* | DctB | DctD | $C_4$ dicarboxylate transport | 48 |
| *Rhizobium meliloti* | FixL | FixJ | Nitrogen fixation | 49 |
| *Escherichia coli* | EnvZ | OmpR | Porin expression | 26-28 |
| | PhoM | PhoB | Phosphate assimilation | 39 |
| | CpxA | SfrA | Aerobiosis, conjugation | 46 |
| | UhpB | UhpA | Sugar phosphate transport | 47 |

associated with the inner membrane of *Agrobacterium*[16,17] and apparently loops into the periplasm.[16] This half of VirA apparently serves as the sensory unit since the carboxy-terminal half has been recently shown to possess histidine kinase activity.[18-20] One or more histidine residues within the C-terminal domain are, therefore, phosphorylated. The histidine residue at amino acid 474 in the octopine VirA[20] and at amino acid 618 in the nopaline VirA[19] are sites of phosphorylation. Within the C-terminal half, a consensus nucleotide binding sequence, Gly-X-Gly-X-X-Gly,[21] is present at amino acid residues 658 to 663[19] in nopaline VirA. This sequence together with others in proper context may be essential for specific interaction with ATP.

How VirA activates VirG remains to be determined. However, phosphorylation of VirG is thought to convert the protein into an specific transcriptional activator of the *vir* genes. Truncated VirA containing the C-terminal half of the protein can still function to activate VirG, presumably through transphosphorylation. Thus, the C-terminal half of VirA was predicted to contain the kinase domain. Recent studies of phosphorylation of VirG confirms the earlier view that the N-terminal half of the protein contains the receiver domain.[2,5,6] In other two-component systems, phosphorylation has been correlated both with transcriptional activation/repression of targeted genes[25,26] and with increased binding of the modulator protein to its DNA target.[27,28] The target of phosphorylation in a counterpart two-component regulatory membrane, CheA/CheY, are aspartyl moieties.[4,23] Likewise, aspartate residue 52 of VirG protein appears to be the principal phosphoreceptor.[24] VirG protein without phosphorylation has been shown to bind nonspecifically to the *virE* regulatory region[29,30] but preferentially to the 5′-nontranscribed region of *virA, virB, virC, virD,* and *virG*.[30] This region contains moderately conserved sequences, often in tandem array, and generally located 50 to 80 bp upstream of the predicted transcription start sites of the accompanying *vir* operon.[31,32] The conserved sequences are known as the *vir* box,[31,32] around which virG appears to bind.[30,33,39] That nonphosphorylated VirG can bind specificially to the *vir* box region *in vitro* is interesting since Powell et al.[29] had postulated earlier that specifically of binding of VirG to the *vir* regulatory region may require phosphorylation. All of the previous studies did not test the sufficiency of the *vir* box as a *cis*-acting site for VirG function. Thus, an assay was developed to uncouple transcriptional activation from DNA binding to study the *in vivo* DNA-binding properties of VirG. Powell and Kado[34] recently showed that VirG binds specifically to the *vir* box sequence *in vivo,* that the C-terminal half of VirG contains the DNA-binding domain, that a threshold concentration of VirG is required for specific binding, and that, once initiated, DNA binding proceeds rapidly with a minimal increase in protein concentration and in a positive cooperative manner. These studies have led to the modification of the current model. Specific binding by VirG around the *vir* box activates the *vir* promoters largely by localized increase in the VirG concentration.

Promoter activation can take place in at least two ways; first, in the absence of signal

**TABLE 2**
**Properties of the *ros* Gene and Its Product**

| Molecule | Property | Reference |
|---|---|---|
| Size of *ros* | 426 bp | 9 |
| Regulation of *ros* | Autoregulated | 9 |
| Other species containing *ros* counterparts | *A. rhizogenes; Rhizobium meliloti* | 9 |
| Size of *ros* protein | 15.5 kDa | 9 |
| pI | 7.13 | 9 |
| No. of zinc fingers | One | 9 |
| Binding affinity to *virC* and *virD* promoters | >20,000-fold excess competitor DNA | 38 |
| *ros* box consensus | TAGATTTCA/TGAAATCTA | 9 |

transduction by VirA, specific binding by VirG does not occur at low concentrations of VirG protein but occurs at high concentrations. High expression of *virG* in the absence of inducer and *virA* will fully activate the transcription of the *vir* operons.[35] This ''high copy effect'' of *vir* gene transcriptional activation by increased copies of VirG has been confirmed by other groups.[30] The second promoter activation comes about through the phosphorylation of VirG protein. As a phosphoprotein, VirG is thought to promote transcription of the *vir* operons at lower concentrations due to high specificity of the *vir* promoter region. The mechanism for transcriptional activation by phosphorylated VirG is not known. Perhaps, phospho-VirG facilitates cooperative binding of itself to the target DNA and, in turn, the protein-DNA complex promotes RNA polymerase activity.[29]

# III. REGULATION OF *VIR* GENES BY CHROMOSOMAL GENES

Besides the two-component VirA/VirG regulation of *vir* genes, chromosomal genes are also involved in controlling *vir* gene expression. Early work on the regulation of *vir* genes led to the premature conclusion that these genes were regulated by only VirG activated by VirA. As stressed by Close et al.,[7] and Rogowsky et al.,[35] transcriptional activation or inhibition can occur by VirG alone and by certain chromosomal genes. The following genes have been identified.

## A. THE *ROS* CHROMOSOMAL GENE

The *ros* gene controls the expression of *virC* and *virD*, two operons which are important in T-intermediate production and transfer. This gene encodes a 15.5-kDa protein whose properties are very similar to a repressor.[8] The Ros protein binds specifically to sequences adjacent to the *vir* boxes situated 5′-upstream of these *vir* operons. The properties of the Ros protein are summarized in Table 2. The *ros* gene is autoregulated. Its promoter contains an inverted repeat that is also found in the promoters of *virC* and *virD*. Also, the *ros* upstream regulatory domain contains sequence stretches that are similar to the consensus sequence of the *vir* box[8] (Table 2). Thus, it is plausible that VirG, which is known to bind to the *vir* box element, competes with Ros protein in *vir* and *ros* regulation. Although the molecular interaction between VirG, Ros, and the *vir* regulatory region has not been fully elucidated, Ros is likely to play a role in modulating the expression of *virC* and *virD*. These operons play a definitive role in the T-intermediate formation, host range, and T-DNA transfer. Thus, a mutation in the *ros* gene will fully derepress these two operons. Such an effect can result in excess production of T-intermediates in induced *A. tumefaciens* cells.[36] The *ros* mutation also widens the host range of *A. tumefaciens*,[9] perhaps because of the high *virC* expression and high double-stranded T-intermediate formation, products which would no longer be limiting in the bacterial cell. It appears, therefore, that *ros* has an important biological role in maintaining certain host-range windows for *A. tumefaciens*. The reasons for this role are not apparent at this time. That *ros* gene sequence is also present in *Rhizobium meliloti*[9] raises the question whether nodulation genes are regulated by a *ros* counterpart.

## B. OTHER CHROMOSOMAL GENES

A number of chromosomal genes termed *chvA, chvB, pscA (exoC* in *Rhizobium)*, and *att*[8] appear to partake in promoting efficient virulence by affecting the attachment of *A. tumefaciens* to plant cells. The *chvD* locus seems to influence the expression for *virG* in *A. tumefaciens* cells starved for phosphate and exposed to low pH.[37] Another locus called *chvE* encodes a periplasmic sugar-binding protein which apparently may interact as protein-sugar complex with periplasmic sites on VirA.[11] In the case of *chvE*, induction of the *vir* genes by sugars may be coordinated with the two-component regulatory system since the synergistic expression of *virB* requires the presence of the *vir* inducer acetosyringone. None of these chromosomal genes, however, directly control the expression of *vir* genes.

There are other *Agrobacterium* chromosomal genes that have ancillary roles such as catalyzing exopolysaccharide production which may indirectly aid virulence. Those genes are *chvA, chvB,* and *pscA,* which appear to promote virulence of *A. tumefaciens* by facilitating bacterial attachment to plant cells.[41,42] Recent studies showed that exopolysaccharide production may be negatively regulated by the *psdA* locus located downstream of *pscA*.[40] *psdA* possibly encodes a repressor. Other chromosomal genes that affect the surface properties of *Agrobacterium* have been identified. A locus called *picA* (for *p*lant-*i*nducible *c*hromosomal), which is inducible by carrot root extracts, is thought to affect bacterial cell surface properties, particularly bacterial agglutination.[43] The inducer has not been identified but it may be acidic polysaccharides, such as polygalacturonic acid, since it is sensitive to pectinase.

As research of chromosomal genes progresses, a number of interesting loci will likely be uncovered, which can affect virulence of *A. tumefaciens* and *A. rhizogenes* one way or another. Our research group had used the transposon Tn4431[44] to generate chromosomal gene fusions to a promoterless *lux* operon in order to study *in planta* inducible genes of *Xanthomonas campestris* pv. *campestris*.[45] Similar studies on *A. tumefaciens* may yield a number of plant-inducible chromosomal genes.

From the growing body of information on plant-microbe interactions, many of the pathogens will likely have chromosomal genes that are responsive to host-generated signals. These signals will likely be diverse and associated with biosynthetic pathways linked to general cell wall biosynthesis and wound-healing processes.

# IV. MOLECULAR COMMUNICATION BETWEEN PLASMID AND CHROMOSOMAL GENES

The positive transcriptional activator VirG binds specifically to *vir* box sequences located in the promoter region of each *vir* operon and *ros*. Like many regulatory proteins, VirG can also behave as a repressor.[34] Thus, VirG may repress the expression of *ros* while activating the expression of the *vir* genes. With decreased concentrations of Ros protein, *virC* and *virD* become fully exposed for transcription. Under noninductive conditions, the Ros protein represses the expression of *virC* and *virD*. Thus, a plasmid gene seems to affect the activity of a chromosomal gene, while a chromosomal gene represses the expression of *vir* genes. The term *"molecular communication"* is coined here to describe this phenomenon. Molecular communication at the level of plant to bacterial interaction is also apparent. In the near future, a growing list of examples of molecular communications will become evident.

# V. SUMMARY AND CONCLUSIONS

The recognition of specific hosts by the pathogen (or symbiont), and conversely the recognition of the pathogen (or symbiont) by the plant are expected initial responses finely honed through evolution. The recognizing mechanism can vary from physical contacts (for example, electrical charge impulses, bacterial attachment to plant cells) to chemical signals (for example, lignin precursors, sugars, sugar-phosphates, sugar-sulfates, sugar-acetates). Currently, a majority of workers have been focusing on chemical signals and the elusive receptor of these signals in the plant. Chemical signals elaborated by the plant have been identified. In *A. tumefaciens* these signals comprise plant phenolic compounds that interact with the product of the *virA* gene. Other compounds such as sugars, sugar-phosphates, and opines are synergistic in promoting *vir* gene expression. In total, signal molecules and synergistically acting compounds as well as chromosomal regulatory genes, such as *ros*, all seem to play key roles in optimizing the infectional event leading to disease manifestation (in this case, crown gall transformation).

The molecular communication as described in this chapter occur at two levels. The intercommunication between plasmid and chromosomal genes is one level. The second level is the intercommunication between bacteria and plants by the transduction of chemical and physical signals to initiate transcription of genes required for pathogenesis on the part of the pathogen, and defense on part of the plant.

## ACKNOWLEDGMENTS

I am indebted to all my co-workers for their stimulating discussions and significant contributions on the two-component regulatory system and on the *ros* gene. Among the recent co-workers are Drs. Bradford Powell, Michael Cooley, Marion D'Souza, Yong Huang, and Patrice Morel. The work summarized here was supported by NIH grants CA-11526-21 and GM-45550-01, and grant 87-CRCR-1-2282 from the Cooperative State Research Service, USDA.

## REFERENCES

1. **Hess, J. F., Bourret, R. B., and Simon, M. I.,** Histidine phosphorylation and phosphoryl group transfer in bacterial chemotaxis, *Nature (London),* 336, 139, 1988.
2. **Stock, J. B., Ninfa, A. J., and Stock, A. M.,** Protein phosphorylation and regulation of adaptive responses in bacteria, *Microbiol. Rev.,* 53, 450, 1989.
3. **Bourret, R. B., Hess, J. F., Borkovich, K. A., Pakula, A. A., and Simon, M. I.,** Protein phosphorylation in chemotaxis and two-component regulatory systems of bacteria, *J. Biol. Chem.,* 264, 7085, 1989.
4. **Bourret, R. B., Hess, J. F., and Simon, M. I.,** Conserved aspartate residues and phosphorylation in signal transduction by the chemotaxis protein CheY, *Proc. Natl. Acad. Sci. U.S.A.,* 87, 41, 1990.
5. **Nixon, B. T., Ronson, W. W., and Ausubel, F. M.,** Two-component regulatory systems responsive to environmental stimuli strongly conserved domains with the nitrogen assimilation regulatory genes *ntrB* and *ntrC, Proc. Natl. Acad. Sci. U.S.A.,* 83, 7850, 1986.
6. **Albright, L. M., Huala, E., and Ausubel, F. M.,** Prokaryotic signal transduction mediated by sensor and regulator protein pairs, *Ann. Rev. Genet.,* 23, 311, 1989.
7. **Close, T. J., Tait, R. C., and Kado, C. I.,** Regulation of Ti plasmid virulence genes by a chromosomal locus of *Agrobacterium tumefaciens, J. Bacteriol.,* 164, 774, 1985.
8. **Matthysse, A. G.,** Characterization of nonattaching mutants of *Agrobacterium tumefaciens, J. Bacteriol.,* 169, 313, 1987.
9. **Cooley, M. B., D'Souza, M. R., and Kado, C. I.,** The *virC and virD* operons of the *Agrobacterium* Ti plasmid are regulated by the *ros* chromosomal gene: analysis of the cloned *ros* gene, *J. Bacteriol.,* in
10. **Shimoda, N., Toyoda-Yamamoto, A., Nagamine, J., Usami, S., Katayama, M., Sakagami, Y., and Machida, Y.,** Control of expression of *Agrobacterium vir* genes by synergistic actions of phenolic signal molecules and monosaccharides. *Proc. Natl. Acad. Sci. U.S.A.,* 87, 6684, 1990.
11. **Cangelosi, G. A., Ankenbauer, R. G., and Nester, E. W.,** Sugars induce the *Agrobacterium* virulence genes through a periplasmic binding protein and a transmembrane signal protein, *Proc. Natl. Acad. Sci. U.S.A.,* 87, 6708, 1990.

12. **Kado, C. I.,** Molecular mechanism of crown gall tumorigenesis, *Critical Reviews in Plant Science,* 10, 1, 1991.
13. **Morris, J. W. and Morris, R. O.,** Identification of an *Agrobacterium tumefaciens* virulence gene inducer from the pinaceous gymnosperm *Pseudotsuga menziesii, Proc. Natl. Acad. Sci. U.S.A.,* 87, 3614, 1990.
14. **Messens, E., Dekeyser, R., and Stachel, S. E.,** A nontransformable *Triticum monococcum* monocotyledonous culture produces the potent *Agrobacterium vir*-inducing compound ethyl ferulate, *Proc. Natl. Acad. Sci. U.S.A.,* 87, 4368, 1990.
15. **Tate, M. E. and Savage, A.,** Chemistry and *vir* induction in agrobacteria. *Fallen Leaf Lake Conf. Abstr.,* p. 42, 1989.
16. **Melchers, L. S., Regensburg-Tuink, T. J. G., Bourret, R. B., Sedee, N. J. A., Schilperoort, R. A., and Hooykaas, P. J. J.,** Membrane topology and functional analysis of the sensory protein VirA of *Agrobacterium tumefaciens, EMBO J.,* 8, 1919, 1989.
17. **Leroux, B., Yanofsky, M. F., Winans, S. C., Ward, J. E., Ziegler, S. F., and Nester, E. W.,** Characterization of the *virA* locus of *Agrobacterium tumefaciens:* a transcriptional regulator and host range determinant, *EMBO J.,* 6, 849, 1987.
18. **Morel, P., Powell, B. S., and Kado, C. I.,** Characterization of three functional domains responsible for a kinase activity in VirA, a transmembrane sensory protein encoded by the Ti plasmid of *Agrobacterium tumefaciens, C. R. Acad. Sci. Paris,* 310, 21, 1989.
19. **Huang, Y., Morel, P., Powell, B., and Kado, C. I.,** VirA, a coregulator of Ti-specified virulence genes, is phosphorylated *in vitro, J. Bacteriol.,* 172, 1142, 1990.
20. **Jin, S., Roitsch, T., Ankenbauer, R. G., Gordon, M. P., and Nester, E. W.,** The VirA protein of *Agrobacterium tumefaciens* is autophosphorylated and is essential for *vir* gene regulation, *J. Bacteriol.,* 172, 525, 1990.
21. **Wierenga, R. K. and Hol, W. G. J.,** Predicted nucleotide-binding properties of p21 protein and its cancer-associated variants, *Nature (London),* 302, 842, 1983.
22. **Veluthambi, K., Krishnan, M., Gould, J. H., Smith, R. H., and Gelvin, S. B.,** Opines stimulate induction of the *vir* genes of the *Agrobacterium tumefaciens* Ti plasmid, *J. Bacteriol.,* 171, 3696, 1989.
23. **Sanders, D. A., Gillece-Castro, B. L., Stock, A. M., Burlingame, A. L., and Koshland, D. E.,** Identification of the site of phosphorylation of the chemotaxis response regulator protein, CheY*, *J. Biol. Chem.,* 264, 21770, 1989.
24. **Jin, S., Prusti, R. K., Roitsch, T., Ankenbauer, R. G., and Nester, E. W.,** Phosphorylation of the VirG protein of *Agrobacterium tumefaciens* by the autophosphorylated VirA protein: essential role in biological activity of VirG, *J. Bacteriol.,* 172, 4945, 1990.
25. **Ninfa, A. J. and Magasanik, B.,** Covalent modification of the *glnG* product, NRI, by the IglnL product, NRII, regulates the transcription of the *glnALG* operon in *Escherichia coli, Proc. Natl. Acad. Sci. U.S.A.,* 83, 5909, 1986.
26. **Forst, S., Delago, J., and Inouye, M.,** DNA-binding properties of the transcription activator (OmpR) for the upstream sequences of *ompF* in *Escherichia coli* are altered by *envZ* mutations and medium osmolarity, *J. Bacteriol.,* 171, 2949, 1989.
27. **Aiba, H., Nakasai, F., Mizushima, S., and Mizuno, T.,** Phosphorylation of a bacterial activator protein, OmpR, by a protein kinase, EnvZ, results in stimulation of its DNA-binding ability, *J. Biochem.,* 106, 5, 1989.
28. **Forst, S., Delgado, J., and Inouye, M.,** Phosphorylation of OmpR by the osmosensor EnvZ modulates expression of the *ompF* and *ompC* genes in *Escherichia coli, Proc. Natl. Acad. Sci. U.S.A.,* 86, 6052, 1989.
29. **Powell, B. S., Rogowsky, P. M., and Kado, C. I.,** *virG* of *Agrobacterium tumefaciens* plasmid pTiC58 encodes a DNA-binding protein, *Mol. Microbiol.,* 3, 411, 1989.
30. **Pazour, G. J. and Das, A.,** *virG,* an *Agrobacterium tumefaciens* transcriptional activator, initiates translation at a UUG codon and is a sequence-specific DNA-binding protein, *J. Bacteriol.,* 172, 1241, 1990.
31. **Steck, T. R., Morel, P., and Kado, C. I.,** Vir box sequences in *Agrobacterium tumefaciens* pTiC58 and A6, *Nucleic Acids Res.,* 16, 8736, 1988.
32. **Winans, S. C., Jin, S., Komari, T., Johnson, K. M., and Nester, E. W.,** The role of virulence regulatory loci in determining *Agrobacterium* host range, in *Plant Molecular Biology,* D. von Wettsein and N.-H. Chua Eds., Plenum Press, NY, 1987, 573.
33. **Jin, S., Roitsch, T., Christie, P. J., Gordon, M. P., and Nester, E. W.,** The regulatory VirG protein specifically binds to a *cis*-acting regulatory sequence involved in transcriptional activation of *Agrobacterium tumefaciens* virulence genes, *J. Bacteriol.,* 172, 531, 1990.
34. **Powell, B. S. and Kado, C. I.,** Specific binding of VirG to the *vir* box requires a C-terminal domain and exhibits a minimum concentration threshold, *Mol. Microbiol.,* 4, 2159, 1990.
35. **Rogowsky, P. M., Close, T. J., Chimera, J. A., Shaw, J. J., and Kado, C. I.,** Regulation of the *vir* genes of *Agrobacterium tumefaciens* plasmid pTiC58, *J. Bacteriol.,* 169, 5101, 1987.
36. **Close, T. J., Rogowsky, P. M., Kado, C. I., Winans, S. C., Yanofsky, M. F., and Nester, E. W.,** Dual control of *Agrobacterium tumefaciens* Ti plasmid virulence genes, *J. Bacteriol.,* 169, 5113, 1987.

37. **Winans, S. C., Kerstetter, R. A., and Nester, E. W.,** Transcriptional regulation of the *virA* and *virG* genes of *Agrobacterium tumefaciens, J. Bacteriol.,* 170, 4047, 1988.
38. **D'Souza, M. R. and Kado, C. I.,** unpublished results, 1990.
39. **Aoyama, T. and Oka, A.,** A common mechanism of transcriptional activation by the three positive regulators, VirG, PhoB, and OmpR., *FEBS Lett.,* 263, 1, 1990.
40. **Kamoun, S., Cooley, M. B., Rogowsky, P. M., and Kado, C. I.,** Two chromosomal loci involved in production of exopolysaccharide in *Agrobacterium tumefaciens, J. Bacteriol.,* 171, 1755, 1989.
41. **Thomashow, M. F., Karlinsey, J. E., Marks, J. R., and Hurlbert, R. E.,** Identification of a new virulence locus in *Agrobacterium tumefaciens* that affects polysaccharide composition and plant cell attachment, *J. Bacteriol.,* 169, 3209, 1987.
42. **Douglas, C. J., Staneloni, R. J., Rubin, R. A., and Nester, E. W.,** Identification and genetic analysis of an *Agrobacterium tumefaciens* chromosomal virulence region, *J. Bacteriol.,* 161, 850, 1985.
43. **Rong, L., Karcher, S. J., O'Neal, K., Hawes, M. C., Yerkes, C. D., Jayaswal, R. K., Hallberg, C. A., and Gelvin, S. B.,** *picA,* a novel plant-inducible locus on the *Agrobacterium tumefaciens* chromosome, *J. Bacteriol.,* 172, 5828, 1990.
44. **Shaw, J. J., Settles, L. G., and Kado, C. I.,** Transposon Tn4431 mutagenesis of *Xanthomonas campestris* pv. *campestris:* characterization of a nonpathogenic mutant and cloning of a locus for pathogenicity, *Mol. Plant-Microbe Interact.,* 1, 39, 1988.
45. **Kamoun, S. and Kado, C. I.,** A plant-inducible gene of *Xanthomonas campestris* pv. *campestris* encodes an exocellular component required for growth in the host and hypersensitivity on nonhosts, *J. Bacteriol.,* 172, 5165, 1990.
46. **Iuchi, S., and Lin, E. C. C.,** *arcA* (dye), a global regulatory gene in *Escherichia coli* mediating repression of enzymes in aerobic pathways, *Proc. Natl. Acad. Sci. U.S.A.,* 85, 1888, 1988.
47. **Weston, L. A. and Kadner, R. J.,** Role of *uhp* genes in expression of the *Escherichia coli* sugar-phosphate transport system, *J. Bacteriol.,* 170, 3375, 1988.
48. **Ronson, C. W., Astwood, P. M., Nixon, B. T., and Ausubel, F. M.,** Deduced products of C4-dicarboxylate transport regulatory genes of *Rhizobium leguminosarum* are homologous to nitrogen regulatory products, *Nucleic Acids Res.,* 15, 7921, 1987.
49. **David, M., Daveran, M. L., Batut, J., Dedieu, A., Domergue, O., Ghai, J., Hertig, C., Boistard, P., and Kahn, D.,** Cascade regulation of *nif* gene expression in *Rhizobium meliloti, Cell,* 54, 671, 1988.

# PHYTOHORMONES AND OLIVE KNOT DISEASE

**Giuseppe Surico and Nicola S. Iacobellis**

## TABLE OF CONTENTS

# I. INTRODUCTION

Virulent strains of *Pseudomonas syringae* ssp. *savastanoi* (*P. savastanoi*) induce excrescences (knots, galls) at the site of bacterial infection. This pathogenic response is characterized both by cell enlargement and cell division and it includes differentiation into xylem and phloem elements arranged in bundles or cyclic nodules.[1,2]

Knots are most frequently seen on the young stems and branches and twigs of olive plants. On ash, *P. savastanoi* causes wartlike excrescences. Recently[3] it has been proposed that the group of bacteria attacking Oleaceae and the Apocynacea, *Nerium oleander,* be classified as a subspecies of *P. syringae* van Hall, and that the variants on ash, olive, and oleander be considered as separate pathovars: pv. *fraxini* (ash isolates), pv. *nerii* (oleander isolates) and pv. *oleae* (olive, privet and jasmine isolates). This classification is not widely accepted but it is a fact that the disease symptoms on ash differ from those on olive, oleander, privet and jasmine, as there are no galls;[4] also, oleander strains can infect olive plants (as well as privet, jasmine and ash), while olive strains, with very few exceptions, are avirulent on oleander but not on the other plant species.[5-7] Last, ash strains are generally avirulent on oleander but can induce swollen lesions on olive. These differences in the behavior of the *P. savastanoi* strains could, in part, be due to differences in bacterial synthesis and in the secretion of the phytohormones indole-3-acetic acid (IAA) and cytokinins which have been recognized as the major factors in the production of knots.

In this chapter the role of the different phytohormones in the development of host-*P. savastanoi* relationships and corresponding disease symptoms will be outlined.

# II. INDOLE-3-ACETIC ACID

## A. IAA PRODUCTION BY *PSEUDOMONAS SAVASTANOI*

*Pseudomonas savastanoi* converts L-tryptophan into indole-3-acetic acid (IAA) in two steps (Figure 1). In the first reaction, catalyzed by tryptophan 2-monooxygenase, L-tryptophan is converted into indole-3-acetamide (IAM); in the second reaction, catalyzed by an indoleacetamide hydrolase, IAM is converted into IAA and ammonia.[8-10] Accumulation of IAA in liquid growth media supplemented with L-tryptophan occurs rapidly over the first 3 to 5 days, during log and stationary growth phases, but declines thereafter.[11-12] The amounts detected vary depending on the strains, with oleander strains generally, although not always, producing less IAA than olive strains. In minimal culture media, *P. savastanoi* virulent strains accumulate, on the average, 1 to 3 mg of IAA/l of culture medium.[12-13] When tryptophan is added there could be a tenfold or even greater increase in IAA accumulation.

## B. ROLE OF IAA IN PATHOGENICITY

The synthesis of IAA by *P. savastanoi* is a major factor in the production of knots: mutants having a reduced ability to synthesize IAA fail to incite knots, or incite either smaller knots than the parental strains or swollen lesions. The involvement of IAA synthesis in knot formation on oleander was demonstrated by experiments with the Californian oleander strains EW2015, ATCC23868, and EW2009. From the first two, Smidt and Kosuge[12] obtained mutants with altered synthesis of IAA by selection for resistance to α-methyltryptophan. In spite of the fact that mutants unable to accumulate IAA in culture grew in host tissue similarly to IAA-producing strains, they failed to incite knot formation. By contrast, mutants which accumulated twice the amount as compared to the wild-type parent strains caused larger knots. These results showed that the level of IAA production by *P. savastanoi* is at least a limiting factor in knot formation.

A confirmation of the role of IAA in knot induction came from the observation that, in oleander strain EW2009, the genetic loci (*iaaM* and *iaaH*) for the two enzymes tryptophan

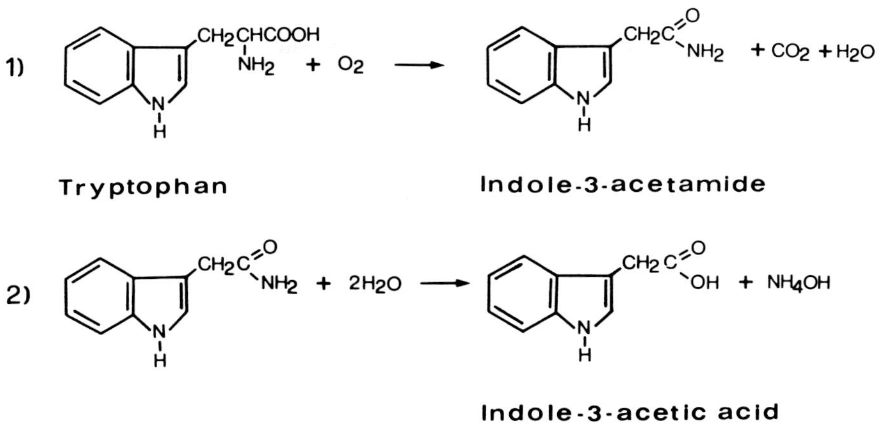


FIGURE 1. Pathway of IAA synthesis in *P. savastanoi*.

2-monooxygenase and indoleacetamide hydrolase are borne on a 52-kb plasmid designated as pIAA1.[14] To demonstrate that plasmid pIAA1 was associated with IAA synthesis, an IAA-deficient mutant of strain EW2009 was transformed with pIAA1. This procedure restored *iaaM* and *iaaH* activities and the mutant reacquired full virulence (i.e., induction of knots). Moreover, the recombinant plasmid pLUC1 was constructed from RSF1010, a vector plasmid, and an *Eco*RI restriction fragment (fragment M) of pIAA1.[15] The introduction of pLUC1 into pIAA1-less strains enabled them to produce IAM and to restore virulence on oleander, but not to the same degree as when pIAA1 was reintroduced. To explain the pathogenic behavior of such a manipulated microorganism, it has been suggested that host enzymes cooperate in bringing about the infection by converting IAM into IAA and that the latter compound causes the knot formation. An alternative hypothesis, excluding host activity, is that IAM of bacterial origin is not further converted into IAA in the plant and that the smaller knots are induced by the IAA synthesized through a chromosomal biosynthetic pathway or more likely through a subsidiary pathway, with indole-3-pyruvic acid as an intermediate, which seems to function in *P. savastanoi*.[17] Working with α-methyltryptophan-resistant mutants of olive and oleander strains, it was observed that in presence of added tryptophan, these mutants, which had lost their *iaaM* and *iaaH* activities, were able to accumulate detectable amounts of IAA in the culture medium.[12,13] When inoculated on olive plants, some of these mutants induced the formation of new tissue which increased with time to yield smaller, greener, and softer knots than those produced by their parental strains. Oleander plants inoculated with IAA-deficient mutants of the oleander strains (these mutants had lost the IAA plasmid) had swollen lesions at the site of inoculation but never any typical knots. However, the strongest evidence for the role of IAA as a virulence factor came from the experiments using insertion mutants performed by Comai and Kosuge.[18] Three spontaneous mutants were isolated with insertions in the monooxygenase gene; these mutants were avirulent and did not produce IAA.

With olive and privet strains it was also possible to isolate α-methyltryptophan-resistant mutants with a reduced capacity for IAA synthesis. The plasmid profile of these mutants was not different from that of the parental strains, but they did not induce knots or induced attenuated symptoms when inoculated on olive or privet plants.[13] Apparently the genes for IAA production are not on a plasmid but on the chromosome.[19]

All these results demonstrated the direct relationship between IAA production and the virulence of *P. savastanoi,* revealed the central role played by the enzyme tryptophan 2-monooxygenase in virulence expression by the bacterium, and stated that the entire IAA operon is necessary for full virulence.

Studies on the epiphytic survival of Iaa$^-$ mutants on olive leaves showed that, besides its direct involvement in symptom production, IAA may confer a selective advantage to the bacterium in nature. In fact, while wild-type strains were able to multiply on olive leaves, their Iaa$^-$ mutants failed to multiply and were either scarcely or not at all detectable a few days after the inoculation on the leaves. Thus, the production of IAA seems to be associated with the epiphytic survival of *P. savastanoi.*[20]

## C. SYNTHESIS AND REGULATION OF IAA: EFFECT ON VIRULENCE

The production and further metabolism of IAA in *P. savastanoi* is subject to a variety of controls. These include: (1) inhibition of anthranilate synthase by L-tryptophan; (2) inhibition of tryptophan 2-monooxygenase by IAM and IAA; and (3) further conversion of IAA into lysine conjugates which have less growth-promoting activity than IAA.

One of the first findings obtained by Kosuge and his co-workers, and confirmed by other researchers, was that the size of the intracellular tryptophan pool regulates the flow of tryptophan for the production of IAA.[16,21] The addition of L-tryptophan to the culture medium greatly enhances production of the auxin by the bacterium even though it does not influence the production of the enzymes responsible for its transformation into IAA. Smidt and Kosuge[12] and Surico et al.[13] found that the addition of 0.5 *M* L-tryptophan to the culture medium resulted in about a three- to tenfold increase in IAA accumulation by some oleander and olive strains over the amounts produced in unsupplemented media.

The size of the cellular tryptophan pool depends upon the feedback inhibition of anthranilate synthase. Strains with a decreased sensitivity to feedback inhibition by tryptophan should accumulate more tryptophan and thus more IAA. However, this does not necessarily mean that super IAA producer mutants are more virulent. Some mutants with a very high capacity for IAA production have been isolated which failed to grow in host tissue and incite knots.[22] Since the primary role in the metabolism of an essential amino acid such as tryptophan is protein synthesis, it seems possible that part of the tryptophan in these mutants is diverted to the synthesis of a secondary metabolite, IAA, to the detriment of protein production and with a consequent decrease in pathogen fitness.[16,22]

From the perspective of the pathogen, an effective means of controlling IAA synthesis is the strong inhibition of tryptophan 2-monooxygenase by IAM ($K_i$ = 7 $\mu M$) and IAA ($K_i$ = 225 $\mu M$) which is reversed by tryptophan. IAA added to cell suspensions of *P. savastanoi* also inhibit IAA synthesis. In this context, a high availability of tryptophan leads to an abnormal synthesis of IAA. This is the case when exogenous tryptophan is added to the bacterial culture medium. By contrast, IAA tends to accumulate in the medium only after the cellular demand for protein synthesis has been satisfied. Thereafter, more tryptophan accumulates, inhibition by IAM and IAA is reversed, and more auxin is produced.[16,23] Of course, any alteration of this balanced metabolic process will affect the virulence.

## D. LYSINE CONJUGATES

IAA is not the end product of the tryptophan metabolism (Figure 2). In oleander, but not in olive strains of *P. savastanoi,* IAA is further converted into indole-3-acetyl-ε-L-lysine (IAA-lysine)[24,25] or, into α-*N*-acetylindole-3-ε-L-lysine (acetyl IAA-lysine)[26] either directly or by way of IAA-lysine. These two related amino acid conjugates are either less active or altogether biologically inactive. When assayed for its growth-promoting activity in the avena coleoptile growth curvature test,[27] IAA-lysine proved to be approximately one third as active as IAA.[28] An additional IAA derivative, though considered as an active storage form of IAA,[29] has recently been isolated from the culture filtrates of an olive *P. savastanoi* strain and identified as IAA-methyl ester.[30]

Considering that the virulence of *P. savastanoi* is directly related to the amount of IAA released in the host tissue, the virulence of the bacterium could conceivably be altered not

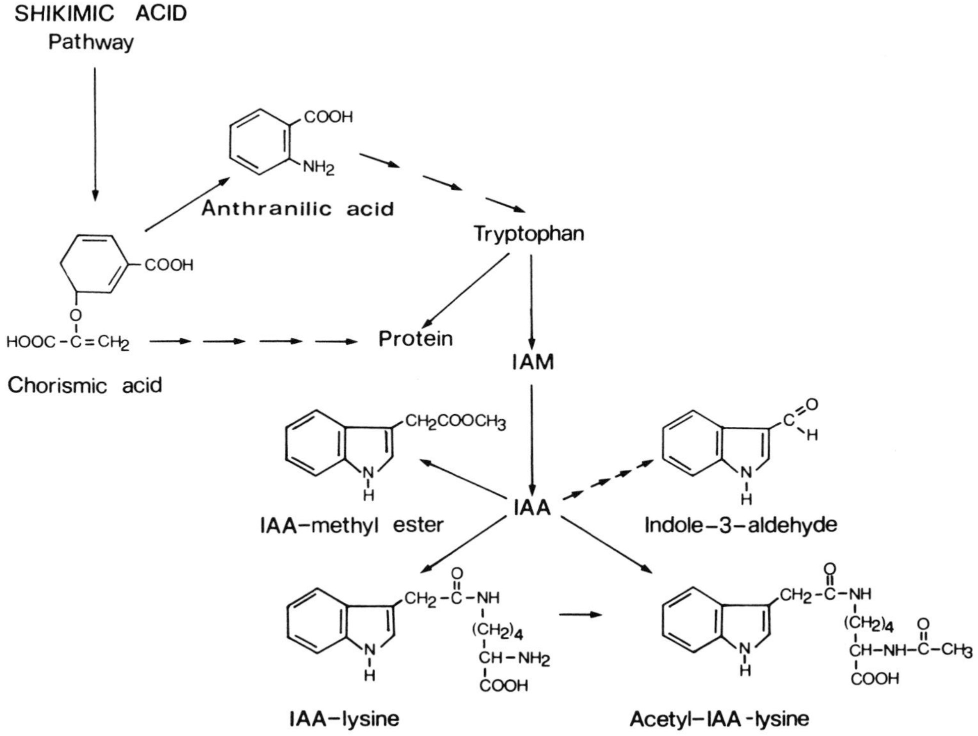


FIGURE 2. Synthesis and transformation of IAA in *P. savastanoi*.

only by the rate of IAA synthesis but also by the rate of the conversion of IAA into IAA-lysine(s). Strains producing IAA-lysine would be expected to release smaller amounts of IAA, while strains lacking IAA-lysine synthase activity would accumulate higher concentrations of IAA. In fact, olive strains, which are unable to convert IAA to IAA-lysine, typically accumulate more IAA in culture than oleander strains,[31] and oleander strains, when inoculated on olive plants, often induce larger knots than olive strains.[32]

### E. RELATIONSHIP BETWEEN IAA-LYSINE SYNTHASE AND VIRULENCE

IAA-lysine synthase is ATP-dependent and also requires a divalent metal cation such as $Mg^+$ or $Mn^+$. The *iaaL* gene for the enzyme is borne on the IAA plasmid but it is not part of the IAA operon since insertion mutations in the *iaaM* gene, the first gene transcribed in the operon, has no effect on the ability of *P. savastanoi* to synthesize IAA-lysine.[31] The *iaaL* locus has been located approximately 2-kb from the IAA operon and is transcribed in the opposite direction as the IAA operon.

Recently,[33] the *iaaL* gene has been sequenced and an open reading frame (ORF2) containing 1188 nucleotides encoding a hydrophilic protein of 396 amino acids with an estimated molecular weight of 44,000 Da has been identified. Another open reading frame (ORF1, 606 nucleotides) encoding a protein of 21,000 Da has been found in the same 4.25-kb *Eco*RI fragment derived from pIAA1 which has been shown to direct the production of IAA-lysine in *P. savastanoi*. ORF1 did not appear to be involved in IAA-lysine synthesis. It could be a component of the acetylation mechanism for IAA-lysine[33] or, alternatively, it might constitute part of an IAA transport mechanism active in *P. savastanoi*.[34]

In order to elucidate the physiological function of IAA-lysine in the pathogen, the *iaaL* gene has been cloned into the vector pUC8 to construct the chimeric plasmids pLG84, pLG137, and others. pLG137 was introduced by triparental mating into the Californian olive strain TK1050. TK1050 transconjugants containing the construct pLG317 accumulated IAA-lysine in culture and the amount of IAA produced was reduced by one third, demonstrating that the conversion of IAA to IAA-lysine can effectively influence IAA pool size in *P. savastanoi*.

There are no data on the virulence of strain TK1050(pLG317) due to the instability of plasmid pLG317 *in planta,*[35] but it was possible to introduce the *iaaL* gene into tobacco plants.[36] Transgenic plants displayed morphological and developmental alterations including epinasty, reduction in root growth, and increased side shoot growth, pointing toward an altered balance of phytohormone levels.

The oleander mutant PB213(*iaaL*::Tn5) accumulated five times as much IAA in culture as its parental strain. However, when the mutant strain was inoculated into oleander plants, it induced only swollen lesions and no typical knots. The authors then investigated the growth of PB213(*iaaL*::Tn5) and PB213 in oleander leaves and observed that the mutant grew more slowly than PB213 and reached a lower population by the time visible symptoms appeared on oleander leaves inoculated with PB213. Thus, the capacity to convert IAA into IAA-lysine may be related to optimal growth and to full expression of the virulence of oleander strains in their hosts.[35]

## F. OXIDATION OF IAA

Another process by which the IAA level could be regulated, besides controlled synthesis and formation of bound auxin, is the enzymic oxidative degradation into inactive compounds. This reaction, widely present in the plant kingdom,[37] also seems to occur in *P. savastanoi*. In fact, indole-3-aldehyde has been isolated[38] from 5-day-old culture filtrates of olive strain ITM317. This compound, which does not seem to be produced by oleander strains, is considered an alternative end product of the oxidation of IAA by peroxidase with indolenine hydroperoxide and indolenine epoxide as intermediates.[39] It is irreversibly formed and probably protects tissues from the accumulation of excessive auxin. Moreover, IAA oxidation has been suggested as being responsible for disease resistance in several host-parasite interactions.[40] However, its role in *P. savastanoi* has yet to be determined. At present, the most interesting hypothesis is that in olive *P. savastanoi* strains IAA oxidation could function like the conversion of IAA to IAA-lysine in oleander strains.

## G. IAA PLASMIDS IN *PSEUDOMONAS SAVASTANOI*

A considerably different situation has been observed in the localization of *iaaM* and *iaaH* genes among strains of *P. savastanoi*. All the oleander strains so far tested harbor a plasmid associated with IAA production but its molecular weight ranges from 52 (pIAA1) to 73 (pIAA2) kb.[14,19,41] In the olive and privet strains thus far examined, the genes for IAA production are chromosome-borne. Also, in oleander strains, copies of the IAA genes seem to be located on the chromosome, but these genes are either not functional or produce little IAA.[42]

Genes *iaaM* and *iaaH* have been localized within a 4-kb *Eco*RI fragment derived from pIAA1 of the oleander strain EW2009. The cloning of the *Eco*RI fragment in vector plasmids RSF1010 and pB328 to obtain the recombinant plasmids pLUC1, pLUC2, and pCP3, respectively, made the physical mapping of the *iaaM* and *iaaH* loci possible, establishing that they are organized as an operon (Figure 3) with *iaaH* being distal to the promoter.[15,43-45]

Sequence analysis of *iaaM* indicated an open reading (ORF frame) encoding a protein of 557 amino acids with a molecular weight of 61,783 Da. This protein corresponds to

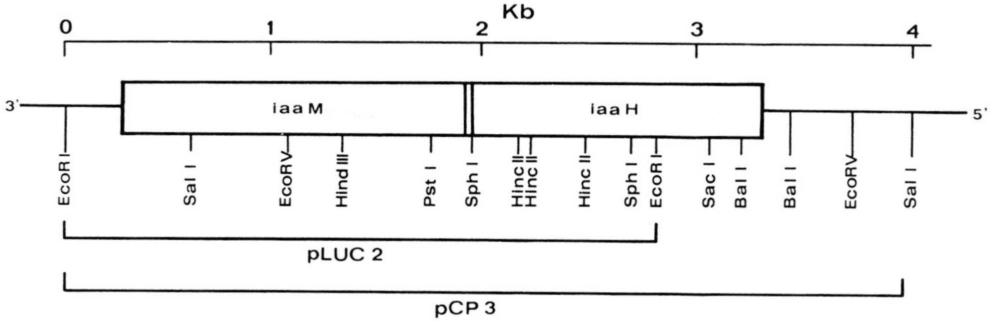


FIGURE 3. Restriction maps of recombinant plasmids pLUC2 and pCP3 containing IAA coding regions of the IAA plasmid. Adapted from Yamada, T., et al., *Proc. Natl. Acad. Sci. U.S.A.*, 82, 6522, 1985.

tryptophan monooxygenase purified to homogeneity by Hutcheson and Kosuge.[46] The nucleotide sequence of the *iaaH* gene revealed the presence of an ORF specifying a protein of 455 amino acids with a molecular weight of 48,515 Da. By utilizing a direct transcription-translation system, it was observed that pCP3 encoded two proteins with molecular weights of 62,000 and 47,000 Da, respectively, in agreement with the deduced size of these proteins. The identity of the two proteins was confirmed by the fact that a cell-free lysate of *Escherichia coli* (pCP3) contains tryptophan monooxygenase and indoleacetamide hydrolase activities. Recent investigations on the promoter for *iaaM* and *iaaH* seem to indicate that transcription starts at −400 bp from the first ATG of *iaaM*. Moreover, an analysis of the nucleotide sequence upstream of the predicted transcriptional start revealed both *Pseudomonas* and *E. coli* promoter consensus sequences.[45,47,48]

The two-step pathway from tryptophan to IAA also operates in *Agrobacterium tumefaciens* which induces crown gall tumors by causing an overproduction of auxin and cytokinins in transformed plant cells.[48] The coding sequences of IAA genes in *P. savastanoi* share a significant homology with the sequences of the analogous genes, *tms-1* (= *iaaM*) and *tms-2* (= *iaaH*), encoding IAA production in the T-DNA of the octopine-type plasmid of *A. tumefaciens*.[44] Strong homology was observed in the apparent FAD-binding site of tryptophan monooxygenase and the central part of the *tms-1* product, suggesting that the two proteins have similar functions. The results proved an overall similarity in the biosynthetic pathway of IAA in the two phytopathogenic bacteria which, typically, cause localized overgrowth in host tissues.

## H. IS ELEMENTS IN *PSEUDOMONAS SAVASTANOI*

In *P. savastanoi,* IAA synthesis does not occur if strains lack pIAA, or if pIAA undergoes deletions involving *iaaM* and *iaaH* genes. Moreover, some $Iaa^-$ mutants of the oleander strains EW2009 and EW2015 have been found with insertions in the monooxygenase genes and a consequent loss of IAA production and virulence. Two transposable elements are found to reside in *P. savastanoi,* IS51 and IS52.[18,49] Although these IS elements share the property of transposition into *iaaM,* but at different sites, no similarities have been found between the nucleotide sequences of the two. However, IS51 and IS52 have similar sizes (1311 and 1219 bp, respectively) and terminal inverted repeats of 20 and 10 bp, respectively.[49]

An interesting homology between a 531-bp region of IS51 and a portion of the central region of T-DNA from *A. tumefaciens* has been observed.[49] These findings and the reported homology between the IAA genes in *P. savastanoi* and in *A. tumefaciens* strengthen the hypothesis that these genes have a common origin. The T-DNA region homologous to IS51 may be a residue of a transpositional event, while IS51 found downstream of the *iaaH* in some *P. savastanoi* strains may be the border of a transposon.[49]

It has been suggested that the deletions of the IAA operon, which occur more frequently than insertional events in the IAA genes, may be mediated by the IS elements.[18,50] However, it is not yet clear how the mechanism works and whether the virulence genes are specific targets of inactivation. Recent observations have, in fact, shown that, in oleander strain PB213, IAA genes are lost by two deletions of 18- and 22-kb.[51] The deletion end points are in a region of repetitive sequences and all the deletions share a common end point upstream the IAA genes. The other end points downstream of the above genes are around a resident copy of IS51, which is interrupted by another insertion element, IS53. This new insertion element has been cloned and sequenced.[51] The transposase gene appears to be regulated by a heat shock promoter.[52] Independent studies using another oleander strain, ITM519, showed that heat shock treatment dramatically enhances the loss of IAA production by the bacterium.[53] At critical temperature (29.5 to 32.5°C) 56% of the treated cells are $Iaa^-$ mutants; all are characterized by a 20-kb specific deletion of the IAA plasmid. These findings confirm that deletions may be mediated by the presence of IS elements. The fact that IS elements may transpose at a high frequency when the bacterium is under stress (high temperature, etc.) indicates that they could be involved in the mechanism of the evolution of *P. savastanoi* as a pathogen.

# III. CYTOKININS

## A. CYTOKININ PRODUCTION BY *PSEUDOMONAS SAVASTANOI*

Initial studies on the cytokinin production ability of *P. savastanoi* were performed with the olive type-strain NCPPB639.[32,54] Ethyl acetate extracts from 3- to 10-day-old cultures stimulated the growth of olive as well as tobacco callus. The TLC fractionation of the crude extracts gave rise to several active bands, the most active of which had the same $R_f$ values as isopentenyladenine (iP) and isopentenyladenosine (iPA). More recently, the study was extended and five cytokinins were isolated from the culture filtrate of oleander strain NCPPB640. Three were identified as zeatin (Z), zeatin riboside (ZR), and iPA by comparing their biological activities, chromatographic behavior, and mass spectra with those of authentic compounds.[55,56] The other two were novel biologically active cytokinins, which were identified by $^1$H- and $^{13}$C-NMR and mass spectrometry as 1′-methylzeatin (1′-MeZ) and 1″-methylzeatin riboside (1″-MeZR).[57-60] These results were confirmed by independent studies using the oleander strain PB213 and the olive strain EW1006.[61] Moreover, three more cytokinins, dihydrozeatin (DHZ), dihydrozeatin riboside (DHZR), and iP were identified by high-performance liquid chromatography, radioimmunoassay, and mass spectrometry (Figure 4).[61,63]

Time course experiments performed with NCPPB640 indicated that cytokinins were already present in 1-day-old cultures and increased with the growth of bacteria (Figure 5).[64] The relative quantities of the four cytokinins, Z, ZR, 1′-MeZ and 1″-MeZR, varied with time. Zeatin, the predominant compound in the culture, reached its highest level of accumulation in 3-day-old cultures (about 11 μmol/l) when the bacteria were still in the growth phase. Thereafter, its level suddenly decreased. In 10-day-old cultures, the most abundant cytokinins were the methylated derivatives.

There is little agreement on the level of cytokinins accumulated in culture by *P. savastanoi,* probably because different strains, culture media, and procedures for quantification were used by different authors. For example, MacDonald et al.[61] found that olive strain EW1006 accumulated up to 0.4 μg/ml of Z and 0.18 μg/ml of 1″-MeZR while oleander strain PB213-2 secreted only 0.08 μg/ml each of Z and ZR; Roberto and Kosuge[47] reported that the level of Z and ZR in culture approached 10 μg/ml; Surico et al.[13] found that the total cytokinin activity as shown by the culture filtrates of several olive and oleander strains varied in a biological assay from 3.0 to 5.3 μg of Z equivalents per milliliter of culture medium. Moreover, the *P. savastanoi* strains so far examined show variation as to the cytokinins synthesized. Zeatin is found in almost all strains, whereas ZR and the methylated derivatives are present only in some strains.[55,65]

All the above data indicate that there are differences in the cytokinin metabolism and enzymic regulation in *P. savastanoi.* Moreover, the level of accumulation of this class of phytohormones in culture is exceptional and much higher than that accumulated by other phytopathogenic bacteria (i.e., *A. tumefaciens* and *Rhodococcus fascians*) inducing hyperplasia and/or growth disorders through cytokinin production.[66,67] Only in the case of *P. amygdali,* the causal agent of hyperplastic canker in almond, are cytokinins produced at a level that is comparable to that accumulated by *P. savastanoi.*[68,69]

## B. CYTOKININ BIOSYNTHESIS

Several studies have found that cytokinins can arise by biosynthesis *de novo* both in higher plants and microorganisms.[70-72] In simple terms, an isopentenyl transferase enzyme catalyzes the transfer of the isoprenoid chain, of mevalonic acid (MVA) origin, from dimethylallylpyrophosphate (iPP) to 5′-adenosine monophosphate (5′-AMP) yielding isopentenylmonophosphate (iPMP). The hydroxylation of the side chain, which apparently occurs either at the nucleotide, nucleoside, or free-base stage, gives rise to Z and/or ZR. Free cytokinins may be further metabolized by an oxidative pathway with the cleavage of the side chain yielding adenine (Ade) and/or adenosine (Ado) or by glucosylation (Figure 6).

A similar biosynthetic pathway of cytokinins may occur in *P. savastanoi.* The fact that there is an isopentenyl transferase activity in the bacterium,[61] that young cultures contain iP and iPA, and that Z can be converted into DHZ, DHZR, 1′-MeZ and 1″-MeZR, as indicated by time course experiments, support this hypothesis. The methylated cytokinins, which proved to be at least as active as Z when assayed on tobacco callus growth, lettuce seed germination, and chlorophyll synthesis in etiolated cucumber cotyledons,[73,74] may, however, also arise by way of a direct biosynthetic route. The isolation of the gene(s) responsible for the methylation might elucidate this point.

## C. CYTOKININ GENES

Studies on the production of cytokinins by plasmid deletion mutants of strains PB213 and EW1006 indicated that the biosynthesis of cytokinins is, at least in part, plasmid-coded.[61] Specifically, the complete loss of the 42-kb plasmid and a 40-kb deletion of the 105-kb plasmid in strains PB213 and EW1006, respectively, abolished the secretion of Z and ZR. The construction of a library of plasmid DNA fragments from wild-type strain EW1006 in the wide-host range vector pRK290 and consequent transformation of *E. coli*(HB101) with the recombinant DNA resulted in one clone, pPS001, which secreted cytokinins in culture.

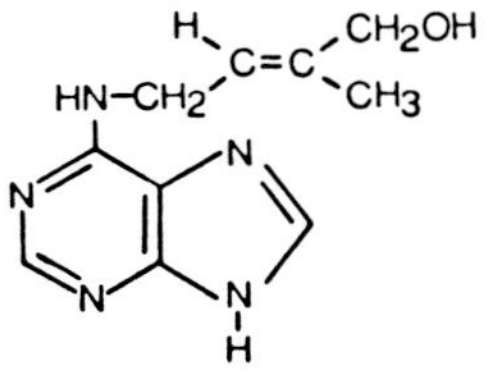

Zeatin

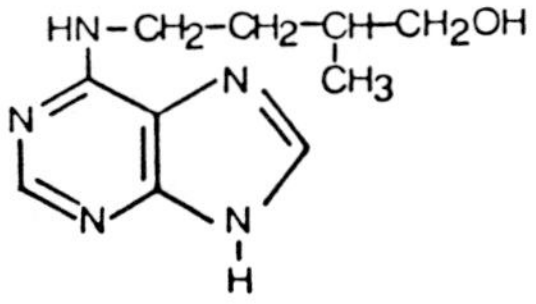

Dihydrozeatin

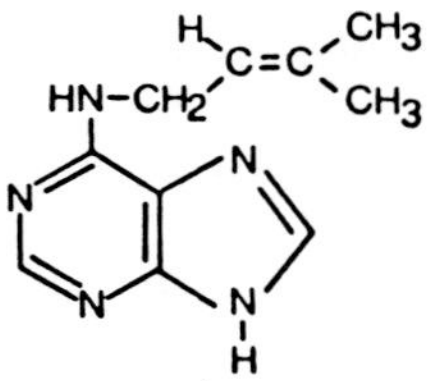

Isopentenyl adenine

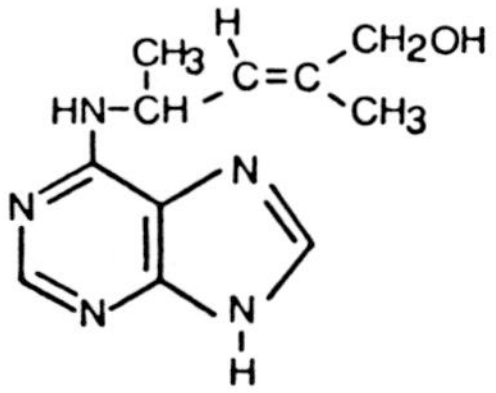

1′-Methylzeatin

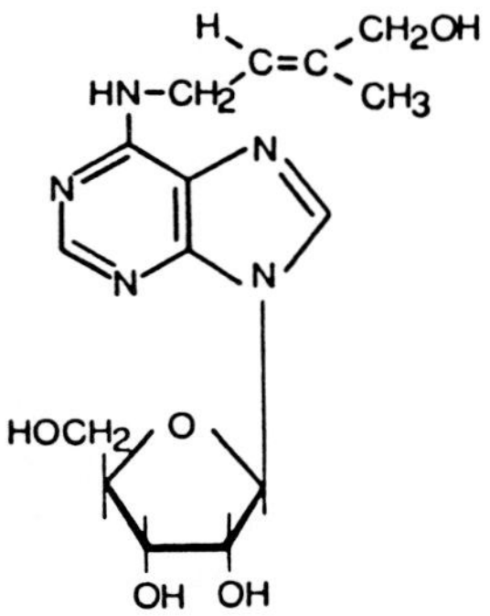

Zeatin riboside

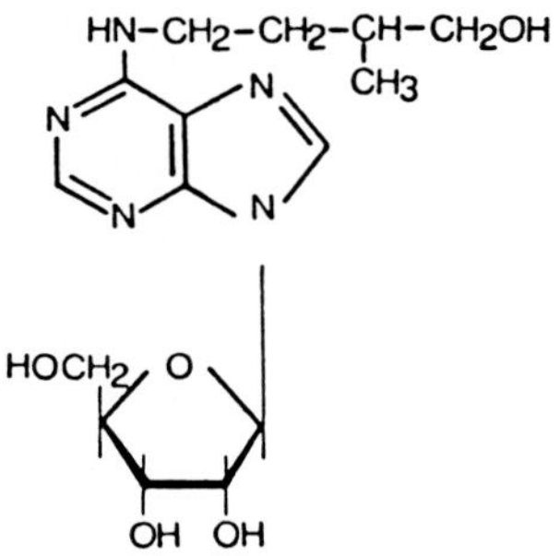

Dihydrozeatin riboside

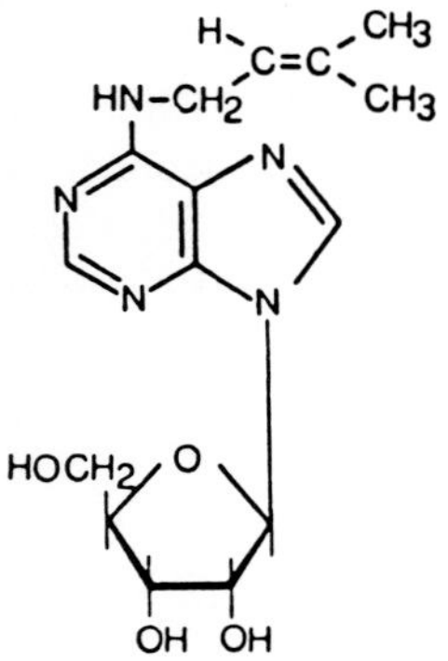

Isopentenyl adenosine

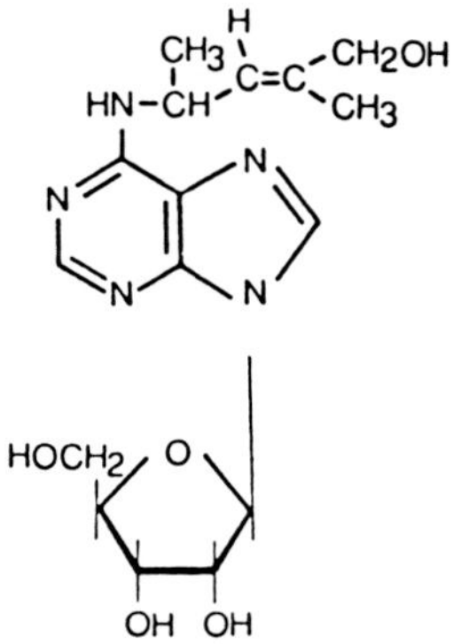

1″-Methylzeatin riboside

FIGURE 4. Structure of the cytokinins synthesized by *P. savastanoi*.

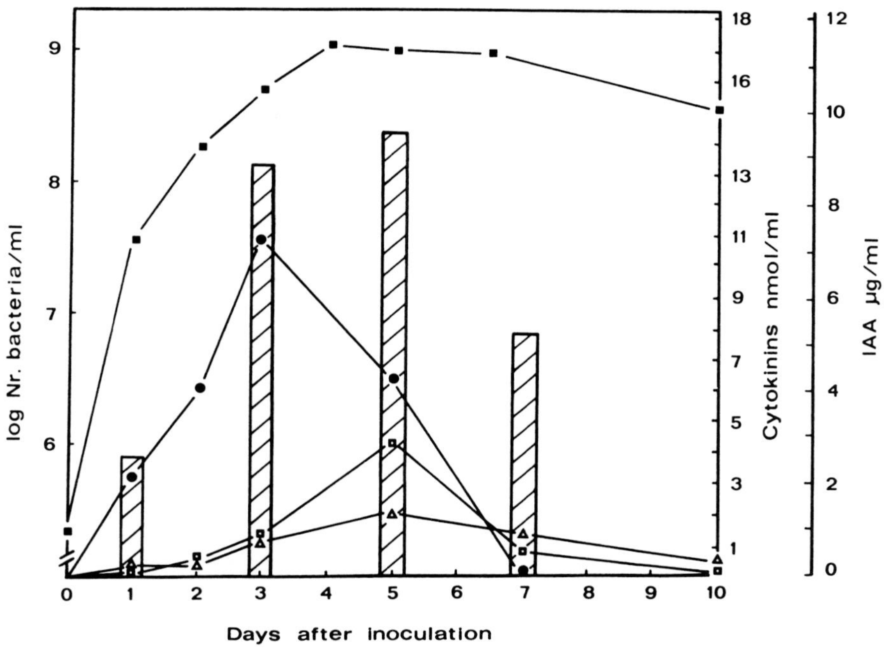


FIGURE 5. Relationship between growth of *P. savastanoi* strain NCPPB640 (-■-) and accumulation in culture of cytokinins and IAA (-▨-).-●-, Z;-□-, ZR;-△-, 1′-MeZ plus 1″-MeZR. Levels of phytohormones were estimated by high-performance liquid chromatography on two mixtures containing Z and ZR and 1′-MeZ and 1″-MeZR, respectively.

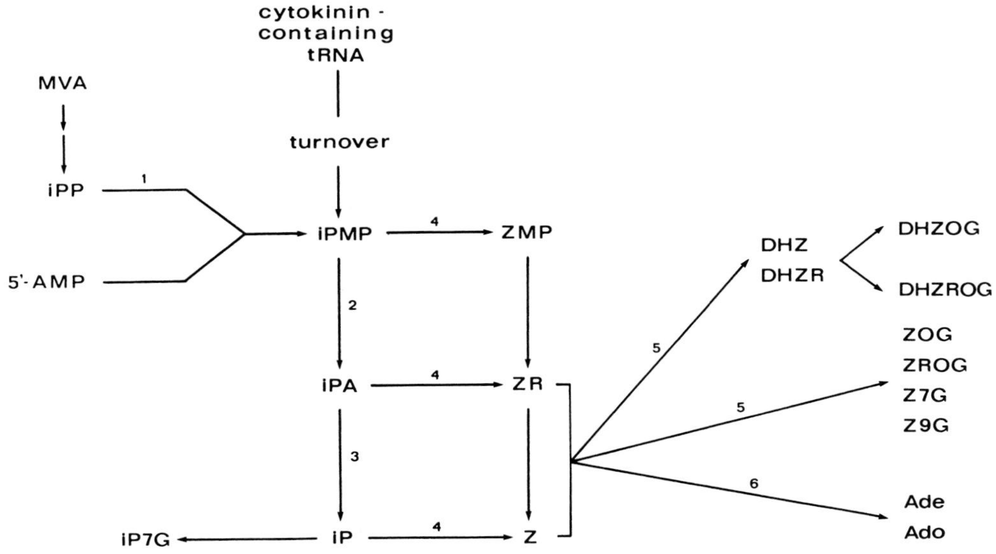


FIGURE 6. Biosynthetic and metabolic pathways of cytokinins. Pathway modified from Morris, O. R. (*Annu. Rev. Plant Physiol.*, 37, 509, 1986) and Chen, C. H. (in *Plant Growth Substances*, Wareing, P. F., Ed., Academic Press, San Diego, 1982, 155. 1, isopentenyl transferase; 2, 5'-nucleotidase; 3, adenosine nucleosidase; 4, cytokinin hydroxylase; 5, cytokinin glucosylation pathway; 6, cytokinin oxidase pathway. Abbreviations: DHZROG, dihydrozeatin riboside-*O*-glucoside; ZOG, zeatin *O*-glucoside; Z7G, zeatin-7-glucoside; Z9G, zeatin-9-glucoside; iP7G, isopentenyladenine-7-glucoside; ZMP, ribosylzeatin 5′-phosphate. For the other abbreviations see the text.

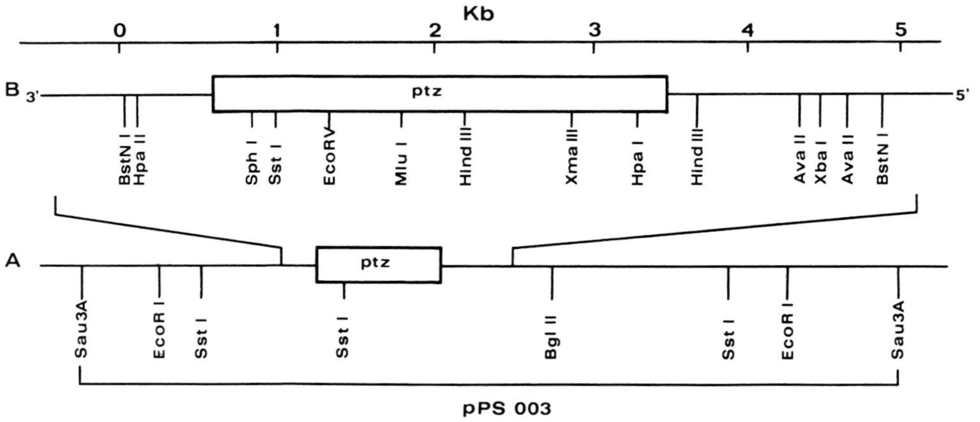


FIGURE 7. (A) Restriction map of 5.2-kb fragment (pSOO3) from *P. savastanoi* plasmid pCK1 containing the *ptz* locus. (B) Expanded map of the *ptz*-containing region. Adapted from Powell, G. K. and Morris, R. O., *Nucleic Acids Res.*, 14, 225, 1986.

A further analysis made it possible to locate the *Pseudomonas trans*-zeatin secretion locus (*ptz*) in the inserted fragment, and to subclone it. The resulting clone, pPS003, yielding a 5.2-kb fragment, was able to secrete Z, ZR, iP, and iPA in culture, and its cell lysate showed isopentenyl transferase activity.[61,75] The cloned fragment was mapped to locate the *ptz* locus (Figure 7) and DNA sequence analysis made it possible to find an ORF of 702 nucleotides, which encodes a protein with a predicted molecular weight of 28,816 Da. The expression of *ptz* by *in vitro* translation produces a protein with a molecular weight of 27,000 Da.[75]

The comparison of nucleotide sequences of *ptz* from *P. savastanoi* with the cytokinin biosynthetic genes, *ipt* and *tzs,* from *A. tumefaciens,* and of their deduced amino acid sequences, showed that the three genes were substantially homologous (at least 50%).

An examination of the sequences upstream of the translation-initiation codon of *ptz* revealed a ribosome binding site and a promoter region almost identical to the consensus sequences of *E. coli.*[75] The *tzs* gene also has a ribosome binding site of *E. coli* associated to it, but it lacks the −10 and −35 *E. coli* promoter consensus sequences. By contrast, the *ipt* gene is characterized by eukaryotic promoter sequences. Downstream from the translation-termination codon of *ptz* there is a 34-kb inverted repeat typical of prokaryotic transcription termination signals. While *tzs* is characterized by inverted repeats like *ptz, ipt* has a polyadenylation site associated to it. In any case, the overall homology of the cytokinin biosynthetic genes from *P. savastanoi* and *A. tumefaciens* suggests that these genes have a common origin, as already suggested for the IAA genes of the two bacteria.[44] A common origin seems to be more appropriate for *tzs* and *ptz* since, besides the homology in their coding region, they also share common prokaryotic control sequences.

The similarity of the phytohormone genes and their products in *P. savastanoi* and *A. tumefaciens* evoke a common basis, for at least some aspects of their tumorigenicity. However, *A. tumefaciens* induces crown gall tumors by transferring oncogenes (*tms-1, tms-2, ipt*) from the tumor-inducing plasmid to plant cells,[76] whereas in *P. savastanoi,* which does not appear to transfer any of its DNA to the host plant, the counterpart genes (*iaaM, iaaH, ptz*) are expressed within the bacterium.

## D. ROLE OF CYTOKININS

The idea that the cytokinins were involved in gall formation was prompted by the studies of Hewitt,[77] Wilson,[1] Wilson and Magie[78] and Surico[2] on the anatomy of olive and oleander galls. In particular, it was observed that around the colony that *P. savastanoi* establishes inside the diseased tissues, abnormal cell enlargement, a proliferation of hypertrophied cells, a differentiation of new cambial zones, and abnormally arranged tracheary elements and sieve tubes occurred.

The succession and pattern of these events clearly indicate that they were induced by factors identifiable with plant growth substances. It was already known that *P. savastanoi* was able to synthesize IAA *in vitro*[79,80] and that, in many systems, in addition to IAA and sucrose, cytokinins are necessary for cambial development and vascular differentiation.[81] In this context, there were enough indications which encouraged the research on cytokinin production by *P. savastanoi*. This research, begun in 1974, led to the identification of eight different cytokinins in the culture filtrate of *P. savastanoi* strains (Figure 4).

With regard to the role of cytokinins, an investigation carried out using six wild-type olive and oleander strains characterized by different levels of IAA and cytokinin production and their relative $Iaa^-$ mutants showed that typical knots formed only when inoculation was with *P. savastanoi* strains that produced cytokinins in addition to IAA.[13] It was found that the length of the incubation period decreases and the rate of knot growth increases as bacterial production of IAA and cytokinins increases. In particular, it was noted that the amount of cytokinins produced *in vitro* was related to knot size.

A study on the multiplication of *P. savastanoi* in the woody tissues of olive plants showed that the number of viable bacteria present at each inoculation site first increased and then declined in a parabolic trend.[82] The size of knots was positively correlated with the maximum values of bacterial populations, but there was no correlation between knot weight gain and the bacterial multiplication rate after inoculation. At first, knots increased very slowly in size while *P. savastanoi* grew exponentially; thereafter, knots increased rapidly while the multiplication rate of bacteria decreased. The rapid increase in knot volume coincided with the proliferating activity of cambial zones, which differentiated inside the parenchymatic tissue of the knots at the time of maximum knot enlargement (about 30 days after inoculation). Therefore, it seems that the meristematic regions form when the concentrations of IAA and cytokinins produced by *P. savastanoi in planta* reach an optimal value. Thus, while the primary role of IAA is the stimulation of preexisting cambia with the consequent proliferation of disorganized tissues, cytokinins, together with IAA and other factors, mainly control the size and the anatomical structure of knots. In fact, as stated above, strains producing elevated levels of cytokinins induce larger knots, but IAA-deficient mutants, though producers of cytokinins, fail to induce the typical symptoms. However, partially contrasting results have been obtained by Roberto and Kosuge.[63] From oleander strain PB213 they obtained a spontaneous mutant, PB213-16, which lacked the 42-kb plasmid previously correlated with cytokinin production. The phytohormone phenotype of this mutant was thus $Iaa^+/Z^-$. When inoculated on oleander shoots, PB213-16 gave rise only to attenuated symptoms, as did an $Iaa^-/Z^+$ mutant, PB213-3, and a mutant which produced neither IAA nor Z, PB213-14. Similar, but less conflicting, results have been obtained by Iacobellis et al.[53] By treating oleander strain ITM519 with heat shock, mutants with different phytohormone phenotypes, including $Iaa^-$/cytokinin$^+$ (ITM519-7), $Iaa^+$/cytokinin$^-$ (ITM519-41), and $Iaa^-$/cytokinin$^-$ (ITM519-6), were obtained. The phenotype of these mutants was investigated by *in vitro* production of phytohormones and DNA homology studies on plasmid preparations, using tryptophan monooxygenase (*iaaM*) and isopentenyl transferase (*ipt*) genes as probes. Virulence was assessed by inoculation on oleander shoots and leaves (where *P. savastanoi* particularly induces typical knots). On the shoots, only the wild-type strain

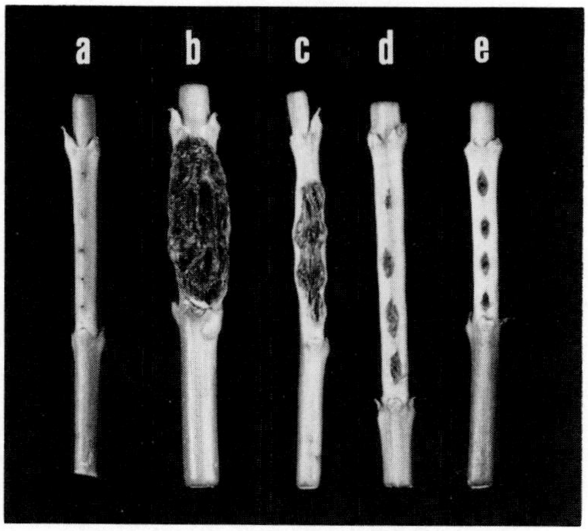


FIGURE 8. Symptoms on oleander shoots 35 days after inoculation with *P. savastanoi* ITM519 and mutants defective in phytohormone production. Plants were inoculated by injecting into growing oleander shoots 100 μl of a $10^8$ bacterial suspension of the indicated strains: (a) sterile water; (b) ITM519; (c) ITM519-41; (d) ITM519-7; (e) ITM519-6. Plants inoculated with ITM519-41 ($Iaa^+$/$cytokinin^-$) induced attenuated symptoms. ITM519-7 ($Iaa^-$/$cytokinin^+$) and ITM519-6 ($Iaa^-$/$cytokinin^-$) induced, respectively, necrotic swellings and necrosis at the inoculation sites.

ITM519 induced hyperplastic canker at the site of inoculation (Figure 8); ITM519-7 and ITM519-41 induced, respectively, necrotic swellings and attenuated hyperplastic cankers. On the other hand, ITM519-6 was completely avirulent inducing only the necrosis of the inoculated tissue. Surprisingly, when inoculated on oleander leaves, mutant ITM519-41 induced galls whose size did not apparently differ from those induced by ITM519 while the other two mutants induced only necrotic symptoms. These results confirm the previous observation regarding the secondary role of cytokinins in knot formation, but they also suggest that the plant host interacts with the pathogen to influence symptom development.

## IV. PHYTOHORMONES IN OLIVE KNOTS

There has not been a great deal of research on the examination of phytohormone content of olive and oleander knots. Beltrà[83] reported that knots on olive plants contained high concentrations of phytohormones since ether extracts of diseased tissues led to the bifurcation of the midrib of olive and privet leaves and increased their surface area. Studies of the knot cytokinin level are at a similar early stage. Iacobellis and Surico[84] found that extracts from naturally occurring olive knots showed high cytokinin activity compared with the extracts from the bark of healthy branches: 100 ng vs 1 ng of benzyladenine equivalents per gram of tissue. Cytokinin-like active substances were tentatively identified as dihydrozeatin-*O*-glucoside (DHZOG) and zeatin riboside-*O*-glucoside (ZROG). No free cytokinins were found.

# V. CONCLUDING REMARKS

In the last 10 years, remarkable progress has been made in developing an understanding of the *P. savastanoi* infection process. There is now strong evidence that IAA and cytokinins are the major determinants of virulence in the interaction between *P. savastanoi* and the host plant. Mutants that fail to produce IAA do not produce knots. On the other hand, a failure to produce cytokinins mainly leads to an attenuation of the symptoms. Full-sized knots are only formed by strains which produce both IAA and cytokinins. However, while the bacterium undoubtedly produces phytohormones in culture, we do not yet have clear evidence that it does so in the host tissue. It is also surprising that in spite of the existence in nature of strains producing different IAA:cytokinins ratios, knots with different morphologies have never been observed on any host plant. Probably the host-plant interacts with the pathogen by locally regulating the IAA:cytokinin ratio, thus avoiding root or shoot differentiation.

Another point that still needs clarification is why oleander is resistant to infection by strains isolated from olive. In our pluriennal experience, only the Californian olive strain EW1017 proved to be able to induce symptoms on oleander. However, a physiological and genetic analysis of EW1017 revealed that this strain actually was an oleander strain which, under natural conditions, spreads from oleander to olive plants from whose knots was then isolated.[7] Olive and oleander strains differ from each other in a number of traits, including amount of IAA produced in culture, conversion of IAA to less active or inactive derivatives, location of IAA genes, production of bacteriocines,[85,86] fatty acid profile,[87,88] and utilization of sucrose[89] but no one trait appears to be important enough to justify a different host range. Nevertheless, olive strains are able to multiply in oleander leaves. Evidently, olive strains do not produce IAA in oleander or, if they do, IAA cannot express its action. On the other hand, we cannot rule out the possibility that the mere production of IAA is not enough for knots to develop. Several plant pathogenic pseudomonads and xanthomonads produce IAA and/or cytokinins,[49,90,92] but in very few of them can phytohormone production be related to generalized or localized growth disorders. On the other hand, we observed that IAA may influence the epiphytic survival of *P. savastanoi*. Moreover, Silverstone et al.[93] in tests with isogenic bacterial strains, which differed from the wild-type parent strains only by their IAA production, showed that IAA enhances survival of *P. savastanoi* in oleander tissue, and that pIAA encodes, not only IAA production, but also some, as yet unknown, functions which contribute to the ecological fitness of the pathogen.

Remarkable similarities have emerged between *P. savastanoi* and *A. tumefaciens,* both in phytohormone production and the nucleotide sequences of the genes encoding their production. Also in *P. savastanoi* the wounding of the potential host could be necessary for plant infection and knot induction.[94] In the future we should gain a greater understanding of the contribution of host plant factors to the process of knot formation.

## REFERENCES

1. **Wilson, E. E.,** Pathological histogenesis in oleander tumors induced by *Pseudomonas savastanoi, Phytopathology,* 55, 1244, 1965.
2. **Surico, G.,** Osservazioni istologiche sui tubercoli della rogna dell'olivo, *Phytopathol. Mediterr.,* 16, 109, 1977.
3. **Janse, J. D.,** *Pseudomonas synringae* subsp. *savastanoi* (ex Smith) subsp. nov., nom. rev., the bacterium causing excrescences on Oleaceae and *Nerium oleander* L., *Int. J. System. Bacteriol.,* 32, 166, 1982.
4. **Janse, J. D.,** The bacterial disease of ash *(Fraxinus excelsior),* caused by *Pseudomonas syringae* subsp. *savastanoi* pv. *fraxini.* III. Pathogenesis, *Eur. J. For. Pathol.,* 12, 218, 1982.
5. **Sutic, D. and Dowson, W. J.,** The reactions of olive, oleander and ash, cross-inoculated with some strains and forms of *Pseudomonas savastanoi* (Smith) Stevens, *Phytopathol. Z.,* 46, 305, 1963.
6. **Janse, J. D.,** The bacterial disease of ash *(Fraxinus excelsior),* caused by *Pseudomonas syringae* subsp. *savastanoi* pv. *fraxini.* II. Etiology and taxonomic considerations, *Eur. J. For. Pathol.,* 11, 425, 1981.
7. **Surico, G., Comai, L., and Kosuge, T.,** Pathogenicity of strains of *Pseudomonas syringae* pv. *savastanoi* and their indoleacetic acid-deficient mutants on olive and oleander, *Phytopathology,* 74, 490, 1984.
8. **Magie, A. R., Wilson, E. E., and Kosuge, T.,** Indoleacetamide as an intermediate in the synthesis of indoleacetic acid in *Pseudomonas savastanoi, Science,* 141, 1281, 1963.
9. **Kosuge, T., Heskett, M. G., and Wilson, E. E.,** Microbial synthesis and degradation of indole-3-acetic acid. I. The conversion of L-tryptophan to indole-3-acetamide by an enzyme system from *Pseudomonas savastanoi, J. Biol. Chem.,* 241, 3738, 1966.
10. **Hutzinger, O. and Kosuge, T.,** Microbial synthesis and degradation of indole-3-acetic acid. II. The source of oxygen in the conversion of L-tryptophan to indole-3-acetamide, *Biochim. Biophys. Acta,* 136, 389, 1967.
11. **Kuo, T. T. and Kosuge, T.,** Factors influencing the production and further metabolism of indole-3-acetic acid by *Pseudomonas savastanoi, J. Gen. Appl. Microbiol.,* 15, 51, 1969.
12. **Smidt, M. and Kosuge, T.,** The role of indole-3-acetic acid accumulation by alpha methyl tryptophan-resistant mutants of *Pseudomonas savastanoi* in gall formation on oleanders, *Physiol. Plant Pathol.,* 13, 203, 1978.
13. **Surico, G., Iacobellis, N. S., and Sisto, A.,** Studies on the role of indole-3-acetic acid and cytokinins in the formation of knots on olive and oleander plants by *Pseudomonas syringae* pv. *savastanoi, Physiol. Plant Pathol.,* 26, 309, 1985.
14. **Comai, L. and Kosuge, T.,** Involvement of plasmid deoxyribonucleic acid in indoleacetic acid synthesis in *Pseudomonas savastanoi, J. Bacteriol.,* 143, 950, 1980.
15. **Comai, L., and Kosuge, T.,** Cloning and characterization of *iaaM,* a virulence determinant of *Pseudomonas savastanoi, J. Bacteriol.,* 149, 40, 1982.
16. **Kosuge, T. and Comai, L.,** Metabolic regulation in plant pathogen interactions from the perspective of the pathogen, in *Plant Infection: The Physiological and Biochemical Basis,* Asada, Y., et al., Eds., Japanese Scientific Society, Press, Tokyo, Springer-Verlag, Berlin, 1982, 175.
17. **Beltrà, R.,** Tryptophan metabolism of *Pseudomonas savastanoi, Microbiol. Esp.,* 17, 123, 1964.
18. **Comai, L. and Kosuge, T.,** Transposable element that causes mutations in a plant pathogenic *Pseudomonas* sp., *J. Bacteriol.,* 154, 1162, 1983.
19. **Comai, L., Surico, G., and Kosuge, T.,** Relation of plasmid DNA to indoleacetic acid production in different strains of *Pseudomonas syringae* pv. *savastanoi, J. Gen. Microbiol.,* 128, 2157, 1982.
20. **Varvaro, L. and Surico, G.,** Epiphytic survival of wild types of *Pseudomonas syringae* pv. *savastanoi* and their Iaa$^-$ mutants on olive leaves, in *Proc. 2nd Int. Working Group on Pseudomonas syringae pvs.,* Greece, p. 20, 1984.
21. **Kosuge, T. and Yamada, T.,** Virulence determinants in plant pathogen interactions, in *Molecular Determinants of Plant Diseases,* Nishimura, S., et al., Eds., Japanese Scientific Society Press, Tokyo, Springer-Verlag, Berlin, 1987, 147.
22. **Kosuge, T. and Sanger, M.,** Indoleacetic acid, its synthesis and regulation: a basis for tumorigenicity in plant disease, *Recent Adv. Phytochem.,* 20, 147, 1987.
23. **Kosuge, T., Comai, L., and Glass, N. L.,** Virulence determinants in plant-pathogen interactions, in *Plant Molecular Biology,* Goldberg, R., Ed., Alan R. Liss, Inc., NY, 1983, 167.
24. **Hutzinger, O. and Kosuge, T.,** Microbial synthesis and degradation of indole-3-acetic acid. II. The isolation and characterization of indole-3-acetyl-ϵ-L-lysine, *Biochemistry,* 7, 601, 1968.
25. **Hutzinger, O. and Kosuge, T.,** 3-Indoleacetyl-L-lysine, a new conjugate of 3-indoleacetic acid produced by *Pseudomonas savastanoi,* in *Biochemistry and Physiology of Plant Growth Substances,* Wightman, F. and Setterfield, G., Eds., The Runge Press, Ottawa, 1968, 183.
26. **Evidente, A., Surico, G., Iacobellis, N. S., and Randazzo, G.,** α-*N*-Acetylindole-3-acetyl-ϵ-L-lysine: a metabolite of indole-3-acetic acid from *Pseudomonas syringae* pv. *savastanoi, Phytochemistry,* 25, 125, 1986.

27. **Nitsch, J. P. and Nitsch, C.,** Studies on the growth of coleoptile and first internode sections. A new sensitive, straight-growth test for auxin, *Plant Physiol.*, 31, 94, 1956.
28. **Kosuge, T. and Kuo, T. T.,** Metabolism of indole-3-acetic acid by the phytopathogen, *Pseudomonas savastanoi,* in *Biochemical Regulation in Diseased Plants or Injury,* Phytopathological Society of Japan, Tokyo, 1968, 313.
29. **Cohen, J. D. and Bandurski, R. S.,** Chemistry and physiology of the bound auxins, *Annu. Rev. Plant Physiol.*, 33, 403, 1982.
30. **Surico, G., Iacobellis, N. S., and Sisto, A.,** unpublished data, 1989.
31. **Glass, N. L. and Kosuge, T.,** Cloning of the gene for indoleacetic acid-lysine synthetase from *Pseudomonas syringae* subsp. *savastanoi, J. Bacteriol.*, 166, 598, 1986.
32. **Surico, G., Sparapano, L., Lerario, P., Durbin, R. D., and Iacobellis, N. S.,** Studies on growth-promoting substances produced by *Pseudomonas savastanoi* (E. F. Smith) Stevens, *Proc. 4th Congr. Mediterr. Phytopathol. Union,* Yugoslavia, 1975, 449.
33. **Roberto, F. F., Klee, H., White, F., Nordeen, R., and Kosuge, T.,** Expression and fine structure of the gene encoding $N^{\epsilon}$-(indole-3-acetyl)-L-lysine synthetase from *Pseudomonas savastanoi, Proc. Natl. Acad. Sci. U.S.A.*, 87, 5797, 1990.
34. **Marlow, J. L. and Kosuge, T.,** Tryptophan and indoleacetic acid transport in the olive and oleander knot organism *Pseudomonas savastanoi* (E. F. Smith) Stevens, *J. Gen. Microbiol.*, 72, 211, 1972.
35. **Glass, N. L. and Kosuge, T.,** Role of indoleacetic acid-lysine synthetase in regulation of indoleacetic acid pool size and virulence of *Pseudomonas syringae* subsp. *savastanoi, J. Bacteriol.*, 170, 211, 1972.
36. **Spena, A. and Schulze, S. C.,** Chimaeras and transgenic plant mosaics: a new tool in plant biology, in *Advances in Molecular Genetics of Plant-Microbe Interactions,* Vol. 1, *Proc. 5th Int. Symp. on Mol. Genet. Plant-Microbe Interact.*, Switzerland, 1990, Henneche, H. and Verma, D. P. S., Eds., Kluwer Academic Publishers, Dordrecht, The Netherlands, 1991, 352.
37. **Moore, T. C.,** *Biochemistry and Physiology of Plant Hormones,* Springer-Verlag, NY, 1979, chap. 2.
38. **Evidente, A. and Surico, G.,** Isolation of indole-3-aldehyde from *Pseudomonas syringae* pv. *savastanoi, J. Nat. Prod.*, 49, 938, 1986.
39. **Hinman, R. L. and Lang, J.,** Peroxidase-catalyzed oxidation of indole-3-acetic acid, *Biochemistry,* 4, 144, 1965.
40. **Elstner, E. F.,** Hormones and metabolic regulation in disease, in *Biochemical Plant Pathology,* Callow, J. A., Ed., John Wiley & Sons, 1983, 415.
41. **Surico, G.,** Indoleacetic acid and cytokinins in the olive knot disease. An overview of their role and their genetic determinants, in *Biology and Molecular Biology of Plant-Pathogen Interactions,* NATO ASI Ser. Vol. H1, Bailey, J., Ed., Springer-Verlag, Berlin, 1986, 315.
42. **Ziegler, S. F., White, F. F., and Nester, E. W.,** Genes involved in indoleacetic acid production in plant pathogenic bacteria, in *Plant Pathogenic Bacteria, Proc. 6th Int. Conf. on Plant Pathogenic Bacteria,* Maryland, 1985, Civerolo, E. L., et al., Eds., Martinus Nijhoff Publishers, Dordrecht, The Netherlands, 1987, 18.
43. **Comai, L. and Kosuge, T.,** The genetics of indoleacetic acid production and virulence in *Pseudomonas savastanoi,* in *Molecular Genetics of the Bacteria-Plant Interactions,* Puhler, A., Ed., Springer-Verlag, Berlin, 1983, 363.
44. **Yamada, T., Palm, C. J., Brooks, B., and Kosuge, T.,** Nucleotide sequences of the *Pseudomonas savastanoi* indoleacetic acid genes show homology with *Agrobacterium tumefaciens* T-DNA, *Proc. Natl. Acad. Sci. U.S.A.*, 82, 6522, 1985.
45. **Palm, C. J., Gaffney, I., and Kosuge, T.,** Cotranscription of genes encoding indoleacetic acid production in *Pseudomonas syringae* subsp. *savastanoi, J. Bacteriol.*, 171, 1002, 1989.
46. **Hutcheson, S. and Kosuge, T.,** Regulation of 3-indoleacetic acid production in *Pseudomonas syringae* pv. *savastanoi.* Purification and properties of tryptophan 2-monooxygenase, *J. Biol. Chem.*, 260, 6281, 1985.
47. **Roberto, F. F. and Kosuge, T.,** Aspects of phytohormone metabolism in *Pseudomonas savastanoi,* in *Physiology and Biochemistry of Plant-Microbial Interactions,* Keen, N. T., Kosuge, T., and Walling, L. L., Eds., The American Society of Plant Physiologists, 1988, 31.
48. **Binns, A. N. and Thomashow, M. F.,** Cell biology of *Agrobacterium tumefaciens* infection and transformation of plants, *Annu. Rev. Microbiol.*, 42, 575, 1988.
49. **Yamada, T., Lee, P.-D., and Kosuge, T.,** Insertion sequence elements of *Pseudomonas savastanoi:* nucleotide sequence and homology with *Agrobacterium tumefaciens* transfer DNA, *Proc. Natl. Acad. Sci. U.S.A.*, 83, 8263, 1986.
50. **Starlinger, P.,** IS elements and transposons, *Plasmid,* 3, 241, 1980.
51. **Soby, S., Kirkpatrick, B., and Kosuge, T.,** High frequency deletion of the iaa operon in *Pseudomonas syringae* pv. *savastanoi, Proc. 5th Int. Symp. Mol. Genet. Plant-Microbe Interact.*, Switzerland, 1990, 56 (abstract).
52. **Soby, S.,** personal communication, 1990.
53. **Iacobellis, N. S., Morea, M., Sisto, A., and Caponero, A.,** unpublished results, 1990.

54. **Surico, G., Sparapano, L., Lerario, P., Durbin, R. D., and Iacobellis, N. S.,** Cytokinin-like activity in extracts from culture filtrates of *Pseudomonas savastanoi, Experientia,* 31, 929, 1975.
55. **Surico, G., Evidente, A., Iacobellis, N. S., and Randazzo, G.,** Cytokinins: identification of compounds isolated from *Pseudomonas syringae* pv. *savastanoi* and diseased olive tissues, in *Phytobacteriology and Plant Bacterial Diseases of Quarantine Significance,* Italian-US Workshop/Seminar "Plant Pathology and Quarantine," Quacquarelli, A. and Casano, F. J., Eds., Tipolitografia Mondazzi, Roma, 1984, 49.
56. **Iacobellis, N. S., Surico, G., Evidente, A., Iasiello, I., and Randazzo, G.,** Note preliminari sulla identificazione di alcune citocinine estratte dai liquidi colturali di *Pseudomonas syringae* (Smith) Young et al. pv. *savastanoi, Phytopathol. Mediterr.,* 24, 315, 1985.
57. **Surico, G., Evidente, A., Iacobellis, N. S., and Randazzo, G.,** A cytokinin from the culture filtrate of *Pseudomonas syringae* pv. *savastanoi, Phytochemistry,* 24, 1499, 1985.
58. **Evidente, A., Surico, G., Iacobellis, N. S., and Randazzo, G.,** 1′-Methyl zeatin, an additional cytokinin from *Pseudomonas syringae* pv. *savastanoi, Phytochemistry,* 25, 525, 1986.
59. **Itaya, T., Fujii, T., Evidente, A., Randazzo, G., Surico, G., and Iacobellis, N. S.,** Syntheses and absolute configurations of the cytokinins 1′-methyl zeatin and its 9-riboside, *Tetrahedron Lett.,* 27, 6349, 1986.
60. **Ohba, M., Fujii, T., Evidente, A., Surico, G., and Iacobellis, N. S.,** 1′-methyl zeatin and its 9-β-D-ribofuranoside: the carbon-13 nuclear magnetic resonance spectra revisited, *Heterocycles,* 31, 599, 1990.
61. **MacDonald, E. M. S., Powell, G. K., Regier, D. A., Glass, N. L., Roberto, F., Kosuge, T., and Morris, R. O.,** Secretion of zeatin, ribosylzeatin, and ribosyl-1″-methylzeatin by *Pseudomonas savastanoi, Plant Physiol.,* 82, 742, 1986.
62. **Morris, R. O., Powell, G. K., Beaty, J. S., Durley, R. C., Hommes, N. G., Lica, L., and MacDonald, E. M. S.,** Cytokinin biosynthetic genes and enzymes from *Agrobacterium tumefaciens* and other plant-associated prokaryotes, in *Plant Growth Substances,* Bopp, M., Ed., Springer-Verlag, Berlin, 1986, 185.
63. **Roberto, F. F. and Kosuge, T.,** Phytohormone metabolism in *Pseudomonas syringae* subsp. *savastanoi,* in *Molecular Biology of Plant Growth Control,* Fox, E. W. and Jacobs, M., Eds., Alan R. Liss, Inc., NY, 1987, 371.
64. **Surico, G., Evidente, A., Iacobellis, N. S., and Randazzo, G.,** On the presence and level of different cytokinins in culture filtrate of *Pseudomonas syringae* pv. *savastanoi,* in *Plant Pathogenic Bacteria, Proc. 6th Int. Conf. Plant Pathogenic Bacteria,* Maryland, 1985, Civerolo, E. L., et al., Eds., Martinus Nijhoff, Publishers, Dordrecht, The Netherlands, 1987, 566.
65. **Morris, O. R.,** Genes specifying auxin and cytokinin biosynthesis in phytopathogens, *Annu. Rev. Plant. Physiol.,* 37, 509, 1986.
66. **Regier, D. A. and Morris, R. O.,** Secretion of *trans*-zeatin by *Agrobacterium tumefaciens:* a function determined by the nopaline Ti plasmid, *Biochem. Biophys. Res. Commun.,* 104, 1560, 1982.
67. **Murai, N., Skoog, F., Doyle, M. E., and Hanson, R. S.,** Relationship between cytokinin production, presence of plasmids, and fasciation caused by strains of *Corynebacterium fascians, Proc. Natl. Acad. Sci. U.S.A.,* 77, 619, 1980.
68. **Iacobellis, N. S., Evidente, A., and Surico, G.,** Isolation of plant growth regulators from *Pseudomonas amygdali, Experientia,* 44, 70, 1988.
69. **Iacobellis, N. S., Evidente, A., Surico, G., Sisto, A., and Gammaldi, G.,** Production of phytohormones by *Pseudomonas amygdali* and their role in the hyperplastic bacterial canker of almond, *J. Phytopathol.,* 129, 177, 1990.
70. **Chen, C. M.,** Cytokinin biosynthesis in cell-free systems, in *Plant Growth Substances,* Wareing, P. F., Ed., Academic Press, San Diego, 1982, 155.
71. **Letham, D. S. and Palni, L. M. S.,** The biosynthesis and metabolism of cytokinins, *Annu. Rev. Plant Physiol.,* 34, 163, 1983.
72. **Taya, Y., Tanaka, Y., and Nishimura, S.,** 5′-AMP is a direct precursor of cytokinin in *Dictyostelium discoideum,* Nature (London), 271, 545, 1978.
73. **Fujii, T., Itaya, T., and Matsubara, S.,** Purines. XXXIII. Syntheses and cytokinin activities of both enantiomers of 1′-methylzeatin and their 9-β-D-ribofuranosides, *Chem. Pharm. Bull.,* 37, 1758, 1989.
74. **Evidente, A., Fujii, T., Iacobellis, N. S., Riva, S., Sisto, A., and Surico, G.,** Structure-activity relationships of zeatin cytokinins produced by plant pathogenic Pseudomonades, *Phytochemistry,* 1991, in press.
75. **Powell, G. K. and Morris, R. O.,** Nucleotide sequence and expression of a *Pseudomonas savastanoi* cytokinin biosynthetic gene: homology with *Agrobacterium tumefaciens tmr* and *tzs* loci, *Nucleic Acids Res.,* 14, 225, 1986.
76. **Ream, W.,** *Agrobacterium tumefaciens* an interkingdom genetic exchange, *Annu. Rev. Phytopathol.,* 27, 583, 1989.
77. **Hewitt, W. B.,** Leaf-scar infection in relation to the olive-knot disease, *Hilgardia,* 12, 41, 1938.
78. **Wilson, E. E. and Magie, A. R.,** Systemic invasion of the host plant by the tumor inducing bacterium *Pseudomonas savastanoi, Phytopathology,* 54, 576, 1964.

79. **Beltrà, R.**, Relacion entre la concentration de triptofano en los tallos de olivo y la localization de los tumores bacterianos, *Microbiol. Esp.*, 11, 401, 1958.
80. **Beltrà, R.**, El acido beta-indoleacetic y los tumores vegetales de origen bacteriano, *Rev. Latinoamer. Microbiol.*, 2, 23, 1959.
81. **Butcher, D. N. and Ingram, D. S.**, *Plant Tissue Culture*, E. Arnold, London, 1976, chap. 3.
82. **Varvaro, L. and Surico, G.**, Moltiplicazione di *Pseudomonas savastanoi* (E. F. Smith) Stevens nei tessuti dell'Olivo, *(Olea europaea L.), Phytopathol. Mediterr.*, 17, 179, 1978.
83. **Beltrà, R.**, Efecto morfogenetico observado en los extractos hormonales de los tumores del olivo, *Microbiol. Esp.*, 14, 177, 1961.
84. **Iacobellis, N. S. and Surico, G.**, Occurrence of cytokinins in knot tissues induced by *Pseudomonas syringae* pv. *savastanoi*, in *Proc. 2nd Int. Working Group on Pseudomonas syringae pvs.*, Greece, p. 32, 1984.
85. **Varvaro, L. and Surico, G.**, Multiplication of wild types of *Pseudomonas syringae* pv. *savastanoi* (Smith) Young et al. and their indoleacetic acid-deficient mutants in olive tissues, in *Plant Pathogenic Bacteria, Proc. 6th Int. Conf. Plant Pathogenic Bacteria*, Maryland, 1985, Civerolo, E. L. et al., Eds., Martinus Nijhoff Publishers, Dordrecht, The Netherlands, 1987, 556.
86. **Iacobellis, N. S., Contesini, A. M., and Surico, G.**, unpublished data, 1990.
87. **Varvaro, L. and Sasser, M.**, Fatty acid profiles of *Pseudomonas syringae* pv. *savastanoi*, paper presented at 3rd Int. Working Group on *Pseudomonas syringae* pvs., Portugal, 1987.
88. **Wells, J. M., Casano, F. J., and Surico, G.**, unpublished data, 1990.
89. **Wilson, E. E., Heskett, M. G., Johnson, M. L., and Kosuge, T.**, Metabolic behaviour of *Pseudomonas savastanoi* isolates from olive and oleander on certain carbohydrate and amino substrates, *Phytopathology*, 62, 349, 1972.
90. **Green, E. M.**, Cytokinin production by microorganism, *Botan. Rev.*, 46, 25, 1980.
91. **Fett, W. F., Osman, S. F., and Dunn, M. F.**, Auxin, production by plant pathogenic *pseudomonads* and *xanthomonads, Appl. Environ. Microbiol.*, 53, 1839, 1987.
92. **Akiyoshi, D. E., Regier, D. A., and Gordon, M. P.**, Cytokinin production by *Agrobacterium tumefaciens* and *Pseudomonas* spp., *J. Bacteriol.*, 169, 4242, 1987.
93. **Silverstone, S. F., Bostock, R. M., Gilchrist, D. G., and Kosuge, T.**, Effect of IAA production on *in planta* survival of *Pseudomonas savastanoi*, in Abstr. APS/CPS Annu. Meeting, Grand Rapids, Michigan, 1990, A112.
94. **Surico, G.**, Observations on olive leaves infection by *Pseudomonas syringae* pv. *savastanoi*, in *Plant Pathogenic Bacteria, Proc. 7th Int. Conf. Plant Pathogenic Bacteria, Budapest, Hungary*, Klement, Z., Ed., Akademiai Kiado, 1989, 41.

# IRON: A MODULATOR IN BACTERIAL VIRULENCE AND SYMBIOTIC NITROGEN-FIXATION

**Dominique Expert and Paul R. Gill, Jr.**

## TABLE OF CONTENTS

# I. THE BIOLOGY OF IRON: ESSENTIALITY, TOXICITY AND ACQUISITION

## A. THE IMPORTANCE OF IRON IN BIOLOGICAL SYSTEMS

One cannot miss the vital importance of iron in the basic biochemical mechanisms of virtually every living cell — in animals, plants or microbes. Some of the iron-containing metalloproteins play key roles in such fundamental reactions as ribonucleotide reduction, heme-dependent hydroxylation systems, and respiratory electron transfer.[1] Indeed, besides cytochromes and Fe/S proteins involved in redox reactions, metalloflavoproteins, various oxidases and hydrogenases, and enzymes involved in the detoxification of oxidizing radicals, the importance of reversible oxygen binding proteins, such as hemoglobin, hemoglobin-like proteins, and leghemoglobin present in root nodules of nitrogen-fixing plants, should also be acknowledged.

Why is iron involved in so many metabolic processes? The iron atom itself displays a powerful catalytic activity because of its electronic structure, which can undergo reversible changes through several oxidation states differing by one electron.[2] Iron may have become essential among other available resources very early in the evolution of life in that it is found to transfer a single electron in a wide variety of biochemically important redox reactions.[3] The electron transfer ability of iron can be modified by its ligand environment, and these redox reactions also differ widely in their reduction potentials, spanning almost the entire range of biochemical redox reactions to nearly 1 v.[4] Although iron is an essential element, it is toxic in excessive quantities. The toxicity of the free element is borne in its ability to catalyze the single electron reduction of oxygen to form superoxide, hydrogen peroxide, and ·OH. An overabundance of this metal in humans can provoke and exacerbate disease.[5] The extreme reactivity of the iron atom in the presence of oxygen results in the formation of oxidizing radicals which can be very damaging to biomolecules. Therefore, when present as the free element, iron severely damages living tissues, for instance, by the peroxidation of polyunsaturated lipids, one of the major constituents of biological membranes. In humans, iron is transported in the bloodstream as a transferrin complex; it is also stored in ferritins. In animals, ferritins are particularly abundant in liver.[6] Iron is associated with the chloroplast in plants.[7] There is evidence that ferritins also exist in bacteria.[8]

Acknowledged to be the fourth most abundant element in the earth's crust and widely distributed in soils and rocks as ferric oxides, such as goethite, hematite, ferrihydrite,[9] iron might seem to be readily available to satisfy any biological requirement. Its availability, however, may have been greatly compromised with the advent of the photosynthetic adaptation that results in the formation of oxygen, i.e., photosystem II. It is believed that this adaptation switched the planet from an anaerobic to an aerobic environment. The resulting consequences for the acquisition of iron in the words of J.B. Neilands: "a crisis on a grand scale." The relatively soluble $Fe^{2+}$ (ferrous) form is oxidized to the very insoluble $Fe^{3+}$ (ferric) form in the presence of oxygen. The $K_{sp}$ of $Fe(OH)_3$ is $10^{-38}$. Thus, at pH 7, $Fe^{3+}$ is available at $10^{-17}$ *M*, which is far below that required for microbial growth which is usually about $10^{-6}$ *M*. As conditions become more alkaline, the solubility of iron decreases three orders of magnititude per pH unit.

A world full of microoganisms and all the available iron has turned into rust! To deal with this crisis, several strategies appear to have been adopted, for example, retreating to anaerobic environments or shifting to other cofactors as may have been done by members of *Lactobacillus*.[10] The absolute requirement for this element, however, seems to have provoked an evolutionary trend among microorganisms when it became less available: when grown under iron-limiting conditions bacteria and fungi generally produce and excrete into their environment low-molecular weight ferric-specific chelating (iron-solubilizing) compounds (i.e., siderophores) the ferric complex of which is recognized and transported into

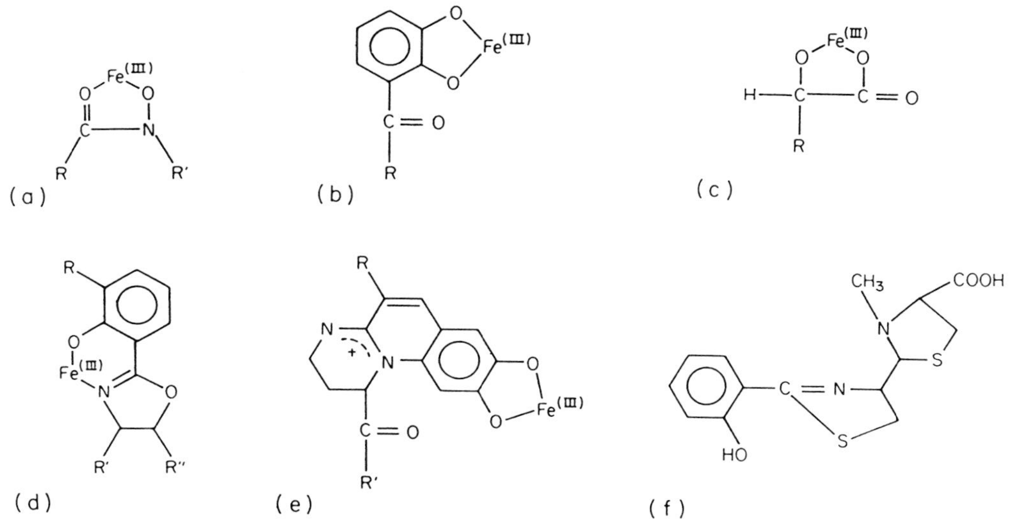


FIGURE 1. Bidentate ligand systems of siderophores: (a) Hydroxamate (ferrichromes, rhodotorulic acid, other hydroxamates); (b) catechol (enterobactin, other catechols); (c) α-hydroxyacid (citrate-containing siderophores, pseudobactin); (d) 2-(2-hydroxyphenyl)-oxazoline (mycobactins, agrobactin, parabactin, vibriobactin); (e) fluorescent quinolinyl chromophore (pseudobactins); (f) 2-(2-*o*-hydroxyphenyl-2-thiazolin-4-yl)-3-methylthiazolidine-4-carboxylic acid. Reproduced from *Microbiol. Sci.*, with permission from Blackwell Scientific, Oxford.

the microbe that produced it. A functional siderophore thus has two roles: it must solubilize iron from the environment, i.e., it must be capable of sequestering iron from $OH^-$, and the ferric complex must be recognized and transported into the microbe.

Although siderophores differ widely in their overall structure, which accounts for the specific recognition and uptake by a given microbe, the iron-chelating functional groups are much more conserved.[4] To form a stable hexadentate coordination complex with ferric iron, microbial siderophores have, for example, been observed to contain one to three catechol (2,3-dihydroxybenzoic acid) groups, generally produced by bacteria. Three such groups are found in enterobactin produced by *Escherichia coli*. Hydroxamate siderophores are generally produced by fungi; three hydroxamate groups are found in the fungal siderophore, ferrichrome. More exotic iron-chelating moieties are found in the pseudobactins, various quinoline groups, and an α-hydroxylcarboxylic acid in pseudobactin (see below), or a cyclic threonine to yield an oxazoline ring in agrobactin and parabactin (see Figure 1). Besides the high-affinity ferric-specific siderophores, many organic compounds (e.g., organic acids) can form a weak complex with iron. The ferric cation solubilized with citric acid, a weak but respectable iron-chelator, is an iron source for a *Bradyrhizobium* strain.[11] Perhaps it will prove to be an iron source for this bacterium *in planta* since iron is well known to be solubilized as ferric:citrate in plants.[12] Citric acid also serves as a carbon source, but not an iron source, for *Salmonella typhimurium;* it serves as an iron source but not a carbon source for either *E. coli*,[13] or *Bradyrhyzobium* sp.[11]

Recognition and transport of a given ferric:siderophore complex by bacteria has been shown to require several genes. One class of these genes, encoding outer membrane proteins, specifically recognize the ferric complex of known siderophores, the chirality of which is also specific.[14] To make matters more complicated, various microbes have learned to utilize, as an iron source, the ferric complexes of siderophores they do not produce, but which are produced by other organisms. Ferric-ferrichrome, a fungal siderophore, abundant in many cheese preparations, is readily transported by *E. coli* as an iron source. The *tonA* locus, so carefully studied by Luria and Delbrück in 1943[15] was later found to encode the outer membrane receptor FhuA for ferric:ferrichrome.

Plants can also use microbial siderophores as iron sources. The $^{59}Fe$ complex of agrobactin, a catechol-type siderophore from *Agrobacterium tumefaciens,* was shown to serve as an iron supply for bean and pea.[16] In a similar study, ferrioxamine B, produced by several fungi, was shown to serve as an iron supply for oat.[17] A new siderophore from *Rhizopus arrhizus,* whose structure has recently been partially determined, appeared to be very efficient in providing iron to corn and cotton plants.[18] Pseudobactin from *Pseudomonas* strain B10, however, could not serve as an iron source for corn and pea plants.[19] The mechanisms by which iron is mobilized to the plant is not yet clearly understood. In roots of dicots and non-grass monocots, a specific process is induced by iron stress. This involves a set of reductases responsible for the conversion of Fe(III) to Fe(II) at the plasmalemma of similarly low iron-induced rhizodermal transfer cells which appear to be closely associated with an extruding flux of protons and reducing agents. Whether such enzymes participate in the reduction of microbial siderophores is currently under investigation. Grasses use another strategy which, like microbes, involves the production and utilization of the ferric complexes of phytosiderophores.[20]

The iron requirements of Gram-negative bacteria generally range from 0.36 to 1.8 $\mu M$, but higher concentrations have been reported for strongly aerobic organisms from several genera including *Pseudomonas, Azotobacter, Corynebacterium,* and *Mycobacterium.*[21] Recently, Mössbauer spectrometry studies showed that such needs correspond to intracellular levels that may vary from 25 to 250 ppm iron per gram of dry weight, depending on iron availability in the environment.[22] In plants, the iron contents varies in the same range, being from about 60 to 300 ppm; deficient plants may have 10 to 30 ppm, while excess conditions can give levels of 400 to 1000 ppm. The concentration of chelated iron for optimal plant growth under controlled conditions in solution cultures is on the order of 0.1 to 10 $\mu M$, depending on various factors such as genotype, phosphate concentration, type of chelate, and pH of the solution.[21] By comparison, the human body contains about 4 g of iron, a content which is maintained through a rigid control over absorption-to-balance loss.

## B. IRON IN MICROBIAL COMPETITION AND INFECTION

Conditions of iron limitation enhance the development of chlorosis which is common in plants grown in calcareous soils. Because of its insolubility, particularly in alkaline soils, iron deficiency-induced chlorosis of agricultural plants has been observed to be an important factor in crop damage worldwide.[23] As a limiting factor, iron is sometimes the stake in an ardent and complex competition between various members of a particular ecological niche or habitat. For instance, in the rhizosphere, bacteria and fungi can counter the lack of soluble iron by excreting siderophores that capture iron from natural chelates and transport the ferric complexes into the cell via a specific transport system. A number of rhizosphere pseudomonads produce a fluorescent siderophore, termed pseudobactins or pyoverdins. Members of this family of siderophores consist of, in addition to a fluorescent quinoline moiety with various modifications, an amino acid side chain of up to eight amino acids, two of which are modified for iron chelation, and which together form a hexadentate complex with $Fe^{3+}$. Depending on the producing organism, the composition and sequence of the amino acid side chain and the amino acid modifications for iron chelation is variable.[24] The utilization of a given ferric:pseudobactin as an iron source by a given fluorescent rhizosphere pseudomonad is also variable. The growth of some strains can be inhibited by a pseudobactin produced by another pseudomonad on an iron-limiting medium, whereas others can utilize siderophores they do not produce.[25] *Pseudomonas* sp. WCS358, can utilize as an iron source virtually any pseudobactin-like siderophore produced by another pseudomonad. The plant growth-promoting ability of this species has been shown to rely on its ability to capture iron in the rhizosphere via its siderophore system.[26,27]

Perhaps the most remarkable, and certainly the most studied, example of competition for iron resides in pathogenic microorganisms that invade vertebrate hosts. The topic of iron and infection originated almost 50 years ago, when Schade and Caroline[28] discovered the presence of specific iron-binding proteins in blood and in egg whites that were thought to inhibit the growth of certain bacteria. These investigators observed that raw hen egg white had bacteriostatic properties and that this effect could be reversed with added iron. Even though this finding was initially ignored, it has now been nearly three decades since particular attention has been drawn to iron as a clue in the puzzle of bacterial virulence and animal host defense. Indeed, sequestration of the metal by the host, and the consequent lack of its availability for the pathogen represent the central issues. The major iron-binding glycoproteins present in body fluids, lactoferrin in secretions and transferrin in serum, can withhold iron from many microorganisms, and thus show a broad-spectrum antimicrobial activity. Lactoferrin binds iron efficiently at low pH and thus is a good iron scavenger in areas of infection. In addition, transferrin not only delivers iron to host cells but also acts as a defense protein. In response to microbial invasion, the host lowers its iron level by saturation of transferrin and temporarily by storing iron in ferritin.[29] Also, since some pathogenic bacteria can use transferrin,[30] heme or hemoglobin[31,32] as an iron source in wounds, vertebrates produce haptoglobulin which can combine with hemoglobin to render the iron unavailable to the invading microorganisms.[33] Furthermore, the immune system may produce antibodies directed specifically against certain components of the microbial iron uptake systems, thereby blocking their function.[33] To counter iron retention by the host, one alternative for the pathogen is to produce a siderophore whose ability to function in an infection arena like the human bloodstream, greatly exceeds that found when the pathogen is free-living. Aerobactin, for instance, originally detected in *Aerobacter aerogenes*,[34] is produced by several pathogenic strains of *E. coli* and other enteric organisms through the acquisition of plasmids and transposons as a part of systems that enhance their virulence.[35] Aerobactin can remove iron from tranferrin in serum, whereas enterobactin, the chromosomally encoded siderophore of *E. coli,* and whose affinity for iron is much higher, is rapidly inactivated by lipophilic serum proteins.[36]

The observation that iron is both essential and toxic suggests that the uptake of this element is carefully controlled. The regulation of iron uptake by *E. coli* has been studied in detail. The *fur* (*f*erric *u*ptake *r*egulation) gene product, a classical repressor, that uses $Fe^{2+}$ as a corepressor (a signal for iron sufficient conditions), controls the uptake of this metal by this bacterium.[37] Not only is this gene subject to catabolite repression, but it is also autoregulated in a way that decreases its production more slowly than siderophore production when available iron is not completely limiting growth. In *E. coli,* chromosomal- and plasmid-encoded high-affinity iron uptake systems are controlled by the *fur* gene product. A palindromic sequence of 18 nucleotides (the DNA "iron box" ) has been identified as a consensus operator, and binding site for the $Fe^{2+}$:Fur protein complex in these high-iron repressed loci.[38] Thus, under iron-sufficient conditions, siderophore biosynthesis is turned off.

Under iron-limiting conditions in the human bloodstream, bacterial virulence determinants (seeming unrelated to iron transport) are likewise derepressed. Transcription of the Shiga-like toxin type I in *E. coli* is repressed by the Fur protein in the presence of iron.[39] The Fur protein can also function as an iron-dependent repressor of the diphtheria toxin promote cloned from corynephage β of *Corynebacterium diphtheriae*,[40] as well as of the hemolysin operon promote of *Serratia*[41] and *Vibrio*.[32] From this standpoint, regulation by iron in human and animal infectious diseases illustrates the remarkable ability of pathogenic bacteria to respond to variable environmental conditions as they move throughout the host. When the host is healthy, free iron is unavailable and toxins are produced. When the host is debilitated, there is no point in producing more toxins that have already done their job. The nutritional requirement for this metal in an infection arena may have provoked a second evolutionary trend among pathogens, in that recognition of an iron-limiting environment is correlated with with the "*entry*" and the need to grow within a host.

## II. IRON IN PLANT BACTERIAL PATHOGENESIS

The problem of iron availability for microbial plant pathogens and rhizobial symbionts has only been raised recently. In light of the data obtained in animal infections, J. B. Neilands, who had focused on the biochemistry of microbial iron transport systems, opened the debate about the production of siderophores by plant pathogenic bacteria as a potential virulence-enhancing factor. Subsequently, interest was turned to symbiotic *Rhizobium* whose needs for iron seemed to be critical, requiring large amounts of this metal in the enzymic reduction of atmospheric nitrogen. In the following sections we will describe the recent developments on these topics, with particular attention to *Erwinia* and *Rhizobium,* the studies of which have undoubtedly contributed the most to broaden the concept of iron as a modulator in plant-microbe interactions.

To unravel the question of iron acquisition in plant pathogenesis, Leong and Neilands first surveyed a set of economically important pathogenic bacterial genera for production of siderophores.[42] In response to iron deficiency, most of the microorganisms examined appeared to produce siderophores. Several strains of *Agrobacterium tumefaciens,* as well as the saprophytic bacterium, *A. radiobacter* 84, produce catechol type siderophores. Catechols were also found in low iron cultures of *Erwinia carotovora* 78, and *Pseudomonas syringae* pv. *phaseolicola* G50 was shown to synthesize a siderophore closely related to those isolated from *P. fluorescens* spp. In *E. amylovora* 178 and 213, a siderophore activity was detected by bioassay, but neither catechol- nor hydroxamate-type siderophores were found.

Agrobactin, the siderophore of *A. tumefaciens* B6, was characterized as a threonyl peptide of spermidine acylated with three residues of 2,3-dihydroxybenzoic acid.[43] In response to iron stress, the production of agrobactin appeared to be associated with the induction of several outer membrane proteins with apparent molecular weights of approximately 80,000. Mutants unable to produce or to utilize agrobactin were isolated after chemical mutagenesis, but they were still capable of initiating tumors in plants. Therefore, agrobactin does not seem to be required for tumor formation.[44] Considering the extensive knowledge that has accumulated about this pathogen over the last decade, this result is not surprising. Indeed, in such a case, the plant disorder results from the transfer of bacterial genetic material (T-DNA) to plant cells via a limited number of bacteria whose survival in the plant environment may not be absolutely required.[45] However, in some circumstances *A. tumefaciens* is capable of generating secondary tumors at a distance from the initial infection site as a possible consequence of bacterial spreading within the plant vessels.[45a] From this standpoint, iron acquisition may be important to the survival of this pathogen in the plant where iron has commonly been found to be transported by organic acids, mostly citrate and malate.[12] Although *A. tumefaciens* can use ferric citrate as an iron source, production of agrobactin may be advantageous in some conditions as well as for the survival of the bacterium in soil.

The importance of iron acquisition in the pathogenicity of *P. syringae* has recently been assessed. In most *P. syringae* pathovars, the release of phytotoxins seems to be of significance to plant pathogenesis. Syringomycin disrupts physiological functions in the plasma membrane of host cells, and is capable of eliciting necrosis that resembles the natural disease syndrome. Furthermore, a study of $Tox^-$ mutants has shown the toxin to contribute efficiently to the virulence of the strains tested in an immature cherry fruit assay.[46] The biosynthesis of this toxin is stimulated by high concentrations of iron and the authors postulated that the pathogen might need to acquire sufficient quantities of iron from the environment to produce syringomycin.[47] A pyoverdin-like (pseudobactin-like), siderophore was identified,[48] but using siderophore biosynthetic and transport mutants, Cody and Gross showed that iron uptake did not appear to play a role in the virulence of this bacterium.[49]

In fact, the assumption that iron assimilation plays a role in plant infection was experimentally shown by with work on the pathogenicity of *E. chrysanthemi,* a pectinolytic bacteria that causes soft rot on a wide range of plants. These bacteria induce maceration of parenchymatous tissue mainly by degrading the plant cell walls as a consequence of production of extracellular bacterial enzymes.[50,51] Besides the secretion of pectin-degrading enzymes, Expert and Toussaint suggested that other determinants, such as the production of siderophores might be important in the infection of saintpaulia plants (african violets) by *E. chrysanthemi* 3937.[52] Indeed, bacteriocin-resistant mutants defective in the synthesis of one, two, or all three of the low iron-regulated outer membrane proteins with apparent molecular weights of approximately 80,000, which were still able to secrete the lytic enzymes, were all shown to produce only a localized, necrotic response in the infected saintpaulia plants. Normally, these bacteria induce a systemic soft rot. Further studies showed that induction of the three proteins in the wild-type strain was associated with production of a catechol-type siderophore called chrysobactin, whose structure was identified as *N*-[$N^2$-(2,3-dihydroxybenzoyl)-D-lysyl]-L-serine.[53] Mutants which no longer produced chrysobactin (defective at different stages in the biosynthetic pathway of this siderophore or which failed to take up the chrysobactin ferric:complex) were recovered by insertion mutagenesis with a phage mini-Mu derivative. In particular, several mutants unable to utilize ferric:chrysobactin lack the 82,000-Da low iron-regulated outer membrane protein,[54] providing additional evidence that this protein serves as the receptor for the ferric:chrysobactin complex. These mutants, unlike the parental strain, were unable to invade the host plant although most of them could still cause a localized necrotic response. Restoration of these functions as a result of precise excision of the prophage responsible for the mutation or from genetic complementation by recombinant plasmids restored virulence of the mutants. The behavior of the mutants *in planta* led the authors to suggest that competition for iron exists between the invading bacteria and the host. Bacterial growth conditions during infection were assessed from analysis of the intercellular fluid of healthy saintpaulia plant leaves and their petioles. The iron level in this fluid appeared to be low enough for the bacterial cells to induce the production of chrysobactin and its uptake system.[55] Furthermore, the three low iron-regulated outer membrane proteins appeared to be induced when analyzed from the infecting bacteria grown *in planta*.[56] This finding is particularly relevant since this *Erwinia* strain cannot use ferric citrate as an iron source[55] (see below).

Genetics of this high-affinity iron transport system is currently under investigation. The functions involved (Cbs, Cbu, Cbx, and Fct for: *c*hryso*b*actin biosynthesis, *u*ptake, e*x*cretion, and *f*erri-*c*hrysobactin *t*ransport, respectively) are encoded by a genomic region of 55 kb that has recently been localized on the genetic linkage map of *E. chrysanthemi* 3937.[57] Cloning and complementation tests, using *E. coli* mutants defective in the production of enterobactin, a siderophore very similar to chrysobactin, were used to identify an operon of about 8 kb in length necessary for the conversion of chorismate to the catechol intermediate, the biosynthetic locus involved in catechol activation and the *fct* gene encoding the cognate receptor for the chrysobactin ferric:chelate.[58] Interestingly, the first gene of this operon to be transcribed is the *fct* uptake gene. This system was found to be negatively regulated by iron, through the Cbr (*c*hryso*b*actin *r*egulation) gene product which acts as a transcriptional repressor in the presence of iron. From this standpoint, this regulator appears to be functionally analogous to the *E. coli* Fur protein, even though there is apparently no sequence homology or functional complementation between these two loci. Further investigations will determine whether these two regulatory proteins bind DNA via the same consensus operator. Recent data indicates that iron through the *cbr* regulatory locus controls directly or indirectly other virulence functions not related per se to iron transport. Indeed, the pectinolytic activity resulting from five pectate lyase isoenzymes in strain 3937 was found to be stimulated up

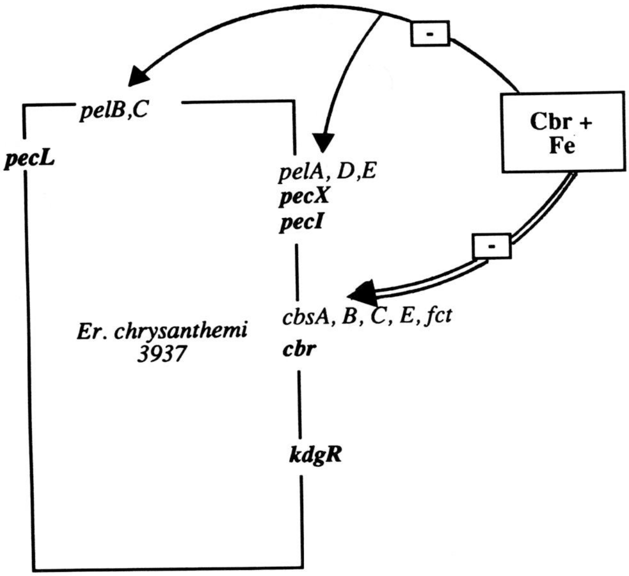


FIGURE 2. Genes controlled by iron via the *cbr* network in *Erwinia chrysanthemi* 3937. The loci related to iron correspond to the phenotypes described in the text. The loci *pecL, X, I* and *kdg* R refer to regulatory functions specific for the *pelA* to *pelE* pectate lyase encoding genes. (-) Negative regulation. Whether or not *cbr* directly acts at level of transcription of *pel* genes is still unknown.

to threefold by iron starvation, and this was correlated with a proportional increase in mRNA levels[59] (see Figure 2).

A second example of plant-bacterial pathogenesis where iron has been implicated as a critical factor for virulence involves another member of *Erwinia, (E. amylovora),* the causal agent of the fire blight disease. The strain 1430 is particularly destructive to pear and apple trees, generating a rapid systemic necrosis throughout the plant. In this strain, iron deficiency induces the production of an hydroxamate-type siderophore which was correlated with the appearance of two new polypeptides migrating in the 80,000 Da range, in the bacterial outer membrane.[60] Mutants, isolated by insertion mutagenesis with a phage Mu derivative, displayed attenuated virulence on several plant hosts.[61] One of the six mutants failed to grow in the presence of an iron chelator, and accordingly failed to produce any functional siderophore, thus indicating that iron uptake could be involved in the pathogenicity. R′ clones that were capable of complementing the hydroxamate biosynthetic defect were recovered by using the *in vivo* cloning vector RP4::miniMu. Whether or not the presence of these R′ plasmids restores virulence of the mutant is currently under investigation.[62]

## III. IRON IN SYMBIOTIC NITROGEN FIXATION

The reduction or fixation of atmospheric nitrogen to ammonia as a biochemical conversion has been found only to occur in bacteria. A large number of nitrogen-fixing bacteria have been discovered representing several diverse genera.[63, 64] Although, bacteria such as *Anabaena* sp., and *Clostridium pasteurianum, Methanococcus voltae,* and *Thiobacillus ferrooxidans* can fix nitrogen, others are the either enterics or obligate aerobes. The nitrogen-fixing free-living saprotrophs such as *Klebsiella pneumoniae* and *Azotobacter* species and those that form a symbiotic relationship with plants, such as various species of the rhizobial group, i.e., *Rhizobium, Bradyrhizobium,* and *Azorhizobium,* have been the focus of many studies relating to nitrogen fixation.[65] The rhizobial group induce nodule formation, enter the plant, differentiate into nitrogen-fixing bacteroids, and reside in the newly formed nodules. The genes required for nitrogen fixation have also been extensively studied in *Klebsiella, Azotobacter* and the rhizobial strains. For *K. pneumoniae,* these genes have been designated as *nif* genes. In rhizobial strains similar *nif* genes have been identified, as well as an additional class of genes, termed *fix* genes.[66] The *fix* genes are specifically required for nitrogen fixation *in planta*. The *nif* genes, whose functions have been identified, encode the enzymes of the dinitrogenase-dinitrogenase reductase complex as well as those required for the biosynthesis of the nitrogenase cofactors. DNA sequence analysis indicates that a number of rhizobial *fix* genes share sequence homologies with bacterial ferredoxins as well as other genes required for respiratory electron transport,[65] suggesting that new such systems are required in the nitrogen-fixing nodule function. Other *fix* genes have been shown to have a regulatory function.[66]

Biochemical analyses of functional nodules as well as their molecular components provide strong evidence that the requirement for iron would be greater for the symbiosis than for the host plant alone. Verma and Nadler estimated the lower limit to be 0.5 m*M* based on the iron content of the bacteroids, the turnover number of the nitrogenase enzyme complex, bacteroid volume, and the daily fluctuation in nitrogenase activity.[67] The production of iron-containing heme proteins, i.e., leghemoglobins, occurs during nodular development and can comprise >20% of the total soluble protein in the nodule.[68] In addition, the dinitrogenase-dinitrogenase reductase enzyme complex, nonheme iron proteins, constitutes about 11% of the total protein in the bacteroid.[69] Each dinitrogenase contains 24 to 36 atoms of iron, and each dinitrogenase reductase contains 4 atoms of iron.[70] An additional amount of iron may be required for the formation of novel respiratory systems in the bacteroids of rhizobia.

## IV. RHIZOBIAL AND PLANT IRON NUTRITION

The focus of this section is to describe both bacterial and plant iron nutrition in symbiotic nitrogen fixation. The mineral requirements for the growth of higher plants has been studied in detail.[71] For a mineral nutrient to be considered essential, several criteria have been proposed. The plant must not be able to complete its life cycle in the absence of the mineral element. The mineral element must not be replaceable with another element, and it must be involved in some component of plant metabolism such as part of an enzyme cofactor. Of the known essential micronutrients, iron was the first to be discovered.

Several criteria have also been proposed to assess the mineral constraints inherent to symbiotic nitrogen fixation. There should be a higher requirement for a given nutrient for symbiotic nitrogen fixation than for the host plant alone. There should also be a "negative interaction" between the nutrient and combined nitrogen for growth of the symbiotic legume.[72] The term "negative interaction" in this case means that the addition of the mineral in amounts that satisfy the symbiosis are not enhanced further with the addition of combined nitrogen. Furthermore, when the symbiotic legume is deficient in the mineral, the addition of combined nitrogen does not improve the development, growth, or nitrogen-fixing capability of the symbiosis.

It has been shown that the application of iron to iron-stressed legumes increased nodule number and plant mass.[73-77] This observation does not indicate whether the application of iron has an indirect effect on the iron metabolism of the plant or on the development of functional nodules. It was demonstrated, however, that iron deficiency in peanut plants (*Arachis hypogaea* L.) specifically limited bradyrhizobial nodule development and the development of a functional nitrogen-fixing apparatus.[78] This study was undertaken because it was shown that very iron-deficient peanut plants grown in alkaline field soils failed to nodulate after inoculation unless foliar applications of iron were administered. In glasshouse tests, an increase in the number of nodules, bacteroids (up to 200-fold increases), leghemoglobin production, and nitrogenase activity occurred when iron-stressed plants were given foliar applications of iron. Interestingly, of the two peanut plant cultivars tested, one showed a more dramatic response using the same bacterial inoculum. They further demonstrated that foliar applications of iron resulted in an increase in the iron content of roots and nodules indicating that the iron had been translocated. The initial formation and development of nodules induced by rhizobia can be monitored by first removing the outer cortex of the root tissue. After staining, the nodule "initials" can then be counted with the aid of a dissecting microscope. The number of nodule "initials" was about the same for treated and untreated plants suggesting that nodule development rather than nodule formation was limited by available iron in this system. The bradyrhizobia used in this study were capable of surviving in the soil even under iron-limiting conditions. The above study provided the first evidence that iron deficiency in legume symbiosis specifically limits nodular development and function.

In a related study, the role of plant iron nutrition on bradyrhizobial nodulation and nitrogen fixation in lupines (*Lupinus angustifolius* L.) was investigated.[79] To address the criteria for the essentiality of a mineral nutrient in legume symbiosis, combined nitrogen was added to inoculated plants grown under very-limiting, moderately limiting, and high iron conditions. Under very-limiting iron conditions, the addition of combined nitrogen did not improve the symbiosis. However, under moderately limiting iron conditions, the addition of combined nitrogen improved shoot growth, suggesting that iron deficiency was operationally nitrogen deficiency. A negative interaction was observed for the addition of combined nitrogen and high concentrations of iron. In this system it was noted that the increased levels of available iron led to increases in the number of nodule "initials," suggesting that nodule initiation is directly affected by the availability of iron. Also, the external iron concentration required for maximal nodule initiation was higher than that required for maximal chlorophyll biosynthesis. Taken together, these two studies indicate that the essentiality of iron in symbiosis is at the level of functional nodule development.

Although, there is wide variation among plants in their ability to acquire iron from their natural environment, they are generally termed as being either iron-efficient or -inefficient depending on whether or not they can acquire iron from a given soil. In a study comparing the sensitivity of two lupine species and one pea species to iron deficiency, it was shown that whereas one lupine species was very sensitive the other was only moderately sensitive. The pea species, however, was unaffected under the same conditions.[80] A further complication is that the efficiency of iron acquisition by various rhizobia also varies, which might affect their ability to form functional nodules.[81-83] It should be clear from such studies that the precise legume-*Rhizobium* combination, as well as the soil type used for plant growth, may affect the symbiosis.

A number of rhizobial strains have been examined for their ability to acquire iron under iron-limiting laboratory conditions. Siderophore production has been shown for a variety of strains, and there is a large variation in the chemical nature of rhizobial siderophores. Both hydroxamate-like and catechol-like siderophores as well as others with unknown iron-binding functional groups are produced by different strains and, as stated above, the ferric complex of citric acid is used by yet another.[11,84-92] Interestingly, anthranilic acid serves as a siderophore for a *R. leguminosarum* strain.[90] The complete structure has been determined for a novel siderophore produced by *R. meliloti,* namely, DM-4.[91,92] This siderophore, termed rhizobactin, contains an ethylenediamine group for iron chelation. The use of such a functional group had not been previously seen in microbial siderophores. Although the structure has not been completed, the siderophore produced by *R. meliloti* 1021 may contain other unique iron-binding moieties.[93] In a study of a number of *R. meliloti* strains, it was found that, like the pseudomonad strains discussed above, there is variation in the ability or inability of a given rhizobial strain to use a siderophore produced by another strain as an iron source.[91,93]

Presumably, these diverse siderophores can function to acquire iron for the given rhizobial strain in their free-living phase of growth. Of particular interest is the role of such iron transport systems for nodular growth and development. Are such systems functional for bacterial iron acquisition once the bacterium is growing *in planta*? After saturation mutagenesis of *R. meliloti* 1021 with Tn5, a number of single-site Tn5 insertion mutants defective in either the production, the transport of the ferric:complex, or the negative regulation of the siderophore, rhizobactin 1021, produced by the strain were isolated.[94] All of the siderophore auxotrophs were complemented to wild-type levels of siderophore production with a cloned genomic region of about 30-kb. This result suggests that the genes required for rhizobactin 1021 biosynthesis are confined to this 30-kb region.[94] None of the transport mutants were defective in the production of iron-regulated outer membrane proteins, an expected component of the system, and may represent pleiotropic mutations defective in some general membrane function. The regulatory mutants could not be complemented in *trans,* and may represent "operator" mutations.

These mutants were analyzed for their ability to produce nodules, fix nitrogen, and induce an increase in plant growth (alfalfa).[95] Thirty days after inoculation, the plant weight and nodule numbers were about the same for wild-type and mutant inoculations. The nitrogenase activity, however, was two to three times lower for plants inoculated with the mutants compared to wild-type inoculated plants. Seventy days after inoculation, a dramatic difference in nitrogenase activity was seen comparing plants inoculated with the wild-type and those inoculated with many of the mutants. The nitrogenase activity was an average of 13.5 times greater for plants inoculated with the wild-type compared to that of the mutants strains: 54, 20, 21, 22, 25, and 30 (see Figure 3). Strains 62 and 54 are transport mutants, strain 63 is a regulatory mutant, and strains 20 to 30 are siderophore production mutants. When grown in a liquid medium containing a non-utilizable synthetic iron chelator, the growth of the mutants was attenuated compared to the growth of the wild-type. Except for mutants 27 and 29, the degree of growth attenuation was correlated with the decrease in nitrogenase activity relative to the wild-type *in planta*. Inoculation with strains 27 and 29 resulted in the formation of very large nodules as well as nitrogenase activity that exceeded that of the wild-type, which cannot be readily interpreted. Perhaps the plant has somehow compensated for the inability of these mutants to acquire iron. Taken together, however, these results indicate that a high-affinity iron transport system can be required for efficient nitrogen fixation, presumably to provide nutritional iron to the bacteroids.[95]

In another recent study the role of bacterial iron nutrition on heme biosynthesis and symbiotic nitrogen fixation in *R. leguminosarum* was investigated.[96] One chemically-induced mutant of this bacterium was shown to form ineffective nodules, and was arrested at a late stage of nodular development, i.e., after bacteroid development. The mutant was also defective in heme production which may be related to its defect in iron acquisition. Interestingly, it failed to use its own siderophore, anthranilic acid, but still retained some ability to use other weak chelators that appear to be used as an iron source by this strain, like citrate, suggesting that the mutation in iron acquisition may be in a transport function. The ability of this mutant to acquire iron like the wild type and form effective nodules was restored with a cosmid suggesting, at the least, a very close link with symbiotic nitrogen fixation and iron acquisition. A molecular dissection of this cosmid, as well as the one described for *R. meliloti*,[94] should provide much needed information for the precise role and mechanisms of rhizobial iron transport systems in symbiotic nitrogen fixation.

## V. CONCLUDING REMARKS

We have outlined evidence indicating that the presence of a bacterial iron-transport

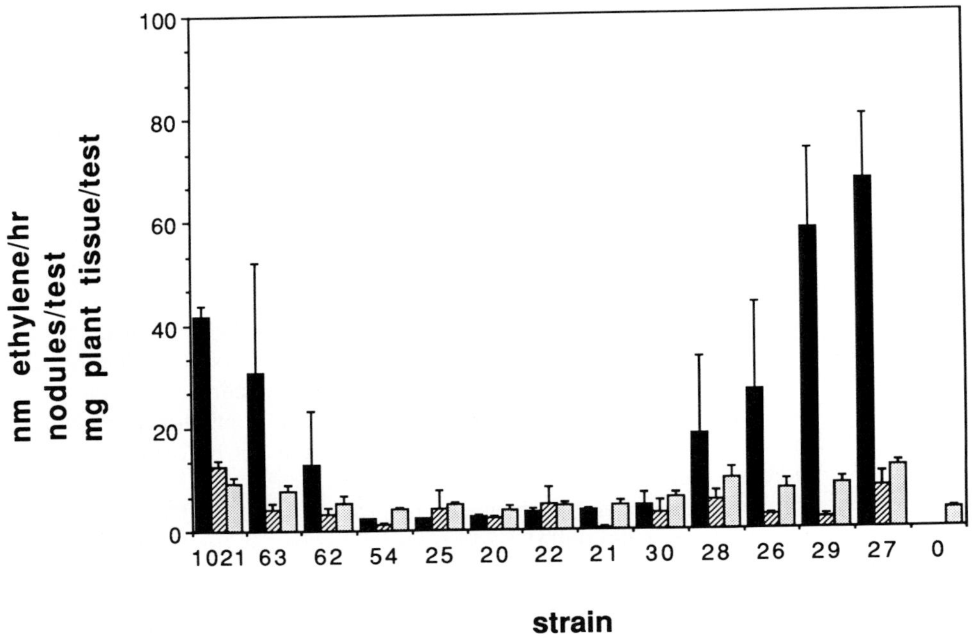


FIGURE 3. Nitrogenase activity (■), number of nodules per plant (▨), and plant tissue, mg dry weight, (▩) for plants grown in plant nodulation medium. Results with bacterial inoculations are indicated by strain numbers: 1021 for Rm1021 and 20 to 63 for the PRR Tn5 mutants described in the text; 0 indicates no bacterial inoculation. The results are based on an average of 6 plants for bacterial inoculations and 16 plants for uninoculated plants. The standard error for each analysis is indicated. Reproduced from *Plant and Soil,* with permission from Kluwer Academic Publishers.

system in a plant-invading bacterium can significantly influence its ability to complete its life cycle *in planta*. Presumably, the role of such a system is to provide nutritional iron to the bacterium necessary for its growth and development. In the plant-bacterial systems that we have discussed, i.e., systemic soft rot incited by *Erwinia* and symbiotic nitrogen fixation by *Rhizobium* and *Bradyrhizobium,* substantial growth and movement or development of the bacterium is required once it is in the plant. The importance of bacterial siderophore systems did not appear to be required in at least two other plant-bacterial pathogen interaction systems in which a localized disease is incited by the bacterium. The requirement for bacterial iron transport systems in bacterial plant diseases may be restricted to those which incite vascular or systemic diseases. The iron requirement for rhizobia to form nitrogen-fixing nodules may also be a reflection of the increase in iron required to form the nitrogen-fixing apparatus. One other possible beneficial role for the pathogen to retain the capacity to form iron-binding compounds is to reduce the level of free iron that might be released from damaged tissue. An increase in the level of free iron, which could be released from broken chloroplasts, might serve to increase the levels of the bactericidal oxidizing radicals that are produced during infection.

The response by animals to withhold iron from invading pathogens, i.e., the *i*ron *w*ithholding *r*esponse (IWR), is well-documented as described above. Analogous studies of plant responses related to the IWR are still in their infancy. It is known, for example, that the localization of iron-containing phytoferritin is altered in infected tissue compared to uninfected tissue.[97] Studies related to the IWR are certainly needed, and would help to begin to unravel the biochemical mechanisms of why siderophore auxotrophs of *Erwinia* and *Rhizobium* are unable to complete their life cycles *in planta*. Perhaps plant-invading bacteria are dosed with high levels of iron, and they need siderophore production to sequester the toxic metal; it is still truly a "black box." From a purely nutritional point of view, however, various plant-derived iron chelators have been described which might serve as an iron source for invading pathogens.[98] Interestingly, since the vascular fluid of plants contains iron solubilized mostly as ferric:citrate, the *Erwinia* strain described in this review is unable to use ferric:citrate as an iron source. It would be informative to determine whether a given vascular phytopathogen that does not produce a siderophore, can instead use a plant iron chelator, such as ferric:citrate, as an iron source, as might be the case for some rhizobia during their symbiotic phase of growth. Like certain animal pathogens, the ability to utilize a host-derived iron chelator as an iron source, could provide an economical advantage to an invading pathogen, in that only an uptake system would be required.

One additional observation deserves attention. The *fnr* gene product monitors the redox state of *E. coli* perhaps by noting the redox state of iron,[99] using $Fe^{2+}$ as a cofactor. Recently, the *fnr* gene product was shown to have homology with both *nif* and *fix* gene products, whose functions are known to be sensitive to the oxygen status of the cell. A metal ion has recently been implicated to be required for the proper functioning of the *nif*A gene products of *B. japonicum*[100] and of *K. pneumonia*.[101] The *fix*K gene of *R. meliloti* was shown to have homology with the *fnr* gene,[102] and both of the gene products act as positive and negative transcriptional regulators. From such studies one might suggest a possible regulatory role for this metal in yet other systems with no obvious relation to nutritional iron transport.

## REFERENCES


1. **Griffiths, E.,** Iron in biological systems, in *Iron and Infection,* Bullen, J. J. and Griffiths, E., Eds., John Wiley & Sons, NY, 1987, 1.
2. **Walsh, C.,** Hemoprotein oxidases, monooxygenases, and reductases, *Enzymatic Reaction Mechanisms,* Freeman, San Francisco, 1979, 464.
3. **Segel, I. H.,** Standard reduction potentials of some oxidation-reduction half-reactions, *Biochemical Calculations,* 2nd ed., John Wiley & Sons, NY, 1975, 414.
4. **Neilands, J. B.,** Siderophores of bacteria and fungi, *Microbiol. Sci.,* 1, 9, 1984.
5. **Halliwell, B. and Gutteridge, J. M. C.,** Oxygen is poisonous—an introduction to oxygen toxicity and free radicals, in *Free Radicals in Biology and Medicine,* Halliwell, B. and Gutteridge, J. M. C., Eds., Oxford University Press, New York, 1987, 1.
6. **Theil, E. C. and Aisen, P.,** The storage and transport of iron in animal cells, in *Iron Transport in Microbes, Plants and Animals,* Winkelmann, G., van der Helm, D., and Neilands, J. B., Eds., VCH Verlagsgesellschaft mbH, Weinheim, FRG, 1987, 491.
7. **Briat, J.-F., Laulhère, J.-P., Labouré, A.-M., Proudhon, D., and Lescure, A.-M.,** Structure, function, and synthesis of plant ferritins, in *Metal Ion Homeostasis: Molecular Biology and Chemistry,* Winge, D. and Hamer, D., Eds., Alan R. Liss, NY, 1989, 137.
8. **Andrews, S. C., Harrison, P. M., and Guest, J. R.,** Cloning, sequencing, and mapping of the bacterioferritin gene *(bfr)* of *Escherichia coli* K-12, *J. Bacteriol.,* 171, 3940, 1989.
9. **Schwertmann, U.,** Solubility and dissolution rate of iron oxides as influenced by type, crystallinity and isomorphous substitution, presented at 5th Int. Symp. Iron Nutrition and Interact. in Plants, *Jerusalem, Israel, June 11-17, 1989.*
10. **Archibald, F.,** *Lactobacillus plantarum,* an organism not requiring iron, *FEMS Microbiol. Lett.,* 19, 29, 1983.
11. **Guerinot, M. L., Meidl, E. J., and Plessner, O.,** Citrate as a siderophore in *Bradyrhizobium japonicum, J. Bacteriol.,* 172, 3298, 1990.
12. **Tiffin, L. O.,** Iron translocation. I. plant culture, exudate sampling, iron citrate analysis, *Plant Physiol.,* 41, 510, 1966.
13. **Braun, V., Hantke, K., Eick-Helmerich, K., Köster, W., Preßler, Sauer, M., Schäffer, S., Schöffler, H., Staudenmaier, H., and Zimmermann, L.,** Iron transport systems in *Escherichia coli, in Iron Transport in Microbes, Plants and Animals,* Winkelmann, G., van der Helm, D. and Neilands, J. B., Eds., VCH Verlagsgesellschaft mbH, Weinheim, FRG, 1987, 35.

14. **Neilands, J. B.,** Microbial envelope proteins related to iron, *Annu. Rev. Microbiol.,* 36, 285, 1982.
15. **Luria, S. E. and Delbrück, M.,** Mutations of bacteria from virus sensitivity to virus resistance, *Genetics,* 28, 491, 1943.
16. **Becker, J. O., Messens, E., and Hedges, R. W.,** The influence of agrobactin, on the uptake of ferric ion by plants, *FEMS Microbiol. Ecol.,* 31, 171, 1985.
17. **Reid, C. P. P., Crowley, D. E., Kim, H. J., Powel, P. E., and Szaniszlo, P. J.,** Utilization of iron when supplied as ferrated synthetic chelate or as ferrated hydroxamate siderophore, *J. Plant Nutr.,* 7, 437, 1984.
18. **Halmann, M., Sery, T., Chen, Y., Bar-ness, E., Gilan, S., and Hadar, Y.,** A novel siderophore from *Rhizopus arrhizus* and the utilization of its iron complex by plants, presented at 2nd Int. Symp. Iron Transport, Storage, and Metabolism, July 20-22, 1990, University of Texas, Austin.
19. **Becker, J. O., Hedges, R. W., and Messens, E.,** Inhibitory effect of pseudobactin on the uptake of iron by higher plants, *Appl. Environ. Microbiol.,* 49, 1090, 1985.
20. **Bienfait, H. F.,** Mechanisms in Fe-efficiency reactions of higher plants, *J. Plant Nutr.,* 11, 605, 1988.
21. **Lankford, C. E.,** Bacterial iron assimilation, *Crit. Rev. Microbiol.,* 2, 273, 1973.
22. **Matzanke, B. F., Bill, A. X., Trutwein, and Winkelmann, G.,** Main components of iron metabolism in enterobacteriaceae-An in vivo Mössbauer study, presented at 2nd Int. Symp. Iron Transport, Storage, and Metabolism, *July 20-22, 1990, University of Texas, Austin, abstracts.*
23. **Vose, P. B.,** Iron nutrition in plants: a world overview, *J. Plant Nutr.,* 5, 233, 1982.
24. **Leong, J.,** Siderophores: their biochemistry and possible role in the biocontrol of plant pathogens, *Annu. Rev. Phytopathol.,* 24, 187, 1986.
25. **Buyer, J. S. and Leong, J.,** Iron transport-mediated antagonism between plant-growth promoting and plant-deleterious *Pseudomonas* strains, *J. Biol. Chem.,* 261, 791, 1986.
26. **Schippers, B., Geels, F. P., Bakker, P. A. H. M., Bakker, A. W., and Weisbeek, P.,** Methods of studying plant growth-promoting pseudomonads-problems and progress, in *NATO Advanced Research Workshop: Iron, Siderophores and Plant Disease,* Swinburne, T. R., Ed., Plenum Press, NY, 1986, 149.
27. **Schippers, B., Bakker, A. W., and Bakker, P. A. H. M.,** Interactions of deleterious and beneficial rhizosphere microorganisms and the effect of cropping practices, *Annu. Rev. Phytopathol.,* 25, 339, 1987.
28. **Schade, A. L. and Caroline, L.,** Raw egg white and the role of iron in growth inhibition of *Shigella dysenteriae, Staphylococcus aureus, Escherichia coli* and *Saccharomyces cerevisiae, Science,* 100, 14, 1944.
29. **Weinberg, E. D.,** Iron withholding: a defense against infection and neoplasia, *Physiol. Rev.,* 64, 65, 1984.
30. **Mickelsen, P. A. and Sparling, P. F.,** Ability of *Neisseria gonorrhoeae, Neisseria meningitidis,* and commensal *Neisseria* species to obtain iron from transferrin and iron compounds, *Infect. Immun.,* 33, 555, 1981.
31. **Ludwig, A., Jarchau, T., Benz, R., and Goebel, W.,** The repeat domain of *Escherichia coli* HlyA is responsible for its $Ca^{2+}$-dependent binding to erythrocytes, *Mol. Gen. Genet.,* 214, 553, 1988.
32. **Stoebner, J. A. and Payne, S. M.,** Iron-regulated hemolysin production and utilization of heme and hemoglobin by *Vibrio cholerae, Infect. Immun.,* 56, 2891, 1988.
33. **Griffiths, E.,** The iron-uptake systems of pathogenic bacteria, in *Iron and Infection,* Bullen, J. J. and Griffiths, E., Eds., John Wiley & Sons, NY, 1987, 69.
34. **Gibson, F. and Magrath, D. J.,** The isolation and characterization of a hydroxamic acid (aerobactin) formed by *Aerobacter aerogenes, Biochim. Biophys. Acta,* 192, 175, 1969.
35. **Waters, V. L. and Crosa, J. H.,** DNA environment of the aerobactin iron uptake system genes in prototypic ColV plasmids, *J. Bacteriol.,* 167, 647, 1986.
36. **Konopka, K. and Neilands, J. B.,** Effect of serum albumin on siderophore-mediated utilization of transferrin iron, *Biochemistry,* 23, 2122, 1984.
37. **Bagg, A. and Neilands, J. B.,** Molecular mechanism of regulation of siderophore mediated iron assimilation, *Microbiol. Rev.,* 51, 509, 1987.
38. **deLorenzo, V., Giovannini, F., Herrero, M., and Neilands, J. B.,** Metal ion regulation of gene expression: Fur repressor-operator interaction at the promoter region of the aerobactin system of pColV-K30, *J. Mol. Biol.,* 203, 875, 1988.
39. **Calderwood, S. and Mekalanos, J. J.,** Iron regulation of Shiga-like toxin expression in *Escherichia coli* is mediated by the *fur* locus, *J. Bacteriol.,* 169, 4759, 1987.
40. **Tai, S.-P.S. and Holmes, R. K.,** Iron regulation of the cloned diphtheria toxin promoter in *Escherichia coli, Infect. Immun.,* 56, 2430, 1988.
41. **Poole, K. and Braun, V.,** Iron regulation of *Serratia marcescens* hemolysin gene expression, *Infect. Immun.,* 56, 2967, 1988.
42. **Leong, S. A. and Neilands, J. B.,** Siderophore production by phytopathogenic microbial species, *Arch. Biochem. Biophys.,* 218, 351, 1982.
43. **Ong, S. A., Peterson, T., and Neilands, J. B.,** Agrobactin, a siderophore of *Agrobacterium tumefaciens, J. Biol. Chem.,* 254, 1860, 1979.

44. **Leong, S. A. and Neilands, J. B.,** Relationship of siderophore mediated iron assimilation to virulence in crown gall disease, *J. Bacteriol.,* 141, 482, 1981.
45. **Chilton, M. D., Drummond, M. H., Merlo, D. J., Sciaky, D., Montoya, A., Gordon, M. P., and Nester, E. W.,** Stable incorporation of plasmid DNA into higher plant cells: the molecular basis of crown-gall tumorigenesis, *Cell,* 11, 263, 1977.
45a. **Tempé, J.,** personal communication, 1990.
46. **Xu, G.-W. and Gross, D. G.,** Evaluation of the role of syringomycin in plant pathogenesis by using Tn5 mutants of *Pseudomonas syringae* pv. *syringae* defective in syringomycin production, *Appl. Environ. Microbiol.,* 54, 1345, 1988.
47. **Gross, D. C.,** Evidence that syringomycin synthesis in *Pseudomonas syringae* pv. *syringae* and defined conditions for its production, *J. Appl. Bacteriol.,* 58, 167, 1985.
48. **Cody, Y. S. and Gross, D. C.,** Characterization of pyoverdin, the fluorescent siderophore of *Pseudomonas syringae* pv. *syringae, Appl. Environ. Microbiol.,* 53, 928, 1987.
49. **Cody, Y. S. and Gross, D. C.,** Outer membrane protein mediating iron uptake via pyoverdin, the fluorescent siderophore produced by *Pseudomonas syringae* pv. *syringae, J. Bacteriol.,* 169, 2207, 1987.
50. **Collmer, A. and Keen, N. T.,** The role of pectic enzymes in plant pathogenesis, *Annu. Rev. Phytopathol.,* 24, 383, 1986.
51. **Kotoujansky, A.,** Molecular genetics of pathogenesis by soft rot erwinias, *Annu. Rev. Phytopathol.,* 25, 405, 1987.
52. **Expert, D. and Toussaint, A.,** Bacteriocin-resistant mutants of *Erwinia chrysanthemi:* possible involvement of iron acquisition in phytopathogenicity, *J. Bacteriol.,* 163, 221, 1985.
53. **Persmark, M., Expert, D., and Neilands, J. B.,** Isolation, characterization and synthesis of chrysobactin, a compound with siderophore activity from *Erwinia chrysanthemi, J. Biol. Chem.,* 264, 3187, 1989.
54. **Enard, C., Diolez, A., and Expert, D.,** Systemic virulence of *Erwinia chrysanthemi* 3937 requires a functional iron assimilation system, *J. Bacteriol.,* 170, 2919, 1988.
55. **Enard, C., Franza, T., Neema, C., Gill, P. R., Persmark, M., Neilands, J. B., and Expert, D.,** The requirement of chrysobactin dependent iron transport for virulence incited by *Erwinia chrysanthemi on Saintpaulia ionantha, Plant and Soil,* 130, 263, 1991.
56. **Neema, C. and Expert, D.,** unpublished data, 1990.
57. **Franza, T., Enard, C., Van Gijsegem, F., and Expert, D.,** Genetic analysis of the *Erwinia Chrysanthemi,* 3937 chrysobactin iron-transport system: characterization of a gene cluster involved in uptake and biosynthetic pathways, 5, 1319, 1991.
58. **Franza, T. and Expert, D.,** The virulence associated chrysobactin iron uptake system of *Erwinia chrysanthemi* 3937 involves an operon encoding transport and biosynthetic functions, *J. Bacteriol.,* 1991, in press.
59. **Sauvage, C., Franza, T., and Expert, D.,** Iron as a modulator of pathogenicity of *Erwinia chrysanthemi* 3937 on *Saintpaulia ionantha, in Proc. 5th Int. Symp. Mol. Genet. Plant-Microbe Interact.,* 1990, in press.
60. **Vanneste, J. L. and Expert, D.,** Detection and characterization of an iron uptake system in *Erwinia amylovora, Acta Hortic.,* Fire Blight, 273, 249, 1990.
61. **Vanneste, J. L., Paulin, J. P., and Expert, D.,** Bacteriophage Mu as a genetic tool to study *Erwinia amylovora* pathogenicity and hypersensitive reaction on tobacco, *J. Bacteriol.,* 172, 932, 1990.
62. **Vanneste, J. L. and Expert, D.,** unpublished data, 1990.
63. **Werner, D.,** Stickstoff ($N^2$)-Fixierung und Produktionsbiologie, *Angew. Bot.,* 54, 67, 1980.
64. **Bothe, H. and Newton, W. E., Eds.,** *Proc. 7th Int. Symp. Nitrogen Fixation,* Gustav Fischer-Verlag, Stuttgart, 1988.
65. **Triplett, E. W., Roberts, G. P., Ludden, P. W., and Handelsman, J.,** What's new in nitrogen fixation, *ASM News,* 55, 15, 1989.
66. **Long, S. R.,** *Rhizobium* genetics, *Annu. Rev. Genet.,* 23, 483, 1989.
67. **Verma, D. P. S. and Nadler, K. D.,** The *Rhizobium*-legume symbiosis: the host's point of view, in *Plant Gene Research I:* Genes Involved in Microbe-Plant Interactions, Verma, D. P. S. and Hohn, T., Eds., Springer-Verlag, Berlin, 1984, 58.
68. **Nadler, K. D. and Avissar, Y.,** Heme biosynthesis in soybean root nodules. 1. On the role of bacteroid δ-aminolevulinic synthase and δ-aminolevulinic acid dehydratase in the synthesis of heme of leghemoglobin, *Plant Physiol.,* 60, 433, 1977.
69. **Verma, D. P. S. and Long, S. R.,** The molecular biology of the *Rhizobium*-legume symbiosis, *Int. Rev. Cytol. Suppl.,* 14, 211, 1983.
70. **Bothe, H., Yates, M. G., and Cannon, F. C.,** Physiology, biochemistry, and genetic dinitrogen fixation, in *Encyclopedia of Plant Physiology,* New Series, Vol. 15A, Lauchli, A. and Bieleski, R. L., Eds., Springer-Verlag, NY, 1983, 241.
71. **Marschner, H.,** Introduction, definition, and classification of mineral nutrients, *Mineral Nutrition in Higher Plants,* Academic Press, San Diego, 1986, 3.
72. **O'Hara, G. W., Boonkerd, N., and Dilworth, M. J.,** Mineral constraints to nitrogen fixation, *Plant Soil,* 108, 93, 1988.

73. **Rai, R., Singh, S. N., and Prasad, V.,** Effect of pressmud amend pyrite on symbiotic $N_2$-fixation, active iron contents of nodules, grain yield and quality of chickpea *(Cicer arietinum* Linn) genotypes in calcareous soil, *J. Plant Nutr.*, 5, 905, 1982.
74. **Hartzook, A.,** The problem of iron deficiency in peanuts *(Arachis hypogaea* L.) on basic and calcareous soils in Israel, *J. Plant Nutr.*, 5, 923, 1982.
75. **Rai, R., Prasad, V., Choudhury, S. K., and Sinha, N. P.,** Iron nutrition and symbiotic $N_2$-fixation of lentil *(Lens culinaris)* genotypes in calcareous soil, *J. Plant Nutr.*, 7, 399, 1984.
76. **Hemantarajan, A. and Garg, O. K.,** Introduction of nitrogen-fixing nodules through iron and zinc fertilization in the non-nodule-forming French bean *(Phaseolus vulgaris* L.), *J. Plant Nutr.*, 9, 281, 1986.
77. **Hemantarajan, A.,** Iron fertilization in relation to nodulation and nitrogen fixation in French bean *(Phaseolus vulgaris* L.), *J. Plant Nutr.*, 11, 829, 1988.
78. **O'Hara, G. W., Dilworth, M. J., Boonkerd, and Parkpian, P.,** Iron-deficiency specifically limits nodule development in peanut inoculated with *Bradyrhizobium* sp., *New Phytol.*, 108, 51, 1988.
79. **Tang, C., Robson, A. D., and Dilworth, M. J.,** The role of iron in nodulation and nitrogen fixation in *Lupinus angustifolius* L., *New Phytol.*, 114, 173, 1990.
80. **White, P. F. and Robson, A. D.,** Lupin species vary widely in their sensitivity to Fe deficiency, *Aust. J. Agric. Res.*, 40, 539, 1989.
81. **O'Hara, G. W., Hartzook, A., Bell, R., and Loneragan, J. F.,** Response to *Bradyrhizobium* strain of peanut cultivars grown under iron stress, *J. Plant Nutr.*, 11, 843, 1988.
82. **Reigh, G. and O'Connell, M.,** Siderophore production is strain specific in Rhizobium, in Nitrogen Fixation: Hundred Years After, Bothe, H., DeBruijn, F. J., and Newton, W. E., Eds., Gustav Fischer, NY, 1988, 826.
83. **Ames-Gottfred, N. P., Christie, B. R., and Jordan, D. C.,** Use of the chrome azurol S agar plate technique to differentiate and field isolates of *Rhizobium leguminosarum* biovar *trifolii, Appl. Environ. Microbiol.*, 55, 707, 1989.
84. **Carrillo-Castaneda, G. and Peralta, J. R. V.,** Siderophore-like activities in *Rhizobium phaesoli, J. Plant Nutr.*, 11, 935, 1988.
85. **Modi, M., Shah, K. S., and Modi, V. V.,** Isolation and characterization of catechol-like siderophore from cowpea *Rhizobium* RA-1, *Arch. Microbiol.*, 141, 156, 1985.
86. **Nambiar, P. T. C. and Sivaramakrishnan, S.,** Detection and assay of siderophores in cowpea rhizobia (Bradyrhizobium) using radioactive Fe ($^{59}$Fe), *Lett. Appl. Microbiol.*, 4, 37, 1987.
87. **Patel, H. N., Chakraborty, R. N., and Desai, S. B.,** Isolation and partial characterization of phenolate siderophore form *Rhizobium leguminosarum* IARI 102, *FEMS Microbiol. Lett.*, 56, 131, 1988.
88. **Skorupska, A., Choma, A., Derylo, M., and Lorkiewicz, Z.,** Siderophore containing 2,3-dihydroxybenzoic acid and threonine formed by *Rhizobium trifolii, Acta Biochim. Pol.*, 35, 119, 1988.
89. **Skorupska, A., Derylo, M., and Lorkiewicz, Z.,** Siderophore production and utilization by *Rhizobium trifolii, Biol. Metals,* 2, 45, 1989.
90. **Rioux, C. R., Jordan, D. C., and Rattray, J. B. M.,** Iron requirement of *Rhizobium leguminosarum* and secretion of anthranilic acid during growth on an iron-deficient medium, *Arch. Biochem. Biophys.*, 248, 175, 1986.
91. **Smith, M. J. and Neilands, J. B.,** Rhizobactin, a siderophore from *Rhizobium meliloti, J. Plant Nutr.*, 7, 449, 1984.
92. **Smith, M. J., Shoolery, J. N., Schwyn, B., Holden, I., and Neilands, J. B.,** Rhizobactin, a structurally novel siderophore from *Rhizobium meliloti, J. Am. Chem. Soc.*, 107, 1739, 1985.
93. **Schwyn, B. and Neilands, J. B.,** Siderophores from agronomically important species of the Rhizobiaceae, Comments *Agric. Food Chem.*, 1, 95, 1987.
94. **Gill, P. R. and Neilands, J. B.,** Cloning of genomic region required for a high-affinity iron-uptake system in *Rhizobium meliloti* 1021, *Mol. Microbiol.*, 3, 1183, 1989.
95. **Gill, P. R., Barton, L. L., Scoble, M. D., and Neilands, J. B.,** A high-affinity iron transport system of *Rhizobium meliloti* may be required for efficient nitrogen fixation *in planta,* presented at Proc. 4th Int. Symp. Iron Nutrition Interact. Plants, Jerusalem, Israel, 1990; *Plant Soil,* 130, 211-217, 1990.
96. **Nadler, K. D., Johnston, A. W. B., Chen, J.-W., and John, T. R.,** A *Rhizobium leguminosarum* mutant defective in symbiotic iron acquisition, *J. Bacteriol.*, 172, 670, 1990.
97. **Maramorosch, K. and Hirumi, H.,** Phytoferritin accumulation in leaves of diseased coconut palms, *Protoplasma,* 78, 175, 1973.
98. **Marschner, H.,** Functions of mineral nutrients: micronutrients, *Mineral Nutrition in Higher Plants,* Academic Press, San Diego, 1986, 269.
99. **Spiro, S., Roberts, R. E., and Guest, J. R.,** FNR-dependent repression of the *ndh* gene of *Escherichia coli* and metal ion requirement for FNR-regulated gene expression, *Mol. Microbiol.*, 3, 601, 1989.
100. **Fischer, H.-M., Bruderer, T., and Hennecke, H.,** Essential and non-essential domains in the *Bradyrhizobium japonicum* NifA protein: identification of indispensable cysteine residues potentially involved in redox reactivity and/or metal binding, *Nucleic Acids Res.*, 16, 2207, 1988.

101. **Henderson, N., Austin, S., and Dixon, R. A.,** Role of metal ions in negative regulation of nitrogen fixation by the *nif*L gene product from *Klebsiella pneumoniae, Mol. Gen. Genet.*, 216, 484, 1989.
102. **Batut, J., Daveran-Mingot, M.-L., David, M., Jacobs, J., Gamerone, A. M., and Kahn, D.,** *fix*K, a gene homologous with *fnr* and *crp* from Escherichia coli, regulates nitrogen fixation genes both positively and negatively in *Rhizobium meliloti, EMBO J.*, 8, 1279, 1989.

# BACTERIAL PHYTOTOXIN SYRINGOMYCIN AND ITS INTERACTION WITH HOST MEMBRANES

**Jon Y. Takemoto**

## TABLE OF CONTENTS

# I. INTRODUCTION

Syringomycin is one of several phytotoxins produced by the pathovars of *Pseudomonas syringae*. Reviews on the *P. syringae* toxins have appeared periodically during the last 10 years.[1-3] The majority of these toxins are peptides or amino acids and are secreted by the bacteria. Typically, each pathovar produces a single kind of toxin molecule, although the toxins themselves are not specific to a given host. For example, *P. syringae* pv. *tabaci, P. syringae* pv. *phaseolicola,* and certain isolates of *P. syringae* pv. *syringae* produce tabtoxin, phaseolotoxin, and syringomycin, respectively. An exception is coronatine which is synthesized by several different *P. syringae* pathovars. With *P. syringae* pv. *syringae* isolates, there is further diversity in the type of toxin produced. For example, citrus fruit isolates produce syringotoxin, but stone fruit isolates generate syringomycin. Syringomycin-like toxins, the syringostatins, are produced from a lilac isolate of *P. syringae* pv. *syringae.*[4] The structures of syringomycin, syringotoxin, and the syringostatins are strikingly similar, and together they form a distinct family of bacterial phytotoxins.

The mode of action of syringomycin is fundamentally different from that of other *P. syringae* toxins. Tabtoxin, phaseolotoxin, and coronatine inhibit the activities of host intracellular enzymes. These effects lead to chloroses in host plants. Tabtoxin inhibits glutamine synthetase,[5] and phaseolotoxin inhibits ornithine carbamoyltransferase.[6,7] Coronatine may directly inhibit chlorophyll formation by impeding δ-aminolevulinic acid metabolism,[8] but an effect on ethylene biosynthesis has also been suggested.[9] In contrast, syringomycin action leads to necroses that most likely result from disturbances of membrane functions. The mechanisms of action of the other *P. syringae* toxins are either vague or completely unknown.

This chapter presents an overview of several aspects of syringomycin including the work in the author's laboratory on the mechanism of action of this toxin. The body of knowledge on syringomycin has grown significantly and is beginning to contribute to principles about plant-microbe interactions. A recent major advance was the determination of the structure of syringomycin. Other contributions have been in the areas of its role, biosynthesis, and mechanism of action.

# II. ROLE OF SYRINGOMYCIN IN PLANT DISEASE

J. E. DeVay and his colleagues first described the role of syringomycin in plant disease.[10-14] They observed a correlation between the pathogenicity of *P. syringae* isolates on peach trees and syringomycin production. Application of partially purified preparations of the toxin to young peach trees gave typical disease symptoms. Several studies suggested that syringomycin is a significant contributor to a variety of plant diseases. These diseases include necroses of stone fruits and cowpea and holcus spot disease of maize.[15]

The above studies implied a role for syringomycin in certain plant diseases. However, the question remained whether syringomycin is necessary for disease as a pathogenicity factor or instead increases the severity of disease as a virulence factor. Recent work by Xu and Gross[16] helped clarify this issue. These investigators created and isolated nontoxigenic ($tox^-$) strains of *P. syringae* by transposon Tn5 mutagenesis. Fifteen isolates did not produce syringomycin as judged by their inability to inhibit growth of the fungus, *Geotricum candidum* (used in bioassay). In tests for pathogenicity on cherry fruits, two distinct mutant groups were observed. Eight strains were pathogenic, but the disease index was 17 to 66% lower than for the toxigenic parent strain. The remaining seven strains were nonpathogenic. These results show that syringomycin is not essential for pathogenicity, but contributes to virulence. However, the large fraction (7 of 15) of nonpathogenic, $tox^-$ strains strongly suggests a link between syringomycin production and pathogenicity. Earlier studies also suggested such a link.[10,12] Finally, the *in planta* population dynamics of $tox^-$ and nonpathogenic strains were the same as the toxigenic parent strain. This observation shows that bacterial growth within the invaded plant tissues does not require syringomycin.

Although described as a phytotoxin, syringomycin is also an antifungal agent.[13] Indeed, fungi and yeasts such as *G. candidum, Rhodotorula pilimanae,* and *Saccharomyces cerevisiae* are extremely sensitive to the toxin and conveniently used for its bioassay. Bacteria are less susceptible or not inhibited by this toxin,[13] and there is no documentation about its effects on animals.

The antifungal properties of syringomycin may be important in the survival of *P. syringae* pv. *syringae* in nature. These bacteria are epiphytes and plant pathogens. As epiphytes on surfaces of leaves and stems, they must successfully compete with other microbes for scarce nutrients. Conceivably, syringomycin production plays a role in controlling fungal populations to allow bacterial survival. Studies on this topic are lacking. The population dynamics studies of Xu and Gross[16] showed no *in planta* growth differences between syringomycin-producing and $tox^-$ strains. However, these experiments were conducted over a period of 72 h after stab inoculation of sweet cherry fruits. Consequently, they do not concern the long-term survival capabilities of $tox^-$ strains under conditions where they must compete with other microbes for nutrients.

Syringomycin could help bacterial survival in another way as it may play a role in localized bacterial colonization of the host. Using scanning electron microscopy, Mansvelt and Hattingh[17] observed that 6 days after spraying apple leaves with *P. syringae* pv. *syringae,* the cells concentrated within substomatal chambers. They concluded that the cells directly colonized the chambers by entering through the stomata. Presumably, the chambers provided protected sites for bacterial colonization. Syringomycin may aid colonization within the stomatal chambers. Mott and Takemoto[18] recently showed that low concentrations of syringomycin transported through the stems of *Xanthium strumarium* L. closed stomata in leaves. Conceivably, once colonization begins, a protected environment within the chambers is established with syringomycin induction of stomata closure. Certain observations suggest mechanisms by which syringomycin causes stomata closure.[18] The kinetics of the effect of syringomycin paralleled those of abscisic acid. Abscisic acid closes stomata by turgor-induced swelling of guard cells, and this may be the most likely mechanism for syringomycin as well (see Section V).

## III. STRUCTURE OF SYRINGOMYCIN

Despite wide interest in syringomycin for over 20 years, its structure was not known until recently. Early studies by J. E. DeVay and his colleagues indicated that syringomycin is a peptide.[10,11,19] A collaborative effort by the groups of A. Ballio and J. Takemoto culminated in the first publication on the structure of syringomycin.[20] The work centered on syringomycin-E purified from stone fruit isolates of *P. syringae* pv. *syringae.* Syringomycin-E is the major form with several copurifying minor forms.[21] This report was quickly followed by publication of papers from the laboratories of A. Isogai and A. Ballio on the structures of syringomycin from a sugar cane isolate,[22] two newly identified syringomycin-like toxins, syringostatin A and B, from a lilac isolate,[23] and syringotoxin.[24] All of these molecules are cyclic lipodepsinonapeptides (Figure 1). They contain a hydroxylated fatty acid, serine, diaminobutyric acid, dehydroaminobutyric acid, hydroxyaspartic acid, and chlorothreonine, and four other amino acids which differ between syringomycin, the syringostatins, and syringotoxin. At the C-terminal end, the carboxyl group of chlorothreonine closes a macrocyclic ring with the OH group of the *N*-terminal serine. The chlorothreonine may be unique to these molecules.

Initially, the reports on syringomycin conflicted on the relative order of the chlorothreonyl and hydroxyaspartyl residues on the C-terminal end. There was also disagreement on whether

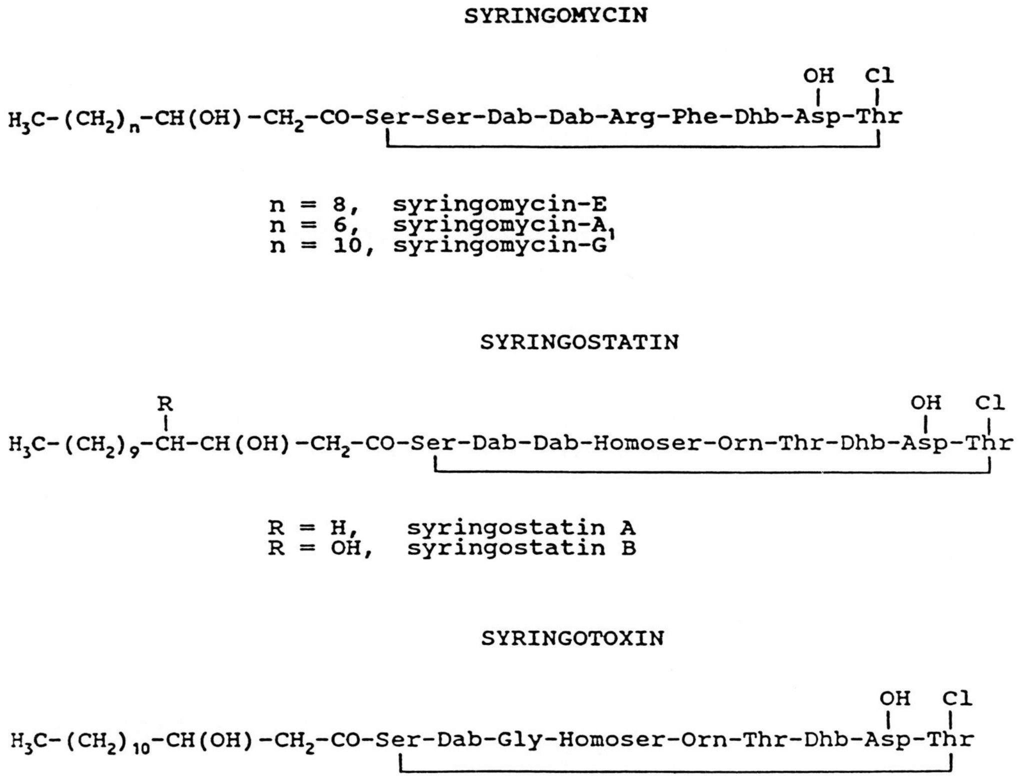


FIGURE 1. The structures of syringomycin, syringostatin, and syringotoxin. Dab,1,4 diaminobutyric acid; Dhb,2,3-dehydroaminobutyric acid.

the peptide linkage of the hydroxyaspartyl residue is alpha or beta. Fukuchi et al.[22] presented strong arguments for the order of the C-terminal residues shown in Figure 1 and for alpha linkage of the hydroxyaspartyl residue. More recent analyses on syringomycin-E confirm the structure shown in Figure 1.[24]

Two minor forms of syringomycin ($A_1$ and G) vary from the E form and each other in the length of the fatty acid moiety. The $A_1$ form has 3-hydroxydecanoic acid instead of 3-hydroxydodecanoic acid; the G form has 3-hydroxytetradecanoic acid. Similarly, syringostatin A and B differ in the fatty acid, having 3-hydroxytetradecanoic acid and 3,4-dihydroxytetradecanoic acid, respectively.

Syringotoxin closely resembles syringostatin A. In syringotoxin, glycine replaces the second diaminobutyryl group from the N-terminal end of syringostatin A.

The structures of all of these toxins are strikingly similar. The one identical feature is the dehydroaminobutyric acid-hydroxyaspartic acid-chlorothreonine-o-serine sequence. This conserved feature may be important when considering the evolution, biosyntheses, and mechanisms of action of these toxins.

## IV. BIOSYNTHESIS OF SYRINGOMYCIN

How syringomycin is synthesized is largely unknown. The determination of its structure, however, provides clues about possible biosynthetic mechanisms. By analogy to other bacterial cyclic peptides, it is likely that a large multifunctional peptide synthetase and not ribosomes catalyzes its synthesis.[25]

An early question on syringomycin biosynthesis centered on whether or not the genes were plasmid-encoded. Correlations between the loss of syringomycin production and the curing of a plasmid in *P. syringae* pv. *syringae* strain HS191 suggested plasmid involvement.[26] Further studies began to show that plasmids had no role in either toxin production or pathogenicity.[27,28] With the identification of two chromosomal genes, *syrA* and *syrB*, implicated in syringomycin synthesis[29] (see below), it is now quite clear that plasmids are not involved. The only *pseudomonad* phytotoxin known to be plasmid-associated is coronatine. This is consistent with coronatine production by several pathovars of *P. syringae*, each having apparently gained the appropriate plasmid.

Recent genetic findings of Xu and Gross[29] begin to provide some insight about syringomycin synthesis. DNA fragments containing two genes, *syrA* and *syrB*, necessary for syringomycin production were isolated. They were identified by complementation of Tn5 tox$^-$ mutants with a cosmid gene bank. The bank was generated from the chromosome of a toxigenic stone fruit isolate, *P. syringae* pv. *syringae* strain B301D. Subcloning and *Tn5* insertion analysis led to size estimates of 2.3 to 2.8 kb and 2.4 to 3.3 kb, respectively, for *syrA* and *syrB*. Strain W4S770, with a Tn5 insert in *syrA*, was pleiotropically defective in syringomycin production and pathogenicity, leading to speculation that *syrA* may be regulatory for several pathogenicity genes. On the other hand, *syrB* was not associated with pathogenicity, and could be a structural gene for an enzyme directly involved in syringomycin synthesis. If so, it probably codes for a subunit of a larger synthetic enzyme complex. The size of *syrB* showed coding for a protein no larger than 100 kDa; but, cyclic peptide synthetases are normally large complexes of about 500 kDa. Electrophoretic gels of the *syrA* and *syrB* mutants revealed that protein fractions with sizes of about 130 kDa and 350 kDa found in the toxigenic parent strain were missing. Conceivably, these represent the monomeric and trimeric forms of *syrB*.

There were similar findings for genes of syringotoxin biosynthesis.[30] Analyses of Tn5 insertion mutants revealed two high $M_r$ proteins, ST1 (470 kDa) and ST2 (435 kDa) which correlated with syringotoxin production. It is noteworthy that ST1 and ST2 gene segments showed significant sequence similarities to genomic DNA fragments from syringomycin-producing *P. syringae* pv. *syringae* strains. This is not unexpected considering the structural similarities of these toxins.

Several factors regulate syringomycin synthesis. Iron, phosphate, histidine, and temperature influence the amounts of syringomycin produced.[31] These are probably important determinants for establishing disease by *P. syringae* pv. *syringae*. Although there is some information about iron and its relationship to toxin production, the regulatory mechanisms at large are unknown.

*Pseudomonas syringae* pv. *syringae* strain B301D did not produce toxin in iron-free potato-dextrose broth, although this medium did not affect growth.[31] When grown in deferrated medium, the cells lacked the 130- and 350-kDa proteins which were also absent in *syrA* and *syrB* mutants.[29] These findings suggest that the synthesis of one or more enzymes necessary for syringomycin production requires iron. Iron influences syringotoxin production in the same way. Morgan and Chatterjee[30] showed that syringotoxin levels were directly proportional to iron concentrations in the growth medium. Furthermore, the amounts of the putative biosynthetic enzymes, ST1 and ST2, showed corresponding increases.

The studies described above provide the base for future work which should lead to an understanding of syringomycin synthesis. An understanding will require determination of the function of *syrB* and its relationship to the putative syringomycin synthetase. To accomplish this, the synthetase itself will need to be isolated and characterized. Other biosynthetic genes and their products will need to be identified. For example, it will be necessary to learn about the components which distinguish the constituents of syringomycin, such as chlorothreonine and the fatty acid moiety. Finally, studies on the biosynthetic mechanisms leading to the consensus dehydroaminobutyric acid-hydroxyaspartic acid-chlorothreonine-*o*-serine sequence will be important. It is likely that all toxin-producing *P. syringae* pv. *syringae* isolates use the same mechanism to produce this sequence.

# V. MECHANISM OF ACTION

Even the earliest studies[12,32] suggested that syringomycin acts on host membranes. It is now clear that one primary site of action for this toxin is the plasma membrane. However, only a few mechanistic details of effects on this membrane are known.

## A. EFFECTS ON PLASMA MEMBRANE ELECTRICAL POTENTIAL AND PH GRADIENTS

A series of studies with the yeasts, *R. pilimanae* and *S. cerevisiae,* showed that a target site for syringomycin is the plasma membrane.[33-37] Toxin concentrations as low as 0.5 μg/ml inhibited the growth of these organisms. At these levels, the electrical potential (cell interior, negative) and pH gradient (interior, alkaline) across the plasma membrane were enhanced.[33,36] These parameters were measured by the cellular uptake of the membrane permeable ions, tetraphenylphosphonium (TPP) and dimethyloxazolidine (DMO), respectively. Since the uptake of these ions into cells required an intact membrane, a detergent effect resulting in membrane dissolution did not occur. With *S. cerevisiae,* TPP and DMO uptake were dependent upon the presence of glucose, showing that these stimulated effects were also energy-dependent.[36]

How $H^+$ ions move across the plasma membrane in response to syringomycin is complex and not well understood. The DMO uptake studies suggested that syringomycin increased the pH gradient across the plasma membrane to make the cytoplasm more alkaline. However, [$^{31}$P] NMR spectral analyses showed that the pH of the cytoplasm decreased.[35] pH measurements of the cell suspension medium showed that the toxin also transiently decreased the cellular extrusion of $H^+$ ions.[34] Thus, syringomycin administration appeared to decrease the $H^+$ ion concentration in both the cytoplasm and the cell exterior. When the electrical potential was partially collapsed with high amounts of TPP ($>$25 μ*M*), syringomycin-treated cells acidified the external medium.[34] One possible scenario to explain these effects is the following: syringomycin initially enhances both $H^+$ influx and efflux across the plasma membrane and also $H^+$ influx into an intracellular compartment (e.g., the vacuole). The inward movement of $H^+$ across the plasma membrane is accompanied by extrusion of one or more cations (e.g. $K^+$, as discussed below), which contributes to the electrical potential (interior, negative). With a small electrical potential, the $H^+$ efflux is prominent, but as it becomes greater, $H^+$ efflux is repressed.

## B. EFFECT ON $K^+$ EFFLUX

As mentioned above, syringomycin causes the efflux of $K^+$ in yeast.[34,37] $K^+$ efflux occurs simultaneously with electrical potential and pH changes,[34] and may be a major contributor to the syringomycin-induced electrical potential increase. In experiments with whole cells, the $K^+$ efflux did not resemble the action of well-known $K^+$ ionophores, such as valinomycin and nigericin, and it was not affected by dicyclocarbodiimide, an inhibitor of the $H^+$-ATPase.[37] Therefore, $K^+$ efflux occurs independently of $H^+$-ATPase activity. This is consistent with the suggestion made above that syringomycin activates a cation exchange system (i.e., a $H^+/K^+$ antiporter). Such a system could account for the $K^+$ efflux and accompanying $H^+$ influx.

### C. $K^+$ EFFLUX AND THE CLOSING OF LEAF STOMATA

Syringomycin also causes $K^+$ efflux in plant cells.[34] Because $K^+$ transport is a major contributor to the opening and closing of leaf stomata, the influence of syringomycin on these processes was investigated.[18] Stomata open because of the swelling of guard cells caused by the influx of $K^+$. Conversely, they close with the efflux of $K^+$. Normally, these events occur in response to environmental (light, $O_2$, and humidity) or hormonal (abscisic acid) stimuli. Cut leaves of *X. strumarium* were perfused via the petiole with syringomycin solutions, and gas exchange was used to measure stomatal opening and closing. Syringomycin was also applied directly to isolated epiderm layers of *Vicia faba* leaves and the stomata observed with a microscope. In both cases, low concentrations of syringomycin closed stomata. Significantly, fusicoccin, a plant toxin that causes $K^+$ influx,[38] reversed the effect of syringomycin. The reversal by fusicoccin lends support to the argument that syringomycin does not permanently disrupt the plasma membrane. The biological significance of this ability to close stomata is discussed above.

### D. EFFECT ON THE PLASMA MEMBRANE $H^+$-ATPASE

Because syringomycin caused pH gradient changes across the plasma membrane, its effects on the plasma membrane $H^+$-ATPase were investigated. This enzyme is similar in fungi and plants,[39] and catalyzes ATP-dependent electrogenic $H^+$ transport across the plasma membrane. The $H^+$-ATPase plays a pivotal role in the function and regulation of plasma membrane ion transport systems in fungi and plants.

The plasma membrane $H^+$-ATPase activities of both *R. pilimanae* and red beet storage tissue were stimulated by syringomycin.[33,40,41] With the red beet enzyme, 0.1% (w/v) deoxycholate inhibited the effect. Stimulation did not occur with partially purified enzyme. This indicated that syringomycin did not act directly on the enzyme. Also, with membranes, several uncouplers and ionophores showed no effect on the stimulatory ability of syringomycin. Therefore, the stimulated activity was not a response to the collapse of the electrical potential in closed vesicles. At the concentrations used (2 to 10 μg/ml), syringomycin most likely did not have a detergent lysis effect. Based on these observations, it was suggested that syringomycin indirectly stimulates the $H^+$-ATPase and thus electrogenic cellular $H^+$ efflux. As discussed above (Section V.A), it is predicted that $H^+$ efflux occurs only when the electrical potential is sufficiently small. Therefore, in leaky plasma membrane vesicles or when the electrical potential is small, the $H^+$-ATPase operates in an activated state with syringomycin.

### E. PROTEIN PHOSPHORYLATION OF PLASMA MEMBRANE POLYPEPTIDES

Protein phosphorylation regulates many plasma membrane functions such as ion transport. Thus, it is reasonable to speculate that one or more protein kinases mediate the effect of syringomycin. The following observations also suggest this possibility:

1. Deoxycholate (0.1%, w/v) eliminated the ability of syringomycin to activate the red beet $H^+$-ATPase.[41] Treatment of plant plasma membranes with 0.1% deoxycholate also eliminated endogenous protein kinase activity including the phosphorylation of a 100-kDa protein presumed to be the $H^+$-ATPase.[41,42,46]
2. Syringomycin did not affect the activity of a partially purified $H^+$-ATPase preparation.[41] This suggests that in the membrane the toxin stimulates the activity indirectly by a regulatory mechanism that involves other components.
3. The syringomycin effects resembled processes in yeast that are mediated by protein kinases. For example, adding phorbol esters or glucose to cells of *S. cerevisiae* activated plasma membrane $H^+$-ATPase activity and caused the phosphorylation of a 100-kDa membrane polypeptide presumed to be the ATPase.[43,44]

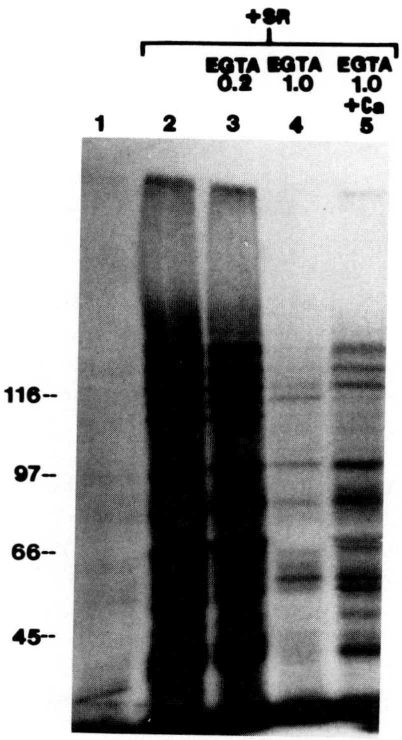


FIGURE 2. Gel autoradiogram showing the effects of syringomycin and $Ca^{2+}$ on the phosphorylation of red beet storage tissue plasma membrane proteins. Purified plasma membranes (400 μg protein) were incubated with ATP (135 m*M* and containing 25 μCi [γ-$^{32}$P]ATP) in a volume of 1 ml, and solubilized and electrophoresed as described by Bidwai and Takemoto.[46] Phosphorylations were conducted for 10 min without syringomycin (lane 1) or with syringomycin (10 μg per 25 μg protein) (lanes 2-5). The additions of EGTA and $Ca^{2+}$ are designated. $Ca^{2+}$ was added at a concentration of 2 m*M*. Reproduced with permission from Bidwai, A. P., and Takemoto, J. Y., *Proc. Natl. Acad. Sci. U.S.A.*, 85, 1408, 1988.

4. The activity of a polypeptide-dependent protein kinase (PK-P) from membranes of *S. cerevisiae* was stimulated by syringomycin.[45]

The protein kinase activity in red beet plasma membranes and its response to syringomycin were investigated in some detail.[40,46] Syringomycin (10 μg/ml) caused a rapid and large increase (4- to 15-fold) in endogenous protein kinase activity. The activity was inhibited by 0.1% deoxycholate. Many plasma membrane polypeptides were phosphorylated including a 100-kDa polypeptide which corresponds in size to the $H^+$-ATPase[46] (Figure 2). The phosphate bonds to most of these polypeptides were not cleaved with hydroxylamine indicating that monophosphoester linkage occurred via serine, threonine, or tyrosine groups characteristic of protein kinase activity. The phosphorylation of most of the polypeptides

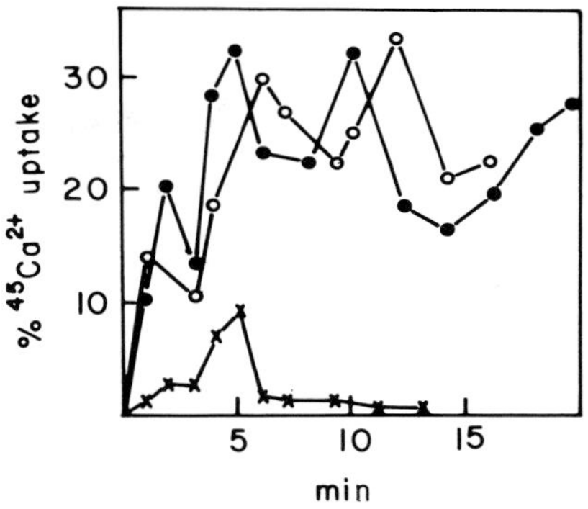


FIGURE 3. $^{45}Ca^{2+}$ uptake into red beet storage tissue slices treated with 5 μg/ml syringomycin (●) or 10 μ*M* A23187 (○). Syringomycin was also added to slices pretreated with EGTA (X). Reproduced with permission from Takemoto, J. Y., Giannini, J. L., Vassey, T., and Briskin, D. P., in *Phytotoxins and Plant Pathogenesis,* Graniti, A., Durbin, R. D., and Ballio, A., Eds., Springer-Verlag, Berlin, 1989, 167.

(including the 100-kDa polypeptide) appeared to be $Ca^{2+}$-dependent. Addition of the $Ca^{2+}$ chelator, EGTA, to the phosphorylation reaction mixture eliminated the phosphorylation of these polypeptides. EGTA added together with a slight excess of $Ca^{2+}$ restored the phosphorylations. Cyclic AMP had no effects on the stimulation of phosphorylation by syringomycin. Significantly, these results may be the first indication of the regulation of the plant $H^+$-ATPase by protein phosphorylation, although further work on this possibility is required.[39] Syringomycin effects on phosphorylation of the yeast plasma membrane proteins have not been reported.

### F. $Ca^{2+}$ TRANSPORT

$Ca^{2+}$ has a potentially important role in the response to syringomycin. A strong hint for this role is the requirement of $Ca^{2+}$ for the syringomycin-stimulated phosphorylation of red beet plasma membrane proteins.[46] Cellular $Ca^{2+}$ is an important second messenger since it is a mediator of many biochemical events.[47-49] As a result, cells spend a great deal of energy optimizing the levels and distribution of $Ca^{2+}$ within the various cell compartments.[50] $Ca^{2+}$ transport across the plasma and other membranes are coordinated to achieve this. Most significantly, sudden $Ca^{2+}$ flux changes accompany cellular responses to external signals which are initially perceived and processed at the plasma membrane. Often involved are changes in plasma membrane ion transport and modulation of protein kinases. As described above, such changes are characteristic of the syringomycin response.

To investigate the role of $Ca^{2+}$, the effects of syringomycin on $Ca^{2+}$ transport were studied. Syringomycin-treated red beet storage tissue slices showed rapid and transient fluxes of $^{45}Ca^{2+}$ with an overall increase in uptake[51] (Figure 3). Sealed inside-out vesicles of the red beet plasma membrane capable of ATP-dependent uptake of $^{45}Ca^{2+}$ were prepared. $^{45}Ca^{2+}$ uptake by these vesicles was inhibited with low concentrations of syringomycin. When the inside-out vesicles were preloaded with $^{45}Ca^{2+}$, and then treated with syringomycin, a rapid efflux of label from the vesicles was observed. Artificial liposomes preloaded with $^{45}Ca^{2+}$ did not show $^{45}Ca^{2+}$ efflux with syringomycin. However, treatment of these liposomes with the $Ca^{2+}$ ionophore, A23187, did cause efflux. Thus, exposure to syringomycin results in $Ca^{2+}$ flux changes across the plasma membrane that are not due to an

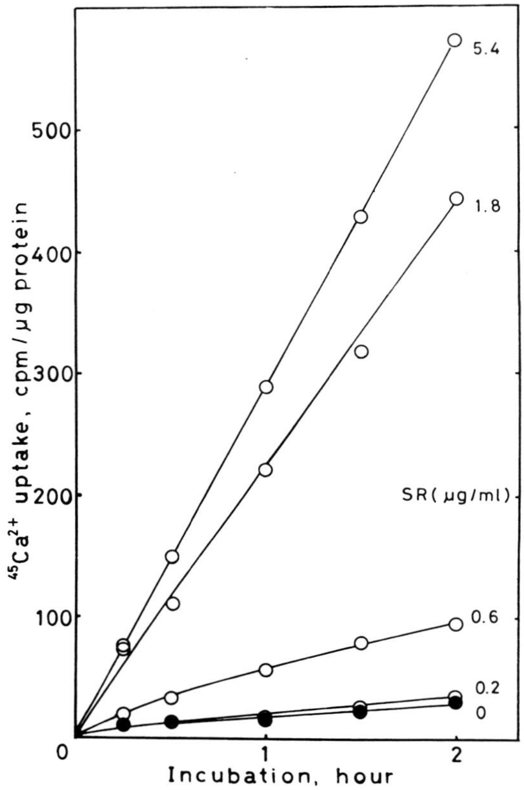


FIGURE 4. $^{45}Ca^{2+}$ uptake by the yeast, *S. cerevisiae*, treated with various amounts of syringomycin. Cells were grown in medium containing 1% (w/v) yeast extract, 2% (w/v) peptone, and 2% (w/v) dextrose, washed, and suspended ($10^8$ cells/ml) in the same medium containing $^{45}Ca^{2+}$ (4 μCi/ml) and either with or without syringomycin. At various times, the cells were centrifuged, washed, and the cell radioactivity determined. Unpublished data from N. Taguchi and T. Miyakawa, Hiroshima University, Japan.

ionophoric effect. Importantly, the observations strongly suggest that the toxin affects $Ca^{2+}$ transport processes commonly involved in signaling.

The inhibition of energy-dependent $Ca^{2+}$ transport into the red beet plasma membrane vesicles could arise by either of two mechanisms: (1) $Ca^{2+}$-ATPases which function in pumping $Ca^{2+}$ out of the cell across the plasma membrane are inhibited, or (2) plasma membrane $Ca^{2+}$ channels are opened allowing diffusion of this ion. Learning which event occurs should provide significant insight into the mechanism of action of syringomycin. The $Ca^{2+}$-ATPase pump of the red beet plasma membrane has been partly characterized,[52] but nothing is known of the $Ca^{2+}$ channels in this system.

Recently, it was observed that syringomycin also caused the rapid uptake of $Ca^{2+}$ into *S. cerevisiae* (unpublished data) (Figure 4). *Saccharomyces cerevisiae* has well-known experimental advantages, and it provides attractive opportunities to study the role of $Ca^{2+}$ transport in the syringomycin response. Some facts are known about $Ca^{2+}$transport in this organism, and a few associated genes have been identified.[53-59] Investigating the roles of the products of these genes in the syringomycin response could reveal mechanisms of toxin action that are coupled to $Ca^{2+}$ transport.

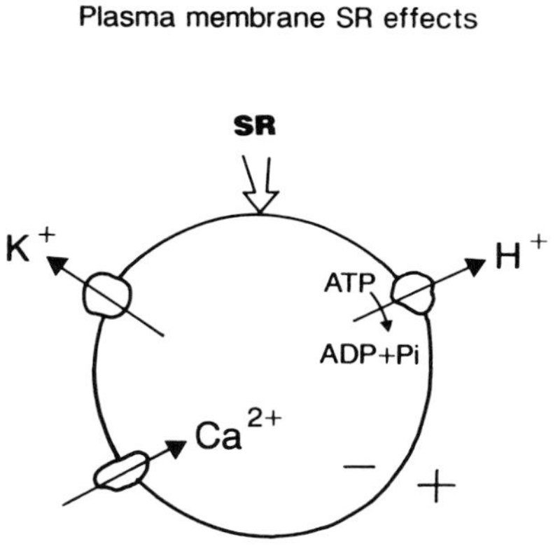


FIGURE 5. Summary of syringomycin (SR) effects on ion transport and electrical potential of host plasma membranes.

### G. MITOCHONDRIA AND UNCOUPLER ACTION

Host mitochondria have been suggested as sites of action for syringomycin and syringotoxin.[60] Both toxins showed uncoupler effects on mitochondria isolated from maize. Syringotoxin was more effective than syringomycin. There was no correlation, however, between these effects and resistance to the toxin-producing *P. syringae* pv. *syringae* strains. Other findings indicate that mitochondria are not required for syringomycin action. A respiratory mutant (petite) strain of *S. cerevisiae* lacking mitochondrial function responded identically to syringomycin as the parent wild-type strain.[36] Also, in some respects, syringomycin does not behave as an uncoupler. With red beet plasma membrane vesicles, well-known uncouplers such as valinomycin, nigericin, gramicidin, and carbonylcyanide *p*-trifluoromethoxyphenylhydrazone did not stimulate the $H^+$-ATPase or did they cause characteristic $K^+$ efflux or membrane potential changes in yeast as did syringomycin.[37,41] It seems that while syringomycin affects isolated mitochondria, action on this organelle and uncoupler effects are not necessary for the toxin response in intact cells and tissues.

The situation with syringotoxin is different. Studies with black lipid membranes and artificial liposomes show that syringotoxin could act as an uncoupler.[61,62] Low concentrations of syringotoxin formed conductive ion channels which were ideally selective for chloride ions.[63] Similar artificial membrane studies have not been reported for syringomycin.

Clearly, more work is needed to determine if mitochondrial uncoupling and ion channel formation play critical roles in the actions of syringomycin and syringotoxin. Conceivably, the two toxins differ in this respect. Syringotoxin is a more effective uncoupler with isolated mitochondria and can form membrane channels. Perhaps the structures of syringomycin and syringotoxin are sufficiently different to impart distinct mechanisms of action.

### H. SUMMARY OF MECHANISM OF ACTION AND FUTURE RESEARCH

The best known effects of syringomycin on host plasma membrane ion transport are schematically summarized in Figure 5. It is possible to speculate how these various events might be related. For example, it may be postulated that the toxin's primary effect is to open $Ca^{2+}$ channels which, in turn, raises the cytoplasmic levels of this ion. A $Ca^{2+}$-dependent protein kinase is then activated which phosphorylates, and thereby activates, the $H^+$ pumps and $K^+$ channels. The resultant ionic imbalance leads to cell death. This sequence of events, however, is only a conjecture. Despite efforts to kinetically order the events,[34] the sequential relations between them remain unknown.

To progress in solving the mechanism of action of syringomycin, new experimental strategies are needed. Two approaches are promising. One is the use of patch clamp techniques.[63] These techniques will allow measurements of single membrane ion channel activities in response to syringomycin. A second is the use of molecular genetic techniques in *S. cerevisiae*. *S. cerevisiae* mutants with altered responses to syringomycin may be isolated and characterized. Widely used molecular genetic methods can then be applied to identify the genes which are altered and to determine their encoded proteins. Once identified, the order of function of these gene products can be determined. Such an undertaking with yeast is justified since it is highly sensitive to the toxin and plants and yeasts show similar responses to the toxin. Eventually, however, the findings with yeast must be tested for relevance to plant disease.

## REFERENCES

1. **Gross, D. C. and Cody, Y. S.,** Mechanisms of plant pathogenesis by *Pseudomonas* species, *Can. J. Microbiol.,* 31, 403, 1985.
2. **Mitchell, R. E.,** Structure: Bacterial, in *Toxins in Plant Disease,* Durbin, R. D., Ed., Academic Press, San Diego, 1981, 259.
3. **Mitchell, R. E.,** The relevance of non-host-specific toxins in the expression of virulence by pathogens, *Annu. Rev. Phytopathol.,* 22, 215, 1984.
4. **Isogai, A., Fukuchi, N., Yamashita, S., Suyama, K., and Suzuki, A.,** Syringostatins, novel phytotoxins produced by *Pseudomonas syringae* pv. *syringae, Agr. Biol. Chem.,* 53, 3117, 1989.
5. **Thomas, M. D., Langston-Unkefer, P. J., Uchytil, T. F., and Durbin, R. D.,** Inhibition of glutamine synthetase from pea by tabtoxinine-B-lactam, *Plant Physiol.,* 71, 912, 1983.
6. **Tam, L. Q. and Patil, S. S.,** Mode of action of the toxin from *Pseudomonas phaseolicola* II. Mechanism of inhibition of bean ornithine carbamoyltransferase, *Plant Physiol.,* 49, 808, 1972.
7. **Patil, S. S. and Tam, L. Q.,** Mode of action of the toxin from *Pseudomonas phaseolicola* I. Toxin specificity, chlorosis, and ornithine accumulation, *Plant Physiol.,* 49, 803, 1972.
8. **Gulya, T. J. R. and Dunleavy, J. M.,** Inhibition of chlorophyll synthesis by *Pseudomonas glycinea, Crop Sci.,* 19, 261, 1979.
9. **Ferguson, I. B. and Mitchell, R. E.,** Stimulation of ethylene production in bean leaf discs by the pseudomonad phytotoxin coronatine, *Plant Physiol.,* 77, 969, 1985.
10. **De Vay, J. E., Lukezic, F. L., Sinden, S. L., English, H., and Coplin, D. L.,** A biocide produced by pathogenic isolates of *Pseudomonas syringae* and its possible role in the bacterial canker disease of peach trees, *Phytopathology,* 58, 95, 1968.
11. **Sinden, S. L., De Vay, J. E., and Backman, P. A.,** Properties of syringomycin, a wide spectrum antibiotic and phytotoxin produced by *Pseudomonas syringae,* and its role in the bacterial canker disease of peach trees, *Physiol. Plant Pathol.,* 1, 199, 1971.
12. **Backman, P. A. and De Vay, J. E.,** Studies on the mode of action and biogenesis of the phytotoxin syringomycin, *Physiol. Pathol.,* 1, 215, 1971.
13. **De Vay, J. E., Gonzalez, C. F., and Wakeman, R. J.,** Comparison of the biocidal activities of syringomycin and syringotoxin and the characterization of isolates of *Pseudomonas syringae* from citrus hosts, *Proc. 4th Int. Conf. Plant Pathol. Bacteriol.,* 4, 643, 1978.
14. **Gross, D. C. and De Vay, J. E.,** Production and purification of syringomycin, a phytotoxin produced by *Pseudomonas syringae, Physiol. Plant Pathol.,* 11, 13, 1977.
15. **Gross, D. C. and De Vay, J. E.,** Role of syringomycin in holcus spot of maize and systemic necrosis of cowpea caused by *Pseudomonas syringae, Physiol. Plant Pathol.,* 11, 1, 1977.
16. **Xu, G.-W. and Gross, D.,** Evaluation of the role of syringomycin in plant pathogenesis by using Tn5 mutants of *Pseudomonas syringae* pv. *syringae* defective in syringomycin production, *Appl. Environ. Microbiol.,* 54, 1345, 1988.
17. **Mansvelt, E. L. and Hattingh, M. J.,** Scanning electron microscopy of invasion of apple leaves and blossoms by *Pseudomonas syringae* pv. *syringae, Appl. Environ. Microbiol.,* 55, 533, 1989.
18. **Mott, K. A. and Takemoto, J. Y.,** Syringomycin, a bacterial phytotoxin, closes stomata, *Plant Physiol.,* 90, 1435, 1989.
19. **Gross, D. C., De Vay, J. E., and Stadtman, F. H.,** Chemical properties of syringomycin and syringotoxin: toxigenic peptides produced by *Pseudomonas syringae, J. Appl. Bacteriol.,* 43, 453, 1977.

20. **Segre, A., Bachmann, R. C., Ballio, A., Bossa, F., Grgurina, I., Iacobellis, N. S., Pucci, P., Simmaco, M., and Takemoto, J. Y.,** The structure of syringomycins A1, E and G., *FEBS Lett.*, 255, 27, 1989.
21. **Ballio, A., Barra, D., Bossa, F., De Vay, J. E., Grgunina, I., Iacobellis, N. S., Marino, G., Pucci, P., Simmaco, M., and Surico, G.,** Multiple forms of syringomycin, *Physiol. Mol. Plant Pathol.*, 33, 493, 1988.
22. **Fukuchi, N., Isogai, A., Yamashita, S., Suyama, K., Takemoto, J. Y., and Suzuki, A.,** Structure of phytotoxin syringomycin produced by a sugar cane isolate of *Pseudomonas syringae* pv. *syringae*, *Tetrahedron Lett.*, 31, 1589, 1990.
23. **Isogai, A., Fukuchi, N., Yamashita, S., Suyama, K., and Suzuki, A.,** Structures of syringostatins A and B, novel phytotoxins produced by *Pseudomonas syringae* pv. *syringae* isolated from lilac blights, *Tetrahedron Lett.*, 31, 695, 1990.
24. **Ballio, A., Bossa, F., Collina, A., Gallo, M., Iacobellis, N. S., Paci, M., Pucci, P., Scaloni, A., Segre, A., and Simmaco, M.,** Structure of syringotoxin, a bioactive metabolite of *Pseudomonas syringae pv. syringae*, *FEBS Lett.*, 269, 377, 1990.
25. **Kleinkauf, H., and von Dohren, H.,** Biosynthesis of peptide antibiotics, *Annu. Rev. Microbiol.*, 41, 259, 1987.
26. **Gonzalez, C. F. and Vidaver, A. K.,** Syringomycin production and holcus spot disease of maize: plasmid-associated properties in *Pseudomonas syringae*, *Curr. Microbiol.*, 2, 75, 1979.
27. **Currier, T. C. and Morgan, M. K.,** Plasmids of *Pseudomonas syringae:* no evidence of a role in toxin production or pathogenicity, *Can. J. Microbiol.*, 29, 84, 1983.
28. **Gonzalez, C. F., Layher, S. K., Vidaver, A. K., and Olsen, R. H.,** Transfer, mapping, and cloning of *Pseudomonas syringae* pv. *syringae* plasmid pCG131 and assessment of its role in virulence, *Phytopathology*, 74, 1245, 1984.
29. **Xu, G.-W. and Gross, D.,** Physical and functional analyses of the syrA and syrB genes involved in syringomycin production by *Pseudomonas syringae* pv. *syringae*, *J. Bacteriol.*, 170, 5680, 1988.
30. **Morgan, M. K. and Chatterjee, A. K.,** Genetic organization and regulation of proteins associated with production of syringotoxin by *Pseudomonas syringae* pv. *syringae*, *J. Bacteriol.*, 170, 5689, 1988.
31. **Gross, D. C.,** Regulation of syringomycin synthesis in *Pseudomonas syringae* pv. *syringae* and defined conditions for its production, *J. Appl. Bacteriol.*, 58, 167, 1985.
32. **Paynter, V. A. and Alconero, R.,** A specific fluorescent antibody for detection of syringomycin in infected peach tree tissues, *Phytopathology*, 69, 493, 1979.
33. **Zhang, L. and Takemoto, J. Y.,** Effects of *Pseudomonas syringae* phytotoxin, syringomycin, on plasma membrane functions of *Rhodotorula pilimanae*, *Phytopathology*, 77, 297, 1987.
34. **Reidl, H. H. and Takemoto, J. Y.,** Mechanism of action of bacterial phytotoxin, syringomycin. Simultaneous measurement of early responses in yeast and maize, *Biochim. Biophys. Acta*, 898, 59, 1987.
35. **Reidl, H. H., Grover, T. A., and Takemoto, J. Y.,** $^{31}P$-NMR evidence for cytoplasmic acidification and phosphate extrusion in syringomycin-treated cells of *Rhodotorula pilimanae*, *Biochim. Biophys. Acta*, 1010, 325, 1989.
36. **Zhang, L. and Takemoto, J. Y.,** Mechanism of action of *Pseudomonas syringae* phytotoxin, syringomycin. Interaction with the plasma membrane of wild-type and respiratory-deficient strains of *Saccharomyces cerevisiae*, *Biochim. Biophys. Acta*, 861, 201, 1986.
37. **Zhang, L. and Takemoto, J. Y.,** Syringomycin stimulation of potassium efflux by yeast cells, *Biochim. Biophys. Acta*, 987, 171, 1989.
38. **Marre, E.,** Fusicoccin: a tool in plant physiology, *Annu. Rev. Plant Physiol.*, 30, 273, 1979.
39. **Serrano, R.,** Structure and function of proton translocating ATPase in plasma membranes of plants and fungi, *Biochim. Biophys. Acta*, 947, 1, 1988.
40. **Bidwai, A. P. and Takemoto, J. Y.,** *Pseudomonas syringae* phytotoxin, syringomycin, stimulates the phosphorylation of red beet plasma membrane proteins, in *Plant Membranes: Structure, Function, Biogenesis*, Leaver, C. J. and Sze, H., Eds., Alan R. Liss, NY, 1987, 383.
41. **Bidwai, A. P., Zhang, L., Bachmann, R. C., and Takemoto, J. Y.,** Mechanism of action of *Pseudomonas syringae* phytotoxin, syringomycin. Stimulation of red beet plasma membrane ATPase activity, *Plant Physiol.*, 83, 39, 1987.
42. **Briskin, D. P. and Leonard, R. T.,** Phosphorylation of the adenosine triphosphatase in a deoxycholate-treated plasma membrane fraction from corn roots, *Plant Physiol.*, 70, 1459, 1982.
43. **Portillo, F. and Mazon, M. J.,** Activation of yeast plasma membrane ATPase by phorbol ester, *FEBS Lett.*, 192, 95, 1985.
44. **McDonough, J. P. and Mahler, H. P.,** Covalent phosphorylation of the $Mg^{2+}$-dependent ATPase of yeast plasma membranes, *J. Biol. Chem.*, 257, 14579, 1982.
45. **Abdel-Ghany, M., Raden, D., Racker, E., and Katchalski-Katzir, E.,** Phosphorylation of synthetic random polypeptides by protein kinase P and other protein-serine (threonine) kinases and stimulation or inhibition of kinase activities by microbial toxins, *Proc. Natl. Acad. Sci. U.S.A.*, 85, 1408, 1988.

46. **Bidwai, A. P. and Takemoto, J. Y.,** Bacterial phytotoxin, syringomycin, induces a protein kinase-mediated phosphorylation of red beet plasma membrane polypeptides, *Proc. Natl. Acad. Sci. U.S.A.,* 84, 6755, 1987.
47. **Kauss, H.,** Some aspects of calcium-dependent regulation in plant metabolism, *Annu. Rev. Plant. Physiol.,* 38, 47, 1987.
48. **Carafoli, E. and Penniston, T.,** The calcium signal, *Sci. Amer.,* 253, 70, 1985.
49. **Gilroy, S., Blowers, D. P., and Trewavas, A. J.,** Calcium: a regulation system emerges in plant cells, *Development,* 100, 181, 1987.
50. **Carafoli, E.,** Intracellular calcium homeostasis, *Annu. Rev. Biochem.,* 56, 395, 1987.
51. **Takemoto, J. Y., Giannini, J. L., Vassey, T., and Briskin, D. P.,** Syringomycin effects on plasma membrane $Ca^{2+}$ transport, in *Phytotoxins and Plant Pathogenesis,* Graniti, A., Durbin, R. D., and Ballio, A., Eds., Springer-Verlag, Berlin, 1989, 167.
52. **Giannini, J. L., Ruiz-cristin, J., and Briskin, D. P.,** Calcium transport in sealed vesicles from red beet (beta vulgaris L.) storage tissue. II. Characterization of $^{45}Ca^{2+}$ uptake into plasma membrane vesicles, *Plant Physiol.,* 85, 1137, 1987.
53. **Ohya, Y., Ohsumi, Y., and Anraku, Y.,** Genetic study of the role of calcium ions in the cell division cycle of *Saccharomyces cerevisiae:* a calcium-dependent mutant and its trifluoperazine-dependent pseudorevertants, *Mol. Gen. Genet.,* 193, 389, 1984.
54. **Ohya, Y., Miyamoto, S., Ohsumi, Y., and Anraku, Y.,** Calcium-sensitive cls4 mutant of *Saccharomyces cerevisiae* with a defect in bud formation, *J. Bacteriol.,* 165, 28, 1986.
55. **Ohya, Y., Ohsumi, Y., and Anraku, Y.,** Isolation and characterization of Ca-sensitive mutants of *Saccharomyces cerevisiae, J. Gen. Microbiol.,* 132, 979, 1986.
56. **Miyamoto, S., Ohya, Y., Ohsumi, Y., and Anraku, Y.,** Nucleotide sequence of the CLS4 (CDC24) gene of *Saccharomyces cerevisiae, Gene,* 54, 125, 1987.
57. **Schmitt, H. D., Puzicha, M., and Gallwitz, D.,** Study of a temperature-sensitive mutant of the ras-related YPT1 gene product in yeast suggests a role in the regulation of intracellular calcium, *Cell,* 53, 635, 1988.
58. **Shih, C.-K., Wagner, R., Feinstein, S., Kanik-Ennulat, C., and Neff, N.,** A dominant trifluoperazine resistance gene from *Saccharomyces cerevisiae* has homology with $F_0F_1$ATP synthetase and confers calcium-sensitive growth, *Mol. Cell. Biol.,* 8, 3094, 1988.
59. **Rudolph, H. K., Fink, G. R., Buckley, C. M., Antebi, A., Dorman, T. E., Le Vitre, J., Davidow, L. S., Mao, J., and Moir, D. T.,** The yeast secretory pathway is perturbed by mutations in PMR1, a member of a $Ca^{2+}$ ATPase family, *Cell,* 58, 133, 1989.
60. **Surico, G. and De Vay, J. E.,** Effect of syringomycin and syringotoxin produced by *Pseudomonas syringae* pv. *syringae* on structure and function of mitochondria isolated from holcus spot resistance and susceptible maize lines, *Physiol. Plant Pathol.,* 21, 39, 1982.
61. **Ziegler, W., Pavlovkin, J., and Pokorny, J.,** Effect of syringotoxin on the permeability of bilayer lipid membranes, *Biologia,* 39, 693, 1984.
62. **Pokorny, J. and Ziegler, W.,** Is the syringotoxin-channel permeable to $Pr3^+$ ions?, *Biologia,* 39, 701, 1984.
63. **Hedrich, R. and Schroeder, J. I.,** The physiology of ion channels and electrogenic pumps in higher plants, *Annu. Rev. Plant Physiol.,* 40, 539, 1989.

# BACTERIALLY DELIVERED TOXINS FOR STUDYING PLANT METABOLISM

Thomas J. Knight and Pat J. Langston-Unkefer

## TABLE OF CONTENTS

# I. INTRODUCTION

Plant-microbe interactions in the rhizosphere, although poorly understood, have crucial effects on plant growth.[1] Thousands of soil bacteria and other microorganisms live in the rhizosphere and utilize the plant as a source of nutrients, thus forming highly complex relationships. These relationships range from simple beneficial or detrimental associations affecting only one member of the association to mutualistic relationships affecting multiple organisms. There are notable relationships in which plants utilize bacterial functions to increase their nutrient uptake and supply: bacterial siderophores enhance iron uptake by plants; microbial solubilization of phosphate increases the supply of phosphorous available for plant uptake; nonsymbiotic nitrogen-fixing associations in the rhizosphere increase the supply of nitrogen to the host plant.[2,3] A well-recognized rhizosphere interaction is the signaling between a legume and its nitrogen-fixing symbiont that allows, initiates, or specifies nodulation. Another type of plant-microbe interaction is the bacterial release of toxins, or inhibitors, and the consequences of the action of these inhibitors. These bacteria are often recognized as plant pathogens that, under the proper conditions, release toxins with specific metabolic targets. These bacteria include: *Pseudomonas andropogonis* and certain strains of *Bradyrhizobium japonicum* which produce a cystathionine β-lyase inhibitor, rhizobitoxine; *P. syringae* pv. *phaseolicola* which produces an ornithine carbamoyltransferase inhibitor, phaseolotoxin; and *P. syringae* pv. *tabaci* which produces a glutamine synthetase inhibitor, tabtoxinine-β-lactam. In their natural environment, these plant-microbe interactions either go unnoticed because they are functioning at suboptimum levels or manifest themselves with highly deleterious effects. The magnitude of these interactions is usually regulated by the environment in which the organisms are functioning.

Inactivation of specific enzymes in metabolic pathways has been very useful in identifying biosynthetic intermediates. This approach was used very effectively in characterizing amino acid biosynthesis in microorganisms where specific enzyme inactivation was accomplished by either mutation or treatment with an inhibitor. However, mutations are not as readily produced and characterized in plants because their generation time is much longer and many plants are polyploid. Thus, using an inhibitor to specifically impair or block a particular step in plants is attractive. Although we understand many of the metabolic routes in plants, we do not fully understand the coordination of metabolism or the relationship between metabolism and regulation of plant gene expression. Specific inhibition of key metabolic steps can lead to insights into these questions and allow examination of the relative metabolic flux through competing pathways. A powerful plant research tool is the combination of a selective metabolic inhibitor with a long-term bacterial delivery system for this inhibitor. A unique research tool is created by careful selection of the host plant, the bacterial toxin and its specific target, and the culture conditions for the plants and bacteria.

In this chapter, we will discuss specific advantages and disadvantages of these new tools and outline how they are developed. We will describe several bacterial delivery systems whose toxins have been used or are potentially useful as biochemical research tools. We will describe the utility of one such system which uses *P. syringae* pv. *tabaci* to release a glutamine synthetase inhibitor, tabtoxinine-β-lactam. This system is being used to study nitrogen metabolism in legumes and cereals.

# II. ADVANTAGES OF MICROBIALLY DELIVERED INHIBITORS AS RESEARCH TOOLS

Bacterial toxins are often produced by plant pathogens that have evolved strategies to optimize their desired effect on the host plant.[2] These toxins are integral parts of the phytopathogenic strategies and frequently target metabolically important enzymes or cellular

**TABLE 1**
**Bacterially Delivered Toxins as Research Tools**

**Advantages**

1. Toxins target important enzymes or functions.
2. Bacterial synthesis of toxins is cheap.
3. Toxins are easy to apply.
4. Toxins are specific for target enzymes.
5. Radiolabeled toxin can be biosynthesized.
6. Toxins can be selective for isozymes of target enzymes.
7. Toxins are an effective method for long-term treatments.
8. Long-term perturbation of important steps gives insight to regulation of processes.
9. Treatment is reversible by removing the bacteria.
10. Investigations of plant metabolism and plant-microbe interactions are possible.

**Disadvantages**

1. Requires understanding and management of bacterial growth and culturing.
2. Requires understanding and management of toxin production.
3. Requires ruling out other bacterial factors when interpreting results.

functions (Table 1). Thus, these toxins are often natural inhibitors of key enzymes in central metabolism and cellular functions. These toxins can potentially provide the researcher with an inhibitor of an enzyme acting at a metabolic regulatory step. Perturbation of such important steps in metabolism leads to fundamental changes in the host plant and opens new avenues of investigation. An example of such a perturbation is the relocation of ammonia assimilation from the roots to the leaves of oat plants, accomplished by inhibition of oat root glutamine synthetase by the action of the toxin of *P. syringae* pv. *tabaci*, tabtoxinine-β-lactam. In this case, the perturbation is beneficial because it is accompanied by increased leaf protein. A second example of a beneficial perturbation is the increased growth and nodulation observed in pv. *tabaci*-infested alfalfa plants.[5]

Another advantage of using a bacterial delivery system for a metabolic inhibitor is the length of time the inhibitor treatment can be carried out effectively. Establishing a viable population of the bacterium in the plant's rhizosphere can provide toxin (or inhibitor) for several weeks; during this time the effects of long-term enzyme inhibition becomes well developed and can be studied. An important property of these systems is the possibility of reversing the perturbation. By inoculating, and later removing the pathogen, the system can be allowed to return to a state with normal function of the target enzyme. However, these systems may not return to a completely normal state because the initial perturbation may have induced irreversible changes in the plant. Identification of these changes can provide insight to the impact of different metabolic systems on gene regulation.

Bacterial synthesis of metabolic inhibitors is reasonably inexpensive, especially when the bacterium is used to produce the inhibitor *in situ,* i.e., in the plant's rhizosphere or leaves. The ease and efficiency with which these inhibitors can be generated and applied to large numbers of plants allow selection of plants resistant to the inhibitor; this strategy can be used to obtain toxin-resistant target enzymes.[4,5] Biosynthesis of these toxins, which are sometimes the only known inhibitor of a particular enzyme, is an efficient way to obtain useful quantities of them. Another advantage of bacterial synthesis of an inhibitor is the relative ease with which radiolabeled inhibitor can be obtained in large quantities.

These toxins can be selective inhibitors of isoforms of their target enzymes. Such selectivity can be of great utility in metabolic studies for several applications: partitioning metabolic pathways between tissues, understanding specific functions of different isoforms of enzymes, or partially inhibiting an enzyme activity or function. These toxins can be

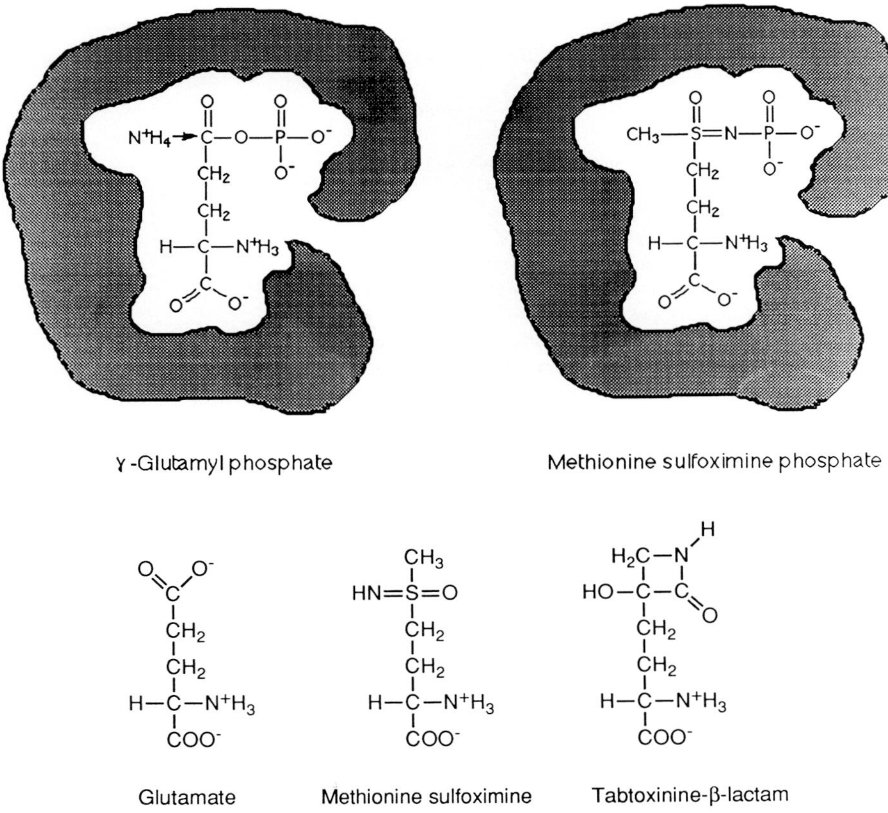


FIGURE 1. Glutamine synthetase active site represented with the γ-glutamyl phosphate intermediate and with methionine sulfoximine phosphate bound. Structures of glutamate, methionine sulfoximine, and tabtoxinine-β-lactam are shown to allow ready comparison of their structures.

selective inhibitors of isoforms of the target enzyme, because they are often not perfectly constructed inhibitors of their target enzymes. Specifically, these toxins are not always good analogs of the transition state. The transition state is the most sterically acceptable form of the substrate bound by the enzyme. A transition state analog has the conformation and structure of the substrate after the substrate has been activated for catalysis in the enzyme active site. Such analogs often mimic the binding of several of the substrates bound at an active site. For example, methionine sulfoximine binds in the glutamate-binding site of glutamine synthetase and becomes phosphorylated during the inactivation (Figure 1). Methionine sulfoximine phosphate, the true inhibitor, is formed; it occupies the binding sites for glutamate and the phosphate group of the normal reaction intermediate, g-glutamyl phosphate (Figure 1).[6-8] Such inhibitors are irreversible because they fit the active site extremely well and are thus very strongly bound. Inhibitors with bulky substituents, such as tabtoxinine-β-lactam (Figure 1), simply do not fit the active site as well and are either not bound as strongly or not bound at all in a particular active site. If a toxin is sterically excluded from the active site of one isoform of a target enzyme, this isoform is toxin-resistant; if this toxin fits into and is bound in the active site of another isoform of the enzyme, this form is toxin-sensitive. Thus, inhibitor selectivity depends upon the spatial definition of the active site; i.e., what structures can be bound and which structures can not be bound in the active site. Selective inhibition between isoforms of a target enzyme can thus be accomplished. Such selectivity exists in the action of tabtoxinine-β-lactam toward the nodule-specific (GSns) and root (GSr) isoforms of glutamine synthetase (Figure 2).[5] A second example of enzyme isoform selectivity is the action of phaseolotoxin on L-ornithine carbamoyltransferase from different sources.[9-11] Such selectivity is in marked contrast to the failure of many dedicated efforts to find forms of glutamine synthetase that are fully functional and truly resistant to the popular and commonly used GS inhibitor, methionine sulfoximine.[6-8,12]

## III. DISADVANTAGES OF MICROBIALLY DELIVERED INHIBITORS AS RESEARCH TOOLS

The chief disadvantage of this research tool is that bacterial growth and function in the rhizosphere is influenced by several factors, all of which must be considered when using bacteria to produce and deliver inhibitors. Because good experimental protocol requires consistent production and release of a toxin inhibitor, the factors influencing toxin synthesis and release into the rhizosphere must be identified and controlled.[13] This control can be difficult. Toxins, like many other extracellular metabolites, can be produced in response to some stress, the most common being nutrient stress. Therefore, the experimental design must consider nutrient sources for the plant and the bacterium, amount of bacterial inoculum, consistent culture of the pathogen, and population of the pathogen in the rhizosphere. Keeping these factors invariant in the experiments alleviates many complications in experimental interpretation. Consistent population dynamics of the pathogen in the rhizosphere is most readily maintained in sterilized soil or sand to avoid competitive interactions from other soil microorganisms. Limiting the amount of inoculum is an easy and efficient means of controlling the initial bacterial population. The population after inoculation depends upon temperature, pH, supplemental nutrients, and the interaction between the host plant and the pathogen.

The least understood parameter is the interaction between the plant and the pathogen that influences toxin delivery. An example is the effect of compounds released by the plant into the rhizosphere through its root exudate in moderating toxin delivery in the *P. syringae* pv. *tabaci* system.[14,15] Recent data obtained in our laboratory indicate that reduced nitrogen, in the form of ammonia and certain amino acids, in root exudate inhibits toxin production.[14,15] The composition of root exudate thus influences toxin production. As the interaction between pv. *tabaci* and alfalfa matures, and glutamine synthetase is impaired, ammonia is released into the rhizosphere, and the composition of the exudate changes as the plant begins to fix and assimilate more nitrogen. Clearly, the relationship between the pv. *tabaci* and alfalfa is dynamic, not static. A complication of bacterial delivery of low levels of toxin is the difficulty in determining the concentration of the toxin in the rhizosphere. In practice, the best indication that an effective concentration of toxin is being delivered to the plant is loss of the target enzyme activity.

Presently, there are only a few reasonably well-characterized phytopathogenic bacteria that release enzyme inhibitors, although many others probably exist in nature; as plant pathologists continue their work, other toxin-releasing bacteria will be found and characterized.[2,3] Genetic engineering of inhibitors or herbicides can also be expected to provide additional microbially delivered inhibitors.

## IV. DEVELOPMENT OF RESEARCH SYSTEMS USING *IN SITU* PRODUCTION AND RELEASE OF BACTERIAL INHIBITORS

Although development and use of a bacterium-toxin system requires a significant commitment of time and resources, it provides the researcher with a unique approach to the investigation of the target enzyme and the metabolic system in which it functions. It also

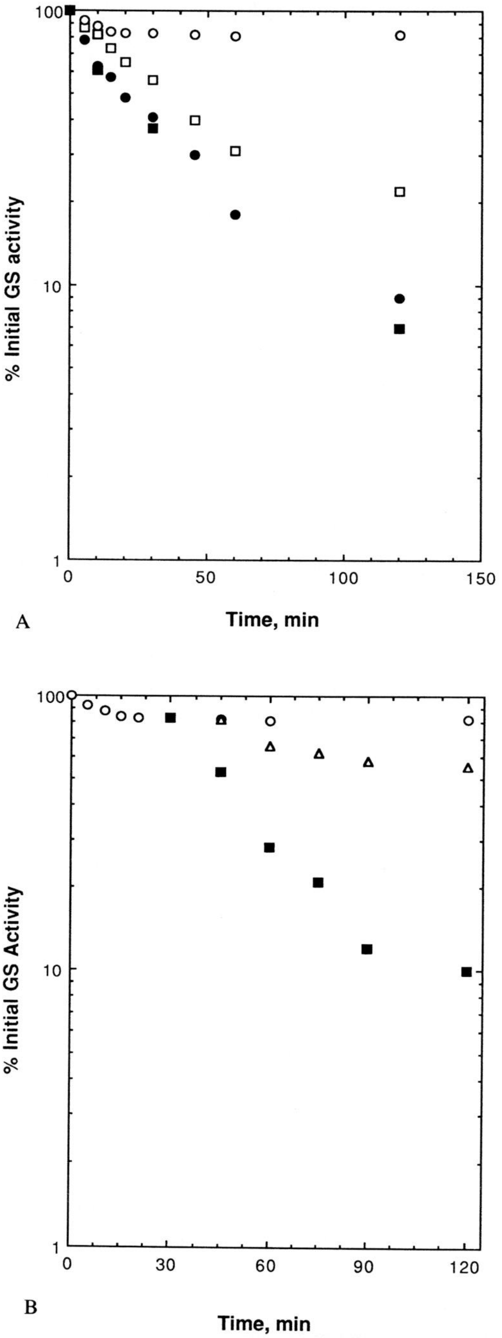


FIGURE 2. A. Selective inactivation glutamine synthetase isoforms by tabtoxinine-β-lactam. Root (squares) and nodule specific (circles) glutamine synthetase were incubated with 100 μ*M* inhibitor and the progress of the inactivations monitored. Tabtoxinine-β-lactam (open symbols) and methionine sulfoximine (closed symbols) were used to demonstrate the selective action of tabtoxinine-β-lactam as compared with nonselective inactivation of both of these enzyme forms by methionine sulfoximine. B. Partial inactivation of nodule-specific glutamine synthetase by tabtoxinine-β-lactam. The initial concentration of tabtoxinine-β-lactam was 100 μ*M*. At 30 min and 45 min, samples were withdrawn and either 100 μ*M* methionine sulfoximine (squares) or 1 mM tabtoxinine-β-lactam (triangles) added and the progress of these inactivations monitored.

**TABLE 2**
**The Optimal Bacterial Toxin Delivery System**

| Toxin | Bacterium |
|---|---|
| 1. Specific target | 1. Appropriate host range |
| 2. Targets desired metabolism | 2. Colonizer of rhizosphere |
| 3. Selectivity between different forms of the target | 3. Consistent toxin production |
| 4. Quantitative assay | 4. Understanding of plant-microbe interaction |
| 5. Chemically stable | |
| 6. Readily purified | |

rewards the user with a second avenue of investigation — plant-microbe interactions and their ultimate development for practical uses. The desirable properties of such a system are summarized in Table 2.

Once a suitable pathogen inhibitor target system is found, the next step is to find appropriate plant systems. Ideally, the pathogen should not exhibit strong host specificity with respect to plant rhizosphere colonization and release of its toxin; this lack of host specficity will allow more flexibility in choosing the plant for study. The plant systems chosen will probably be the ones in which the inhibitor acts selectively toward the different forms of the target enzyme. If the target enzyme is encoded by multiple genes in the host plant, the chances of finding selective action is greatly enhanced. Thus, it is probably an advantage if the enzyme is present in several different tissues. However, polyploidy may be of little utility in enhancing the chances of finding selective action by the toxin. In some cases, an appropriate initial screen for a plant system will be plant survivability. For example, nodulated alfalfa survives pv. *tabaci* and immediately reveals its increased growth and vigor; oats are sensitive to pv. *tabaci* infestation and most of them die rapidly after inoculation, but a few oat plants are variants, and these oats survive the inoculation. Subsequent examination of these variant oats revealed that when their rhizosphere was infested with pv. *tabaci,* they contained no root glutamine synthetase activity but had elevated levels of leaf glutamine synthetase activity. This variant oat leaf glutamine synthetase was resistant to inactivation by the toxin of pv. *tabaci*.[4] In these two plant systems, oat and alfalfa, the toxin acts selectively to inactivate only the root form of the target enzyme; thus these systems are valuable for studying the benefits of conducting ammonia assimilation exclusively in leaves. As tools for the study of nitrogen assimilation, they are complementary systems, allowing investigation in both a legume and a cereal. The considerable utility of this pair of systems suggests that other types of investigations will benefit from examining and developing more than one type of plant.

After the basic system is designed, several important technical parameters must be considered. Of primary importance is culturing the pathogen in such a way as to assure that it consistently produces toxin in culture. The correct means of culturing a plant pathogen is specific to a particular pathogen. Next, steps must be taken to assure that the toxin is produced consistently in the rhizosphere. In addition to the consideration of the population of the pathogen in the rhizosphere, we must consider the need to control or alter the soil media, the pH, temperature, and nutrients supplied to the plant. As discussed earlier, the root exudate may influence toxin production; this influence can be either positive or negative. Once toxin biosynthesis is consistent in the laboratory, the utility of producing and purifying radiolabeled toxin should be considered. Radiolabeled toxin will allow the investigator to confirm the site of the toxin's action, e.g., if the toxin is an irreversible enzyme inhibitor, and establish the toxin's stoichiometry with its target.[16] Radiolabeling will also allow the researcher to examine the toxin's uptake by the roots and its translocation in the plant and will allow estimation of toxin concentrations in plant tissues. Monitoring the fate of the radiolabeled toxin in the plant will alert one to the action of detoxification mechanisms in the plant or allow one to easily verify that the inhibitor is intact in a toxin-resistant plant identified by screening.

A very important component of such a research system is a $tox^-$ mutant or isolate of the pathogen. This $tox^-$ mutant will allow identification of any effects observed in pathogen-infested plants that are not due to the toxin and will also allow confirmation of the effects that are toxin-dependent. We consider this an absolute requirement, because of the complexity of plant-microbe interactions in the rhizosphere.

Legumes form a dynamic tripartite association — the plant, its symbiont, and the pathogen. In this case, the impact of the pathogen upon the symbiont must also be considered. If the symbiont contains the target of the toxin, it will also need a protection or detoxification mechanism to assure its survival in the presence of the pathogen. In the case of pv. *tabaci* and *Rhizobium meliloti,* coexistence is possible because some isolates of *R. meliloti* contain a lactamase activity that inactivates the tabtoxinine-β-lactam.[15,17-19]

## V. INVESTIGATION OF PLANT METABOLISM WITH BACTERIAL TOXINS

At the outset of the project, the investigation will probably proceed in a more or less straightforward manner as the obvious studies are done on the effects upon the relevant metabolite pools and the activities of other enzymes in the pathway. However, long-term perturbations of metabolism in our systems have been accompanied by changes in gene expression for closely related metabolic systems. Such changes in gene expression will be responsible for added complexity in the system and will generate effects not necessarily readily anticipated from knowledge about the action of an inhibitor. These secondary effects are also an investigative advantage for probing and understanding the more global regulation mechanisms functioning to control plant metabolism and physiology. These larger scale regulation mechanisms are very poorly understood and are an area in which much work needs to be done. Clearly, the investigations with these systems are multidisciplinary, utilizing biochemistry, physiology, plant pathology, and molecular biology.

### A. TOXINS USED WITHOUT MICROBIAL DELIVERY

Several bacterially produced inhibitors are discussed, and their potential to be developed as a bacterial delivery system are mentioned. These systems are summarized in Table 3.

#### 1. Ethylene Methionine Biosynthesis Studies with Rhizobitoxine-Producing Bacteria

Rhizobitoxine and several other ether-containing amino acids are irreversible inhibitors of methionine biosynthesis, and as a consequence of this, have been used to block ethylene biosynthesis from methionine (Figure 3).[20,21] Several plant-colonizing bacteria produce these inhibitors, analogs of which can have different effects in different systems.[21,22] Rhizobitoxine is produced by one group of *Bradyrhizobium japonicum,*[23] and by *P. andropogonis,*[24] while the other inhibitors are produced by isolates of *Pseudomonas aeruginosa* and some species of *Streptomyces. Bradyrhizobium japonicum* is an effective colonizer of the rhizosphere of many plants. Isolates of *P. aeruginosa,* while not normally recognized as major plant pathogens, have been obtained from several crops including tobacco, onion, bulbs, and palms.[2,3] It has been associated with chlorosis in several of these plant species. *Pseudomonas andropogonis* pv. *andropogonis* is a pathogen of sorghum, corn, and clover, and *P. a.* pv. *stizolobii* is a pathogen of velvet bean.[2,3] These bacteria are expected to colonize the plant rhizosphere under experimental conditions. They could provide an excellent opportunity to examine in detail the effects of prolonged inhibition of ethylene biosynthesis or methionine biosynthesis in a number of different types of plants and could be useful in developmental studies.

**TABLE 3**
**Bacterial Toxins as Research Tools**

| Toxin | Bacterium | Host plants | Specific target | Metabolism affected |
|---|---|---|---|---|
| Rhizobitoxine and analogs | *B. japonicum, P. andropogonis, P. aeruginosa, Strepto-myces* sp. | Soybean, clover, corn, sorghum, tobacco, bean, onion, palms | Cystathionine β-lyase (EC 4.4.1.8) | Ethylene biosynthesis |
| Phaseolotoxin | *P. syringae* pv. *phaseolicola* | Beans | Ornithine carbamoyl-transferase (EC 2.1.3.3) | Arginine and poly-amine biosynthesis |
| Tagetitoxin | *P. syringae*pv. *tagetis* | Marigolds, other Asteraceae | RNA polymerase III | Transcription of 7 SK and 7 SL RNAs, tRNA, 5 S RNA, U6snRNA |
| Phosphinothricyl-alanyl-alanine (pre-cursor of phosphin-othricin) | *Streptomyces viri-dochromogenes* | ? Soil microbe | Glutamine synthetase (EC 6.3.1.2) | Ammonia assimilation |
| (*N*-Phosphono) methionine-sulfoxi-minyl-alanyl-alanine (precursor of MSX phosphate) | *Streptomyces* sp. | ? Soil microbe | Glutamine synthetase | Ammonia assimila-tion |
| Tabtoxin (precursor of tabtoxinine-β-lactam) | *P. syringae* pv. *tabaci* | Tobacco, oats, alfalfa, soybean, clover, bean, pea, tomato | Glutamine synthetase | Ammonia assimilation |

Rhizobitoxine has been shown to be an active site-directed irreversible inhibitor of cystathionine β-lyase (EC 4.4.1.8).[25,26] This enzyme catalyzes the conversion of cystathionine to homocysteine, which is the penultimate step in methionine biosynthesis. Purified rhizobitoxine has been used effectively to inhibit the spinach enzyme both *in vivo* and *in vitro* during studies of ethylene biosynthesis.[25,27] Yu and Yang also used rhizobitoxine as a tool to study ethylene biosynthesis.[28] As a result of these and other studies, an analog of rhizobitoxine was brought into practical application as an apple tree spray to delay ripening, retard immature fruit drop, and increase the force needed to remove the fruit.[29]

## 2. Arginine Biosynthesis/Function Studies with Phaseolotoxin

Phaseolotoxin is a reversible competitive inhibitor of ornithine carbamoyltransferase (EC 2.1.3.3) (Figure 3) .[30] This inhibitor is synthesized as a tripeptide,[30] but the active form of the inhibitor is a single novel amino acid that is released by hydrolysis of the tripeptide.[11] The name, phaseolicola, is often used interchangeably to mean both the tripeptide and the active toxic moiety of the tripeptide. Phaseolotoxin is, produced and released by *Pseudomonas syringae* pv. *phaseolicola,* a chlorosis-inducing pathogen of bean leaves.[31-33] *Pseudomonas syringae* pv. *phaseolicola* is associated with seeds and probably lives on the roots of these plants, especially under controlled experimental conditions. This inhibitor could be of use in studies involving possible feedback regulation of arginine biosynthesis by arginine in plants.[34] It could also be used to impair, not abolish, the biosynthesis of the arginine used as a precursor of polyamines and then to study any of the many effects of polyamines. Polyamines are positive contributors to tolerance of high levels of salt. Polyamines are also known to stimulate protein synthesis, are concerned with DNA synthesis in the nuclei, and may be involved in membrane stabilization. In plants, polyamines contribute to control of cellular pH and are also involved in nicotine biosynthesis.

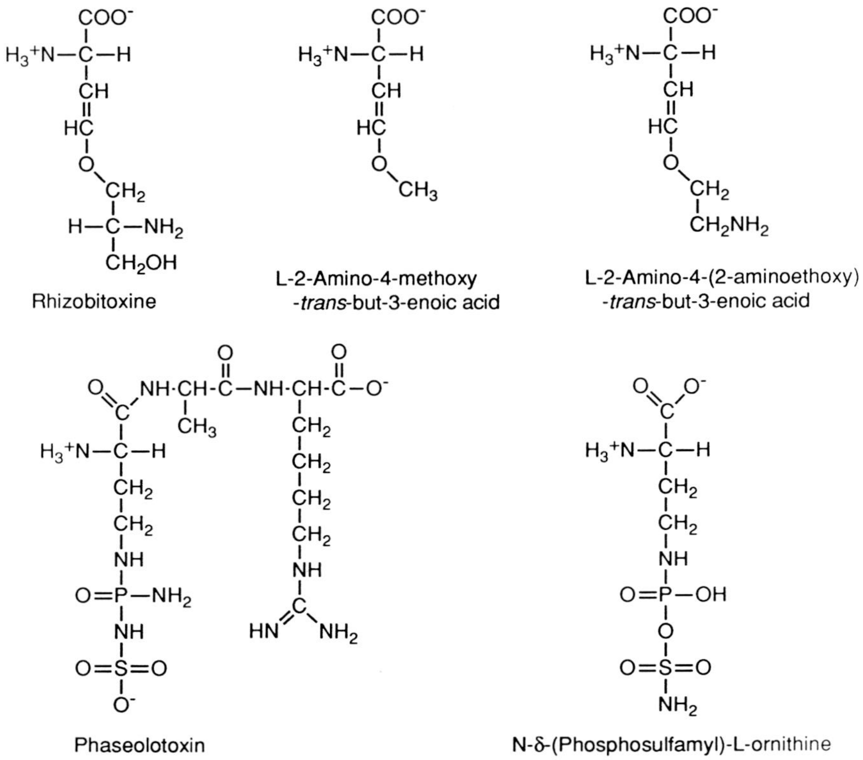


FIGURE 3. Structures of rhizobitoxine and related ether-containing inhibitors and phaseolotoxin, the dipeptide precursor, and the active enzyme inhibitor released from phaseolotoxin.

## 3. Systems Encoded by Chloroplast DNA with Tagetitoxin-Producing Bacteria

Tagetitoxin is a chlorosis-inducing phytotoxin produced by *P. syringae* pv. *tagetis*. It is a specific inhibitor of RNA polymerase III, which is active in plastids in plants.[35,36] RNA polymerase III transcribes genes for 5 S RNA, tRNAs, 7 SK and 7 SL RNA, U6snRNA and other small stable RNAs.[37] Because of its target, tagetitoxin stops all *de novo* synthesis in chloroplasts; thus, it is an appropriate tool to consider using in studies of chloroplast-encoded functions. Although nothing is known about its effect on root plastids, these appear to be an attractive targets for investigation using tagitoxin. A natural pathogen of marigold, *P. syringae* pv. *tagetis* can also infect a wide range of Asteraceae following inoculation.[2,3] This toxin is effective only in developing tissues and perhaps under conditions in which chloroplast genes are being induced. It is also expected to grow in the rhizosphere of plants under controlled experimental conditions.

## 4. Ammonia Assimilation with Phosphinothricin-Producing Bacteria

Phosphinothricyl-alanyl-alanine is a tripeptide (Figure 4) produced by *Streptomyces viridochromogenes*.[38] It is enzymically degraded to phosphinothricin, an irreversible inhibitor of glutamine synthetase.[38-40] *Streptomyces viridochromogenes* is a soil microorganism and is probably able to colonize plant rhizospheres when competing bacteria are controlled. Phosphinothricin is an effective inhibitor, although its utility as a selective inhibitor is probably quite limited; no such selectivity has been reported. Phosphinothricin is, like methionine sulfoximine, a very good analog of the transition state formed by glutamine synthetase,[40] which is the reason for its success as a broad spectrum herbicide.

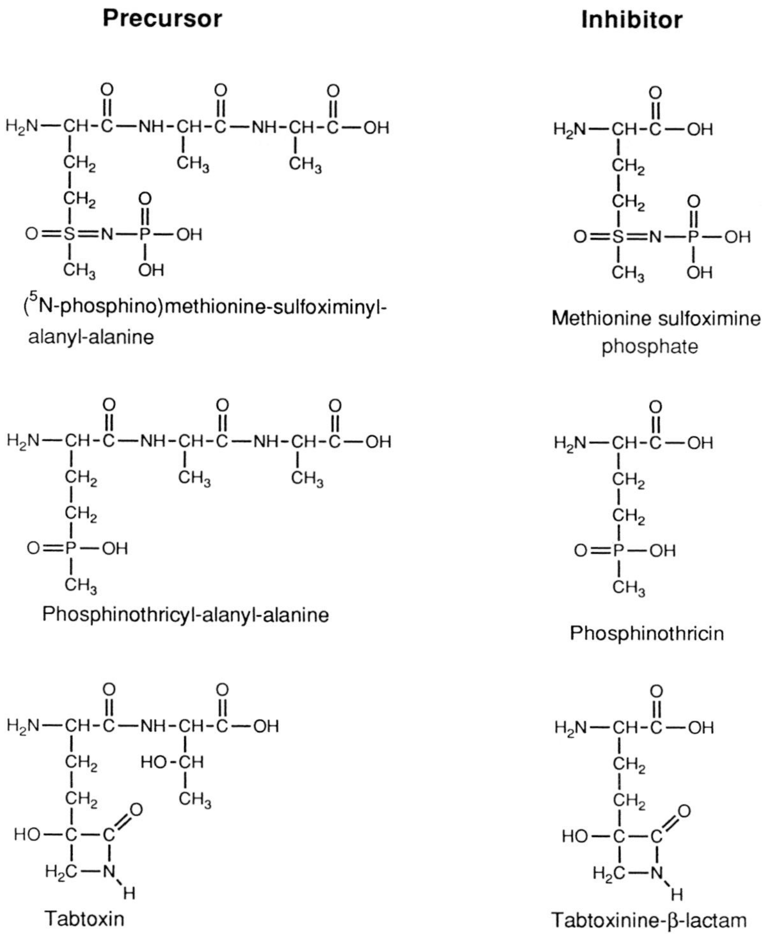


FIGURE 4. Structures of several glutamine synthetase inhibitors that are biosynthesized as small peptides and the structures of the actual enzyme inhibitors released from them.

## 5. Ammonia Assimilation with Methionine Sulfoximine Phosphate-Producing Bacteria

L-($N^5$-Phosphono)methionine-(*S*)-sulfoximinyl-L-alanyl-L-alanine is a tripeptide (Figure 4), produced by species of soil- borne *Streptomyces*.[41] It can be enzymatically hydrolyzed to release methionine sulfoximine phosphate, an inhibitor of glutamine synthetase. Methionine sulfoximine phosphate is also an excellent analog of the transition state formed during catalysis by glutamine synthetase.[6-8] In addition to its microbial synthesis, it is formed during the inactivation of glutamine synthetase by methionine sulfoximine.[6-8] Like phosphinothricin, no selectivity toward isoforms of glutamine synthetase has been reported, to our knowledge, for either methionine sulfoximine or methionine sulfoximine phosphate. A methionine sulfoximine-resistant glutamine synthetase was reported in *Salmonella typhimurium,* but this enzyme functioned poorly and the organism grew slowly.[12] For applications requiring selective action against enzyme isoforms, not simply specific targeting of an enzyme activity, it will probably be of lesser utility than the inhibitor discussed below.

## B. BACTERIAL DELIVERY OF A SELECTIVE GLUTAMINE SYNTHETASE-INHIBITING TOXIN: *PSEUDOMONAS SYRINGAE* PV. *TABACI*-PLANT SYSTEMS

We have developed two systems for studying plant nitrogen metabolism using the unusual effects produced by infesting the rhizosphere of plants with *P. syringae* pv. *tabaci*. These systems include both legumes and cereals:[4,5] alfalfa and oats. These two systems embody both selective action of the bacterial toxin tool and long-term delivery of the inhibitor, allowing investigation of the consequences of long-term perturbations. Alfalfa survives infestation of its rhizosphere by pv. *tabaci* partly because its nodule-specific form of glutamine synthetase (GSns) is resistant to inactivation by the toxin of pv. *tabaci*, tabtoxinine-β-lactam (TβL). On the other hand, wild-type oats are very sensitive to pv. *tabaci*, and all of the glutamine synthetase isoforms in the oat are rapidly inactivated by TβL.[16] A research system has been developed in oats by screening for their tolerance of the infestation. These tolerant oats contain leaf glutamine synthetases that are resistant to inactivation by TβL; however, their root glutamine synthetase, GSr, is rapidly inactivated by TβL. These oats breed true for their resistance to TβL. These infested, tolerant oat plants are extremely vigorous, healthy plants,[4] an observation that supports the hypothesis that reduction of nitrate in leaves, near the site of photosynthesis, is desirable.

There are several properties of *P. syringae* pv. *tabaci* that are most relevant to its use as a research tool. It colonizes the rhizosphere of most plants, although it is commonly recognized as a leaf chlorosis-inducing pathogen of tobacco. TβL, the toxin released by pv. *tabac*i, is an active site-directed, irreversible inhibitor of glutamine synthetase.[44] It binds in 1:1 stoichiometry to the glutamine synthetase subunits and remains tightly bound. The inactivation of glutamine synthetase requires ATP and proceeds in multiple kinetic phases. In all of these properties, TβL is similar to methionine sulfoximine, another commonly used inhibitor of glutamine synthetase.[44] TβL-resistant, fully functional forms of glutamine synthetase have been identified, which is in contrast to the case for methionine sulfoximine for which no fully functional, resistant forms of glutamine synthetase are known. Because pv. *tabaci* also uses glutamine synthetase to assimilate ammonia, it requires self-protection mechanisms of which there are at least three: one is the presence of a β-lactamase enzyme that converts TβL to an inactive form; the second is a shift of its metabolism away from dependence upon the target enzyme; the third is its biosynthesis of TβL as a chemically blocked dipetide form of the toxin, tabtoxin (Figure 4).[17-19] Tabtoxin is a dipeptide of TβL and either serine or threonine;[45] plants obligingly cleave the tabtoxin peptide bond and release the active toxin, TβL.[46] The isolate of *R. meliloti,* the alfalfa symbiont, also contains glutamine synthetases, but has a β-lactamase that can cleave TβL to an inactive form; this β-lactamase is probably a factor in its survival in the alfalfa rhizosphere in the presence of a significant population of pv. *tabaci.*[15,47]

### 1. The Legume/pv. *tabaci* System

Alfalfa plants grown and cultured with *P. syringae* pv. *tabaci* following a prescribed time table grow vigorously. The most important feature is the time allowed for the alfalfa to nodulate before inoculating with pv. *tabaci*. Plants inoculated simultaneously with *Rhizobium* and pv. *tabaci* are seriously stressed for several days after the coinoculation. Typical harvest times are also noted on the time table. When compared with alfalfa plants not infested with *P. syringae* pv. *tabaci,* the infested plants have greater above and below ground fresh weights; these fresh weights are approximately twice the weights of these tissues in the noninfested control plants. The *P. syringae* pv. *tabaci*-infested plants also contain approximately twice the control amount of total nitrogen, although the concentration of the nitrogen is the same as that found in control plants. These plants also have significantly greater total nitrogen-fixing capability;[5] the increased nitrogen-fixing capability results from an increase

**TABLE 4**
**Effects of Infestation of the Alfalfa Rhizosphere on Total Plant Nitrogen, Acetylene Reduction, and Nitrogenase Protein**

| Parameter | Control | pv. *tabaci*-infested |
|---|---|---|
| Total assimilated N (mg/plant) | 1.37 (100%) | 2.85 (208%) |
| Nitrogenase (μmol acetyl red/g nod/h) | 13.6 (100%) | 26.4 (194%) |
| Nitrogenase protein (relative #/mg nod) | 1.0 (100%) | 1.8 (180%) |

[a] Plant age is 45 days, grown and treated as described previously.[5]
[b] Acetyl red, acetylene reduction; nod, nodules.

in both nodule numbers and nodule fresh weights. One quite striking characteristic of the roots and nodules of the *P. syringae* pv. *tabac*i-infested plants is the lack of aged or even older-appearing roots and nodules. The nodules of these plants are very healthy and do not appear to develop the senescent regions common in normal alfalfa nodules. In other gross physiological parameters, the roots and nodules of these plants appear normal.

We compared the total nitrogenase protein in infested and noninfested nodules and found that the pv. *tabac*i-infested nodules contained as much as 1.8 times nitrogenase protein as the control nodules, based on grams per fresh weight of nodule tissue (Table 4). Equal weights of nodules were harvested from noninfested and pv. *tabaci*-infested plants (45 days). The nodules were ground in buffer (50 m*M* imidazole, pH 7.5, 10 m*M* MgC12) and incubated (24 h) with an excess of antibodies to nitrogenase (*Azotobactor vinelandi*). The nitrogenase-antibody complex was harvested as the precipitated protein by centrifugation (12,000 $g$ × 15 min). The protein was resuspended and the total immunoprecipitated protein determined. This increase in total nitrogenase protein could be explained by either of two mechanisms; either more nitrogenase was synthesized in each bacteroid in the nodules or more bacteroids developed in these nodules. These two mechanisms are being investigated.

Infestation by pv. *tabaci* of the rhizosphere of nodulated soybeans is also accompanied by increased nodulation, nitrogen fixation, and total plant growth.[49] A few initial tests for the growth enhancement in other legumes are also promising.

The activity of glutamine synthetase, the target of the TβL released by pv. *tabaci,* was examined in the roots, nodules, and leaves of the infested alfalfa plants and compared with that in noninfested plants. GSr activity in the roots was completely absent in the infested plants.[5] In the nodule plant tissues, two glutamine synthetase forms, GSr and GSns, were present; however, which form or forms of glutamine synthetase assimilated the ammonia produced by fixation is not yet clear. GSr activity was absent in the nodules of these plants; only GSns was active in the plant fraction of these nodules. Partially purified GSr was very rapidly and completely inactivated by TβL, whereas partially purified GSns was essentially resistant to inactivation by TβL (Figure 2). The purified GSns was only partially inactivated (20%) by TβL and the remaining activity was resistant to further inactivation by TβL (Figure 2).[5,15] Clearly, TβL acted selectively upon the GSr form in the roots and nodules of these plants. As expected in the pv. *tabaci*-infested nodules, ammonia levels were elevated, and glutamine and glutamate levels were altered when compared with the noninfested control plants.

The glutamine synthetase forms in alfalfa leaves are not impaired in infested plants. Although this finding suggested that the alfalfa leaf glutamine synthetase was insensitive to inactivation by TβL, subsequent investigation showed that the leaf glutamine synthetase was sensitive to inactivation by TβL.[15] The explanation for the survival of this TβL-sensitive glutamine synthetase in the alfalfa leaves appears to rest in the host plant's regulation of the bacterial synthesis and release of TβL into the rhizosphere. As discussed above, recent data indicates that the reduced nitrogen in the rhizosphere exudate of nitrogen-fixing legumes is very effective in inhibiting TβL biosynthesis. Oat seedlings deprived of nitrogen are very sensitive to infestation of their rhizosphere by pv. *tabaci;* however, oat plants that are well supplied with nitrogen survive the infestation by pv. *tabaci*. Root exudate from these well-nourished oats also strongly inhibits TβL biosynthesis, whereas, root exudate from oat seedlings deprived of nitrogen is a poor inhibitor of TβL production. We conclude from these studies that pv. *tabaci* probably releases less TβL into the rhizosphere of a nitrogen-fixing legume or a cereal that is well supplied with nitrogen than it releases into the rhizosphere of a plant with a suboptimal nitrogen supply. We propose that the reduced amount of TβL released into the rhizosphere of the nitrogen-supplied plants is not sufficient to allow effective concentrations to be accumulated in all of the tissues in the plant; in other words, the leaf simply does not receive enough toxin for it to be effective. We might speculate that TβL is produced and released by pv. *tabaci* as a means of assuring itself an abundant supply of reduced nitrogen, although, this interpretation may be entirely too simplistic or even irrelevant in the real world of the rhizosphere.

We returned our focus to the survival of pv. *tabaci*-infested alfalfa plants. We expected to find that the root glutamine synthetase activity would be lost when TβL was provided to the plants. Furthermore, we expected to observe loss of the leaf GS only when sufficient TβL was supplied that it could be distributed throughout the plant. This was precisely what we observed in the pv. *tabaci*-infested alfalfa plants. We could not measure TβL in the rhizosphere because it was not present in sufficient quantities to allow its quantification; instead, we had to rely upon measurements of glutamine synthetase activity to monitor the action of TβL. We have also shown that alfalfa plants readily translocate TβL to their foliar regions, and therefore their survival is not a function of an inability to move the toxin to the leaves. Thus, the influence that the reduced nitrogen released into the rhizosphere has upon TβL production appears to be one of the complex interactions in this tripartite association.

Glutamine synthetase forms in *R. meliloti* must be considered with respect to TβL action. Although these glutamine synthetases were sensitive to the toxin, they were present in approximately normal levels in the bacteroids of infested plants. *Rhizobium meliloti* also survived the infestation of the rhizosphere by pv. *tabaci*. We found that the isolate of *R. meliloti* we have been using contained a β-lactamase activity that hydrolyzes the β-lactam ring of TβL;[15] thus, this *R. meliloti* contained one of the protection mechanisms used by pv. *tabaci* against TβL.[17,19]

## 2. The Cereal/pv. *tabaci* System

As mentioned above, most oat plants are very sensitive to infestation of their rhizosphere by pv. *tabaci* and these pathogen-sensitive plants die rapidly following the complete loss of their glutamine synthetase activity.[16] A few oat plants are tolerant of *P. syringae* pv. *tabaci* infestation of their rhizosphere.[4] They readily take up the toxin and do not detoxify or degrade it; pv. *tabaci* establishes and maintains a significant population in their rhizosphere. In this case, their tolerance of pv. *tabaci* is a result of the presence in their leaves of glutamine synthetases that are resistant to inactivation by TβL. Their root glutamine synthetase, on the other hand, is rapidly and completely inactivated by TβL. When their rhizosphere is infested with pv. *tabaci,* these tolerant oat plants grow vigorously and are quite fertile, producing good seed yields.[4] These infested oat plants contain about twice the normal specific activity of glutamine synthetase in their leaves when expressed per unit weight of protein and even greater amount when based on fresh weight.[4] They also have approximately twice the normal concentration of total leaf protein.[4]

Normally, when oats are growing on nitrate, they reduce a significant amount of nitrate to ammonia and assimilate this ammonia to glutamine and glutamate in their roots. This ammonia assimilation cannot occur in the root of pv. *tabaci*-infested, tolerant plants because of the absence of root glutamine synthetase activity.[4] The reduction of nitrate to ammonia in their roots is improbable because they would be unable to assimilate it in the roots and would need to translocate it to their leaves for assimilation. Plants do not, as a matter of course, translocate significant amounts of ammonia because of its damaging effects on chloroplast membrane structures. The amino acid pools have also changed in the roots and leaves of these plants as is to be expected if there is high glutamine synthetase activity in their leaves and there is no glutamine synthetase activity in their roots.[4] These infested plants appear to be assimilating ammonia in their leaves. The vigor with which these plants grow is consistent with the prediction that plants would benefit from reducing and assimilating nitrate in their leaves, near the site of photosynthate production. This relocation of nitrogen reduction and assimilation to the leaves removes any possible limitation of these activities resulting from a limitation of movement of carbon to their roots.

Bacterial toxins have played an important role in various metabolic studies. Their role in the future, however, could be expanded to include studies of the effects of long-term metabolic perturbations focusing on understanding the regulation or interaction of multiple metabolic systems and the resulting consequences on plant gene expression. Natural toxins were discussed in this chapter, but they need not be the only inhibitors used; as we develop microbial delivery systems for engineered, biosynthesized pesticides, a new group of inhibitors will become available. While the use of these possible research systems will undoubtedly be limited to specific applications, they provide unique experimental advantages over conventional investigative methods.

## REFERENCES


1. **Curl, E. A. and Truelove, B.,** *The Rhizosphere,* Springer-Verlag, Berlin, 1986.
2. **Durbin, R. D.,** *Toxins in Plant Disease,* Academic Press, San Diego, 1980.
3. **Fahy, P. C. and Persley, G. J.,** *Plant Bacterial Diseases,* Academic Press, San Diego, 1983.
4. **Knight, T. J., Bush, D. R., and Langston-Unkefer, P. J.,** Oats tolerant of *Pseudomonas syringae* pv. *tabaci* contain tabtoxinine-β-lactam-insensitive leaf glutamine synthetases, *Plant Physiol.,* 88, 33, 1988.
5. **Knight, T. J. and Langston-Unkefer, P. J.,** Enhancement of symbiotic dinitrogen fixation by a toxin-releasing plant pathogen, *Science,* 241, 951, 1988.
6. **Meister, A.,** Catalytic Mechanisms of glutamine synthetase; overview of glutamine metabolism, in *Glutamine: Metabolism, Enzymology, and Regulation,* Mora, J. and Palacios, R., Eds., Academic Press, San Diego, 1980, 1.
7. **Ronzio, R. A., Rowe, W. B., and Meister, A.,** Studies on the mechanism of inhibition of glutamine synthetase by methionine sulfoximine, *Biochemistry,* 8, 1066, 1969.
8. **Weisbrod, R. E. and Meister, A.,** Studies on glutamine synthetase from *Escherichia coli, J. Biol. Chem.,* 248, 3997, 1973.
9. **Ferguson, A. R., Johnston, J. S., and Mitchell, R. E.,** Resistance of *Pseudomonas syringae* pv. *phaseolicola* to its toxin, phaseolotoxin, *FEMS Microbiol. Lett.,* 7, 123, 1980.
10. **Staskawicz, B. J., Panopoulos, N. J., and Hoogenraad, N. J.,** Phaseolotoxin-insensitive ornithine carbamoyltransferase of *Pseudomonas syringae* pv. *phaseolicola:* Basis for immunity to phaseolotoxin, *J. Bacteriol.,* 142, 720, 1980.
11. **Templeton, M. D., Sullivan, P. A., and Shepherd, M. G.,** Phaseolotoxin-insensitive L-ornithine transcarbamoylase from *Pseudomonas syringae* pv. *phaseolicola, Physiol. Mol. Plant Pathol.,* 29, 393, 1987.
12. **Miller, E. S. and Brenchley, J. E.,** L-Methionine SR-sulfoximine-resistant glutamine synthetase from mutants of *Salmonella typhimurium, J. Biol. Chem.,* 256, 11307, 1981.

13. **Shaw, P. D.,** Production and isolation, in *Toxins in Plant Disease,* Durbin, R. D., Ed., Academic Press, San Diego, 1980, 21.
14. **Ghosh, S., Knight, T. J., and Langston-Unkefer, P. J.,** Influence of reduced nitrogen on production of tabtoxinine-β-lactam by *Pseudomonas syringae* pv. *tabaci,* in preparation.
15. **Knight, T. J., Gosh, S., Dickstein, R., and Langston-Unkefer, P. J.,** Tabtoxinine-β-lactam related interactions in the tripartite association between alfalfa, *Rhizobium meliloti,* and *Pseudomonas syringae* pv. *tabaci,* in preparation.
16. **Knight, T. J., Durbin, R. D., and Langston-Unkefer, P. J.,** Effects of tabtoxinine-β-lactam on nitrogen metabolism in *Avena sativa* L. roots, *Plant Physiol.,* 82, 1045, 1986.
17. **Knight, T. J., Durbin, R. D., and Langston-Unkefer, P. J.,** Self-protection of *Pseudomonas syringae* pv. *tabaci* from its toxin, tabtoxinine-β-lactam, *J. Bacteriol.,* 169, 1954, 1987.
18. **Knight, T. J., Durbin, R. D., and Langston-Unkefer, P. J.,** Role of glutamine synthetase adenylylation in the self-protection of *Pseudomonas syringae* subsp. *tabaci* from its toxin, tabtoxinine-β-lactam, *J. Bacteriol.,* 166, 224, 1986.
19. **Durbin, R. D. and Langston-Unkefer, P. J.,** The mechanisms for self-protection against bacterial phytotoxins, *Annu. Rev. Phytopathol.,* 26, 313, 1988.
20. **Owens, L. D., Lieberman, M., and Kunishi, A.,** Inhibition of ethylene production by rhizobitoxine, *Plant Physiol.,* 48, 1, 1971.
21. **Lieberman, M.,** Biosynthesis and action of ethylene, *Annu. Rev. Plant Physiol.,* 30, 533, 1979.
22. **Wang, C. Y. and Mellenthin, W. M.,** Effect of aminoethoxy analog of rhizobitoxine on ripening of pears, *Plant Physiol.,* 59, 546, 1977.
23. **Minamisawa, K.,** Division of rhizobitoxine-producing and hydrogen-uptake positive strains of *Bradyrhizobium japonicum* by nifDKE sequence divergence, *Plant Cell Physiol.,* 31, 81, 1990.
24. **Mitchell, R. E.,** Biosynthesis of rhizobitoxine from L-aspartic acid and L-threo-hydroxythreonine by *Pseudomonas andropogonis, Phytochemistry,* 28, 1617, 1989.
25. **Giovanelli, J., Owens, L. D., and Mudd, S. H.,** Mechanism of inhibition of spinach β-cystathionase by rhizobitoxine, *Biochim. Biophys. Acta,* 227, 671, 1971.
26. **Owens, L. D., Guggenheim, S., and Hilton, J. L.,** Rhizobium-synthesized phytotoxin: an inhibitor of S-cystathionase in *Salmonella typhimurium, Biochim. Biophys. Acta,* 158, 219, 1968.
27. **Giovanelli, J., Owens, L. D., and Mudd, S. H.,** β-Cystathionase: *In vivo* inactivation by rhizobitoxine and role of the enzyme in methionine biosynthesis in corn seedlings, *Plant Physiol.,* 51, 492, 1972.
28. **Yu, Y-B. and Yang, S. F.,** Auxin-induced ethylene production and its inhibition by aminoethoxyvinylglycine and cobalt ion, *Plant Physiol.,* 64, 1074, 1979.
29. **Bangerth, F.,** The effect of a substituted amino acid on ethylene biosynthesis, respiration, ripening and preharvest drop of apple fruits, *J. Am. Hortic. Sci.,* 103, 401, 1978.
30. **Kamdar, H. V., Rowley, K. B., Clements, D., and Patil, S. S.,** *Pseudomonas syringae* pv. *phaseolicola* genomic clones harboring heterologous DNA sequences suppress the same phaseolotoxin-deficient mutants, *J. Bacteriol.,* 173, 1073, 1991.
31. **Patil, S. S., Tam, L. Q., and Sakai, W. S.,** Mode of action of the toxin from *Pseudomonas phaseolicola* 1. Toxin specificity, chlorosis, and ornithine accumulation, *Plant Physiol.,* 49, 803, 1972.
32. **Kwok, O. C. H. and Patil, S. S.,** Activation of a chlorosis-inducing toxin of *Pseudomonas syringae* pv. *phaseolicola* by leucine aminopeptidase, *FEMS Microbiol. Lett.,* 14, 247, 1982.
33. **Turner, J. G. and Mitchell, R. E.,** Association between symptom development and inhibition of ornithine carbamoyltransferase in bean leaves treated with phaseolotoxin, *Plant Physiol.,* 79, 468, 1985.
34. **Turner, J. G.,** Effect of phaseolotoxin on the synthesis of arginine and protein, *Plant Physiol.,* 80, 760, 1986.
35. **Mathews, D. E. and Durbin, R. D.,** Tagetitoxin inhibits RNA synthesis directed by RNA polymerases form chloroplasts and *Escherichia coli, J. Biol. Chem.,* 265, 493, 1990.
36. **Steinberg, T. H., Mathews, D. E., Durbin, R. D., and Burgess, R. R.,** Tagetitoxin: A new inhibitor of eukaryotic transcription by RNA polymerase III, *J. Biol. Chem.,* 265, 499, 1990.
37. **Geiduschek, E. P. and Tocchini-Valentini, G. P.,** Transcription by RNA polymerase III, *Annu. Rev. Biochem.,* 57, 873, 1988.
38. **Bayer, E., Gugel, K. H., Hagele, K., Hagenmaier, H., Esipov, S. E., Konig, W. A., and Zaehner, H.,** Metabolic products of microorganisms. 98. Phosphinothricin and phosphinothricylalanylalanine, *Helv. Chim. Acta,* 55, 224, 1972.
39. **Manderscheid, R. and Wild, A.,** Studies on the mechanism of inhibition of phosphinothricin of glutamine synthetase isolated from *Triticum aestivum* L., *J. Plant Physiol.,* 123, 135, 1986.
40. **Logusch, E. W., Walker, D. M., McDonald, J. F., and Franz, J. E.,** Substrate variability as a factor in enzyme inhibitor design: inhibition of ovine brain glutamine synthetase by α-γ-substituted phosphinothricins, *Biochemistry,* 28, 3043, 1989.
41. **Pruess, D. L., Scannell, J. P., Ax, H. A., Kellet, M., Wiess, F., Demny, T. C., and Stempel, A.,** Antimetabolites produced by microorganisms. VII. L-($N^5$-phosphono)methionine-*S*-sulfoximinyl-L-alanyl-L-alanine, *J. Antibiot.,* 26, 26, 1973.

42. **Zollner, H.,** *Handbook of Enzyme Inhibitors,* VCH, NY, 1989, 95.
43. **Valleau, W. D., Johnson, E. M., and Diachun, S.,** Association of tobacco leaf spot bacteria with roots of crop plants, *Science,* 96, 164, 1942.
44. **Langston-Unkefer, P. J., Robinson, A. C., Knight, T. J., and Durbin, R. D.,** Inactivation of pea seed glutamine synthetase by the toxin, tabtoxinine-β-lactam, *J. Biol. Chem.,* 262, 1608, 1987.
45. **Stewart, W. W.,** Isolation and proof of structure of wildfire toxin, *Nature (London),* 229, 174, 1971.
46. **Uchytil, T. F. and Durbin, R. D.,** Hydrolysis of tabtoxins by plant and bacterial enzymes, *Experientia,* 36, 301, 1980.
47. **Krieg, N. R., Ed.,** *Bergey's Manual of Systematic Bacteriology,* Williams & Wilkins, Baltimore, MD, 1984.
48. **Ta, T. C., Macdowall, F. D. H., and Faris, M. A.,** Excretion of nitrogen assimilated from $N_2$ fixed by nodulated roots of alfalfa *(Medicago sativa), Can. J. Bot.,* 64, 2063, 1986.
49. **Langston-Unkefer, P. J., Knight, T. J., and Sengupta-Gopalan, C.,** Beneficial consequences of a selective glutamine synthetase inhibitor in oats and legumes, in *Biological Nitrogen Fixation Associated With Rice Production,* Dutla, S. K. and Sloger, C., Eds., Howard University Press, Washington, D.C., 1991, 47.

# *Section III: Bacterial-Plant Interactions (Symbiotic)*

# ATTACHMENT, LECTIN AND INITIATION OF INFECTION IN (*BRADY)RHIZOBIUM*-LEGUME INTERACTIONS

**Jan W. Kijne, Ben J. J. Lugtenberg, and Gerrit Smit**

## TABLE OF CONTENTS

# I. INTRODUCTION

Under conditions of nitrogen limitation *(Brady)rhizobium* bacteria form symbiotic root nodules on leguminous as well as on a few nonleguminous plants. In these nodules the bacteria, in the form of bacteroids, convert atmospheric nitrogen into ammonia, which is used by the plant as a nitrogen source. Nodulation is a host plant-specific process: of the various species and biovars of *(Brady)rhizobium* only a few are able to nodulate a particular plant and fix nitrogen efficiently.

Nodulation presumably starts with chemotaxis of rhizobia towards one or more root exudate components. Once the bacteria are close to the root, they attach to the tips of root hairs,[1] to epidermal cell junctions[2] or at wound sites[3] where they are optimally exposed to flavonoids, inducers of *nod*(nodulation) genes,[4-7] that are secreted by the plant root. Among the products of these *nod* genes are various enzymes that are involved in the synthesis of a series of lipo-oligosaccharides[8,9] which induce a variety of processes in the host plant like root hair deformation, root hair curling, cell division, secretion of additional *nod* gene inducers,[10] and initiation of infection threads.

For many rhizobia, primary target sites for infection are young growing root hairs. It is generally assumed, but not proven, that attachment of rhizobia to the root hair surface is an essential step in tight ("marked") curling of legume root hairs prior to root infection. Moreover, attachment has been proposed to be a determinant of host plant specificity (e.g., Reference 11). In this chapter, attention will be focused on the processes of attachment and initiation of infection.

Attachment of rhizobia is a nonhost-specific two-step process. The first attachment step is probably mediated by a bacterial surface-located $Ca^{2+}$-binding protein, designated as rhicadhesin, and an unknown, but common, plant root surface component.[12] The nature of the molecules involved in the additional attachment step depends on the rhizobial species and its growth conditions. It has been described that cellulose fibrils,[13] proteinaceous fimbriae,[14,15] and plant lectins[16,17] can be involved in this step.

Attachment of rhizobia to growing root hair tips can be followed by curling of the root hair. Tight curling is indicated as the phenotype Hac. In homologous interactions Hac causes root hair tip growth activity to be concentrated at one site, the infection focus, resulting in redirection of tip growth toward the root cortex, i.e., infection thread formation (phenotype Inf).[18] Lipo-oligosaccharides play a role in hair deformation and curling.[8,9]

The plant's lectin plays a major role in the next nodulation step, initiation of infection. Its role, as well as that of its putative bacterial counterparts lipo-oligosaccharides, extracellular polysaccharides (EPS) and lipopolysaccharides (LPS) in this process will be discussed.

# II. ATTACHMENT

## A. BACTERIAL GROWTH CONDITIONS

Rhizobia may attach to different types of cells at the host root surface, e.g., root epidermal cells, root hairs, and root cap cells. For associations of rhizobia with legumes such as pea, clover, and soybean, just emerging root hairs constitute the primary sites of infection. In pea, tips of growing root hairs are preferred sites of attachment.[1] Studying the molecular mechanism of rhizobial attachment to pea root hairs, Smit et al.[1,12,13,19,20] and Kijne et al.[16,21] found that bacterial growth conditions, and especially nutrient limitations, strongly affect the characteristics of attachment. Carbon limitation of *R. leguminosarum* bv. *viciae* results in a nonhost-specific attachment mechanism.[1,13] However, manganese limitation as well as other limitations (e.g., for oxygen and nitrogen) appear to result in a faster attachment process, in which, among other components, host-plant specific molecules are involved, i.e., root lectin.[16,21] Furthermore, manganese-limited cells are significantly more infective than carbon-limited cells, demonstrating the profound influence of bacterial growth conditions on rhizobial attachment and infectivity.[16] It would be interesting to learn whether nutrient limitation is a prerequisite for attachment and nodulation under field conditions. If so, the root nodule symbiosis might be viewed as a survival mechanism by which the bacteria and the host plant avoid nutrient limitation. Considering the composition of plant root exudates,[22,23] a carbon source is probably not a growth-limiting nutrient in the rhizosphere.

### B. THE LECTIN RECOGNITION HYPOTHESIS

Initially, it was hypothesized that host specificity, as found in the *(Brady)rhizobium*-legume symbiosis, is a consequence of specific binding of the homologous rhizobial partner to the root hair cell surface of the host plant. The molecular basis for the recognition mechanism was proposed to be binding of a host plant lectin to the bacterial surface.[11,24] Legume lectins constitute a homologous group of sugar-binding glycoproteins from which each monomer contains a specific sugar-binding site and a conserved hydrophobic pocket.[25] The lectin recognition hypothesis was tested by examining the attachment ability of nodulating or nonnodulating rhizobial strains in the absence or presence of lectin-haptenic monosaccharides.[26] Lectin-mediated attachment of *R. leguminosarum* bv. *trifolii* to clover root hair tips could be observed, and appeared to be dependent on the presence of the Sym plasmid.[11,26-29] However, a low level of nonhost-specific attachment was also observed, which could not be prevented by addition of lectin haptens. Host-specific attachment as well as lectin binding appeared to be dependent on the growth phase of the bacteria.[30-32] Data supporting the lectin-recognition hypothesis were also reported for other *(Brady)rhizobium*-legume combinations.[17,33]

In contrast to these data, several studies were unable to demonstrate specific lectin-mediated binding of homologous rhizobia to legume root hairs.[1,13-15,34-36] Heterologous rhizobia appeared to adhere equally well to host and nonhost plant surfaces in comparison with homologous rhizobia. Moreover, absence of the Sym plasmid did not affect rhizobial attachment behavior.[1,37]

The recent information showing that bacterial growth conditions largely influence bacterium-plant adhesion, and the fact that various research groups use different culture and assay conditions, offer an explanation for these data. In addition, it is well established that culture age affects the expression of lectin receptors on the rhizobial cell surface.[30,32,38] Therefore, lectin may be involved in adhesion, but only under specific growth conditions.

### C. RHICADHESIN AND CALCIUM

Calcium limitation is the only limitation tested that does not lead to optimal attachment of rhizobia to pea root hairs. Growth of rhizobia under low $Ca^{2+}$ conditions strongly reduces the initial direct attachment of rhizobia to pea root hair surfaces, suggesting the involvement of a $Ca^{2+}$-dependent adhesin (Figure 1).[13] Although growth under low $Ca^{2+}$ conditions affects a number of cell surface characteristics (e.g., results in loss of flagella and the O-antigenic part of the LPS), these characteristics are not responsible for the attachment-minus phenotype.[20,39] Instead, it appeared that under low $Ca^{2+}$ conditions a rhizobial $Ca^{2+}$-dependent adhesin, rhicadhesin, was released from the bacterial surface into the growth medium.[39]

Rhicadhesin can be isolated from the surface of *R. leguminosarum* bv. *viciae* grown under normal $Ca^{2+}$ conditions, and has been purified using its ability to inhibit attachment of both carbon- and manganese-limited bacteria to the surface of pea root hairs.[12,20,39] This protein has the following properties: (1) it is heat-labile and sensitive to protease treatment,[12,20] (2) it has an isoelectric point of 5.1 and a molecular weight of 14,000, (3) it is a $Ca^{2+}$-binding protein,[12] (4) $Ca^{2+}$ is required for anchoring of rhicadhesin to the bacterium,

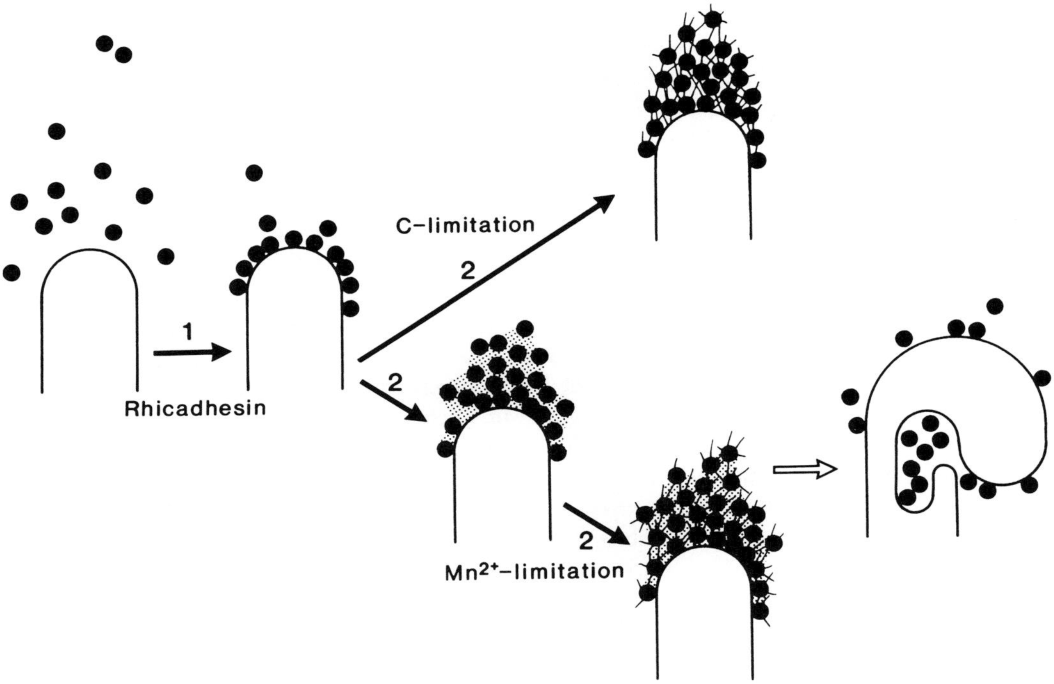


FIGURE 1. A detailed model for attachment of rhizobial cells to pea root hair tips. Step 1 attachment is mediated by rhicadhesin and leads to the attachment of single rhizobial cells to the surface of the root hair tip. The mechanism of step 2 attachment depends on the growth conditions of the rhizobia and results in the formation of aggregates of bacteria on the tip of the root hair. In the case of carbon-limited bacteria, rhizobial cellulose fibrils are involved in the second step of attachment, whereas in the case of manganese-limited rhizobia, host plant lectins are also involved. In the latter case, the aggregates are formed within a shorter period of time. Lectin-mediated accumulation is correlated with optimal root infection, as indicated by the open arrow (see also refs. 12 and 16).

not to the plant, and (5) it is released in the growth medium when the rhizobia grow under low $Ca^{2+}$ conditions.[39] Rhicadhesin is common among all genera of the family *Rhizobiaceae,* including *Bradyrhizobium* and *Agrobacterium,* and is also present in *Azotomonas* spp. Adhesin activity is not associated with bacterial species tested that represent a number of other genera.[12] Its expression was found to be independent from the rhizobial Sym plasmid or the Ti plasmid of *Agrobacterium.*[12]

*Rhizobium* mutants lacking rhicadhesin have not yet been found. Such mutants will be essential to answer the question whether rhicadhesin-mediated attachment is a prerequisite for nodulation. Interestingly, rhicadhesin appeared to play an essential role in tumorigenicity of *Agrobacterium,* since both attachment and virulence of *A. tumefaciens chvB* (chromosomal virulence) mutants[40,41] can be restored by treatment with purified rhicadhesin isolated from either wild-type *A. tumefaciens* or *R. leguminosarum* bv. *viciae.*[19,42]

Rhicadhesin-mediated binding of *Rhizobiaceae* cells to root surfaces was not only found with pea, but with all tested plants, including monocotyledonous species.[12] Therefore, it seems that this mechanism is common for the first attachment step in *Rhizobiaceae*-plant interactions and is based on binding to a common plant surface component. The identity of the plant receptor remains to be determined.

## D. RHIZOBIAL AGGREGATION AT ROOT HAIR TIPS

Following binding to the root hair surface, rhizobia may form aggregates on root hair tips. For *R. leguminosarum* bv. *viciae,* extracellular fibrils were found to mediate aggregation (cap formation; see Figure 1).[1,13] These fibrils were purified and chemically characterized as cellulose fibrils. Fibril-overproducing mutants showed greatly increased cap-forming ability, whereas fibril-negative mutants lost this ability completely. Both types of mutants showed normal nodulation properties, indicating that both cellulose fibrils and fibril-mediated cap formation are not a prerequisite for successful nodulation under the conditions used.[13] It cannot be excluded, however, that cellulose fibrils are important colonization factors under field conditions (e.g., by being involved in competition for nodulation sites).

Pea lectin (Psl) was found to accelerate accumulation of manganese-limited *R. leguminosarum* bv. *viciae* cells at pea root hair tips.[16,43] Psl is produced in pea seeds and roots and is secreted into the rhizosphere.[44-46] Involvement of Sym plasmid-located genes in lectin-enhanced accumulation of *R. leguminosarum* bv *viciae* could not be demonstrated.[16] This is consistent with the observation that several *Rhizobium* species are able to bind Psl[47] and that Psl can precipitate EPS of various fast-growing rhizobia.[48]

Interestingly, clover lectin, trifoliin A,[49] binds specifically to homologous *R. leguminosarum* bv. *trifolii* cells (for a review, see Dazzo et al.[28]). Like Psl in pea, trifoliin A is synthesized in clover seeds and roots and is secreted into the rhizosphere.[50,51] As a result, trifoliin A specifically mediates accumulation of *R. leguminosarum* bv. *trifolii* cells at clover root hair tips.[27-29] This accumulation is dependent on the presence of *R. leguminosarum* bv *trifolii* Sym plasmid-located *nod* genes.[29,52] In contrast to Psl in pea, trifoliin A is present on the outer cell wall surface of growing clover root hairs where it may also mediate direct binding of *R. leguminosarum* bv. *trifolii* cells to the root hair surface.[28]

These observations show that under appropriate culture conditions legume lectins may contribute to the presence of an increased number of rhizobia at primary infection sites. The precise role of certain *nod* genes in rhizobial accumulation remains to be established.

## E. RHIZOBIAL ATTACHMENT IS A TWO-STEP PROCESS

From the available data it can be concluded that (1) homologous as well as heterologous rhizobia are able to attach to (host) plant root hairs, (2) the presence of a Sym plasmid is not a prerequisite for nonhost-specific attachment, (3) more than one type of binding occurs in *(Brady)rhizobium*-plant root hair associations, and (4) under certain conditions host plant lectins enhance accumulation of (homologous) rhizobia at root hair tips.

In all cases studied, attachment appeared to be a two-step process with an initial binding phase followed by firm adherence and accumulation (see also Reference 28). Fibrillous appendages of rhizobia appear to be involved in the second step of attachment leading to bacterial aggregation and anchoring of bacteria to the plant cell surface. Also, in case of the closely related bacterium *A. tumefaciens,* there are indications that cellulose fibrils enable anchoring of the bacteria which might be advantageous under certain conditions (e.g., flushing of attachment sites by water).[53] For *B. japonicum,* fimbriae were reported to mediate firm attachment.[14,15] This type of binding is specifically inhibited in the presence of galactose, suggesting the presence of a fimbriae-associated lectin on the bacterial cell surface. Recently, Ho et al.[54,55] isolated a 38-kDa lactose/galactose-binding protein from *B. japonicum,* designated Bj38. These authors consider it likely that this protein is associated with bradyrhizobial fimbriae, although its localization at the bacterial surface has not (yet) been demonstrated directly. Similar to mutants defective in production of fimbriae, mutants not producing Bj38 show reduced firm attachment ability. However, both types of mutants are still able to attach and to nodulate,[56,57] indicating that firm attachment is not essential for infection of soybean under laboratory conditions. Similarly, cellulose fibrils are not a prerequisite for nodulation of pea by *R. leguminosarum* bv *viciae.*[13]

These observations suggest that the initial weak attachment step is essential in root infection. However, the second fibril-mediated attachment step may not be unimportant. Bacterial appendages, such as fimbriae and cellulose fibrils, affect the physicochemical properties of the cell surface[58] and, therefore, may have general effects in cell-to-cell interactions. For instance, fimbriae-mediated adhesion likely involves hydrophobic interactions.[14] A number of studies have shown that adhesion to root cells is positively correlated with adhesion to abiotic surfaces.[1,14,20] Possibly, certain physicochemical interactions between the plant and the bacterial surface promote weak adherence of bacteria to plant cells under natural conditions.

## III. INITIATION OF INFECTION

### A. INFECTION STEPS

After attachment of homologous rhizobia, a growing legume root hair may respond by curling and infection thread initiation.[59,60] For induction of tight curling, presence of living rhizobia at the root hair surface is essential.[61] In pea, up to 25% of growing root hairs curl upon inoculation with *R. leguminosarum* bv. *viciae*.[59] Extensive curling of pea root hairs (supercurling; phenotype Hac$^{++}$ according to Van Brussel et al.[62]) is found in presence of the heterologous *R. leguminosarum* bv. *trifolii*.[59,63] However, other heterologous rhizobia are unable to induce significant root hair curling in pea but generate various root hair deformations (phenotype Had). Apparently, although not widely recognized, induction of tight root hair curling has a high degree of host plant specificity.[59,61] In contrast to the situation with the homologous *R. leguminosarum* bv. *viciae,* curling of pea root hairs by the heterologous *R. leguminosarum* bv. *trifolii* is not followed by infection thread formation.

An infection thread in a legume root hair is a tip-growing tubular structure containing the invading rhizobia. Infection thread initiation coincides with an arrest of tip growth in the infected root hair. By using computer simulation, Van Batenburg et al.[18] predicted that tight curling of a root hair results from induction and redirection of tip growth by attached rhizobia. Concentration of the plant cell tip growth machinery at the site of the enclosed rhizobia may subsequently result in infection thread initiation. Growth of an infection thread is likely to require, in addition to a tip growth area, a growth pressure like the cell turgor functions for a root hair. Entrapment of rhizobia may be essential for generating this growth pressure. Thus, according to the model of Van Batenburg et al., initiation of rhizobial infection includes at least four essential elements: (1) attachment of bacteria, (2) induction/redirection of plant cell tip growth, (3) concentration of the tip growth activity at one site (the infection focus), and (4) generation of a growth pressure (Figure 2). Elements (1) and (2) enable root hair curling, whereas subsequently, elements (3) and (4) enable infection thread growth. Since growth of the infection thread is accompanied by rhizobial cell divisions, attachment of most enclosed and invading rhizobia is apparently not essential and may be even disadvantageous for elements (3) and (4).

### B. ROOT HAIR CURLING AND SIGNAL MOLECULES

In the preceding part of this chapter, it has been shown that rhizobia may attach to plant roots in absence of *nod* genes. In contrast, both common and host-range *nod* genes are required for tight curling and infection of legume root hairs (see chapter by Kondorosi et al., this volume). Molecules (most) probably involved in root hair curling and infection include low-molecular weight signals, surface polysaccharides produced by rhizobia, and lectin produced by the host plant. Both common and host-specific *nod* genes can be involved in production of rhizobial signal molecules able to influence growth of legume root hairs (see chapter by Dénarié et al., this volume). Lerouge et al.[8] were the first to chemically identify a rhizobial *nod* gene-related signal molecule. This molecule, produced by *R. meliloti* and called NodRm-1, is a sulfated lipo-oligosaccharide that induces root hair deformation and branching (new growth areas) as well as nodule primordium initiation (phenotype Noi) in the host plant alfalfa (see chapter by Dénarié et al., this volume). According to Lerouge

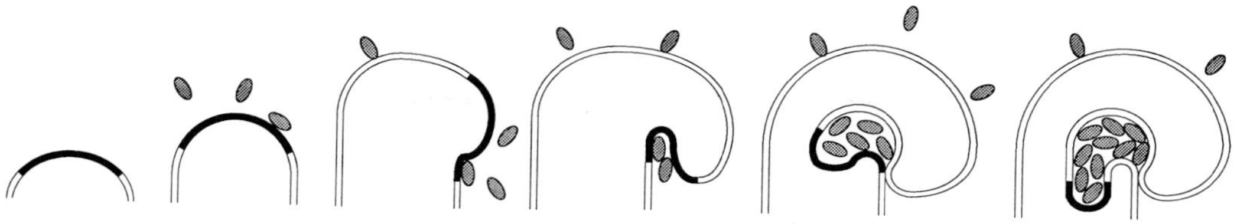

FIGURE 2. Successive steps in legume root hair infection by rhizobia. The tip growth area of the plant cell is represented in black (based on ref. 18).

et al.,[64] the heterologous host plant *Vicia* (vetch) does not respond to this molecule unless the sulfate group is absent (NodRm-1 without the sulfate group is called NodRm-2, and is produced by *R. meliloti* mutants affected in the host-range gene *nodH* [see Dénarié et al., this volume]). This shows that NodRm-1 (1) is able to induce/redirect tip growth in host plant cells, (2) induces a host plant-specific response; and (3) contains a structure (i.e., a sulfate group) that prevents functional recognition by heterologous plants. According to the model of Van Batenburg et al., attachment of *R. meliloti* and production of NodRm-1 may be sufficient for induction of root hair curling in alfalfa. Similar signal molecules have been found for *R. leguminosarum* bv. *viciae* and bv. *trifolii*.[9,10] The involvement of host-range *nod* genes in signal production, as shown for example with *R. meliloti,* explains host plant specificity of Hac.

Induction of *nod* genes by proper flavonoids is a prerequisite for production of substantial amounts of *nod* gene-related signals by rhizobia (see chapter by Kondorosi et al., this volume). However, a basic level of *nod* gene expression may be sufficient for induction of hair curling. Spaink et al.[65] found that *R. leguminosarum* bv. *trifolii nodD* mutants carrying a heterologous *nodD* gene are able to induce hair curling in *Trifolium pratense* (red clover). Since flavonoids from red clover induce *nod* genes from *R. leguminosarum* bv. *trifolii* only in concert with the homologous *nodD* gene,[65,66] flavonoid-induced derepression of *nod* genes may be necessary for infection thread initiation whereas the basic expression level may already be sufficient for root hair curling.

### C. INFECTION THREAD INITIATION

Attachment, production of a tip growth signal, and *nod* gene induction are required but not sufficient for infection thread initiation. *Rhizobium leguminosarum* bv. *viciae* is able to attach to white clover root hairs,[27] induces supercurling, and experiences *nod* gene induction by white clover exudate,[66] but is unable to initiate infection thread formation. Similarly, *R. leguminosarum* bv. *trifolii* is able to attach to pea root hairs,[1] and induces supercurling but is unable to infect host plants of *R. leguminosarum* bv. *viciae*. According to the model of Van Batenburg et al., supercurling results from the inability of the rhizobia to generate an infection focus. The residual tip growth activity of the root hair is responsible for this extensive curling. The difference in host plant specificity between the *R. leguminosarum* bv. *viciae* and bv. *trifolii* is primarily determined by the host-range gene *nodE*.[67] This gene is involved in production of a signal molecule similar, but not identical, to NodRm-1.[9] Interestingly, *nodE* mutants of these biovars induce supercurling in root hairs of their homologous hosts vetch and white clover, respectively,[68] similar to the response to a *R. leguminosarum* strain lacking host-range *nod* genes.[62] Apparently, *nod* gene-related signals produced in the absence of a functional *nodE* gene or in presence of a heterologous *nodE* gene are able to induce supercurling but are not able to contribute to attraction of all tip growth activity into an infection focus, i.e., unable to induce tight curling. Consequently, the root hair continues the curling process and cell infection is not initiated. These observations suggest that in nodulation of vetch and white clover *nodE* is involved in the generation of an infection focus, presumably by its role in the modification of a lipo-oligosaccharide signal molecule.[9]

In addition to functional host plant-specific signals, surface-located polysaccharides of rhizobia also play a role in root hair infection. The evidence primarily results from the study of the nodulation behavior of polysaccharide mutants (see chapter by Gray et al., this volume). *R. leguminosarum* produces a high-molecular weight EPS that is more or less soluble and either can be recovered from the growth medium or can be washed off from the cells.[52,69-71] Several EPS-deficient mutants (*exo* mutants) of the *R. leguminosarum* bv. *viciae* and bv. *trifolii* do not nodulate their host plants because of abortive infection thread formation.[71-75] Similarly, *exo* mutants of *R. meliloti* induce abortive infection threads in alfalfa,[76,77] in spite of being able to produce host plant-specific bioactive signal molecules like NodRm-1.[8] Since both *R. leguminosarum* and *R. meliloti* infect host root hairs via broad infection threads containing an infection thread matrix (for a review, see Kijne[78]), EPS may play a role in the synthesis of the thread matrix. By its putative hydrophilic spongelike nature, this matrix may function in attracting water for rhizobial cell proliferation and/or in generation of pressure for infection thread growth. Since aborted infection threads must have been initiated, EPS of these mutants apparently is not essential for generation of an infection focus.

Many genes control EPS biosynthesis and production (see chapter by Gray et al., this volume), and not all *exo* mutants of *R. leguminosarum* and *R. meliloti* yield the aborted Inf phenotype. Some of them are normally nodulating[71,79] suggesting that production of a large amount of EPS under laboratory conditions is not a prerequisite for successful root infection. Other *exo* mutants are unable to induce root hair curling and are noninfective.[71] An interesting example of the latter class of mutants is *R. leguminosarum* bv. *viciae* RBL5515 *exo4*::Tn5 pRL1JI (in short, RBL5515 *exo4*).[71] In this mutant, Tn5 is located in the middle of the open reading frame (ORF) of gene *pssA*, which is involved in EPS synthesis.[74,80] RBL5515 *exo4* carrying a *R. leguminosarum* bv. *viciae* Sym plasmid is $Hac^-$ on vetch but still able to induce root hair deformations. However, if this plasmid is replaced by the *R. leguminosarum* bv. *trifolii* Sym plasmid pSym5, this mutant is able to normally nodulate white clover.[71] Analysis of the low amount of EPS produced by RBL5515 *exo4* revealed its structure to be similar to that of the parental strain but lacking acetyl groups esterified to the glucosyl residue carrying the EPS side chain. These observations suggest that the structure rather than the amount of EPS is relevant for root hair infection, and point at a possible role of chromosomally encoded EPS substitutions in determination of host plant-specific nodulation by *R. leguminosarum* biovars. This suggestion is consistent with the results of Philip-Hollingsworth et al.[81] showing different glycosyl substitution patterns for EPSs of *R. leguminosarum* bv. *viciae* and bv. *trifolii* strains. At the species level, differences in (brady)-rhizobial EPS structure are more obvious.[70] Whether *nod* genes play a role in the modification of the structure of EPS is still a matter of debate.[52,71] Furthermore, production of functional *nod* gene-related signals by RBL5515 *exo4* has yet to be tested.

Interestingly, addition of $Exo^+$ $Nod^-$ helper strains[82] or purified homologous EPS[83,84] can restore the nodulation ability of some *exo* mutants. This suggests a possible signal function for EPS in infection initiation rather than a structural role during infection thread formation. When testing this hypothesis, one should be aware of possible contamination of EPS with low-molecular weight signal molecules.[85]

Acidic EPS does not seem to be essential for infection of bean and soybean.[74,75,86-89] Instead, complete LPS (LPSI) appears to be required for infection thread formation in these plants.[90,91] This difference is probably correlated with the different type of infection threads (narrow threads without matrix) induced by these rhizobia.[78,92]

## D. LEGUME LECTIN

Root lectin was found to be also involved in root hair infection by rhizobia. In pea, the pattern of location of Psl on the surface of the root entirely corresponds with the susceptibility of root epidermal cells to infection by *R. leguminosarum* bv. *viciae*.[43,45] Root hair-forming cells represent the majority of Psl-positive epidermal cells. Psl is encoded by a small gene family of which only one gene appears to be functional.[93,94] Introduction of the *psl* gene into white clover (hairy) roots enabled these roots to be infected and nodulated by *R. leguminosarum* bv. *viciae,* although nodulation was limited and delayed in time.[46] Specific ligand localization showed that in these transgenic roots Psl is present at sites similar to those on pea roots.[43,95] Since *R. leguminosarum* bv. *viciae* induces supercurling in normal white clover roots, these results indicate that lectin contributes to root hair infection at the level of formation of an infection focus. Such observations, together with the fact that *R. leguminosarum* bv. *viciae* does not bind the clover lectin trifoliin A,[49] also indicate that lectin is not necessarily involved in attachment and induction of supercurling.

The clover host specificity barrier for *R. leguminosarum* bv. *viciae* can also be broken by exchanging the *R. leguminosarum* bv. *viciae nodE* gene for that of *R. leguminosarum* bv. *trifolii*. In this case, tight curling and normal nodulation result.[67] Since *nodE* is involved in synthesis of an extracellular low-molecular weight lipo-oligosaccharide signal molecule,[9] it is tempting to suggest that lectin is involved in reception of rhizobial infection signals. However, it cannot be excluded that the NodE protein has an additional role, e.g., in modification of other rhizobial saccharides.

In clover, lectin is probably also involved in root infection. Trifoliin A is present at clover root hair tips (e.g., Reference 96), and both capsular polysaccharides (CPS) and LPS of *R. leguminosarum* bv. *trifolii* bind at these sites.[28,97] Addition of microgram quantities of these polysaccharides to clover roots results in significant enhancement of root hair infections by *R. leguminosarum* bv. *trifolii,* in contrast to addition of certain related non-trifoliin A-binding polysaccharides.[28,98] A similar biological activity is displayed by trifoliin A-binding oligosaccharide fragments of *R. leguminosarum* bv. *trifolii* EPS and CPS. *In vivo,* such fragments may be generated by enzymes present in clover root exudate.[99,100]

Soybean roots produce and secrete a lectin similar, but probably not identical, to the soybean seed lectin SBL.[101] Pretreatment of *B. japonicum* cells with SBL or the related root lectin specifically enhances infection and nodulation of soybean by low inocula of wild-type bacteria or by normal inocula of a mutant strain exhibiting a delayed-nodulation phenotype.[102-104] Since this mutant strain shows normal attachment and hair curling ability, the lectin apparently induces a response in bradyrhizobia that results in enhancement of an infection step beyond curling, i.e., infection thread initiation. The molecular mechanism of this effect is unknown. The possibility that such responses also play a role in other rhizobia-legume interactions is still open (e.g., Mody and Modi[105]).

Taken together, these observations strongly suggest that legume root lectin plays a role in root hair infection, by either recognizing rhizobial lipo-oligosaccharide signals, (fragments of) surface polysaccharides or their precursor molecules. Either the sugar-binding sites or the hydrophobic pockets present in lectin molecules may be involved. The relationship between lectin (ligands) and plant cell tip growth is still unclear.

## IV. CONCLUSIONS

Various molecules, including rhicadhesin, cellulose fibrils, fimbriae, lipo-oligosaccharides, EPS, and LPS at the rhizobial side, and flavonoids and lectin at the host plant's side (see above) are now known to participate in early interactions of some, if not all, rhizobia-legume associations. The role of chromosomal and Sym plasmid-encoded *nod* genes in synthesis of rhizobial extracellular molecules is presently the subject of considerable research efforts. The study of rhizobial attachment has shown that growth conditions of the bacteria are of prime importance in determining the molecular mechanism of the attachment process (see above). Growth conditions may also influence the synthesis of molecules involved in root hair curling and infection thread formation. For these molecules, an essential site of action is a root hair curl, a poorly characterized micro environment. Characterization of rhizobial growth conditions at, on, and in legume root hairs (e.g., pH, nutrient concentrations, $pO_2$) will be necessary to put genetical and biochemical research on rhizobial infectivity into its proper perspective. At the plant's side, several putative components obviously relevant for root hair infection still await characterization, e.g., common receptor molecules for rhicadhesin and specific receptor molecules for lipo-oligosaccharides. Moreover, legume mutants are required for testing the roles of flavonoids and lectins in various nodulation steps.

## REFERENCES

1. **Smit, G., Kijne, J. W., and Lugtenberg, B. J. J.,** Correlation between extracellular fibrils and attachment of *Rhizobium leguminosarum* to pea root hair tips, *J. Bacteriol.,* 168, 821, 1986.
2. **Turgeon, B. G. and Bauer, W. D.,** Ultrastructure of infection thread development during infection of soybean by *Rhizobium japonicum, Planta,* 163, 328, 1985.
3. **Chandler, M. R.,** Some observations on infection of *Arachis hypogaea* L. by *Rhizobium, J. Exp. Bot.,* 29, 749, 1978.
4. **Redmond, J. W., Batley, M., Djordjevic, M. A., Innes, R. W., Kuempel, P. L., and Rolfe, B. G.,** Flavones induce expression of nodulation genes in *Rhizobium, Nature (London),* 323, 632, 1986.
5. **Firmin, J. L., Wilson, K. E., Rossen, L., and Johnston, A. W. B.,** Flavonoid activation of nodulation genes in *Rhizobium* reversed by other compounds present in plants, *Nature (London),* 324, 90, 1986.
6. **Peters, N. K., Frost, J. W., and Long, S. R.,** A plant flavone, luteolin, induces expression of *Rhizobium meliloti* nodulation genes, *Science,* 233, 917, 1986.
7. **Wijffelman, C., Zaat, B., Spaink, H., Mulders, I., Van Brussel, T., Okker, R., Pees, E., De Maagd, R., and Lugtenberg, B.,** Induction of *Rhizobium nod* genes by flavonoids: differential adaptation of promoter, *nodD* gene and inducers for various cross-inoculation groups, in *Recognition in Microbe-Plant Symbiotic and Pathogenic Interactions,* NATO ASI Ser., Vol. H4, Lugtenberg, B., Ed., Springer-Verlag, Berlin, 1986, 123.
8. **Lerouge, P., Roche, P., Faucher, C., Maillet, F., Truchet, G., Promé, J. C., and Dénarié, J.,** Symbiotic host-specificity of *Rhizobium meliloti* is determined by a sulfated and acetylated glucosamine oligosaccharide signal, *Nature (London),* 344, 781, 1990.
9. **Spaink, H. P., Geiger, O., Sheeley, D. M., Van Brussel, A. A. N., York, W. S., Reinhold, V. N., Lugtenberg, B. J. J., and Kennedy, E. P.,** The biochemical function of the *Rhizobium leguminosarum* proteins involved in the production of host specific signal molecules, in *Advances in Molecular Genetics of Plant-Microbe Interactions, Vol. 1,* Hennecke, H. and Verma, D. P. S., Eds., Dordrecht, The Netherlands, 142, 1991.
10. **Van Brussel, A. A. N., Recourt, K., Pees, E., Spaink, H. P., Tak, T., Wijffelman, C. A., Kijne, J. W., and Lugtenberg, B. J. J.,** An extracellular biovar-specific signal of *Rhizobium leguminosarum* biovar *viciae* induces increased *nod* gene-inducing activity in root exudate of *Vicia sativa* ssp. *nigra, J. Bacteriol.,* 172, 5394, 1990.
11. **Dazzo, F. B. and Hubbell, D. H.,** Cross-reactive antigens and lectins as determinants of symbiotic specificity in the *Rhizobium*-clover association, *Appl. Microbiol.,* 30, 1018, 1975.
12. **Smit, G., Logman, T. J. J., Boerrigter, M. E. T. I., Kijne, J. W., and Lugtenberg, B. J. J.,** Purification and partial characterization of the $Ca^{2+}$-dependent adhesin from *Rhizobium leguminosarum* biovar *viciae,* which mediates the first step in attachment of Rhizobiaceae cells to plant root hair tips, *J. Bacteriol.,* 171, 4054, 1989.
13. **Smit, G., Kijne, J. W., and Lugtenberg, B. J. J.,** Both cellulose fibrils and a $Ca^{2+}$-dependent adhesin are involved in the attachment of *Rhizobium leguminosarum* to pea root hair tips, *J. Bacteriol.,* 169, 4294, 1987.
14. **Vesper, S. J. and Bauer, W. D.,** Role of pili (fimbriae) in attachment of *Bradyrhizobium japonicum* to soybean roots, *Appl. Environ. Microbiol.,* 52, 134, 1986.
15. **Vesper, S. J., Malik, N. S. A., and Bauer, W. D.,** Transposon mutants of *Bradyrhizobium japonicum* altered in attachment to host roots, *Appl. Environ. Microbiol.,* 53, 1959, 1987.

16. **Kijne, J. W., Smit, G., Díaz, C. L., and Lugtenberg, B. J. J.,** Lectin-enhanced accumulation of manganese-limited *Rhizobium leguminosarum* cells on pea root hair tips, *J. Bacteriol.,* 170, 2994, 1988.
17. **Stacey, G., Paau, A. S., and Brill, W. J.,** Host recognition in the *Rhizobium*-soybean symbiosis, *Plant Physiol.,* 66, 609, 1980.
18. **Van Batenburg, F. H. D., Jonker, R., and Kijne, J. W.,** *Rhizobium* induces marked root hair curling by redirection of tip growth, a computer simulation, *Physiol. Plant.,* 66, 476, 1986.
19. **Smit, G.,** Adhesins from *Rhizobiaceae* and Their Role in Plant-Bacterium Interactions, Ph.D. Thesis, Leiden University, The Netherlands, 1988.
20. **Smit, G., Kijne, J. W., and Lugtenberg, B. J. J.,** Roles of flagella, lipopolysaccharide and a $Ca^{2+}$-dependent cell surface protein in attachment of *Rhizobium leguminosarum* to pea root hair tips, *J. Bacteriol.,* 171, 569, 1989.
21. **Kijne, J. W., Smit, G., Díaz, C. L., and Lugtenberg, B. J. J.,** Attachment of *Rhizobium leguminosarum* to pea root hair tips, in *Recognition in Microbe-Plant Symbiotic and Pathogenic Interactions,* NATO ASI Ser. Vol. H4, Lugtenberg, B., Ed., Springer-Verlag, Berlin, 1986, 101.
22. **Boulter, D., Jeremy, J. J., and Wilding, M.,** Amino acids liberated into the culture medium by pea seedling roots, *Plant Soil,* 24, 121, 1966.
23. **Lipton, D. S., Blanchar, R. W., and Blevins, D. G.,** Citrate, malate, and succinate concentrations in exudates from P-sufficient and P-stressed *Medicago sativa* L. seedlings, *Plant Physiol.,* 85, 315, 1987.
24. **Bohlool, B. B. and Schmidt, E. L.,** Lectins: a possible basis for specificity in the *Rhizobium*-legume root nodule symbiosis, *Science,* 185, 269, 1974.
25. **Van Driessche, E.,** Structure and function of leguminous lectins, in *Advances in Lectin Research,* Vol. 1, Franz, H., Ed., Springer-Verlag, Berlin, 1988, 73.
26. **Dazzo, F. B.,** Bacterial attachment as related to cellular recognition in the *Rhizobium*-legume symbiosis, *J. Supramol. Struct. Cell. Biochem.,* 16, 29, 1981.
27. **Dazzo, F. B., Napoli, C. A., and Hubbell, D.,** Adsorption of bacteria to roots as related to host specificity in the *Rhizobium*-clover symbiosis, *Appl. Environ. Microbiol.,* 32, 166, 1976.
28. **Dazzo, F. B., Truchet, G. L., Sherwood, J. E., Hrabak, E. M., Abe, M., and Pankratz, S. H.,** Specific phases of root hair attachment in the *Rhizobium trifolii* clover symbiosis, *Appl. Environ. Microbiol.,* 48, 1140, 1984.
29. **Zurkowski, W.,** Specific adsorption of bacteria to clover root hairs, related to the presence of plasmid pWZ2 in cells of *Rhizobium trifolii, Microbios,* 27, 27, 1980.
30. **Dazzo, F. B. and Brill, W. J.,** Bacterial polysaccharide which binds *Rhizobium trifolii* to clover root hairs, *J. Bacteriol.,* 137, 1362, 1979.
31. **Dazzo, F. B., Urbano, M. R., and Brill, W. J.,** Transient appearance of lectin receptors on *Rhizobium trifolii, Curr. Microbiol.,* 2, 15, 1979.
32. **Mort, A. J. and Bauer, W. D.,** Composition of the capsular and extracellular polysaccharides of *Rhizobium japonicum:* changes with culture age and correlation with binding of soybean seed lectin to the bacteria, *Plant Physiol.,* 66, 158, 1980.
33. **Kato, G., Maruyama, Y., and Nakamura, M.,** Role of bacterial polysaccharides in the adsorption process of the *Rhizobium*-pea symbiosis, *Agric. Biol. Chem.,* 44, 2843, 1980.
34. **Badenoch-Jones, J., Flanders, D. J., and Rolfe, B. G.,** Association of *Rhizobium* strains with roots of *Trifolium repens, Appl. Environ. Microbiol.,* 49, 1511, 1985.
35. **Mills, K. M. and Bauer, W. D.,** *Rhizobium* attachment to clover roots, *J. Cell Sci. Suppl.,* 2, 333, 1985.
36. **Pueppke, S. G.,** Adsorption of slow- and fast-growing rhizobia to soybean and cowpea roots, *Plant Physiol.,* 75, 924, 1984.
37. **Menzel, G., Uhig, H., and Weischael, G.,** Settling of rhizobia and other soil bacteria on the roots of some legumes and non-legumes, *Zentralbl. Bakteriol. Parasitenk. Infektionskr. Hyg.,* 127, 348, 1972.
38. **Mort, A. J. and Bauer, W. D.,** Structure of the capsular and extracellular polysaccharides of *Rhizobium japonicum* that bind soybean lectin. Application of two new methods for cleavage of polysaccharides into specific oligosaccharide fragments, *J. Biol. Chem.,* 257, 1870, 1981.
39. **Smit, G., Tubbing, D. M. J., Kijne, J. W., and Lugtenberg, B. J. J.,** Role of $Ca^{2+}$ in the activity of rhicadhesin from *Rhizobium leguminosarum* biovar *viciae,* which mediates the first step in attachment of *Rhizobiaceae* cells to plant root hair tips, *Arch. Microbiol.,* 155, 278, 1991.
40. **Douglas, C., Halperin, W., and Nester, E. W.,** *Agrobacterium tumefaciens* affected in attachment to plant cells, *J. Bacteriol.,* 152, 1265, 1982.
41. **Douglas, C. J., Staneloni, R. J., Rubin, R. A., and Nester, E. W.,** Identification and genetic analysis of an *Agrobacterium tumefaciens* chromosomal virulence region, *J. Bacteriol.,* 161, 850, 1985.
42. **Lugtenberg, B. J. J., Smit, G., Díaz, C. L., and Kijne, J. W.,** Role of attachment of *Rhizobium leguminosarum* cells to pea root hair tips in targeting signals for early symbiotic steps, in *Molecular Signals in Microbe-Plant Symbiotic and Pathogenic Systems,* NATO ASI Ser. Vol. H36, Lugtenberg, B. J. J., Ed., Springer-Verlag, Berlin, 1989, 129.
43. **Díaz, C. L.,** Root Lectin as a Determinant of Host-Plant Specificity in the *Rhizobium*-Legume Symbiosis, Ph.D. Thesis, Leiden University, The Netherlands, 1989.

44. **Díaz, C. L., Lems-Van Kan, P., Van der Schaal, I. A. M., and Kijne, J. W.,** Determination of pea (*Pisum sativum* L.) root lectin using an enzyme-linked immunoassay, *Planta,* 161, 302, 1984.
45. **Díaz, C. L., Van Spronsen, P. C., Bakhuizen, R., Logman, G. J. J., Lugtenberg, E. J. J., and Kijne, J. W.,** Correlation between infection by *Rhizobium leguminosarum* and lectin on the surface of *Pisum sativum* L. roots, *Planta,* 168, 350, 1986.
46. **Díaz, C. L., Melchers, L. S., Hooykaas, P. J. J., Lugtenberg, B. J. J., and Kijne, J. W.,** Root lectin as a determinant of host specificity in the *Rhizobium*-legume symbiosis, *Nature (London),* 338, 579, 1989.
47. **Van der Schaal, I. A. M., Kijne, J. W., Díaz, C. L., and Van Iren, F.,** Pea lectin binding by *Rhizobium,* in *Lectins, Biology, Biochemistry, Clinical Biochemistry,* Vol. 3, Bøg-Hansen, T. C. and Spengler, G. A., Eds., Walter de Gruyter, Berlin, 1983, 531.
48. **Kamberger, W.,** An ouchterlony double diffusion study on the interaction between legume lectins and rhizobial cell surface antigens, *Arch. Microbiol.,* 121, 83, 1979.
49. **Dazzo, F. B., Yanke, W. E., and Brill, W. J.,** Trifoliin: a *Rhizobium* recognition protein from white clover, *Biochim. Biophys. Acta,* 539, 276, 1978.
50. **Dazzo, F. B. and Hrabak, E. M.,** Presence of trifoliin A, a *Rhizobium*-binding lectin, in clover root exudate, *J. Supramol. Struct. Cell. Biochem.,* 16, 133, 1981.
51. **Sherwood, J. E., Truchet, G. L., and Dazzo, F. B.,** Effect of nitrate supply on the *in vivo* synthesis and distribution of trifoliin A, a *Rhizobium trifolii*-binding lectin, in *Trifolium repens* seedlings, *Planta,* 162, 540, 1984.
52. **Philip-Hollingsworth, S., Hollingsworth, R. I., Dazzo, F. B., Djordjevic, M. A., and Rolfe, B. G.,** The effect of interspecies transfer of *Rhizobium* host-specific nodulation genes on acidic polysaccharide structure and *in situ* binding by host lectin, *J. Biol. Chem.,* 264, 5710, 1989.
53. **Matthysse, A. G.,** Role of bacterial cellulose fibrils in *Agrobacterium tumefaciens* infections, *J. Bacteriol.,* 154, 906, 1983.
54. **Ho, S-C., Wang, J. L., and Schindler, M.,** Carbohydrate binding activities of *Bradyrhizobium japonicum.* I. Saccharide specific inhibition of homotypic and heterotypic adhesion, *J. Cell Biol.,* 111, 1631, 1990.
55. **Ho, S-C., Schindler, M., and Wang, J. L.,** Carbohydrate binding activities of *Bradyrhizobium japonicum.* II. Isolation and characterization of a galactose-specific lectin, *J. Cell Biol.,* 111, 1639, 1990.
56. **Vesper, S. J. and Bhuvaneswari, T. V.,** Nodulation of soybean roots by an isolate of *Bradyrhizobium japonicum* with reduced firm attachment capability, *Arch. Microbiol.,* 150, 15, 1988.
57. **Ho, S-C.,** personal communication, 1990.
58. **Irvin, R. T.,** Hydrophobicity of proteins and bacterial fimbriae, in *Microbial Cell Surface Hydrophobicity,* Doyle, R. J. and Rosenberg, M., Eds., American Society Microbiology, Washington, DC, 1990, 137.
59. **Haack, A.,** Über den Einfluss der Knöllchenbakterien auf die Wurzelhaare von Leguminosen und Nichtleguminosen, *Zentrabl. Bakteriol. Parasitenk. Infektionskr. Hyg. Abt. II,* 117, 343, 1964.
60. **Newcomb, W., Sipell, D., and Peterson, R. L.,** The early morphogenesis of *Glycine max* and *Pisum sativum* root nodules, *Can. J. Bot.,* 57, 2603, 1979.
61. **Yao, P. Y. and Vincent, J. M.,** Host specificity in the root hair curling factor of *Rhizobium* spp., *Aust. J. Biol. Sci.,* 22, 413, 1969.
62. **Van Brussel, A. A. N., Pees, E., Spaink, H. P., Tak, T., Wijffelman, C. A., Okker, R. J. H., Truchet, G., and Lugtenberg, B. J. J.,** Correlation between *Rhizobium leguminosarum nod* genes and nodulation phenotypes on *Vicia,* in *Nitrogen Fixation: Hundred Years After,* Bothe, H., de Bruijn, F., and Newton, W. E., Eds., Gustav Fischer, Stuttgart, 1988, 483.
63. **Van Brussel, A. A. N. and Wijffelman, C. A.,** personal communication, 1990.
64. **Lerouge, P., Roche, P., Promé, J-C., Faucher, C., Vasse, J., Maillet, F., Camut, S., de Billy, F., Barker, O. G., Dénaríe, J., and Truchet, G.,** *Rhizobium meliloti* nodulation genes specify the product of an alfalfa-specific sulphated lipo-oligo saccharide signal, in *Nitrogen Fixation: Achievements and Objectives,* Gresshoff, P. M., Roth, L. E., Stacey, G., and Newton, W. E., Eds., Chapman and Hall, New York, 1990, 177.
65. **Spaink, H. P., Okker, R. J. H., Wijffelman, C. A., Tak, T., Goosen-de Roo, L., Pees, E., Van Brussel, A. A. N., and Lugtenberg, B. J. J.,** Symbiotic properties of rhizobia containing a flavonoid-independent hybrid *nodD* product, *J. Bacteriol.,* 171, 4045, 1989.
66. **Spaink, H. P., Wijffelman, C. A., Pees, E., Okker, R. J. H., and Lugtenberg, B. J. J.,** *Rhizobium* nodulation gene *nodD* as a determinant of host specificity, *Nature (London),* 328, 337, 1987.
67. **Spaink, H. P., Weinman, J., Djordjevic, M. A., Wijffelman, C. A., Okker, R. H. J., and Lugtenberg, B. J. J.,** Genetic analysis and cellular localization of the *Rhizobium* host-specificity determining NodE protein, *EMBO J.,* 8, 2811, 1989.
68. **Van Brussel, A. A. N.,** personal communication, 1990.
69. **Robertsen, B. K., Aman, P., Darvill, A. G., McNeil, M., Albersheim, P.,** Host-symbiont interactions. V. The structure of acidic extracellular polysaccharides secreted by *Rhizobium leguminosarum* and *Rhizobium trifolii, Plant Physiol.,* 67, 389, 1981.
70. **Carlson, R. W.,** Surface chemistry, in *Ecology of Nitrogen Fixation,* Vol. 2. *Rhizobium,* Broughton, W., Ed., Oxford University Press, NY, 1982, 199.

71. **Canter Cremers, H. C. J.,** Role of Exopolysaccharide in Nodulation by *Rhizobium leguminosarum* biovar *viciae,* Ph.D. Thesis, Leiden University, The Netherlands, 1990.
72. **Napoli, C. and Albersheim, P.,** *Rhizobium leguminosarum* mutants incapable of normal extracellular polysaccharide production, *J. Bacteriol.,* 141, 1454, 1980.
73. **Chakravorty, A. K., Zurkowski, W., Shine, J., and Rolfe, B. G.,** Symbiotic nitrogen fixation: molecular cloning of *Rhizobium* genes involved in exopolysaccharide synthesis and effective nodulation, *Mol. Gen. Genet.,* 192, 459, 1982.
74. **Borthakur, D., Barber, C. E., Lamb, J. W., Daniels, M. J., Downie, J. A., and Johnston, A. W. B.,** A mutation that blocks exopolysaccharide synthesis prevents nodulation of peas by *Rhizobium leguminosarum* but not of beans by *Rhizobium phaseoli* and is corrected by cloned DNA from *Rhizobium* or the phytopathogen *Xanthomonas, Mol. Gen. Genet.,* 203, 320, 1986.
75. **Diebold, R. and Noel, K. D.,** *Rhizobium leguminosarum* exopolysaccharide mutants: biochemical and genetic analyses and symbiotic behavior on three hosts, *J. Bacteriol.,* 171, 4821, 1989.
76. **Leigh, J. A., Signer, E. R., and Walker, G. C.,** Exopolysaccharide deficient mutants of *Rhizobium meliloti* that form ineffective nodules, *Proc. Natl. Acad. Sci. U.S.A.,* 82, 6231, 1985.
77. **Leigh, J. A., Reed, J. W., Hanks, A. M., Hirsch, A. M., and Walker, G. C.,** *Rhizobium meliloti* mutants that fail to succinylate their calcofluor-binding exopolysaccharide are defective in nodule invasion, *Cell,* 51, 579, 1987.
78. **Kijne, J. W.,** The infection process, in *Biological Nitrogen Fixation,* Stacey, G., Burris, R. H., and Evans, H. J., Eds., Chapman and Hall, NY, in press, 1991.
79. **Sanders, R., Raleigh, E., and Signer, E.,** Lack of correlation between extracellular polysaccharide and nodulation ability in *Rhizobium, Nature (London),* 292, 148, 1981.
80. **Borthakur, D., Barker, R. F., Latchford, J. W., Rossen, L., and Johnston, A. W. B.,** Analysis of *pss* genes in *Rhizobium leguminosarum* required for exopolysaccharide synthesis and nodulation of peas: Their primary structure and their interaction with *psi* and other nodulation genes, *Mol. Gen. Genet.,* 213, 155, 1988.
81. **Philip-Hollingsworth, S., Hollingsworth, R. I., and Dazzo, F. B.,** Host-range related structural features of the acidic extracellular polysaccharides of *Rhizobium trifolii* and *Rhizobium leguminosarum, J. Biol. Chem.,* 264, 1461, 1989.
82. **Klein, S., Hirsch, A. M., Smith, C. A., and Signer, E. R.,** Interaction of *nod* and *exo* of *Rhizobium meliloti* in alfalfa nodulation, *Mol. Plant-Microbe Interact.,* 1, 94, 1988.
83. **Djordjevic, S. P., Chen, H., Batley, M., Redmond, J. W., and Rolfe, B. G.,** Nitrogen-fixing ability of exopolysaccharide synthesis mutants of *Rhizobium* sp. strain NGR234 and *Rhizobium trifolii* is restored by the addition of homologous exopolysaccharides, *J. Bacteriol.,* 169, 53, 1987.
84. **Pühler, A., Enenkel, B., Hilleman, A., Kapp, D., Keller, M., Muller, P., Niehaus, K., Priefer, U. B., Quandt, J., and Schmidt, C.,** *Rhizobium meliloti* and *Rhizobium leguminosarum* mutants defective in surface polysaccharide synthesis and root nodule development, in *Nitrogen Fixation: Hundred Years After,* Bothe, H., de Bruijn, F., and Newton, W. E., Eds., Gustav Fischer, Stuttgart, 1988, 483.
85. **Rolfe, B. G., Sargent, C. L., Weinman, J. J., Djordjevic, M. A., McIver, J., Redmond, J. W., Batley, M., Yuan, D. C., and Sutherland, M. W.,** Signal exchange between *Rhizobium trifolii* and clovers, in *Molecular Signals in Microbe-Plant Symbiotic and Pathogenic Systems,* NATO ASI Ser. Vol. H36, Lugtenberg, B. J. J., Ed., Springer-Verlag, Berlin, 1989, 129.
86. **Law, I. J., Yamamoto, Y., Mort, A. J., and Bauer, W. D.,** Nodulation of soybean by *Rhizobium japonicum* mutants with altered capsule synthesis, *Planta,* 154, 100, 1982.
87. **Kim, C-H., Tully, R. E., and Keister, D. L.,** Exopolysaccharide-deficient mutants of *Rhizobium fredii* HH303 which are symbiotically effective, *Appl. Environ. Microbiol.,* 55, 1852, 1989.
88. **Ko, Y. H. and Gayda, R.,** Nodule formation in soybeans by exopolysaccharide mutants of *Rhizobium fredii* USDA191, *J. Gen. Microbiol.,* 136, 105, 1990.
89. **O'Connell, K. P., Araujo, R. S., and Handelsman, J.,** Exopolysaccharide-deficient mutants of *Rhizobium* sp. strain CIAT899 induce chlorosis in common bean *(Phaseolus vulgaris), Mol. Plant-Microbe Interact.,* 3, 424, 1990.
90. **Noel, K. D., VandenBosch, K. A., and Kulpaca, B.,** Mutations in *Rhizobium phaseoli* that lead to arrested development of infection threads, *J. Bacteriol.,* 168, 1392, 1986.
91. **Stacey, G.,** personal communication, 1990.
92. **Rae, A. and N. J. Brewin,** personal communication, 1989.
93. **Kaminski, P. A., Buffard, D., and Strosberg, A. D.,** The pea lectin gene family contains only one functional lectin gene, *Plant Mol. Biol.,* 9, 497, 1987.
94. **De Pater, B. S.,** unpublished data, 1990.
95. **Díaz, C. L., Hooykaas, P. J. J., Lugtenberg, B. J. J., and Kijne, J. W.,** Functional expression of pea lectin on root hair tips of transgenic clover hairy roots, in *Nitrogen Fixation: Achievements and Objectives,* Gresshoff, P. M., Roth, L. E., Stacey, G., and Newton, W. E., Eds., Chapman and Hall, NY, 1990, 733.
96. **Truchet, G. L., Sherwood, J. E., Pankratz, H. S., and Dazzo, F. B.,** Clover root exudate contains a particulate form of the lectin trifoliin A which binds *Rhizobium trifolii, Physiol. Plant.,* 66, 575, 1986.

97. **Dazzo, F. B. and Brill, W. J.,** Bacterial polysaccharide which binds *Rhizobium trifolii* to clover root hairs, *J. Bacteriol.,* 137, 1362, 1979.
98. **Abe, M., Sherwood, J. E., Hollingsworth, R. I., and Dazzo, F. B.,** Stimulation of clover root hair infection by lectin-binding oligosaccharides from the capsular and extracellular polysaccharides of *Rhizobium trifolii, J. Bacteriol.,* 160, 517, 1984.
99. **Dazzo, F. B., Truchet, G. L., Sherwood, J. E., Hrabak, E. M., and Gardiol, A. E.,** Alteration of the trifoliin A-binding capsule of *Rhizobium trifolii* 0403 by enzymes released from clover roots, *Appl. Environ. Microbiol.,* 44, 478, 1982.
100. **Bhuvaneswari, T. V. and Solheim, B.,** Root hair deformations in white clover/*Rhizobium trifolii* symbiosis, *Physiol. Plant.,* 68, 1144, 1985.
101. **Vodkin, L. O. and Raikhel, N. V.,** Soybean lectin and related proteins in seeds and roots of $Le^+$ and $Le^-$ soybean varieties, *Plant Physiol.,* 81, 558, 1986.
102. **Stacey, G., Halverson, L. J., Nieuwkoop, T., Banfalvi, Z., Schell, M. G., Gerhold, D., Deshmane, N., So, J. S., and Sirotkin, K. M.,** Nodulation of soybean: *Bradyrhizobium japonicum* physiology and genetics, in *Recognition in Microbe-Plant Symbiotic and Pathogenic Interactions,* NATO ASI Ser. Vol. H4, Lugtenberg, B. J. J., Ed., Springer-Verlag, Berlin, 1986, 87.
103. **Halverson, L. J. and Stacey, G.,** Host recognition in the *Rhizobium*-soybean symbiosis. Evidence for the involvement of lectin in nodulation, *Plant Physiol.,* 77, 621, 1985.
104. **Halverson, L. J. and Stacey, G.,** Effect of lectin on nodulation by wild-type *Bradyrhizobium japonicum* and a nodulation-defective mutant, *Appl. Environ. Microbiol.,* 51, 753, 1986.
105. **Mody, B. and Modi, V. V.,** Characterization of peanut agglutinin from tissues of *Arachis hypogaea* and determination of a novel biological function of purified seed lectin in legume-*Rhizobium* interaction, *Ind. J. Exp. Biol.,* 27, 519, 1989.

# *RHIZOBIUM* NODULATION SIGNALS

**Jean Dénarié and Philippe Roche**

## TABLE OF CONTENTS

In addition to their agricultural importance, due to symbiotic nitrogen fixation, the associations between legumes and nodule bacteria provide good experimental systems to investigate several problems of great interest in plant sciences, such as (1) the specificity of recognition between plants and microbes, (2) the exchange of information between two symbiotic organisms to coordinate their development and functioning, and (3) the induction of a major plant developmental switch, the nodule organogenesis. Work during recent years has shown us that during the early steps of the symbiosis each partner influences the expression of specific genes of the other by an exchange of small signal molecules.[1,2] A. Kondorosi has reviewed in another chapter of this volume how leguminous hosts signal to their bacterial partners to regulate the expression of the rhizobial nodulation (*nod*) genes. In this chapter we will describe how *Rhizobium* signals back to the plant by producing lipo-oligosaccharidic nodulation factors. In addition to the various chapters of this volume dedicated to the *Rhizobium*-legume symbiosis, a number of reviews have recently been published on the molecular biology of the symbiosis,[1-4] the genetics of *Rhizobium*,[5,6] and the *Rhizobium* nodulation genes.[7-9] This chapter will therefore, not be a comprehensive review of the structure and function of *nod* genes but will detail our present knowledge about their role in the production of lipo-oligosaccharidic symbiotic signals. In addition, we will describe the plant reactions to these signals and discuss the role of these bacterial molecules in determining host specificity, infection, and nodulation.

# I. INTRODUCTORY REMARKS

## A. THE RHIZOBIA: A GREAT VARIETY OF NODULE BACTERIA

Bacteria which are able to elicit nodule formation on legumes have recently been classified into four genera, *Rhizobium, Bradyrhizobium, Sinorhizobium,* and *Azorhizobium,* on the basis of their host specificity and of various taxonomical criteria (Table 1).[3] However, symbiotic relationships from only a very few legumes have been studied so far, and as further symbiosis are examined other new species deserving the status of a genus are likely to be described.[3,10,11]

Systematic studies based on nucleic acid hybridization, have revealed that these four genera of Gram-negative nodule bacteria are very distant phylogenetically.[3] *Rhizobium, Bradyrhizobium* and *Azorhizobium* are much closer to nonsymbiotic relatives than they are to each other.[10,12,13] *Rhizobium* is closely related to *Agrobacterium* which is able to transform dicotyledonous plants and induce tumors, while *Rhodopseudomonas palestris,* a photosynthetic soil bacterium, is the closest known relative of *Bradyrhizobium. Azorhizobium* is most closely related to *Xanthobacter.*[10] The reason why these distant genera have been grouped in the same family, the Rhizobiaceae, is their unique property of establishing a nitrogen-fixing symbiosis with plants of the family Leguminosae. We will refer to these bacteria collectively as "rhizobia."

## B. NODULATION: A VARIETY OF INFECTION AND DEVELOPMENT PATTERNS

The Leguminosae is a very important family containing more than 15,000 species grouped in three subfamilies, the Caesalpinioideae, which is the most ancient group, and the more advanced Mimosoideae and Papilionoideae.[14] Legumes are very diverse in morphology, habit, and ecology, ranging from arctic annual herbaceous plants to tropical trees. Nevertheless the great majority of them are nodulated by rhizobia.[3] Several species of *Parasponia,* in the family Ulmaceae, form nitrogen-fixing nodules with a variety of *Rhizobium* and *Bradyrhizobium* strains most of which can also nodulate some legumes.[15] The Ulmaceae are no more closely related to the Leguminosae than almost any other family of dicotyledons.[16] In both legumes and *Parasponia,* nodules are formed generally on the roots,

**TABLE 1**
**Rhizobia-Plant Associations**

| Rhizobia | Plant |
|---|---|
| *Rhizobium meliloti* | Alfalfa |
| *Rhizobium leguminosarum* | |
| biovar *viciae* | Pea, vetch |
| biovar *trifolii* | Clover |
| biovar *phaseoli* | Bean |
| *Rhizobium* sp. NGR234 | Tropical legumes, *Parasponia* |
| *Rhizobium* loti | Lotus |
| *Bradyrhizobium japonicum* | Soybean |
| *Bradyrhizobium* "cowpea" | Tropical legumes |
| *Sinorhizobium fredii* | Soybean |
| *Azorhizobium caulinodans* | *Sesbania rostrata* |

however some aquatic legumes belonging to the genera *Aeschynomene* and *Sesbania* also exhibit stem nodulation.[17]

The routes of infection are varied and include simple entry through cracks in the epidermis as in the case of peanut or formation of walled infection threads through root hairs as with alfalfa, clover, pea, or vetch.[4] The mode of infection is a characteristic of the host since the same rhizobial strain can penetrate different host species by either root hair threads or epidermis crack entry, and a given species is infected by the same type of process whatever the strain.[3,4,18] Associated with the infection is the induction of cell divisions in the cortex and the formation of nodule primordia.[1,4] Cortical cell division is an early plant reaction and a first round of cell division can be seen in the cortex of alfalfa and soybean within 24 h after inoculation, that is, before infection can be observed in the form of marked hair curling.[19,20]

The nodules which develop from these primordia are genuine differentiated organs, not mere tumors or deformed roots. Nodules of a given plant species have a characteristic ontogeny, anatomy, morphology, and type of development, either determinate, as with soybean, or indeterminate (with a persistent meristem), as with alfalfa, pea, or vetch.[4,21] As for the mode of infection, the structural characteristics of a nitrogen-fixing nodule are specified by the host plant and not by the bacterial strain,[4,18] indicating that the plant possesses the genetic information for symbiotic infection and nodulation and that the role of the bacteria is to turn on the plant genetic program.

## C. A WIDE VARIETY OF HOST RANGES

The symbiosis between root-nodule bacteria and legumes is specific. However, the degree of specificity varies tremendously among rhizobia.[3,7,9,22] Some strains have a narrow host range, for example, *R. leguminosarum* biovar *trifolii* strains elicit nitrogen-fixing nodules only on species of *Trifolium* (clover) and *R. meliloti* strains only on species of the three genera *Medicago, Melilotus,* and *Trigonella.*[22] There are even strains that discriminate between genotypes within a legume species. For example, most isolates of *R. leguminosarum* bv. *viciae* nodulate European pea (*Pisum sativum*) varieties but not certain peas from Afghanistan that require special strains with an extended host range.[23] In contrast, some isolates can have a very broad host range. For example, *Rhizobium* sp. strain NGR234 is currently known to nodulate 35 different legume genera as well as the nonlegume *Parasponia.*[24] Some *Bradyrhizobium* strains also have a broad host range and may nodulate numerous tropical legumes from both the Papilionoideae and Mimosoideae subfamilies and the nonlegume *Parasponia.*[22]

Very frequently, strains have been reported to form effective nodules on one plant species (or genus) and ineffective ones on another, showing that specificity is not limited to nodulation but may also affect the late stages of nodule development and the establishment of a nitrogen-fixing symbiosis. In this chapter, however, we will only consider the specificity of the early steps of symbiosis, infection, and nodulation. In the following text, for the sake of brevity, *R. leguminosarum* bv. *viciae* and bv. *trifolii* will be called *R. leguminosarum* and *R. trifolii,* respectively.

## II. NODULATION GENES

To analyze the mechanisms by which root nodule bacteria determine host specificity, infection, and nodulation, bacterial genetics has proved to be essential. There are many recent reviews on the genetics of nodulation in *Rhizobium* and *Bradyrhizobium.*[1,5-8,25,26] Genetic and molecular studies have led to the identification of a number of nodulation (*nod*) genes which can be grouped in three categories, the regulatory, the common, and the host-specific *nod* genes.

### A. THE *NOD* REGULON

The regulation of *nod* genes in rhizobia is reviewed in a separate chapter of this volume by A. Kondorosi. We will simply summarize here some basic general features. Regulation is governed by *nodD* genes which have been found in all rhizobia studied so far and the *nodD* genes of *Rhizobium, Bradyrhizobium, Sinorhizobium* and *Azorhizobium* are homologous.[3,27,28] There are variations between species in the number of *nodD* copies. For example *R. leguminosarum* and *A. caulinodans* have only one *nodD* copy whose inactivation results in a complete loss of the ability to form nodules.[8,28] In contrast three *nodD* copies, and a *nodD*-like gene *syrM,* have been found in *R. meliloti* and the inactivation of a single copy causes only a delay in nodulation (Figure 1).[35,37,50,51] NodD proteins are positive regulators and activate the transcription of other *nod* operons. They bind upstream of the *nod* operons to promoter regions which contain extended conserved sequences the "*nod* boxes."[52-54] Activation of *nod* gene transcription by the NodD proteins requires the presence of plant signals, flavonoid or other phenolic compounds, which are present in legume root exudates.[1,55] Species-specific variations (1) in the composition of *nod*-inducing compounds in the root exudate on the plant side, and (2) in the NodD protein reactivity to such compounds on the bacterial side, provide a first level of specificity in the legume-rhizobia early interactions.[50,56,57]

Common and host-specific *nod* genes have been shown to be part of the same regulon in *R. leguminosarum* and *R. meliloti.*[34,41,53,54,58] They are under the control of the same regulatory components: *nod* boxes as cis-acting elements, NodD proteins as trans-acting elements, and phenolic compounds as inducers. The presence of a long highly conserved *nod* box upstream of known common and host-specific *nod* operons has suggested a strategy to identify new *nod* genes.[34,52] Synthetic *nod* box probes have been used to identify new *nod* boxes and thus presumptive new *nod* operons.[34,59] This strategy allowed the identification of new *nod* loci in *R. meliloti* (see n4, n5, and n6, Figure 1),[34] in *R. leguminosarum,*[8] *B. japonicum,* and *Rhizobium* sp. NGR234.[59,60] This method is particularly interesting for identifying host-specific *nod* genes which may have been overlooked by standard methods, such as the screening for symbiotic defects after a random insertional mutagenesis, because of the subtlety of the symbiotic behavior changes in the mutant strains.

Presently, in *R. leguminosarum* and *R. meliloti,* an empirical definition of a *nod* gene is based on the type of gene regulation observed (belonging to a *nod* regulon with *nod* box, *nodD* control, flavonoid induction) rather than on the symbiotic phenotype caused by mutations in this gene (alteration of infection and nodulation). For example, the infection and

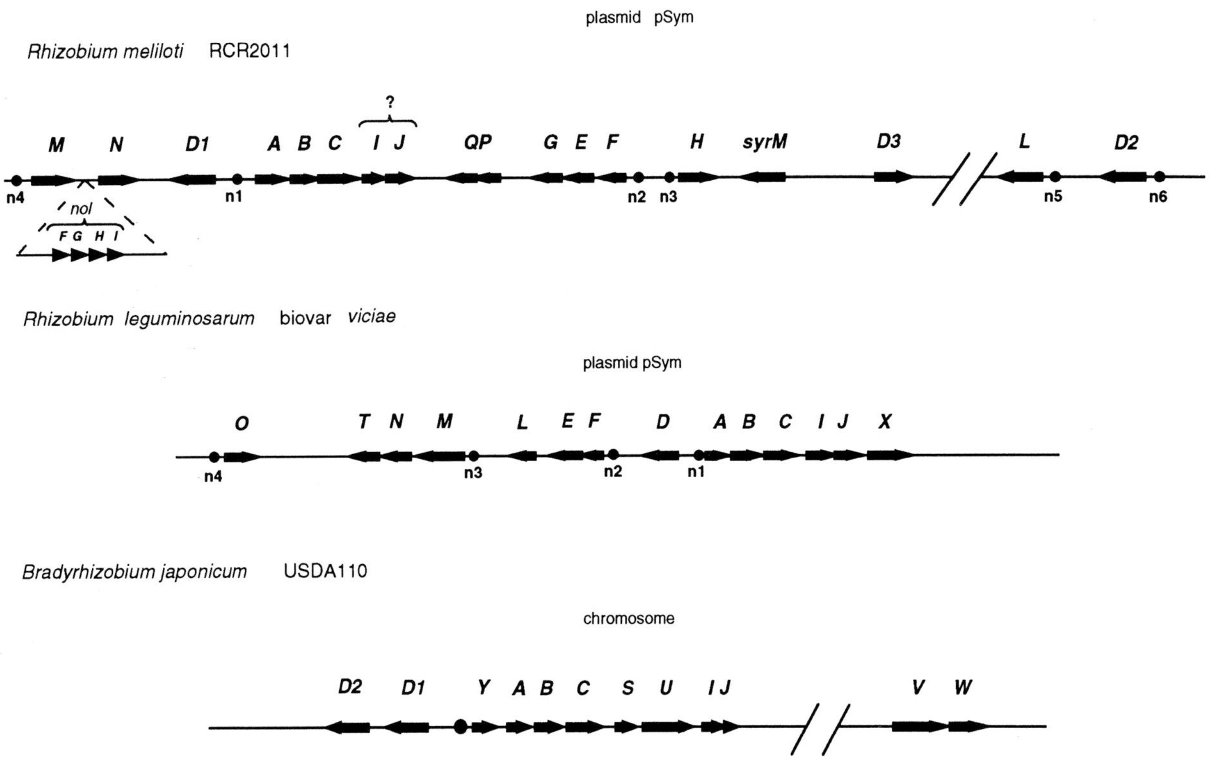


FIGURE 1. Genetic map of nodulation (*nod* or *nol*) genes of *R.meliloti, R.leguminosarum,* and *B.japonicum*. Genes with the same symbol have been shown to be homologous, by hybridization and sequence comparison, across species. Absence of a symbol does not necessarily mean that the gene is absent, as the maps are incomplete. *Rhizobium meliloti* refs., 29-38; *R. leguminosarum* refs., 39-46; *B. japonicum* refs., 47-49.

the nodule development are drastically altered in Exo$^-$ and Ndv$^-$ mutants but the altered genes are not considered as *nod* genes.[61,62] In contrast, mutants in some genes of the *nod* regulon may exhibit no detectable or only slight modifications in infection and nodulation.

The similarity of the basic regulatory components (*nodD* genes, *nod* boxes, phenolic inducers) in *Rhizobium* , *Bradyrhizobium,* and *Azorhizobium,* in spite of their phylogenetic diversity, strongly supports the hypothesis of a common origin of these *nod* genes and suggests possible lateral transfers.

## B. COMMON *NOD* GENES

The *nodABC* genes have been found in all *Rhizobium, Sinorhizobium,* and *Bradyrhizobium* isolates studied so far[6] and have also recently been found in *Azorhizobium caulinodans.*[63] These genes have been called common *nod* genes as they are structurally conserved and, in many cases, are functionally interchangeable between *Rhizobium* and *Bradyrhizobium* species.[6,26] In most species the *nodABC* genes are part of a single operon and are transcribed in this order (Figure 1). However, in *R. phaseoli* biotype I *nodA* is separated by around 20 kb from the *nodBC* genes and in *R. loti* the *nodB* gene is separated from the *nodACIJ* genes.[64,65]

Inactivation of *nodABC* genes abolishes the ability to elicit any symbiotic reaction in the plant, including root hair curling (Hac$^-$), cortical cell divisions, and nodule formation (Nod$^-$).[1] Mutants in *nodABC* genes are always Nod$^-$ regardless of (1) the host, (2) the mode of infection (crack-in or infection thread formation within root hairs), (3) the type of nodule development (determinate or indeterminate), and (4) the nodule location on the plant (root or stem). That three conserved bacterial gene products are controlling very different types of infection and nodulation on various hosts is quite intriguing and may suggest that the plant possesses the genetic program for rhizobial infection and nodulation and that the role of the *nodABC* genes is to trigger the plant program. What could be the function of the *nodABC* gene products? NodA and NodB proteins are cytosolic and their sequences do not show detectable homology with known prokaryotic proteins.[66,67] In contrast, NodC is located in the outer membrane and the proposed structure shows striking similarities with various eukaryotic cell-surface receptors.[67,68]

The *nodIJ* genes are present in *R. leguminosarum, R. trifolii,* and *B. japonicum,*[40,46,48] and partial sequence data suggest their presence in *R. meliloti* and *A. caulinodans.*[30,63] They are located downstream of *nodC* and seem to be part of the same operon as *nodC* (Figure 1). Mutations in *nodIJ* result in a nodulation delay with *R. leguminosarum,* but have no detectable effect with *B. japonicum.* The presence of the nodABCIJ genes in the genetically distant *Rhizobium, Bradyrhizobium,* and *Azorhizobium* isolates suggests a common origin for these genes.

## C. HOST-SPECIFIC *NOD* GENES

Nodulation genes have been identified that have not been found to be functionally or structurally conserved among rhizobia.[69] They have been defined as host-specific *nod* genes based on the following criteria: (1) Mutations in these genes cannot be fully complemented by the introduction of corresponding (and sometimes highly homologous) genes from other rhizobial species or biovars.[69] (2) Mutations result in an alteration or extension of the host range. For example, in *R. trifolii,* in contrast to wild-type strains, *nodFE*$^-$ mutants poorly nodulate white and red clovers but have acquired the ability to infect and nodulate peas.[70] In *R. meliloti, nodH*$^-$ mutants are Hac$^-$ Inf$^-$ Nod$^-$ on the homologous host alfalfa and Hac$^+$ Inf$^+$ Nod$^+$ on heterologous hosts, such as vetch species.[32,71,72] (3) These genes are necessary in interspecific crosses to modify or extend the host range of the recipient strain to include plants nodulated by the donor bacterium.[72-76]

Whereas mutations in the *nodABC* genes totally block nodulation, mutations in host specific *nod* genes do not abolish the ability to nodulate. Instead they cause a nodulation delay or a decrease in the number of nodules formed, alterations which can be associated to changes in the host range. For example, in *R. leguminosarum,* individual mutations in the host-range *nodF, E, L, M, N, O, T* genes do not block nodulation.[77,78] However, the introduction in a *Rhizobium* strain, cured of the pSym plasmid or having a deletion of the whole *nod* region, of a clone carrying only the regulatory *nodD* gene together with the common *nodABCIJ* genes does not allow nodulation.[78] These results indicate that although mutations in the individual host-specific *nod* genes cause leaky phenotypes, the presence of at least some of these genes is essential for nodulation.[77,78]

The variation in the host range which is mediated by host-specific *nod* genes, seems to rely on two types of genetic mechanisms, allelic variability within a gene and nonallelic variability by using different combinations of genes. The *nodFE, nodMN,* and *nodL* genes are common to *R. leguminosarum, R. trifolii,* and *R. meliloti* (Figure 1). In the case of *nodE,* it has been clearly shown that allelic variation is responsible for the host specificity. The *nodE* alleles present in *R. leguminosarum* and *R. trifolii,* respectively, are highly homologous. Nevertheless, this gene is a major determinant of the clear difference in host range between these two biovars,[75] and a region of the NodE protein has been localized which determines its host-specific properties.[75]

Various examples of nonallelic genes have been described which can be characteristic of bacterial species, biovars, or races within a biovar. Some host-specific genes can be present in a particular *Rhizobium* species and not in another one. For example, the *nodG, nodH,* and *nodPQ* genes have been identified in *R. meliloti* but not in *R. leguminosarum* and *R. trifolii* (Figure 1).[31,32,79] Reciprocally, the *nodT* gene present in *R. leguminosarum* and *R. trifolii* has not been identified in *R. meliloti.*[46] The *nodO* gene which has been found in the *R. leguminosarum* biovar *viciae* and not in the closely related biovar *trifolii,*[45] seems to be more specific. Of an even narrower specificity is the *nodX* gene that is present in a particular "race" of *R. leguminosarum* biovar *viciae* and confers the ability to nodulate Afghanistan lines of pea.[42]

Sequence comparisons have been fruitful and significant homologies have been found between some host-specific *nod* gene products and known proteins, suggesting enzymic functions for particular host-range proteins, such as acyl synthesis and transfer (NodFE), alcohol dehydrogenation (NodG), acetyl transfer (NodL), and amidophosphoribosyl transfer (NodM) (Table 2).[6,25] The NodQ protein was found to exhibit a strong homology with GTP-binding proteins.[33,36] Biochemical and immunological studies have allowed protein cellular localization and have shown that NodF is located in the cytoplasm, NodE in the cytoplasmic membrane whereas the *nodO* gene product is a secreted protein.[44,45,75,84] Sequence data suggest that NodT is located in the outer membrane.[46] Sequence homology studies have suggested enzymic functions for the host-specific *nod* genes, raising the question of whether these Nod enzymes are involved in the modification of surface macromolecular bacterial components or in the production of extracellular low-molecular weight signals.

## III. EXTRACELLULAR NOD SIGNALS

### A. PRELUDE

#### 1. Root Hair Curling and Branching Factors

Infection of temperate legumes by rhizobia generally occurs through root hairs and an early and easy detectable plant reaction is the appearance of a marked hair curling. For a very long time (at least nine decades!) the role of the bacteria in eliciting this plant reaction has been studied. At the turn of the century, Hiltner[85] reported that aqueous bacteria-free filtrates from mature *Pisum sativum* nodules contained a substance that induces root hair initiation (Hai) or deformation (Had). Among the deformations observed were curled and

**TABLE 2**
***nod* Gene Products of *Rhizobium***

| Genes | Species biovar[a] | Cellular localization[b] | Homologies[c] | Ref. |
|---|---|---|---|---|
| *nodA* | *Rl; Rt; Rm* | Cytoplasmic | | 29,39,66,67 |
| *nodB* | *Rl; Rt; Rm* | Cytoplasmic | | 29,39 |
| *nodC* | *Rl; Rt; Rm* | Outer Mb | Receptor | 29,30,39,67,68 |
| *nodD* | *Rl; Rt; Rm* | Cytoplasmic Mb | DNA-binding proteins | 35 |
| *nodE* | *Rl; Rt; Rm* | Cytoplasmic Mb | β-Ketoacyl synthases | 75 |
| *nodF* | *Rl; Rt; Rm* | Cytoplasmic | Acyl carrier proteins | 41 |
| *nodG* | *Rm* | | Alcohol dehydrogenases | 31 |
| *nodH* | *Rm* | | (Sulfotransferases) | 31,32,80 |
| *nodI* | *Rl; Rt; (Rm?)* | Cytoplasmic Mb | ATP-binding proteins | 40,46 |
| *nodJ* | *Rl; Rt; (Rm?)* | Cytoplasmic Mb | Hydrophobic domains | 40,46 |
| *nodL* | *Rl; Rt; Rm* | Cytoplasmic Mb | Acetyltransferases | 43,81 |
| *nodM* | *Rl; Rt; Rm* | | Amidophosphoribosyltransferases (D-glucosamine synthase) | 38,43,78,82 |
| *nodN* | *Rl; Rt; Rm* | | | 43 |
| *nodO* | *Rl* | Secreted | Hemolysin; $Ca^{2+}$-binding proteins | 44,45 |
| *nodP* | *Rm* | | (ATP-sulfurylase) | 31,36,83 |
| *nodQ* | *Rm* | | GTP-binding proteins (ATP-sulfurylase) | 31,36,83 |
| *nodT* | *Rl; Rt* | Outer Mb | Transit sequences | 46 |
| *nodX* | *Rl (Tom)* | | | 42 |

[a] Rhizobia species: *Rm, Rhizobium meliloti; Rl, Rhizobium leguminosarum* bv. *viciae; Rt, Rhizobium leguminosarum* bv. *trifolii*.
[b] Mb, membrane.
[c] After determination of the structure of the Nod factors of *R. meliloti*, *nod* sequences were reinvestigated. The new homologies are given in parentheses.

branched root hairs. Since that time the concept of an extracellular bacterial root hair-deforming factor has developed. Whereas the plant reaction is much more pronounced (giving a marked hair curling) in the presence of living bacteria, a hair-deforming activity can nevertheless be observed in sterile filtrates from *Rhizobium* cultures.[86-89] These observations led Yao and Vincent[90] to hypothesize that at least two bacterial factors are involved in root hair deformation: (1) a dialyzable heat-stable substance which is found in culture filtrates and causes branching and slight curling and (2) a nondialyzable substance which demands the close proximity of living invasive bacteria and is required for marked hair curling.

It has also been known for a long time that root hair curling is specific, the maximal effect being observed with the homologous *Rhizobium* bacteria and with degree of specificity being observed with the filtrates as well.[86,90,91] The production of the extracellular factor(s) is stimulated when bacteria are grown in the presence of root exudates.[92] It has been shown that the factor is not the plant hormone indoleacetic acid[93] but discrepancies exist between reports concerning the chemical nature of the curling and branching factors: they were proposed to be a protein,[88] aromatic compounds,[94] or oligosaccharides.[92] These discrepancies could be due to the excretion of various *Rhizobium* metabolites which are able to elicit hair deformations in growth media. Interdisciplinary approaches including bacterial genetics have since addressed the following questions. Are the *nod* genes which control root hair curling, involved in the synthesis of the nondialyzable surface components or of the extracellular factors proposed by Yao and Vincent?[90] If many extracellular root-deforming factors are present, which ones are under the control of *nod* genes?

### 2. Nodule Organogenesis-Inducing Principle

Cytological studies of early steps of infection have shown that rhizobia can elicit at a distance mitotic cell divisions in the inner cortex. Induction of the first rounds of cell division within the cortex of alfalfa or soybean can be observed before the detection of marked hair curling and the initiation of infection thread formation within the root hairs.[19,20] At a later stage, when infection threads are growing through the root hairs and the outer cortex, foci of cell divisions can be seen at a distance from the thread tips.[95]

*Rhizobium meliloti* mutants, auxotrophic for leucine, remain sequestered within the infection threads and cannot be released into the plant cell cytoplasm.[95] Nevertheless, these mutants elicit the formation of nodules with a normal anatomy but having a ''bacteria-free'' central tissue. These observations led to the hypothesis that rhizobia can elicit nodulation at a distance from the target cortical cells by a ''nodule organogenesis-inducing principle'' (NOIP) which can diffuse through the plant cell wall.[95] Other classes of bacterial mutants able to induce the formation of ''empty nodules'' have been reported, such as *Agrobacterium* transconjugants carrying the *R. meliloti nod* genes or *R. meliloti* mutants defective in exopolysaccharide (*exo*) or β-1,2 glucan (*ndv*) synthesis.[61,62] These bacteria, which do not form infection threads and multiply at a reduced level within intercellular spaces of the outer root cortex, nevertheless, elicit the formation of nodules. All these results indicate that rhizobia are signaling to target plant cells from a distance.

More direct evidence that bacteria are signaling to the plant via diffusible molecules was provided by the following experiments. *Bradyrhizobium japonicum* and *R. meliloti* elicit the first rounds of cell division even when the bacteria are separated from the host root by a filter membrane.[19] The mitogenic factors are specific since this induction of cell division is observed only on the homologous host.[19] Moreover, *R. meliloti* is able to elicit the formation of empty nodules when separated from the plant by a filter membrane.[96] Recently, it has been reported that a certain proportion of alfalfa seedlings is able to develop nodules in the absence of rhizobia showing that these plants have the genetic program required for nodule organogenesis.[97] The role of the bacteria is to switch on this program, possibly by a diffusible signal molecule. Factors isolated from *R.trifolii* culture filtrates were reported to elicit foci of cortical cell divisions which resemble nodule primordia.[94] The chemical structure of one such compound BF-5, which also elicits root hair branching, was recently shown to be *N*-acetylglutamic acid.[98]

The induction of cortical cell division and of the formation of a nodule meristem are likely to be associated with modifications to the hormone balance. The exogenous application of cytokinin results in cortical cell divisions in pea root explants[99] and in soybean, cowpea, and alfalfa roots.[19] The addition of auxin-transport inhibitors, such as *N*-(1-naphtyl)phtalamic acid and 2,3,5-triiodobenzoic acid, induces nodule-like outgrowths in alfalfa in which early nodulin genes such as ENOD2 and Nms-30, are expressed.[100] The authors suggested that the auxin-transport inhibitors mimic the activity of compounds produced upon the induction of the *Rhizobium nod* genes. The introduction into *R. meliloti* of a plasmid carrying the *tzs* gene from *Agrobacterium tumefaciens* results in a high production of zeatin. When this *tzs* clone is transferred into a *R. meliloti* strain deleted of the *nod* region the transconjugant recovers the ability to elicit some cortical cell divisions.[101] This finding supports the view that a modification of hormone balance is involved in the mitosis induction and raises the question of the involvement of *nod* genes in bacterial hormone synthesis.

Strains of *Bradyrhizobium* and *Rhizobium* are able to synthesize and excrete cytokinin-active compounds.[102,103] The cytokinin profile of most strains include zeatin and its derivatives. Rhizobia are also able to produce auxins.[104] However, mutations in the *nodABC* genes which result in a $Nod^-$ phenotype do not alter significantly the production of these plant hormones, at least in pure culture.[103,104] More studies are required to learn if, independently of the *nod* regulon, the synthesis of cytokinins and auxins is under the control of plant factors which might modify their production during symbiosis. The production of such bacterial compounds *in planta* might play a role in the modification of hormone balance in the course of nodulation, but genetic data indicate that they are unlikely to be the *nod*-specified NOIP. What is the chemical nature of NOIP? What is the biochemical function of the proteins encoded by the *nod* genes?

### B. NOD FACTORS: BIOLOGICAL EVIDENCE

A series of experiments associating *Rhizobium* genetics to a physiological approach were performed by Ton van Brussel and his colleagues. They had observed that *R. leguminosarum* strains provoke a deformation of the root system, the thick and short root (Tsr) phenotype, on a particular host *Vicia sativa* subsp. *nigra*.[105] Analyzing the role of *Rhizobium* in eliciting this reaction, they showed that the *nodABC* genes were required and that the sterile supernatant of a culture grown in the presence of root exudate had a Tsr activity.[106,107] Root exudate can be replaced by an appropriate flavonoid *nod* inducer such as naringenin.[108] The sterile supernatants of *nod*-induced cultures also have a spectacular effect on the growth of root hairs (Hai phenotype) and on hair deformations (Had).[108] Filtration through molecular sieves indicated that the Tsr/Had factors have a low molecular mass. These experiments have paved the way for the study of the *nod* gene functions by showing that the *nodABC* genes are involved in the production of small extracellular factors, and by providing biological assays to identify the active factors. It has been shown then that the *nodAB* genes of *R. meliloti* specify the synthesis of a Nod factor which stimulates mitosis of cultured plant protoplasts. This factor is heat-stable, partially hydrophobic, and has a low molecular weight.[109]

In contrast to the Tsr assay which can be used only with a limited number of legumes,[105,106,110] the Had assay seems to be of general applicability[92,94,108] and can be used with alfalfa.[110] The sterile supernatant of *R. meliloti* cultures induced by the flavone luteolin elicits hair deformation on alfalfa.[110] Inactivation of *nodABC* genes suppresses this Had activity. Using vetch (Tsr and Had) and alfalfa (Had) bioassays it was possible to show that the biological activity of sterile supernatants is specific and exhibits the same host specificity as observed for nodulation by the bacteria: filtrates from *R. leguminosarum* are active on vetch and not on alfalfa whereas those from *R. meliloti* are active on alfalfa and not on vetch.[110] These results suggested a role of the host-range *nod* genes in extracellular factor production. *Rhizobium meliloti nodH* mutants have an altered host range: they have lost the ability to infect and nodulate the homologous host alfalfa and have acquired the ability to infect and nodulate vetch, an heterologous host.[32,71] The sterile filtrate of a *nodH* mutant is active on vetch and not on alfalfa, implying that the *nodH* gene is modifying the specificity of an extracellular Nod factor.[110]

Similar results are obtained when the *R. meliloti nodABC* and *nodH* genes are introduced into *Escherichia coli:* the *E. coli* transconjugants produce an extracellular alfalfa-specific factor only in the presence of an active *nodH* gene.[111] This result indicates that the metabolites required for the production of an alfalfa-specific Nod factor may be ubiquitous among Gram-negative bacteria.

When the plasmid pGMI515, containing the *R. meliloti* host-range *nodFEG, nodH,* and *nodPQ* genes, is transferred into *R. leguminosarum* the transconjugants gain the ability to nodulate alfalfa and exhibit a strongly reduced nodulation of vetch.[72] Mutations in either *nodH* or *nodQ* restore the normal *R. leguminosarum* specificity, indicating an epistasy (dominance) of these *R. meliloti* genes over the host-specific *nod* genes of *R. leguminosarum*.[72] The specificity of the filtrates of these different strains, as revealed by the Had biological assay, strictly parallels the symbiotic specificity.[72] These results indicate that the *nodQ* gene is also involved in the modification of the Nod factor specificity. A *R. meliloti* derivative with a deletion of the whole nodulation region and having a plasmid carrying only the common *nodABCIJ* genes and the regulatory *nodD1* gene, produces a Nod factor which is not active on alfalfa but active on vetch[72,110] and clover.[111] To interpret all these data the following model was proposed.[72,110] The common *nodABC* genes determine the production of a Nod factor precursors, which can be detected with the vetch bioassays and host-range genes, such as *nodH* and *nodQ* genes in *R. meliloti,* modify this precursors and determine the production of a host-specific signal.[72,110]

## C. NOD FACTORS: CHEMICAL CHARACTERIZATION

### 1. *Rhizobium meliloti*

When the supernatant of a luteolin-induced *R. meliloti* culture is extracted with organic solvents of various polarity most of the Had activity is found in the butanol fraction, suggesting an amphiphilic nature for the Nod factors. HPLC fractionation of the butanol extract, followed by Had bioassays, revealed that the concentration of the active molecule(s) was insufficient for complete structural analysis. It was thus necessary to increase the production of Nod factors.

We hypothesized that the rate of transcription of the *nod* genes might be a major limiting factor. Whereas the regulation of the expression of the common *nod* genes has been extensively studied in *R. meliloti,* the regulation of the host-range gene expression was poorly understood (see chapter by A. Kondorosi in this volume). We thus constructed transcriptional fusions of the *E. coli lacZ* reporter gene to the the *nodFE, nodG,* and *nodH* genes and observed that the host-specific *nod* genes are dependent on the same regulatory circuit as the common *nod* genes.[58] For example, when the regulatory gene couple *syrM/nodD3* is present on an Inc-P1 plasmid, the constitutive level of expression of both the common and the host-specific genes is strongly increased.[58] The recombinant plasmid pGMI149 carries the common *nodABCIJ* genes, the *nodFEG, nodH, nodPQ* host-range genes, and the *nodD1, syrM, nodD3* regulatory genes. The introduction of pGMI149 into *R. meliloti* strains resulted in an increase of more than a 100-fold in the Had activity of the culture supernatant on alfalfa. As a result of this amplification it was possible to observe two far-UV (220 nm) absorbing peaks, comigrating with the Had activity, when the butanol-soluble fraction was chromatographed by HPLC on an analytical $C_{18}$ reverse-phase column.[112] Tn5 insertions into either the *nodA* or *nodC* genes resulted in the simultaneous disappearance of these two UV-absorbing peaks and of the Had activity, indicating that these compounds, which require the *nodABC* genes for their production, were Nod factors.

For structural analysis of the Nod factors, large-scale cultures were required. To facilitate subsequent processing of large volumes of supernatants, an exopolysaccharide-deficient *R. meliloti* derivative was used by virtue of its nonmucoid characteristics. The major Nod factors produced by strains differing only by an *exoB* mutation, were found to be identical as studied by both the biological Had assays and physicochemical criteria (HPLC, NMR, and mass spectrometry). Nod factors were purified by a series of preparative reverse-phase $C_{18}$ HPLC, gel permeation on a Sephadex LH20 column, ion-exchange chromatography on a DEAE column, and analytical reverse-phase $C_{18}$ HPLC. The approximate yield of Nod factors varied from 0.4 to 2.5 mg/l of induced culture.[112,113]

The structure of the major Nod factors corresponding to the two UV-absorbing peaks was established by mass spectrometry, NMR spectroscopy, chemical modification and radioactive labeling and is given in Figure 2. The two peaks correspond to the α and β anomers of the same molecule NodRm-1 (recently renamed NodRm-IV(S), see the nomenclature proposal in the legend of Figure 2.)[112] NodRm-IV(S) is an *N*-acyl-tri-*N*-acetyl-β-1,4-D-glucosamine tetrasaccharide bearing a sulfate group on carbon-6 of the reducing sugar moiety (mol. mass 1102). The acyl moiety, carried by the nonreducing terminal sugar, is a $C_{16}$ chain with two double bonds at positions 2 and 9.[112] The E conjugated double bond at position 2 is responsible for most of the UV-absorbing property of the molecule. A related compound, Ac-NodRm-1 (renamed NodRm-IV(Ac,S); mol. mass 1144), bearing an *O*-

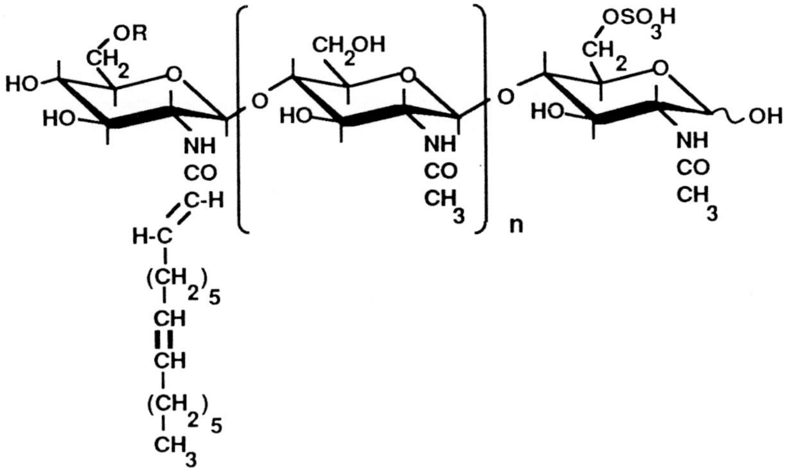


FIGURE 2. Structures of Nod factors of *Rhizobium meliloti*. The substitute at position 6 on the nonreducing end can be a hydrogen (R = H) or an acetyl group (R = $CH_3CO$). The following abbreviated nomenclature is proposed.[80] The oligosaccharidic chain can be tetrameric (with n = 2, NodRm-IV series) or pentameric (n = 3, NodRm-V series). Substitutions on the sugar backbone, other than the *N*-acyl chain, are given between brackets starting from the non-reducing terminal sugar. The abbreviations are "S" for sulfate and "Ac" for *O*-acetate, e.g., the four major Nod factors secreted by *R. meliloti* are NodRm-IV(Ac,S) which was formerly named Ac-NodRm-1, NodRm-IV(S) formerly named NodRm-1, NodRm-V(Ac,S), and NodRm-V(S).[80,113a,114]

acetate group on the carbon-6 of the nonreducing terminal sugar was also found (Figure 2).[113] When, instead of pGMI149, a plasmid amplifying only the regulatory gene couple *syrM*/*nodD3* was used to increase the Nod factor production, the major compound was NodRm-IV(Ac,S) and two new compounds NodRm-V(Ac,S) (mol. mass 1347) and NodRm-V(S) (mol. mass 1305) were detected which were similar to NodRm-IV(Ac,S) and NodRm-IV(S) respectively but contained five glucosamine residues instead of four (Figure 2).[80,113a,114] The proportion of *O*-acetylated molecules and of pentasaccharides excreted by the nonoverproducing wild-type strain remains unknown. The differences observed between the Nod factors produced by overproducing strains containing either the pGMI149 plasmid or the *syrM*/*nodD3* clone indicate that *nod* genes outside of the pGMI149 region (see Figure 1) are involved in the *O*-acetylation and in the control of the number of the glucosamine residues. The presence of pGMI149 results in the amplification of only a subset of the structural *nod* genes and is likely to modify the stoechiometry of the different Nod proteins.[80,113a,114] It is therefore probable that the population of NodRm factors from the overproducing strains which are only engineered for regulatory genes will more closely reflect the Nod factor population produced by the wild-type strain.[80,113a,114] In both types of overproducing constructs other minor compounds were detected which differed slightly from NodRm-IV(S) in the length of the aliphatic chain ($C_{16}$, $C_{18}$) and number and location of double bonds. All these molecules can be considered as NodRm factors since none of them are detectable in the supernatants of *nodA* or *nodC* mutants. Chemical and biological studies of the various NodRm factors produced by different *nod* mutants should indicate whether these minor molecules (1) are intermediates in the biosynthetic pathway of NodRm-IV(S) and NodRm-IV(Ac,S) or (2) if their presence is due to an imperfect specificity of the enzymes involved in the biosynthetic pathway.

**2. Other *Rhizobium* Species**

As discussed previously, legume-rhizobia associations are characterized by varying degrees of symbiotic host specificity, with *Rhizobium* sp. strain NGR234 having an unusually broad host range including nodulation of a large number of different legume genera and the non legume *Parasponia*.[24] The production of extracellular Nod factors by this strain was detected by a root hair deformation test with the tropical legume *Macroptilium*.[115] An assay for the simple detection of sulfated Nod factors was devised by labeling *nod*-induced cultures with $^{35}S$-sulfate, eluting the sterile supernatant through a Sep-Pak $C_{18}$ cartridge, and analyzing the nonpolar fraction by thin-layer chromatography (TLC).[116] Two major sulfated compounds, NodNGR-1 and NodNGR-2, were observed which showed a *Macroptilium* hair-deforming activity and were not detectable in a strain deleted of the *nodABC* genes. A purification strategy similar to the one described for NodRm signals was then used with large-scale cultures to obtain the amounts required for structural studies. Mass and NMR spectrometry revealed that NodNGR-2 has the same general structure as that of the NodRm factors produced by *R. meliloti:* a sulfated lipo-oligosaccharide with *N*-acetyl-D-glucosamine residues and a fatty acid side chain.[116] Interestingly, these molecules also have structural features clearly different from the NodRm factors.

In the case of *R. leguminosarum,* $^{35}S$-sulfate labeling experiments did not detect any Nod factors and, therefore, a different strategy was devised.[78,84] Sequence homology and biochemical studies of the nodFE gene products indicated that they are likely to be involved in the synthesis of fatty acids.[78,84] It was thus hypothesized that Nod metabolites could be efficiently labeled by radioactive precursors of the fatty acid synthetic pathway, such as $^{14}C$-labeled acetate.[78,84] This was indeed the case and growing *R. leguminosarum* wild-type strain in *nod* gene-inducing conditions and in the presence of $^{14}C$-labeled acetate allowed the detection of Nod metabolites.[78,84] After *n*-butanol extraction of the supernatant and reverse-phase $C_{18}$ TLC, five compounds could be detected, three of them being most abundant.[84] Similar results were obtained with *R. trifolii*.[84] Several of the *R. leguminosarum* Nod metabolites have been purified with a combination of column liquid chromatography procedures.[84] FAB mass and NMR spectrometry analysis indicated that the major compounds have a general structure similar to the Nod factors found in *R. meliloti:* they contain a β-linked poly-*N*-acetylglucosamine moiety with an *O*-acetyl substitution and an acyl chain.[84,117] As in *R. meliloti,* compounds were found in which the oligosaccharide backbone contains either four or five sugar residues. But the following differences with *R. meliloti* have been observed: (1) the compounds do not contain a sulfate group; and (2) the major fatty acid chain is mono-unsaturated and contains a double bond at a different position.[84,117]

## D. BIOCHEMICAL FUNCTION OF *NOD* GENE PRODUCTS

What could be the function of the common and specific *nod* genes in the biosynthesis and transport of the Nod factors? To answer this question three approaches have been used: (1) determination of the structure of the molecules produced by *Rhizobium nod* mutant strains, (2) prediction of the function of *nod* gene products by the search for homology with already known enzymes, and (3) physiological or biochemical study of the *nod* gene products.

### 1. The Common *nod* Genes and the Molecule Backbone

In *R. leguminosarum* and *R. meliloti* a Tn5 insertion in the *nodABC* operon suppresses the production of the lipo-oligosaccharidic Nod factors.[84,112] This is not due to a polar effect on the downstream *nodIJ* genes since a Tn5 insertion in the *nodIJ* region does not suppress Nod factor production.[112] Indeed, mutations in *nodABC* are so far the only ones to completely suppress any detectable Nod factor production and symbiotic activity: The proteins encoded by these genes have a crucial role and cannot be functionally replaced by other rhizobial proteins. Biochemical and immunological studies have shown that NodA and NodB are cytosolic whereas NodC is an outer membrane protein.[66,67,68,118] In *R. meliloti,* the *nodAB* genes have been reported to be sufficient for the production of low-molecular weight partially hydrophobic factors which stimulate mitotic cell divisions of plant protoplasts.[109] It was proposed that the NodC protein could rather be involved in transport.[68,109] In contrast, the *nodAB* genes were found not to be sufficient for the production of root hair-deforming factors, the *nodC* gene also being required.[111] $^{14}$C-labeling experiments have shown that in *R. leguminosarum* a mutation in *nodC* results in the absence of detectable amounts of lipo-oligosaccharidic factors not only in the culture medium but also in any of the bacterial cell compartments indicating that the *nodC* gene is involved in the synthesis of the Nod factors.[84]

A *R. leguminosarum* strain containing only the *nodDABC* genes produces four butanol-soluble Nod compounds. These compounds have chromatographic properties which indicate that they have the general lipo-oligosaccharidic structure of the wild-type Nod factors.[84] If we suppose that these Nod metabolites are acylated β-1,4 *N*-acetyl-D-glucosamine oligosaccharides, two hypotheses can be proposed for the functions of the NodABC proteins: They catalyze (1) the formation of acylated oligosaccharides directly or (2) simply the formation of an oligosaccharidic backbone, whose acylation is specified by bacterial "housekeeping" enzymes. The oligosaccharidic backbone of four or five *N*-acetyl-D-glucosamine residues could either be generated by β-1,4 polymerization of *N*-acetyl-D-glucosamine monomers or of β-1,4 dimers of *N*-acetyl-D-glucosamine and *N*-acetyl-D-muramic acid (murein), which are precursors of peptidoglycan synthesis, or by the cleavage of preexisting linear β-1,4-*N*-acetyl-D-glucosamine polymers (chitin). It is worth noting that whereas chitin is a major component of fungal cell walls, the presence of chitin has not yet been reported in bacteria.[119]

Supporting the hypothesis of a polymerization is the recent finding that the *nodM* gene, found in *R. leguminosarum* and *R. meliloti,* shares homology with the glucosamine synthase encoded by the *E. coli glmS* gene (Table 2).[43,78,82,84] Physiological experiments have shown directly that *nodM* codes for a glucosamine-phosphate synthase that is involved in glucosamine synthesis.[38,78] A second glucosamine-phosphate synthase is present, whose synthesis does not require flavonoid induction, and is thus likely a "housekeeping" enzyme.[78] The presence of this second enzyme could explain why *nodM*$^-$ mutants exhibit only a moderate delay in infection and nodulation,[38,43,82] and why neither a qualitative or a quantitative influence of the *nodM* gene on the synthesis of the Nod factors could be detected.[84]

The *nodIJ* genes are commonly present in rhizobia and are likely to be in the same operon as *nodC*. It is thus tempting to hypothesize that they are involved in signal metabolism or transport. The symbiotic phenotype of *nodIJ*$^-$ mutants is a slight delay in infection and nodulation.[40,71,120] NodI has sequence similarity to ATP-binding proteins which play a role in the transport of low-molecular weight compounds (Table 2).[40] NodJ is a membrane protein which may act in conjunction with NodI.[40] A possible role for such a protein couple could be signal transport, but in *R. leguminosarum* no influence of the *nodIJ* genes on the abundance and structure of Nod metabolites in various cell compartments or in the growth medium could be detected.[84]

### 2. Acylation and *O*-Acetylation of the Sugar Backbone

There are both biological and biochemical evidence that *nodFE* and *nodL* genes are involved in the synthesis of extracellular Nod factors. The sterile supernatant from *R. leguminosarum* elicits on a homologous host, *Vicia sativa* subsp. *nigra,* an increase in the *nod* gene-inducing activity (Ini) of the root exudate.[121] The *nodF* gene appears to be required for both the timely appearance and the maximum level of the plant Ini activity and the *nodE* gene is responsible for the biovar specificity of Ini. Moreover, in *R. leguminosarum,* the *nodFE* genes are involved together with the common *nodABC* genes in the production of a host-specific rhizobial signal which activates the transcription of host plant genes.[122] After $^{14}$C-labeling with acetate or glucose of a wild-type strain of *R. leguminosarum*, the culture supernatant is submitted to a preparative and to an analytical $C_{18}$ reverse-phase HPLC. A radioactive peak coelutes with vetch root hair-deforming activity.[78] With strains carrying mutations in *nodE* or *nodL* the elution profile was altered indicating that their gene products modify the root hair-curling molecules.[78] When the butanol extracts from $^{14}C$-labeled acetate cultures of *R. leguminosarum* are fractionated by TLC, differences are observed according to the *nod* gene combinations.[84] With a strain carrying only the *nodABC* genes and the regulatory *nodD* gene, four compounds are detected. If the *nodFE* genes are added, five compounds are detected. The further addition of *nodL* results in a change of mobility for the five compounds.[84] What could be the function of these genes?

The sequence of the *nodL* gene product of *R. leguminosarum* shows homology to the acetyltransferases encoded by the *lacA* and *cysE* genes of *E. coli*.[81] It can thus be hypothesized that *nodL* specifies either the *N*- or the *O*-acetylation of the glucosamine residues. The comparison of the FAB-mass spectra of molecules produced by *R. leguminosarum* in the presence or the absence of *nodL* indicates that *nodL* determines an *O*-acetylation.[84] Recently a *nodL* gene has also been identified in *R. meliloti*.[123] A mutant having a deletion which removes the *nodL* region has been shown to produce NodRm-IV(S) but not NodRm-IV(Ac,S), which indicates that in *R. meliloti* the NodL protein specifies the *O*-acetylation on the carbon-6 of the terminal nonreducing acylated glucosamine residue.[124]

The *nodFE* genes are part of the same operon and the last codon of *nodF* overlaps the first codon of *nodE* in both *R. leguminosarum* and *R. meliloti* suggesting a translational coupling.[31,32,41] The NodF and NodE proteins might thus operate in a coordinate manner in signal synthesis. NodF was found to show homology to acyl carrier proteins.[41] The NodF protein has been purified from *R. leguminosarum* and shown to carry a 4′-phosphopantetheine moiety which suggests that this protein might function as a component of an enzyme system for the biosynthesis of a fatty acid.[84] The NodE protein shares homology with a group of β-ketoacyl synthases such as the *E. coli* condensing enzyme of fatty acid biosynthesis FabB, and those presumed to be involved in the synthesis of β-ketide antibiotics in *Streptomyces* species.[75] It is thus tempting to hypothesize that the *nodFE* genes specify the acylation of the oligosaccharide. However, it is worth noting that a *R. leguminosarum* strain carrying only the *nodDABC* genes (and not the *nodFE* genes) produces Nod factors which are likely to be acylated since they are soluble in butanol and have mobilities on reverse-phase TLC plates which are only slightly different from those of the Nod factors produced by the wild type.[84] Similarly, a *R. meliloti* strain deleted of the *nodFE* genes still produces butanol-extractable molecules which are similar to NodRm-IV(S) and acylated.[125] However, the composition of the lipidic moiety of these factors is clearly more varied than in the case of NodRm-IV(S), as detected by TLC and FAB-MS.[125] This suggests that the *nodFE* genes might determine the host range by acylating the oligosaccharide in a specific way; in the absence of the NodF and NodE proteins "house-keeping" acyl carrier protein(s) and condensing enzymes would acylate the oligosaccharidic backbone less specifically. However, other functions for *nodFE* genes, such as a role in oligosaccharide chain length determination, are suggested by an increased accumulation of five *N*-acetylglucosamine residues in NodFE *R. leguminosarum* overproducing strains. This could be an indication for a transglycosidation activity of the NodE protein.[84] A conclusion on the function of the NodFE proteins should, therefore, await further studies.

In *R. meliloti,* downstream of *nodFE,* is located the *nodG* gene which is not found in *R. leguminosarum*.[32,71] NodG shares homology with alcohol dehydrogenases and might be involved in a specific dehydrogenation of the acyl chain.[31,120]

### 3. Sulfation of the *Rhizobium meliloti* Nod Factors

In *R. meliloti,* the *nodH* and *nodQ* genes determine the alfalfa specificity of Nod factors.[72,108,111] By which mechanism do they operate? *Rhizobium meliloti* strains carrying the pGMI149 plasmid overproduce Nod factors. HPLC and TLC experiments show that strains having a *nodH*::Tn5 insertion in both the pSym and the pGMI149 plasmid do not produce NodRm-IV(S) and NodRm-IV(Ac,S) but excrete more hydrophobic compounds.[126,127] Mass and NMR spectrometry revealed that these compounds were α and β anomers of lipo-oligosaccharides, NodRm-IV and NodRm-IV(Ac), which differ from NodRm-IV(S) and NodRm-IV(Ac,S), respectively, only by the absence of the sulfate group.[80,127] Surprisingly, mutants with a Tn5 insertion in *nodP* or *nodQ* produced a mixture of sulfated and nonsulfated factors.[80,127] This "leaky" phenotype might be due to the presence of a second copy of the *nodPQ* genes which has been found on the second megaplasmid.[36,128] Indeed, double mutants having a Tn5 insertion in both *nodQ* copies excrete only nonsulfated factors.[80] Thus *nodH* and *nodPQ* genes are involved in the sulfation of the lipo-oligosaccharide factors.

In procaryotes as well as in eukaryotes the sulfation of metabolites requires, first, the synthesis of activated forms of sulfate, adenosine 5′-phosphosulfate (APS) and 3′-phosphoadenosine 5′-phosphosulfate (PAPS), and, second, the transfer of the sulfate group onto the target molecule. The *nodP* and *nodQ* gene of *R. meliloti* have strong homologies with the *cysN* and *cysD* genes of *E. coli,* respectively, which code for subunits of an ATP sulfurylase.[83] NodP and NodQ products have been shown to have an ATP sulfurylase activity both *in vitro* and *in vivo* and their function is likely to generate activated sulfate.[83] It is not known whether the subsequent PAPS synthesis is specified by a *nod* gene yet to be discovered or by a housekeeping enzyme. The *nodH* gene product has a significant homology with steroid sulfotransferases from mammals and is likely to specify the transfer of the sulfate group from an activated donor onto lipo-oligosaccharide precursors such as NodRm-IV, NodRm-IV(Ac), NodRm-V and NodRm-V(Ac).[80]

Interesting questions are raised by the study of the Nod factor biosynthetic pathway. First, do the enzymes encoded by the *nod* genes correspond to saprophytic (housekeeping) isoenzymes? If so, how do their characteristics differ from those of the corresponding enzymes of the general metabolism? No sulfated Nod factors can be detected in *nodH*$^-$ mutants which indicate that there is no "housekeeping" sulfotransferase able to specify the transfer of activated sulfate specifically to the lipo-oligosaccharide Nod factors. The situation is puzzling with the *nodPQ* genes. Genetic evidence suggests that there are at least three ATP-sulfurylase systems in *R. meliloti.* One is encoded by the *nodP1Q1* genes and another one is encoded by *P2Q2* genes which are quite homologous to *nodP1Q1* and are located on the second megaplasmid.[83] However, double *Q1*$^-$*Q2*$^-$ mutants are not auxotrophic for sulfur amino acids indicating that a third ATP-sulfurylase exists to generate the APS required for the cell sulfur metabolism.[128] What is intriguing is that the three systems seem to catalyze the same reaction, that is, to generate APS from ATP, yet surprisingly, *nodP1Q1* and *nodP2Q2* genes are efficient to mediate sulfation of Nod factors, whereas the third system does not lead to any detectable Nod factor sulfation.[80] Why is the activated sulfate which is synthesized by the housekeeping enzyme not used by the NodH sulfotransferase? Are there different forms of activated sulfate in *R. meliloti*? Or does the activated sulfate remain bound to a specific protein and this complex has to be recognized by the enzymes utilizing the activated sulfate? Or is the rate of APS production quite different for the three enzymes? Or is there only one form of activated sulfate but present in different cell compartments?

There is a striking correlation among *R. meliloti* strains between the production of sulfated or nonsulfated Nod factors and the host specificity of infection and nodulation.[72,110,126,127] A strain carrying the wild-type *nod* region infects and nodulates alfalfa and produces sulfated lipo-oligosaccharides. A *nodH*$^-$ mutant which exhibits a shift in the host range, and infects and nodulates vetch, excretes nonsulfated signals. A *nodQ*$^-$ mutant which has an extended host range (alfalfa and vetch) produces both types of signals. These correlations indicate that the *nodH* and *nodPQ* genes specify the host range by determining the sulfation of lipo-oligosaccharidic Nod factors.

### 4. Alterations in Nod Factor Synthesis and Symbiotic Phenotypes

Mutations in nodulation genes other than the *nodABC* genes result, in general, in moderate alterations of symbiotic properties, a delay in infection and nodulation, and a change of host specificity. Leakiness of these *nod* mutations could have different causes. These *nod* genes may code for Nod enzymes which have "housekeeping" isoenzymes. In *R. meliloti*, the *nodPQ* mutations are leaky because of the presence of a second copy of homologous genes encoding an ATP-sulfurylase which can partially complement the *nodPQ*-encoded ATP-sulfurylase.[36,83] In *nodFE* mutants the NodFE proteins could be partially and inaccurately replaced by bacterial enzymes having similar but different functions. In such mutants the modified signals would have altered symbiotic activity. Another type of leakiness of *nod* mutations is given by *nodL*. In this case, there is apparently no complementation by other enzymes but the Nod factor which is produced by the *nodL* mutant (with no *O*-acetylation) is still active symbiotically with an efficiency that could vary with hosts. Finally, host-range mutants produce intermediates in the Nod factor biosynthetic pathway which are active on new hosts. This is the case for *nodH* mutants which produce nonsulfated intermediates which are no longer on alfalfa but are active on a nonhomologous host, vetch.[80]

### 5. *NodO*, A Secreted Protein

One example has been described of a *nod* gene product whose function seems not to be the mediation of the synthesis of Nod factors but rather facilitation directly or indirectly of the interaction between a *Rhizobium* Nod factor and a plant receptor. The *nodO* gene is part of the *R. leguminosarum* bv. *viciae nod* regulon: it is preceded by a *nod* box and its transcription is regulated by a nodD protein and plant flavonoid signals.[44,45] This gene is specific for biovar *viciae* and has not been detected in the related biovars *phaseoli* and *trifolii*.[78] A mutation in *nodO* causes a slight delay of nodulation.[45] A *R. leguminosarum* strain in which a deletion has removed all the nodulation region except the *nodDABCIJ* genes is completely Nod$^-$. Such a strain produces Nod factor precursors,[84] but these seem not to be sufficient to allow elicitation of nodule formation. The introduction into this strain of clones carrying either *nodFE, nodFEL* or *nodO* genes partially restores nodulation.[78] *nodFEL* genes are known to specify the modification of the Nod factors and we can thus suppose that their introduction will result in the production of modified factors having a better symbiotic activity. In contrast, the *nodO* gene does not seem to modify quantitatively or qualitatively the Nod factor production.[84] *nodO* codes for a secreted $Ca^{2+}$-binding protein homologous to *E. coli* hemolysin and related proteins.[44,45] By which mechanism could the secretion of this protein partially restore nodulation by a strain producing only the *nodABC*-specified precursors? This could be by facilitating the interaction between these precursors and a plant component.[78] The function of NodO in natural conditions could be to facilitate Nod factor/plant receptor interactions in the rhizosphere.

## E. NOD FACTORS: BIOLOGICAL ACTIVITY

During the last few years it has been shown that the sterile supernatant of *nod*-induced *Rhizobium* cultures or purified Nod factors elicit a great variety of plant reactions affecting root hairs, the root cortex, or the whole root system. Most of the experiments were performed with alfalfa (a *R. meliloti* host), pea and vetch (*R. leguminosarum* hosts).

### 1. Changes in Root Hair Morphology

Sterile supernatants of *Rhizobium* cultures elicit root hair deformations of different kinds and Had bioassays can be potent tools, together with physicochemical methods, for the purification of Nod factors.[112,113] The sensitivity of the bioassay is great and Nod factors elicit root hair deformation at concentrations as low as $10^{-11}$ *M*.[112,113] The ability of the Had assay to discriminate between Nod factors and other *Rhizobium* metabolites varies with the plant. Alfalfa root hairs are not significantly deformed by the filtrates from *R. meliloti* strains deleted of the *nodABC* genes.[112] In contrast, vetch and white clover root hairs can be deformed by the filtrate of Nod$^-$ strains, reacting to various *Rhizobium* excreted metabolites (branching factors) which are not specified by *nod* genes.[94,108] Another present limitation of the Had assay is the difficulty in quantitating the reactions and thus only clearcut differences in activity can be detected. This has been the case for the study of the specificity of Nod factors. The supernatant of *R. meliloti* is clearly more active on alfalfa than on vetch and the supernatant of *R. leguminosarum* is clearly active on vetch and not on alfalfa.[72,110] Specificity is also observed with purified Nod factors. For example, sulfated Nod factors of *R. meliloti,* such as NodRm-IV(S) and NodRm-IV(Ac,S), elicit root hair deformation on alfalfa and not on vetch at concentrations in the range $10^{-8}$ to $10^{-11}$ *M*.[80,112] In contrast, nonsulfated factors produced by a *nodH*$^-$ mutant are Had$^-$ on alfalfa and Had$^+$ on vetch.[80] In *R. meliloti,* an absolute correlation was found between the specificity of infection and nodulation by bacterial cells, on the one hand, and the specificity of root hair deformation by purified Nod factors, on the other hand (see Section III.D.3). This indicates that it is by specifying the production of these lipo-oligosaccharidic Nod factors that the *Rhizobium nod* genes determine the host range.[80,126,127] This correlation also indicates that plant molecules which recognize in a specific manner these Nod signals (signal receptors) are present at the surface or within the root hairs.

The introduction of a pea lectin gene into white clover seedlings results in the transfer of the ability to be infected and nodulated by a pea-specific *Rhizobium.*[129] This finding supports the hypothesis that plant lectins are involved in the *Rhizobium*-legume specific recognition. Root lectin is specially abundant at the tip of root hairs in the root zone susceptible to *Rhizobium* infection and is predominantly present at the external surface of the plasma membrane.[130,131] The Nod signals have structural properties which are characteristic of effective lectin ligands such as hexosamine oligosaccharides.[132] A plant root lectin could thus be part of the Nod signal receptor complex.[131]

### 2. Changes in Root Hair Physiology and Gene Expression

Analysis of the mechanisms of early root hair reactions to Nod factors has been initiated by monitoring the modifications of the transmembrane potential in single root hair cells of alfalfa. The addition of supernatant of wild-type *R. meliloti* or of purified NodRm-IV(S) causes a rapid (within 1 to 2 min) depolarization.[51] The membrane potential recovers after 10 to 15 min and hyperpolarizes slightly. The hair cell is then refractory to further stimulation. This electrophysiological approach is providing both a sensitive quantitative bioassay which should allow the study of the relationship between the structure and the activity of the Nod signals, and a tool to analyze the very first interactions between the signals and the plant.[51]

Another approach addresses the question of the root hair response to Nod signals at the gene expression level. Once the signal has been perceived and transduced to the root hair nucleus what are the host genes whose transcription is specifically induced? Plant genes induced during the early steps of symbiosis have been recently identified and characterized. PsENOD5 (for *Pisum sativum* early nodulin 5) codes for an arabinogalactan-like protein and is expressed in the infected zones of roots, in cells containing infection threads.[133,134] PsENOD12 encodes a proline-rich protein and is also expressed in cells containing infection threads.[122,134] PsENOD5 and PsENOD12 are likely to have a role in the infection process. Molecular assays have been developed using root hair RNA, reverse transcriptase, and PCR amplification to study gene expression in root hairs.[122] These assays revealed that the application of sterile culture filtrates elicit both PsENOD5 and PsENOD12 expression in root hairs.[122,134] The addition of purified Nod factors, in the concentration range $10^{-9}$ to $10^{-12}$ *M*, also elicits the transcription of these two genes.[2,135] A mutation in *nodFE* suppresses this induction, thus providing a molecular evidence that the Nod factor modifications specified by these genes are important for the transcription inducing-activity of the symbiotic signals.[122]

The picomolar concentration at which Nod factors are inducing the transcription of some ENOD genes is at least a 1000-fold lower than the concentration at which classical plant hormones are active.[136] Some fungal oligosaccharidic elicitors trigger host defense mechanisms at nanomolar concentration.[137] ENOD12 gene expression is not a defense reaction and is not induced by infection with a pathogenic fungus.[122] Oligosaccharidic pathogen elicitors and lipo-oligosaccharidic symbiotic signals when activating specific plant gene expression could mimic plant endogenous signaling mechanisms involving oligosaccharides with biological regulatory properties (oligosaccharins).[138]

Two hypotheses can be proposed to explain how Nod signals could be transduced to the root hair nucleus to specifically activate the ENOD gene transcription.[101] The Nod factor could diffuse through the root hair membrane, the cytosol, and bind to a regulatory protein as described for corticoid and steroid receptors in animal systems.[139] Alternatively, the Nod factor could bind to a membrane receptor and the signal could be then tranduced by secondary messengers, as frequently described in animal systems.[139]

### 3. Changes in Root Metabolism

*Rhizobium leguminosarum* elicits on *Vicia sativa* subsp. *nigra* the development of thick and short roots (Tsr).[106,108,140] This particular root reaction can also be caused by sterile filtrates of *nod*-induced cultures and by purified lipo-oligosaccharidic Nod factors.[80,84,106] The addition of the ethylene-inhibitor aminoethoxyvinylglycine (AVG) suppresses the development of the Tsr phenotype and restores a normal nodulation pattern.[140] The fact that an ethylene inhibitor antagonizes a Nod factor-induced effect suggests that Nod factors may modify ethylene metabolism in roots. The Tsr response of *V. sativa* subsp. *nigra* could be due to an overreaction to infection (and to Nod factors) presumably caused by an overproduction of endogenous ethylene.[140]

Flavonoids in root exudate of leguminous plants activate the transcription of *nod* genes (see the chapter by A. Kondorosi in this volume). Inoculation of *V. sativa* subsp. *nigra* with its homologous symbiont *R. leguminosarum* results in an increased *nod* gene-inducing activity (Ini) in root exudate due to a change in the composition of flavonoids.[121] This Ini effect is biovar-specific since it is not induced by nonhomologous bacteria such as *R. meliloti* and *R. trifolii.* The *R. leguminosarum nodDABCFEL* genes are essential for the induction of Ini and the *nodE* gene is responsible for its biovar specificity. Ini can be induced by purified *O*-acetylated *R. leguminosarum* Nod factors but not by the *nodDABC*-specified precursors.[84] Nod factors can thus elicit in a specific way changes in the flavonoid root metabolism and, subsequently, in the flavonoid composition of the root exudate.[121] It is worth noting that the two types of vetch reactions, possibly involving the metabolism of ethylene (Tsr phenotype) and of flavonoids (Ini phenotype), exhibit different requirements for bacterial signal molecules. Whereas the Tsr reaction can be triggered by factors either *O*-acetylated or not, the Ini reaction requires *O*-acetylated signals.[84]

### 4. Induction of Cortical Cell Division

Since the common *nodABC* genes are involved not only in root hair curling but also in the induction of cortical cell division,[20] a logical question to address was the role of these *nod* genes in the production of mitotic agents. In *R. meliloti* the *nodAB* genes are responsible for the synthesis of factors that stimulate cell divisions of plant protoplasts.[109] The addition of NodRm-1 to alfalfa seedlings has clear mitogenic effects.[141] At $10^{-9}$ *M* NodRm-IV(S) induces discrete foci of cell divisions in the inner cortex. At higher concentrations, cell divisions still occur as discrete foci but can also extend to larger zones of the root including the whole cortex.[141] Similarly, Nod factors of *R. leguminosarum* elicit the appearance of cell division foci in the cortex of *Vicia hirsuta* roots.[142] *In situ* hybridization experiments have revealed that the PsENOD12 gene is expressed in these foci of dividing cells, elicited by Nod factors, as it is in similar foci induced in the course of a normal bacterial infection. With a rhizobial infection, the induction of the first rounds of cortical cell division is induced at a distance by the bacteria and is a very rapid plant response which takes place before a marked hair curling can be observed.[20] Is this "remote control" due to a diffusion through the cortical cell layers of the Nod factors themselves or of a cascade of second messengers?

### 5. Induction of Nodule Formation

The addition of sulfated Nod factors of *R. meliloti* to aseptically grown alfalfa plants results in the appearance of root deformations which vary in shape from bumps to round, elongated or multilobate structures.[141] Light microscopy shows that these deformations share ontogenic and anatomic features of rhizobia-induced alfalfa nodules. They have a cortical origin, apical meristems, and peripheral vascular bundles.[141] Moreover, their formation is completely repressed by an excess of nitrate as for bacteria-induced nodules. It can thus be concluded that NodRm-1 elicits the formation of genuine nodules.

To study the structure-function relationships of NodRm factors, chemically modified molecules were prepared and assayed for nodule-inducing ability. Results show that the sulfate group, the reducing function of the sugar backbone, and at least one of the two double bonds of the acyl chain are all important for triggering alfalfa nodule organogenesis.[141] Together, these results indicate that the lipo-oligosaccharide Nod factors correspond to the postulated "nodule organogenesis-inducing principle" excreted by *Rhizobium* during infection.[95]

Whereas oligosaccharides are well known to be involved in the elicitation of plant defense responses,[137] there are only a few reports of their role in the regulation of plant morphogenesis.[138,143] For example, xyloglycan oligosaccharides have anti-auxin activity in pea stem segments at nanomolar concentration[144] and pectin cell wall fragments regulate tobacco thin cell layer explant morphogenesis.[145] However, in these experiments oligosaccharides were not prepared from living plants but were processed from plant cell walls by enzymic or chemical hydrolysis and thus the physiological significance of their effect is still hypothetical. On the contrary, sulfated Nod factors are specified by the bacterial genes whose function is to determine nodule formation and their organogenic activity directly shows that lipo-oligosaccharides can serve as signal molecules in plants and trigger a specific morphogenesis.

### 6. Transformation of Tobacco Plants with *nodA* and *nodB* Genes

To study the biological effects of *nodAB*-specified factors in plants, tobacco has been transformed with the *R. meliloti nodA* and *nodB* genes either singly or in combination.[146] Transgenic plants carrying *nodA* exhibit reduction in the internode distance and an altered leaf morphology, whereas those carrying *nodB* show a strongly reduced growth and a compact inflorescence. Transgenic plants expressing both *nodAB* genes have bifurcated leaves and the formation of two or more stems emerging independently from the leaf axle can be observed.[146] These results indicate that the NodA and NodB proteins alone or in combination are specifying in plants the production of factors which are active in nonlegumes and can affect the phytohormone balance.

## IV. CONCLUSIONS AND PERSPECTIVES

The control of infection is obviously complex and involves, in addition to the *nod* genes, rhizobial genes mediating the synthesis of bacterial surface components, such as exopolysaccharides, lipopolysaccharides or β-1,2-glucan. The role of these genes is discussed in other chapters of this book and we will simply discuss here the role of the *nod* genes.

*Rhizobium, Bradyrhizobium,* and *Azorhizobium* are very distant genetically. For example, *Rhizobium* is reported to be as distant from *Bradyrhizobium* as *E. coli* is from *P. aeruginosa.*[3] Nevertheless, part of the genetic information required to infect and nodulate legumes is conserved among these three distinct genera. First, they have in common a regulatory machinery which allows an adaptation of the *nod* gene expression to the plant signals present in the host rhizosphere. In the three genera, similar NodD regulatory proteins are found, which bind to highly conserved *nod* boxes in the promoters of the *nod* operons, and are activated by phenolic compounds. Second, they have in common *nodABC* genes which play a pivotal role both for infection and nodulation. In several *Rhizobium* species (*R. meliloti, R. leguminosarum,* and a tropical broad host-range strain NGR234) the *nodABC* genes have been shown to be required for the synthesis of extracellular Nod factors of low molecular weight which belong to the same class of molecules. We will now discuss how the discovery of these Nod factors allow us to propose a simple model for the mechanism by which rhizobial *nod* genes determine infection, nodulation, and host specificity and how these findings open new pathways for the molecular dissection of the rhizobia-legume symbiosis.

## A. A SIMPLE MODEL FOR THE BIOLOGICAL ROLE OF *NOD* GENES

### 1. Control of Infection and Nodulation

*Rhizobium meliloti* produces sulfated lipo-oligosaccharidic Nod factors. Inactivation of the *nodABC* genes, which are required for the production of Nod factors, always results in the loss of the bacterial ability to elicit any kind of plant reaction such as hair curling, infection thread formation, cell division, or nodulation on any host plant. These results indicate that *R. meliloti nod* genes specify alfalfa root hair curling, infection, and nodulation via the synthesis of Nod factors.[112,141] This hypothesis is strongly supported by the fact that purified NodRm-IV(S) is sufficient to induce root hair deformation,[112] transcription of infection-related nodulin genes,[2] and mitotic activity in the root cortex followed by the formation of nodules.[141] *Rhizobium leguminosarum*-purified Nod factors also elicit root hair deformation,[84] cortical cell division, and transcription of infection-related nodulins in vetch.[135] NodRm-like factors thus have a pivotal role in the major processes induced by *Rhizobium.*

The Nod factors of the three species studied so far belong to the same class of molecules. They are lipo-oligosaccharides with a sugar moiety composed of β-1,4-*N*-acetyl-D-glucosamine tetra- or pentasaccharides with an acyl chain on the nonreducing terminal sugar. Since the *nodABC* genes are required for their synthesis, and since the *nodABC* genes are conserved among rhizobia, it seems reasonable to predict that *Azorhizobium, Bradyrhizobium,* and *Sinorhizobium* produce Nod factors having this same general structure.

There is a paradoxical situation. Just three conserved *nodABC* genes, of the various rhizobial genera, are required for eliciting very different types of infection and nodulation on quite varied legumes. The simplest hypothesis to explain this is to suppose that the host plants possess the whole genetic program for both infection and nodulation and that the role of rhizobia is to switch on these programs via the Nod factors. Thus the Nod molecules produced by a given strain will trigger different routes of infection and different types of nodule development on different hosts. More generally, this class of lipo-oligosaccharidic Nod factors is hypothesized to be responsible for the rhizobial infection and nodulation of all legumes and *Parasponia.*

### 2. Control of Host Specificity

Recognition of legume phenolic signals by rhizobial regulatory NodD proteins is the first level of specific interaction (see the chapter by A. Kondorosi in this volume). Host-specific *nod* genes are, however, essential determinants of bacterial host specificity.[7,32,72,75] On the basis of biological data, it had been proposed that the common *nodABC* genes determine the synthesis of a Nod factor precursor which is then modified into a plant-specific signal(s) by the host-specific *nod* gene products.[72,110] This model has been since supported by the chemical characterization of the Nod factors. Indeed, the *R. meliloti nodH* and *nodPQ* genes encode the sulfation of precursors to produce alfalfa-specific signals.[80,127] *Rhizobium leguminosarum* and *R. meliloti* host-range *nodFE* and *nodL* genes also modify lipo-oligosaccharide molecules. The perfect correlation observed in *R. meliloti* between the specificity of the Nod factors and the symbiotic specificity of the bacteria supports the hypothesis that the host-range genes operate during symbiosis by encoding the synthesis of specific Nod signals.

Nod factors from different rhizobia are similar but they are diverse and this variety is most likely due to the host-specific genes. Genetic diversity can be caused either by allelic or nonallelic variation. The corresponding biochemical diversity could be due to differences in the composition of the sugar backbone, by structural variations in the lipid chain(s) and by various other substitutions. Interesting structural peculiarities have been found in factors from NGR234 which might suggest the existence in this strain of Nod enzymes and thus of *nod* genes not present in *R. leguminosarum* and *R. meliloti*.[147]

To ensure the diversity of symbiotic signals by modification of the precursor molecules, the host-specific *nod* genes might have evolved from a number of bacterial genes encoding a variety of enzymes. Given the extreme diversity of the host range of individual rhizobia from tropical or temperate areas we can expect that future structural studies of Nod factors from various sources, together with the identification of the *nod* genes specifying their synthesis, will reveal a very great deal of variation around one theme.

## B. BACTERIAL SYNTHESIS OF NOD FACTORS

A combination of *nod* gene sequence comparisons and of the chemical study of the factors accumulating in *nod* mutants has been used to determine the role of the *nodPQ, nodH,* and *nodL* genes in the synthesis of Nod factors. The same approach will be used for other *nod* genes. However additional biochemical, immunological, enzymological, and physiological studies will probably be required to understand precisely the function of each *nod* gene product involved in the biosynthesis and transport of Nod factors. Nod products are located in various cell compartments, in the cytosol (NodAB, NodF), in the cytoplasmic membrane (NodE, NodI), and in the outer membrane (NodC). What are the functional reasons for this partition? Since the Nod factors are essentially extracellular, what is the mechanism of their excretion? In the Nod factor biosynthetic pathways, what are the relationships between the Nod enzymes, on the one hand, and the cellular general metabolism, on the other hand, for the synthesis of the sugar backbone, of the N-acyl chain, of the various substitutions on the sugar backbone and for sulfur metabolism?

The study of the biochemical functions of the key *nodABC* genes might be facilitated by the use of *E. coli.* When introduced into *E. coli,* the *nodABC* genes determine the production of extracellular hair-deforming compounds, which suggests that the metabolites which are modified by the *nodABC* genes to give the Nod precursors are present in this bacterium.[111] *Escherichia coli* might be used as a model to facilitate the study of the Nod precursor biosynthesis because of the existence of a series of mutants (including conditional ones) of the peptidoglycan, lipid, or lipo-polysaccharide synthesis, of acyl carrier proteins, etc.

The host-specific *nodFE, nodH, nodL,* and *nodPQ* genes code for enzymes that have been shown to be involved in the Nod factor biosynthesis. The transfer of host-range genes of *R. trifolii* into *R. leguminosarum* has been reported to be associated with changes in the composition of exopolysaccharides.[148] Are the Nod enzymes only involved in the modification of low-molecular weight signals or could they also be involved in the modification of macromolecular components of the bacterial surface such as lipo-polysaccharides or exopolysaccharides?

In the supernatants of cultures of the *Rhizobium* species studied so far, more than one Nod factor was found — up to five compounds with *R. leguminosarum*. What is the biological meaning of this phenomenon? It is not easy to assay the effects of individual compounds because possible contaminations with minor components of high activity could be misleading. For example NodRm-IV(S) elicits transcription of early nodulin genes at a concentration of $10^{-12}$ *M*! The construction of a series of mutant strains, each producing only one or a reduced number of compounds, could help to answer is question.

## C. FATE AND ACTIVITY OF NOD SIGNALS IN THE PLANT

The Nod factors have first to be recognized and bind at least transiently to a plant component. The amphiphilic nature of the Nod signals does not allow us to predict whether they can diffuse through plant membranes or bind to a membrane receptor. Whatever the location of the receptor(s) the possibility of preparing signal derivatives for radioactive and/or photoaffinity labeling should allow the isolation of Nod factor-binding proteins. Availability of purified Nod factors will also aid an investigation of their possible specific binding to lectins. Various kinds of labeling and the preparation of specific antibodies should also facilitate the cytological study of their adsorption and fate in the plant host. Nod factors are major determinants of the host specificity of bacteria. On the plant side, is the signal recognition based on only one type of molecules (receptors) or is it a more complex phenomenon?

An important question to address is whether *Rhizobium* is mimicking an endogenous plant signaling mechanism, and hence whether lipo-oligosaccharide signals similar to the Nod factors are synthesized in uninfected leguminous or nonleguminous plants. Nod factors share structural analogy with some glycosphingolipids (GSL), a class of molecules which are ubiquitous membrane components of animal cells.[149] GSL are complex lipo-oligosaccharides containing lipid chains linked to oligosaccharides which may contain *N*-acetylglucosamine and *N*-acetyl-galactosamine units. The carbon-6 of a sugar (glucose or *N*-acetyl-D-glucosamine) can be esterified to a sulfate residue.[150,151] GSL can interact with lectins and have been shown to have two different types of functions: (1) modulators of transmembrane signal transducers by influencing protein kinases, and (2) mediators for cell-cell recognition.[149] It might be interesting to look for the presence of such molecules in plant cells. Moreover, the broad knowledge about the mode of action of GSL in animal cells could provide some ideas on how to analyze the Nod factor functions in plants.

Two molecular markers (ENOD5 and ENOD12 genes) are already available to initiate studies of how Nod signals activate the transcription of plant symbiotic genes.[122] However, these two genes are probably part of a large set of plant genes whose expression is regulated by the Nod signals. Exogenous addition of Nod factors may help to identify new plant genes activated either in the root hairs or in the foci of dividing cells in the cortex. What are the molecular events which determine the developmental switch in these cells that leads to the formation of new organs, the nodules? Understanding this switch is not only of interest for plant developmental biology, but could also lead to the genetic engineering of legumes to improve the formation of nodules, the organ of nitrogen fixation.

# ACKNOWLEDGMENTS

We are grateful to our colleagues J. C. Promé, G. Truchet, C. Rosenberg and F. Debellé for stimulating discussions and to W. J. Broughton and N. J. P. Price for providing unpublished informations. We thank J. V. Cullimore for critical reading of the manuscript. This work was supported in part by a grant from the Conseil Régional Midi-Pyrénées.

# REFERENCES


1. **Long, S. R.,** *Rhizobium*-legume nodulation: Life together in the underground, *Cell,* 56, 203, 1989.
2. **Nap, J. P. and Bisseling, T.,** Development biology of a plant-prokaryote symbiosis: the legume root nodule, *Science,* 250, 948, 1990.
3. **Young, J. P. W. and Johnston, A. W. B.,** The evolution of specificity in the legume-*Rhizobium* symbiosis, *Trends Ecol. Evol.,* 4, 331, 1989.
4. **Rolfe, B. G. and Gresshoff, P. M.,** Genetic analysis of legume nodule initiation, *Annu. Rev. Plant Physiol. Plant Mol. Biol.,* 39, 297, 1988.
5. **Long, S. R.,** *Rhizobium* genetics, *Annu. Rev. Genet.,* 23, 483, 1989.
6. **Martinez, E., Romero, D., and Palacios, R.,** The *Rhizobium* genome, *Crit. Rev. Plant Sci.,* 9, 59, 1990.
7. **Djordjevic, M. A., Gabriel, D. W., and Rolfe, B. G.,** *Rhizobium,* the refined parasite of legumes, *Annu. Rev. Phytopathol.,* 25, 145, 1987.
8. **Johnston, A. W. B.,** The symbiosis between *Rhizobium* and legumes, in *Genetics of Bacterial Diversity,* Hopwood, D. A. and Chater, K. F., Eds., Academic Press, San Diego, 1989, 393.
9. **Gabriel, D. W. and Rolfe, B. G.,** Working models of specific recognition in plant-microbe interactions, *Annu. Rev. Phytopathol.,* 28, 365, 1990.
10. **Dreyfus, B., Garcia, J. L., and Gillis, M.,** Characterization of *Azorhizobium caulinaudans* gen. nov., sp. nov., a stem-nodulating nitrogen-fixing bacterium isolated from *Sesbania rostrata, Int. J. Syst. Bacteriol.,* 38, 89, 1988.
11. **Chen, W. X., Yan, G. H., and Li, J. L.,** Numerical taxonomic study of fast-growing soybean rhizobia and a proposal that *Rhizobium fredii* be assigned to *Shiborhizobium* gen. nov, *Int. J. Syst. Bacteriol.,* 38, 392, 1988.
12. **Hennecke, H., Kaluza, K., Thöny, B., Fuhrmann, M., Ludwig, W., and Stackebrandt, E.,** Concurrent evolution of nitrogenase genes and 16S rRNA in *Rhizobium* species and other nitrogen fixing bacteria, *Arch. Microbiol.,* 142, 342, 1985.
13. **Jarvis, B. D. W., Gillis, M., and De Ley, J.,** Intra- and intergeneric similarities between the ribosomal ribonucleic acid cistrons of *Rhizobium* and *Bradyrhizobium* species and some related bacteria, *Int. J. Syst. Bacteriol.,* 36, 129, 1986.
14. **Polhill, R. M. and Raven, P. H.,** *Advances in Legume Systematics,* Royal Botanic Gardens, Kew, Richmond, England, 1978.
15. **Trinick, M. J. and Galbraith, J.,** The *Rhizobium* requirements of the non-legume *Parasponia* in relationship to the cross-inoculation concepts of legumes, *New Phytol.,* 86, 17, 190.
16. **Stebbins, G. L.,** *Flowering Plants: Evolution Above the Species Level,* Harvard University Press, Cambridge, MA, 1974.
17. **Dreyfus, B., Alazard, D., and Dommergues, Y. R.,** Stem-nodulating *Rhizobia,* in *Current Perspectives on Microbial Ecology,* Klug, M. G. and Reddy, C. E., Eds., American Society of Microbiology, Washington, DC, 1984, 161.
18. **Dart, P.,** Infection and development of leguminous nodules, in *A Treatise on Dinitrogen Fixation,* Sect. III, Hardy, R. W. F. and Silver, W. S., Eds., John Wiley & Sons, NY, 1977, 367.
19. **Bauer, W. D., Bhuvaneswari, T. V., Calvert, H. E., Law, I. J., Malik, N. S. A., and Vesper, S. J.,** Recognition and infection by slow-growing rhizobia, in *Nitrogen Fixation Research Progress,* Evans, H. J., Bottomley, P. J., and Newton, W. E., Eds., Martinus Nijhoff, Dordrecht, 1985, 247.
20. **Dudley, M. E., Jacobs, T. W., and Long, S. R.,** Microscopic studies of cell divisions induced in alfalfa root hairs by *Rhizobium meliloti, Planta,* 171, 289, 1987.
21. **Sprent, J. I.,** Evolution, structure and function of nitrogen-fixing root nodules: confession of ignorance, in *Nitrogen Fixation: Achievements and Objectives,* Gresshoff, P. M., Stacey, G., and Newton, W. E., Eds., Chapman and Hall, NY, 1990, 45.
22. **Trinick, M. J.,** *Rhizobium,* in *Nitrogen Fixation,* Vol. 2, Broughton, W. J., Ed., Oxford University Press, NY, 1982, 76.
23. **Lie, T. A.,** Symbiotic specialisation in pea plants: the requirement of specific *Rhizobium* strains for peas from Afghanistan, *Ann. Appl. Biol.,* 88, 462, 1978.
24. **Lewin, A., Rosenberg, C., Meyer, Z. A. H., Wong, C. H., Nelson, L., Manen, J. F., Stanley, J., Dowling, D. N., Dénarié, J., and Broughton, W. J.,** Multiple host-specificity loci of the broad host-range *Rhizobium* sp. NGR234 selected using the widely compatible legume *Vigna unguiculata, Plant Mol. Biol.,* 8, 447, 1987.
25. **Stacey, G.,** Workshop summary: Compilation of the *nod, fix* and *nif* genes of Rhizobia and information concerning their function, in *Nitrogen Fixation: Achievements and Objectives,* Gresshoff, P. M., Roth, L. E., Stacey, G., and Newton, W. E., Eds., Chapman and Hall, NY, 1990, 239.
26. **Barbour, W. M., Wang, S. P., and Stacey, G.,** Molecular genetics of *Bradyrhizobium* symbioses, in *Biological Nitrogen Fixation,* Stacey, G., Burris, R., and Evans, H., Eds., Chapman and Hall, NY, 1991, 645.

27. **Appelbaum, E. R., Thompson, D. V., Idler, K., and Chartrain, N.,** *Bradyrhizobium japonicum* USDA 191 has two *nodD* genes that differ in primary structure and function, *J. Bacteriol.*, 170, 12, 1988.
28. **Goethals, K., Van Den Eede, G., Van Montagu, M., and Holsters, M.,** Identification and characterization of a functional *nodD* gene in *Azorhizobium caulinodans* ORS571, *J. Bacteriol.*, 172, 2658, 1990.
29. **Török, I., Kondorosi, E., Strepkowski, T., Postfai, J., and Kondorosi, A.,** Nucleotide sequence of *Rhizobium meliloti* nodulation genes, *Nucleic Acids Res.*, 12, 9509, 1984.
30. **Jacobs, T. W., Egelhoff, T. T., and Long, S. R.,** Physical and genetic map of a *Rhizobium meliloti* gene region and nucleotide sequence of *nodC, J. Bacteriol.*, 162, 469, 1985.
31. **Debellé, F. and Sharma, S. B.,** Nucleotide sequence of *Rhizobium meliloti* RCR2011 genes involved in host specificity of nodulation, *Nucleic Acids Res.*, 14, 7453, 1986.
32. **Horvath, B., Kondorosi, E., John, M., Schmidt, J., Török, I., Gyorgypal, Z., Barabas, I., Wieneke, U., Schell, J., and Kondorosi, A.,** Organization, structure and symbiotic function of *Rhizobium meliloti* nodulation genes determining host specificity for alfalfa, *Cell*, 46, 335, 1986.
33. **Cervantès, E., Sharma, S. B., Maillet, F., Vasse, J., Truchet, G., and Rosenberg, C.,** The production of host-specific *nodQ* gene of *Rhizobium meliloti* shares homology with translation, elongation and initiation factors, *Mol. Microbiol.*, 3, 745, 1989.
34. **Gerhold, D., Stacey, G., and Kondorosi, A.,** Use of a promoter-specific probe to identify two loci from the *Rhizobium meliloti* nodulation regulon, *Plant Mol. Biol.*, 12, 181, 1989.
35. **Mulligan, J. T. and Long, S. R.,** A family of activator genes regulates expression of *Rhizobium meliloti* nodulation genes, *Genetics*, 122, 7, 1989.
36. **Schwedock, J. and Long, S. R.,** Nucleotide sequence and protein products of two new nodulation genes of *Rhizobium meliloti, nodP* and *nodQ, Mol. Plant-Microbe Interact.*, 2, 181, 1989.
37. **Honma, M. A., Asomaning, M., and Ausubel, F. M.,** *Rhizobium meliloti nodD* genes mediate host-specific activation of *nodABC, J. Bacteriol.*, 172, 901, 1990.
38. **Kondorosi, A.,** Overview on genetics of nodule induction: factors controlling nodule induction by *Rhizobium meliloti*, in *Advances in Molecular Genetics of Plant-Microbe Interactions*, Vol. 1, Hennecke, H. and Verma, D. P. S., Eds., Kluwer Academic, Dordrecht, 1991, 111.
39. **Rossen, L., Johnston, A. W. B., and Downie, J. A.,** DNA sequence of the *Rhizobium leguminosarum* nodulation genes *nodAB* and *C* required for root hair curling, *Nucleic Acids Res.*, 12, 9497, 1984.
40. **Evans, I. and Downie, J. A.,** The NodI product of *Rhizobium leguminosarum* is closely related to ATP-binding bacterial transport proteins: nucleotide sequence of the *nodI* and *nodJ* genes, *Gene*, 43, 95, 1986.
41. **Shearman, C. A., Rossen, L., Johnston, A. W. B., and Downie, J. A.,** The *Rhizobium* gene *nodF* encoded a protein similar to acyl carrier protein and is regulated by *nodD* plus a factor in pea root exudate, *EMBO J.*, 5, 647, 1986.
42. **Davis, E. O., Evans, I. J., and Johnston, A. W. B.,** Identification of *nodX*, a gene that allows *Rhizobium leguminosarum* biovar *viciae* strain TOM to nodulate Afghanistan peas, *Mol. Gen. Genet.*, 212, 531, 1988.
43. **Surin, B. P. and Downie, J. A.,** Characterization of the *Rhizobium leguminosarum* genes *nodLMN* involved in efficient host specific nodulation, *Mol. Microbiol.*, 2, 173, 1988.
44. **de Maagd, R. A., Wijfjes, A. H. M., Spaink, H. P., Ruiz-Sainz, J. E., Wijffelman, C. A., Okker, R. J. H., and Lugtenberg, B. J. J.,** *nodO*, a new *nod* gene of the *Rhizobium leguminosarum* biovar *viciae* Sym plasmid encodes a secreted protein, *J. Bacteriol.*, 171, 6764, 1990.
45. **Economou, A., Hamilton, W. D. O., Johnston, A. W. B., and Downie, J. A.,** The *Rhizobium* nodulation gene *nodO* encodes a $Ca^{2+}$-binding protein that is exported without N-terminal cleavage and is homologous to haemolysin and related proteins, *EMBO J.*, 9, 349, 1990.
46. **Surin, B. P., Watson, J. M., Hamilton, W. D. O., Economou, A., and Downie, J. A.,** Molecular characterization of the nodulation gene, *nodT*, from two biovars of *Rhizobium leguminosarum, Mol. Microbiol.*, 4, 245, 1990.
47. **Lamb, J. W. and Hennecke, H.,** in *Bradyrhizobium japonicum* the common nodulation genes *nodABC* are linked to NifA and FixA, *Mol. Gen. Genet.*, 202, 512, 1986.
48. **Göttert, M., Hitz, S., and Hennecke, H.,** Identification of *nodS* and *nodU*, two inducible genes inserted between the *Bradyrhizobium japonicum nodYABC* and *nodIJ* genes, *Mol. Plant-Microbe Interact.*, 3, 308, 1990.
49. **Göttfert, M., Grob, P., and Hennecke, H.,** Proposed regulatory pathway encoded by the *nodV* and *nodW* genes, determinants of host specificity in *Bradyrhizobium japonicum, Proc. Natl. Acad. Sci. U.S.A.*, 87, 2680, 1990.
50. **Györgypal, Z., Iyer, N., and Kondorosi, A.,** Three regulatory *nodD* alleles of diverged flavonoid-specificity are involved in host-dependent nodulation by *Rhizobium meliloti, Mol. Gen. Genet.*, 212, 85, 1988.
51. **Long, S. R., Fisher, R. F., Ogawa, J., Swanson, J., Ehrhardt, D. W., Atkinson, E. M., and Schwedock, J.,** *Rhizobium meliloti* nodulation gene regulation and molecular signals, in *Advances in Molecular Genetics of Plant-Microbe Interactions*, Vol. 1, Hennecke, H. and Verma, D. P. S., Eds., Kluwer Academic, Dordrecht, 1991, 127.

52. **Rostas, K., Kondorosi, E., Horvath, B., Simoncsits, A., and Kondorosi, A.,** Conservation of extended promoter regions of nodulation genes in *Rhizobium, Proc. Natl. Acad. Sci. U.S.A.,* 83, 1757, 1986.
53. **Fisher, R. F., Egelhoff, T. T., Mulligan, J. T., and Long, S. R.,** Specific binding of proteins from *Rhizobium meliloti* cell-free extracts containing NodD to DNA sequences upstream of inducible nodulation genes, *Genes Dev.,* 2, 282, 1988.
54. **Fisher, R. F. and Long, S. R.,** DNA footprint analysis of the transcriptional activator proteins NodD1 and NodD3 on inducible *nod* genes promoters, *J. Bacteriol.,* 171, 5492, 1989.
55. **Peters, N. K., Frost, J. W., and Long, S. R.,** A plant flavone, luteolin, induces expression of *Rhizobium meliloti* nodulation genes, *Science,* 233, 977, 1986.
56. **Horvath, B., Bachem, C. W. B., Schell, J., and Kondorosi, A.,** Host-specific regulation of nodulation genes in *Rhizobium* is mediated by a plant-signal, interacting with the *nodD* gene product, *EMBO J.,* 6, 841, 1987.
57. **Spaink, H. P., Wijffelman, C. A., Pees, E., Okker, R. J. H., and Lugtenberg, B. J. J.,** *Rhizobium* nodulation gene *nodD* as a determinant of host specificity, *Nature (London),* 328, 337, 1987.
58. **Maillet, F., Debellé, F., and Dénarié, J.,** Role of the *nodD* and *syrM* genes in the activation of the regulatory gene *nodD3,* and of the common and host-specific *nod* genes of *Rhizobium meliloti, Mol. Microbiol.,* 4, 1975, 1990.
59. **Göttfert, M., Lamb, J. W., Gasser, R., Semenza, J., and Hennecke, H.,** Mutational analysis of the *Bradyrhizobium japonicum* common *nod* genes and further *nod* box-linked genomic DNA regions, *Mol. Gen. Genet.,* 215, 407, 1989.
60. **Lewin, A., Cervantès, E., Wong, C. H., and Broughton, W. J.,** *nodSU,* two new *nod* genes of the borad host-range *Rhizobium* strain NGR234 encode host-specific nodulation of the tropical tree Leucaena leucocephala, Mol. Plant-Microbe Interact., 3, 317, 1990.
61. **Finan, T. M., Hirsch, A. M., Leigh, J. A., Johansen, E., Kuldau, G. A., Deegan, S., Walker, G. C., and Signer, E. R.,** Symbiotic mutants of *Rhizobium meliloti* that uncouple plant and bacterial differentiation, *Cell,* 40, 869, 1985.
62. **Stanfield, S., Telpi, L., O'Brochta, D., Helinski, D. R., and Ditta, G. S.,** The *ndvA* gene product of *Rhizobium meliloti* is required for Beta-(1-2)glucan production and has homology to the ATP binding export protein HlyB, *J. Bacteriol.,* 170, 3523, 1988.
63. **Goethals, K., Gao, M., Tomekpe, K., Van Montagu, M., and Holsters, M.,** Common *nodABC* genes in *nod* locus 1 of *Azorhizobium caulinodans:* nucleotide sequence and plant-inducible expression, *Mol. Gen. Genet.,* 219, 289, 1989.
64. **Vasquez, M., Davalos, A., De Las Penas, A., Sanchez, F., and Quinto, C.,** Novel organization of the common nodulation genes in *Rhizobium leguminosarum* bv. *phaseoli* strains, *J. Bacteriol.,* 173, 1250, 1991.
65. **Scott, D. B.,** personal communication, 1990.
66. **Schmidt, J., John, M., Wieneke, U., Krüssmann, H. D., and Schell, J.,** Expression of the nodulation gene *nodA* in *Rhizobium meliloti* and localization of the gene product in the cytosol, *Proc. Natl. Acad. Sci. U.S.A.,* 83, 9581, 1986.
67. **Johnson, D., Roth, E. L., and Stacey, G.,** Immunogold localization of the NodC and NodA proteins of *Rhizobium meliloti, J. Bacteriol.,* 171, 4583, 1989.
68. **John, M., Schmidt, J., Wieneke, U., Krüssmann, H. D., and Schell, J.,** Transmembrane orientation and receptor-like structure of the *Rhizobium meliloti* common nodulation protein NodC, *EMBO J.,* 7, 583, 1988.
69. **Kondorosi, E., Banfalvi, Z., and Knodorosi, A.,** Physical and genetic analysis of a symbiotic region of *Rhizobium meliloti:* identification of nodulation genes, *Mol. Gen. Genet.,* 193, 445, 1984.
70. **Djordjevic, M. A., Schofield, P. R., and Rolfe, B. G.,** Tn5 mutagenesis of *Rhizobium trifolii* host specific nodulation genes result in mutants with altered host range ability, *Mol. Gen. Genet.,* 200, 463, 1985.
71. **Debellé, F., Rosenberg, C., Vasse, J., Maillet, F., Martinez, E., Dénarié, J., and Truchet, G.,** Assignment of symbiotic developmental phenotypes to common and specific nodulation *(nod)* genetic loci of *Rhizobium meliloti, J. Bacteriol.,* 168, 1075, 1986.
72. **Faucher, C., Camut, S., Dénarié, J., and Truchet, G.,** The *nodH* and *nodQ* host range genes of *Rhizobium meliloti* behave as avirulence genes in *R. leguminosarum* bv. *viciae* and determine changes in the production of plant-specific extracellular signals, *Mol. Plant-Microbe Interact.,* 2, 291, 1989.
73. **Djordjevic, M. A., Innes, R. W., Wijffelman, C. A., Schofield, P. R., and Rolfe, B. G.,** Nodulation of specific legumes is controlled by several distinct loci in *Rhizobium trifolii, Plant Mol. Biol.,* 6, 389, 1986.
74. **Debellé, F. Maillet, F., Vasse, J., Rosenberg, C., De Billy, F., Truchet, G., Dénarié, J., and Ausubel, F. M.,** Interference between *Rhizobium meliloti* and *Rhizobium trifolii* nodulation genes: genetic basis of the *R. meliloti* dominance, *J. Bacteriol.,* 170, 5718, 1988.
75. **Spaink, H. P., Weinman, J., Djordjevic, M. A., Wijffelman, C. A., Okker, R. J. H., and Lugtenberg, B. J. J.,** Genetic analysis and cellular localization of the *Rhizobium* host specificity-determining NodE protein, *EMBO J.,* 8, 2811, 1989

76. **Surin, B. P. and Downie, J. A.,** *Rhizobium leguminosarum* genes required for expression and transfer of host specific nodulation, *Plant Mol. Biol.*, 12, 19, 1989.
77. **Downie, J. A. and Surin, B. P.,** Either of two *nod* gene loci can complement the nodulation defect of a *nod* deletion mutant of *Rhizobium leguminosarum* bv. *viciae, Mol. Gen. Genet.*, 222, 81, 1990.
78. **Downie, J. A., Marie, C., Scheu, A. K., Firmin, J. L., Wilson, K. E., Davis, A. E., Cubo, T. M., Mavridou, A., Johnston, A. W. B., and Economou, A.,** Genetic and biochemical studies on the nodulation genes of *Rhizobium leguminosarum* bv. *viciae,* in *Advances in Molecular Genetics of Plant-Microbe Interactions,* Vol. 1., Hennecke, H. and Verma, D. P. S., Eds., Kluwer Academic, Dordrecht, 1991, 134.
79. **Rodrigues-Quinones, F., Banfalvi, Z., Murphy, P., and Kondorosi, A.,** Interspecies homology of nodulation genes in *Rhizobium, Plant Mol. Biol.*, 8, 61, 1987.
80. **Roche, P., Debellé, F., Maillet, F., Lerouge, P., Faucher, C., Truchet, G., Dénarié, J., and Promé, J. C.,** in preparation, 1991.
81. **Downie, J. A.,** The *nodL* gene from *Rhizobium leguminosarum* is homologous to the acetyl transferase encoded by *lacA* and *cysE, Mol. Microbiol.*, 3, 1649, 1989.
82. **Baev, N., Endre, G., Petrovics, G., Banfalvi, Z., and Kondorosi, A.,** Six nodulation genes of *nod* box locus 4 in *Rhizobium meliloti* are involved in nodulation signal production: *nodM* codes for D-glucosamine synthetase, *Mol. Gen. Genet.*, 228, 113, 1991.
83. **Schwedock, J. and Long, S. R.,** ATP sulfurylase activity of the *nodP* and *nodQ* gene products of *Rhizobium meliloti, Nature (London),* 348, 644, 1990.
84. **Spaink, H. P., Geiger, O., Sheeley, D. M., van Brussel, A. A. N., York, W. S., Reinhold, V. N., Lugtenberg, B. J. J., and Kennedy, E. P.,** The biochemical function of the *Rhizobium leguminosarum* proteins involved in the production of host specific signal molecules, in *Advances in Molecular Genetics of Plant-Microbe Interactions,* Vol. 1, Hennecke, H. and Verma, D. P. S., Eds., Kluwer Academic, Dordrecht, 1991, 142.
85. **Hiltner, L.,** Über die Ursachen, welche die Grösse, Zahl, Stellung, und Wirkung der Wurzelknöllchen der Leguminosen bedingen, *Arb. Biol. Reichsanst. Land Forstwirt Berlin Dahlem,* 1, 177, 1900.
86. **McCoy, E.,** Infection by *Bact. radicicola* in relation to the microchemistry of the host's cell walls, *Proc. Roy. Soc. London, Ser B,* 110, 514, 1932.
87. **Yao, P. Y. and Vincent, J. M.,** Host specificity in the root hair "curling factor" of *Rhizobium* spp., *Austr. J. Biol. Sci.*, 22, 413, 1969.
88. **Hubbell, D. H.,** Studies on the root hair "curling factor" of *Rhizobium, Bot. Gaz.*, 131, 337, 1970.
89. **Solheim, B. and Raa, J.,** Characterization of the substances causing the deformation of root hairs of *Trifolium repens* when inoculated with *Rhizobium trifolii, J. Gen. Microbiol.*, 77, 241, 1973.
90. **Yao, P. Y. and Vincent, J. M.,** Factors responsible for the curling and branching of clover root hairs by *Rhizobium, Plant Soil,* 45, 1, 1976.
91. **Vincent, J. M.,** Root-nodule symbiosis with *Rhizobium,* in *The Biology of Nitrogen Fixation,* Quispel, A., Ed., North-Holland, Amsterdam, 1974, 265.
92. **Bhuvaneswari, T. V. and Solheim, B.,** Root hair deformation in the white clover-*Rhizobium trifolii* symbiosis, *Physiol. Plant.*, 63, 25, 1985.
93. **Fahraeus, G. and Lyunggren, J.,** Pre-infection phases of the legume symbiosis, in *The Ecology of Soil Bacteria,* Gray, T. R. G. and Parkinson, D., Eds., Liverpool University Press, Liverpool, 1968, 396.
94. **Hollingsworth, R. I., Squartini, A., Philip-Hollingsworth, S., and Dazzo, F. B.,** Root hair deforming and nodule initiating factors from *Rhizobium trifolii,* in *Signal Molecules in Plant and Plant-Microbe Interactions,* Lugtenberg, B. J. J., Ed., Springer-Verlag, Berlin, 1989, 387.
95. **Truchet, G., Michel, M., and Dénarié, J.,** Sequential analysis of the organogenesis of lucerne *(Medicago sativa)* root nodules using symbiotically-defective mutants of *Rhizobium meliloti, Differentiation,* 16, 163, 1980.
96. **Kapp, D., Niehaus, K., Kuandt, J., Müller, P., and Pühler, A.,** Cooperative action of *Rhizobium meliloti* nodulation and infection mutants during the process of forming mixed infected alfalfa nodules, *Plant Cell,* 2, 139, 1990.
97. **Truchet, G., Barker, D. G., Camut, S., De Billy, F., Vasse, J., and Huguet, T.,** Alfalfa nodulation in the absence of *Rhizobium, Mol. Gen. Genet.*, 219, 65, 1989.
98. **Hollingsworth, R. I., Philip-Hollingsworth, S., and Dazzo, F. B.,** Isolation, characterization, and structural elucidation of a "*nod* signal" excreted by *Rhizobium trifolii* ANU843 which induces root hair branching and nodule-like primordia in axenic white clover seedlings, in *Nitrogen Fixation: Achievements and Objectives,* Gresshoff, P. M., Roth, L. E., Stacey, G., and Newton, W. E., Eds., Chapman and Hall, NY, 1990, 193.
99. **Libbenga, K. R. and Harkes, P. A. A.,** Initial proliferation of cortical cells in the formation of root nodules in *Pisum sativum, Planta,* 114, 17, 1973.
100. **Hirsch, A. M., Bhuvaneswari, T. V., Torrey, J. G., and Bisseling, T.,** Early nodulin genes are induced in alfalfa root outgrowths elicited by auxin transport inhibitors, *Proc. Natl. Acad. Sci. U.S.A.*, 86, 1244, 1989.

101. **Long, S. R. and Cooper, J.,** Overview of symbiosis, in *Molecular Genetics of Plant-Microbe Interactions,* Verma, D. P. S. and Palacios, R., Eds., APS Press, St. Paul, MN, 1988, 163.
102. **Sturtevant, D. B. and Taller, B. J.,** Cytokinin production by *Bradyrhizobium japonicum, Plant Physiol.,* 89, 1247, 1989.
103. **Taller, B. J. and Sturtevant, D. B.,** Cytokinin production by rhizobia, in *Advances in Molecular Genetics of Plant-Microbe Interactions,* Vol. 1, Hennecke, H. and Verma, D. P. S., Eds., Kluwer Academic, Dordrecht, 1991, 215.
104. **Badenoch-Jones, J., Summons, R. E., Djordjevic, M. A., Shine, J., Letham, D. S., and Rolfe, B. G.,** Mass spectrometric quantification of indole-3-acetic acid in *Rhizobium* culture supernatants: relation to root hair curling and nodule initiation, *Appl. Environ. Microbiol.,* 44, 275, 1982.
105. **van Brussel, A. A. N., Tak, T., Wetselaar, A., Pees, E., and Wijffelman, C. A.,** Small leguminosae as test plants for nodulation of *Rhizobium leguminosarum* and other rhizobia and agrobacteria harbouring a leguminosarum plasmid, *Plant Sci. Lett.,* 27, 317, 1982.
106. **van Brussel, A. A. N., Zaat, S. A. J., Canter-Cremers, H. C. J., Wijffelman, C. A., Pees, E., Tak, T., and Lugtenberg, B. J. J.,** Role of plant root exudate and Sym plasmid-localized nodulation genes in the synthesis by *Rhizobium leguminosarum* of Tsr factor, which causes thick and short roots on common vetch, *J. Bacteriol.,* 165, 517, 1986.
107. **Canter-Cremers, H. C. J., van Brussel, A. A. N., Plazinski, J., and Rolfe, B. G.,** Sym plasmid and chromosomal gene products of *Rhizobium trifolii* elicit developmental responses on various legume roots, *J. Plant Physiol.,* 122, 25, 1986.
108. **Zaat, S. A. J., van Brussel, A. A. N., Tak, T., Pees, E., and Lugtenberg, B. J. J.,** Flavonoids induce *Rhizobium leguminosarum* to produce *nodDABC* gene-related factors that cause thick and short roots and root hair responses on common vetch, *J. Bacteriol.,* 169, 3388, 1987.
109. **Schmidt, J., Wingender, R., John, M., Wieneke, U., and Schell, J.,** *Rhizobium meliloti nodA* and *nodB* genes are involved in generating compounds that stimulate mitosis of plant cells, *Proc. Natl. Acad. Sci. U.S.A.,* 85, 8578, 1988.
110. **Faucher, C., Maillet, F., Vasse, J., Rosenberg, C., van Brussel, A. A. N., Truchet, G., and Dénarié, J.,** *Rhizobium meliloti* host range *nodH* gene determines production of an alfalfa-specific extracellular signal, *J. Bacteriol.,* 170, 5489, 1988.
111. **Banfalvi, Z. and Kondorosi, A.,** Production of root hair deformation factors by *Rhizobium meliloti* nodulation genes in *Escherichia coli: HsnD (nodH)* is involved in the plant host-specific modification of the NodABC-factor, *Plant Mol. Biol.,* 13, 1, 1989.
112. **Lerouge, P., Roche, P., Faucher, C., Maillet, F., Truchet, G., Promé, J. C., and Dénarié, J.,** Symbiotic host-specificity of *Rhizobium meliloti* is determined by a sulphated and acylated glucosamine oligosaccharide signal, *Nature (London),* 344, 781, 1990.
113. **Roche, P., Lerouge, P., Ponthus, C., and Promé, J. C.,** Structural determination of bacterial nodulation factors involved in the *Rhizobium meliloti*-alfalfa symbiosis, *J. Biol. Chem.,* 266, 10933, 1991.
113a. **Long, S. R.,** personal communication, 1990.
114. **Ardourel, M. Y., Roche, P., and Promé, J. C.,** personal communication, 1990.
115. **Broughton, W. J., Krause, A., Lewin, A., Perret, X., Price, N. P. J., Relic, B., Rochepeau, P., Wong, C. W., Pueppke, S. G., and Brenner, S.,** Signal exchange mediates host-specific nodulation of tropical legumes by the broad host-range *Rhizobium* species NGR234, in *Advances in Molecular Genetics of Plant-Microbe Interactions,* Vol. 1, Hennecke, H. and Verma, D. P. S., Eds., Kluwer Academic, Dordrecht, 1991, 162.
116. **Price, N. P. J., Lerouge, P., Relic, B., Promé, J. C., and Broughton, W. J.,** personal communication, 1990.
117. **Spaink, H. P.,** personal communication, 1990.
118. **John, M., Schmidt, J., Wieneke, U., Kondorosi, E., Kondorosi, A., and Schell, J.,** Expression of the nodulation gene *nodC* of *Rhizobium meliloti* in *Escherichia coli:* role of the *nodC* gene product in nodulation, *EMBO J.,* 4, 2425, 1985.
119. **Watanabe, T., Oyanagi, W., Suzuki, K., and Tanaka, H.,** Chitinase system of *Bacillus circulans* WL-12 and importance of chitinase A1 in chitin degradation, *J. Bacteriol.,* 172, 4017, 1990.
120. **Swanson, J., Tu, J. K., Ogawa, J. M., Sanga, R., Fisher, R., and Long, S. R.,** Extended region of nodulation genes in *Rhizobium meliloti* 1021. I. Phenotypes of Tn5 insertion mutants, *Genetics,* 117, 181, 1987.
121. **van Brussel, A. A. N., Recourt, K., Pees, E., Spaink, H. P., Tak, T., Wijffelman, C. A., Kijne, J. W., and Lugtenberg, B. J. J.,** A biovar-specific signal of *Rhizobium leguminosarum* bv. *viciae* induces nodulation gene-inducing activity in root exudate of *vicia sativa* subsp. *nigra, J. Bacteriol.,* 172, 5394, 1990.
122. **Scheres, B., Van De Wiel, C., Zalensky, A., Horvath, B., Spaink, H. P., Van Eck, H., Zwartkruis, F., Wolters, A. M., Gloudemans, T., Van Kammen, A., and Bisseling, T.,** The ENOD12 gene product is involved in the infection process during the pea-*Rhizobium* interaction, *Cell,* 60, 281, 1990.

123. **Ardourel, M. Y., Lortet, G., and Rosenberg, C.**, personal communication, 1990.
124. **Ardourel, M. Y., Roche, P., Lortet, G., Promé, J. C., and Rosenberg, C.**, personal communication, 1990.
125. **Roche, P., Debellé, F., and Promé, J. C.**, personal communication, 1990.
126. **Lerouge, P., Roche, P., Promé, J. C., Faucher, C., Vasse, J., Maillet, F., Camut, S., De Billy, F., Barker, D. G., Dénarié, J., and Truchet, G.**, *Rhizobium meliloti* nodulation genes specify the production of an alfalfa-specific sulphated oligosaccharide signal, in *Nitrogen Fixation: Achievements and Objectives*, Gresshoff, P. M., Newton, W. E., Roth, E. L., and Stacey, G., Eds., Chapman-Hall, NY, 1990, 177.
127. **Roche, P., Lerouge, P., Promé, J. C., Faucher, C., Vasse, J., Maillet, F., Camut, S., De Billy, F., Dénarié, J., and Truchet, G.**, NodRm-1, a sulphated lipo-oligosaccharide signal of *Rhizobium meliloti* elicits hair deformation, cortical cell division and nodule organogenesis on alfalfa roots, in *Advances in Molecular Genetics of Plant-Microbe Interactions*, Vol. 1, Hennecke, H. and Verma, D. P. S., Eds., Kluwer Academic, Dordrecht, 1991, 119.
128. **Long, S. R.**, personal communication, 1990.
129. **Diaz, C. L., Melchers, L. S., Hooykaas, P. J. J., Lugtenberg, B. J. J., and Kijne, J. W.**, Root lectin and a determinant of host-plant specificity in the *Rhizobium*-legume symbiosis, *Nature (London)*, 338, 579, 1989.
130. **Diaz, C. L., Van Spronsen, P. C., Bakhuizen, R., Logman, G. J. J., Lugtenberg, B. J. J., and Kijne, J. W.**, Correlation between infection by *Rhizobium leguminosarum* and lectin on the surface of *Pisum sativum* L. roots, *Planta*, 168, 530, 1986.
131. **Lugtenberg, B. J. J., Diaz, C. L., Smit, G., De Pater, S., and Kijne, J. W.**, Roles of the lectin in the *Rhizobium*-legume symbiosis, in *Advances in Molecular Genetics of Plant-Microbe Interactions*, Vol. 1, Hennecke, H. and Verma, D. P. S., Eds., Kluwer Academic, Dordrecht, 1991, 174.
132. **Lis, H. and Sharon, N.**, Lectins as molecules and as tools, *Annu. Rev. Biochem.*, 55, 35, 1986.
133. **Scheres, B., van Engelen, F., van der Knaap, E., Van de Wiel, C., Van Kammen, A., and Bisseling, T.**, Sequential induction of nodulin gene expression in the developing pea nodule, *Plant Cell*, 2, 687, 1990.
134. **Bisseling, T., Franssen, H., Govers, F., Horvath, B., Moerman, M., Scheres, B., Van de Wiel, C., and Yang, W. C.**, Early nodulins in pea and soybean nodule development, in *Advances in Molecular Genetics of Plant-Microbe Interactions*, Vol. 1, Hennecke, H. and Verma, D. P. S., Eds., Kluwer Academic, Dordrecht, 1991, 300.
135. **Horvath, B. and Bisseling, T.**, personal communication, 1990.
136. **Zeroni, M. and Hall, M. A.**, Molecular effects of hormone treatment on tissue, in *Hormonal Regulation of Development*, Vol. 1: Molecular Aspects, MacMillan, J., Ed., Springer-Verlag, Berlin, 1980, 511.
137. **Darvill, A. G. and Albersheim, P.**, Phytoalexins and their elicitors: A defense against microbial infection in plants, *Annu. Rev. Plant. Physiol.*, 35, 243, 1984.
138. **Darvill, A. G., Albersheim, P., Bucheli, P., Doares, S., Doubrava, N., Eberhard, S., Gollin, D. J., Hahn, M. G., Marfa-Riera, V., York, W. S., and Mohnen, D.**, Oligosaccharins: plant regulatory molecules, in *Signal Molecules in Plants and Plant-Microbe Interactions*, NATO ASI Series, Lugtenberg, B. J. J., Ed., Springer-Verlag, Berlin, 1989, 41.
139. **Hahn, M. G.**, Animal receptors. Examples of cellular signal perception molecules, in *Signal Molecules in Plants and Plant-Microbe Interactions*, NATO ASI Series, Lugtenberg, B. J. J., Ed., Springer-Verlag, Berlin, 1989, 1.
140. **Zaat, S. A. J., van Brussel, A. A. N., Tak, T., Lugtenberg, B. J. J., and Kijne, J. W.**, The ethylene-inhibitor aminoethoxyvinylglycine restores normal nodulation by *Rhizobium leguminosarum* biovar *viciae* on *Vicia sativa* subsp. *nigra* by suppressing the "Thick and short roots" phenotype, *Planta*, 177, 141, 1989.
141. **Truchet, G., Roche, P., Lerouge, P., Vasse, J., Camut, S., De Billy, F., Promé, J. C., and Dénarié, J.**, Sulphated lipo-oligosaccharide signals of *Rhizobium meliloti* elicit root nodule organogenesis in alfalfa, *Nature*, 351, 670, 1991.
141a. **Horvath, B. and Bisseling, T.**, personal communication, 1990.
142. **Bisseling, T.**, personal communication.
143. **Tran Than Van, K., Toubart, P., Cousson, A., Darvill, A. G., Gollin, D. J., Chelf, P., and Albersheim, P.**, Manipulation of the morphogenetic pathway of tobacco explants by oligosaccharins, *Nature (London)*, 314, 615, 1985.
144. **McDougall, G. J. and Fry, S. C.**, Xyloglucan oligosaccharides promote growth and activate cellulase: evidence for a role of cellulase in cell expansion, *Plant Physiol.*, 93, 1042, 1990.
145. **Eberhard, S., Doubrava, N., Marfa, V., Mohnen, D., Southwick, A., Darvill, A. G., and Albersheim, P.**, Pectic cell wall fragments regulate tobacco thin-cell-layer explant morphogenesis, *Plant Cell*, 1, 747, 1989.
146. **Schmidt, J., John, M., Wieneke, U., Stacey, G., Röhrig, H., and Schell, J.**, Studies on the function of *Rhizobium meliloti* nodulation genes, in *Advances in Molecular Genetics of Plant-Microbe Interactions*, Vol. 1, Hennecke, H. and Verma, D. P. S., Eds., Kluwer Academic, Dordrecht, 1991, 150.

147. **Price, N. P. J., Lerouge, P., Promé, J. C., and Broughton, W. J.**, personal communication, 1990.

148. **Philip-Hollingsworth, S., Hollingsworth, R. I., Dazzo, F. B., Djordjevic, M. A., and Rolfe, B. G.**, The effect of interspecies transfer of *Rhizobium* host-specific nodulation genes on acidic polysaccharide structure and *in situ* binding by host lectin, *J. Biol. Chem.*, 264, 5710, 1989.

149. **Hakomori, S. I.**, Bifunctional role of glycosphingolipids: Modulators for transmembrane signaling and mediators for cellular interactions, *J. Biol. Chem.*, 265, 18713, 1990.

150. **Slomiany, A., Slomiany, B. L., and Annese, C.**, Sulfated trihexosylceramide from gastric mucosa containing N-acetylglucosamine, *Eur. J. Biochem.*, 109, 471, 1980.

151. **Roberts, D. D.**, Sulfatide-binding proteins, in *Methods Enzymol.*, 138, 4735, 1987.

# REGULATION OF NODULATION GENES IN RHIZOBIA

**Adam Kondorosi**

## TABLE OF CONTENTS

# I. INTRODUCTION

Soil bacteria, referred to as rhizobia belonging to the genera *Rhizobium, Bradyrhizobium,* and *Azorhizobium* have the unique ability to induce nitrogen-fixing nodules on the roots or stems of leguminous plants. Nodule development consists of several stages determined by sets of genes both in the micro- and macrosymbiont. The first observable signs of bacterium-plant interaction is the curling of root hairs and induction of meristematic cell division. A particular rhizobial species can nodulate only appropriate plant hosts indicating the host-specific nature of these interactions. The bacterial genes determining these early events of nodulation are the nodulation (*nod*) genes. Due to the fairly high number of nodulation genes, the terminology has used up all the letters of the alphabet. Therefore, the recently identified *nod* genes were designated as *nol* genes.

It has been observed that rhizobia exposed to plants have an increased ability to nodulate the plant host suggesting that some plant compounds may stimulate the expression of nodulation genes.[1] In 1985, several laboratories reported that, indeed, low-molecular weight compounds in plant root exudates activate the expression of *nod* genes.[2-4] It was found that the only *nod* gene which expresses constitutively, the *nodD,* has a positive regulatory role in *nod* gene expression. Studies of various *nod* promoter regions revealed a highly conserved sequence, the *nod* box, a *cis*-acting element for coordinated *nod* gene activation by NodD and the plant signals, suggesting that the *nod* genes form a regulatory circuit, the *nod* regulon.[5] The inducing compounds were identified as flavonoids,[6-8] acting in conjunction with NodD in a host-specific manner.[9,10] Further studies have led to the demonstration of additional factors controlling *nod* gene expression.[11,12] Below, these intricacies of *nod* gene regulation will be detailed.

Two chapters in this volume deal with nodulation genetics, the function of *nod* genes and the biology of nodule induction. Therefore, these aspects of nodulation will not be dealt with in this chapter. Several recent papers have also reviewed *nod* genetics and *nod* gene regulation.[13-20]

# II. ORGANIZATION OF *NOD* GENES

Currently, numerous *nod* (and *nol*) genes have been identified and sequenced from strains and species of the three rhizobial genera, leading to the description of over thirty distinct nodulation genes.

## A. LOCATION AND CLUSTERING OF *NOD* GENES

In the *Rhizobium* genus (formerly referred to as "fast-growing *Rhizobium* species"), the *nod* genes are localized on large indigenous plasmids (sym plasmids) in the vicinity of nitrogen fixation (*nif*) genes. In *Bradyrhizobium* (formerly "slow-growing *Rhizobium* species") and in *Azorhizobium* the symbiotic genes are located on the chromosome.[19,21,22] In the three rhizobial genera the *nod* genes were found in one or in a few gene clusters and these gene clusters are located on a relatively short DNA segment (Figure 1).[22,23] In a number of strains, however, reiteration of symbiotic genes, including the *nod* genes was detected and the reiterated sequences are more scattered in the genome.[23,24] In some cases each copy of the *nod* gene was shown to be functional, such as the allelic forms of *nodD*.[25-29] The detected reiteration of *nod* genes by hybridization may be due to the evolution of some *nod* genes from certain genes with "housekeeping" functions. For instance, the previously observed reiteration of *nodM* was recently explained by the demonstration that *nodM* encodes glucosamine synthase[30] which is needed for the synthesis of the Nod factor[31] and the reiterated sequence indicates a *glmS* gene required for cell wall synthesis. The *nodM,* therefore, can be considered as a *glmS* under plant signal control.[30] Similarly, the *nodP* and *Q* genes

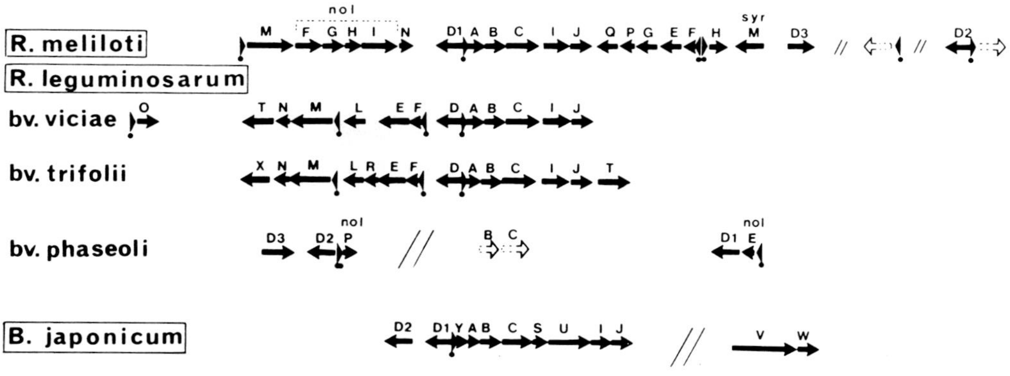


FIGURE 1. Organization of nodulation genes in different *Rhizobium* and *Bradyrhizobium* species and biovars. Full arrows, sequenced genes; open arrows, genes without published sequences; letters, *nod* genes, unless otherwise indicated; full points with triangles above them indicate *nod* box sequences with their orientation.

correspond to *cysD* and *cysN* and have ATP-sulfurylase activity which is needed for the modification of the Nod factor.[32]

## B. COMMON AND HOST-SPECIFIC NODULATION GENES

The *nodABC* and *nodD* genes have been found in all rhizobial strains investigated so far. The *nodABC* genes are structurally and functionally conserved, therefore, they termed as common *nod* genes.[21] Mutations in these genes completely block the early interactions, root hair deformation and curling (Had$^-$ Hac$^-$) and, consequently, nodulation (Nod$^-$). These mutations can be complemented by *nodABC* genes from any other rhizobial strain. The *nodD* is common in that sense that it plays a positive regulatory role in rhizobia, but its activation by the plant signal is specific, i.e., it induces *nod* gene expression in a host-dependent manner.[15,19,33]

The overwhelming majority of the various *nod* genes are present only in certain rhizobial species or genera. Mutations in these genes cause delay of root hair curling or nodulation and cannot be complemented by *nod* genes of any other *Rhizobium*. Therefore, these genes were designated as host-specific nodulation genes (often referred to as *hsn* genes[21]). For some bacterium-plant associations, the presence of a particular *hsn* gene is essential: these *hsn* genes extend the host range to a single plant species[34] or genotype.[35-37] Two chapters of this volume[38,39] describe the *nod* genes in more detail.

## C. THE *NOD* OPERONS

In the course of *nod* gene identification, the *nod* gene regions were delimited by directed transposon mutagenesis[40] followed by sequencing. By this combined genetic and physical analysis the *nod* genes have been identified. The proximity and the same polarity of *nod* genes common to all legumes (common *nod* genes) suggested that they form a single transcriptional unit which was then supported by gene expression studies.

By constructing *nod-lacZ* fusions the expression of *nod* genes has been studied in numerous rhizobial strains. As discussed below, this technique allowed the demonstration that *nod* genes (except *nodD*) are inducible by plant signal molecules in conjunction with NodD; it also helped to assign genes in transcriptional units.

Based on these studies, it was shown that the *nodABC* form one transcriptional unit in most species.[41-51] Similarly, in many cases transcriptional units for other *nod* genes have been defined (Figure 1).[34,52-55]

## D. THE *NOD* BOX

Sequence analysis of the regions upstream of the inducible *nod* genes did not reveal a consensus *E. coli* promoter sequence. At the same time, a highly conserved 47-bp sequence, designated as *nod* box,[5] is present in front of all inducible *nod* transcriptional units (Figure 1). Strains with mutated *nod* box had a phenotype characteristic for mutations in the gene(s) downstream of the *nod* box and abolished *nod* gene inducibility, indicating that the *nod* box is an essential element in *nod* gene expression.[5,56] The mRNA initiation site for several transcriptional units has been determined and was located about 25 to 28 bp downstream of the *nod* box,[57-59] again supporting the concept that the *nod* box is an essential part of the *nod* promoter.

The 47-bp *nod* box sequence consists of four highly conserved domains:[5] two domains are so highly conserved among *nod* box sequences within one strain or among species that a 25-bp synthetic oligomer could be used as a hybridization probe to clone *nod* box-containing DNA regions from the same or from different species.[5,60] By this approach, three additional *nod* transcriptional units were identified in *R. meliloti*[5] and further genetic studies indicated, that downstream from the *nod* boxes, functional plant signal inducible *nod* genes are located that contribute to the nodulation of certain plant hosts.[30,61]

Recently, less conserved *nod* box sequences have been reported in *B. japonicum*[59,60] and in *Azorhizobium caulinodans*.[62] Divergent *nod* boxlike sequences have also been found in front of inducible genes of other members of the one-component prokaryotic regulatory genes of the LysR family.[63,64] Taking into consideration also the divergent *nod* box sequences of different rhizobial strains, Wang and Stacey[59] proposed a slightly different consensus *nod* box than was reported originally by Rostas et al.[5] In their model, the *nod* box can be viewed as a repeat of a 9-bp sequence. In the *nod* boxes also associated with the promoters of *nodD* genes, four repeats are apparent, while in the divergent sequences only two repeats can be recognized. In *nod* boxes with four repeats, two pairs of repeats are separated by 4-bp, indicating that they are on opposite sides of the DNA helix. Promoters regulated by members of the LysR family were reported in several cases to contain four repeats of a 9-bp sequence.[65]

Using gel retardation assays and DNase I footprinting experiments, it was shown that the regulatory NodD protein binds to the *nod* box.[11,66-69] The footprint is fairly extensive, covering the entire *nod* box.[11,68] From the existence of repeats and from footprint data with modified *nod* box sequences it is likely that the *nod* box DNA may bend in order to contact multimeric NodD or two sides of a NodD protein.[5,11,69]

## E. THE *NOD* REGULON

As mentioned above, the *nod* box sequence has been found in the promoter regions of all plant signal-inducible *nod* operons. This *cis*-acting element binds the positive regulatory NodD protein. The *nod* operons are coordinately activated, as observed by a number of laboratories.[3,5,14,58,61,70] Thus, *nodD* and the inducible *nod* genes form a single regulatory circuit, the *nod* regulon, as originally proposed by Rostas et al.[5]

Recently two *nod* genes of *B. japonicum* were described which do not seem to belong to the *nod* regulon.[71] The predicted gene products are members of the family of two-component regulatory systems, suggesting that these genes may regulate yet unknown genes influencing nodulation.

The *nod* regulon is functional in the taxonomically distant *E. coli,* albeit both the constitutive *nodD* and the inducible *nod* genes express at a rather low level.[30] Nevertheless, it was possible to complement the *E. coli glmS*$^-$ mutation by introducing *R. meliloti nodM* and *nodD* genes with their own promoters. In addition, for the identification of *nod* gene products an *in vitro* transcription system was developed for *R. meliloti* and comparison of this system with that of *E. coli* for the production of Nod proteins gave the same protein sizes.[72]

## III. NodD, A POSITIVE REGULATOR

All rhizobial strains tested so far contain one or more copies (2 or 3) of the *nodD* gene.[24] The copy number may be different even within species. NodD proteins have a molecular weight of 34 kDa.[9,25,45,47,55,73] Mutants containing no functional *nodD* are Nod$^-$. The *nodD* sequences are highly conserved but are not identical even within a particular strain.[25,74,75] In strains carrying several *nodD* genes the phenotype of mutations in a particular *nod* gene is dependent also on the plant host.[19] *Rhizobium meliloti* carries three *nodD* alleles and mutations in the *nodD1, nodD2,* or in *nodD3* genes cause delayed nodulation on *Medicago sativa.* In contrast to *nodD1* and *nodD3* mutants, mutations in *nodD2* do not affect the nodulation of another host, *Melilotus albus.*[25-27,76]

In strains harboring single *nodD,* the NodD protein regulates its own transcription.[3,4] There is no evidence for autoregulation of *nodD1* and *nodD2* in *R. meliloti*,[2,27,33] the *nodD3*, however, seems to be also under NodD control.[12,70] The same observation was made for one of the three *nodD* alleles of *R. leguminosarum* bv. *phaseoli.*[75] The expression of the *nodD1* gene of *B. japonicum* and the *nodD3* gene of *R. meliloti* is enhanced upon addition of *nod* gene inducers.[12,73] In front of both genes diverged *nod* box sequences were found[59,77] which are likely to be implicated in the induction of these *nodD* genes.

The role of NodD is to activate the expression of inducible *nod* genes in conjunction with specific plant signal molecules, various flavonoids, or isoflavones. Different *nodD* genes show different inducer specificities and, as a consequence, influence the host specificity of nodulation.[9,10] The specificity of each different NodD occurs at the level of their amino acid sequence. Hybrid *nodD* genes constructed from *nodD* genes of different species revealed that fairly large regions of the protein are involved in determining the signal specificity.[9,78] This region is primarily the carboxy-terminal part of the protein. Point mutations were generated in NodD which abolished the ability to activate *nod* genes or altered the plant signal specificity of NodD or allowed signal-independent activation.[79-81] One must note that specific binding of the flavonoids to NodD has not been shown chemically.

The NodD protein exhibits some resemblance to animal steroid receptors which are also known to interact with some flavonoid ligands. It was found that several steroid hormones or analogs could activate *nod* genes in conjunction with NodD or had synergistic effects with flavonoids in *nod* gene activation.[82] In addition, the consensus ligand-binding domain of the steroid receptors shows homology to the putative flavonoid-binding region of NodD, suggesting that the ligand-binding domain of steroid receptors and NodD had common evolutionary origin.

As discussed in Section II.D., the NodD protein binds to the *nod* box. In the highly conserved N-terminal region, a helix-turn-helix motif, characteristic for DNA-binding sequences, was found.[55] This sequence is conserved in the N-terminal regions of the prokaryotic positive regulatory proteins belonging to the LysR family.[83] Binding of NodD to the *nod* box does not require the presence of plant signal molecules.[11,66,67] It seems that upon addition of the appropriate inducer, the NodD bound to the *nod* box with its DNA-binding domain undergoes a conformational change, resulting in a form which activates transcription of genes lying downstream.

A series of hybrid *nodD* genes from *R. meliloti nodD1* and *R. trifolii nodD* genes were constructed and a hybrid *nodD* gene, consisting of 75% of the *R. meliloti nodD1* at the 5′-end, activated inducible *nod* promoters to maximal levels even in the absence of flavonoids.[78,84] It seems that this hybrid NodD has a conformation which can activate transcription without further modification by plant signals. *Rhizobium leguminosarum* bv. *viciae* or bv. *trifolii* NodD mutants possessing inducer-independent ability to activate *nod* gene expression have also been obtained by nitrosoguanidine mutagenesis.[79-81]

Mutational analysis of *nodD* genes resulted in several different classes of mutants, including those which have altered flavonoid sensitivity, are deficient in autoregulation, or allow a low level of constitutive expression.[79-81] Members of these mutant classes were located at several regions of the NodD protein.

The NodD protein sequence contains several hydrophobic regions but as a whole it is not extremely hydrophobic.[9,55,85] Binding studies between NodD and the *nod* promoter fragments were performed generally with cleared lysates of rhizobia.[11,66,67] Using NodD-specific antiserum, the NodD protein was localized in the cytoplasmic membrane.[85] It was suggested that the NodD is an inner membrane protein with a substantial domain extending into the cytosol. Parallel with these studies, Recourt et al.[86] reported that the flavonoid inducer molecules added to the *Rhizobium* cells are accumulated in the cytoplasmic membrane. It was proposed that the flavonoids bind to NodD within the membrane which results in its conversion into a transcription-activating form.

As discussed in Section II.D., the NodD in its positive regulatory form may be a dimer or multimer. In strains carrying multiple *nodD* alleles, the presence of a *nodD* allele may interfere with the inducing ability of the product of another *nodD* allele, if the inducer can interact only with the latter protein.[81,87-89] The wild-type NodD protein could suppress the inducer-independent activity, indicating that the wild-type NodD binds to the *nod* box in the absence of inducers and acts as a repressor of inducible *nod* genes. Studies with mutant NodD proteins suggested that in the presence of inducer, NodD may become multimeric and in this form can induce *nod* gene expression. On this basis, the inducer-independent mutants were proposed to be locked in the multimeric form.[81] Based on sequence homology, Kofoid and Parkinson[90] identified a putative receiver domain of NodD where several of the inducer-independent mutations mapped.[81] As discussed below, recent work from our laboratory showed interaction of NodD1, NodD2, and NodD3 proteins with another regulatory protein, SyrM, supporting the idea that the transcription-activating form of NodD is not monomeric.[91] The ligand-binding region of the steroid receptors and the putative flavonoid-binding region of NodD consist of two modules and the second module of the hormone receptor is involved in dimerization which might be the case also for NodD.[82]

In some rhizobial strains (*R. fredii,* and *B. japonicum*) containing two *nodD* copies, the homology of the two genes is not higher or even less than among *nodD* genes of different species.[74] The two *nodD* genes of *R. fredii* are functionally different, mutations in these genes do affect nodulation but cause different phenotypes and mutations in *nodD2* also affect EPS synthesis.[74] The *syrM* gene discussed below also shows characteristics of *nodD* but, at the same time, it is rather different in several properties, just like the other members of the LysR family.[92,93]

## IV. PLANT SIGNAL MOLECULES INDUCING *NOD* GENES

In different rhizobial-plant associations, the *nod* gene-inducing molecules were identified as compounds of the phenylpropanoid pathway. For different *Rhizobium* species various flavonoids, e.g., flavones, flavanones, chalcones or isoflavones, have been detected as inducers.[6-8,94-97]

Roots of different plant species exude a fairly broad range of flavonoids which may be recognized by NodD proteins. The signal molecules can act as either *nod* gene inducers or inhibitors (anti-inducers), depending on the signal specificity of a particular NodD protein. In a number of cases, a particular compound acts as inducer on one NodD, as inhibitor on another NodD, or has no effect at all in conjunction with a third type of NodD. The degree of effect of the compound shows also broad variation.[88,98] As expected, NodD proteins of rhizobia with broad host range (*Rhizobium* sp. NGR234, *R. parasponiae, R. fredii, B. japonicum*) are responsive to a wide range of compounds.[88,95,99-101] The NodD proteins from these species also accept certain isoflavones and generally only a hydroxyl moiety at the C-

7 atom of the flavonoid ring is needed for activation. Even certain monocyclic aromatic compounds (vanillin and isovanillin) are strong inducers in combination with the NodD1 of *Rhizobium* sp. NGR234.[101]

In contrast, NodD proteins of the narrow host-range rhizobia, such as *R. leguminosarum* or *R. meliloti,* can accept only a narrow range of flavonoids. In general, isoflavones are inhibitory for *R. meliloti* and substitutions at the C-5 and C-4′ positions are needed for the inducing ability on both species.[17,33,88,98] These observations on flavonoid specificities reflect only general tendencies, since, among the three copies, NodD3 can accept a broader range of flavonoids than NodD1 or NodD2.[88]

The structural features of flavonoids required for *nod* gene induction in combination with different NodD proteins have been determined by using primarily synthetic compounds. For a few plant species the inducers in seed or root exudates have been identified.[6,7,94-98,102,103] The detected inducers can activate *nod* gene expression at very low (often nanomolar) concentrations. In nature, however, the flavonoids are commonly found as *O*-glucosides,[104] which are less potent inducers.

The isoflavonoids often play a role in plant defense against pathogens and are known as phytoalexins.[104] In *Bradyrhizobium* some *nod* gene-inducer isoflavonoids (phytoalexins) are strong antibacterial agents. At higher concentrations, the *nod* gene-inducing flavonoids may also inhibit bacterial growth.

The seeds and roots exude different ranges of flavonoids. It was found, for example, that in *Medicago* seed exudates the major inducer is luteolin,[7] while in roots chalcones and other flavonoids were the most potent inducers.[96] Interactions between *nod* gene inducers from *Medicago* seeds and roots were observed, suggesting that these synergistic interactions may create a zone highly favorable to root nodule formation near the top of the primary root.[102] In *Trifolium,* high levels of *nod* gene expression were noted at the zone of emerging root hairs but inhibition at the root tips was observed.[105] It was suggested that stimulatory and inhibitory flavonoids fluctuate at the different root segments and vary over time, which means that the plant may exert a control over *nod* gene induction in the course of the nodulation process. The levels of flavone nodulation signals in the rhizosphere can limit root nodulation, which in turn can limit the efficiency of symbiotic nitrogen fixation.[106]

Inoculation of the plant with a homologous *Rhizobium* strain results in an increased *nod* gene inducing activity (called Ini) in the root exudate.[107] Inoculation with heterologous strains do not show Ini. Ini is a biovar-specific phenotype and is likely to be evoked by the *Rhizobium* Nod factor.

## V. NEGATIVE CONTROL OF *NOD* GENE EXPRESSION

The first indication for the existence of negative regulation of *nod* genes came from the observation that in the overwhelming majority of *R. meliloti* isolates of different geographical origin the induction of *nod* genes by plant signal molecules was rather low (about twofold increase above background level), in contrast to certain strains used in laboratories for many years where the induction was higher (four- to fivefold above background).[108] Extracts of both types of bacteria were assayed for *trans*-acting factors binding to various *nod* promoter regions.[11] It was found that with extracts of strains allowing high level of *nod* gene expression binding of only NodD was detectable, while with extracts from strains of low *nod* gene inducibility the existence of a second protein binding to the overlapping *nodD1*-*nodA* promoters was demonstrated. A gene library constructed from the DNA of a low expression strain was mated into the high expression strain containing an inducible *nod* promoter-*lacZ* fusion. Among colonies remaining white, the gene region responsible for the production of this novel *trans*-acting factor was found. Further analyses proved that the protein is a repressor which binds to the overlapping *nodD1* and *nodA* promoters and to the *nodD2* promoter.[11]

In DNase I footprint assays 27-bp nucleotides around the transcription initiation site were protected. Comparison of the *nodD1-nodA* overlapping promoter with the *nodD2* promoter indicated a conserved sequence of 21-bp within the protected region. DNA sequencing and genetic studies in our laboratory[109] revealed that the protein is of 14 kDa in size and contains a DNA-binding motif.

It was observed that in repressor-containing strains the *nodD* genes expressed weakly.[110] Mutations in the repressor gene resulted in increased expression of the *nodD* genes as well as of the inducible *nod* genes. The repressor gene does not map in the *nod-nif* region; most likely it is a chromosomal gene.[11] Interestingly, the *nod* repressor-producing strains were more efficient in nodule induction than their isogenic repressor-minus derivatives. This observation is in line with reports that relatively low levels of *nod* gene induction are sufficient for nodulation[2,9,95] and overexpression of *nod* genes may inhibit or decrease nodulation.[88,111] It was proposed that the fine-tuning of *nod* gene expression by both positive and negative *trans*-acting factors results in producing the Nod factor(s) at a concentration which allows optimal expression of the Nod factor-inducible genes in the plant host.[11]

The *nod* repressor was reported only for *R. meliloti*. Several recent publications, however, suggest its existence also in a number of different *Rhizobium* and *Bradyrhzobium* strains where a low level of *nod* gene expression was observed.[19,112,113]

Using antibodies raised against several Nod proteins of *R. leguminosarum* bv. *viciae*, severely reduced levels of these proteins were detected in bacteroids of *Pisum sativum* root nodules.[114] In addition, transcripts of inducible *nod* genes were not found and the amount of *nodD* transcript was also reduced. The data suggest that *nod* gene expression in bacteroids is switched off by a negative transcriptional regulator. Using *nod* promoter-*gusA* fusions a similar observation was made for *R. meliloti* bacteroids.[115]

The *nod* gene expression is also negatively influenced by certain environmental conditions, as discussed in the following section. Moreover, in *Rhizobium* strains where *nodD* regulates its own transcription, this autoregulation ensures fine-tuning of *nod* gene induction which acts against the overproduction of the Nod factors.

## VI. ADDITIONAL FACTORS MODULATING *NOD* GENE EXPRESSION

### A. SyrM

In *R. meliloti* strain 1021 a novel positive regulatory locus, *syrM* (symbiotic regulator) was detected.[87] This locus is located between *nodH*(*hsnD*) and *nodD3* (Figure 1). When the DNA region carrying *nodD3* and *syrM* is carried in *trans* on a plasmid, which is present in several (4 to 6) copies in the cell, the expression of a *nodC-lacZ* fusion was detected in the absence of inducers.[87,93] The loci *syrM* and *nodD3* are required for this high level of constitutive *nod* gene expression. It was observed that strains carrying the plasmid with the *syrM-nodD3* region formed unusually mucoid colonies and Tn5 mutation in *syrM* restored the normal colony morphology. Interestingly, another locus, *SyrA*, on the same DNA region but on the opposite site of *nodD3,* is also involved in the mucoid phenotype but not involved in *nod* gene activation. It was suggested that *syrM* activates *nod* genes via *nodD3* and exopolysaccharide production via *syrA*.[87]

Mutants of *R. meliloti* 1021 indicated that *syrM* is not needed for nodulation.[26] In contrast, in *R. meliloti* strain 41, the *syrM* locus does contribute to nodulation.[93] The *syrM* gene has been sequenced from both strains and several differences, particularly in the 5′- flanking regions were found.[92, 116] The expression of *syrM* in strain 1021 is very low.[87,115] Extracts from the two strains show striking differences also in gel mobility shift assays with the *syrM* promoter.[116] It is possible that in certain *R. meliloti* strains the syrM is expressed at high enough levels to contribute to nodulation of the plant host. The *syrM* expresses at high levels in the nodules suggesting that *syrM* may be involved in the late stage of symbiotic nodule development.[69]

The SyrM sequence revealed that this protein is also member of the LysR family.[92,93] The NodD and SyrM sequences exhibit only about 30% of similarity. The strongest homology was detected at the N-terminal region where a helix-turn-helix motif was found. The SyrM belongs to a subclass of the LysR family together with NodD and the *Pseudomonas putida* NahR, the positive regulator of genes determining the utilization of naphthalene.[117]

SyrM seems to play a role in the activation of the expression of *nodD3*.[87] The *nodD3* gene is plant signal-inducible[12] and high induction of *nodD3* requires also a functional *nodD3* and the other two *nodD* genes.[70,87,93]

Recent data from our laboratory indicate that SyrM in itself can activate *nod* gene expression, albeit at a low level.[20,116] In addition, SyrM is likely to interact with the three NodD proteins. In the *syrM*$^-$ derivative of *R. meliloti* strain 41, the NodD3 is activated by specific plant signal molecules, just like the NodD1 or NodD2 proteins. It was suggested that when SyrM is produced, SyrM and the NodD proteins may form heteromeric complexes which can activate *nod* genes in the absence of inducers.[109,116]

## B. NITROGEN CONTROL

The presence of combined nitrogen in the soil suppresses nodule induction. The role of the plant in this inhibition has been known for a long time[118] and recently a shoot factor was implicated in the nitrogen control of nodulation.[119] Plant mutants that nodulate in the presence of high nitrate concentrations were isolated.[120,121]

Recently it was shown for *R. meliloti* that the bacterial nodulation genes are also under nitrogen ($NH_4^+$) control.[12] In the presence of high ammonia concentration, the expression of inducible *nod* genes (including *nodD3*) remained low. At the same time, no considerable inhibition of the *nodD1* gene was observed, suggesting that the ammonia effect is mediated, at least partly, via *nodD3*. The general nitrogen regulatory system (the *ntrA* and *ntrC* genes) as well as a new chromosomal locus were found to be involved in the control of *nod* gene expression. Mutation in the chromosomal locus abolished the effect of ammonia on *nod* gene activation and, in addition, it rendered the bacterium more efficient and more competitive in nodulation at intermediate $NH_4^+$ concentrations.

The inhibitory effect of ammonia on *nod* gene expression was observed also in *Rhizobium* sp. NGR234[122] and in *B. japonicum*.[123] In all the three species, it was observed that among various combined nitrogen sources only ammonia had a drastic effect. One explanation is that in rhizobia the uptake of $NH_4^+$ is fairly efficient. In NGR234, the involvement of the *ntr* system in *nod* gene regulation was also demonstrated,[122] while in *B. japonicum* the *ntrC* gene does not have a role.[123] Mutation in the sigma factor gene *ntrA* (*rpoN*) of NGR234 affected negatively not only nodulation but also the establishment and maintenance of the peribacteroid membrane.[123] It was proposed that the *ntrA* is a primary coregulator of symbiosis. Further studies are needed to elucidate whether ammonia directly or the combined nitrogen status of the cells exerts the observed negative control on *nod* gene expression.

## C. OTHER FACTORS

There are numerous environmental factors which are known to influence the formation of nodules.[1,118] Particularly, low pH and ion imbalances adversely affect nodulation. These factors also restrict the growth of rhizobia. These factors (pH, Ca, and Al ions) were shown to affect *nod* gene induction indirectly via the physiology of the bacterium, albeit some more direct effect was also observed.[124]

Recent data from our laboratory suggest that in *R. meliloti* the expression of *nodD3*, *syrM*, and the *nod* repressor is controlled by several, yet uncharacterized, factors which is mediated by several chromosomal loci.[116] A chromosomal mutant allowing only a very low

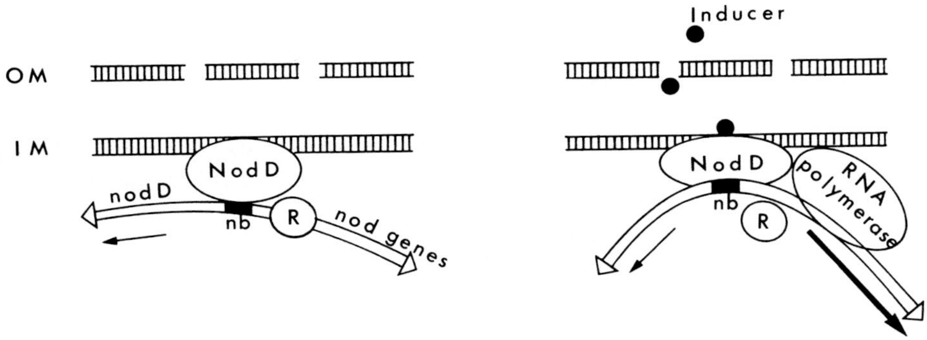


FIGURE 2. Model for *nod* gene regulation. OM, outer membrane; IM, inner membrane; nb, *nod* box; R, *nod* repressor. (From Kondorosi, A., Kondorosi, E., John, M., Schmidt, J., and Schell, J., The role of nodulation genes in bacterium-plant communication, in *Genetic Engineering*, Vol. 13, Setlow, J. K., Ed., Plenum Publishing, New York, 1991, 120. With permission.)

level of *nod* gene expression was found and the mutation seems to affect the stability or activity of NodD.[69] Partial sequencing of the mutated region revealed a sequence homologous to the *E. coli groEL* product and other chaperonin proteins, suggesting that this protein may be involved in the folding or assembly of NodD.

## VII. MODEL OF *NOD* GENE REGULATION

In the absence of the inducer, the constitutively produced NodD protein in its transcriptionally inactive conformation binds to the *nod* box but cannot promote binding of RNA polymerase and/or initiation of transcription (Figure 2). In *nod* repressor-containing strains, the repressor binds to the operator site of the overlapping *nodD*-inducible *nod* promoters and the transcription of *nodD* is also decreased. When the plant signal molecules are supplied by the plant, they cross the outer membrane of the bacterium and accumulate in the inner membrane. At this site, the ligand-binding domain of NodD interacts with the flavonoids. If the appropriate flavonoid is recognized, the NodD undergoes a conformational change and binds the RNA polymerase which competes with the repessor, leading to the induction of the *nodABC* and other inducible *nod* operons. In the repressor-containing strains, upon induction the *nodD* is transcribed at higher levels. In some strains containing autoregulated *nodD,* autoregulation acts against overproduction of *nodD*. In strains with multiple *nodD* alleles, the nonactivated NodD (either binding or not binding noninducing flavonoids) acts as repressor of *nod* genes. Various environmental factors (e.g., ammonia) or physiological conditions, such as pH, further modulate the level of active NodD. All these factors contribute to the fine-tuning of *nod* gene expression.[11]

## VIII. CONCLUSIONS

In the past few years research on the regulation of *nod* genes has achieved considerable progress in the understanding how the plant host can activate the nodulation genes in its appropriate microsymbiont. The interaction of flavonoids with NodD represents an interesting model system for studying the specific reception of signal molecules and its transduction resulting in gene activation. Modulation of the level of *nod* gene induction by a number of internal and external factors may provide the ability for either of the symbiotic partners to fine-tune *nod* gene expression.

The Rhizobium *nod* gene products are involved in the synthesis of specific Nod factors,[31,38,125] oligosaccharins,[31] and these signal molecules were suggested to cause changes in the hormonal balance of the root cells.[18,31] These hormonelike molecules act at very low concentrations,[31] and perhaps, efficient nodulation requires optimal concentrations of the Nod factors.[110,125] Plant hormones produced by *Rhizobium* were found to enhance the effect of Nod factors.[126] Thus, modulation of *nod* gene expression by a number of genes and factors may serve to optimize the concentration of the hormonelike Nod signal molecules. Identification of these factors and of their mode of action may elucidate the hierarchy of the various regulatory genes and the biological significance of fine-tuning of *nod* gene control.

## ACKNOWLEDGMENTS

I thank all of my colleagues who sent reprints, preprints, or unpublished material, E. Kondorosi for suggestions on the manuscript and for preparing the figures and C. Deforeit for typing.

## REFERENCES

1. **Halverson, L. J. and Stacey, G.,** Signal exchange in plant-microbe interactions, *Microbiol. Rev.,* 50, 193, 1986.
2. **Mulligan, J. T. and Long, S. R.,** Induction of *Rhizobium meliloti nodC* expression by plant exudate requires *nodD, Proc. Natl. Acad. Sci. U.S.A.,* 82, 6609, 1985.
3. **Innes, R. W., Kuempel, P. L., Plazinski, J., Canter-Cremers, H., Rolfe, G. B., and Djordjevic, M. A.,** Plant factors induce expression of nodulation and host-range genes in *Rhizobium trifolii, Mol. Gen. Genet.,* 201, 426, 1985.
4. **Rossen, L., Shearman, C. A., and Johnston, A. W. B., and Downie, J. A.,** The *nodD* gene of *Rhizobium leguminosarum* is autoregulatory and in the presence of plant exudate induces the *nodA,B,C* genes, *EMBO J.,* 4, 3369, 1985.
5. **Rostas, K., Kondorosi, E., Horvath, B., Simoncsits, A., and Kondorosi, A.,** Conservation of extended promoter regions of nodulation genes in *Rhizobium, Proc. Natl. Acad. Sci. U.S.A.,* 873, 1757, 1986.
6. **Redmond, J. W., Batley, M., Djordjevic, M. A., Innes, R. W., Kuempel, P. L., and Rolfe, B. G.,** Flavones induce expression of nodulation genes in *Rhizobium, Nature (London),* 323, 632, 1986.
7. **Peters, N. K., Frost, J. W., and Long, S. R.,** A plant flavone, luteolin, induces expression of *Rhizobium meliloti* nodulation genes, *Science,* 233, 977, 1986.
8. **Firmin, J. L., Wilson, K. E., Rossen, L., and Johnston, A. W. B.,** Flavonoid activation of nodulation genes in *Rhizobium* reversed by other compounds present in plants, *Nature (London),* 324, 90, 1986.
9. **Horvath, B., Bachem, C., Schell, J., and Kondorosi, A.,** Host-specific regulation of nodulation genes in *Rhizobium* is mediated by a plant-signal interacting with the *nodD* gene product, *EMBO J.,* 6, 841, 1987.
10. **Spaink, H. P., Wijffelman, C. A., Pees, E., Okker, R. J. H., and Lugtenberg, B. J. J.,** *Rhizobium* nodulation gene *nodD* as a determinant of host specificity, *Nature (London),* 328, 337, 1987.
11. **Kondorosi, E., Gyuris, J., Schmidt, J., John, M., Duda, E., Hoffmann, B., Schell, J., and Kondorosi, A.,** Positive and negative control of *nod* gene expression in *Rhizobium meliloti* is required for optimal nodulation, *EMBO J.,* 8, 1331, 1989.
12. **Dusha, I., Bakos, A., Kondorosi, A., de Bruijn, F., and Schell, J.,** The *Rhizobium meliloti* early nodulation genes *(nodABC)* are nitrogen-regulated: Isolation of a mutant strain with efficient nodulation capacity on alfalfa in the presence of ammonium, *Mol. Gen. Genet.,* 219, 89, 1989.
13. **Kondorosi, E. and Kondorosi, A.,** Nodule induction on plant roots by *Rhizobium, Trends Biochem. Sci.,* 11, 296, 1986.
14. **Rossen, L., Davies, E. O., and Johnston, A. W. B.,** Plant-induced expression of *Rhizobium* genes involved in host specificity and early stages of nodulation, *Trends Biochem. Sci.,* 12, 430, 1987.
15. **Long, S. R.,** *Rhizobium*-legume nodulation: life together in the underground, *Cell,* 56, 203, 1989.
16. **Winsor, B. A. T.,** A *nod* at differentiation: the *nodD* gene product and initiation of *Rhizobium* nodulation, *Trends Genet.,* 5, 199, 1989.
17. **Appelbaum, E.,** The *Rhizobium/Bradyrhizobium*-legume symbiosis, in *Molecular Biology of Symbiotic Nitrogen Fixation,* Gresshoff, P. M., Ed., CRC Press, Boca Raton, FL, 1990, 131.

18. **Kondorosi, A., Kondorosi, E., John, M., Schmidt, J., and Schell, J.,** The role of nodulation genes in bacterium-plant communication, in *Genetic Engineering,* Vol. 13, Setlow, J. K., Ed., Plenum Press, New York, 1991, 115.
19. **Kondorosi, A.,** *Rhizobium*-legume interactions: nodulation genes, in *Plant-Microbe Interactions, Vol. 3,* Kosuge, T. and Nester, E. W., Eds., McGraw-Hill, NY, 1989, 383.
20. **Kondorosi, A.,** Overview on genetics of nodule induction: factors controlling nodule induction by *Rhizobium meliloti,* in *Advances in Molecular Genetics of Plant-Microbe Interactions,* Vol. 1, Hennecke, H. and Verma, D. P. S., Eds., Kluwer Academic, Dordrecht, 1991, 111.
21. **Kondorosi, E., Banfalvi, Z., and Kondorosi, A.,** Physical and genetic analysis of a symbiotic region of *R. meliloti:* identification of nodulation genes, *Mol. Gen. Genet.,* 193, 445, 1984.
22. **Putnoky, P. and Kondorosi, A.,** Two gene clusters of *Rhizobium meliloti* code for early essential nodulation functions, a third influences nodulation efficiency, *J. Bacteriol.,* 167, 881, 1986.
23. **Martinez, E., Romero, D., and Palacios, R.,** The *Rhizobium* Genome, *Plant Sci.,* 9, 59, 1990.
24. **Rodriguez-Quinones, F., Banfalvi, Z., Murphy, P., and Kondorosi, A.,** Interspecies homology of nodulation genes in *Rhizobium, Plant Mol. Biol.,* 8, 61, 1987.
25. **Göttfert, M., Horvath, B., Kondorosi, E., Putnoky, P., Rodriguez-Quinones, F., and Kondorosi, A.,** At least two *nodD* genes are necessary for efficient nodulation of alfalfa by *Rhizobium meliloti, J. Mol. Biol.,* 191, 411, 1986.
26. **Honma, M. A. and Ausubel, F. M.,** *Rhizobium meliloti* has three functional copies of the *nodD* symbiotic regulatory gene, *Proc. Natl. Acad. Sci. U.S.A.,* 84, 8558, 1987.
27. **Györgypal, Z., Iyer, N., and Kondorosi, A.,** Three regulatory *nodD* alleles of diverged flavonoid-specificity are involved in host-dependent nodulation by *Rhizobium meliloti, Mol. Gen. Genet.,* 212, 85, 1988.
28. **Rodriguez-Quinones, F., Fernandez-Burriel, M., Banfalvi, Z., Megias, M., and Kondorosi, A.,** Identification of a conserved, reiterated DNA region that influences the efficiency of nodulation in strain RS1051 of *Rhizobium leguminosarum* bv. *trifolii, Mol. Plant-Microbe Interact.,* 2, 75, 1989.
29. **Davis, E. O. and Johnston, A. W. B.,** Analysis of three *nodD* genes in *Rhizobium leguminosarum* biovar phaseoli: *nodD1* is preceded by *nolE,* a gene whose product is secreted from the cytoplasm, *Mol. Microbiol.,* 4, 921, 1990.
30. **Baev, N., Endre, G., Petrovics, G., Banfalvi, Z., and Kondorosi, A.,** Six nodulation genes of *nod* box locus 4 in *Rhizobium meliloti* are involved in nodulation signal production: *nodM* codes for D-glucosamine synthetase, *Mol. Gen. Genet.,* in press, 1991.
31. **Lerouge, P., Roche, P., Faucher, C., Maillet, F., Truchet, G., Promé, J. C., and Dénarié, J.,** Symbiotic host-specificity of *Rhizobium meliloti* is determined by a sulphated and acylated glucosamine oligosaccharide signal, *Nature (London),* 344, 781, 1990.
32. **Schwedock, J. and Long, S. R.,** ATP sulfurylase activity of *Rhizobium meliloti nodP and nodQ, Nature (London),* 348, 644, 1990.
33. **Györgypal, Z., Kiss, G. B., and Kondorosi, A.,** Transduction of plant signal molecules by the *Rhizobium* NodD protein, *Bioassays,* in press, 1990.
34. **Horvath, B., Kondorosi, E., John, M., Schmidt, J., Torok, I., Györgypal, Z., Barabas, I., Wieneke, U., Schell, J., and Kondorosi, A.,** Organization, structure and symbiotic function of *Rhizobium meliloti* nodulation genes determining host-specificity for alfalfa, *Cell,* 46, 335, 1986.
35. **Gotz, R., Evans, I. J., Downie, J. A., and Johnston, A. W. B.,** Identification of the host-range DNA which allows *Rhizobium leguminosarum* strain TOM to nodulate Afghanistan peas, *Mol. Gen. Genet.,* 212, 85, 1985.
36. **Surin, B. P., Watson, J. M., Hamilton, W. D. O., Economou, A., and Downie, J. A.,** Molecular characterization of the nodulation gene, *nodT,* from two biovars of *Rhizobium leguminosarum, Mol. Microbiol.,* 4, 245, 1990.
37. **Sadowsky, M. J., Cregan, P. B., Göttfert, M., Sharma, A., Gerhold, D., Rodriguez-Quinones, F., Keyser, H. H., Hennecke, H., and Stacey, G.,** The *Bradyrhizobium japonicum nolA* gene and its involvement in the genotype-specific nodulation of soybeans, *Proc. Natl. Acad. Sci. U.S.A.,* in press, 1990.
38. **Dénarié, J. and Roche, P.,** *Rhizobium* nodulation signals, in *Molecular Signals in Plant Microbe Communications,* Verma, D. P. S., Ed., CRC Press, Boca Raton, FL, 1991.
39. **Kijne, J. W., Lugtenberg, B. J. J., and Smit, G.,** Attachment, lectin, and initiation of infection in (brady) rhizobium-legume interactions, in *Molecular Signals in Plant-Microbe Communications,* Verma, D. P. S., Ed., CRC Press, Boca Raton, FL, 1991.
40. **Ruvkun, G. B. and Ausubel, F. M.,** A general method for site directed mutagenesis in prokaryotes, *Nature (London),* 289, 85, 1981.
41. **Török, I., Kondorosi, E., Stepkowski, T., Posfai, J., and Kondorosi, A.,** Nucleotide sequence of *R. meliloti* nodulation genes, *Nucleic Acids Res.,* 12, 9509, 1984.
42. **Rossen, L., Johnston, A. W. B., and Downie, J. A.,** DNA sequence of the *R. leguminosarum* nodulation genes *nodA, B* and *C* required for root hair curling, *Nucleic Acids Res.,* 12, 9497, 1984.

43. **Djordjevic, M. A., Schofield, P. R., and Rolfe, B. G.,** Tn5 mutagenesis of *Rhizobium trifolii* host-specific nodulation genes result in mutants with altered host-range ability, *Mol. Gen. Genet.*, 200, 263, 1985.

44. **Downie, J. A., Knight, C. D., Johnston, A. W. B., and Rossen, L.,** Identification of genes and gene products involved in the nodulation of peas by *Rhizobium leguminosarum, Mol. Gen. Genet.*, 198, 255, 1985.

45. **Egelhoff, T. T., Fisher, R. F., Jacobs, T. W., Mulligan, J. T., and Long, S. R.,** Nucleotide sequence of *Rhizobium meliloti* 1021 nodulation genes: *nodD* is read divergently from *nodABC, DNA,* 4, 424, 1985.

46. **Wijffelman, C. A., Pees, E., Van Brussel, A. A. N., Okker, R. J. H., and Lugtenberg, B. J. J.,** Genetic and functional analysis of the nodulation region of the *Rhizobium leguminosarum* symplasmid pRL1JI, *Arch. Microbiol.*, 143, 225, 1985.

47. **Scott, K. F.,** Conserved nodulation genes from the nonlegume symbiont *Bradyrhizobium* sp. *(Parasponia), Nucleic Acids Res.*, 14, 2891, 1986.

48. **Bachem, C., Kondorosi, E., Banfalvi, Z., Horvath, B., Kondorosi, A., and Schell, J.,** Identification and cloning of nodulation genes from a wide host range *Rhizobium* strain MPIK3030, *Mol. Gen. Genet.*, 199, 271, 1985.

49. **Noti, J. F., Dudas, B., and Szalay, A. A.,** Isolation and characterization of nodulation genes from *Bradyrhizobium* sp. strain IRC78, *Proc. Natl. Acad. Sci. U.S.A.*, 82, 7379, 1985.

50. **Lamb, J. W. and Hennecke, H.,** In *Bradyrhizobium japonicum* the common nodulation genes, *nodABC*, are linked to *nifA* and *fixA, Mol. Gen. Genet.*, 202, 512, 1986.

51. **Nieuwkoop, A. J., Banfalvi, Z., Deshmane, N., Gerhold, D., Schell, M. G., Sirotkin, K. M., and Stacey, G.,** A locus encoding host range is linked to the common nodulation genes of *Bradyrhizobium japonicum, J. Bacteriol.*, 169, 2631, 1987.

52. **Djordjevic, M. A., Innes, R. W., Wijffelman, C. A., Schofield, P. R., and Rolfe, B. G.,** Nodulation of legumes is controlled by several distinct loci in *Rhizobium trifolii, Plant Mol. Biol.*, 6, 389, 1986.

53. **Evans, I. J. and Downie, J. A.,** The *nodI* gene product of *Rhizobium leguminosarum* is closely related to ATP-binding bacterial transport proteins: nucleotide sequence analysis of the *nodI* and *nodJ* genes, *Gene*, 43, 95, 198.

54. **Schofield, P. R. and Watson, J. M.,** DNA sequence of *Rhizobium trifolii* nodulation genes reveals a reiterated and potentially regulatory sequence preceding *nodABC* and *nodFE, Nucleic Acids Res.*, 14, 2891, 1986.

55. **Shearman, C. A., Rossen, L., Johnston, A. W. B., and Downie, J. A.,** The *Rhizobium leguminosarum* nodulation gene *nodF* encodes a protein similar to acyl-carrier protein and it is regulated by *nodD* plus a factor in pea root exudate, *EMBO J.*, 5, 647, 1986.

56. **Spaink, H. P., Okker, R. J. H., Wijffelman, C. A., Pees, E., and Lugtenberg, B. J. J.,** Promoters in the nodulation region of the *Rhizobium leguminosarum* Sym plasmid pRL1JI, *Plant Mol. Biol.*, 9, 27, 1987.

57. **Fisher, R. F., Brierley, H. L., Mulligan, J. T., and Long, S. R.,** Transcription of *Rhizobium meliloti* nodulation genes, *J. Biol. Chem.*, 262, 6849, 1987.

58. **Fisher, R. F., Swanson, J. A., Mulligan, J. T., and Long, S. R.,** Extended region of nodulation genes in *Rhizobium meliloti* 1201. II. Nucleotide sequence, transcription start sites, and protein products, *Genetics*, 117, 191, 1987.

59. **Wang, S. P. and Stacey, G.,** Inducible expression of the *Bradyrhizobium japonicum nodD1* gene is dependent on a divergent *nod* box, submitted, 1991.

60. **Göttfert, M., Lamb, J. W., Gasser, R., Semenza, J., and Hennecke, H.,** Mutational analysis of the *Bradyrhizobium japonicum* common *nod* genes and further *nod* box-linked genomic DNA regions, *Mol. Gen. Genet.*, 215, 407, 1989.

61. **Gerhold, D., Stacey, G., and Kondorosi, A.,** Use of a promoter specific probe to identify two loci from the *Rhizobium meliloti* nodulation regulon, *Plant Mol. Biol.*, 12, 181, 1989.

62. **Goethals, K., Gao, M., Geelen, D., Van den Eede, G., Van Montagu, M., and Holsters, M.,** *Nod* box-related sequences in *Azorhizobium caulinodans* strain ORS571, in *Nitrogen Fixation: Achievements and Objectives,* Gresshoff, P. M., Roth, E., Stacey, G., and Newton, W. E., Eds., Chapman and Hall, NY, 1990, 536.

63. **Brunel, F. and Davison, J.,** Cloning and sequencing of *Pseudomonas* genes encoding vanillate demethylase, *J. Bacteriol.*, 170, 4924, 1988.

64. **Venturi, V.,** personal communication, 1989.

65. **Bolker, M. and Kahmann, R.,** The *Escherichia coli* regulatory protein OxyR discriminates between methylated and unmethylated states of the phage Mu *mom* promoter, *EMBO J.*, 8, 2403, 1989.

66. **Hong, G. F., Burn, J. E., and Johnston, A. W. B.,** Evidence that DNA involved in the expression of nodulation *(nod)* genes in *Rhizobium* binds to the product of the regulatory gene *nodD, Nucleic Acids Res.*, 15, 9677, 1987.

67. **Fisher, R. F., Egelhoff, T. T., Mulligan, J. T., and Long, S. R.,** Specific binding of proteins from *Rhizobium meliloti* cell-free extracts containing NodD to DNA sequences upstream of inducible nodulation genes, *Genes Dev.,* 2, 282, 1988.
68. **Fisher, R. F. and Long, S. R.,** DNA footprint analysis of the transcriptional activator proteins NodD1 and NodD3 on inducible *nod* gene promoters, *J. Bacteriol.,* 171, 5492, 1989.
69. **Long, S. R., Fisher, R. F., Ogawa, J., Swanson, J., Ehrhardt, D. W., Atkison, E. M., and Schwedock, J. S.,** *Rhizobium meliloti* nodulation gene regulation and molecular signals, in *Advances in Molecular Genetics of Plant-Microbe Interactions,* Vol. 1, Hennecke, H. and Verma, D. P. S., Eds., Kluver Academic, Dordrecht, 1991, 127.
70. **Maillet, F., Debellé, F., and Dénarié, J.,** Role of the *nodD* and *syrM* genes in the activation of the regulatory gene *nodD3*, and of the common and host-specific *nod* genes of *Rhizobium meliloti, Mol. Microbiol.,* 4, 1990.
71. **Göttfert, M., Grob, P., and Hennecke, H.,** Proposed regulatory pathway encoded by the *nodV* and *nodW* genes, determinants of host specificity in *Bradyrhizobium japonicum, Proc. Natl. Acad. Sci. U.S.A.,* 87, 2680, 1990.
72. **Dusha, I., Schröder, J., Putnoky, P., Banfalvi, Z., and Kondorosi, A.,** A cell-free system for *Rhizobium meliloti* to study the specific expression of nodulation genes, *Eur. J. Biochem.,* 160, 69, 1986.
73. **Banfalvi, Z., Nieuwkoop, A., Schell, M., Besl, L., and Stacey, G.,** Regulation of *nod* gene expression in *Bradyrhizobium japonicum, Mol. Gen. Genet.,* 214, 420, 1988.
74. **Appelbaum, E. R., Thompson, D. V., Idler, K., and Chartrain, N.,** *Rhizobium japonicum* USDA 191 has two *nodD* genes that differ in primary structure and function, *J. Bacteriol.,* 170, 12, 1988.
75. **Davis, E. O. and Johnston, A. W. B.,** Regulatory functions of the three *nodD* genes of *Rhizobium leguminosarum* biovar *phaseoli, Mol. Microbiol.,* 4, 933, 1990.
76. **Honma, M. A., Asomaning, M., and Ausubel, F. M.,** *Rhizobium meliloti nodD* genes mediate host-specific activation of *nodABC, J. Bacteriol.,* 172, 9801, 1990.
77. **Iyer, N. and Kondorosi, A.,** unpublished data, 1990.
78. **Spaink, H. P., Wijffelman, C. A., Okker, R. J. H., and Lugtenberg, B. E. J.,** Localization of functional regions of the *Rhizobium nodD* product using hybrid *nodD* genes, *Plant Mol. Biol.,* 12, 59, 1989.
79. **Burn, J. Rossen, L., and Johnston, A. W. B.,** Four classes of mutations in the *nod* gene of *Rhizobium leguminosarum* biovar *viciae* that affect its ability to autoregulate and/or activate other *nod* genes in the presence of flavonoid inducers, *Genes Dev.,* 1, 456, 1987.
80. **Burn, J. E., Hamilton, W. D., Wootton, J. C., and Johnston, A. W. B.,** Single and multiple mutations affecting properties of the regulatory gene *nodD* of *Rhizobium, Mol. Microbiol.,* 3, 1567, 1989.
81. **McIver, J., Djordjevic, M. A., Weinman, J. J., Bender, G. L., and Rolfe, B. G.,** Extension of host range of *Rhizobium leguminosarum* bv. *trifolii* caused by point mutations in *nodD* that result in alterations in regulatory function and recognition of inducer molecules, *Mol. Plant-Microbe Interact.,* 2, 97, 1989.
82. **Györgypal, Z. and Kondorosi, A.,** Homology of the ligand-binding regions of *Rhizobium* symbiotic regulatory protein NodD and vertebrate nuclear receptors, *Mol. Gen. Genet.,* 226, 337, 1990.
83. **Henikoff, S., Haughn, G. W., Calvo, J. M., and Wallace, J. C.,** A large family of bacterial activator proteins, *Proc. Natl. Acad. Sci. U.S.A.,* 85, 6602, 1988.
84. **Spaink, H. P. S., Okker, R. J. H., Wijffelman, C. A., Tak, T., Goosen-de Roo, L., Pees, E., Van Brussel, A. A. N., and Lugtenberg, B. J. J.,** Symbiotic properties of rhizobia containing a flavonoid-independent hybrid *nodD* product, *J. Bacteriol.,* 171, 4045, 1989.
85. **Schlaman, H. R. M., Spaink, H. P., Okker, R. J. H., and Lugtenberg, B. J. J.,** Subcellular localization of the *nodD* gene product in *Rhizobium leguminosarum, J. Bacteriol.,* 171, 4686, 1989.
86. **Recourt, K., Van Brussel, A. A. N., Driessen, A. J. M., and Lugtenberg, B. J. J.,** Accumulation of *nod* gene inducer, the flavonoid naringenin, in the cytoplasmic membrane of *Rhizobium leguminosarum* biovar *viciae* is caused by the pH-dependent hydrophobicity of naringenin, *J. Bacteriol.,* 171, 4370, 1989.
87. **Mulligan, J. T. and Long, S. R.,** A family of activator genes regulates expression of *Rhizobium meliloti* nodulation genes, *Genetics,* 122, 7, 1989.
88. **Györgypal, Z., Kondorosi, E., and Kondorosi, A.,** Diverse signal sensitivity of NodD protein homologs from narrow and broad host range rhizobia, *Mol. Plant-Microbe Interact.,* 4, 356, 1991.
89. **Hartwig, U. A., Maxwell, C. A., Joseph, C. M., and Phillips, D. A.,** Effects of alfalfa *nod* gene-inducing flavonoids on *nodABC* transcription in *Rhizobium meliloti* strains containing different *nodD* genes, *J. Bacteriol.,* 172, 1990.
90. **Kofoid, E. C. and Parkinson, J. S.,** Transmitter and receiver modules in bacterial signaling proteins, *Proc. Natl. Acad. Sci. U.S.A.,* 85, 4981, 1988.
91. **Kondorosi, E.,** unpublished results, 1990.
92. **Barnett, M. J. and Long, S. R.,** DNA sequence and translational product of a new nodulation-regulatory locus: SyrM has sequence similarity to NodD proteins, *J. Bacteriol.,* 172, 3695, 1990.

93. **Kondorosi, E., Györgypal, Z., Dusha, I., Baev, N., Pierre, M., Hoffmann, B., Himmelbach, A., Banfalvi, Z., and Kondorosi, A.,** *Rhizobium meliloti* nodulation genes and their regulation, in *Nitrogen Fixation: Achievements and Objectives,* Gresshoff, P., Roth, E., Stacey, G., and Newton, W. E., Eds., Chapman and Hall, NY, 1990, 207.
94. **Zaat, S. A., Wijffelman, C. A., Spaink, H. P., Van Brussel, A. A. N., Okker, R. J. H., and Lugtenberg, B. J. J.,** Induction of the *nodA* promoter of *Rhizobium leguminosarum* Sym plasmid pRL1JI by plant flavonones and flavones, *J. Bacteriol.,* 169, 198, 1987.
95. **Kosslak, R., Bookland, R., Barkei, J., Paaren, H. E., and Appelbaum, E. R.,** Induction of *Bradyrhizobium japonicum* common *nod* genes by isoflavones isolated from *Glycine max, Proc. Natl. Acad. Sci. U.S.A.,* 84, 7428, 1987.
96. **Maxwell, C. A., Hartwig, U. A., Joseph, C. M., and Phillips, D. A.,** A chalcone and two related flavonoids released from alfalfa roots induce *nod* genes of *Rhizobium meliloti, Plant Physiol.,* 91, 842, 1989.
97. **Sadowsky, M. J., Olson, E. R., Foster, V. E., Kosslak, R. M., and Verma, D. P. S.,** Two host-inducible genes of *Rhizobium fredii* and characterization of the inducing compound, *J. Bacteriol.,* 170, 171, 1988.
98. **Zaat, S. A. J., Schripsema, J., Wijffelman, C. A., Van Brussel, A. A. N., and Lugtenberg, B. J. J.,** Analysis of the major inducers of the *Rhizobium nodA* promoter from *Vicia sativa* root exudate and their activity with different *nodD* genes, *Plant Mol. Biol.,* 13, 175, 1989.
99. **Bender, G. L., Nayudu, M., Le Strange, K. K., and Rolfe, B. G.,** The *nodD1* Gene from *Rhizobium* strain NGR234 is a key determinant in the extension of host range to the nonlegume *Parasponia, Mol. Plant-Microbe Interact.,* 1, 259, 1988.
100. **Bassam, J. B., Djordjevic, M. A., Redmond, J. W., Batley, M., and Rolfe, B. G.,** Identification of a *nodD*-dependent locus in the *Rhizobium* strain NGR234 activated by phenolic factors secreted by soybeans and other legumes, *Mol. Plant-Microbe Interact.,* 1, 161, 1988.
101. **Le Strange, K. K., Bender, G. L., Djordjevic, M. A., Rolfe, B. G., and Redmond, J. W.,** The *Rhizobium* strain NGR234 *nodD1* gene product responds to activation by the simple phenolic compounds vanillin and isovanillin present in wheat seedling extracts, *Mol. Plant-Microbe Interact.,* 3, 214, 1990.
102. **Hartwig, U. A., Maxwell, C. A., Joseph, C. M., and Phillips, D. A.,** Interactions among flavonoid *nod* gene inducers released from alfalfa seeds and roots, *Plant Physiol.,* 91, 1138, 1989.
103. **Hartwig, U. A., Maxwell, C. A., Joseph, C. M., and Phillips, D. A.,** Chrysoeriol and luteolin released from alfalfa seeds induce *nod* genes in *Rhizobium meliloti, Plant Physiol.,* 92, 116, 1990.
104. **Ingham, J.,** Phytoalexins from the Leguminosae, in *Phytoalexins,* Bailey, J. A. and Mansfield, J. W., Eds., John Wiley & Sons, NY, 1982, 21.
105. **Djordjevic, M. A., Redmond, J. W., Batley, M., and Rolfe, B. G.,** Clovers secrete specific phenolic compounds which either stimulate or repress *nod* gene expression in *Rhizobium trifolii, EMBO J.,* 6, 1173, 1987.
106. **Kapulnik, Y., Joseph, C. M., and Phillips, D. A.,** Flavone limitations to root nodulation and symbiotic nitrogen fixation in alfalfa, *Plant Physiol.,* 84, 1193, 1987.
107. **Van Brussel, A. A. N., Recourt, K., Pees, E., Spaink, H. P., Tak, T., Wijffelman, C. A., Kijne, J. W., and Lugtenberg, B. J. J.,** A biovar-specific signal of *Rhizobium leguminosarum* bv. *viciae* induces increased nodulation gene-inducing activity in root exudate of *Vicia sativa* subsp. *nigra, J. Bacteriol.,* 172, 5394, 1990.
108. **Kondorosi, E., Gyuris, J., Schmidt, J., John, M., Duda, E., Schell, J., and Kondorosi, A.,** Positive and negative control of nodulation genes in *Rhizobium meliloti* strain 41, in *Molecular Plant-Microbe Interactions,* Palacios, R. and Verma, D. P. S., Eds., APS Press, St. Paul, MN, 1988, 106.
109. **Kondorosi, E., Buiré, M., Cren, M., Iyer, N., Hoffman, B., Kondorosi, A.,** Involvement of the *nyrM* and *nodD3* genes of *Rhizobium meliloti* in *nod* gene activation and in optimal nodulation of the plant host, *Mol. Microbiol.,* in press, 1991.
110. **Kondorosi, A, Kondorosi, E., Dusha, I., Banfalvi, Z., Gyuris, J., John, M., Schmidt, J., Bakos, A., Hoffmann, B., Duda, E., de Bruijn, F., and Schell, J.,** Factors controlling root nodule induction by Rhizobium meliloti, in *Molecular Signals in Microbe-Plant Symbiotic and Pathogenic Systems,* NATO ASI Ser., Vol. 36, Lugtenberg, B., Ed., Springer-Verlag, Berlin, 1989, 319.
111. **Knight, C. D., Rossen, L., Robertson, J. G., Wells, B., and Downie, J. A.,** Nodulation inhibition by *Rhizobium leguminosarum* multicopy *nodABC* genes and analysis of early stages of plant infection, *J. Bacteriol.,* 166, 552, 1986.
112. **Stacey, G.,** personal communication, 1990.
113. **Schlaman, H. R. M., Okker, R. J. H., and Lugtenberg, B. J. J.,** Subcellular localization of the *Rhizobium leguminosarum nodI* gene product, *J. Bacteriol.,* 172, 5486, 1990.
114. **Schlaman, H. R. M., Okker, R. J. H., and Lugtenberg, B. J. J.,** Expression of nodulation genes of *Rhizobium leguminosarum bv. viciae* is suppressed in nodules, submitted, 1990.
115. **Sharma, S. B. and Signer, E. R.,** Temporal and spatial regulation of the symbiotic genes of *Rhizobium meliloti* in planta revealed by transposon Tn5-*gusA*, *Genes Dev.,* 4, 344, 1990.

116. **Kondorosi, E. and Hoffmann, B.,** unpublished results, 1990.
117. **Schell, M. A. and Sukordhaman, J.,** Evidence that the transcription activator encoded by the *Pseudomonas putida nahR* gene is evolutionarily related to the transcription activator encoded by the *Rhizobium nodD genes, J. Bacteriol.,* 171, 1952, 1989.
118. **Quispel, A.,** *The Biology of Nitrogen Fixation,* North-Holland, Amsterdam, 1974.
119. **Delves, D. A., Mathews, A., Day, D. A., Carter, A. S., Carroll, B. J., and Gresshoff, P. M.,** Regulation of the soybean-*Rhizobium* symbiosis by shoot and root factors, *Plant Physiol.,* 82, 588, 1986.
120. **Jacobsen, E. and Feenstra, W. J.,** A new pea mutant with efficent nodulation in the presence of nitrate, *Plant Sci. Lett.,* 33, 337, 1984.
121. **Carroll, B. J., McNeil, D. L., and Gresshoff, P. M.,** Isolation and properties of soybean *(Glycine max* (L.) *Merr)* mutants that nodulate in the presence of high nitrate concentrations, *Proc. Natl. Acad. Sci. U.S.A.,* 82, 4162, 1985.
122. **Van Slooten, J. C., Cervantes, E., Broughton, W. J., Wong, C. H., and Stanley, J.,** Sequence and analysis of the *rpoN* sigma factor gene of *Rhizobium* sp. strain NGR234, a primary coregulator of symbiosis, *J. Bacteriol.,* 172, 5563, 1990.
123. **Wang, S. P. and Stacey, G.,** Ammonia regulation of *nod* genes in *Bradyrhizobium japonicum, Mol. Gen. Genet.,* 223, 329, 1990.
124. **Richardson, A. E., Simpson, R. J., Djordjevic, M. A., and Rolfe, B. G.,** Expression of nodulation genes in *Rhizobium leguminosarum* biovar *trifolii* is affected by low pH and by Ca and Al ions, *Environ. Microbiol.,* 54, 2541, 1988.
125. **Banfalvi, Z. and Kondorosi, A.,** Production of root hair deformation factors by *Rhizobium meliloti* nodulation genes in *Escherichia coli;* HsnD (NodH) is involved in the plant host specific modification of the NodABC factors, *Plant Mol. Biol.,* 13, 1, 1989.
126. **Pardo, M. A. and Martinez, E.,** A *Rhizobium* mutant in the flavonoid-inducible indole-acetic acid production has reduced nodulation efficiency, in *Abstr. Proc. 5th Int. Symp. Mol. Genet. Plant-Microbe Interact.,* Interlaken, Switzerland, p. 102, 1990.

# RHIZOBIAL POLYSACCHARIDES REQUIRED IN SYMBIOSES WITH LEGUMES

**K. Dale Noel**

## TABLE OF CONTENTS

# I. INTRODUCTION

Bacterial cells exhibit various types of surfaces. Often they are surrounded by capsules and/or layers of slime, chiefly composed of acidic and neutral polysaccharides (or sometimes acidic polypeptides).[1] Internal to such layers are the outermost permanent surfaces of the bacterial cell, the thick peptidoglycan wall of Gram-positive eubacteria or the outer membrane of Gram-negative eubacteria.[2] The latter membrane has a lipid bilayer, whose outer leaflet is composed mainly of abundant lipopolysaccharides (LPSs). The outer membrane also contains abundant integral proteins, but access to the majority of these proteins is restricted by the polysaccharide portion of the LPS molecules.[3] In addition, both Gram-positive and Gram-negative eubacteria often contain protein structures — flagella, pili, and fimbriae — and cellulose fibrils that project well beyond the outermost permanent layer.

In general, the most pervasive bacterial surface components are polysaccharides. Surface polysaccharides apparently are crucial in many pathogenic and commensal relationships of bacteria with animals[1,2] and plants.[74] In the mutualistic *Rhizobium*-legume symbiosis, as well, a variety of bacterial surface polysaccharides are required, as revealed by recent studies with polysaccharide-deficient rhizobial mutants.

The reader is referred to other chapters in this volume for details of the *Rhizobium*-legume symbiosis. The following points are important in understanding the phenotypes discussed in this chapter. The bacteria induce cell division in the cortex of immature regions of legume roots. As these dividing cells develop into a root nodule, bacteria enter by an infection thread whose wall is laid down by the plant. Bacteria are released endocytotically from the infection thread into interior cells of the developing nodule, where they proliferate and become nitrogen-fixing "bacteroids." Clover, alfalfa, pea, *Vicia* spp., and *Leucaena* spp. all exhibit *indeterminate* nodule development,[4] i.e., cell division and infection continue indefinitely in the mature nodule. On the other hand, soybean and bean nodules are *determinate* — cell division and infection cease as these nodules mature. Determinate and indeterminate nodulation differ also in earlier stages of nodule development, such as where cell division arises in the cortex. Each of these developmental steps requires compatibility between the symbiotic partners. For example, each rhizobial strain can nodulate only certain hosts, although the host range can be rather broad.

One generalization that will emerge from considering rhizobial mutants is that, when important in symbiosis, surface polysaccharides are usually required for infection (infection thread formation and release of bacteria into infected plant cells) rather than for eliciting nodule formation. Probably because of the infection defects, mutants deficient in these polysaccharides are $Ndv^-$; i.e., the nodules they induce are incompletely developed. Because infection is absent or incomplete, such mutants also are $Fix^-$; i.e., the elicited nodules fix little or no nitrogen. Since $Fix^-$ applies also to bacteria that lack nitrogen-fixing ability but elicit fully developed nodules, the $Ndv^-$ designation is more revealing, but it still covers a wide range of specific developmental defects.[16,87]

# II. LIPOPOLYSACCHARIDE (LPS)

## A. STRUCTURE

Lipopolysaccharide (LPS) structure is divided conceptually into two portions, the lipid A moiety and an attached polysaccharide (Figure 1). Lipid A is the predominant lipid of the outer leaflet of the outer membrane in well-known enteric bacteria, such as *Escherichia coli*. It consists of a "backbone" sugar or disaccharide to which fatty acids are attached. Among rhizobia, three types of backbone are known: glucosamine (*R. meliloti* and *Sinorhizobium fredii*), a glucosamine-galacturonic acid disaccharide (*R. leguminosarum*), and 2,3-diamino-2,3-dideoxyglucose alone or in combination with glucosamine (certain *Bradyrhi-*

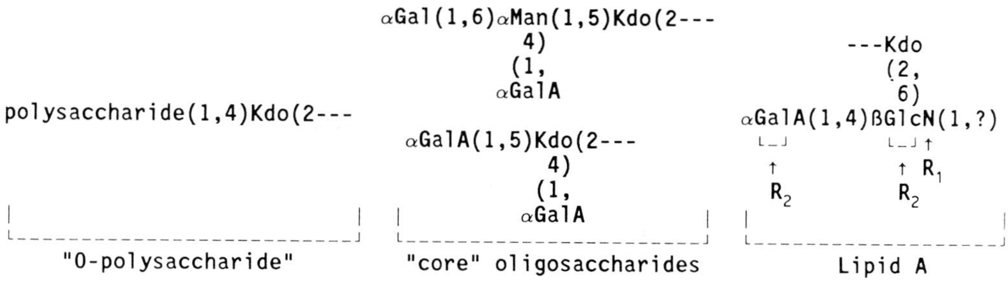


FIGURE 1. Structural features of the LPS of *R. leguminosarum.*[15,36,105] Studies have been based on fragments released by chemical cleavage at the ketosidic bonds of Kdo residues. Therefore, the residues to which the Kdo residues are attached in the intact LPS molecules are unknown, as indicated by dashed lines (---). The "core" oligosaccharide structures are conserved among the various strains that have been studied, whereas the *O*-polysaccharide is strain-dependent. The core oligosaccharide portion apparently is attached to lipid A through a Kdo residue attached to O-6 of GlcN. $R_1$, fatty acids 27-OH-28:0 and 27-O-(27-OH-28:0)-28:0 amide-linked to GlcN. $R_2$, hydroxyfatty acids, mainly 3-OH-14:0 and 27-OH-28:0, ester-linked to GalA and GlcN. An unknown substituent (?) is attached at the reducing end of GlcN. Abbreviations: GalA, galacturonic acid; Man, mannose; Gal, galactose; GlcN, glucosamine; Kdo, D-*manno*-3-deoxyoctulonic acid.

*zobium* species).[5] The major fatty acids of lipid A include the recently discovered 27-hydroxyoctacosanoic acid,[6] (Figure 1), which is long enough to span the bilayer of the outer membrane, and shorter 3-hydroxy fatty acids typical of lipid A in enteric bacteria. The long fatty acid appears to be a consequence of phylogeny.[5] Its role, if any, in symbiosis is unclear.

The polysaccharide portion of the LPS projects outward from the bacterial cell. The LPS polysaccharide is further subdivided conceptually into two regions, the (inner) core oligosaccharide(s) and the (outer, strain-dependent) O-antigen. Although compositions of the LPS polysaccharide portions of *R. meliloti, Bradyrhizobium japonicum,* and *S. fredii* strains have been determined,[11-14] the only rhizobial species for which there is much information regarding overall structure is *R. leguminosarum*[7,15] (Figure 1). The LPS "core" of *R. leguminosarum* differs substantially from the classic LPS core structure of enteric bacteria. There are no heptose residues, and mild acid hydrolysis releases two oligosaccharides (Figure 1) rather than one.[7] The strain-dependent immunodominant *O*-polysaccharides of *R. leguminosarum* also are released by mild hydrolysis of LPS, as longer fragments that terminate in D-*manno*-3-deoxyoctulonic acid (Kdo) but do not contain either "core" structure (Figure 1).[8]

A convenient technique for analyzing the LPS content of a bacterial strain is sodium dodecyl sulfate polyacrylamide gel electrophoresis (SDS-PAGE).[9,10] When *R. leguminosarum* LPS is subjected to this technique, two forms of LPS are separated. The more slowly migrating bands (LPS I) contain *O*-polysaccharide,[8] whereas the faster migrating bands (LPS II) have only core oligosaccharides[8] (Figure 2).

## B. GENETICS

The first systematic studies of rhizobial *lps* genetics grew out of the realization that an abundant class of Ndv$^-$ mutants of *R. leguminosarum* CFN42, obtained by random transposon mutagenesis, was defective in the polysaccharide portion of the LPS.[16] Eventually, it was shown that mutations affecting LPS were located in five genetic (*lps*) regions defined by cosmid cloning of the DNA of strain CFN42.[17,18] One of these regions is located on an indigenous plasmid other than the Sym ("symbiotic") plasmid in strain CFN42 and in 6 of 10 other *R. leguminosarum* strains surveyed.[19,20] Mutations in this *lps* region result in LPS molecules that lack *O*-polysaccharide and have truncated core oligosaccharide.[19,21] In a much more extensive *lps* region located on the chromosome of strain CFN42, nine transcriptional units (at least nine genes) have been identified by complementation analysis[22] (Table 1).

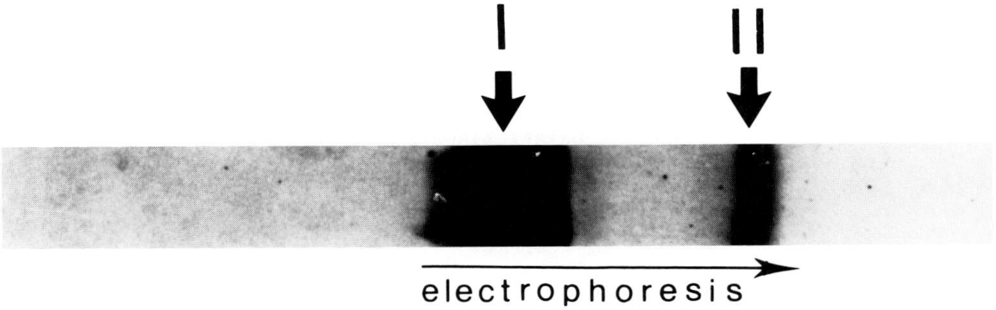


FIGURE 2. Purified LPS of *R. leguminosarum* CFN42 separated on SDS-PAGE.[16] One lane of a slab gel polymerized from 15% acrylamide is shown. The arrows indicate bands LPS I and LPS II, as revealed by periodate-silver staining.[10] Anti-LPS antisera raised against cells or total LPS of strain CFN42 react almost exclusively with LPS I.[8] This antigenicity is correlated with the presence in LPS I, but not in LPS II, of an "*O*-polysaccharide" (Figure 1) released from *R. leguminosarum* LPS by mild acid hydrolysis.[8]

**TABLE 1**
**Genes for Polysaccharide Synthesis**

| Polysaccharide | Strain[a] | Genes/loci[b] | Function | Ref. |
|---|---|---|---|---|
| LPS | Rl CFN42 | *lpsα*[c] *(lpsABCDEFGHI)* | Core, OPS[d] | 7,22 |
| | | *lpsβ*[c], *lpsγ*[c] | Core | 17,19 |
| | Rl VF39 | *lpsα* (>5 genes), *lpsβ* | OPS, core[e] | 20,24 |
| | Rl ANU843 | *lpsα*, *lpsβ* | OPS, core[e] | 19,21 |
| EPS I | Rm SU47 | *exo (PMALFOB)C* | Synthesis[f] | 59,60 |
| | | *exo (NKJG)D* | Quan./qual.[g] | 59,60 |
| | | *exoRS(FX)* | Regulation | 59,67 |
| | | *exoH* | Succinylation | 51 |
| EPS II | Rm SU47 | *exp(ACDEFG), exoBC* | Synthesis[f] | 49 |
| | | *expR, mucR* | Regulation | 49,50 |
| EPS | NGR234 | *exoBDYCG* | Synthesis[f] | 60 |
| | | *exoXY* | Regulation | 66 |
| EPS | Rl 8002 | *pss-1, pss-2* | Synthesis | 61 |
| | | ***psi, psr***,*pss-2* | Regulation | 61,65 |
| | Rl CFN42 | *exoA1,A2* | Synthesis[f] | 18 |
| β-1,2-Glucan | Rm 102F34 | *ndvB, exoC* | Synthesis[f] | 86,88 |
| | | *ndvA* | Export | 86,88 |

[a] Abbreviations: Rm, *R. meliloti;* Rl, *R. leguminosarum.*

[b] Only a few of the loci listed have been defined as genes. Most are transcriptional units defined by transposon mutagenesis. Underlining means that the loci are found on a plasmid other than the Sym (nod-nif) plasmid. **Bold** type is used to indicate loci *(psi, psr)* located on the Sym plasmid.

[c] *lps* regions α, β, and γ refer to *lps* DNA cloned on recombinant plasmids pCOS109.11, pCOS126, and pCOS309, respectively.[17]

[d] Mutations in these genes affect LPS core oligosaccharides or O-polysaccharide (OPS).

[e] Presumed necessary for the same portions of LPS as the corresponding genetic regions of strain CFN42.

[f] Mutations in these genes abolish synthesis of the EPS.

[g] Mutations in these genes affect the quantity or some quality of the EPS.

Mutations in this region result in changes in the *O*-polysaccharide with unaltered core[17,23] or alterations in core accompanied by loss of *O*-polysaccharide.[7,15] A gene located elsewhere[17] is also important for core synthesis.[7,15] A mutation in a fourth region results in fewer LPS I molecules than normal.[17] Mutations in a fifth region affect both exopolysaccharide (EPS) and LPS structure.[18]

*R. leguminosarum* VF39 has an *lps* region with at least five genes[24] that may correspond to the extensive chromosomal *lps* region of CFN42. *R. meliloti lpsB, lpsX, lpsY, lpsZ,* and *fix-23* mutations affect bacteriophage sensitivity and polysaccharide(s) that may be LPS, but the structural effects have not been defined.[25-28] There have been reports of mutants affected in LPS structure in other rhizobial species, but the mutated genes have not been mapped or cloned.

## C. SYMBIOTIC PHENOTYPES OF LPS MUTANTS

*Rhizobium leguminosarum* mutants that lack LPS I (*O*-polysaccharide-containing molecules) generally are defective in the infection process and exhibit the Ndv$^-$ phenotype. Such mutants derived from *R. leguminosarum* CFN42 (whose host is bean) elicit nodules on bean plants in which infection threads become strangely bloated and abort, mainly within infected root hair cells.[16] These nodules exhibit normal meristematic activity during the first few days, but later differentiation events do not occur. The nodules remain small, white, and devoid of leghemoglobin.[16] Although the occurrence of meristematic activity signifies that the bacterial *nod* genes are functional, nodules clearly are elicited at lower efficiency, because they are more widely spaced on the root than normal.[16]

LPS I is also required for complete infection by *R. leguminosarum* strains capable of nodulating clover and *Vicia hirsuta*.[21,24,29] However, on these indeterminate hosts, mutants without LPS I infect sufficiently to be deposited in low numbers into some nodule cells. The resulting nodule development is incomplete, and nitrogenase activity is barely detectable.

Lps mutants of *Bradyrhizobium japonicum,* whose host range includes soybean, are also symbiotically defective. One mutant derived from *B. japonicum* 61A76 was reported to lack certain LPS sugars and to elicit soybean nodules that developed very slowly.[30] Two mutants derived from *B. japonicum* 110 have been reported to have a truncated LPS. One, perhaps more severely truncated, does not elicit any visible nodulation.[13] The other is Ndv$^-$ in a manner similar to the *R. leguminosarum* Lps mutants on bean hosts.[31]

Since severely truncated LPS in enteric bacteria leads to pleiotropic changes in the outer membrane,[2] Lps mutants of *R. leguminosarum* have been investigated for this possibility. However, no consistent differences in outer membrane protein content, or release of outer membrane proteins into the medium could be detected, even of mutants lacking *O*-polysaccharide and one of the oligosaccharides.[16] The only obvious change consistently associated with these mutants *ex planta* is that the cells clump together in liquid culture and on soft agar ("swarm") plates. The latter observation has been interpreted to indicate deficiency in motility,[24,29] but microscopic observations indicate that such mutants derived from *R.leguminosarum* CFN42 are motile. Using a mutant in which *lps* genes were deleted, it was possible to isolate a suppressor mutant that lacked LPS I but did not clump in liquid or swarm plates.[78] However, this double mutant still had the Ndv$^-$ phenotype, indicating that this clumping or "motility" defect was not the cause of the symbiotic defect.

## D. POSSIBLE SYMBIOTIC FUNCTIONS

Why is complete LPS (e.g., LPS I of *R. leguminosarum*) necessary? One possibility is that it serves as a signal compound to trigger developmental events important for rhizobial infection. In this role, LPS traditionally is considered a fixed receptor on the bacterial outer membrane, but it also might serve as a diffusible elicitor. Release of LPS and outer membrane fragments from bacterial cells is common,[32] and LPS I molecules are present in cell-free supernatants of *R. leguminosarum* cultured *ex planta*.[18,33,34]

The phenotype of one Lps mutant of *R.leguminosarum* CFN42, in particular, seems relevant to this possible role. This mutant appears to have the wild-type LPS I structure but at approximately one third the normal concentration.[17] This mutant is as deficient in infection as mutants that lack LPS I completely. If the core oligosaccharides and *O*-polysaccharide of this mutant have the normal structure, as indicated by preliminary data,[17,23] then the implication is that the normal *abundance* of LPS I is critical. Such a requirement would not be predicted if LPS serves only as a fixed surface receptor.

Other observations suggest that there is great latitude in *O*-polysaccharide structures that are symbiotically effective, thereby placing certain restrictions on how LPS I could serve as a signal. Whereas *R. leguminosarum* LPS core oligosaccharide structures are conserved between strains of different biovars (host ranges),[7,15] *O*-polysaccharides vary as much within biovars as they do between biovars.[36] By genetic manipulation it has been possible to eliminate the LPS I of *R. leguminosarum* ANU843 and replace it with an LPS I whose composition is almost identical with that of *R. leguminosarum* CFN42.[21] The resulting strain has the host range of strain ANU843, is Fix$^+$, and nodulates as well as strain ANU843. Therefore, although *O*-polysaccharide must be present, its specific structure may be irrelevant. However, although the ANU843 and CFN42 LPS I molecules do not cross-react immunologically, it is conceivable that they share a structural feature that is critical in a host-independent manner. Another alternative is that a signal-promoting structure is in the core or lipid A part of LPS and the role of *O*-polysaccharide is merely to allow aqueous diffusion.

Other possible functions are predicated on the notion that the bacterial outer membrane is sheathed in an almost continuous layer of LPS *O*-polysaccharide moieties. This layer could act as a hydrophilic shield to prevent access of toxic hydrophobic molecules that otherwise would diffuse into or through the bacterial outer membrane. The LPS *O*-antigens of animal pathogens are often viewed similarly as shields against hydrophobic toxins, phagocytosis, and complement action.[2] The rhizobial *O*-polysaccharide does apparently compart hydrophilicity, as shown by the increased hydrophobicity of Lps$^-$ *R. leguminosarum* mutants.[29] However, if the role is merely to maximize hydrophilic surface character, it is puzzling that so many of the *O*-polysaccharide residues are *O*-methylated deoxysugars.[8,36] Another protective role might be to mask elicitors of host defense responses or to otherwise preempt the host defense cascade.[35] A related type of role is to provide an appropriate bacterial surface for endocytosis or other potential surface-surface interactions during infection,[29] by analogy to LPS action during cellular invasion by certain bacterial endoparasites. This function might require only certain "hydrophilic" or "charge" qualities,[29] or it might require specific structures.

On indeterminate hosts of *R. leguminosarum,* infection by Lps mutants proceeds much further than infection of soybean or bean by such mutants. This may indicate that LPS has at least two distinct functions and only the second one is necessary on the indeterminate hosts. In keeping with the phenotype exhibited on these latter hosts, the second role may be to provide a compatible bacterial surface for interaction with plant membranes at the time of endocytosis or bacteroid proliferation. The first role may be a very nonspecific one that can be carried out by either LPS or EPS, but is adopted by LPS on determinate hosts because (possibly) EPS synthesis is repressed.

## E. CHANGES IN LPS STRUCTURE DURING SYMBIOSIS

Bacterial pathogens of animals very often exhibit changed surfaces during symbiosis, presumably in response to the change in environment and in order to cope with animal immune systems.[37] A few studies have suggested that *R. leguminosarum* also changes its surface during symbiosis, particularly the LPS content and structure.[38-41] Pea bacteroids of *R. leguminosarum* 300 have an LPS I antibody epitope not detected in the bacteria grown under traditional conditions in the laboratory.[39,41] This epitope is observed, however, when bacteria are grown at low pH (5.3 or lower) or under low oxygen tensions.[42] Therefore, this LPS epitope is not symbiosis-specific. Bean bacteroids also exhibit antigenic features that can be partially reproduced by unusual growth conditions *ex planta*.[34,40] An antigenic change

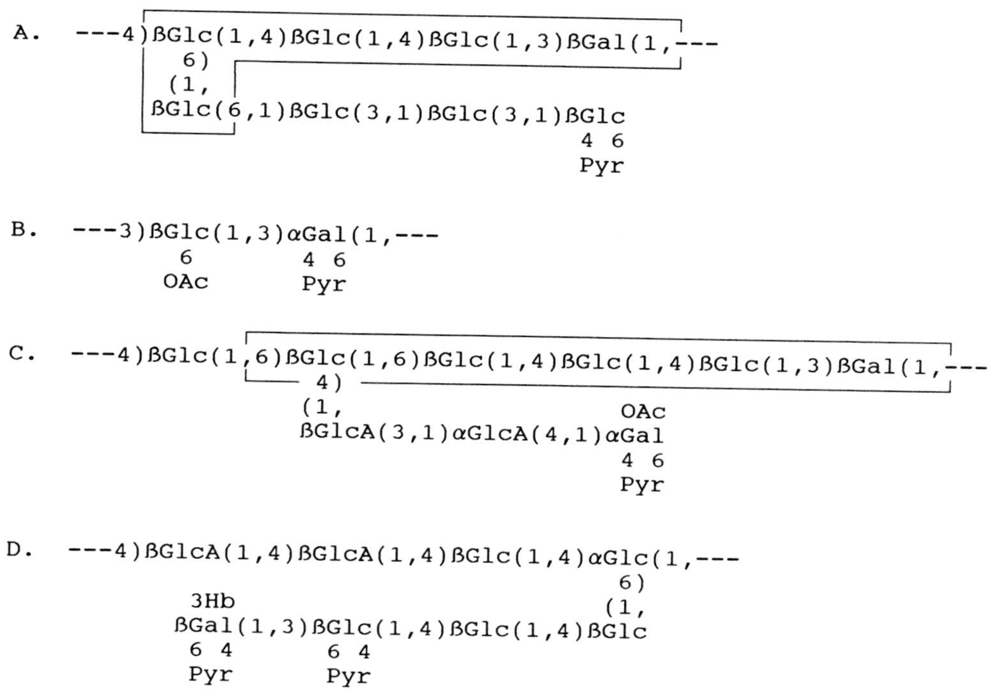


FIGURE 3. EPS repeat units. (A) EPS I of *R. meliloti*.[104] Acetyl and succinyl groups occur at unknown positions. (B) EPS II of *R. meliloti*.[49] (C) EPS of strain NGR234.[52] (D) The most common EPS of *R. leguminosarum* strains.[45,54] *O*-acetyl substitutions occur at variable positions in structure (D).[94] The boxed areas in (A) and (C) indicate the portions shared by these structures.[60] Abbreviations: Glc, glucose; Pyr, pyruvate; OAc, *O*-acetate; GlcA, glucuronic acid; 3Hb, 3-hydroxybutyrate; Gal, galactose.

induced by exposure of *R. leguminosarum* CFN42 to bean root exudate is controlled by a gene(s) in the *nod* region of the Sym plasmid.[34] This is the first indication that any gene in the *nod* region has influence on LPS structure. The significance of these changes in LPS is still unclear. In particular, it is not known whether these modifications are related to the symbiotic requirement for LPS I. It would be helpful to isolate mutations that prevent these epitope changes without affecting the gross LPS structure. One could then test whether such mutations affect symbiosis and/or limit bacterial growth at low pH or oxygen concentration.

## III. EXTRACELLULAR POLYSACCHARIDE (EPS)

Most or all rhizobia are thought to exhibit capsules under certain conditions. The presence of capsules surrounding *R. leguminosarum* 0403 has been particularly well-documented.[43,44] Most strains also elaborate slimy (mucoid) colonies on solid agar and, during late stages of growth in liquid or agar, release particular acidic exopolysaccharides (EPSs) that dominate the composition of extracellular material. The structures of many acidic EPSs have been elucidated (Figure 3), and mutants deficient in their synthesis have been isolated.

The chemical and cytological relationships between capsule, slime, and EPS are uncertain. The major acidic EPS is often designated as "capsular,"[43-46] but without proof that it is responsible for the capsule seen by microscopy. Indeed, there is another candidate for the capsular polysaccharide.[106] Study of the available nonmucoid mutants might settle this issue.

### A. EPS-DEFICIENT MUTANTS

Mutants deficient in a succinoglycan (EPS I of Figure 3) have been isolated from *R. meliloti* SU47 by selecting for nonfluorescent colonies in the presence of the polysaccharide-binding fabric whitener Calcofluor.[47] These ($Exo^-$) mutants induce nodules that lack infection threads or have very attenuated infections on *Medicago, Melilotus,* and *Trigonella* hosts of strain SU47.[48] This $Ndv^-$ phenotype on alfalfa (*Medicago*) can be suppressed completely by mutations that allow the synthesis of a normally cryptic EPS (EPS II of Figure 3).[49,50] In other words, either EPS I or EPS II is sufficient to allow symbiotic proficiency of strain SU47. Other than acid character, the only obvious feature common to the structures of EPS I and EPS II is a glucose-β(1,3)-galactose disaccharide motif (Figure 3). In both cases the glucose of this disaccharide may be acetylated. Acidity itself may be the key feature, since mutants in which EPS I lacks the succinyl moiety are $Ndv^-$ as well, despite having abundant EPS of normal molecular weight.[51]

Another *R. meliloti* wild isolate, strain 41, appears to have at least one other redundant component for the infection role carried out by EPS I or EPS II. *exoB* mutations in this strain prevent EPS I and EPS II production but do not prevent the induction of well-developed, nitrogen-fixing nodules.[26,27] However, *R. meliloti* 41 *exoB* mutants do become $Ndv^-$ if also mutated in one of the following loci: *lpsX, lpsY, lpsZ, or fix-23*.[26-28] Either *exoB*$^+$ or *fix-23*$^+$ completely suppresses the $Ndv^-$ phenotype of *exoB fix-23* double mutations.[27] Moreover, transfer of *lpsZ*$^+$ into *R. meliloti* SU47 (which lacks this gene) converts *exoB*$^-$ mutants into $Ndv^+$ $Fix^+$ strains.[26] As mentioned before, these suppressing loci may specify LPS structure; *fix-23* mutations clearly affect a Kdo-containing polymer(s).[27] *lpsZ* and *fix-23* are obviously distinct genetically, since *lpsZ* is located on an indigenous plasmid whereas *fix-23, lpsX,* and *lpsY* are chromosomal.[26-28]

The major EPS of wide-host-range strain NGR234 (Figure 3)[52] also has been studied extensively. The symbiotic phenotype of nonmucoid mutants that lack this EPS depends on the host. At least some of these mutants are $Ndv^+Fix^+$ on hosts that form determinate nodules, but on *Leucaena leucocephala,* which forms indeterminate nodules, all nonmucoid mutants exhibit an $Ndv^-$ noninfective phenotype very similar to that of $Exo^-$ *R. meliloti* SU47 mutants.[53]

*Rhizobium leguminosarum* has a major acidic EPS whose structure is basically the same in most strains (Figure 3), irrespective of host range (biovar).[54] Nonmucoid ($Exo^-$) mutants are deficient in this EPS.[18] $Exo^-$ biovar *viciae* strains are $Nod^-$ on pea, i.e., no apparent nodule tissue is induced.[18,55] $Exo^-$ biovar *trifolii* strains are $Ndv^-$ on clover.[18,56] $Exo^-$ biovar *phaseoli* strains are $Ndv^+Fix^+$ on bean.[18,55] Once again, the indeterminate hosts (pea and clover) require $Exo^+$ microsymbionts, whereas a determinate host (bean) does not.

Since it may turn out that *R. meliloti* LPS and EPS can substitute for each other in carrying out symbiotic functions, it is important to stress that there is no evidence for this possibility in *R. leguminosarum*. $Exo^-$ *R. leguminosarum* strains are symbiotically defective on clover and pea, regardless of having normal LPS structure.[18] Likewise, mutants lacking LPS I are defective on all hosts, whether or not they still produce the normal EPS.[57] Even if the *lpsBXYZ* and *fix-23* mutations affect *R. meliloti* LPS, it is quite possible that the resulting defects are not comparable to the absence of LPS I from *R. leguminosarum*.

*Sinorhizobium fredii* HH303 is a microsymbiont of determinate-nodulating soybean. Mutants of this strain that lack the major acidic EPS are $Nod^+Fix^+$ on soybeans.[58]

### B. GENETICS AND REGULATION

Genetic loci required for the synthesis or regulation of *R. meliloti* EPS I and EPS II, the major EPS of strain NGR234, and the acidic EPS of *R. leguminosarum* are listed in Table 1 (see earlier). These loci have been defined by mutation, genetic complementation, and, in a few cases, by nucleotide sequence.[18,59-61] The most complete genetic analysis has been done with *R. meliloti*.

Only two of the protein products of these genes have been assigned functions; *exoC* of *R. meliloti* specifies phosphoglucomutase,[62] and the *exoB* gene product of *R. meliloti* appears to be UDPgalactose 4′-epimerase.[63,64] Cross-complementation and genetic hybridization between strain NGR234 and *R. meliloti* suggest that five *exo* genes (*exoFMALP* of *R. meliloti*) may encode glycosyl transferases for five consecutive residues and linkages that are shared by the EPSs of these bacteria (Figure 3).[60]

*Rhizobium meliloti* SU47, strain NGR234, and *R.leguminosarum* 8002 each have a gene that inhibits EPS production when present in multiple copies relative to one or more counterbalancing genes.[65-68] The mechanism of inhibition remains unresolved; one hypothesis proposes that it occurs by interaction of the *exo* gene products at the cytoplasmic membrane.[68] Mutations in the inhibitory gene (*psi*) of *R.leguminosarum* 8002 cause an Ndv$^-$ phenotype on bean,[65,69] whereas mutation of the corresponding gene in *R. meliloti* has no effect.[67] Since bean tolerates microsymbionts incapable of producing EPS, it is tempting to speculate that EPS is actually deleterious on this host. Hence, *psi* may be required during symbiosis to prevent EPS production during infection.

An indication that rhizobia have recruited EPSs as symbiotic determinants is that there are genes on Sym plasmids that regulate EPS abundance. Aside from *psi,* examples include *syrA, syrM, and nodD3 of R. meliloti,*[70] and *nodD2* of *S. fredii.*[71]

### C. POSSIBLE SYMBIOTIC ROLES AND THE EFFECTS OF ADDED EPS

As with LPS, one difficulty in defining the role of acidic EPS is that the severity of the infection block associated with the EPS-deficient condition depends upon the particular symbiosis: Exo$^-$ mutants can be Nod$^-$, noninfective Ndv$^-$, partially infective Ndv$^-$, or Ndv$^+$Fix$^+$. It is possible then that EPS plays different roles in different host-microsymbiont combinations, and more than one role on a given host.

Since EPSs are diffusible, they may be acting as signal molecules to elicit necessary plant responses[72] or as suppressors of deleterious plant reactions.[35] Perhaps processing of EPS generates "oligosaccharin"-type hormones.[73] Another possibility is that acidic EPSs bind to and neutralize proteins or smaller cationic molecules from the plant that are toxic to the bacteria. There is evidence that this may indeed be an EPS function for plant pathogen *Erwinia amylovora.*[74] Likewise, EPS may prevent bacterial toxins or diffusible bacterial elicitors of host-defense responses from reaching the plant.[75] Other possibilities include the contrasting functions of facilitating adherence to plant surfaces or preventing deleterious adherence during infection.

Addition of purified EPS and/or EPS fragments prior to, during, or after inoculation of plants with Exo$^-$ mutants of *R.leguminosarum* ANU843, *R. meliloti* SU47, and strain NGR234 has been reported to partially reverse the infection defects of these mutants.[76,77] In each case, only EPS isolated from the wild-type microsymbiont of the host was effective; i.e., the EPSs exhibited the same host-specificities as the strains that produced them. The requirement for specific structures, however, is not supported by experiments in which strain NGR234 was manipulated genetically to produce EPS I of *R. meliloti* instead of the NGR234 EPS.[108] Such hybrid strains infect *L. leucocephala,* and NGR234 host that requires EPS for infection. Note that this result is analogous to the result of changing the *O*-polysaccharides of *R. leguminosarum* strains.[21] On the other hand, in both of these cases[21,108] the heterologous polysaccharide still shares substantial structure with the polysaccharide normally present, and this shared structure could be responsible for providing the symbiotic function. One Exo$^-$ *R. leguminosarum* strain was restored to normal symbiotic proficiency when *exo* DNA from *Xanthomonas campestris* was transferred into it.[55] However, there was no analysis of whether the recombinant produced *Xanthomonas*-type EPS or *R. leguminosarum*-type EPS.

Attempts to restore infection to Lps$^-$ mutants of *R.leguminosarum* CFN42 by adding purified wild-type LPS have failed.[78] However, the LPS may not have been isolated from cells in the proper physiological state. It has been reported that prior incubation of clover roots with low concentrations of LPS from *R. leguminosarum* strains increases the frequency of infection threads. Such enhancement by LPS from *R. leguminosarum* 0403 is restricted to LPS isolated from bacteria at a specific stage of growth.[46]

## IV. β-GLUCANS

Rhizobia produce various types of β-glucans. *Rhizobium* and *Agrobacterium* spp. make cyclic β-1,2-glucans[79,80] that reside mainly in the periplasm but also are secreted beyond the outer membrane. These molecules contain 15 to 30 glucose residues and their synthesis appears to be regulated in response to the osmolarity of the growth medium.[81] *Bradyrhizobium* spp. also have cyclic glucans, albeit with β-1,3- and β-1,6-linkages, whose function may be similar.[82] Long β-1,4-glucans (cellulose) are found on the surfaces of at least *R. leguminosarum.*[83]

Mutations in genes *chvA* and *chvB* of the tumor-inducing pathogen *A. tumefaciens* result in cyclic β-1,2-glucan deficiency.[84,85] The mutants are avirulent and defective in assays of attachment to host cells.

The homologous genes of *R. meliloti* 102F34 are designated as *ndvA* and *ndvB.*[86] Nodules induced by *ndv* mutants are incompletely developed and devoid of bacteria.[86] Certain mutations suppress *ndv* mutations such that normally developed nodules are elicited on alfalfa seemingly without β-1,2-glucan production.[88] The target(s) of the suppressor mutations is not yet known. It may be that the suppressor mutations induce the production of another polysaccharide that compensates for the lack of β-1,2-glucan. These pseudorevertants are still deficient in adaptation to low osmolar conditions *ex planta* and are only marginally less deficient in root hair attachment and infection thread formation than the *ndv* mutant parents.[88] Therefore, the symbiotic role of β-1,2-glucan may not be related to osmolarity adaptation, and even its role in attachment and infection thread initiation seems unimportant.

Cellulose mutants also were isolated first from *A. tumefaciens.*[89] Such mutants are fully virulent unless the plant is flooded with water after inoculation. Matthysse has suggested that cellulose acts in a second phase of attachment, after which the bacteria are very tightly bound to the plant surfaces.[89] Mutants incapable of synthesizing cellulose fibrils also have been isolated from an *R. leguminosarum* strain.[83] These mutants are incapable of a type of attachment that yields large aggregates on pea root tips, but they nodulate in a normal manner under laboratory conditions. There is no report of whether these mutants nodulate after washing the inoculum from roots, or of whether they are deficient under field conditions.

## V. LECTINS, HOST SPECIFICITY, AND ATTACHMENT

One of the recurring ideas about capsules is that they facilitate attachment of bacterial colonies to surfaces.[1,90] Consideration of this possibility with respect to rhizobia has been skewed by a provocative hypothesis concerning the mechanism of host specificity of nodule formation. Almost two decades ago it was proposed that host lectins (carbohydrate-binding proteins) were responsible for the type of attachment of rhizobia to root surfaces that could lead to infection.[91] Further, it was speculated that the selectivity of the lectins for specific bacterial surface polysaccharides was responsible for the host range of the bacteria.[92,93] A great deal of subsequent work has been performed to test this hypothesis, and several versions of the hypothesis have been proposed.[43,46,72]

Two recent studies involving the transfer of recombinant DNA bolster the view that lectins are involved in host-specific nodulation of clover and pea by *R. leguminosarum.* In one study, the pea microsymbiont *R. leguminosarum* 300 received the *nod* genes specifying the host range of clover microsymbiont *R. leguminosarum* ANU843.[94] The resulting strain was capable of nodulating clover, and its binding of clover lectin trifoliin A in an *in situ* assay increased dramatically, from nil to a level comparable to that of clover symbiont ANU843. In the other study, white clover roots were transformed with pea lectin genes and could then be nodulated by the pea microsymbiont *R. leguminosarum* 248.[95] Most of these nodules were of the Ndv$^-$ type but some were developed and infected sufficiently to fix nitrogen. In this latter experiment, the presence of pea lectin was not actually measured, nor was the effect of the hapten for pea lectin tested. In neither experiment was there any attempt to determine which bacterial polysaccharide (if any) was acting as the lectin substrate.

Most work has been based on the idea that a polysaccharide or polysaccharide fragment is the lectin substrate. Based on indirect assays, clover lectin trifoliin A has been inferred to bind to both capsular polysaccharide and LPS, but only to the versions of these molecules produced by *R. leguminosarum* bv. *trifolii* strains.[43,46] However, this idea has not been tested by direct assays of the binding of purified polysaccharides to purified lectins, nor have *exo* and *lps* mutants of *R. leguminosarum* bv. *trifolii* been tested for lectin binding in published studies.

In principle, polysaccharides need not be the substrates for the lectins, nor must a host-determining step involve attachment. A very recent idea is that lectins might recognize the lipo-oligosaccharides that appear to be specified by the *Rhizobium nod* genes.[96] Although these *nod* factors may serve as fixed receptors (e.g., for attachment) on the bacterial surfaces, they are active in the absence of the bacteria and are found in culture fluids.[96] Therefore, they most likely function as secreted molecules.

Two types of assays have been used to measure rhizobial attachment to roots. One type involves microscopic examination of bacterial aggregation to and colonization of the root surface.[97] An advantage of using the microscope is that it is possible to distinguish the topology and location of attachment. However, quantitation is very difficult. The other type of assay uses roots or root segments incubated in bacterial suspensions.[98] This allows quantitation and analysis of kinetics but is not limited to microniches, such as developing root hairs, that may be the most important, although it is possible to sample segments only from infectible regions of the root. Controversy centers not only on how attachment should be measured but also on whether it is host specific[43] or nonspecific.[99,100]

Whether lectins and bacterial polysaccharides are required for attachment is also subject to debate. Recent observations with *R. leguminosarum* were interpreted to indicate that a calcium-dependent bacterial protein (adhesin), bacterial cellulose fibrils, as well as pea lectin are involved in the attachment to root hairs.[99] Only the adhesin protein is involved under all conditions, and neither it nor cellulose seem to be host-specific in action. Assay of Lps mutants indicated that LPS *O*-polysaccharide was not required for attachment in the microscope assay used.[101]

## VI. POLYSACCHARIDE MUTANTS AS TOOLS TO STUDY NODULE DEVELOPMENT

Mutants deficient in rhizobial polysaccharides have confirmed the idea that nodule meristematic activity (cell division) can be induced without rhizobial infection. In this context *exo, lps,* and *ndv* are bonafide determinants for infection, whereas the role of *nod* genes in infection (if any) has not been satisfactorily dissected from their roles in meristem induction.

Regardless of which polysaccharide is deficient, whenever nodules are induced without complete infection, the Ndv$^-$ phenotype results. Therefore, it appears that normal bacterial release from infection threads into plant cells is required for nodule differentiation and completion of anatomical development. For instance, such nodules do not have the class of nodule-specific plant proteins known as ''late nodulins.''[102,107] In view of the various degrees of truncation of the infection process in the different kinds of polysaccharide mutants, these mutants have been greatly underexploited in investigating how infection elicits the various nodulins and other differentiation events.

Nodule development elicited by $Ndv^-$ mutants appears to be insufficient to trigger the feedback mechanism that normally suppresses nodulation after the initial clusters of nodules appear.[103] The $Ndv^-$ nodules often appear in larger numbers, and, unlike the normal distribution, are scattered throughout the root system. Investigation of the regulation of nodulation may be greatly aided by these mutants.

## VII. FINAL COMMENTS

Nonsymbiotic bacteria have LPS, EPS, and glucan polymers analogous to those of rhizobia. Obviously, these molecules are not solely important in symbiosis. LPS, in particular, is difficult to establish as symbiotically important, in addition to being intrinsically important as an integral component of the bacterial outer membrane. However, one should not assume that LPS molecules with intact *O*-polysaccharides are required for membrane integrity. There is a widespread and mistaken impression that studies with *E. coli* and *Salmonella* spp. have shown that mutants lacking the *O*-polysaccharide have defective outer membranes which leak proteins and exhibit generally altered permeability. The truth is that only mutants with severe truncations of the core oligosaccharide have this problem.[2]

Pathogenic symbioses of bacteria with plant and animal hosts often, but not always, also require these polysaccharides.[1,2,37,74,90] In some cases, the hypothetical roles are extensions of their nonsymbiotic roles, e.g., facilitating adherence of colonies to surfaces and protection against toxic substances. However, many proposed roles apply only to certain types of hosts or pathogenesis, such as warding against mammalian immunity mechanisms with capsules (and LPS) or occlusion of plant vessels by the EPS of wilt pathogens. The latter example illustrates a cautionary lesson that obviously applies to studies of rhizobia. Although most EPS-deficient mutants of wilt or blight pathogens are avirulent, there are exceptions.[74] Hence, one must be cautious in interpreting mutant phenotypes to mean that only a particular polysaccharide can provide the required function. Despite current paradoxes and functions limited to particular interactions, paradigms that apply to symbiosis in general should emerge as studies of these various systems continue.

There is no compelling evidence from genetic studies that the symbiotic functions of *R. leguminosarum* LPS I require specific structures. However, experiments with purified EPS indicate that in some cases the symbiotic roles of EPS might require host-specific EPS structures. Nevertheless, it is unlikely that EPS is required for the host specificity of nodule formation, the type of host specificity that has occasioned the most study. For one thing, nodules appear on the proper hosts after inoculation by $Exo^-$ mutants. Moreover, there is evidence that the *nod* genes acknowledged to determine host specificity direct the synthesis of lipooligosaccharides[96] that have structures quite distinct from those of the well-studied EPSs. Perhaps, then, the putative host specificity of EPS structure is expressed during infection or later developmental events. However, the reader should be aware of arguments that the EPS of *R. leguminosarum* undergoes structural modifications which determine or help determine the host range for nodule formation.[94]

Many of the speculated functions seemingly could be carried out by either EPS or LPS. However, EPS and LPS differ tremendously at the molecular level and mainly reside in different surface layers. Indeed, it is clear that LPS and EPS do not substitute for each other in all symbioses. Only with *R. meliloti* is there evidence that they might perform the same role(s), and it has not yet been proved that LPS is the third alternative to EPS I and EPS II. In *R. leguminosarum* strains, EPS and LPS clearly have somewhat distinct roles because both are required on indeterminate hosts, whereas on a determinate host (bean) only LPS will serve.

One key area of future analysis is the complete determination of LPS structures in representative strains, particularly of *R. leguminosarum* and *R. meliloti*. Possible symbiosis-specific structural changes in EPS and LPS should be considered as well.

Perhaps the greatest challenge for researchers in this area is to develop experimental systems to test the various alternative hypotheses. Aside from experiments based on the lectin hypothesis, there have been very few attempts to provide experimental models for the roles of polysaccharides in symbiosis. More will be required than merely suppressing mutant phenotypes by adding purified polysaccharides, although this approach is useful at least for defining structural requirements.

## ACKNOWLEDGMENTS

I am grateful to Russell Carlson and Ramadas Bhat for unpublished information and their insights regarding polysaccharide structures.

## REFERENCES

1. **Costerton, J. W. and Irvin, R. T.,** The bacterial glycocalyx in nature and disease, *Annu. Rev. Microbiol.,* 35, 299, 1981.
2. **Nikaido, H. and Vaara, M.,** Molecular basis of bacterial outer membrane permeability, *Microbiol. Rev.,* 49, 1, 1985.
3. **Bentley, A. T. and Klebba, P. E.,** Effect of lipopolysaccharide structure on reactivity of antiporin monoclonal antibodies with the bacterial cell surface, *J. Bacteriol.,* 170, 1063, 1988.
4. **Sprent, J. I.,** Root nodule anatomy, type of export product, and evolutionary origin in some Leguminosae, *Plant Cell Environ.,* 3, 35, 1980.
5. **Bhat, U. R., Mayer, H., Yokota, A., Hollingsworth, R. I., and Carlson, R. W.,** Occurrence of lipid A variants with 27-hydroxyoctacosanoic acid in lipopolysaccharides from members of the family *Rhizobiaceae, J. Bacteriol.,* 173, 2155, 1991.
6. **Hollingsworth, R. I. and Carlson, R. W.,** 27-Hydroxyoctacosanoic acid is a major structural fatty acyl component of the lipopolysaccharide of *Rhizobium trifolii* ANU843, *J. Biol. Chem.,* 246, 9300, 1989.
7. **Carlson, R. W., Garcia, F., Noel, D., and Hollingsworth, R.,** The structures of the lipopolysaccharide core components from *Rhizobium leguminosarum* biovar *phaseoli* CE3 and the two of its symbiotic mutants, CE109 and CE309, *Carbohydr. Res.,* 195, 101, 1989.
8. **Carlson, R. W., Kalembasa, S., Turowski, D., Pachori, P., and Noel, K. D.,** Characterization of the lipopolysaccharide from a *Rhizobium phaseoli* mutant that is defective in infection thread development, *J. Bacteriol.,* 169, 4923, 1987.
9. **Jann, B., Reske, K., and Jann, K.,** Heterogeneity of lipopolysaccharides. Analysis of polysaccharide chain lengths by sodium dodecyl sulfate-polyacrylamide gel electrophoresis, *Eur. J. Biochem.,* 60, 239, 1975.
10. **Tsai, C. and Frasch, C. E.,** A sensitive silver stain for detecting lipopolysaccharides in polyacrylamide gels, *Anal. Biochem.,* 119, 115, 1981.
11. **Zevenhuizen, L. P. T. M., Scholten-Koerselman, I., and Posthumus, M. A.,** Lipopolysaccharides of *Rhizobium, Arch. Microbiol.,* 125, 1, 1980.
12. **Carrion, M., Bhat, U. R., Reuhs, B., and Carlson, R. W.,** Isolation and characterization of the lipopolysaccharides from *Bradyrhizobium japonicum, J. Bacteriol.,* 172, 1725, 1990.
13. **Puvanesarajah, V., Schell, F. M., Gerhold, D., and Stacey, G.,** Cell surface polysaccharides from *Bradyrhizobium japonicum* and a nonnodulating mutant, *J. Bacteriol.,* 169, 137, 1987.
14. **Carlson, R. W. and Yadav, M.,** Isolation and partial characterization of the extracellular polysaccharides and lipopolysaccharides from fast-growing *Rhizobium japonicum* USDA 205 and its nod$^-$ mutant, HC205, which lacks the symbiotic plasmid, *Appl. Environ. Microbiol.,* 50, 1219, 1985.
15. **Bhat, U. R., Bhagyalakshmi, S. K., Hollingsworth, R. I., and Carlson, R. W.,** Structural investigation of KDO region of core components from the lipopolysaccharide of *Rhizobium leguminosarum,* submitted.
16. **Noel, K. D., VandenBosch, K. A., and Kulpaca, B.,** Mutations in *Rhizobium phaseoli* that lead to arrested development in infection threads, *J. Bacteriol.,* 168, 1392, 1986.

17. **Cava, J. R., Elias, P. M., Turowski, D. A., and Noel, K. D.,** *Rhizobium leguminosarum* CFN42 genetic regions encoding lipopolysaccharide structures essential for complete nodule development on bean plants, *J. Bacteriol.*, 171, 8, 1989.
18. **Diebold, R. and Noel, K. D.,** *Rhizobium leguminosarum* exopolysaccharide mutants: biochemical and genetic analyses and symbiotic behavior on three hosts, *J. Bacteriol.*, 171, 4821, 1989.
19. **Brink, B. A.,** Genetic and functional comparison of LPS biosynthetic loci in *Rhizobium leguminosarum* strains of different host ranges, Ph.D. dissertation, Marquette University, Milwaukee, WI, 1989.
20. **Hynes, M. F. and McGregor, N. F.,** Two plasmids other than the nodulation plasmid are necessary for formation of nitrogen-fixing nodules by *Rhizobium leguminosarum, Mol. Microbiol.*, 4, 567, 1990.
21. **Brink, B. A., Miller, J., Carlson, R. W., and Noel, K. D.,** Expression of *Rhizobium leguminosarum* CFN42 genes for lipopolysaccharide in strains derived from different *R. leguminosarum* soil isolates, *J. Bacteriol.*, 172, 548, 1990.
22. **Cava, J. R., Tao, H., and Noel, K. D.,** Mapping of complementation groups within a *Rhizobium leguminosarum* CFN42 chromosomal region required for lipopolysaccharide synthesis, *Mol. Gen. Genet.*, 221, 125, 1990.
23. **Turowski, D. A. and Noel, K. D.,** unpublished data, 1989.
24. **Priefer, U. B.,** Genes involved in lipopolysaccharide production and symbiosis are clustered on the chromosome of *Rhizobium leguminosarum* biovar *viciae* VF39, *J. Bacteriol.*, 171, 6161, 1989.
25. **Clover, R., Kieber, J., and Signer, E. R.,** Lipopolysaccharide mutants of *Rhizobium meliloti* are not defective in symbiosis, *J. Bacteriol.*, 171, 3961, 1989.
26. **Williams, M. N. V., Hollingsworth, R. I., Klein, S., and Signer, E. R.,** The symbiotic defect of *Rhizobium meliloti* exopolysaccharide mutants is suppressed by *lpsZ*, a gene involved in lipopolysaccharide biosynthesis, *J. Bacteriol.*, 172, 2622, 1990.
27. **Putnoky, P., Petrovics, G., Kereszt, A., Grosskopf, E., Ha, D. T. C., Banfalvi, Z., and Kondorosi, A.,** *Rhizobium meliloti* lipopolysaccharide and exopolysaccharide can have the same function in the plant-bacterium interaction, *J. Bacteriol.*, 172, 5450, 1990.
28. **Williams, M. N. V., Hollingsworth, R. I., Brzoska, P. M., and Signer, E. R.,** *Rhizobium meliloti* chromosomal loci required for suppression of exopolysaccharide mutations by lipopolysaccharide, *J. Bacteriol.*, 172, 6596, 1990.
29. **deMaagd, R. A., Rao, A. S., Mulders, I. H. M., Leentje, G. R., Loosdrecht, M. C. M., Wijiffelman, C. A., and Lugtenberg, G. J. J.,** Isolation and characterization of mutants of *Rhizobium leguminosarum* bv. *viciae* 248 with altered lipopolysaccharides: possible role of surface charge or hydrophobicity in bacterial release from the infection thread, *J. Bacteriol.*, 171, 1143, 1989.
30. **Maier, R. J. and Brill, W. J.,** Involvement of *Rhizobium japonicum* O antigen in soybean nodulation, *J. Bacteriol.*, 133, 1295, 1978.
31. **Carlson, R. W., Lakshmi, B., Bhat, U. R., and Stacey, G.,** Isolation and analysis of the lipopolysaccharides from symbiotic mutants of *B. japonicum,* in *Nitrogen Fixation: Achievements and Objectives,* Gresshoff, P. M., Roth, L. E., Stacey, G., and Newton, W. E., Eds., Chapman and Hall, NY, in press.
32. **Mayrand, D. and Grenier, D.,** Biological activities of outer membrane vesicles, *Can. J. Microbiol.*, 35, 607, 1989.
33. **Carlson, R. W. and Lee, R. P.,** A comparison of the surface polysaccharides from *R. leguminosarum* 128C53 $str^r$ $rif^r$ with the surface polysaccharides from its $Exo^-$ mutant, *Plant Physiol.*, 71, 223, 1983.
34. **Tao, H. and Noel, K. D.,** unpublished data, 1990.
35. **Djordjevic, M. A., Gabriel, D. W., and Rolfe, B. G.,** *Rhizobium*–the refined parasite of legumes, *Annu. Rev. Phytopathol.*, 25, 145, 1987b.
36. **Carlson, R. W.,** Heterogeneity of *Rhizobium* lipopolysaccharides, *J. Bacteriol.*, 158, 1012, 1984.
37. **DiRita, V. J. and Mekalanos, J. J.,** Genetic regulation of bacterial virulence, *Annu. Rev. Genet.*, 23, 455, 1989.
38. **Van Brussel, A. A. N., Planqué, K., and Quispel, A.,** The wall of *Rhizobium leguminosarum* in bacteroid and free-living forms, *J. Gen. Microbiol.*, 101, 51, 1977.
39. **VandenBosch, K. A., Brewin, N. J., and Kannenberg, E. L.,** Developmental regulation of a *Rhizobium* cell surface antigen during growth of pea root nodules, *J. Bacteriol.*, 171, 4537, 1989.
40. **Sindhu, S. S., Brewin, N. J., and Kannenberg, E. L.,** Immunochemical analysis of lipopolysaccharides from free-living and endosymbiotic forms of *Rhizobium leguminosarum, J. Bacteriol.*, 172, 1804, 1990.
41. **Wood, E. A., Butcher, G. W., Brewin, N. J., and Kannenberg, E. L.,** Genetic derepression of a developmentally regulated lipopolysaccharide antigen from *Rhizobium leguminosarum* 3841, *J. Bacteriol.*, 171, 4549, 1989.
42. **Kannenberg, E. L. and Brewin, N. J.,** Expression of a cell surface antigen from *Rhizobium leguminosarum* 3841 is regulated by oxygen and pH, *J. Bacteriol.*, 171, 4543, 1989.
43. **Dazzo, F. B. and Truchet, G. L.,** Interactions of lectins and their saccharide receptors in the *Rhizobium*-legume symbiosis, *J. Membrane Biol.*, 73, 1, 1983.

44. **Sherwood, J. E., Vasse, J., Dazzo, F. B., and Truchet, G. L.,** Development and trifoliin A-binding ability of the capsule of *Rhizobium trifolii, J. Bacteriol.,* 159, 145, 1984.
45. **Hollingsworth, R. I., Dazzo, F. B., Hallenga, K., and Musselman, B.,** The complete structure of the trifoliin A lectin-binding capsular polysaccharide of *Rhizobium trifolii* 843, *Carbohydr. Res.,* 172, 97, 1988.
46. **Dazzo, F. B. and Hollingsworth, R. E.,** Trifoliin A and carbohydrate receptors as mediators of cellular recognition in the *Rhizobium trifolii*-clover symbiosis, *Biol. Cell,* 51, 267, 1984.
47. **Leigh, J. A., Signer, E. R., and Walker, G. C.,** Exopolysaccharide-deficient mutants of *Rhizobium meliloti* that form ineffective nodules, *Proc. Natl. Acad. Sci. U.S.A.,* 82, 6231, 1985.
48. **Finan, T. M., Hirsch, A. M., Leigh, J. A., Johansen, E., Kuldau, G. A., Deegan, S., Walker, G. C., and Signer, E. R.,** Symbiotic mutants of *Rhizobium meliloti* that uncouple plant from bacterial differentiation, *Cell,* 40, 869, 1985.
49. **Glazebrook, D. W. and Walker, G. C.,** A novel exopolysaccharide can function in place of the calcofluor-binding exopolysaccharide in nodulation of alfalfa by *Rhizobium meliloti, Cell,* 56, 661, 1989.
50. **Zhan, H., Levery, S. B., Lee, C. C., and Leigh, J. A.,** A second exopolysaccharide of *Rhizobium meliloti* strain SU47 that can function in root nodule invasion, *Proc. Natl. Acad. Sci. U.S.A.,* 86, 3055, 1989.
51. **Leigh, J. A., Reed, J. W., Hanks, J. F., Hirsch, A. M., and Walker, G. C.,** *Rhizobium meliloti* mutants that fail to succinylate their calcofluor-binding exopolysaccharide are defective in nodule invasion, *Cell,* 51, 579, 1987.
52. **Djordjevic, S. P., Rolfe, B. G., Batley, M., and Redmond, J. W.,** The structure of the exopolysaccharide from *Rhizobium* sp. strain ANU280 (NGR234), *Carbohydr. Res.,* 148, 87, 1986.
53. **Chen, H., Batley, M., Redmond, J., and Rolfe, B. G.,** Alteration of the effective nodulation properties of a fast-growing broad host range *Rhizobium* due to changes in exopolysaccharide synthesis, *J. Plant Physiol.,* 120, 331, 1985.
54. **McNeil, M., Darvill, J., Darvill, A. G., and Albersheim, P.,** The discernible structural features of the acidic polysaccharides secreted by different *Rhizobium* species are the same, *Carbohydr. Res.,* 126, 307, 1986.
55. **Borthakur, D., Barber, C. E., Lamb, J. W., Daniels, M. J., Downie, J. A., and Johnston, A. W. B.,** A mutation that blocks exopolysaccharide synthesis prevents nodulation of peas by *Rhizobium leguminosarum* but not of beans by *Rhizobium phaseoli* and is corrected by cloned DNA from *Rhizobium* or the phytopathogen *Xanthomonas, Mol. Gen. Genet.,* 203, 320, 1986.
56. **Chakravorty, A. K., Zurkowski, W., Shine, J., and Rolfe, B. G.,** Symbiotic nitrogen fixation: molecular cloning of *Rhizobium* genes involved in exopolysaccharide synthesis and effective nodulation, *J. Mol. Appl. Genet.,* 1, 585, 1982.
57. **Noel, K. D. and Carlson, R. W.,** unpublished, 1989.
58. **Kim, C-H., Tully, R. E., and Keister, D. L.,** Exopolysaccharide-deficient mutants of *Rhizobium fredii* HH303 which are symbiotically effective, *Appl. Environ. Microbiol.,* 55, 1852, 1989.
59. **Long, S. R., Reed, J. W., Himawan, J., and Walker, G. C.,** Genetic analysis of a cluster of genes required for synthesis of the calcofluor-binding exopolysaccharide of *Rhizobium meliloti, J. Bacteriol.,* 170, 4239, 1988.
60. **Zhan, H., Gray, J. X., Levery, S. B., Rolfe, B. G., and Leigh, J. A.,** Functional and evolutionary relatedness of genes for exopolysaccharide synthesis in *Rhizobium meliloti* and *Rhizobium* sp. strain NGR234, *J. Bacteriol.,* 172, 5245, 1990.
61. **Borthakur, D., Barker, R. F., Latchford, J. W., Rossen, L., and Johnston, A. W. B.,** Analysis of *pss* genes of *Rhizobium leguminosarum* required for exopolysaccharide synthesis and nodulation of peas: their primary structure and their interaction with *psi* and other nodulation genes, *Mol. Gen. Genet.,* 213, 155, 1988.
62. **Uttaro, A. D., Cangelosi, G. A., Geremia, R. A., Nester, E. W., and Ugalde, R. A.,** Biochemical characterization of avirulent *exoC* mutants of *Agrobacterium tumefaciens, J. Bacteriol.,* 172, 1640, 1990.
63. **Canter Cremers, H. C. J., Batley, M., Redmond, J. W., Eydems, L., Breedveld, M. W., Zevenhuisen, L. P. T. M., Pees, E., Wijffelman, C. A., and Lugtenberg, B. J. J.,** *Rhizobium leguminosarum exoB* mutants are deficient in the synthesis of UDPglucose 4′-epimerase, *J. Biol. Chem.,* 265, 21122, 1990.
64. **Buendia, A., Enenkel, B., Koplin, R., Niehaus, K., Arnold, W., and Puhler, A.,** Structural and functional analysis of the *Rhizobium meliloti exoZ/exoB* fragment of megaplasmid 2: *exoB* encodes a UDPgalactose 4-epimerase, presented at Int. Symp. Mol. Genet. Plant-Microbe Interact., Sept. 11-16, 1990, Abstr. P145.
65. **Borthakur, D. and Johnston, A. W. B.,** Sequence of *psi,* a gene of the symbiotic plasmid of *Rhizobium phaseoli* which inhibits exopolysaccharide synthesis and nodulation, and demonstration that its transcription is inhibited by *psr,* another gene on the symbiotic plasmid, *Mol. Gen. Genet.,* 207, 149, 1987.
66. **Gray, J. X., Djordjevic, M. A., and Rolfe, B. G.,** Two genes that regulate exopolysaccharide production in *Rhizobium* sp. strain NGR234: DNA sequences and resultant phenotypes, *J. Bacteriol.,* 172, 193, 1990.

67. **Zhan, H. and Leigh, J. A.,** Two genes that regulate exopolysaccharide production in *Rhizobium meliloti, J. Bacteriol.,* 172, 5254, 1990.
68. **Gray, J. X. and Rolfe, B. G.,** Exopolysaccharide production in *Rhizobium* and its role in invasion, *Mol. Microbiol.,* 4, 1425, 1990.
69. **Johnston, A. W. B.,** personal communication, 1989.
70. **Mulligan, J. T. and Long, S. R.,** A family of activator genes regulates expression of *Rhizobium meliloti* nodulation genes, *Genetics,* 122, 7, 1989.
71. **Appelbaum, E. R., Thompson, D. V., Idler, K., and Chartrain, N.,** *Rhizobium japonicum* USDA 191 has two *nodD* genes that differ in primary structure and function, *J. Bacteriol.,* 170, 12, 1988.
72. **Halverson, L. J. and Stacey, G.,** Signal exchange in plant-microbe interactions, *Microbiol. Rev.,* 50, 193, 1986.
73. **Darvill, A. G. and Albersheim, P.,** Phytoalexins and their elicitors—defense against microbial infection in plants, *Annu. Rev. Plant Physiol.,* 35, 243, 1984.
74. **Coplin, D. L. and Cook, D.,** Molecular genetics of extracellular polysaccharide biosynthesis in vascular phytopathogenic bacteria, *Mol. Plant-Microbe Interact.,* 3, 271, 1990.
75. **O'Connell, K. P., Araujo, R. S., and Handelsman, J.,** Exopolysaccharide-deficient mutants of *Rhizobium* sp. strain CIAT899 induce chlorosis in common bean *(Phaseolus vulgaris), Mol. Plant-Microbe Interact.,* 3, 424, 1990.
76. **Djordjevic, S. P., Chen, H., Batley, M., Redmond, J. W., and Rolfe, G. G.,** Nitrogen fixation ability of exopolysaccharide synthesis mutants of *Rhizobium* sp. strain NGR234 and *Rhizobium trifolii* is restored by the addition of homologous exopolysaccharides, *J. Bacteriol.,* 169, 53, 1987.
77. **Battista, L., Lee, C. C. and Leigh, J. A.,** Specific forms of exopolysaccharides from *Rhizobium meliloti* restore symbiotic effectiveness to *R. meliloti* Exo$^-$ mutants, in *Nitrogen Fixation: Achievements and Objectives,* Gresshoff, P. M., Roth, L. E., Stacey, G., and Newton, W. E., Eds., Chapman and Hall, NY, in press.
78. **Flowers, K. and Noel, K. D.,** unpublished data, 1990.
79. **Abe, M., Amemura, A., and Higashi, S.,** Studies of cyclic beta-1,2-glucan obtained from the periplasmic space of *Rhizobium trifolii* cells, *Plant Soil,* 64, 315, 1982.
80. **Dell, A.,** The cyclic structure of beta-D-(1—2)-linked D-glucans secreted by rhizobia and agrobacteria, *Carbohydr. Res.,* 117, 185, 1983.
81. **Miller, K. J., Kennedy, E. P., and Reinhold, V. N.,** Osmotic adaptation by Gram-negative bacteria: possible role for periplasmic oligosaccharides, *Science,* 231, 48, 1986.
82. **Miller, K. J., Gore, R. S., Johnson, R., Benesi, A. J., and Reinhold, V. N.,** Cell-associated oligosaccharides of *Bradyrhizobium* spp, *J. Bacteriol.,* 172, 136, 1990.
83. **Smit, G., Kijne, J. W., and Lugtenberg, B. J. J.,** Involvement of both cellulose fibrils and a $Ca^{2+}$-dependent adhesin in the attachment of *Rhizobium leguminosarum* to pea root hair tips, *J. Bacteriol.,* 169, 4294, 1987.
84. **Douglas, C. J., Halperin, W., and Nester, E. W.,** Identification and genetic analysis of a *Agrobacterium tumefaciens* mutants affected in attachment to plant cells, *J. Bacteriol.,* 161, 850, 1985.
85. **Puvanesarajah, V., Schell, F. M., Stacey, G., Douglas, C. J., and Nester, E. W.,** Role for 2-linked-beta-D-glucan in the virulence of *Agrobacterium tumefaciens, J. Bacteriol.,* 164, 102, 1985.
86. **Dylan, T., Ielpi, L., Stanfield, S., Kashyap, L., Douglas, C., Yanofsky, M., Nester, E. W., Helinski, D. R., and Ditta, G.,** *Rhizobium meliloti* genes required for nodule development are related to chromosomal virulence genes in *Agrobacterium tumefaciens, Proc. Natl. Acad. Sci. U.S.A.,* 83, 4403, 1986.
87. **Vandenbosch, K. A., Noel, K. D., Kaneko, J., and Newcomb, E. H.,** Nodule initiation elicited by noninfective mutants of *Rhizobium phaseoli, J. Bacteriol.,* 162, 950, 1985.
88. **Dylan, T., Nagpal, P., Helinski, D. R., and Ditta, G. S.,** Symbiotic pseudorevertants of *Rhizobium meliloti ndv* mutants, *J. Bacteriol.,* 172, 1409, 1990.
89. **Matthysse, A. G.,** Role of bacterial cellulose fibrils in *Agrobacterium tumefaciens* infection, *J. Bacteriol.,* 154, 906, 1983.
90. **Sutherland, I. W.,** Bacterial surface polysaccharides: structure and function, *Int. Rev. Cytol.,* 113, 187, 1988.
91. **Hamblin, J. and Kent, S. P.,** Possible role of phytohaemagglutinin in *Phaseolus vulgaris, Nature New Biol.,* 245, 28, 1973.
92. **Bohlool, B. and Schmidt, E.,** Lectins: a possible basis for specificity in the *Rhizobium*-legume root nodule symbiosis, *Science,* 185, 269, 1974.
93. **Dazzo, F. and Hubbell, D.,** Cross-reactive antigens and lectin as determinants of symbiotic specificity in the *Rhizobium*-clover association, *Appl. Microbiol.,* 30, 1017, 1975.
94. **Philip-Hollingsworth, S., Hollingsworth, R. I., Dazzo, F. B., Djordjevic, M. A., and Rolfe, B. G.,** The effect of interspecies transfer of *Rhizobium* host-specific nodulation genes on acidic polysaccharide structure and *in situ* binding by host lectin, *J. Biol. Chem.,* 264, 5710, 1989.

95. **Diaz, C. L., Melchers, L. S., Hooykaas, P. J. J., Lugtenberg, B. J. J., and Kijne, J. W.,** Root lectin as a determinant of host-plant specificity in the *Rhizobium* legume symbiosis, *Nature (London),* 338, 579, 1989.

96. **Lerouge, P., Roche, P., Faucher, C., Maillet, F., Truchet, G., Prome, J. C., and Denarie, J.,** Symbiotic host-specificity of *Rhizobium meliloti* is determined by a sulphated and acylated glucosamine oligosaccharide signal, *Nature (London),* 344, 781, 1990.

97. **Dazzo, F. B., Truchet, G. L., Sherwood, J. E., Hrabak, E. R., Abe, M., and Pankratz, S. H.,** Specific phases of root hair attachment in the *Rhizobium trifolii*-clover symbiosis, *Appl. Environ. Microbiol.,* 48, 1140, 1984.

98. **Vesper, S. J. and Bauer, W. D.,** Role of pili (fimbriae) in attachment of *Bradyrhizobium japonicum* to soybean roots, *Appl. Environ. Microbiol.,* 52, 134, 1986.

99. **Smit, G., Logman, T. J. J., Boerrigter, M. E. T. I., Kijne, J. W., and Lugtenberg, B. J. J.,** Purification and partial characterization of the *Rhizobium leguminosarum* biovar *viciae* $Ca^{2+}$-dependent adhesin, which mediates the first step in attachment of cells of the family Rhizobiaceae to plant root hair tips, *J. Bacteriol.,* 171, 4054, 1989.

100. **Badenoch-Jones, J., Flanders, D. J., Rolfe, B. G.,** Association of *Rhizobium* strains with roots of *Trifolium repens, Appl. Environ. Microbiol.,* 49, 1511, 1985.

101. **Smit, G., Kijne, J. W., and Lugtenberg, B. J. J.,** Roles of flagella, lipopolysaccharide, and a $Ca^{2+}$-dependent adhesin in the attachment of *Rhizobium leguminosarum* biovar *viciae* to pea root hair tips, *J. Bacteriol.,* 171, 569, 1989.

102. **Nap, J. P. and Bisseling, T.,** Developmental biology of a plant-prokaryote symbiosis: the legume root nodule, *Science,* 250, 948, 1990.

103. **Calvert, H. E., Pence, M. K., Pierce, M., Malik, N. S. A., and Bauer, W. D.,** Anatomical analysis of the development and distribution of *Rhizobium* infections in soybean roots, *Can. J. Bot.,* 62, 2375, 1984.

104. **Aman, P., McNeil, M., Franzen, L., Darvill, A. G., and Albersheim, P.,** Structural elucidation, using HPLC-MS and GLC-MS of the acidic polysaccharide secreted by *Rhizobium meliloti* strain 1021, *Carbohydr. Res.,* 95, 263, 1981.

105. **Carlson, R. W.,** personal communication, 1991.

106. **Zevenhuizen, L. P. T. M. and Van Neerven, A. R. W.,** Gel-forming capsular polysaccharide of *Rhizobium leguminosarum* and *Rhizobium trifolii, Carbohydr. Res.,* 124, 166, 1983.

107. **Verma, D. P. S., Delauney, A. J., Guida, M., Hirel, B., Schafer, R., and Koh, S.,** Control of expression of nodulin genes, in *Molecular Genetics of Plant-Microbe Interactions 1988,* Palacios, R. and Verma, D. P. S., Eds., APS Press, St. Paul, MN, 1988, 315.

108. **Gray, J. X., Zhan, H., Levery, S. B., Battisti, L., Rolfe, B. G., and Leigh, J. A.,** Heterologous exopolysaccharide production in *Rhizobium* sp. strain NGR234 and consequences for nodule development, *J. Bacteriol.,* 173, 3066, 1991.

# THE ROLE OF THE *RHIZOBIUM* CELL SURFACE DURING SYMBIOSIS

**James X. Gray, Ruud A. de Maagd, Barry G. Rolfe, Andrew W. B. Johnston, and Ben J. J. Lugtenberg**

## TABLE OF CONTENTS

# I. INTRODUCTION

As described elsewhere in this book and in recent reviews,[1-5] the symbiotic interaction between legumes and *Rhizobium* is complex and multifaceted, involving coupled differentiation and development in both partners. In this interaction there is intimate contact between both partners throughout the infection process. There is recognition at the very earliest stages when the bacteria first make contact with the root hairs of its host all the way through to the point at which the bacteroids are fixing nitrogen and are in close, perhaps direct, contact with the plant-specified peribacteroid membrane which surrounds them.

It should, therefore, come as no surprise that structures at the cell surface of both partners should play an important role in the establishment of nitrogen-fixing nodules. At least in the case of the bacterial symbiont, genetic evidence has amply supported this statement. By the isolation of various classes of mutants and the demonstration that these are affected in plant cell invasion, nodulation, and nitrogen fixation, it has been concluded that several genes whose protein products themselves are likely to be at the cell surface or which are exported from the cell are important for these processes. These gene products play key roles in the establishment of effective nitrogen-fixing nodules on a range of leguminous plants.

# II. THE *RHIZOBIUM* CELL SURFACE

The cell surface of *Rhizobium,* as in other Gram-negative bacteria, consists of a cytoplasmic membrane and an outer membrane, separated by the periplasmic space containing the peptidoglycan.[6] The development of methods for separation of the various cell compartments of *Rhizobium* has allowed the subcellular localization of molecules involved in symbiosis.[7] The outer membrane consists of proteins, lipopolysaccharides, and phospholipids. The protein profiles of the outer membranes of *R. leguminosarum* strains are very similar and most proteins belong to basically four groups of protein antigens defined by cross-reacting antibodies.[8,9] Some outer membrane proteins of *R. leguminosarum* are extremely heat and SDS-resistant, calcium-stabilized oligomeric proteins, with covalent links to the peptidoglycan layer;[10] which is uncommon for Gram-negative bacteria. It has been shown that the composition of the *Rhizobium* cell envelope undergoes extensive changes during the differentiation from the free-living to the bacteroid state, i.e., some of the outer membrane protein antigens almost completely disappear in bacteroids.[9] In addition, the lipopolysaccharides undergo changes (see below and the chapter by Noel et al. in this volume). One of the symbiotically repressed outer membrane protein antigens was recently cloned.[11] The question of whether these developmental changes are essential in establishing a successful symbiosis and what the mechanisms are for this regulation, is yet to be resolved.

Loosely attached to the outer membrane of *Rhizobium* cells are a number of surface structures, some of which appear to play a role in the early steps of symbiosis, particularly in attachment. *R. leguminosarum* flagella, although present and functioning in cell motility, appear not to be essential for the establishment of symbiosis, at least not under laboratory conditions.[12] However, this does not exclude the likely possibility that motility, and hence flagella, are important for nodulation in the soil.[13,14]

Rhicadhesin is a small 14-kDa surface protein involved in the initial attachment of the bacterium to the host root hair tips.[15] Microfibrils present on the surface of *R. leguminosarum,* in amounts varying from strain to strain consist of cellulose ($\alpha$-1,4-glucan).[16,17] Fimbriae, proteinaceous filaments extending from the bacterial cell surface, have been identified in *Bradyrhizobium japonicum.*[18] The role of the latter surface structures in attachment is reviewed in another chapter in this volume (Kijne et al.).

# III. *NOD* GENE PRODUCTS AND THE *RHIZOBIUM* CELL SURFACE

In several species of *Rhizobium,* clusters of *nod* genes that determine the ability of the bacteria to participate in the early steps of infection and which are of key importance in determining host range for a given strain, have been identified. Several Nod proteins have been shown directly, by subcellular localization or inferred by analysis of the deduced polypeptide products (obtained from DNA sequence determination), to be associated with one or other of the bacterial membranes, the periplasm, or to be exported from the cell.

The *nodC* gene is absolutely required for nodulation and, indeed, all the steps of infection. The product of *nodC* is located in the bacterial cell envelope.[19-21] John et al.[20] proposed that its C-terminus is located in the inner membrane and an N-terminal region is located in the outer membrane. Such a localization would be unique among bacterial cell envelope proteins. The function of *nodC* combined with those of *nodA* and *nodB,* is the production of an exported tetrasaccharide signal molecule, which induces the earliest observable steps in the symbiosis.[22] The transcription of several operons of nodulation genes are coordinately regulated by the product of the *nodD* gene.[23-27] The product of *nodD* is a DNA-binding protein,[28-30] which is associated with the inner membrane of *Rhizobium* cells.[31] The NodD protein responds to flavonoid inducer molecules,[32-35] and it is significant that upon incubation of subcellular fractions of *Rhizobium* cells with flavonoid inducers, flavonoids appear to have a high affinity for the cytoplasmic membrane, probably inserting in its phospholipid bilayer.[36] The combined data suggests that there may be some complex interaction between the inner membrane, the inducers, *nod* boxes, and NodD.

Other nodulation genes such as *nodE, nodI, nodJ, nodX,* and *nolE,* all encode polypeptides with hydrophobic regions at their N-terminus,[37-40] suggesting that they are either located within a membrane or transported through the membrane. The products of several of these genes have been localized after cell fractionation. The product of *nodE* was shown to be present in the inner membrane and its deduced protein sequence is similar to polyketide synthetase and fatty acid synthetases.[37] The *nodI* gene product, which was also shown to be present in the inner membrane,[41] encodes a protein that is similar in sequence to a family of bacterial proteins that are involved in the transport of a variety of low-molecular weight compounds across the inner membrane.[38]

Transport of protein through the inner membrane may also be inferred from studies of translational fusions of the *E. coli phoA* gene to the particular gene under study. For example, alkaline phosphatase activity of such a fusion to the *nolE* gene confirmed that part of the protein is exposed to the periplasm.[40]

The *R. leguminosarum* bv. *viciae nodO* gene encodes a secreted protein, homologous to a family of exported bacterial proteins, that lack a conventional N-terminal signal sequence for export.[42-44] A region of the NodO polypeptide has significant homology to the calcium-binding domain of *E. coli* hemolysin, which is involved in binding to the target cell surface and therefore suggests that the NodO protein also binds to a (plant) cell surface receptor.[43,44] A further inference based upon the homology to the above mentioned family of secreted proteins, is that specialized export machinery for NodO secretion exists in the *Rhizobium* cell envelope.

# IV. TRANSPORT SYSTEMS INVOLVED IN NODULATION

Given the trafficking of various molecules that must occur between the two partners during symbiosis, it would not be surprising if bacterial mutants, defective in certain transport systems, were also affected in their symbiotic capabilities. This has turned out to be the case for two systems, namely those involved in the transport of $Fe^{2+}$ or $Fe^{3+}$ (iron) ions and dicarboxylic acids. Mutants of *R. leguminosarum* bv. *viciae* that are unable to take up ferric or ferrous ions from the medium and overaccumulate protoporphyrin (a precursor of heme), could successfully nodulate peas, but were unable to fix nitrogen.[45] Nadler et al.[45] proposed that the bacteria provide the heme moiety of leghemoglobin, which is essential for nitrogen-fixation.

Bacteroids rapidly take up dicarboxylic acids *in vitro*.[46] Mutants defective in the uptake of succinate were shown to induce nonnitrogen-fixing nodules.[47-49] This is consistent with the observation by Udvardi et al.,[50] who showed that the peribacteroid membrane, which surrounds the bacteroids, has a dicarboxylate anion transporter with a high specificity for succinate. Therefore, it is generally accepted that *Rhizobium* bacteroids make use of certain dicarboxylic acids provided by the plant, as a source of carbon skeletons and energy. Furthermore, the product of the *dctC* gene, which is required for the uptake of dicarboxylic acids is periplasmically located. The product of another gene, involved in *dct* tregulation, is located in the inner membrane where it ''senses'' the presence of succinate in the medium.[51]

## V. *RHIZOBIUM* SURFACE POLYSACCHARIDES

Wild-type *Rhizobium* cells often produce large quantities of exopolysaccharides (EPSs) and form very mucoid colonies on laboratory media. The major types of polysaccharide molecules found loosely attached to the cell surface are the neutral β-1,2-glucans and the acidic EPSs. The lipopolysaccharides (LPSs) which form part of the outer membrane and contain the somatic or O-antigens, make up the other major group of rhizobial polysaccharides. The acidic EPSs are high molecular weight complex heteropolysaccharide polymers of a linked repeating unit that consists of 7,8, or 9 sugar residues. These oligosaccharide repeat units contain sugars (usually hexoses) linked by various α and β linkages in either a linear order or with branched side chains and quite often the sugars contain the noncarbohydrate substituents of succinate, pyruvate, acetate, and hydroxybutanoate. The acidic nature comes from the uronic acids, pyruvate, and succinyl groups. Each *Rhizobium* species usually synthesizes only one structural form at a time for each of its three polysaccharides. Thus, it has been proposed that this uniqueness in structure of the EPS and LPS of a *Rhizobium* strain may play a role in its host-specific interaction with plants.[52] In this respect, it may be relevant that some plant defensive genes are activated by small oligosaccharides that are not covalently attached to proteins. Oligosaccharide molecules derived from degraded cell walls of fungi and plants have been identified as signals and regulatory molecules affecting expression of certain plant genes.[53,54] However, with regard to a possible role for polysaccharides in host specificity, a correlation between polysaccharide structure and host range has not yet been conclusively demonstrated, either for EPS or LPS (see below).

## VI. ACIDIC EPS AND THE ESTABLISHMENT OF NODULATION

Extensive evidence implicates EPS molecules in many *Rhizobium*-plant interactions. Acidic EPS appears to be essential for the successful symbiosis between *R. meliloti* and alfalfa,[55] *R. leguminosarum* bv. *viciae* and pea or vetch,[56,57] *R. leguminosarum* bv. *trifolii* and clover,[58] which all form indeterminate nodules. Most EPS mutants of *R. leguminosarum* bv. *viciae* are completely $Nod^-$ and fail to induce marked root hair curling.[57]

Contrary to this, $Exo^-$ (defective in EPS synthesis) mutants of *R. leguminosarum* bv. *phaseoli* and *R. fredii* were shown to be unaffected with regard to their symbiotic ability on *Phaseolus* beans and soybeans.[56,59] This was further expanded upon,[60] when two classes of $Exo^-$ mutants of *R. leguminosarum* bv. *phaseoli* were isolated and found to be still symbiotically effective on beans; and that the identical mutations in *R. leguminosarum* bv. *viciae* and *R. leguminosarum* bv. *trifolii* were also $Exo^-$, but formed defective nodules on peas and clover, respectively. These observations seem to suggest that EPS is required for indeterminate nodules, but it is not necessary for determinate ones.

On the whole, those $Exo^-$ mutants of *Rhizobium* that are unable to form functional nodules, can be functionally complemented by $Exo^+Nod^-$ mutants of the same wild-type strain when coinoculated on the host. A $Nod^-$ mutant of *Rhizobium* sp. NGR234 that had been cured of its symbiotic plasmid and, consequently lacked the essential *nod* genes, was coinoculated with $Exo^-$ mutants of *Rhizobium* sp. NGR234 that were able to curl root hairs, but were incapable of forming nodules; together they could form nitrogen-fixing nodules on *Leucaena leucocephala.*[61] Similar coinoculation experiments between $Exo^-Nod^+$ and $Exo^+Nod^-$ mutants have resulted in functional nodules on clover by *R. trifolii,*[62] on peas by *R. leguminosarum* bv. *viciae,*[63] and on alfalfa by *R. meliloti.*[64,65] In the case of pea nodules, the majority of bacteroids originated from the $Exo^+$ coinoculant.[63] Klein et al.[64] conducted a detailed examination of the occupants from alfalfa nodules resulting from similar coinoculations of defective, but complementary, strains. Their results were quite interesting in that $Exo^-Fix^+$ mutants could be assisted into forming a $Fix^+$ nodule by a $Exo^+Nod^-Fix^-$ helper strain, but not by a $Exo^+Nod^+Fix^-$ helper or even the wild type. For an effective combination, one of the coinoculants must provide the $Nod^+$ phenotype and the other the $Exo^+$ phenotype. If both of these functions can be provided by a single strain, then the $Exo^-$ participant is excluded from the infection, perhaps by an inability to compete due to the necessity to acquire one of its prerequisites in *trans.*[64] The ability of $Exo^-$ mutants of *R. leguminosarum* bv. *trifolii, Rhizobium* sp. NGR234, and *R. meliloti* to induce nitrogen-fixing nodules on clover, *L. leucocephala,* and alfalfa, respectively, was restored by the accompaniment of EPS or oligosaccharide purified from their respective wild-type *Rhizobium* parental strains.[66,67] Addition of heterologous EPS did not correct the defective symbiotic phenotype for any tested $Exo^-$ mutant. The $Exo^-$ mutant is still able to provide early nodulation signals, but for indeterminate nodules EPS is clearly necessary for the continual progression of the infection thread.

The specificity of action of polysaccharides indicates that they have more than a simple passive or structural role on the *Rhizobium* cell surface. For example, the second EPS (EPS II or EPSb) from *R. meliloti* strain SU47 is sufficient for symbiosis with alfalfa, but not with four other plants that are normally hosts for *R. meliloti.*[68,69] In addition, *exoH* mutants of *R. meliloti* produce only large molecular weight (MW) polymers of the acidic EPS, which is not succinylated, and they are defective in nodule invasion and the growth of infection threads.[70] The lack of succinyl groups is correlated with the absence of low MW acidic EPS, suggesting that succinylation is important for cleavage of the high MW EPS polymer into the tetramer signal molecules. Further evidence of the specificity of *Rhizobium* EPS was found in studies of transconjugants made with *Rhizobium* sp. NGR234 mutants and cosmids carrying *R. meliloti exo* genes. If the *exoY* and the biosynthetic *exo* genes of NGR234 were deleted and replaced with *R. meliloti exo* genes, then $Exo^+$ colonies were formed with EPS of the *R. meliloti* succinoglycan type, but these hybrid transconjugant strains were still defective in their nodulation of *L. leucocephala.*[71] Thus, the production of the homologous EPS is necessary for the successful $Nod^+Fix^+$ phenotypes induced by *Rhizobium* sp. NGR234. One possible explanation for this proposed specific biological activity of the oligosaccharides, is that the different oligosaccharide ligands bind to specific membrane receptors at or near the root hair surface and are involved in a crucial early step, e.g., initiating infection threads. Different receptor molecules would be present in each plant and hence this would contribute to *Rhizobium* host range.

## VII. GENETICS OF EPS BIOSYNTHESIS

Our current knowledge on EPS biosynthesis of *Rhizobium* bacteria comes largely from the genetic studies of three different fast-growing *Rhizobium* species over the last 5 years:

*R. leguminosarum* bv. *viciae* and bv. *phaseoli*, *R. meliloti* strain Rm1021, and *Rhizobium* sp. strain NGR234 (the broadest host range *Rhizobium*, nodulating tropical legumes and the non-legume Parasponia). *Rhizobium meliloti* has a cluster of 13 genes on the second megaplasmid, pRmeSU47b,[72] and several other genes located on the chromosome for EPS synthesis (*exo*).[73] Similarly, a cluster of at least ten *exo* genes has been identified in *Rhizobium* sp.NGR234 and most of these are functionally homologous to the cluster of *R. meliloti exo* genes.[74] The second type of EPS (EPS II) synthesized by *R. meliloti* involves an additional six genes on pRmeSU47b and another on the chromosome.[68,69]

Both in *R. leguminosarum* as well as in *R. meliloti*, the *exoB* gene plays a central role in the synthesis not only of EPS but also of LPS and β-1,2-glucan,[75,76] which encodes a UDP-glucose 4′-epimerase.[76] The pleiotropic nature of these mutants shows that one should be cautious in assuming a role in nodulation for the EPS molecules themselves without considering other possible defects (see also below).

Recent studies on EPS mutants of *R. leguminosarum* strain RBL5515 questions the presumed essential role for EPS in nodulation of vetch by this strain.[57] Two Tn5-induced $Exo^-$ mutants of this strain both produce only 1% of the normal amount of wild-type EPS. However, only one of the two mutants was unable to nodulate normally and this mutant was shown to have an insertion in a gene homologous to the earlier described *pss2* gene,[63] designated *pssA*. The $Nod^+Exo^-$ mutant had an insertion in the promoter region of the same gene, possibly allowing some residual expression. These results suggest that it is not the production of EPS that is essential for nodulation, but that some other function of the *pssA* gene is needed. When the Sym plasmid is exchanged for one allowing nodulation of the clover cross-inoculation group, both mutants, although still producing low amounts of EPS, nodulate white and red clovers normally, showing that the *pssA* gene is not essential for the nodulation of clover.[57]

The first inhibitor of EPS synthesis to be characterized in some detail was *psi* from *R. leguminosarum* bv. *phaseoli*.[77,78] Genes homologous to *psi* have since been discovered in several other bacterial species: *psdA* in *A. tumefaciens*,[79] *exoX* in *Rhizobium* sp. NGR234,[80] and *exoX* in *R. meliloti*.[81] The common characteristic of these genes is that when they are cloned on multicopy plasmids and transferred into their respective wild-type ($Exo^+$) *Rhizobium* strains, the transconjugants have $Exo^-$ phenotypes. Mutants that carry Tn5 insertions within the genomic copies of *psi*, *exoX* or *psdA* produce approximately normal levels of EPS. Thus, there is no absolute requirement of these genes for EPS production, and their total absence does not result in deregulated overproduction of EPS. DNA sequence analysis of *exoX* (from *Rhizobium* sp. NGR234) and *psi* showed that the two genes encode proteins of similar size, 96 and 86 amino acids, respectively. In addition, the hydrophobicity plots are very similar, but at the primary sequence level, there is only an 18-amino acid domain that has 10 residues identical between the two putative proteins. The symbiotic phenotypes associated with this class of mutants are not the same—*R. leguminosarum* bv. *phaseoli psi* mutants had a $Nod^+Fix^-$ phenotype on beans, whereas the *exoX*::Tn5 mutation in *R. meliloti* did not affect symbiosis with alfalfa.

The inhibitory effects of *psi* and *exoX* could be opposed by another class of genes: *pss2* in *R. leguminosarum* bv. *phaseoli*,[63] *exoY* in *Rhizobium* sp. NGR234,[80] and and an equivalent gene in *R. meliloti*.[81] One common property of these genes (*pss2*/*exoY*) is that the inhibition of EPS caused by multicopy *psi*/*exoX* is overcome when the copy number of their counterpart (*pss2*/*exoY*) is at a comparable level. For example, normal EPS production occurs in *Rhizobium* sp. NGR234 transconjugants when both *exoY* and *exoX* are carried together on a plasmid, but inhibition of EPS production occurs when only *exoX* is plasmid borne. Mutations within the *pss2* and *exoY* loci all result in an $Exo^-$ phenotype. Only in the case of the *exoY* mutants has it been established that no oligosaccharide repeat unit is made; presumably, this is also true for the other mutated loci. The nucleotide sequences for *exoY* and *pss2* have been determined and they are clearly homologs. ExoY (226 amino acids) and Pss2 (200 amino acids) are 52% similar with 32% of the homology due to exact matches and their hydrophobicity plots are virtually superimposable. In addition to the loss of EPS synthesis, *Rhizobium* strains mutated in the *pss* or *exoY* genes are unable to nodulate their respective hosts (peas, *L. leucocephala* or alfalfa). These hosts are all examples of legumes that develop an indeterminate nodule. In marked contrast, the Exo$^-$ *R. l* bv. *phaseoli pss2*::Tn5 mutants still form nitrogen-fixing nodules on *Phaseolus* beans, which use a different ontogeny for the growth of a determinate nodule.

The *psi* and *exoX* genes do not inhibit transcription of *pss2* and *exoY*, respectively, or vice versa.[63,80] The sequence analysis of all four genes demonstrates that their encoded proteins all have significant stretches of hydrophobic amino acids and thus argues that they may form a posttranslational complex associated with the membrane.[3,63] It was proposed that the product of *exoX* reversibly binds to a catalytic biosynthetic complex involving products of at least *exoY*.[3]

## VIII. EPS PRODUCTION BY BACTEROIDS

The transformation of *Rhizobium* from free-living to the bacteroid form is marked by a number of morphological, physiological, and biochemical changes. Bacteroids cease cell division, are much larger, show changes in cytochrome composition, and, most importantly, fix atmospheric nitrogen.[82] The bacteroids also become enclosed within a membrane of plant origin,[83] which has the potential to regulate nutrient exchanges between bacteroid and host. The peribacteroid membrane is readily permeable to dicarboxylic acids, but is poorly permeable to hexose sugars.[84]

Several laboratories have investigated the expression of *exo* genes during symbiosis. Expression from two neighboring transcriptional units involved in EPS production by *R. meliloti,* using fusions of the *exo* promoters to the *E. coli lacZ* gene, demonstrated that the DNA regions were highly expressed in both free-living rhizobia and in bacteroids isolated from nitrogen-fixing nodules. However, a fusion of the promoter from the *R. leguminosarum* bv. *phaseoli psi* gene to the *E. coli uidA* (β-glucuronidase) gene indicated that it was expressed at a high level by bacteroids, but only at a very low level by free-living bacteria.[85] Latchford and Johnston[85] propose that since *psi* is involved in the inhibition of EPS synthesis,[77] then perhaps EPS is not produced by bacteroids during symbiosis. In *R. meliloti,* an operon with genes definitely involved in EPS biosynthesis, *exoY* and *exoF,* was also fused to the *E. coli phoA* gene.[86] This promoter was found to be expressed by *Rhizobium* cells in the invasion zone at the front of the infection thread just behind the nodule meristem and it was not active in the bacteroid nitrogen-fixing zone.[86] Collectively, the results of Reuber et al.[86] and Latchford and Johnston[85] support a model that EPS, which is synthesized by products of the *exoY* operon, is regulated by *psi* (*exoX*). In addition, EPS is required at the point where *Rhizobium* cells are penetrating plant cells via the infection thread, but that EPS synthesis ceases when the rhizobia differentiate into bacteroids.

Electron microscopic studies of the infection process of alfalfa has shown electron dense material within the infection thread matrix, which is believed to be EPS.[87] However, this material is absent around mature bacteroids.[87] It was presumed that transformation into bacteroids was accompanied by the cessation of EPS production and that the conditions of the nodule may provide an environmental stimulus. The possibility that oxygen is a regulator of *exo* genes has been explored. It has been known for some time that the central tissue of a nodule is essentially anaerobic,[88] and that a low $O_2$ tension is imperative for synthesis and activity of the nitrogenase enzyme.[89] More recently, the central regulatory component of *nif* gene expression, *nifA,* from *R. meliloti* is induced when the oxygen concentration is reduced to microaerobic levels.[90] Electron microscopy was used to examine cells from core samples taken through colonies growing on soft agar medium.[91] It was found that cells at the aerobic surface were surrounded by EPS, but the presence of EPS diminished as the cells analyzed came from deeper within the colony and at the anaerobic bottom EPS was absent.

An investigation into EPS synthesis in *B. japonicum* under anaerobic free-living conditions and during symbiosis was conducted by Tully and Terry.[92] Their wild-type strain, RT2, produced copious amounts of EPS during aerobic growth in liquid culture, but their symbiotically effective Exo$^-$ mutant, RT176-1, produced about 12% of the RT2 level of EPS production. The amounts of EPS recovered from nodules induced by both of these strains on soybean were both approximately equal and both very low compared to the amount synthesized under free-living conditions. Furthermore, gas chromatographic analysis of the recovered EPS demonstrated that the sugar composition resembled that of EPS recovered from uninoculated root tissue and was very different from the chemical composition of EPS isolated from free-living liquid cultures. Their conclusion was that the host plant tissue was the source of the EPS recovered from the nodules.[92] EPS production by *B. japonicum* was shown to decrease by 92% when liquid cultures were grown under anaerobic compared with aerobic growth conditions. Finally, it was demonstrated that cells failing to synthesize adequate quantities of EPS were also inefficient at binding soybean lectin.[92]

*Rhizobium* mutants that overproduce their EPS have been generated by Tn5 mutagenesis for *R. meliloti,*[65,93] and *Rhizobium* sp. NGR234.[58] It is believed that this Exo$^{++}$ mutant phenotype would result when a repressor or negative regulator is inactivated leading to uncontrolled overproduction of EPS. In general the Exo$^{++}$ mutants were not as effective in symbiosis as the wild type, but in many cases they were still capable of functional symbiosis. One of the loci involved in overproduction of EPS in *R. meliloti,* termed *exoR,* affects transcription from the *exoY,F* promoter and its own expression is influenced by ammonia availability.[93] The level of EPS synthesized by wild-type *R. meliloti* is greatly increased when the bacteria are starved for ammonia. The *exoR*::Tn5 mutant, however, produced excessive amounts of the EPS regardless of whether ammonia was limiting or not. Furthermore, EPS production by transconjugants of the mutant carrying the wild-type allele on a plasmid, was once again under the influence of ammonia levels. Thus, under conditions where the ammonia was not limiting, the *exoR* repressor of EPS synthesis would be active in the wild type. However, when ammonia is absent, EPS synthesis becomes deregulated, indicating that expression of *exoR* is inhibited under conditions of essentially zero ammonia.[93] One might expect similar conditions to exist within the nodule, where fixed nitrogen in the form of ammonia would be a precious commodity and would be transported as quickly as it is made to developing plant tissue. However, this does present the conflicting argument that EPS production is increased by *Rhizobium* within nodules. Of course, these observations by Doherty et al.[93] involved free-living *R. meliloti* and a great many changes take place when the *Rhizobium* differentiate into bacteroids.

## IX. NEUTRAL β-1,2-GLUCAN

Strains of *Rhizobium* and *Agrobacterium* make a β-1,2-glucan,[94,95] which is present in the periplasm and is exported from the bacteria. This polysaccharide is a homopolymer consisting of about 20 residues of β-(1→2)-linked glucose sugars, possibly in a cyclic configuration.[96] Although it can be modified in various ways,[97-99] there is less scope for variation with this polysaccharide as compared to EPS or LPS. *Bradyrhizobium japonicum* does not contain β-1,2-glucan but instead produces smaller cell-associated oligomers of β-1,3- and β-1,6-linked glucose residues, which may have similar functions.[100]

Neutral β-1,2-glucans may have a universal role in the interactions of both *Rhizobium* and *Agrobacterium* with their host plants. Mutations in two genes, termed *ndvA* and *ndvB,*[101] result in an absence of β-1,2-glucan synthesis in *R. meliloti,* without affecting EPS synthesis.[102] These mutants form ineffective nodules on alfalfa.[102,103] In addition, mutations at homologous loci in *A. tumefaciens* (*chvA* and *chvB*) result in avirulent, attachment-defective mutants.[104] Mutations in *chvA* and *chvB* also block nodulation by the strain *A. tumefaciens* (pSym *R. leguminosarum* bv. *phaseoli*), which normally nodulates *Phaseolus* beans.[105] It was shown that *chv* mutants of *A. tumefaciens* failed to bind tightly to plant cells.[106] Similarly, Dylan et al.[107] demonstrated that *ndv* mutants of *R. meliloti* are severely impaired in their binding to the roots of alfalfa seedlings. Analysis of the deduced sequence of the *ndvA* and *chvA* gene products showed that they are related to a family of membrane-associated proteins involved in the transmembrane transport of a variety of molecules, suggesting that *ndvA*/*chvA* is required for export of glucans.[108-110] This is supported by the observation that *chvA* mutants still make β-1,2-glucan, but that it is confined to the cytoplasm. In contrast, mutations in *ndvB* or *chvB* do not make the molecule. Both *ndvB* and *chvB*, code for an inner membrane protein, appearing in SDS-PAGE as a 235-kDa protein, which is thought to be involved in the synthesis as well as acting as a carrier of intermediates of glucan synthesis.[111,112]

The production of β-1,2-glucans is increased at low osmolarity and appears to be essential for optimal osmoadaptation,[113,114] a property which might play a role in nodulation. Also, during infection, the glucans may act to "mop up" (or concentrate) low-molecular weight signals or defense compounds that are generated by the plant, in such a way that the molecules act as a protectant against plant defense reactions. Significant in this respect is the enhancement of infection thread formation and nodulation by *R. leguminosarum* bv. *trifolii* on clover by addition of isolated β-1,2-glucans.[115]

Although it is clear from the studies above, that mutations which abolish β-1,2-glucan production severely compromise the ability to form normal nodules, it is still uncertain that the loss of the glucan itself is responsible for these defects. The *chv* and *ndv* mutations are strongly pleiotropic, also affecting a number of other surface features of the bacterium.[107] Pseudorevertants could be isolated from pink alfalfa nodules that arose at low frequency after inoculation with an *ndv* mutant of *R. meliloti*.[107] Although these pseudorevertants were restored for their symbiotic nitrogen-fixing ability, they were not restored for their ability to produce the glucan, or for osmoadaptation.

## X. LIPOPOLYSACCHARIDES

In contrast to the other polysaccharides, lipopolysaccharide (LPS) are an integral part of the outer membrane. LPS molecules can be divided into three functional regions: lipid A, core oligosaccharide, and O-antigen. Lipid A is the hydrophobic lipid moiety that anchors the entire LPS structure in the outer membrane by embedding in the outer leaflet. Covalently attached to the lipid A is the core oligosaccharide, which consists of a small number of sugar moieties. Two core oligosaccharide structures have been identified in *R. leguminosarum* bv. *trifolii*. One is a trisaccharide consisting of two galacturonic acid residues and one KDO (3-deoxy-D-manno-2-octulosonic acid),[116] and the other is a tetrasaccharide consisting of galactose, galacturonic acid, mannose and KDO.[117] Attached to the core polysaccharide is the polymorphic region of the LPS (O-antigen), which projects outward from the cell surface.

Genetic studies have shown that LPS is also required for normal nodulation and nitrogen-fixation between some combinations of *Rhizobium* and legume hosts. Mutants of *R. leguminosarum* bv. *phaseoli*, defective in their production of LPS and EPS or LPS alone, were now incapable of normal determinate nodulation of *Phaseolus* beans.[60,118-120] Similarly, mutants of *B. japonicum* that produce an altered LPS molecule, but with apparently unaltered EPS, fail to elicit any visible nodule development on soybeans.[121] Microscopic studies of nodulation elicited by some O-antigen deficient mutants showed that induction of meristematic activity and initiation of infection thread formation still occurs, but infection thread formation was soon aborted.[118] On the other hand, mutants of *R. meliloti*, *R. leguminosarum* bv. *trifolii,* and *R. leguminosarum* bv. *viciae* that produce an altered O-antigen structure in their LPS, were still capable of forming normal nitrogen-fixing nodules with alfalfa, clover, and vetch, respectively.[120,122,123] However, mutants of *R. meliloti, R. leguminosarum* bv. *trifolii,* and *R. leguminosarum* bv. *viciae* that produced very little to none of the O-antigen were very abnormal in nodule development, indicating that there is still a requirement for the presence of an O-antigen component of the LPS in cases of indeterminate nodulation, although the structure does not seem to be crucial.[120,122,123]

The genes involved in LPS production in *R. leguminosarum* bv. *viciae* appear to be clustered on the chromosome.[118,119] The antigenicity of LPS from *R. leguminosarum* bv. *viciae* was shown to vary as the cells differentiated from the free-living to the bacteroid form.[124,125] An LPS antigen found on the surface of *R. leguminosarum* bv. *viciae* bacteroids, but not on the surface of free-living cells, could be expressed by free-living cells when cultures were grown at a neutral pH with the oxygen concentration less than 7.5% in the gas phase or when cultures were grown aerobically at a pH less than 5.3.[126] The low oxygen conditions of the nodule,[88] may be the environmental stimulus which activates the developmental LPS gene(s) involved. Of course, another possibility is acidity, but little is known about the pH of the peribacteroid units. Kannenberg and Brewin[126] suggest that it might be acidic due to the proton-ATPases and transport systems for dicarboxylic acids,[50,127] but they also discuss that the uptake of succinate and excretion of $NH_3$ would tend to raise the pH within the peribacteroid space. Whether the pH effect is relevant or not is unclear. Wood et al.[128] isolated mutants of *R. leguminosarum* bv. *viciae* which constitutively expressed the LPS that was normally expressed only in the bacteroids or in free-living bacteria grown at low pH or low $pO_2$. These mutations would appear to be in the gene that responds to the environmental conditions and which normally represses expression of the corresponding LPS genes.

There is one example, reported recently, of a *R. meliloti* strain where the functions of acidic EPS appear to be substituted for by modified LPS molecules.[129] Strain AK631 is a spontaneous $Exo^-$ derivative of the wild-type isolate *R. meliloti* strain Rm41, which is different in several respects to the wild-type strain SU47, including the expression of an *lpsZ* gene in strain Rm41, which is apparently absent in strain SU47. The mutation in strain AK631 appears to be within the *exoB* gene, but instead of having an $Inf^-Fix^-$ phenotype on alfalfa, which is normally associated with *exoB* mutants of strain SU47,[55] strain AK631 is instead $Fix^+$ on alfalfa. It was proposed by Williams et al.[129] that a strain Rm41 gene, *lpsZ,* which is not expressed in strain SU47, was responsible for the production of altered LPS molecules that substitute for the absence of acidic EPS during the infection of alfalfa by strain AK631 *exoB* mutants. However, the structure of these proposed LPS signal molecules has not been determined as yet.

## XI. PLANT LECTIN-*RHIZOBIUM* INTERACTIONS

There is now much evidence in support of the hypothesis that host plant lectins have stringent specificity for *Rhizobium* cells within their cross-inoculation groups.[130] For a number of *Rhizobium* species, evidence has been presented that surface polysaccharides may be involved in this interactions. Trifoliin A is a clover lectin isolated from root hairs that specifically binds to the LPS and EPS of *R. leguminosarum* bv. *trifolii.*[131] Trifoliin A also binds to certain carbohydrate structures on clover and may function as a cross-bridge to link *R. leguminosarum* bv. *trifolii* cells to the clover root hair.[131] Root lectins that specifically recognize their cosymbiont *Rhizobium* cells have also been demonstrated for soybean, sweet clover, and pea.[130] The gene encoding the pea root hair lectin was cloned and transferred to white clover using *Agrobacterium rhizogenes* as the vector.[132] Transgenic clover roots were now susceptible to infection by *R. leguminosarum* bv. *viciae* and, in addition, a relatively low level of delayed nodulation by *R. leguminosarum* bv. *viciae* was also observed.[132] Nodulation of these transgenic clover roots by *R. leguminosarum* bv. *trifolii* was not affected. The majority of the *R. leguminosarum* bv. *viciae*-generated nodules were abnormal, but electron microscopic examination revealed bacteria within intracellular spaces of the cortex, aborted infection threads, and degenerated nodule meristems. Those nodules that did develop normally and fix atmospheric nitrogen had normal nodule ultrastructures as determined by electron microscopy. Diaz et al.[132] concluded that the host range in the *Rhizobium*-legume symbiosis is at least partially determined by symbiont-root-lectin interactions. It should be noted that no evidence has been provided that this interaction is based on a recognition of a bacterial polysaccharide by the plant's lectin. In contrast, based on preliminary experimental evidence, it is conceivable that bacterial lipo-oligosaccharide signal molecules are responsible for the host-specific interaction with lectin.[133,134]

Alterations of the *R. leguminosarum* bv. *viciae* EPS structure by the addition of donor *R. leguminosarum* bv. *trifolii* host-range genes and the subsequent alteration in host range was described by Philip-Hollingsworth et al.[135] The structure of the oligosaccharide repeat units for *R. leguminosarum* bv. *viciae* and the donor *R. leguminosarum* bv. *trifolii* were concluded to differ only in the degree of substitution of 3-hydroxybutyrate and acetate.[136] When a plasmid carrying cloned *R. leguminosarum* bv. *trifolii* host-specific nodulation (*hsn*) genes *nodFERLMN,* was tranferred into *R. leguminosarum* bv. *viciae,* the resulting hybrid was now able to nodulate clover and bind the clover lectin trifoliin A.[135] In contrast, other laboratories in which the influence of induced Sym plasmid-localized *nod* genes on EPS structure was studied, concluded that the EPS structure was not significantly altered by exchange of Sym plasmids, which are known to determine the host range.[137,138]

## XII. CONCLUSIONS

It is clear that various macromolecules on the surface of *Rhizobium* cells are required for sustained invasion of host plant cells that finally result in a nitrogen-fixing nodule. Many of the *nod* genes appear to encode proteins that are located within or transported through the cell membranes, but for many of the genes this is yet to be conclusively demonstrated. In addition, with the exception of *nodD,* many of the precise biochemical functions of individual *nod* genes are yet to be determined. However, it has been shown that induction of the *Rhizobium nod* genes does result in the production of extracellular bacterial lipo-oligosaccharide signals that cause extensive root hair distortion on host plants.[22] Root cortical cell division is also induced by this signal,[139] as is the induction of some early nodulin genes.[140]

Copious quantities of various polysaccharides are also found on the surface of *Rhizobium* cells and while studies of their biological roles are still far from complete, they clearly are important for the development of nitrogen-fixing nodules. General observations seem to suggest that symbiotic partners that develop an indeterminate nodule structure have an absolute requirement for acidic EPS with specific structures and tightly regulated production.[55-58,62] On the other hand, lack of EPS production does not seem to affect development of determinate nodules.[60] The development of both determinate as well as indeterminate nodules requires the O-antigen of the bacterial LPS,[119-123] but a requirement for specific structures of the O-antigen has not yet been found.[141] However, a specific LPS structure, conferred by the *lpsZ* gene in *R. meliloti* can, in some cases, substitute for EPS function in nodule development.[129] Mutants which lack β-1,2-glucans are affected in both indeterminate as well as in determinate nodule development,[102,105] but it is possible that a secondary defect in these mutants, and not the lack of the glucan itself is responsible for these phenotypes.[107]

Whereas the presence of those surface components discussed above is needed during nodule development, some of these components undergo extensive changes during this development or disappear altogether. LPS epitopes and outer membrane protein antigens disappear in bacteroids, while new ones appear.[9,124,125,142] Also, probably EPS synthesis ceases (see above), and it has been shown that transcription of the *nod* genes, with the exception of *nodD,* is very low in bacteroids.[143,144] It will be interesting to see whether these changes occur at the same developmental stage and are caused by a common environmental signal (e.g., low oxygen pressure) and whether all these changes are essential for normal bacteroid development.

If host range is, indeed, determined by rhizobial polysaccharides (see above) or oligosaccharides,[133] the root hair surfaces of various host plants may contain a variety of receptors. Different receptor molecules may be present in each plant and this could contribute to *Rhizobium* host range. Lectins, which are known to bind specific oligosaccharide ligands, may form part of this receptor system. Using transgenic technology, a gene encoding a pea lectin was introduced into white clover roots; the exciting result was an altered host range demonstrated by a low level of delayed nodulation by *R. leguminosarum* bv. *viciae.*[132] Without a doubt, the surfaces of both *Rhizobium* and plant cells contain receptors and contain or release signals involved in the recognition and cellular communication between these two partners.

## REFERENCES


1. **Downie, J. A. and Johnston, A. W. B.,** Nodulation of legumes by *Rhizobium, Plant Cell Environ.,* 11, 403, 1988.
2. **Rolfe, B. G. and Gresshoff, P. M.,** Genetic analysis of legume nodule initiation, *Annu. Rev. Plant Physiol. Plant Mol. Biol.,* 39, 297, 1988.
3. **Gray, J. X. and Rolfe, B. G.,** Exopolysaccharide production in *Rhizobium* and its role in invasion, *Mol. Microbiol.,* 4, 1425, 1990.
4. **Kijne, J. W.,** The infection process, in *Biological Nitrogen Fixation,* Stacey, G., Burris, R. H., and Evans, H. J., Eds., Chapman and Hall, NY, 1991, in press.
5. **Lugtenberg, B. J. J., Kijne, J. W., Smit, G., Spaink, H. P., and Zaat, S. A. J.,** Early stages of *Rhizobium* plant interactions: fundamental research and prospects for applications, *Agro-Ind. Hi-Tech,* 1, 40, 1990.
6. **Lugtenberg, B. J. J. and Van Alphen, L.,** Molecular architecture and functioning of the outer membrane of *Escherichia coli* and other Gram-negative bacteria, *Biochim. Biophys. Acta,* 737, 51, 1983.
7. **De Maagd, R. A. and Lugtenberg, B. J. J.,** Fractionation of *Rhizobium leguminosarum* cells into outer membrane, cytoplasmic membrane, periplasmic and cytoplasmic components, *J. Bacteriol.,* 167, 1083, 1986.
8. **De Maagd, R. A., Van Rossum, C., and Lugtenberg, B. J. J.,** Recognition of individual strains of fast-growing rhizobia by using profiles of membrane proteins and lipopolysaccharides, *J. Bacteriol.,* 170, 3782, 1988.
9. **De Maagd, R. A., de Rijk, R., Mulders, I. H. M., and Lugtenberg, B. J. J.,** Immunological Characterization of *Rhizobium leguminosarum* outer membrane antigens by use of polyclonal and monoclonal antibodies, *J. Bacteriol.,* 171, 1136, 1989.
10. **De Maagd, R. A., Wientjes, F. B., and Lugtenberg, B. J. J.,** Evidence for divalent cation ($Ca^{2+}$)-stabilized oligomeric proteins and covalently bound protein-peptidoglycan complexes in the outer membrane of *Rhizobium leguminosarum, J. Bacteriol.,* 171, 3989, 1989.
11. **De Maagd, R. A., Canter Cremers, H. C. J., and Lugtenberg, B. J. J.,** Cloning and characterization of a gene for production of a symbiotically repressed outer membrane protein of *Rhizobium leguminosarum* biovar *viciae,* in *Proc. 5th Int. Symp. Mol. Genet. Plant-Microbe Interact.,* Göttfert, M., Hennecke, H., and Paul, H., Eds., Mikrobiologisches Institut, Eidgenössische Technische Hochschule Zürich, Switzerland, 1990, 91.
12. **Smit, G., Kijne, J. W., and Lugtenberg, B. J. J.,** Roles of flagella, lipopolysaccharide, and a $Ca^{2+}$-dependent cell surface protein in attachment of *Rhizobium leguminosarum* biovar *viciae* to pea root hair tips, *J. Bacteriol.,* 171, 569, 1989.

13. **El-Haloui, N. E., Ochin, D., and Taillez, R.**, Competitivité pour l'infection entre souches de *Rhizobium meliloti:* role de la mobilité, *Plant Soil,* 95, 337, 1986.
14. **Hunter, W. J. and Fahring, C. J.**, Movement by *Rhizobium* and nodulation of legumes, *Soil Biol. Biochem.*, 12, 537, 1980.
15. **Smit, G., Logman, T. J. J., Boerrigter, M. E. T. I., Kijne, J. W., and Lugtenberg, B. J. J.**, Purification and partial characterization of the *Rhizobium leguminosarum* biovar *viciae* $Ca^{2+}$-dependent adhesion, which mediates the first step in attachment of cells of the family Rhizobiaceae to plant root hair tips, *J. Bacteriol.*, 171, 4054, 1989.
16. **Deinema, M. H. and Zevenhuizen, L. P. T. M.**, Formation of cellulose fibrils by Gram-negative bacteria and their role in bacterial flocculation, *Arch. Microbiol.*, 78, 42, 1971.
17. **Smit, G., Kijne, J. W., and Lugtenberg, B. J. J.**, Involvement of both cellulose fibrils and a $Ca^{2+}$-dependent adhesion in the attachment of *Rhizobium leguminosarum* to pea root hair tips, *J. Bacteriol.*, 169, 4294, 1987.
18. **Vesper, S. J. and Bauer, W. D.**, Role of pili (fimbriae) in attachment of *Bradyrhizobium japonicum* to soybean roots, *Appl. Environ. Microbiol.*, 52, 134, 1986.
19. **John, M., Schmidt, J., Wieneke, Kondorosi, E., Kondorosi, A., and Schell, J.**, Expression of the nodulation gene *nodC* of *Rhizobium meliloti* in *Escherichia coli:* role of the *nodC* gene product in nodulation, *EMBO J.*, 4, 2425, 1985.
20. **John, M., Schmidt, J., Wieneke, U., Krüssmann, H.-D., and Schell, J.**, Transmembrane orientation and receptor-like structure of the *Rhizobium meliloti* common nodulation protein NodC, *EMBO J.*, 7, 583, 1988.
21. **Johnson, D., Roth, L. E., and Stacey, G.**, Immunogold localization of the NodC and NodA proteins, *J. Bacteriol.*, 171, 4583, 1989.
22. **Lerouge, P., Roche, P., Faucher, C., Maillet, F., Truchet, G., Promé, J. C., and Dénarié, J.**, Symbiotic host-specificity of *Rhizobium meliloti* is determined by a sulphated and acylated glucosamine oligosaccharide signal, *Nature (London),* 344, 781, 1990.
23. **Innes, R. W., Kuempel, P. L., Plazinski, J., Canter-Cramers, H., Rolfe, B. G., and Djordjevic, M. A.**, Plant factors induce expression of nodulation and host range genes in *Rhizobium trifolii, Mol. Gen. Genet.*, 201, 426, 1985.
24. **Mulligan, J. T. and Long, S. R.**, Induction of *Rhizobium meliloti nodC* expression by plant exudate requires nodD, *Proc. Natl. Acad. Sci. U.S.A.*, 82, 6609, 1985.
25. **Rossen, L., Shearman, C. A., Johnston, A. W. B., and Downie, J. A.**, The *nodD* gene from *Rhizobium leguminosarum* is autoregulatory and in the presence of plant root exudate induces the *nodABC* genes, *EMBO J.*, 4, 3369, 1985.
26. **Shearman, C. A., Rossen, L., Johnston, A. W. B., and Downie, J. A.**, The *Rhizobium leguminosarum* gene *nodF* encodes a protein similar to acyl carrier protein and is regulated by *nodD* plus a factor in pea root exudate, *EMBO J.*, 5, 647, 1986.
27. **Gerhold, D., Stacey, G., and Kondorosi, A.**, Use of a promoter-specific probe to identify two loci from the *Rhizobium meliloti* nodulation region, *Plant Mol. Biol.*, 12, 181, 1989.
28. **Hong, G.-F., Burn, J., and Johnston, A. W. B.**, Evidence that DNA involved in the expression of nodulation *(nod)* genes in *Rhizobium* binds to the product of the regulatory gene *nodD, Nucleic Acids Res.*, 15, 9677, 1987.
29. **Fisher, R. F., Egelhoff, T. T., Mulligan, J. T., and Long, S. R.**, Specific binding of proteins from *Rhizobium meliloti* cell free extracts containing NodD to DNA sequences upstream of inducible nodulation genes, *Genes Dev.*, 2, 282, 1988.
30. **Kondorosi, E., Gyuris, J., Schmidt, J., John, M., Duda, E., Hoffman, B., Schell, J., and Kondorosi, A.**, Positive and negative control of *nod* gene expression in *Rhizobium meliloti* is required for optimal nodulation, *EMBO J.*, 8, 1331, 1989.
31. **Schlaman, H. R. M., Spaink, H. P., Okker, R. J. H., and Lugtenberg, B. J. J.**, Subcellular localization of the *nodD* gene product in *Rhizobium leguminosarum, J. Bacteriol.*, 171, 4686, 1989.
32. **Redmond, J. W., Batley, M., Djordjevic, M. A., Innes, R. W., Kuempel, P. L., and Rolfe, B. G.**, Flavones induce expression of nodulation genes in *Rhizobium, Nature (London),* 323, 632, 1986.
33. **Peters, N. K., Frost, J. W., and Long, S. R.**, A plant flavone, luteolin, induces expression of *Rhizobium meliloti* nodulation genes, *Science,* 223, 977, 1986.
34. **Firmin, J. L., Wilson, K. E., Rossen, L., Johnston, A. W. B.**, Flavanoid activation of nodulation genes in *Rhizobium* reversed by other compounds present in plants, *Nature (London),* 324, 90, 1986.
35. **Zaat, S. A. J., Wijffelman, C. A., Spaink, H. P., Van Brussel, A. A. N., Okker, R. J. H., and Lugtenberg, B. J. J.**, Induction of the *nodA* promoter of *Rhizobium leguminosarum* Sym plasmid pRL1JI by plant flavanones and flavones, *J. Bacteriol.*, 169, 198, 1987.
36. **Recourt, K., van Brussel, A. A. N., Driessen, A. J. M., and Lugtenberg, B. J. J.**, Accumulation of a *nod* gene inducer, the flavonoid naringenin in the cytoplasmic membrane of *Rhizobium leguminosarum* biovar *viciae* is caused by the pH-dependent hydrophobicity of naringenin, *J. Bacteriol.*, 171, 4370, 1989.

37. **Spaink, H. P., Weinman, J., Djordjevic, M. A., Wijffelman, C. A., Okker, R. J. H., and Lugtenberg, B. J. J.,** Genetic analysis and cellular localization of the *Rhizobium* host specificity-determining NodE protein, *EMBO J.*, 8, 2811, 1989.
38. **Evans, I. J. and Downie, J. A.,** The *nodI* product of *Rhizobium leguminosarum* is closely related to ATP-binding bacterial transport proteins: nucleotide sequence of *nodI* and *nodJ* genes, *Gene*, 43, 95, 1986.
39. **Davis, E. O., Evans, I. J., and Johnston, A. W. B.,** Identification of *nodX*, a gene that allows *Rhizobium leguminosarum* biovar *viciae* strain TOM to nodulate Afghanistan peas, *Mol. Gen. Genet.*, 212, 531, 1988.
40. **Davis, E. O. and Johnston, A. W. B.,** Analysis of three *nodD* genes in *Rhizobium leguminosarum* biovar *phaseoli; nodD1* is preceded by *nolE*, a gene whose product is secreted from the cytoplasm, *Mol. Microbiol.*, 4, 921, 1990.
41. **Schlaman, H. R. M., Okker, R. J. H., and Lugtenberg, B. J. J.,** Subcellular localization of the *Rhizobium leguminosarum nodI* gene product, *J. Bacteriol.*, 172, 5486, 1990.
42. **De Maagd, R. A., Spaink, H. P., Pees, E., Mulders, I. H. M., Wijfjes, A., Wijffelman, C. A., Okker, R. J. H., and Lugtenberg, B. J. J.,** Localization and symbiotic function of a region on the *Rhizobium leguminosarum* Sym plasmid pRL1JI responsible for a secreted flavonoid-inducible 50 kD protein, *J. Bacteriol.*, 171, 1151, 1989.
43. **De Maagd, R. A., Wijfjes, A. H. M., Spaink, H. P., Ruiz-Sainz, J. E., Wijffelman, C. A., Okker, R. J. H., and Lugtenberg, B. J. J.,** *NodO*, a new *nod* gene of the *Rhizobium leguminosarum* biovar *viciae* sym plasmid pRL1JI, encodes a secreted protein, *J. Bacteriol.*, 171, 6764, 1989.
44. **Economou, A., Hamilton, W. D. O., Johnston, A. W. B., and Downie, J. A.,** The *Rhizobium* nodulation gene *nodO* encodes a $Ca^{2+}$-binding protein that is exported without N-terminal cleavage and is homologous to haemolysin and related proteins, *EMBO J.*, 9, 349, 1990.
45. **Nadler, K. D., Johnston, A. W. B., Chen, J.-W., and John, T. R.,** A *Rhizobium leguminosarum* mutant defective in symbiotic iron acquisition, *J. Bacteriol.*, 172, 670, 1990.
46. **Reibach, P. H. and Streeter, J. G.,** Evaluation of active versus passive uptake of metabolites by *Rhizobium japonicum* bacteroids, *J. Bacteriol.*, 159, 47, 1984.
47. **Bolton, E., Higgisson, B., Harrington, A., and O'Gara, F.,** Dicarboxylic acid transport in *Rhizobium meliloti:* isolation of mutants and cloning of dicarboxylic acid transport genes, *Arch. Microbiol.*, 144, 142, 1986.
48. **Finan, T. M., Wood, J. M., and Jordan, D. C.,** Symbiotic properties of $C_4$-dicarboxylic acid transport mutants of *Rhizobium leguminosarum, J. Bacteriol.*, 154, 1403, 1983.
49. **Ronson, C. W., Lyttleton, P., and Robertson, J. G.,** $C_4$-dicarboxylate transport mutants of *Rhizobium trifolii* form ineffective nodules on *Trifolium repens, Proc. Natl. Acad. Sci. U.S.A.*, 78, 4284, 1981.
50. **Udvardi, M. K., Price, D. G., Gresshoff, P. M., and Day, D. A.,** A dicarboxylate transporter on the peribacteroid membrane of soybean nodules, *FEBS Lett.*, 231, 36, 1988.
51. **Jiang, J., Gu, B., Albright, L. M., and Nixon, B. T.,** Conservation between coding and regulatory elements of *Rhizobium meliloti* and *Rhizobium leguminosarum dct* genes, *J. Bacteriol.*, 171, 5244, 1989.
52. **Dudman, W. F.,** The polysaccharides and oligosaccharides of *Rhizobium* and their role in the infection process, in *Advances in Nitrogen Fixation Research,* Veeger, C. and Newton, W. E., Eds., Martinus Nijhoff, The Hague, 1984, 397.
53. **Albersheim, P. and Darvill, A. G.,** Oligosaccharides, *Sci. Am.*, 253, 44, 1985.
54. **Ryan, C. A.,** Oligosaccharides as recognition signals for the expression of defensive genes in plants, *Biochemistry*, 27, 8879, 1988.
55. **Leigh, J. A., Signer, E. R., and Walker, G. C.,** Exopolysaccharide deficient mutants of *Rhizobium meliloti* that form ineffective nodules, *Proc. Natl. Acad. Sci. U.S.A.*, 82, 6231, 1985.
56. **Borthakur, D., Barber, C. E., Lamb, J. W., Daniels, M. J., Downie, J. A., and Johnston, A. W. B.,** A mutation that blocks exopolysaccharide synthesis prevents nodulation of peas by *Rhizobium leguminosarum* but not of beans by *R. phaseoli* and is corrected by cloned DNA from *Rhizobium* or the phytopathogen *Xanthomonas, Mol. Gen. Genet.*, 203, 320, 1986.
57. **Canter Cremers, H. C. J.,** Role of exopolysaccharide in nodulation by *Rhizobium leguminosarum* bv. *viciae,* Ph.D. thesis, Leiden University, The Netherlands, 1990.
58. **Chen, H., Batley, M., Redmond, J., and Rolfe, B. G.,** Alteration of the effective nodulation properties of a fast-growing broad host range *Rhizobium* due to changes in exopolysaccharide synthesis, *J. Plant Physiol.*, 120, 331, 1985.
59. **Kim, C.-H., Tully, R. E., and Keister, D. L.,** Exopolysaccharide-deficient mutants of *Rhizobium fredii* HH303 which are symbiotically effective, *Appl. Environ. Microbiol.*, 55, 1852, 1989.
60. **Diebold, R. and Noel, K. D.,** *Rhizobium leguminosarum* exopolysaccharide mutants: biochemical and genetic analyses and symbiotic behavior on three hosts, *J. Bacteriol.*, 171, 4821, 1989.
61. **Chen, H. and Rolfe, B. J.,** Cooperativity between *Rhizobium* mutant strains: Induction of nitrogen-fixing nodules on the tropical legume *Leucaena leucocephala, J. Plant Physiol.*, 127, 307, 1987.

62. **Chakravorty, A. K., Zurkowski, W., Shine, J., and Rolfe, B. J.,** Symbiotic nitrogen fixation: molecular cloning of *Rhizobium* genes involved in exopolysaccharide synthesis and effective nodulation, *J. Mol. Appl. Genet.*, 1, 585, 1982.
63. **Borthakur, D., Barker, R. F., Latchford, J. W., Rossen, L., and Johnston, A. W. B.,** Analysis of *pss* genes of *Rhizobium leguminosarum* required for exopolysaccharide synthesis and nodulation of peas: Their primary structure and interaction with *psi* and other nodulation genes, *Mol. Gen. Genet.*, 213, 155, 1988.
64. **Klein, S., Hirsch, A. M., Smith, C. A., and Signer, E. R.,** Interaction of *nod* and *exo Rhizobium meliloti* in alfalfa nodulation, *Mol. Plant-Microbe Interact.*, 2, 94, 1988.
65. **Müller, P., Hynes, M., Kapp, D., Niehaus, K., and Pühler, A.,** Two classes of *Rhizobium meliloti* infection mutants differ in exopolysaccharide production and in coinoculation properties with nodulation mutants, *Mol. Gen. Genet.*, 211, 17, 1988.
66. **Djordjevic, S. P., Chen, H., Batley, M., Redmond, J. W., and Rolfe, B. G.,** Nitrogen-fixing ability of exopolysaccharide synthesis mutants of *Rhizobium* sp. strains NGR234 and *R. trifolii* is restored by the addition of homologous exopolysaccharide, *J. Bacteriol.*, 169, 53, 1987.
67. **Battisti, L., Lee, C. C., and Leigh, J. A.,** Specific forms of exopolysaccharides from *Rhizobium meliloti* restore symbiotic effectiveness to *R. meliloti exo mutants, in Proc. 5th Int. Symp. Mol. Genet. Plant-Microbe Interact.*, Göttfert, M., Hennecke, H., and Paul, H., Eds., Mikrobiologisches Institut, Eidgenössische Technische Hochschule Zürich, Switzerland, 1990, 117.
68. **Glazebrook J. and Walker, G. C.,** A novel exopolysaccharide can function in place of the calcofluor-binding exopolysaccharide in nodulation of alfalfa by *Rhizobium meliloti, Cell,* 56, 661, 1989.
69. **Zhan, H., Levery, S. B., Lee, C. C., and Leigh, J. A.,** A second exopolysaccharide of *Rhizobium meliloti* strain SU47 that can function in root nodule invasion, *Proc. Natl. Acad. Sci. U.S.A.*, 86, 3055, 1989.
70. **Leigh, J. A., Reed, J. W., Hanks, J. F., Hirsch, A. M., and Walker, G. C.,** *Rhizobium meliloti* mutants that fail to succinylate their calcofluor-binding exopolysaccharide are defective in nodule invasion, *Cell,* 51, 579, 1987.
71. **Gray, J. X., Zhan, H., Levery, S. B., Rolfe, B. G., and Leigh, J. A.,** Heterologous exopolysaccharide production by chimeric *Rhizobium* sp. strain NGR234 and *R. meliloti* and demonstration of host-symbiont specificity in promotions of nodule development and invasion by exopolysaccharide in *Proc. 5th Int. Symp. Mol. Genet. Plant-Microbe Interact.*, Göttfert, M., Hennecke, H., and Paul, H., Eds., Mikrobiologisches Institut, Eidgenössische Technische Hochschule Zürich, Switzerland, 1990, p116.
72. **Long, S., Reed, J. W., Himawan, J., and Walker, G. C.,** Genetic analysis of a cluster of genes required for the synthesis of the calcofluor-binding exopolysaccharide of *Rhizobium meliloti, J. Bacteriol.*, 170, 4239, 1988.
73. **Finan, T. M., Kunkel, B., De Vos, G. F., and Signer, E. R.,** Second symbiotic megaplasmid in *Rhizobium meliloti* carrying exopolysaccharide and thiamine synthesis genes, *J. Bacteriol.*, 167, 66, 1986.
74. **Zhan, H., Gray, J. X., Levery, S. B., Rolfe, B. G., and Leigh, J. A.,** Functional and evolutionary relatedness of genes for exopolysaccharide synthesis in *Rhizobium meliloti* and *Rhizobium* sp. strain NGR234, *J. Bacteriol.*, 172, 5245, 1990.
75. **Leigh, J. A. and Lee, C. C.,** Characterization of polysaccharides of *Rhizobium meliloti exo* mutants that form ineffective nodules, *J. Bacteriol.*, 170, 3327, 1988.
76. **Canter Cremers, H. C. J., Batley, M., Redmond, J. W., Eydems, L., Breedveld, M., Zevenhuizen, L. P. T. M., Pees, E., Wijffelman, C. A., and Lugtenberg, B. J. J.,** *Rhizobium leguminosarum exoB* mutants are deficient in the synthesis of UDP-glucose 4′-epimerase, *J. Biol. Chem.*, 265, 21122, 1990.
77. **Borthakur, D., Downie, J. A., Johnston, A. W. B., and Lamb, J. W.,** *psi,* a plasmid-linked *Rhizobium phaseoli* gene that inhibits exopolysaccharide production and which is required for symbiotic nitrogen fixation, *Mol. Gen. Genet.*, 200, 278, 1985.
78. **Borthakur, D. and Johnston, A. W. B.,** Sequence of *psi,* a gene on the symbiotic plasmid of *Rhizobium phaseoli* which inhibits exopolysaccharide synthesis and nodulation and demonstration that its transcription is inhibited by *psr,* another gene on the symbiotic plasmid, *Mol. Gen. Genet.*, 207, 149, 1987.
79. **Kamoun, S., Cooley, M. B., Rogowsky, P. M., and Kado, C. I.,** Two chromosomal loci involved in production of exopolysaccharide in *Agrobacterium tumefaciens, J. Bacteriol.*, 171, 1755, 1989.
80. **Gray, J. X., Djordjevic, M. A., and Rolfe, B. G.,** Two genes that regulate exopolysaccharide production in *Rhizobium* sp. strain NGR234: DNA sequences and resultant phenotypes, *J. Bacteriol.*, 172:193, 1990.
81. **Zhan, H. and Leigh, J. A.,** Two genes that regulate exopolysaccharide production in *Rhizobium meliloti, J. Bacteriol.*, 172, 5254, 1990.
82. **Bergersen, F. J.,** *Root Nodules of Legumes: Structure and Functions,* Research Studies Press, Brisbane, 1982.
83. **Paau, A. S., Cowles, J. R., and Raveed, D.,** Development of bacteroids in alfalfa *(Medicago sativa)* nodules, *Plant Physiol.*, 62, 526, 1978.

84. **Price, G. D., Day, D. A., and Gresshoff, P. M.**, Rapid isolation of intact peribacteroid envelopes from soybean nodules and demonstration of selective permeability to metabolites, *J. Plant Physiol.*, 130, 157, 1987.
85. **Latchford, J. W. and Johnston, A. W. B.**, Molecular genetics of exopolysaccharide production by *Rhizobium* bv. *viciae, leguminosarum*, in *Proc. 5th Int. Symp. Mol. Genet. Plant-Microbe Interact.*, Göttfert, M., Hennecke, H., and Paul, H., Eds., Mikrobiologisches Institut, Eidgenössische Technische Hochschule Zürich, Switzerland, 1990, 110.
86. **Reuber, T. L., Long, S., and Walker, G. C.**, Regulation of *Rhizobium meliloti exo* genes in free-living cells and in planta examined by using Tn*phoA* fusions, *J. Bacteriol.*, 173, 426, 1991.
87. **Jordan, D. C., Grinyer, I., and Coulter, W. H.**, Electron microscopy of infection threads and bacteria in young root nodules of *Medicago sativa, J. Bacteriol.*, 86, 125, 1963.
88. **Tjepkema, J. D. and Yocum, C. S.**, Measurement of oxygen partial pressure within soybean nodules by oxygen microelectrodes, *Planta*, 119, 351, 1974.
89. **Bergersen, F. J., Turner, G. L., Gibson, A. H., and Dudman, W. F.**, Nitrogenase activity and respiration of cultures of *Rhizobium* spp. with special reference to the concentration of dissolved oxygen, *Biochim. Biophys. Acta*, 444, 164, 1976.
90. **Ditta, G., Virts, E., Palomares, A., and Kim, C.-H.**, The *nifA* gene of *Rhizobium meliloti* is oxygen regulated, *J. Bacteriol.*, 169, 3217, 1987.
91. **Pankurst, C. E. and Craig, A. S.**, Effect of oxygen concentration, temperature and combined nitrogen on the morphology and nitrogenase activity of *Rhizobium* sp. strain 32H1 in agar culture, *J. Gen. Microbiol.*, 106, 207, 1978.
92. **Tully, R. E. and Terry, M. E.**, Decreased exopolysaccharide synthesis by anaerobic and symbiotic cells of *Bradyrhizobium japonicum, Plant Physiol.*, 79, 445, 1985.
93. **Doherty, D., Leigh, J. A., Glazebrook, J., and Walker, G. C.**, *Rhizobium meliloti* mutants that overproduce the *R. meliloti* acidic calcofluor-binding exopolysaccharide, *J. Bacteriol.*, 170, 4249, 1988.
94. **York, W. S., McNeil, M., Darvill, A. G., and Albersheim, P.**, Beta-2-linked glucans secreted by fast-growing species of *Rhizobium, J. Bacteriol.*, 142, 243, 1980.
95. **Zevenhuizen, L. P. T. M. and Scholten-Koerselman, H. J.**, Surface carbohydrates of Rhizobium I. β-1,2-glucans, *Antonie van Leeuwenhoek, J. Microbiol. Serol.*, 45, 165, 1979.
96. **Dell, A., York, W. S., McNeil, M., Darvill, A. G., and Albersheim, P.**, The cyclic structure of β-D-(1→2)-linked glucans secreted by rhizobia and agrobacteria, *Carbohydr. Res.*, 117, 185, 1983.
97. **Batley, M., Redmond, J. W., Djordjevic, S. P., and Rolfe, B. G.**, Characterization of glycerophosphorylated cyclic β-1,2-glucans from a fast-growing *Rhizobium* species, *Biochim. Biophys. Acta*, 901, 112, 1987.
98. **Hisamatsu, M., Yamada, T., Higashiura, T., and Ikeda, M.**, The production of acidic, *O*-acylated cyclosophorans [cyclic (1→2)-β-D-glucan] by *Agrobacterium* and *Rhizobium* species, *Carbohydr. Res.*, 163, 115, 1987.
99. **Miller, K. J., Reinhold, V. N., Weissborn, A. C., and Kennedy, E. P.**, Cyclic glucans produced by *Agrobacterium tumefaciens* are substituted with *sn*-1-phosphoglycerol residues, *Biochim. Biophys. Acta*, 901, 112, 1987.
100. **Miller, K. J., Gore, R. S., Johnson, R., Benesi, A. J., and Reinhold, V. N.**, Cell-associated oligosaccharides of *Bradyrhizobium* spp., *J. Bacteriol.*, 172, 136, 1990.
101. **Dylan, T., Ielpi, L., Stanfield, S., Kashyap, L., Douglas, C., Yanofsky, M., Nestor, E., Helinski, D. R., and Ditta, G.**, *Rhizobium meliloti* genes required for nodulation are related to chromosomal virulence genes in *Agrobacterium tumefaciens, Proc. Natl. Acad. Sci. U.S.A.*, 83, 4403, 1986.
102. **Cangelosi, G. A., Hung, L., Puvanesarajah, V., Stacey, G., Ozga, D. A., Leigh, J. A., and Nester, E. W.**, Common loci for *Agrobacterium tumefaciens* and *Rhizobium meliloti* exopolysaccharide synthesis and their roles in plant interactions, *J. Bacteriol.*, 169, 2086, 1987.
103. **Geremia, R. A., Cavaignac, S., Zorreguieta, A., Toro, N., Olivares, J., and Ugalde, R. A.**, A *Rhizobium meliloti* mutant that forms ineffective pseudonodules in alfalfa produces normal exopolysaccharide but fails to form β-(1-2) glucan, *J. Bacteriol.*, 169, 880, 1987.
104. **Puvanesarajah, V., Schell, F. M., Stacey, G., Douglas, C. J., and Nester, E. W.**, Role for 2-linked-β-D-glucan in the virulence of *Agrobacterium tumefaciens, J. Bacteriol.*, 164, 102, 1985.
105. **Van Veen, R. J. M., Dulk-Ras, H., Schilperoort, R. A., and Hooykaas, P. J. J.**, Chromosomal nodulation genes: Sym plasmid containing *Agrobacterium* strains need chromosomal virulence genes (*chvA* and *chvB*) for nodulation, *Plant Mol. Biol.*, 8, 105, 1987.
106. **Douglas, C. J., Halperin, W., and Nester, E. W.**, *Agrobacterium tumefaciens* mutants affected in attachment to plant cells, *J. Bacteriol.*, 152, 1265, 1982.
107. **Dylan, T., Nagpal, P., Helinski, D. R., and Ditta, G. S.**, Symbiotic Pseudorevertants of *Rhizobium meliloti ndv* mutants, *J. Bacteriol.*, 172, 1409, 1990.

108. **Stanfield, S. W., Ielpi, L., O'Brochta, D., Helinski, D. R., and Ditta, G. S.**, The *ndvA* gene product of *Rhizobium meliloti* is required for β-(1→2)-glucan production and has homology to the ATP-binding export protein HlyB, *J. Bacteriol.*, 170, 3523, 1988.
109. **Cangelosi, G. A., Martinetti, G., Leigh, J. A., Lee, C. C., Theines, C., and Nester, E. W.**, Role of *Agrobacterium tumefaciens* ChvA protein in export of β-1,2-glucan, *J. Bacteriol.*, 171, 1609, 1989.
110. **De Iannino, N. I., and Ugalde, R. A.**, Biochemical characterization of avirulent *Agrobacterium tumefaciens chvA* mutants: synthesis and excretion of β[1—2]-glucan, *J. Bacteriol.*, 171, 2842, 1989.
111. **Zorreguieta, A., Geremia, R. A., Cavaignac, S., Cangelosi, C. A., and Nester, E. W.**, Identification of the product of an *Agrobacterium tumefaciens* chromosomal virulence gene, *Mol. Plant-Microbe Interact.*, 1, 121, 1988.
112. **Ielpi, L., Dylan, T., Ditta, G. S., Helinski, D. R., and Stanfield, S. W.**, The *ndvB* locus of *Rhizobium meliloti* encodes a 319 kDa protein involved in the production of beta-(1,2)-glucan, *J. Biol. Chem.*, 265, 2843, 1990.
113. **Zorreguieta, A., Cavaignac, S., Geremia, R. A., and Ugalde, R. A.**, Osmotic regulation of beta-(1-2)-glucan synthesis in members of the family Rhizobiaceae, *J. Bacteriol.*, 172, 4701, 1990.
114. **Dylan, T., Helinski, D. R., and Ditta, G. S.**, Hypoosmotic adaptation in *Rhizobium meliloti* requires β-(1→2)-glucan, *J. Bacteriol.*, 172, 1400, 1990.
115. **Abe, M., Amemura, A., and Higashi, S.**, Studies on cyclic β-1,2-glucan obtained from periplasmic space of *Rhizobium trifolii* cells, *Plant Soil,* 64, 315, 1982.
116. **Carlson, R. W., Hollingsworth, R. I., and Dazzo, F. B.**, A core oligosaccharide component from the lipopolysaccharide of *Rhizobium trifolii* ANU843, *Carbohydr. Res.*, 176, 127, 1988.
117. **Hollingsworth, R. I., Carlson, R. W., Garcia, F., and Gage, D. A.**, A new core tetrasaccharide component from the lipopolysaccharide of *Rhizobium trifolii* ANU843, *J. Biol. Chem.*, 264, 9294, 1989.
118. **Cava, J. R., Elias, P. M., Turowski, D. A., and Noel, K. D.**, *Rhizobium leguminosarum* CFN42 genetic regions encoding lipopolysaccharide structures essential for complete nodule development on bean plants, *J. Bacteriol.*, 171, 8, 1989.
119. **Priefer, U. B.**, Genes involved in lipopolysaccharide production and symbiosis are clustered on the chromosome of *Rhizobium leguminosarum* biovar *viciae* VF39, *J. Bacteriol.*, 171, 6161, 1989.
120. **Brink, B. A., Miller, J., Carlson, R. W., and Noel, K. D.**, Expression of *Rhizobium leguminosarum* CFN42 genes for lipopolysaccharide in strains derived from different *R. leguminosarum* field isolates, *J. Bacteriol.*, 172, 548, 1990.
121. **Puvanesarajah, V., Schell, F. M., Gerhold, D., and Stacey, G.**, Cell surface polysaccharides from *Bradyrhizobium japonicum* and a nonnodulating mutant, *J. Bacteriol.*, 169, 137, 1987.
122. **De Maagd, R. A., Rao, A. S., Mulders, I. H. M., Goosen-de Roo, L., van Loosdrecht, M., Wijffelman, C. A., and Lugtenberg, B. J. J.**, Isolation and characterization of mutants of *Rhizobium leguminosarum* biovar *viciae* strain 248 with altered lipopolysaccharides: possible role of surface charge or hydrophobicity in bacterial release from the infection thread, *J. Bacteriol.*, 171, 1143, 1989.
123. **Clover, R. H., Kieber, J., and Signer, E. R.**, Lipopolysaccharide mutants of *Rhizobium meliloti* are not defective in symbiosis, *J. Bacteriol.*, 171, 3961, 1989.
124. **VandenBosch, K., Brewin, N. J., and Kannenberg, E. L.**, Developmental regulation of a *Rhizobium* cell surface antigen during growth in pea root nodules, *J. Bacteriol.*, 171, 4537, 1989.
125. **Goosen-de Roo, L. and de Maagd, R. A.**, Antigenic changes in lipopolysaccharide (LPS-I) and an outer membrane protein of *Rhizobium leguminosarum* bv. *viciae* 248 during the differentiation of bacteria to bacteroids, in *Proc. 8th Int. Congr. Nitrogen Fixation,* Stacey, G., Ed., University of Tennessee, Knoxville, TN, 1990, B20.
126. **Kanneberg, E. L. and Brewin, N. J.**, Expression of a cell surface antigen from *Rhizobium leguminosarum* 3841 is regulated by oxygen and pH, *J. Bacteriol.*, 171, 4543, 1989.
127. **Blumwald, E., Fortin, M. G., Rea, P. A., and Verma, D. P. S.**, Presence of host plasma membrane type $H^+$/ATPase in the membrane envelope enclosing the bacteroids in soybean root nodules, *Plant Physiol.*, 78, 665, 1985.
128. **Wood, E. A., Butcher, G. W., Brewin, N. J., and Kannenberg, E. L.**, Genetic derepression of a developmentally regulated lipopolysaccharide antigen from *Rhizobium leguminosarum* 3841, *J. Bacteriol.*, 171, 4549, 1989.
129. **Williams, M. N. V., Hollingsworth, R. I., Klein, S., and Signer, E. R.**, The symbiotic defect of *Rhizobium meliloti* is suppressed by *lpsZ*$^+$, a gene involved in lipopolysaccharide biosynthesis, *J. Bacteriol.*, 172, 2622, 1990.
130. **Halverson, L. J. and Stacey, G.**, Signal exchange in plant-microbe interactions, *Microbiol. Rev.*, 50, 193, 1986.
131. **Dazzo, F. B. and Truchet, G. L.**, Interactions of lectins and their saccharide receptors in the *Rhizobium*-legume symbiosis, *J. Membr. Biol.*, 73, 1, 1983.

132. **Diaz, C. L., Melchers, L. S., Hooykaas, P. J. J., Lugtenberg, B. J. J., and Kijne, J. W.,** Root lectin as a determinant of host-plant specificity in the *Rhizobium*-legume symbiosis, *Nature (London),* 338, 579, 1989.
133. **Spaink, H. P., Geiger, O., Reinhold, V., Lugtenberg, B. J. J., and Kennedy, E. P.,** The biochemical function of the *Rhizobium leguminosarum* proteins involved in the production of host specific signal molecules, in *Proc. 5th Int. Symp. Mol. Genet. Plant-Microbe Interact.,* Göttfert, M., Hennecke, H., and Paul, H., Eds., Mikrobiologisches Institut, Eidgenössische Technische Hochschule Zürich, Switzerland, 1990, 17.
134. **Lugtenberg, B. J. J., Diaz, C., Smit, G., De Pater, S., and Kijne, J. W.,** Roles of lectin in the *Rhizobium*-legume symbiosis, in *Proc. 5th Int. Symp. Mol. Genet. Plant-Microbe Interact.,* Göttfert, M., Hennecke, H., and Paul, H., Eds., Mikrobiologisches Institut, Eidgenössische Technische Hochschule Zürich, Switzerland, 1990, 8.
135. **Philip-Hollingsworth, S., Hollingsworth, R. I., Dazzo, F. B., Djordjevic, M. A., and Rolfe, B. G.,** The effect of interspecies transfer of Rhizobium host-specific nodulation genes on acidic polysaccharide structure and *in situ* binding by host lectin, *J. Biol. Chem.,* 264, 5710, 1989.
136. **Philip-Hollingsworth, S., Hollingsworth, R., and Dazzo, F.,** Host-range related structural features of the acidic extracellular polysaccharides of *Rhizobium trifolii* and *Rhizobium leguminosarum*, *J. Biol. Chem.,* 264, 1461, 1989.
137. **Canter Cremers, H. C. J., Batley, M., Redmond, J. W., Lugtenberg, B. J. J., Wijfjes, A., and Wijffelman, C. A.,** Distribution of *O*-acetyl group in the exopolysaccharide synthesized by *Rhizobium leguminosarum* is not determined by the Sym plasmid, *J. Biol. Chem.,* in press.
138. **O'Neill, M. A. and Darvill, A. G.,** The degree of esterification and points of substitution by *O*-acetyl and *O*-(3-hydroxybutanoyl) groups in the acidic extracellular polysaccharides secreted by *Rhizobium leguminosarum* biovar *viciae, trifolii,* and *phaseoli* are not related to host range, *J. Biol. Chem.,* in press.
139. **Roche, P., Lerouge, P., Faucher, C., Vasse, J., Prome, J. C., Dénarié, J. and Truchet, G.,** *Rhizobium meliloti* extracellular signals, in *Proc. 5th Int. Symp. Mol. Genet. Plant-Microbe Interact.,* Göttfert, M., Hennecke, H., and Paul, H., Eds., Mikrobiologisches Institut, Eidgenössische Technische Hochschule Zürich, Switzerland, 1990, 7.
140. **Scheres, B., Van de Wiel, C., Zalensky, A., Horvath, B., Spaink, H., Van Eck, H., Zwartkruis, F., Wolters, A. M., Gloudemans, T., Van Kammen, A., and Bisseling, T.,** The ENOD12 gene product is involved in the infection process during the pea-*Rhizobium* interaction, *Cell,* 60, 281, 1990.
141. **De Maagd, R. A. and Lugtenberg, B. J. J.,** Lipopolysaccharide: a signal in the establishment of the *Rhizobium* legume symbiosis, in *Signal Molecules in Plant-Microbe Interactions,* NATO ASI Ser., Lugtenberg, B. J. J., Ed., Springer-Verlag, Berlin, 1989, 337.
142. **Sindhu, S. S. Brewin, N. J., and Kannenberg, E. L.,** Immunochemical analysis of lipopolysaccharides from free-living and endosymbiotic forms of *Rhizobium leguminosarum, J. Bacteriol.,* 172, 1804, 1990.
143. **Sharma, S. B. and Signer, E. R.,** Temporal and spatial regulation of the symbiotic genes of *Rhizobium meliloti* in planta revealed by transposon *Tn5-gusA, Genes Dev.,* 4, 344, 1990.
144. **Schlaman, H. R. M., Horvath, B., Vijgenboom, E., Okker, R. J. H., and Lugtenberg, B. J. J.,** Evidence for a new negative regulation mechanism involved in the suppression of nodulation gene expression of *Rhizobium leguminosarum* bv. *viciae, J. Bacteriol.,* in press.

# RHIZOPINES IN THE LEGUME-*RHIZOBIUM* SYMBIOSIS

**Peter J. Murphy and Christopher P. Saint**

## TABLE OF CONTENTS

# I. INTRODUCTION

Members of the bacterial genera *Rhizobium, Bradyrhizobium, Azorhizobium,* and *Sinorhizobium* form symbiotic associations with the plant family Leguminosae culminating in the conversion of atmospheric nitrogen into an organic form which the plant can utilize. This process, termed symbiotic nitrogen fixation, is very significant in agriculture and in the recycling of nitrogen in the biosphere.

The development of symbiosis involves the rhizobia invading root hairs and entering into plant cells via infection threads, this process setting in play a complex series of interactions resulting in the formation of specialized structures on the roots called nodules. In the nodule, as bacteria are released from infection threads, they are encapsulated by the peribacteroid membrane and enter into plant cells. Here the bacteria terminally differentiate into bacteroids within which nitrogen is reduced to ammonia (reviews[1,2]). Neither the bacteroids nor the free-living bacteria remaining in the infection threads can utilize this fixed nitrogen.

The legume-*Rhizobium* symbiosis is considered to be a mutualistic association with both partners benefiting from the interaction.[3] It is generally accepted that the advantage to the bacteria is a general enhanced growth in the rhizosphere (the region of the soil immediately surrounding the roots of a plant, together with the root surfaces)[3] as a result of a greater input of nutrients into this region.[4] In this environment there are many microorganisms competing for these nutrients. Hence it would be prudent if free-living members of a rhizobial population, both in infection threads within the nodule and in the rhizosphere, could obtain a specific nutritional advantage from the symbiotic interaction. Rhizopines, nodule-specific opine-like compounds, which provide a selective growth substrate for the nodule-inducing rhizobia, may well confer such an advantage. Such compounds are described as opine-like by analogy with *Agrobacterium* opines since the free-living bacteria which induce the production of the compound can also catabolize it, but in contrast to opines, which are produced in the plant, rhizopines are synthesized in the bacteroid within the nodule.

Here we review the discovery and the many unique molecular features of the suite of genes involved in the catabolism and synthesis of the first rhizopine isolated — that which is found in nodules on *Medicago sativa* (alfalfa) induced by *Rhizobium meliloti* strain L5-30. We describe the molecular features which enable the bacteria to take advantage of this selective growth substrate and we provide physiological evidence that rhizopine-utilizing rhizobia are at an advantage in the nodule. We discuss these features with regard to more recently discovered rhizopines.

# II. DISCOVERY OF RHIZOPINES

The discovery of rhizopines is a relatively recent event and owes much to the opine concept developed in the crown gall *Agrobacterium* plant pathogenic interaction (see chapter by Tempé in this volume). This concept states that opines are the chemical mediators of parasitism, with the bacteria redirecting plant metabolites to produce strain-specific, gall-specific compounds which the inducing bacteria can utilize for its propagation.[5] The mechanism by which bacteria achieve this, which involves bacterial opine synthetic and catabolic genes with the former being transferred via the T-DNA to the plant during infection, has been well reviewed.[6,7] This process has been termed genetic colonization.[8]

To test the validity of the opine concept Tempé set about demonstrating its generality.[5] After showing that all *Agrobacterium tumefaciens* strains produced opines,[9] and extending the concept to *Agrobacterium rhizogenes,*[10] he looked at other plant-microbial interactions. Initially, the *Rhizobium*-legume symbiosis was chosen for a number of reasons. First, *Rhizobium* is quite closely related to *Agrobacterium.*[11] Second, the pathogenic and symbiotic

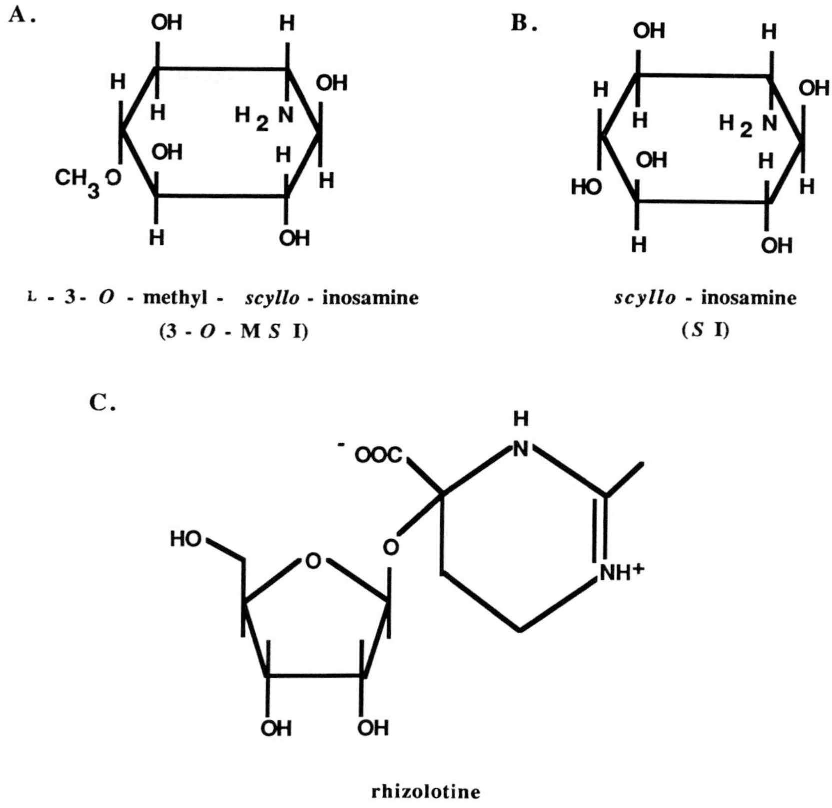


FIGURE 1. The structures of (A) L-3-*O*-methyl-*scyllo*-inosamine (3-*O*-M*S*I), (B) *scyllo*-inosamine (*S*I), and (C) rhizolotine.

states induced by these two organisms are quite similar; indeed, *Rhizobium* has been described as a refined parasite.[12] Third, both interactions involve plasmids (which in the case of *Agrobacterium* are fundamental to the opine concept).[5] The *Rhizobium* system proved to be a wise choice, since in alfalfa (*Medicago sativa*) nodules induced by *R. meliloti* strain L5-30 a nodule-specific "opine-like" compound was recently isolated.[13] The compound was identified as L-3-*O*-methyl-*scyllo*-inosamine (3-*O*-M*S*I, see Figure 1A)[14] and was described as "opine-like" since, of approximately 20 strains tested, only the strain which induced its synthesis in the nodule could catabolize it as a growth substrate.

After studying the genes involved in the synthesis and catabolism of 3-*O*-M*S*I it became evident that although it resembled opines in its physiological aspects the mechanism by which it functioned was quite distinct. This, along with the discovery of other similar compounds (see Section IV), led us to adopt the generic name of rhizopines for such compounds.[14] The name rhizopine was chosen to retain a link to the analogy with *Agrobacterium* opines but to distinguish them from implications of T-DNA transfer, which is an integral part of the opine concept.

## III. CHARACTERIZATION OF RHIZOPINE GENES

### A. ISOLATION AND ANALYSIS OF RHIZOPINE CATABOLIC (*MOC*) GENES

We first directed our attention to the isolation of the catabolic (*moc*) genes as they were expressed in the free-living rhizobia, and we could isolate sufficient quantities of 3-*O*-M*S*I from nodules to be used in catabolic studies.[14] Accordingly, a clone bank of total DNA from strain L5-30 was prepared in the broad host-range cosmid vector pLAFR1[15] and used to complement a $Moc^-$ strain of *R. meliloti*. Transconjugants were screened by growth in minimal medium supplemented with the rhizopine as a sole carbon and/or nitrogen source. When *moc* genes were expressed, 3-*O*-M*SI* disappeared from the growth medium; this is easily detected by high-voltage paper electrophoresis of the culture supernatant followed by staining with $AgNO_3$. 3-*0*-M*SI* proved to be a better carbon than nitrogen source and minimal medium was supplemented with $(NH_4)_2SO_4$ in subsequent experiments. Octopine has also recently been shown, using a chemostat, to be a better carbon than nitrogen source.[16]

On the basis of the above procedure and subsequent subcloning, a 15-kb DNA fragment derived from L5-30 was found to be essential for *moc* activity. This region was further defined by Tn*5* mutagenesis and deletion analysis and two functional catabolic regions were identified.[14] One region, spanned by a 5.4-kb *Eco*RI fragment at one end of the 15-kb fragment, appears to be essential for *moc* activity, since Tn*5* mutants in this region completely abolish 3-*0*-M*SI* catabolism. At the other end of the *moc* region is a 2.4-kb *Eco*RI-*Kpn*I fragment, the removal of which results in the accumulation of *scyllo*-inosamine (*SI*, Figure 1B). Interestingly, Tn5 mutants between these two regions do not effect *moc* activity. *SI* is the result of demethylation of 3-*O*-M*S*l. Thus, it is likely that at least part of the 5.4-kb fragment is involved in this process and the 2.4-kb fragment is required for further catabolism of *SI*. It is also plausible that breakdown products of 3-*O*-M*SI* could then enter common bacterial ''housekeeping'' pathways for further metabolism. Besides catabolic genes, the 5.4-kb *moc* fragment may also contain additional genes, such as those involved in rhizopine uptake. The linkage of uptake genes with catabolic genes has been demonstrated for the opine nopaline.[17]

### B. REGULATION OF *MOC* GENES

The *moc* genes may be regulated by common bacterial regulatory genes. However, they are not regulated by the *ntrC* or *nifA* systems as the *moc* genes are expressed in strains with mutations in these genes.[18] Perhaps this is not surprising since the *nifA* regulon controls symbiotic functions, not processes in free-living bacteria, and the *ntrC* system regulates nitrogen metabolism and 3-*0*-M*SI* is a poor nitrogen source. The possibility exists that the common *ntrA* bacterial regulatory system may control the *moc* genes as this system is involved in the regulation of a wide variety of bacterial genes.[19] This could easily be determined by testing for *moc* activity in a $ntrA^-$ strain. NtrA acts as a sigma factor, at approximately −20 bp upstream from the start of transcription, in concert with upstream regions.[20] If NtrA is regulating the *moc* genes it will be interesting to see what other upstream regulatory sequences it interacts with.

The identification of two distinct regions involved in 3-*0*-M*SI* catabolism raises the question of how these regions are coordinately regulated. Some clue to this may be obtained from earlier studies which indicated that when catabolism of *SI* was blocked, its accumulation prevented the complete conversion of 3-*O*-M*SI* to *SI*, suggesting there is feedback inhibition functioning between the two separate catabolic regions. In addition, expression of the *moc* genes may be subject to substrate regulation. The use of *lacZ* fusions[21] could be used to resolve this. Thus, the overall regulation of the *moc* genes may well be complex, involving several levels of regulation.

### C. ISOLATION OF RHIZOPINE SYNTHESIS (*MOS*) GENES

3-*O*-M*SI* is only found in bacteroids in nodules and not in other parts of the plant or in free-living bacteria. By analogy with *Agrobacterium* we assumed that the primary genes involved in the synthesis of 3-*O*-M*SI* were bacterial. Evidence suggested that this was indeed the case. First, strains of *R. meliloti* other than L5-30 did not induce the rhizopine when inoculated onto the same plant host. Second, a derivative of L5-30 which had a large deletion in the Sym plasmid, but had restored nodulation functions by the addition of an R-prime containing *nod* genes, did not induce rhizopine production during endosymbiosis. Further, as many related bacterial genes are closely linked, we considered it possible that the *moc-mos* genes might also be linked.

Based on these assumptions an R-prime was constructed. This 150-kb plasmid consisted of R68.45 bearing a region encompassing the *moc* genes and surrounding DNA. This R-prime was used to complement a $Mos^-$ strain and nodules induced by this strain produced the rhizopine. The *mos* locus was delineated to a 5-kb fragment by further subcloning and deletion analysis.[14]

## D. *MOC-MOS* LOCI ARE CLOSELY LINKED

One of the early significant discoveries regarding the L5-30 rhizopine was the close linkage of the *moc* and *mos* loci. It was shown that both loci could be isolated on a 25-kb insert in a single cosmid, and are, in fact, only 4.5 kb apart.[14] This was important in establishing an integral relationship between synthesis and catabolism of the rhizopine. Prior to this, the possibility always existed that these loci were unrelated with the rhizopine being catabolized only by general "housekeeping" catabolic processes of the bacteria. The close linkage of these genes strongly suggests this is unlikely and implies that these loci have coevolved as a self-contained synthetic/catabolic functional unit. Together these loci are referred to as the rhizopine cassette.

## E. *MOC-MOS* LOCI ARE ON THE *NOD-NIF* SYM PLASMID

Many genes involved in nodulation and nitrogen fixation are on large plasmids in rhizobia.[22] In many *R. meliloti* strains, two extremely large plasmids (mega plasmids) which comigrate *in situ* lysis gels have been identified.[23] These plasmids have been estimated to be as large as 1600 kb.[24] One of these plasmids has *nod, nif,* and many *fix* functions (p*Syma*), but both plasmids are required for effective nodulation.[23] *Rhizobium* strain L5-30 is no exception and also has two large plasmids. By hybridization studies and mobilizing individual plasmids using Tn*5*-mob[25] the rhizopine cassette was shown to be on the *nod-nif* Sym plasmid of this strain.[14] Attempts were made to link the rhizopine cassette to the *nod-nif* genes of L5-30, but despite using cosmid clones to walk approximately 25-kb, in one direction, and approximately 45-kb in the other, no linkage could be found. Nevertheless, all these genes are missing from L5-22, a derivative of L5-30 which has a deletion in the Sym plasmid estimated to be in the order of several hundred kilobases. The observation that the rhizopine cassette was on the *nod-nif* Sym plasmid was significant as this was the first piece of evidence, beside the physiological analogy with the *Agrobacterium* opines, that the rhizopine may be important in symbiosis.

## F. ANALYSIS OF *MOS* GENES

### 1. Regulation of *mos* Genes

To study the regulation of the *mos* genes we started from the premise that if they play a role in symbiosis they could be regulated by the symbiotic regulatory circuitry. Many genes involved in symbiotic nitrogen fixation are regulated by the *nifA* gene[20,26] so this seemed an appropriate starting point to investigate the regulation of the *mos* genes. The *nifA* gene, which in *R. meliloti* is in turn regulated by low levels of oxygen,[27,28] acts in concert with a more diverse regulatory gene *ntrA* (*rpoN*) to transcriptionally regulate genes involved in nitrogen fixation. In *R. meliloti* both the *nifHDK* operon, encoding the nitrogenase complex[29,30] and the *fixABCX* operon, encoding genes essential for symbiotic nitrogen fixation, which may be involved in electron transport,[31,32] are regulated in this manner. The 5′-regions of these two operons have 160 bp of highly conserved DNA sequence[33] which is considered to constitute a symbiotic promoter. Within this region, at approximately −10 to −27 bp from the start of transcription, is a consensus sequence CTGGYAYR-$N_4$-TTGCA[26,34,35] which is required for *ntrA* regulation.[19] Further upstream, at approximately −120 bp from the start of transcription, is another consensus sequence TGT-$N_{10}$-ACA which is required for NifA regulation in the free living bacteria *Klebsiella pneumoniae.* This consensus sequence is also conserved in *R. meliloti,*[36] although its role here has been questioned.[37] *Rhizobium nifA* regulation is independent of a further regulatory gene *ntrC,* which is involved in the regulation of nitrogen metabolism[38] and is, therefore, distinct from nitrogen fixation in free-living diazotrophs such as *K. pneumoniae,* which require *ntrC* (reviews[26,20]).

To determine if the *mos* genes are symbiotically regulated, we[39] conjugated a plasmid containing the *mos* locus into $NifA^-$ and $NtrC^-$mutants of *R. meliloti*[40,41] and analyzed nodules produced by these strains for the presence of the rhizopine. Since the $NtrC^-$transconjugant induced normal levels of the rhizopine, but none was detected in nodules induced by the $NifA^-$ transconjugant, it was concluded that the *mos* locus was regulated by the symbiotic *nifA* regulatory gene. This was indicative of either the *nifA* gene directly regulating *mos* genes via a NifA-regulated promoter, or there could be a requirement for NifA-regulated functions, such as a requirement for a symbiotically effective nodule for 3-*O*-MSI production. Further work showed that there was no requirement for symbiotically fixed nitrogen since a $NifH^-$ transconjugant containing the *mos* genes induced normal levels of the rhizopine. This suggested that the *mos* genes were directly regulated by a NifA activated promoter. To test this, we prepared a probe from the promoter region of a *R. meliloti fixA* gene. This probe has an extended homology with other NifA regulated genes, such as those of the *nif* operon, and can be considered to be a symbiotic promoter probe. A number of bands from L5-30 DNA hybridized to this probe, among which was a 1.0-kb fragment which could be localized to the *mos* locus. This DNA fragment was sequenced and revealed a striking homology with the *nif* operon promoter: 75% homology over a 132-bp region and the two consensus sequences for NifA and NtrA recognition were conserved. This region is essential for *mos* expression as its removal results in a $Mos^-$ phenotype. Further, a specific primer was prepared to a region downstream of the promoter and used to map the start of transcription by the primer extension method. The start of transcription was almost identical, being one base further downstream to that reported for the *nif* operon in *R. meliloti* strain 102F34.[33]

On the basis of these studies it was concluded[39] that the *mos* locus is regulated by the symbiotic nitrogen fixation regulatory gene *nifA,* and this regulation is directly via a NifA-regulated promoter and is not a secondary effect of other genes regulated by the *nifA* gene. The presence of this promoter ensures that the *mos* genes are transcriptionally regulated in a similar fashion to genes involved in symbiotic nitrogen fixation. Several other rhizobial genes, other than those directly involved in nitrogen fixation, have been reported to be NifA-regulated suggesting they too may have a symbiotic function. These are the *melA* gene, which codes for melanin production,[42] and the *nfe* region, which is involved in efficiency of nodule formation.[43]

Regulatory sequences controlling both *nod* and *nif* genes have been reported to be reiterated in rhizobia.[33,34] It has been suggested that these sequences may control genes that are not essential for symbiosis, but are nevertheless important in the symbiotic interaction. As most symbiotic assays are done under controlled conditions in test tubes, not all of these nonessential symbiotic functions may be immediately obvious. However, although there may be as yet undetermined symbiotic functions, it is also unlikely that all sequences isolated by homology with promoter probes control symbiotic functions. In support of this it has been found by Better et al.[33] that in *R. meliloti* strain 102F34 a small sequence between *nif* and *fix* genes has homology with the *nifH* promoter and the first 47-bp of the *nifH* gene.

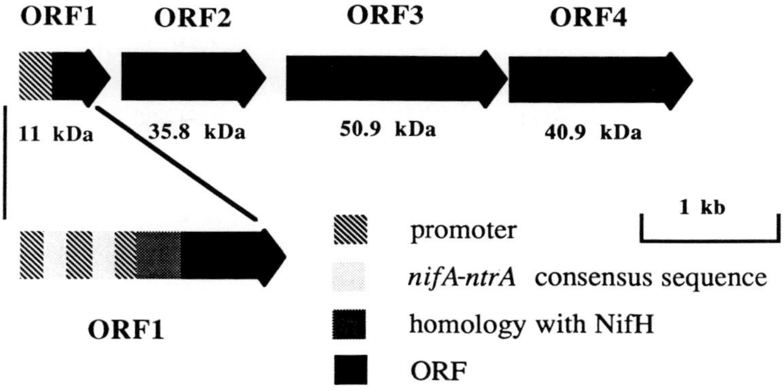


FIGURE 2. Rhizopine synthesis (*mos*) locus from *R. meliloti* strain L5-30.

The open reading frame (ORF) terminates 16-bp downstream from the point where the homology with *nifH* ends. This region is transcribed in the nodule. In addition, approximately 550 bp upstream of the *mos* locus we have found a sequence with extended homology to *fixA* and its promoter. The nucleotide sequence extending past the ATG for 102 bp is highly homologous to the *fixA* sequence. This same ORF then continues for a further 60 bp coding for a small protein of 6170 Da. We do not know with any certainty whether this region encodes a functional protein.

Often associated with these reiterated promoters is a portion of a duplicated symbiotic gene, *nifH* and *fixA* in the cases mentioned above. A reiterated *fixA* sequence, although not the promoter, has also been found approximately 500 bp upstream of *fixABCX* in strain *R. meliloti* 1021.[31] It would seem that the region around symbiotic genes, including the promoter, may be "hot spots" for rearrangements and a number of transcriptional "false starts" may be generated.

Since the *mos* genes have a 5′-region homologous to the *nif* operon, including leader sequence and part of the *nifH* gene (see next section), we investigated the origin of this region. The *mos* locus was probed with a *nifHD* probe and no homology up to 33-kb downstream of the *mos* operon was found.[18] Therefore, it does not appear that *mos* genes have integrated into a duplicated *nif* gene. However, it is possible that if such an integration occurred some time ago any nonfunctional downstream sequence has been subsequently lost.

### 2. Sequence of *mos* Genes

The size of the *mos* locus has been determined by deletion analysis to be 5-kb. Within this region there are four ORF's all reading in the same direction and there are no obvious additional putative promoters, suggesting this region is covered by one transcript (Figure 2).[45] Analysis of the 5′-region revealed some quite remarkable features. Not only did the promoter region show extended homology with the promoters of the *nif* and *fix* operons but the *mos* 5′-untranslated leader sequence shared 99% homology with an equivalent region of the *nif* operon. Perhaps even more surprising is that the 5′-region of the first ORF (ORF 1) of the *mos* locus has extensive homology with the *nifH* gene, which codes for the Fe-S dinitrogen reductase (component II) of the nitrogenase enzyme complex.[46] There is 87% homology between the first 57 bp of *mos* ORF 1 and the *nifH* coding region. The sequences then diverge and ORF 1 continues for a further 249 bp with *mos* ORF 2 starting a further 187 bp downstream.

The conservation between the *mos* ORF 1 and the *nifH* gene is an intriguing observation and may represent a domain important for the functioning of both these genes. Alternatively, this region may not code for a functional portion of a Mos protein but may be merely a consequence of the fusion of the *nif* operon 5′-regulatory region to the *mos* genes, thus enabling these genes to be symbiotically regulated. The overall homology extending over the entire 5′-region, including the promoter and untranslated leader, would suggest that this latter explanation is the correct one. If ORF 1 is part of a fusion, the purpose of which is to ensure that rhizopine production is coordinately regulated with other symbiotic genes, then it may not code for a functional protein. To test this we constructed a frame shift mutant in ORF 1. With this construct, the NifA-regulated transcript would still be produced but ORF 1 would terminate approximately one third the way down its sequence. When this clone was mated into *Rhizobium* and nodules induced, normal levels of rhizopine were obtained. This data supports the model that ORF 1 is the result of a fortuitous fusion ensuring advantageous transcriptional control and ORFs 2,3, and 4 code for functional *mos* proteins. The structure of the *mos* locus is consistent with this hypothesis. *mos* ORF 2 starts a further 187 bp downstream from ORF 1. This is a large gap for an intergenic region in a transcriptional unit. Nevertheless, larger gaps have been reported, as in the case of the *hsnB* and *hsnC* genes in *R. meliloti* strain AK631 where there is a gap of 493 bp.[47] In front of ORF 2 is a potential Shine-Dalgarno sequence[48] which would enable ribosomes to reinitiate and produce Mos proteins 2, 3, and 4.

#### 3. Function of Individual *mos* Genes

When the sequence of ORF 2 was scanned against the Genbank and EMBL libraries, highly significant homology with the product of the *dapA* gene[49] of *E. coli* was observed.[50] The *dapA* gene encodes dihydrodipicolinate synthase which is the first enzyme of the diaminopimelate and lysine pathway.[51] This enzyme condenses pyruvate and aspartic semialdehyde to 2,3-dihydrodipicolinic acid. During the production of the rhizopine, the product of ORF 2 may be involved in the condensation of linear plant carbon skeletons into the circular form. Neither ORF 3 nor ORF 4 has significant homology to other sequenced genes either at the nucleic acid or protein level. However, the protein predicted from ORF 4 is very hydrophobic and likely to be a membrane protein. Such a protein could conceivably be involved in the transport of rhizopine precursors from the plant into the bacteroid. Alternatively, it could be involved in the transport of the rhizopine out of the bacteroid.

## IV. ARE RHIZOPINES UBIQUITOUS?

### A. HOW WIDESPREAD IS THE RHIZOPINE 3-*O*-MSI AND RELATED COMPOUNDS?

Of 20 other *R. meliloti* strains tested for the ability to catabolize the L5-30 rhizopine, 3-*O*-M*S*I, only one other could do so.[52] On further investigation this strain (Rm220-3) was found to synthesize *scyllo*-inosamine (*S*I) in the nodule, a compound which it could also catabolize. *S*I is structurally related to 3-*O*-M*S*I (see Figure 1) and, by definition, is a rhizopine.

We have recently isolated the rhizopine genes from strain Rm220-3 and found some remarkable similarities with those genes from strain L5-30. In strain Rm220-3 rhizopine synthesis and catabolism genes are closely linked on the Sym plasmid and synthesis genes bear a NifA-regulated promoter.[53] This is totally analogous to the situation in strain L5-30. Rhizopine genes for the production and catabolism of 3-*O*-M*S*I in a strain of *R. leguminosarum* bv. *viciae* are also being investigated in our laboratory and preliminary evidence suggests the synthetic and catabolic genes are linked in this strain. Now that probes for the *moc* and *mos* genes are available it would be prudent to screen strain collections to see how widespread the catabolism and synthesis genes for rhizopines are.

### B. ARE THERE OTHER RHIZOPINES?

The rhizobia studied to date all induce rhizopines belonging to the inositol class of compounds. The definition of a rhizopine implies it must be produced exclusively in the nodule (although plant photosynthates are used) and the bacteria must bear the complete genetic suite for its synthesis and catabolism. As such, the structure of the compound *per se* is not important and compounds with vastly different structures to 3-*O*-M*S*I or *S*I can be classified as rhizopines. In this context, a compound, rhizolotine, has been found in nodules induced by *R. loti* (Figure 1C).[54,55] Genetic evidence suggests that this compound is synthesized in the bacteroid as is 3-*O*-M*S*I. However, although the inducing strain can catabolize the compound, it does not appear to be able to utilize it as a growth substrate. It is, therefore, yet to be determined whether rhizolotine is a rhizopine. Isolation and study of rhizolotine genes may help in determining this.

To find other rhizopines it is necessary to inoculate plants with a chosen rhizobial strain, harvest the nodules induced, homogenize these, and remove nonspecific carbon and nitrogen sources by incubation of the extract with a bacteria other than *Rhizobium*.[13] The *Rhizobium* strain which induced the nodule is then inoculated into the biologically purified extract and if growth is observed it indicates the presence of a specific carbon or nitrogen growth substrate. This procedure is far from ideal as potential rhizopines may well be catabolized by the strain of bacteria used for biological purification of extracts. Therefore, in the search for new rhizopines a variety of different extraction methods may be required.

### C. DO ALL RHIZOBIA HAVE RHIZOPINES?

Opines have proved to be remarkably widespread in *Agrobacterium* and this raises the question as to how widespread are rhizopines. As discussed above not all strains have inositol-based rhizopines so there may be a broad range of structurally different rhizopines. Certainly there are a large range of different compounds in nodules.[56]

If rhizopines act as selective growth substrates (see Section VI) then they may be a particular refinement that some rhizobia have developed to survive in the rhizosphere. If many rhizobia or other microorganisms in the soil could utilize the same rhizopine, there would be little advantage to be gained. The advantage of rhizopines may be that they are not generally catabolized by other strains. There could, however, be a wide variety of different rhizopines. This is distinct from opines which fall into a few catabolic classes that can be utilized by a wide range of *Agrobacterium* strains, and also by a range of other bacteria.[57,58]

## V. WHAT IS THE FUNCTION OF RHIZOPINES?

As rhizopines were found by analogy with *Agrobacterium* opines, and as a limited number of strains can utilize rhizopines as a selective growth substrate, we have worked on the premise that rhizopines supply the bacteria with a selective growth substrate. Furthermore, we have not detected any effect on nitrogen fixation or nodulation efficiency using mutants with inactivated *moc* and *mos* genes[59] indicating that rhizopines do not act on these functions. We have designed a number of experiments in an endeavor to test the role of rhizopines in supplying the bacteria with a competitive growth advantage. To date no system investigating competition has been completely tested in the field.[60] Indeed, it was because of the perceived difficulties that Tempé tested the validity of the opine concept by its generality, rather than testing it in the field. As a starting point we have investigated the role of rhizopines in survival under controlled conditions.

An L5-30::Moc$^-$ strain has been constructed using Tn*5* and coinoculated with the wild-type L5-30 strain onto plants in test tubes. After several months, nodules were harvested and bacteria isolated from the infection threads of these nodules. Using appropriate antibiotic markers, the ratios of the bacteria occupying the nodules were determined. Such experiments have been repeated many times and have consistently shown that in the presence of the rhizopine the strain which can utilize it preferentially occupies the nodule. In these experiments, up to 95% of the nodules contain only the rhizopine catabolizing *Rhizobium*.[61] This could be explained by either the rhizopine-catabolizing strain having an advantage of infection or this strain has a growth advantage in a doubly infected nodule. Support for the latter explanation is provided by a time course experiment where nodules analyzed after a longer period following infection showed a higher frequency of the rhizopine-catabolizing strain than nodules analyzed soon after inoculation.[61] The presence of doubly infected nodules is established both under experimental conditions and in nature.[62,63]

It would seem that under our experimental conditions the rhizopine is selectively feeding the free-living bacteria in the infection threads in the nodule. We envisage that this principle would work at a broader level over a longer period of time. Senescence of nodules would release the rhizopine into the rhizosphere, enabling strains with the capacity to catabolize it to flourish, particularly when other nutrient sources are limiting. Hence, the rhizopine utilizing strain could persist in the rhizosphere. This strain could nodulate the plant during a subsequent growing season. Competition in the soil is a very complex issue[60,64] involving many factors. The role of rhizopines in this arena would be a useful addition to enhance the survival of strains efficient in nitrogen fixation and nodulation.

A number of other systems involving secondary plant metabolites implicated in nutritional relationships between plants and soil bacteria have been reported. The plant morning glory (*Calystegia sepium*) produces a compound, calystegin, which *R. meliloti* strain 41 can catabolize as a sole carbon and nitrogen source but 43 other strains tested could not.[65] Another plant secondary plant metabolite, trigonelline, often abundant in legumes, can be catabolized as an energy source by *R. meliloti* strain RCR 2011. The trigonelline catabolic genes of this strain are on the Sym plasmid close to the *nifHDK* genes and appear to be widespread in *R. meliloti*.[66,67] These genes are induced in the rhizosphere of alfalfa and in the infection threads. What separates these two compounds from rhizopines is that they are produced by plants and their production is not under the control of the catabolizing rhizobial strain.

## VI. CONCLUSION

This review has concentrated on the discovery of the first rhizopine L3-*O*-methyl-*scyllo*-inosamine (3-*O*-MSI) and the molecular biology of its synthesis in the nodule and catabolism by *R. meliloti* strain L5-30. We have also described a similar system operating in *R. meliloti* strain Rm220-3 for the production and catabolism of *scyllo*-inosamine (SI), a related rhizopine. In addition, this similarity may well extend to a rhizopine in *R. leguminosarum* bv. *viciae*. These rhizopines belong to an inositol class of compounds. With the exception of rhizolotine, which is yet to be conclusively classed as a rhizopine, no other rhizopines have been found; although we predict that with a thorough search more will be discovered.

Molecular evidence, Sym plasmid location, and NifA regulation of the rhizopine synthesis genes strongly suggest a symbiotic role for rhizopines. Initial studies with a mutant defective in rhizopine catabolism attest to this symbiotic role as being the provision of a selective growth substrate to the free-living bacterial sister cells, as is the presumed case with *Agrobacterium* opines. We envisage the rhizopine cassette may well be important in agriculture by contributing to the persistence of beneficial bacteria in the soil.

One of the unique features of rhizopines is the fact that their synthetic and catabolic genes are linked but are expressed in different developmental stages of the bacterium. The synthetic genes are expressed in the bacteroid within the nodule and the catabolic genes function in the free-living bacteria. This enables the bacteria to take advantage of the input of photosynthates from the plant by sequestering these to produce a specialized growth substrate. By analogy with the opine concept, this process is now referred to as the rhizopine concept (see Figure 3).

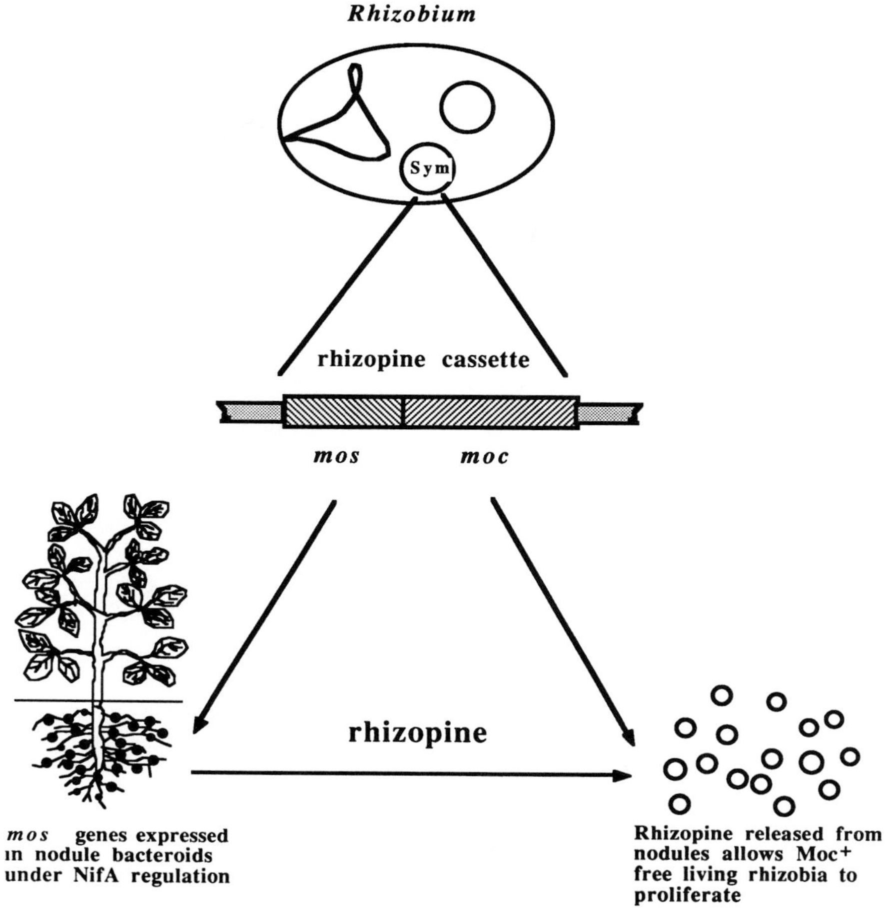


FIGURE 3. The rhizopine concept.

Our studies on the inositol class of rhizopines have indicated common function at both the physiological and genetic levels. Physiologically, specific rhizobia are able to induce the synthesis of, and catabolize, a selective growth substrate. Genetically, synthesis and catabolism of the substrate is controlled by closely linked genes on the Sym plasmid and synthesis is NifA regulated. Whether subsequent rhizopines can be likewise defined awaits further research. It is likely, just as the genetic colonization scheme described by Schell[8] explains the mechanism by which the opine concept[13] functions, that the *moc-mos* system may well prove a model for a unifying theory of rhizopine gene function.

## ACKNOWLEDGMENTS

We would like to thank D. Dowling and M. Wexler for critically reading this review. CPS is supported by a grant (UAD16) from the Australian Wool Corporation.

## REFERENCES

1. **Verma, D. P. S. and Long, S. R.,** The molecular biology of *Rhizobium*-legume symbiosis, *Int. Rev. Cytol. Suppl.,* 14, 211, 1983.
2. **Sprent, J. I.,** Which steps are essential for the formation of functional legume nodules?, *New Phytol.,* 111, 128, 1989.
3. **Stanier, R. Y., Adelberg, E. A., and Ingraham, J. L.,** *General Microbiology,* 4th ed., Macmillan, NY, 1976, 747.
4. **Beringer, J. E., Brewin, N., Johnston, A. W. B., Schulman, H. M., and Hopwood, D. A.,** The *Rhizobium*-legume symbiosis, *Proc. R. Soc. London Ser. B,* 204, 219, 1979.
5. **Tempé, J. Petit, A., and Farrand, S. K.,** Induction of cell proliferation by *Agrobacterium tumefaciens* and *A. rhizogenes:* A parasite's point of view, in *Genes Involved in Microbe-Plant Interactions, Plant Gene Research,* Vol. 1, Verma, D. P. S. and Höhn, T., Eds., Springer-Verlag, Berlin, 1984, 271.
6. **Gheysen, G., Dhaese, P., Van Montagu, M., and Schell, J.,** in *Genetic Flux in Plants,* Hohn, B. and Dennis, E. S., Eds., Springer-Verlag, Berlin, 1985, 11.
7. **Zambryski, P.,** this volume.
8. **Schell, J., Van Montagu, M., De Beuckeler, M., De Block, M., Depicker, A., De Wilde, M., Engler, G., Genetello, C., Hernalsteens, J. P., Holsters, M., Seurinck, J., Silva, B., Van Vliet, F., and Villarroel, R.,** Interactions and DNA transfer between *Agrobacterium tumefaciens,* the Ti-plasmid and the plant host, *Proc. R. Soc. London Ser. B,* 204, 251, 1979.
9. **Guyon, P., Chilton, M-D., Petit, A., and Tempé, J.,** Agropine in "null-type" crown gall tumors: evidence for generality of the opine concept, *Proc. Natl. Acad. Sci. U.S.A.,* 77, 2693, 1980.
10. **Petit, A., David, C., Dahl, G. A., Ellis, J. G., Guyon, P., Casse-Delbart, F., and Tempé, J.,** Further extension of the opine concept: plasmids in *Agrobacterium rhizogenes* cooperate for opine degradation, *Mol. Gen. Genet.,* 180, 204, 1983.
11. **De Ley, J., Bernaerts, M., Rassel, A., and Guilmont, J.,** Approach to an improved taxonomy of the genus, *Agrobacterium, J. Gen. Microbiol.,* 43, 7, 1960.
12. **Djordjevic, M. A., Gabriel, D. W., and Rolfe, B. G.,** *Rhizobium*—the refined parasite of legumes, *Annu. Rev. Phytopathol.,* 25, 145, 1987.
13. **Tempé, J. and Petit, A.,** Les Pisté des Opines, in *Molecular Genetics of the Bacteria Plant Interaction,* Pühler, A., Ed., Springer-Verlag, Berlin, 1983, 14.
14. **Murphy, P. J., Heycke, N., Banfalvi, Z., Tate, M. E. de Bruijn, F. J., Kondorosi, A., Tempé, J., and Schell, J.,** Genes for the catabolism and synthesis of an opine-like compounds in *Rhizobium meliloti* are closely linked and on the Sym plasmid, *Proc. Natl. Acad. Sci. U.S.A.,* 84, 493, 1987.
15. **Friedman, A. M., Long, S. R., Brown, S. E., Buikema, W. J., and Ausubel, F. M.,** Construction of a broad host range cosmid cloning vector and its use in the genetic analysis of *Rhizobium* mutants, *Gene,* 18, 289, 1982.
16. **Bell, C. R.,** Growth of *Agrobacterium tumefaciens* under octopine limitation in chemostats, *Appl. Environ. Microbiol.,* 56, 1775, 1990.
17. **Schardl, C. L. and Kado, C. I.,** A functional map of the nopaline catabolism genes on the Ti plasmid of *Agrobacterium tumefaciens* C58, *Mol. Gen. Genet.,* 191, 10, 1983.
18. **Saint, C. P. and Murphy, P. J.,** unpublished data.
19. **Ronson, C. W., Nixon, B. T., Albright, L. M., and Ausubel, F. M.,** *Rhizobium meliloti ntrA (rpoN)* gene is required for diverse metabolic functions, *J. Bacteriol.,* 169, 2424, 1987.
20. **Gussin, G. N., Ronson, C. W., and Ausubel, F. M.,** Regulation of nitrogen fixation genes, *Annu. Rev. Genet.,* 20, 567, 1986.
21. **Ratet, P., Schell, J., and de Bruijn, F. J.,** Mini-Mu *lac* transposons with broad-host-range origins of conjugal transfer and replication designed for gene regulation in Rhizobiaceae, *Gene,* 63, 41, 1988.
22. **Beringer, J. E., Johnston, A. W. B., and Kondorosi, A.,** Genetic maps of *Rhizobium meliloti* and *R. leguminosarum* biovars: *phaseoli, trifolii* and *viciae,* in *Genetic Maps,* Vol. 4, O'Brien, S. J., Ed., Cold Spring Harbor Laboratory, Cold Spring Harbor, NY, 1987, 245.
23. **Hynes, M. F., Simon, R., Müller, P., Niehaus, K., Labes, M., and Pühler, A.,** The two megaplasmids of *Rhizobium meliloti* are involved in the effective nodulation of alfalfa, *Mol. Gen. Genet.,* 202, 356, 1986.
24. **Charles, C. T. and Finan, T. M.,** Genetic map of *Rhizobium meliloti* megaplasmid pRmeSU47b, *J. Bacteriol.,* 172, 2469, 1990.
25. **Simon, R.,** High frequency mobilization of gram-negative bacterial replicons by the *in vitro* constructed Tn*5*-Mob transposon, *Mol. Gen. Genet.,* 196, 413, 1984.
26. **Ausubel, F. M.,** Regulation of nitrogen fixation genes, *Cell,* 37, 5, 1984.
27. **Ditta, G., Virts, E., Palomores, A., and Kim, C.-H.,** The *nifA* gene of *Rhizobium meliloti* is oxygen regulated, *J. Bacteriol.,* 168, 3217, 1987.

28. **Virts, E. L., Stanfield, S. W., Helinski, D. R., and Ditta, G. S.,** Common regulatory elements control symbiotic and microaerobic induction of *nifA* in *Rhizobium meliloti, Proc. Natl. Acad. Sci. U.S.A.,* 85, 3062, 1988.
29. **Ruvkun, G. B., Sundaresan, V., and Ausubel, F. M.,** Directed Transposon Tn*5* mutagenesis and complementation analysis of *Rhizobium meliloti* symbiotic nitrogen fixation genes, *Cell,* 29, 551, 1982.
30. **Corbin, D., Barran, L., and Ditta, G.,** Organization and expression of *Rhizobium meliloti* nitrogen fixation genes, *Proc. Natl. Acad. Sci. U.S.A.,* 80, 3005, 1983.
31. **Long, S. R.,** *Rhizobium* genetics, *Annu. Rev. Genet.,* 23, 483, 1989.
32. **Earl, C. D., Ronson, C. W., and Ausubel, F. M.,** Genetic and structural analysis of the *Rhizobium meliloti fixA, fixB, fixC* and *fixX* genes, *J. Bacteriol.,* 169, 1127, 1987.
33. **Better, M., Lewis, B., Corbin, D., Ditta, G., and Helinski, D. R.,** Structural relationships among *Rhizobium meliloti* symbiotic promoters, *Cell,* 35, 479, 1983.
34. **Beynon, J. L., Cannon, M. C., Buchanan-Wollaston, V., and Cannon, F. C.,** The *nif* promoters of *Klebsiella pneumoniae* have a characteristic primary structure, *Cell,* 34, 665, 1983.
35. **Ow, D. W., Sundaresan, S., Rothstein, D. M., Brown, S. E., and Ausubel, F. M.,** Promoters regulated by the *glnG (ntrC)* and *nifA* gene products share a heptameric consensus sequence in the − 15 region, *Proc. Natl. Acad. Sci. U.S.A.,* 80, 2524, 1983.
36. **Buck, M., Miller, S., Drummond, M., and Dixon, R.,** Upstream activator sequences are present in the promoters of nitrogen fixation genes, *Nature (London),* 320, 374, 1986.
37. **Better, M., Ditta, G., and Helinski, D. R.,** Deletion analysis of *Rhizobium meliloti* symbiotic promoters, *EMBO J.,* 4, 2419, 1985.
38. **Magasanik, B.,** Genetic control of nitrogen assimilation in bacteria, *Annu. Rev. Genet.,* 16, 135, 1982.
39. **Murphy, P. J., Heycke, N., Trenz, S., Ratet, P., de Bruijn, F. J., and Schell, J.,** Synthesis of an opine-like compound—a rhizopine—in alfalfa nodules is symbiotically regulated, *Proc. Natl. Acad. Sci. U.S.A.,* 85, 9133, 1988.
40. **Szeto, W. W., Zimmerman, J. L., Sundaresan, V., and Ausubel, F. M.,** A *Rhizobium meliloti* symbiotic regulatory gene, *Cell,* 36, 1035, 1984.
41. **Szeto, W. W., Nixon, B. T., Ronson, C. W., and Ausubel, F. M.,** Identification and characterization of the *Rhizobium meliloti ntrC* gene: *R. meliloti* has separate regulatory pathways for activation of nitrogen fixation genes in free-living and symbiotic cells, *J. Bacteriol.,* 169, 1423, 1987.
42. **Hawkins, F. K. L. and Johnston, A. W. B.,** Transcription of a *Rhizobium leguminosarum* biovar *phaseoli* gene needed for melanin synthesis is activated by *nifA* of *Rhizobium* and *Klebsiella pneumoniae, Mol. Microbiol.,* 2, 331, 1988.
43. **San Juan, J. and Olivares, J.,** Implication of *nifA* in regulation of genes located on a *Rhizobium meliloti* cryptic plasmid that affect nodulation efficiency, *J. Bacteriol.,* 171, 4154, 1989.
44. **Rostas, K., Kondorosi, E., Horvath, B., Simoncsits, A., and Kondorosi, A.,** Conservation of extended promoter regions of nodulation genes in *Rhizobium, Proc. Natl. Acad. Sci. U.S.A.,* 83, 1757, 1986.
45. **Trenz, S. and Murphy, P. J.,** unpublished data.
46. **Hageman, R. V. and Burris, R. H.,** Nitrogenase and nitrogenase reductase associate and dissociate with each catalytic cycle, *Proc. Natl. Acad. Sci. U.S.A.,* 75, 2699, 1978.
47. **Horvath, B., Kondorosi, E., John, M., Schmidt, J., Torok, I., Györgypal, Z., Barabas, I., Wieneke, U., Schell, J., and Kondorosi, A.,** Organization, structure and symbiotic function of *Rhizobium meliloti* nodulation genes determining host specificity for alfalfa, *Cell,* 46, 335, 1986.
48. **Shine, J. and Dalgarno, L.,** The 3′-terminal sequence of *Escherichia coli* 16 S ribosomal RNA: complementarity to nonsense triplets and ribosome binding sites, *Proc. Natl. Acad. Sci. U.S.A.,* 71, 1342, 1974.
49. **Richaud, F., Richaud, C., Ratet, P., and Putte, J-C.,** Chromosomal location and nucleotide sequence of the *Escherichia coli dapA* gene, *J. Bacteriol.,* 166, 297, 1986.
50. **Murphy, P. J.,** unpublished data.
51. **Shedlarski, J. G. and Gilvarg, L.,** The pyruvate-aspartic semialdehyde condensing enzyme of *Escherichia coli, J. Biol. Chem.,* 245, 1362, 1970.
52. **Tempé, J.,** unpublished data.
53. **Saint, C. P., Wexler, M., Murphy, P. J., and Murphy, P. J.,** Characterization of genes for synthesis and catabolism of a novel rhizopine induced in nodules by *Rhizobium meliloti* Rm220-3: extension of the rhizopine concept, *J. Bacteriol.,* submitted, 1991.
54. **Scott, D. B., Wilson, R., Shaw, G. J., Petit, A., and Tempé, J.,** Biosynthesis and degradation of nodule specific *Rhizobium loti* compounds in *Lotus* nodules, *J. Bacteriol.,* 169, 278, 1987.
55. **Shaw, G. J., Wilson, R. D., Lane, G. A., Kennedy, L. D., Scott, D. B., and Gainsford, G. J.,** Structure of rhizolotine, a novel opine-like metabolite from *Lotus tenuis* nodules, *J. Chem. Soc. Chem. Commun.,* 180, 1986.
56. **Streeter, J. G.,** Identification and distribution of ononitol in nodules of *Pisum sativum* and *Glycine max, Phytochemistry,* 24, 174, 1985.
57. **Rossignol, G. and Dion, P.,** Octopine, nopaline and octopinic acid utilization in *Pseudomonas, Can. J. Microbiol.,* 31, 68, 1985.

58. **Tremblay, G., Gagliardo, R., Chilton, W. S., and Dion, P.,** Diversity among opine utilizing bacteria: identification of coryneform isolates, *Appl. Environ. Microbiol.*, 53, 1519, 1987.
59. **Heycke, N. and Murphy, P. J.,** unpublished data.
60. **Triplett, E. W.,** The molecular genetics of nodulation competitiveness in *Rhizobium* and *Bradyrhizobium*, *Mol. Plant-Microbe Interact.*, 3, 199, 1990.
61. **Putnoky, P., Kondorosi, A., and Murphy, P. J.,** unpublished data.
62. **Johnston, A. W. B. and Beringer, J. E.,** Mixed inoculations with effective and ineffective strains of *Rhizobium leguminosarum*, *J. Appl. Bacteriol.*, 40, 375, 1976.
63. **Pretorius-Guth, I.-G., Pühler, A., and Simon, R.,** Conjugal transfer of megaplasmid 2 between *Rhizobium meliloti* strains in alfalfa nodules, *Appl. Environ. Microbiol.*, 56, 2354, 1990.
64. **Dowling, D. N. and Broughton, W. J.,** Competition for nodulation of legumes, *Annu. Rev. Microbiol.*, 40, 131, 1986.
65. **Tepfer, D., Goldmann, A., Fleury, V., Maille, M., Message, B., Pamboukdjian, N., Boivin, C., Dénarié, J., Rosenberg, C., Lallemand, J. Y., Descoins, C., Charpin, I., and Amarger, N.,** Calystegins, nutritional mediators in plant-microbe interactions, in *Molecular Genetics of Plant-Microbe Interaction,* Palacios, R. and Verma, D. P. S., Eds., APS Press, MN, 139, 1988.
66. **Goldmann, A., Boivin, C., Fleury, V., Message, B., Lecoeur, L., Maille, M., and Tepfer, D.,** Catabolism of secondary metabolites from legumes by *Rhizobium meliloti:* genes essential for betaine utilization are located on pSym in the symbiotic region, in *Abstr. Proc. 5th Int. Symp. Mol. Genet. Plant-Microbe Interact., Interlaken, Switzerland,* p. 98, 1990.
67. **Boivin, C., Barran, L. R., Malpica, C. A., Camut, S., Truchet, G., and Rosenberg, C.,** *Rhizobium meliloti* genes encoding catabolism of trigonelline, a legume secondary metabolite, are induced under symbiotic conditions, in *Abstr. Proc. 5th Int. Symp. Mol. Genet. Plant-Microbe Interact., Interlaken, Switzerland, p. 99, 1990.*

# *Section IV: Interaction of Other Microorganisms with Plants*

# *ARABIDOPSIS THALIANA* AS A MODEL HOST FOR STUDYING PLANT-PATHOGEN INTERACTIONS

**Keith R. Davis**

## TABLE OF CONTENTS

# I. INTRODUCTION

Although plants are continuously in contact with a variety of potentially pathogenic microorganisms, successful infections are rare. Studies by many workers using biochemical and molecular approaches have demonstrated that a number of diverse responses are induced during a resistance reaction that are not induced, or are induced less rapidly, in a susceptible response.[1-5] However, since the overall response of the plant to microbial attack is very complex, it has been difficult to clearly demonstrate by these correlative approaches that a particular response is required for resistance. Moreover, although significant progress has been made in identifying specific genes that are activated during attempted infection, little is known about the mechanisms involved in the activation and coordinated regulation of these putative defense genes.

Studies using biochemical and molecular approaches combined with genetic analyses of mutant plants with altered disease resistance would provide a very powerful method for dissecting the complex pattern of gene expression that is associated with plant disease resistance. However, the majority of plants that have been used to date for molecular studies of disease resistance are not amenable to genetic or molecular genetic approaches. Recent efforts by several groups have established that *Arabidopsis thaliana* is an excellent host for several bacterial, fungal, and viral phytopathogens. These initial studies clearly indicate that *A. thaliana* offers tremendous potential as a model host suitable for the application of genetic approaches to address fundamental questions concerning the molecular basis of plant disease resistance. This chapter summarizes current knowledge about pathogens of *A. thaliana* and describes several genetic approaches for studying the regulation of plant defense responses.

# II. ADVANTAGES OF *ARABIDOPSIS THALIANA* AS A HOST PLANT

*Arabidopsis thaliana* is a small crucifer which offers many experimental advantages for conducting classical and molecular genetic studies.[6-8] It is a true diploid, has a rapid generation time (6 to 8 weeks), is self-compatible, and produces large numbers of small seeds (as much as 10,000 per plant). It has one of the smallest genomes known in a higher plant (approximately 70 Mb/haploid nucleus),[9,10] and can easily be genetically engineered using standard *Agrobacterium tumefaciens*-mediated transformation protocols.[11] Additional advantages include the ability to grow large numbers of plants in a small area, either in soil or on sterile medium, and the availability of independent isolates of *A. thaliana* (land races or ecotypes),[12] which can be screened for natural variability in resistance to infection by phytopathogens. A critical advantage that *A. thaliana* offers is the excellent genetic tools available for mutant analysis. A number of groups have constructed restriction fragment length polymorphism (RFLP) and physical maps of the *A. thaliana* genome and have correlated these maps with the previously developed genetic map based on morphologic mutants.[13-15] These tools will allow the cloning of specific genes which have been identified by mutation via chromosome walking methods.[8,16,17] In addition, recent work has demonstrated the feasibility of cloning genes that have been mutated by T-DNA insertion,[18-20] and using genomic subtraction to directly clone genes identified by deletion mutations.[21,22]

Genetic approaches have been used very successfully to dissect complex biosynthetic and regulatory pathways in a number of prokaryotic and eukaryotic organisms. Recent studies concerning the biosynthesis of lipids,[23] amino acids,[24] and secondary metabolites such as glucosinolates,[25] and flavonoid derivatives,[26] demonstrate that *A. thaliana* is amenable to the types of molecular and biochemical genetic methods that would be ideal for investigating the complex phenotype of disease resistance.

**TABLE 1**
**Virulence of *Pseudomonas* Species on *A. thaliana*[a]**

| *Pseudomonas* species[b] | Strain | Virulence on *A. thaliana*[c] | Other hosts |
|---|---|---|---|
| *P. s.* pv. *maculicola* | 4326 | + + + | *Brassica,* tomato, radish |
| *P. s.* pv. *maculicola* | 795 | + | *Brassica,* tomato |
| *P. s.* pv. *maculicola* | 2744 | + + | *Brassica,* tomato |
| *P. s.* pv. *maculicola* | 4981 | + + | *Brassica,* tomato |
| *P. s.* pv. *tomato* | DC3000 | + + + | *Brassica,* tomato |
| *P. s.* pv. *tomato* | 3435 | + + + | tomato |
| *P. s.* pv. *tomato* | 3455 | + + + | tomato |
| *P. s.* pv. *tomato* | 5034 | + + + | *Brassica,* tomato |
| *P. s.* pv. *tomato* | 1065 | — | *Brassica,* tomato |
| *P. s.* pv. *tomato* | T1 | — | *Brassica,* tomato |
| *P. cichorii* | 83-1 | — | celery, lettuce |

[a] Compiled from data in ref. 30-32.
[b] *P. s., Pseudomonas syringae.*
[c] Virulence on land race Columbia.

# III. PATHOGENS OF *ARABIDOPSIS THALIANA*

Since *A. thaliana* is of no agronomic importance, very little had been published on its infection by phytopathogens prior to the recent efforts summarized in this chapter. Earlier reports, summarized by Koch and Slusarenko,[27] indicated that *A. thaliana* could be infected by the fungal pathogens *Peronospora parasitica, Puccinia thlaspeos, Albugo candida,* and *Plasmodiophora brassicae. Sclerotinia* has also been reported to infect natural populations of *A. thaliana.*[28] More recently, with the renewed interest in *A. thaliana* as a model system suitable for molecular genetic studies, a few groups initiated systematic screens for phytopathogens that would infect *A. thaliana.* Most of these studies utilized pathogens that were known to infect other crucifers, and thus, would be likely to infect *A. thaliana.* The following sections describe some of the progress that has been made in developing *A. thaliana* infection systems using bacterial, fungal, and viral pathogens.

## A. BACTERIAL PATHOGENS OF *ARABIDOPSIS THALIANA*

### 1. *Pseudomonas* Species

Several groups have established infection of *A. thaliana* with *Pseudomonas* species as a model system for investigating resistance to bacterial phytopathogens.[29-32] One of the more promising systems is the infection of *A. thaliana* with *P. syringae* pv. *tomato* and the closely related *P. syringae* pv. *maculicola,* the causal organisms of bacterial speck disease on tomato and *Brassica* species. A number of *P. syringae* pv. *tomato* and *P. syringae* pv. *maculicola* strains that are either virulent or avirulent when infiltrated into leaves of *A. thaliana* have been identified and characterized in some detail (Table 1). The virulent *P. syringae* pv. *maculicola* and *P. syringae* pv. *tomato* strains cause water-soaked lesions that are often associated with chlorosis of the surrounding tissue. In some cases, these lesions develop necrotic centers (Plate 1*). Disease symptoms can be routinely obtained at doses of $10^4$ bacteria/ml (approximately $10^3$ bacteria per leaf) and become visible 2 to 3 days after infiltration. These symptoms are similar to those observed when these strains are inoculated on other susceptible plant hosts such as radish and tomato. The development of disease symptoms in leaves inoculated with virulent strains of *P. syringae* pv. *maculicola* and *P. syringae* pv. *tomato* is correlated with significant bacterial growth (Figure 1). The virulent

* Plate 1 follows page 402.

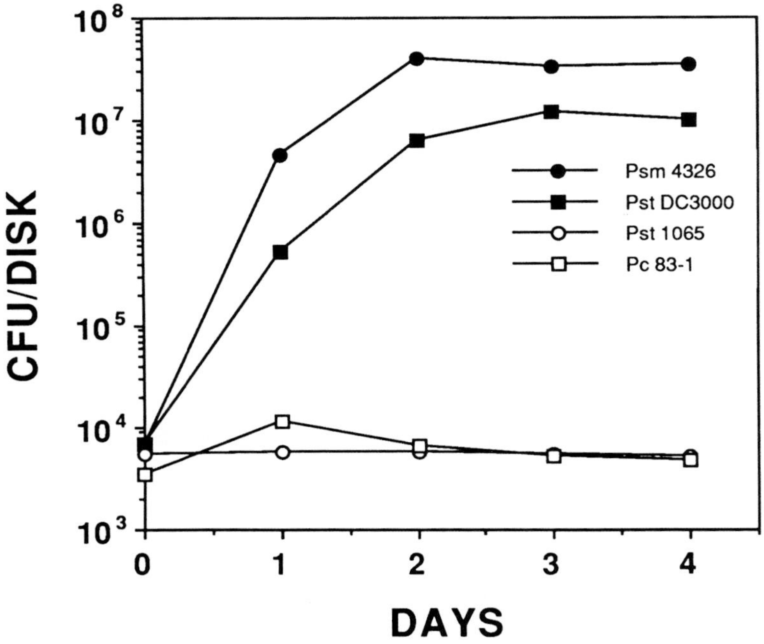


FIGURE 1. *In planta* growth of *Pseudomonas* strains in *A. thaliana*. Bacteria were infiltrated into leaves of *A. thaliana* (land race Columbia) at a dose of $10^5$ CFU/ml. Bacterial growth was measured by plating serial dilutions of homogenates prepared from disks isolated from infiltrated leaves. The data presented are the averages obtained from duplicate disks in a representative experiment. *Pst, P. syringae* pv. *tomato; Psm, P. syringae* pv. *maculicola; Pc, P. cichorii.*

strains *P. syringae* pv. *maculicola* 4326, *P. syringae* pv. *tomato* 5034, and *P. syringae* pv. *tomato* DC3000 multiply $10^3$ to $10^4$-fold during the first several days after infection. In contrast, avirulent strains either caused no visible symptoms, or small necrotic lesions typical of a mild hypersensitive response (Plate 1) and did not multiply significantly *in planta* (Figure 1).[30-32]

Previous work with phytopathogenic pseudomonads has demonstrated that, in some cases, the interaction of specific strains of bacteria with specific cultivars of a host plant are governed by single, dominant avirulence (*avr*) genes in the pathogen and single, dominant resistance genes in the host.[33,34] The bacterial genes are specified as *avr* genes since their presence in a strain will make that strain avirulent when inoculated on host cultivars that contain the corresponding resistance gene. One interpretation of these genetic data is that the *avr* gene product (or some signal molecule produced by the *avr* gene product) is recognized by the gene product of the resistance gene. This recognition event causes the timely and efficient induction of the host defense responses, resulting in a restriction of bacterial growth.

Further studies of the infection of *A. thaliana* with *P. syringae* pv. *tomato* strains indicate that this interaction may also be modulated by single *avr* genes in the bacteria and single resistance genes in the host. Several putative *avr* genes have been isolated from the avirulent *P. syringae* pv. *tomato* strains 1065 and T1.[30,31,35,36] When these *avr* genes are introduced into virulent *P. syringae* pv. *tomato* or *P. syringae* pv. *maculicola* strains, an attenuation of disease symptoms is observed concomitant with decreased multiplication of the bacteria. In complementary studies, evidence for the presence of resistance genes in *A. thaliana* was obtained by screening a number of the available land races of *A. thaliana* for resistance to several of the *P. syringae* pv. *maculicola* and *P. syringae* pv. *tomato* strains which are virulent on land race Columbia. Although no land races were identified that were resistant to the *P. syringae* pv. *maculicola* strains tested,[31,37] several land races resistant to *P. syringae* pv. *tomato* strains have been identified.[31] Similar studies have demonstrated that several *A. thaliana* land races exhibit differential responses to specific cloned *avr* genes.[31,36] In these studies, land races were inoculated with virulent *P. syringae* pv. *tomato* or *P. syringae* pv. *maculicola* strains containing cloned *avr* genes which were known to induce resistance in land race Columbia. Several land races susceptible to these strains were identified, suggesting that these land races do not contain a corresponding resistance gene. Genetic crosses between resistant and susceptible land races indicate that, in some cases, resistance is conferred by the presence of a single, dominant gene.[35] These results suggest that gene-for-gene interactions[38] occur in some *A. thaliana-Pseudomonas* combinations and that this system may be useful for further defining the molecular basis of "race-specific" interactions.

An interaction between *A. thaliana* and a phytopathogenic pseudomonad which results in a hypersensitive response[39] has been described by Davis et al.[29,32] Infection of *A. thaliana* with the avirulent *P. cichorii* strain 83-1 causes rapid tissue collapse (12 to 24 h) after infiltration at doses of $10^6$ bacteria/ml ($10^5$ bacteria per leaf) or higher. These lesions are limited to the infiltrated area and are not associated with chlorosis (Plate 1). The development of the hypersensitive response is correlated with a two to ten-fold increase in bacterial populations during the first 2 days after infiltration (Figure 1). These responses were also observed when *P. cichorii* 83-1 was infiltrated into leaves of tobacco and radish.[40] Further evidence that *P. cichorii* induces a hypersensitive response has been obtained using cultured *Arabidopsis* cells. Previous studies with cultured tobacco cells demonstrated that *Pseudomonas* strains that induce a hypersensitive response in intact plant tissue cause the activation of $K^+/H^+$ exchange across the plasma membrane, resulting in an increase of the pH of the culture medium.[41] Davis et al.[32] have shown that *P. cichorii* 83-1 causes a similar increase in the pH of cultured medium when added to cultured *Arabidopsis* cells. The interaction of *P. cichorii* with *A. thaliana* is most likely an example of nonhost resistance.[34] This is supported by the observations that screens of a number of *A. thaliana* land races with several different *P. cichorii* strains did not identify a strain virulent on any land race and that attempts to clone an *avr* gene from *P. cichorii* 83-1 have been unsuccessful.[37]

## 2. *Xanthomonas campestris* pv. *campestris*

*Xanthomonas campestris* pv. *campestris* is the most economically important pathogen of *Brassica* sp. and causes black rot disease. Several groups have established that many strains of *X. campestris* pv. *campestris* are virulent on *A. thaliana* when inoculated either by leaf infiltration or more "natural" methods of inoculation such as misting leaves with inoculum.[32,42-45] Virulent strains of *X. campestris* pv. *campestris* infiltrated into leaves of *A. thaliana* cause chlorotic lesions which develop within 2 to 3 days after infection. In many cases, these lesions develop necrotic centers which can spread over the entire infected leaf. Symptom development is associated with a $10^2$ to $10^4$-fold increase in bacterial population. These symptoms are not typical of the black rot disease, and it has been suggested that this method of infection is not suitable for studies with *X. campestris* pv. *campestris*.[46] Tsuji and Somerville[42] and Simpson and Johnson[43] have demonstrated that inoculation of *A. thaliana* with *X. campestris* pv. *campestris* by wounding or spray methods results in typical black rot symptoms and thus may be more appropriate methods of inoculation.

Preliminary studies have suggested that the interaction of *A. thaliana* with *X. campestris* pv. *campestris* may also be modulated by specific genes present in the bacterium and in the host plant. Simpson and Johnson[43] reported that certain land races of *A. thaliana* are significantly more resistant than others to infection by specific strains of *X. campestris* pv. *campestris* as measured by symptom expression and bacterial growth *in planta*. Further genetic studies will be required to define the number of host genes conferring this resistance phenotype. Similar results have been described by Tsuji et al.[44] These investigators observed that land race Pr-0 appeared more susceptible to infection by *X. campestris* pv. *campestris* when compared to land race Columbia. Reciprocal crosses of Pr-0 and Columbia revealed that resistance phenotype was conferred by a single, dominant nuclear gene. In addition, Daniels and co-workers have found an *X. campestris* pv. *campestris* strain that is avirulent on land race Columbia. A putative *avr* gene has been cloned from this strain and preliminary genetic studies indicate that this *avr* gene may interact with a resistance gene present in *A. thaliana* land race Columbia.[45]

## B. FUNGAL PATHOGENS OF *ARABIDOPSIS THALIANA*

Recent work on fungal pathogens has extended earlier observations that *Peronospora parasitica,* the causal agent of downy mildew on crucifers, is virulent on *A. thaliana*. Koch and Slusarenko[47] collected field isolates of *A. thaliana* infected with *P. parasitica* and conducted a detailed laboratory analysis of the infection process. It was demonstrated that disease development in *A. thaliana* was very similar to that of *P. parasitica* infection of *Brassica* species. Of particular significance, these authors identified a land race of *A. thaliana* which was resistant to this particular isolate of *P. parasitica*. More recent genetic crosses between the susceptible and resistant land races indicate that the resistant phenotype is dominant.[48] Further studies are required to determine if this resistance is due to a single "resistance" gene. A current limitation of using *P. parasitica* for studies of plant-fungal interactions is that this fungus is an obligate biotroph and thus cannot be grown in defined medium. Also, the sexual cycle of *P. parasitica* is not well characterized and methods for conducting genetic crosses or for introducing cloned genes have not been developed. However, this system does offer some promise for identifying host genes that are involved in resistance to fungal pathogens, both by identifying natural variability in resistance in different land races and by isolation of mutants of *A. thaliana* altered in their resistance phenotype.

Other studies by Koch and Slusarenko have demonstrated that another biotrophic fungus, *Erysiphe cruciferarum* (powdery mildew of crucifers), and the broad host-range phytopathogenic fungi *Rhizoctonia solani* and *Botrytis cinerea* are also pathogens of *A. thaliana*.[27] These initial studies indicate that the infection of *A. thaliana* by these fungi is similar to that observed in infections of other host plants, and thus, represent true compatible interactions. Experiments are currently underway to determine if there is any natural variability in the susceptibility of various land races of *A. thaliana* to infection with these fungi. Other fungal pathogens of *A. thaliana* that have recently been described include *Leptosphaeria maculans*,[49] *Plasmodiophora brassicae*,[50] and *Pythium* species.[48] Further studies of the interaction of these fungi with *A. thaliana* will be required to determine the usefulness of these systems.

## C. VIRAL PATHOGENS OF *ARABIDOPSIS THALIANA*

### 1. Cauliflower Mosaic Virus (CaMV)

Cauliflower mosaic virus (CaMV), a dsDNA caulimovirus that infects a number of crucifers,[51] was one of the first viral pathogens of *A. thaliana* described. In the initial studies by Balàz and Lebeurier,[52] it was noted that two different CaMV isolates were virulent on five different land races of *A. thaliana*. Symptoms included the development of yellow spots on infected leaves approximately 1 week after inoculation, followed a week later by a systemic infection characterized by the development of mosaic symptoms and stunting. Slight differences in the severity of the symptoms expressed in the different land races were observed, however, all plants were clearly susceptible. Inclusion bodies typical of CaMV infection were observed in systemically infected leaves and infectious virus was isolated from inoculated plants.

Infection of *A. thaliana* by CaMV was later confirmed by Melcher.[53] In these studies, systemically infected leaves of land race Columbia developed chlorotic spots, chlorotic mosaic patterns, and, in some cases, complete chlorosis. Other symptoms included vein clearing and the development of necrotic lesions that, occasionally, led to the death of the infected plant. Symptoms observed on stem and flower structures included chlorosis, vein clearing, and stunting. Differences were noted in the pattern of symptoms caused by different CaMV isolates, however, all isolates tested were virulent. It remains to be determined if other land races of *A. thaliana* are resistant to any of the isolates shown to be virulent on land race Columbia.

### 2. Turnip Crinkle Virus (TCV)

Turnip crinkle virus (TCV) is a ssRNA virus that contains a single component genome in association with various satellite RNAs.[54,55] TCV is known to infect several *Brassica* species and is transmitted by beetle vectors.[56] Recent studies by Simon and co-workers[57] have demonstrated that TCV infects *A. thaliana.* Typical symptoms, which include stunting and the eventual browning of leaves, become evident 1 week after infection. The strain TCV-M, a TCV isolate containing the sat-RNA C,[55] is significantly more virulent than strains which do not contain this satellite RNA (Plate 2*). TCV-M was found to infect most land races of *A. thaliana,* however, one land race was found to be resistant. The resistant land race does not develop symptoms and there is little or no virus accumulation. Preliminary genetic analysis of crosses between susceptible and resistant land races indicate that resistance is due to a single gene and efforts are currently underway to map and clone this putative resistance locus.[57]

### 3. Beet Curly Top Virus (BCTV)

Beet curly top virus (BCTV) is a leafhopper-transmitted geminivirus that contains a single-component ssDNA genome.[58] BCTV has a broad host range in comparison to other geminiviruses and is known to infect a number of crucifers.[59] Infectious DNA clones of BCTV have been constructed and *Agrobacterium*-mediated inoculation methods have been developed that allow efficient infection of plants without the insect vector.[60] Recent studies have demonstrated that several strains of BCTV are virulent on *A. thaliana.*[61] Infected plants are stunted and develop abnormal floral structures and, in some cases, the leaves become chlorotic and crinkled. Differences in symptom expression in different land races of *A. thaliana* caused by specific BCTV strains have been noted (Plate 2). Current studies are focused on determining whether land races of *A. thaliana* exhibit differential sensitivity to several BCTV strains.

## IV. INDUCTION OF DEFENSE GENES IN *ARABIDOPSIS THALIANA*

The results of a large number of studies indicate that a number of diverse responses are induced during a resistance reaction that are not induced, or are induced less rapidly, in a susceptible response. Biochemical responses often associated with resistance include the synthesis of polyphenolic lignins and hydroxyproline-rich glycoproteins (HRGP), the synthesis of antimicrobial hydrolases, such as β-1,3-glucanase and chitinase; and the synthesis and accumulation of low-molecular weight antimicrobial compounds called phytoalexins.[1-5] Detailed analyses of the expression of specific genes involved in the elaboration of these putative defense responses have revealed that there are groups of coordinately expressed genes that, for the most part, are regulated at the transcriptional level.[3-5]

Initial studies of putative defense genes in *A. thaliana* utilized a simplified system based

* Plate 2 follows page 402.

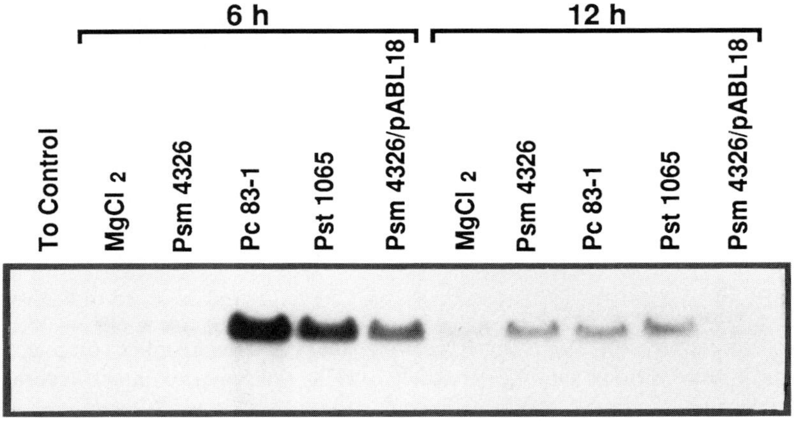


FIGURE 2. Induction of PAL mRNA in *A. thaliana* leaves infiltrated with avirulent and virulent *Pseudomonas* strains. Leaves of 4 week old *A. thaliana* plants (land race Columbia) were infiltrated with the *Pseudomonas* strains at a dose of $10^5$ bacteria per leaf. Leaves were harvested at 6 or 12 h after infiltration. Total RNA was isolated and 10 μg samples analyzed by standard RNA blot hybridization methods using a fragment derived from a genomic *A. thaliana* PAL clone. $T_0$ represents RNA from untreated leaves; $MgCl_2$ represents RNA from leaves infiltrated with 10 m*M* $MgCl_2$, which is the solution used to suspend the bacteria prior to infiltration; *Psm, P.s.* pv. *maculicola; Pst, P.s.* pv. *tomato; Pc, P. cichorii;* pABL18 is a pLAFR3 derivative containing a cloned *avr* gene from *P.s.* pv. *tomato* 1065. The size of the hybridizing mRNA is approximately 2.8 kb, the expected size for a PAL mRNA.

on the induction of defense responses in cultured *Arabidopsis* cells treated with elicitors.[62] These studies demonstrated that the bacterial elicitor, α-1,4-endopolygalacturonic acid lyase (PGA lyase), induced increased activities of several enzymes involved in phenylpropanoid biosynthesis including phenylalanine ammonia-lyase (PAL), 4-coumarate:CoA ligase (4CL), caffeic acid *O*-methyltransferase (CMT), and peroxidase (PO). PAL and 4CL enzyme activities were transiently induced, with maximum induction occurring approximately 8 to 10 h after elicitor treatment. CMT and PO were induced more slowly, reaching maximum levels approximately 24 h after elicitor treatment. The transient increase in PAL and 4CL enzyme activities was preceded by transient increases of PAL and 4CL mRNA levels which reached maximum levels approximately 3 h after elicitor treatment. Studies of elicitor-treated *Arabidopsis* cell cultures demonstrated that PGA lyase also induced the accumulation of β-1,3-glucanase mRNA.[62]

Further studies demonstrated that similar patterns of gene expression were observed in *A. thaliana* leaves infiltrated with phytopathogenic bacteria. Davis et al.[32] and Dong et al.[30] demonstrated that the induction of PAL mRNA was more rapid and reached significantly higher levels in leaves infiltrated with avirulent *Pseudomonas* strains when compared to leaves infiltrated with virulent strains. PAL mRNA was transiently induced 15- to 30-fold in leaves infiltrated with avirulent bacterial strains with maximum levels observed approximately 6 h after infiltration. In contrast, leaves infiltrated with virulent strains exhibited only a two- to tenfold induction of PAL mRNA which reached maximum levels approximately 12 h after infiltration. An important result obtained during these studies was that the introduction of a single *avr* gene from the avirulent strain into the virulent strain caused the transconjugant to induce PAL mRNA in a manner very similar to that observed in leaves infiltrated with the avirulent strain (Figure 2).[30,63] These results are consistent with previous observations indicating that PAL has a role in disease resistance.[64-66] Concurrent studies demonstrated that the expression of β-1,3-glucanase mRNA was distinct from that observed for PAL mRNA.[30] β-1,3-glucanase mRNA was induced to higher levels in leaves infiltrated with virulent *Pseudomonas* strains when compared to avirulent strains. Interestingly, the presence of an *avr* gene in the virulent strain suppressed the induction of β-1,3-glucanase mRNA.

Studies of the organization of PAL and β-1,3-glucanase genes in the *A. thaliana* genome have demonstrated that both of these defense genes are encoded by small multigene families. The *A. thaliana* PAL gene family is composed of three to four members, only one of which has been characterized in any detail.[29,32,67] It remains to be determined how similar these different PAL genes are in primary structure and their patterns of expression during a defense response. Three distinct β-1,3-glucanase genes have been isolated from *A. thaliana*.[30] These three genes were found to be present within a 12-kb region and have 65 to 80% nucleotide sequence similarity with each other. Studies utilizing gene-specific probes indicated that all three of these β-1,3-glucanase genes exhibit similar patterns of expression in leaves infiltrated with *Pseudomonas* strains.[30]

Genes encoding another putative defense protein, chitinase, have also been isolated from *A. thaliana*.[68] Genes for both the acidic and basic forms of chitinase were obtained and it appears that these are single-copy genes in *A. thaliana*. The *A. thaliana* basic chitinase was found to have 73% amino acid sequence similarity with the corresponding tobacco chitinase while the acidic *A. thaliana* chitinase had 60% amino acid sequence similarity with an acidic chitinase from cucumber. No data were presented concerning the expression of the chitinase genes in response to infection with phytopathogens, however, it was shown that these genes were differentially regulated during development and in response to ethylene and salicylate.

# V. GENETIC APPROACHES FOR STUDYING DISEASE RESISTANCE

The major advantage that *A. thaliana* offers as a model system for studying plant disease resistance is the opportunity to use a combination of biochemical, molecular, and genetic approaches to identify specific plant genes involved in the elaboration of a defense response. Genetic studies will be particularly useful for identifying genes whose products play an important role in the regulation of a defense response without dramatic quantitative changes in their amounts and/or activities. Examples of such constitutively expressed components would be receptors involved in microbial recognition and components of specific second messenger systems involved in transducing external signals to the genome. Genetic approaches will also be useful for identifying defense genes encoding proteins that may be extremely difficult to identify using biochemical approaches due to their low abundance or instability. The following sections highlight two complementary genetic approaches to define specific genes and regulatory mechanisms involved in plant disease resistance.

## A. TRADITIONAL GENETIC APPROACHES

A more traditional genetic approach is to screen a mutagenized population of *A. thaliana* for mutants that are altered in their response to infection with either virulent or avirulent phytopathogens. The two major classes of mutants that may be identified in this type of screening are: (1) mutants that become susceptible to an avirulent strain, and (2) mutants that become more resistant to a virulent strain. Mutants in the first class are likely to include plants with defects in the recognition of signal molecules involved in microbial recognition or in components of signal transduction pathways. Mutants in this class may also include plants defective for the biosynthesis of defense compounds. Mutants in the second class may include plants in which defense responses have become constitutively expressed or that have gained the ability to recognize and respond to a normally pathogenic strain.

A potential difficulty with this approach may be that it is not possible to obtain a single mutation which would cause a phenotypic change in the inoculated plants. This might occur if the avirulent strain used was not capable of causing disease symptoms even in the absence of host recognition or defense reactions. Also, it may be difficult to identify mutant plants that have become resistant to a virulent strain if the pathogen used was "hypervirulent" due to the production of large amounts of a toxin or other virulence factors. These problems are most easily overcome by careful analysis of the strains being used so that it is clear that avirulent strains are pathogenic on other related plant hosts (and thus have the genetic ability to infect plant tissues) and virulent strains are shown to be avirulent on other hosts (thus providing some evidence that they are not "hypervirulent"). The ideal situation would be to use strains that are pathogenic on some land races of *A. thaliana* and avirulent on others. Bacterial strains exhibiting differential virulence characteristics have been identified, and moreover, nearly isogenic strains that differ only by the presence of a single *avr* gene have been constructed.[30,31] These strains will be invaluable for isolating mutants which are specifically altered in defense gene regulation.

### B. MOLECULAR GENETIC APPROACHES

Another genetic approach for characterizing the regulation of defense responses that is currently being developed is a molecular-genetic strategy for identifying mutants altered in the regulation of specific genes thought to be involved in disease resistance (Figure 3). This strategy involves constructing transgenic *A. thaliana* plants containing fusions of defense-related gene promoters to an easily screenable reporter gene, β-glucuronidase (GUS).[69] These fusions are then introduced back into wild-type plants and the resulting transformants screened for the appropriate expression of the fusion. The seeds from this line are used to generate a large number of seeds which can be mutagenized. The M2 progeny of the mutagenized population are then screened for aberrent expression of the fusion. Two major classes of mutants in *trans*-acting loci are expected from this screen; either constitutive high level expression of the fusion or loss of inducibility by an appropriate signal. The first class should include mutations in factors involved in repressing gene expression while the second class should include mutations in positive regulatory factors.

This approach should provide a very powerful method for analyzing complex processes such as disease resistance since it allows one to identify mutants that are altered in the expression of specific genes without requiring a visible phenotypic change. By constructing gene fusions containing promoters from several defense genes that exhibit distinct patterns of expression during a resistance response, it will be possible to identify mutations that will allow the characterization of the mechanisms involved in the coordinated regulation of genes involved in disease resistance. Through the analysis of mutants altered in expression of specific promoters, it will be possible to determine if different defense genes are regulated by common mechanisms and whether there are factors specific for each gene. Once mutations in these regulatory factors are identified, it will be possible to identify other components of the regulatory pathway by screening for mutations that suppress the original phenotype. By carefully defining complementation groups and screening for second-site suppressors, it will ultimately be feasible to define the regulatory machinery involved in microbial recognition and the expression of a defense response.

## VI. CONCLUSIONS AND FUTURE PROSPECTS

The results summarized in this chapter offer strong support to the proposition that *A. thaliana* offers an excellent system for studying plant disease resistance.[29,70] The identification of pathogens of *A. thaliana,* a critical first step in developing this system, has been successfully completed. The initial characterization of the interaction of *A. thaliana* with these pathogens indicates that the types of responses observed in infected *A. thaliana,* both at the phenotypic and the molecular levels, are very similar to those described in other plant

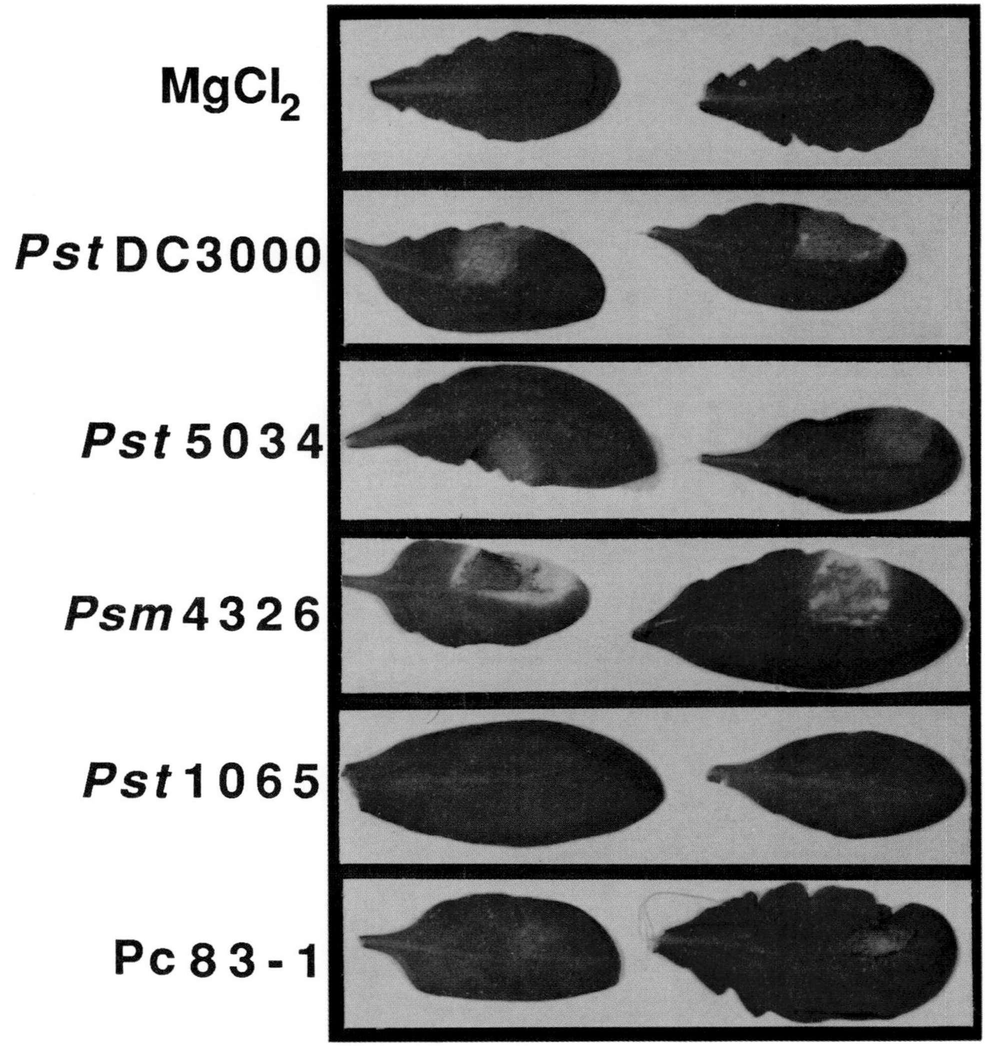


PLATE 1. Phenotypes caused by phytopathogenic *Pseudomonas* strains infiltrated into leaves of *A. thaliana*. Bacterial strains were infiltrated into leaves of *A. thaliana* (land race Columbia) at a dose of $10^6$ CFU/ml (approximately $10^4$ CFU/leaf) and photographed 2 days after infiltration. $MgCl_2$ represents leaves infiltrated with 10 m*M* $MgCl_2$, which is the solution used to suspend the bacteria prior to infiltration. *Pst, P. syringae* pv. *tomato; Psm, P. syringae* pv. *maculicola; Pc, P. cichorii.*

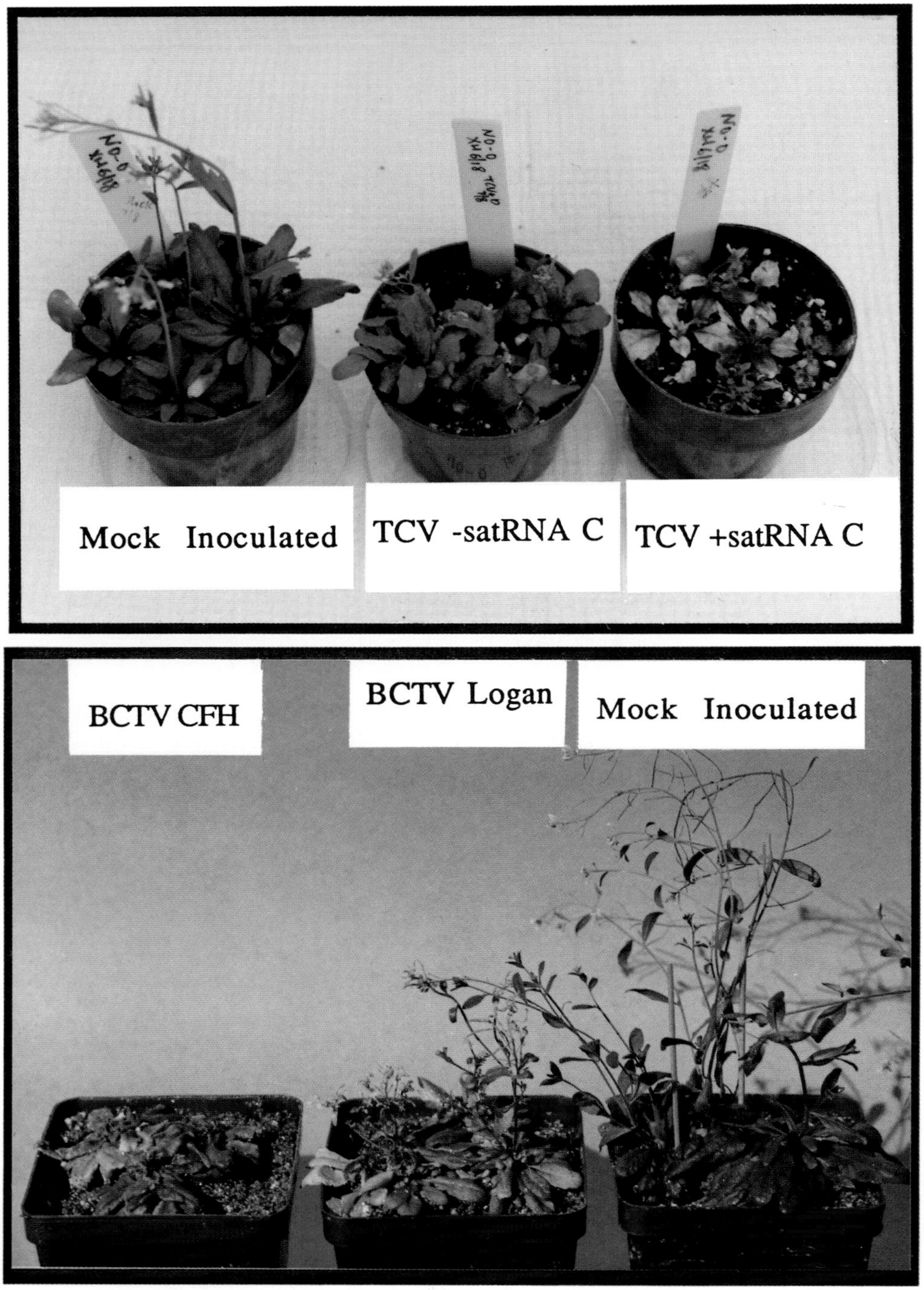


PLATE 2. Symptoms caused by viral infection of *A. thaliana*. Top panel: Symptoms caused by inoculation of turnip crinkle virus (TCV) isolates without (middle) or with (right) satellite RNA (sat RNA) C onto leaves of *A. thaliana* (land race No-0). The plants on the left were mock inoculated. Plants were photographed 13 days after inoculation (photo provided by A. Simon, University of Massachusetts, Amherst). Bottom panel: Symptoms caused by introduction of cloned beet curly top virus (BCTV) into *A. thaliana* (land race Columbia) via *Agrobacterium*-mediated inoculation; left to right, BCTV strain CFH, BCTV strain Logan, and mock-inoculated controls. Plants were photographed 3 weeks after inoculation.

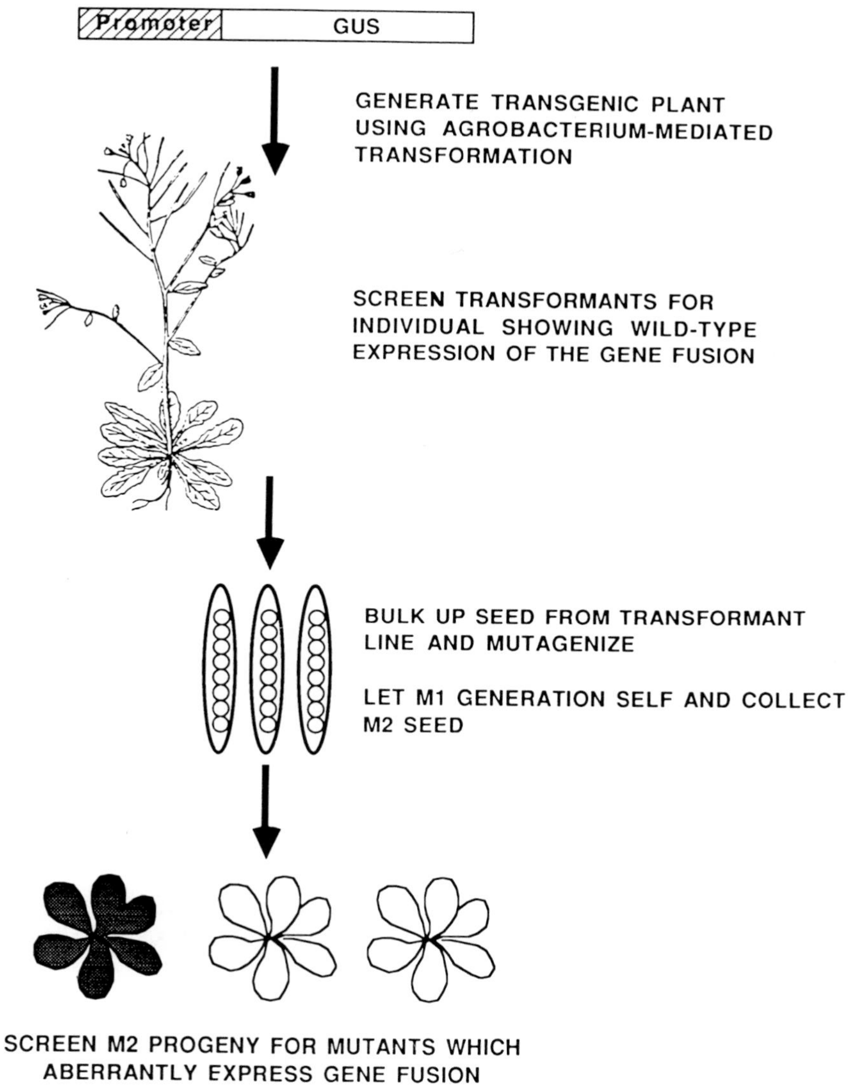


FIGURE 3. A molecular genetic approach for identifying mutations affecting defense gene regulation.

hosts. These similarities appear to extend to the initiation of a resistance response by the interaction of single dominant genes in *A. thaliana* with single avirulence genes in the pathogen. Similar interactions between host resistance genes and pathogen avirulence genes have been described in detail in a number of plant-pathogen interactions.[33,34] Thus, studies using *A. thaliana* may provide some insight into the nature of plant resistance genes and the mechanisms by which resistance genes interact with avirulence genes in the pathogen. Moreover, it is feasible that resistance genes isolated from *A. thaliana* may be transferred to other plants via genetic engineering to generate plants with disease resistances that could not be created using standard breeding techniques.

The identification of pathogens of *A. thaliana* provides infection systems that can be used to bring the power of genetic analysis to the study of the molecular basis of disease resistance in plants. Such genetic studies have been extremely useful in identifying genes in the pathogens which are involved with host interactions and providing some insight into the regulation of these genes. We are now in position to begin similar studies in a host plant.


## ACKNOWLEDGMENTS

I thank my colleagues who provided manuscripts and experimental results prior to their publication and David Bisaro for critically reading this manuscript. I am particularly grateful to Fred Ausubel for his previous and continuing support of my work. The author's unpublished work was supported by funds provided by The Ohio State University.


## REFERENCES


1. **Bell, A. A.,** Biochemical mechanisms of disease resistance, *Annu. Rev. Plant Physiol.*, 32, 21, 1981.
2. **Bailey, J. A. and Mansfield, J. W., Eds.,** *Phytoalexins,* Blackie & Son. London, 1982.
3. **Collinge, D. B. and Slusarenko, A. J.,** Plant gene expression in response to pathogens, *Plant Mol. Biol.*, 9, 389, 1987.
4. **Hahlbrock, K. and Scheel, D.,** Physiology and molecular biology of phenylpropanoid metabolism, *Annu. Rev. Plant Physiol. Plant Mol. Biol.*, 40, 347, 1989.
5. **Dixon, R. A. and Harrison, M. J.,** Activation, structure and organization of genes involved in microbial defense in plants, *Adv. Genet.*, 28, 165, 1990.
6. **Redei, G. P.,** *Arabidopsis* as a genetic tool, *Annu. Rev. Genet.*, 9, 111, 1975.
7. **McCourt, P., and Somerville, C. R.,** The use of mutants for the study of plant metabolism, in *The Biochemistry of Plants,* Vol. 13, Davies, D. D., Ed., Academic Press, San Diego, 1987, 34.
8. **Meyerowitz, E. M.,** *Arabidopsis thaliana, Annu. Rev. Genet.*, 21, 93.
9. **Leutwiler, L. S., Meyerowitz, E. M., and Tobin, E. M.,** The DNA of *Arabidopsis thaliana, Mol. Gen. Genet.*, 194, 15, 1986.
10. **Pruit, R. E. and Meyerowitz, E. M.,** Characterization of the genome of *Arabidopsis thaliana. J. Mol. Biol.*, 187, 169, 1986.
11. **Valvekens, D., Van Montagu, M., and Van Lijsebettens, M.,** *Agrobacterium tumefaciens*-mediated transformation of *Arabidopsis thaliana* root explants by using kanamycin selection, *Proc. Natl. Acad. Sci. U.S.A.*, 85, 5536, 1988.
12. **Kranz, A. R. and Kirchheim, B.,** Genetic resources in *Arabidopsis, Arabidopsis Inf. Serv.*, 24, 3.2.1, 1987.
13. **Koornneef, M.,** Linkage map of *Arabidopsis thaliana* (2n = 10), in *Genetic Maps,* O'Brien, S. J., Ed., Cold Spring Harbor Laboratory, Cold Spring Harbor, NY, 1987, 742.
14. **Chang, C., Bowman, J. L., DeJohn, A. W., Lander, E. S., and Meyerowitz, E. M.,** Restriction fragment length polymorphism linkage map for *Arabidopsis thaliana, Proc. Natl. Acad. Sci. U.S.A.*, 85, 6856, 1988.
15. **Nam, H-G., Giraudat, J., Boer, B., Moonan, F., Loos, W. D. B., Hauge, B. M., and Goodman, H. M.,** Restriction fragment length polymorphism linkage map of *Arabidopsis thaliana, Plant Cell,* 1, 699, 1989.
16. **Somerville, C.,** *Arabidopsis* blooms, *Plant Cell,* 1, 1131, 1989.
17. **Hauge, B. M., Giraudat, J., Hanley, S., and Goodman, H. M.,** Physical mapping of the *Arabidopsis* genome and its applications, in *Plant Molecular Biology,* Herm, R. G., Ed., Plenum, NY, 1991, in press.
18. **Feldmann, K. A., Marks, M. D., Christianson, M. L., and Quatrano, R. S.,** A dwarf mutant of *Arabidopsis* generated by T-DNA insertion mutagenesis, *Science,* 243, 1351, 1989.
19. **Marks, M. D. and Feldmann, K. A.,** Trichome development in *Arabidopsis thaliana.* I. T-DNA tagging of the GLABROUS1 gene, *Plant Cell,* 1, 1043, 1989.
20. **Koncz, C., Martini, N., Mayerhofer, R., Konczkalman, Z., Korber, H., Redei, G. P., and Schell, J.,** High-frequency T-DNA-mediated gene tagging in plants, *Proc. Natl. Acad. Sci. U.S.A.*, 86, 8467, 1989.
21. **Straus, D. and Ausubel, F. M.,** Genomic substration for cloning DNA corresponding to deletion mutations, *Proc. Natl. Acad. Sci. U.S.A.*, 87, 1889, 1990.
22. **Ausubel, F. M.,** personal communication, 1991.

23. **Somerville, C. R. and Browse, J.,** Genetic manipulation of the fatty acid composition of plant lipids, in *Opportunities for Phytochemistry in Plant Biotechnology, Recent Advances in Phytochemistry,* Vol 22, Conn, E. E., Ed., Plenum, NY, 1988, 19.
24. **Last, R. L. and Fink, G. R.,** Tryptophan-requiring mutants of the plant *Arabidopsis thaliana, Science,* 240, 305, 1988.
25. **Haughn, G., Davin, L., Giblin, M., and Underhill, E. W.,** Biochemical genetics of plant secondary metabolism in *Arabidopsis thaliana:* the glucosinolates, submitted, 1991.
26. **Graham, T. L.,** A rapid, high resolution HPLC profiling procedure for plant and microbial aromatic secondary metabolites, *Plant Physiol.,* 95, 584, 1991.
27. **Koch, E. and Slusarenko, A. J.,** Fungal pathogens of *Arabidopsis thaliana* (L.) Heyhn, *Bot. Helvetica,* 100, 257, 1991.
28. **Morgan, I. D.,** A study of jour weed hosts of *Sclerotinia* species in alfalfa fields, *Plant Dis. Rep.,* 55, 1087, 1971.
29. **Davis, K. R., Schott, E., Dong, X., and Ausubel, F. M.,** *Arabidopsis thaliana* as a model system for studying plant-pathogen interactions, in *Signal Molecules in Plants and Plant-Microbe Interactions,* Lugtenberg, B. J. J., Ed., Springer-Verlag, Berlin, 1989, 99.
30. **Dong, X., Mindrinos, M., Davis, K. R., and Ausubel, F. M.,** Induction of *Arabidopsis thaliana* defense genes by virulent and avirulent *Pseudomonas syringae* strains and by a cloned avirulence gene, *Plant Cell,* 3, 61, 1991.
31. **Whalen, M., Innes, R., Bent, A., and Staskawicz, B.,** Identification of *Pseudomonas syringae* pathogens of *Arabidopsis thaliana* and a bacterial gene determining avirulence on both *Arabidopsis* and soybean, *Plant Cell,* 3, 49, 1991.
32. **Davis, K. R., Schott, E., and Ausubel, F. M.,** Virulence of selected phytopathogenic pseudomonads in *Arabidopsis thaliana Mol. Plant-Microbe Interact.,* in press, 1991.
33. **Keen, N. and Staskawicz, B.,** Host range determinants in plant pathogens and symbionts, *Annu. Rev. Microbiol.,* 42, 421, 1988.
34. **Gabriel, D. W. and Rolfe, B. G.,** Working models of specific recognition in plant-microbe interactions, *Annu. Rev. Phytopathol.,* 28, 365, 1990.
35. **Bent, A., Innes, R., Whalen, M., and Staskawicz, B.,** personal communication, 1991.
36. **Mindrinos, M. and Ausubel, F. M.,** personal communication, 1991.
37. **Schott, E., Davis, K. R., and Ausubel, F. M.,** unpublished data, 1990.
38. **Flor, H.,** Current status of the gene-for-gene concept, *Annu. Rev. Phytopathol.,* 9, 275, 1971.
39. **Klement, Z.,** Hypersensitivity, *Phytopath. Prokaryotes,* 2, 149, 1982.
40. **Schott, E. and Ausubel, F. M.,** unpublished data, 1990.
41. **Atkinson, M. M., Huang, J. S., and Knopp, J. A.,** The hypersensitive reaction of tobacco to *Pseudomonas syringae* pv. *pisi:* Activation of a plasmalemma $K^+/H^+$ exchange in tobacco, *Phytopathology,* 77, 843, 1985.
42. **Tsuji, J. and Somerville, S. C.,** *Xanthomonas campestris* pv. *campestris*-induced chlorosis in *Arabidopsis, Arabidopsis Inf. Serv.,* 26, 1, 1988.
43. **Simpson, R. B. and Johnson, L. J.,** *Arabidopsis* as a host for *Xanthomonas campestris* pv. *campestris, Mol. Plant-Microbe Interact.,* 3, 233, 1990.
44. **Tsuji, J., Somerville, S. C., and Hammerschmidt, R.,** Identification of a gene in *Arabidopsis thaliana,* that controls resistance to *Xanthomonas campestris* pv. *campestris, Physiol. Mol. Plant Pathol.,* 38, 57, 1991.
45. **Daniels, M. J., Fan, M-J., Barber, C. E., Clarke, B. R., and Parker, J. E.,** Interactions between *Arabidopsis thaliana* and *Xanthomonas campestris,* in *Advances in Molecular Genetics of Plant-Microbe Interactions,* Vol. 1, Hennecke, H. and Verma, D. P. S., Eds., Kluwer Academic, Dordrecht, 1991, 84.
46. **Shaw, J. J. and Kado, C. I.,** Whole plant wound inoculation for consistent reproduction of black rot of crucifers, *Phytopathology,* 78, 981, 1988.
47. **Koch, E. and Slusarenko, A.,** *Arabidopsis* is susceptible to infection by a downy mildew fungus, *Plant Cell,* 2, 437, 1990.
48. **Muach-Mani, B. and Slusarenko, A.,** unpublished results, 1990.
49. **Sjödin, C. and Glimelius, K.,** Screening for resistance to blackleg *Phoma lingam* (Tode ex Fr.) ''Desm. within Brassicaceae, *J. Phytopathol.,* 123, 322, 1988.
50. **Koch, E. and Williams, P. H.,** unpublished results, cited in 27, 1990.
51. **Maule, A.,** Replication of caulimoviruses in plants and protoplasts, in *Molecular Plant Virology,* Vol. 2, Davies, J. W., Ed., CRC Press, Boca Raton, FL, 1985, 161.
52. **Balàzs, E. and Lebeurier, G.,** *Arabidopsis* as a host of cauliflower mosaic virus, *Arabidopsis Inf. Serv.,* 18, 130, 1981.
53. **Melcher, U.,** Symptoms of cauliflower mosaic virus infection in *Arabidopsis thaliana* and turnip, *Bot. Gaz. (Chicago),* 150, 139, 1989.

54. **Altenbach, S. B. and Howell, S. H.,** Identification of a satellite RNA associated with turnip crinkle virus, *Virology,* 112, 25, 1981.
55. **Simon, A. E. and Howell, S. H.,** The virulent satellite RNA of turnip crinkle virus has a major domain homologous to the 3′ end of the helper virus genome, *EMBO J.,* 5, 3423, 1986.
56. **Morris, T. J. and Carrington, J. C.,** Carnation mottle virus and viruses with similar properites, in *The Plant Viruses,* Vol. 3, Koenig, R., Ed., Plenum Press, NY, 1988, 73.
57. **Simon, A. E.,** personal communication, 1990.
58. **Stanley, J., Markham, P. G., Callis, R. J., and Pinner, M. S.,** The nucleotide sequence of an infectious clone of the geminivirus beet curly top virus, *EMBO J.,* 5, 1761, 1986.
59. **Bennett, C. W.,** The curly top disease of sugarbeet and other plants, *Monogr. Am. Phytopathol. Soc.,* 7, 26, 1971.
60. **Briddon, R. W., Watts, J., Markham, P. G., and Stanley, J.,** The coat protein of beet curly top virus is essential for infectivity, *Virology,* 172, 628, 1989.
61. **Stenger, D., Bisaro, D., and Davis, K. R.,** unpublished data, 1990.
62. **Davis, K. R. and Ausubel, F. M.,** Characterization of elicitor-induced defense responses in suspension-cultured cells of *Arabidopsis, Mol. Plant-Microbe Interact.,* 2, 363, 1989.
63. **Davis, K. R., Bent, A., Innes, R., and Staskawicz, B.,** unpublished results, 1990.
64. **Bell, J. N., Ryder, T. B., Wingate, V. P. M., Bailey, J. A., and Lamb, C. J.,** Differential accumulation of plant defense gene transcripts in a compatible and incompatible plant-pathogen interaction, *Mol. Cell. Biol.,* 6, 1615, 1986.
65. **Cuypers, B., Schmelzer, E., and Hahlbrock, K.,** *In situ* localization of rapidly accumulated phenylalanine ammonia-lyase mRNA around penetration sites of *Phytophthora infestans* in potato leaves, *Mol. Plant-Microbe Interact.,* 1, 157, 1988.
66. **Habereder, H., Schroder, G., and Ebel, J.,** Rapid induction of phenylalanine ammonia-lyase and chalcone synthase mRNAs during fungus infection of soybean *(Glycine max,* L.) roots or elicitor treatment of soybean cell cultures at the onset of phytoalexin synthesis, *Planta,* 177, 58, 1989.
67. **Ohl, S., Hedrick, S. A., Chory, J., and Lamb, C. J.,** Functional properties of a phenylalanine ammonia-lyase promoter from *Arabidopsis, Plant Cell,* 2, 837, 1990.
68. **Samac, D. A., Hironaka, C. M., Yallaly, P. E., and Shah, D. M.,** Isolation and characterization of the genes encoding and acidic chitinase in *Arabidopsis thaliana, Plant Physiol.,* 93, 907, 1990.
69. **Jefferson, R. A., Kavanagh, T. A., and Bevan, M. W.,** GUS fusions: Beta-glucuronidase as a sensitive and versatile gene fusion marker in higher plants, *EMBO J.,* 6, 3901, 1987.
70. **Schott, E. J., Davis, K. R., Dong, X., Mindrinos, M., Guevara, P., and Ausubel, F. M.,** *Pseudomonas syringae* infection of *Arabidopsis thaliana* as a model system for studying plant-bacterial interactions, in *Pseudomonas: Biotransformation, Pathogenesis, and Evolving Biotechnology,* Silver, S., Chakrabarty, A. M., Iglewski, B., and Kaplan, S., Eds., American Society for Microbiology, Washington, D.C., 1990, 82.

# HYDROXYPROLINE-RICH GLYCOPROTEINS IN PLANT-MICROBE INTERACTIONS AND DEVELOPMENT

**Keith L. Wycoff, Richard A. Dixon, and Christopher J. Lamb**

## TABLE OF CONTENTS

# I. INTRODUCTION

The cell wall is often the first line of defense against plant pathogenic microorganisms. It is a structural barrier that few microbes are equipped to breach by force alone without the aid of cell wall-degrading enzymes.[1] In addition, plants have the ability to modify their cell walls and often do so during or following infection. One modification is an increase in the deposition of certain proteins into the cell wall. Proteins with a high content of hydroxyproline constitute a major portion of the protein component of plant cell walls and have been implicated in various aspects of cell growth and development, response to wounding and infection, and cell-cell recognition (recent reviews[2-9]). The purpose of this chapter is both to examine the evidence that plant cell wall proteins, particularly those rich in the amino acids hydroxyproline (Hyp) and proline (Pro), play an important role in the interactions of plants with microorganisms, and to bring together what is known about the signals and signal transduction mechanisms that operate in these interactions. The expression and potential function of these proteins in normal development will also be discussed.

Hydroxyproline is an unusual amino acid, resulting from the posttranslational hydroxylation of peptidyl proline residues. It occurs in relatively few proteins in plants, but is quite abundant in the proteins of plant cell walls.[10,11] Proteins containing hydroxyproline are typically highly glycosylated, hence as a group they are often referred to as hydroxyproline-rich glycoproteins (HRGPs). The HRGPs have been traditionally divided into three groups: (1) extensins; (2) arabinogalactan proteins; and (3) solanaceous lectins. A fourth, and probably related, group of cell wall proteins that have recently begun to be characterized are the hydroxyproline-proline-rich proteins (H/PRPs). The H/PRPs are probably not glycosylated and are distinguished from the other groups by their characteristic repeat sequences. In most cases their high proline content has been deduced from cDNA or genomic nucleotide sequences and they contain an unknown quantity of hydroxyproline; in those H/PRPs whose actual amino acid content has been analyzed, about one half of the prolines are hydroxylated to hydroxyproline.

# II. EXTENSINS

## A. BIOCHEMISTRY, STRUCTURE, AND LOCALIZATION

The extensins, often referred to as cell wall HRGPs, are the most well characterized of the HRGPs. Initially identified as highly insoluble proteins located in the cell wall, subsequent studies showed that they exist as soluble proteins at certain stages of wall development and eventually become insolubilized.[12-14] They are very basic, with isoelectric points between 10 and 12, and hydroxyproline + proline can make up 30 to 55% of their amino acid composition. They are also rich in serine and one or more of the following amino acids: valine, tyrosine, lysine, and histidine. Their sequences are highly repetitive, usually containing the repeating pentapeptide sequence Ser(Hyp)$_4$ within a higher-order repeat such as Ser(Pro)$_4$-Ser-Pro-Ser-(Pro)$_4$-(Tyr)$_3$-Lys.[15] The higher-order repeats can be used to distinguish different classes of cell wall HRGPs (see Showalter and Rumeau,[7] Table IV). Recently, however, proteins and/or cDNA clones have been isolated from maize and sugarbeet that have many of the characteristics of extensins (i.e., basic isoelectric points, high Hyp/Pro content, repeat structure), but have only one or no Ser(Hyp)$_4$ repeats.[16-19]

The extensins that have been purified are generally 40 to 60% carbohydrate, primarily in the form of short oligoarabinosides attached *O*-glycosidically to the hydroxyproline residues. The carbohydrate appears to be important in maintaining extensin structure as linear, rodlike molecules.[20] It is thought that extensin molecules are cross-linked via their tyrosine residues to form an insoluble matrix woven between the structural carbohydrates of the wall (although intermolecular iso-dityrosine crosslinks have not yet been found *in vivo*).[9,14,21]

After translation, the peptide precursors of cell wall HRGPs follow the typical endomembrane route for extracellular proteins. A hydrophobic leader sequence has been identified in many of the cloned genomic and cDNA sequences. Newly translated HRGP monomers are inserted into the rough endoplasmic reticulum (ER) where most of the proline residues become hydroxylated.[22] In monocots, the rough ER also appears to be the site of glycosylation. In dicots glycosylation takes place in the golgi apparatus.[5] After passing through the ER and the golgi apparatus, the glycoproteins are inserted into the cell wall.

The localization of extensin in mature carrot roots was studied using anti-extensin-1 antibodies, followed by protein-A colloidal gold.[23] Extensin-1 was distributed more or less uniformly across the primary cell walls of phloem parenchyma cells. It was present at reduced levels in the middle lamella between cells and was absent from the expanded middle lamella between three or more cells. Cell walls of young seedlings were not labeled. Infrequently, a dead cell surrounded by living cells was seen, and in each case antibody labeling occurred only over the half of the cell wall adjacent to a living cell. These results suggest that extensin-1 is added to the wall by intussusception after the rest of the wall is laid down, and that extensin cannot cross the middle lamella.

Cell wall HRGPs do not accumulate to equal levels in all tissues and cell types. They are more abundant in roots, with lower levels in stems and much lower levels in leaves. Extensin mRNA species from oilseed rape were expressed in roots at levels of at least 400 times those in other organs.[24] Cell wall HRGPs are particularly concentrated in soybean seed coats, where Hyp can make up 30% of the total cell wall protein.[25]

## B. RESPONSE TO INFECTION

It is now well established that cell wall HRGPs play some role in the repertoire of plant defenses against pathogens. The accumulation of cell wall HRGP (or HRGP mRNA) has been observed in a number of different plants following infection by pathogens, including fungi,[15,26-32] bacteria,[30,33] viruses,[33,34] and viroids.[35] In addition, it has been shown that HRGPs accumulate more rapidly in incompatible (resistant) than compatible (susceptible) interactions,[28,32] and that artificially increasing (by heat shock or ethylene) or decreasing (by hypoxia or free hydroxyproline) HRGP levels leads, respectively, to increased or decreased resistance to fungal infection.[36-38] Cell walls of plants with artificially increased HRGP levels were more resistant to degradation by extracellular enzymes from *Cladosporium cucumerinum* than were untreated plants.[37]

Interestingly, all of the plants in which HRGP or HRGP mRNA are infection-inducible are dicots. The monocots that have been examined have much lower levels of cell wall HRGP, and these do not increase following infection.[33]

Recently a number of investigators have taken an immunocytochemical approach to locate the accumulation of cell wall HRGPs during infection at the level of individual cells. An antiserum against melon HRGP, which cross-reacted with purified tomato HRGP, was used to study the subcellular distribution of HRGPs in tomato root tissues infected by *Fusarium oxysporum radicis-lycopersici,* a fungal strain that caused disease on the tomato cultivar used.[26] Cell walls of infected tissue contained considerably more HRGP than uninfected tissue. Labeling was especially heavy over areas where walls were thickened by fibrillar material and over wall appositions resembling papillae. There was also a gradual occlusion of intercellular spaces by a material that was, at first, fibrillar, then increasingly dense and amorphous, and that reacted specifically with the anti-HRGP antibodies. Labeling occurred only in the walls of cells that were in contact with or had been invaded by fungal hyphae and appeared to occur as a result of cell damage. This suggests that, at least in this compatible interaction, expression of these HRGPs occurs too late to affect the initial stages of fungal colonization.

Anti-HRGP antibodies were used to probe bean and melon tissue infected with incompatible (host resistant) races of *Colletotrichum lindemuthianum* (the anthracnose fungus) or *Pseudomonas syringae* pv. *phaseolicola* (the halo blight bacterium). Immunogold labeling showed that HRGPs accumulated in walls and paramural papillae of living plant cells adjoining dead hypersensitive cells. The hypersensitive cells were sites where the growth of bacteria and fungi were arrested, supporting the involvement of these proteins in disease resistance. In bean and melon leaves inoculated with *Pseudomonas fluorescens,* a saprophytic bacterium, HRGP was detected in amorphous material that encapsulated the bacterial cells. Interestingly, the increased labeling of the cell wall was confined to cells contiguous to a hypersensitive cell. Cell walls further from the site of infection became labeled to the same degree as uninoculated control tissue.[30] Unfortunately, it is difficult to make any generalizations about differences in HRGP accumulation between compatible and incompatible interactions by comparing this study to the previously cited study, as both the plants and the pathogens used were different.

Cell wall HRGPs have also been detected in necrotic lesions produced on leaves of *Nicotiana tabacum* infected by tobacco mosaic virus (TMV), using antibodies raised to melon HRGP. In the necrotic core of the lesions there was substantial labeling over hemispherical excrescences formed along the cell walls and over material filling intercellular spaces. In the halo surrounding the core of the lesion there were fewer hemispherical excrescences, but these labeled with the antibody, as did an amorphous material that filled some vascular parenchyma cells, and opaque intertwined fibrils seen in xylem elements. In the green tissue 1.5 mm to 3 mm around the halo, anti-HRGP antibodies only labeled fibrillar material in the xylem vessels. Outside of this region, no labeling was seen.[34]

Two points should be noted from these observations. First, an increase in the apparent amount of HRGP was limited to cells immediately surrounding the infections. This is somewhat at odds with reports of systemic, or at least nonlocalized induction of HRGP expression (see examples below).[15,31,32,39] Second, in contrast to the observations in normal carrot tissue,[23] HRGP seems able to cross the middle lamella and accumulate in intercellular spaces and (dead) xylem elements. In general, these results are consistent with the proposed function of cell wall HRGPs as a physical barrier, immobilizing and preventing further spread of pathogenic organisms in plant tissue. However, it is impossible to say whether HRGP accumulation contributed to the arrest of pathogen growth or merely served as a barrier to potential secondary infections.

The technique of *in situ* hybridization has been useful in localizing the accumulation of HRGP transcripts. Radiolabeled RNA probes corresponding to the noncoding strand of the bean HRGP4.1 cDNA[15] were used to follow the accumulation of HRGP transcripts in bean following infection with virulent and avirulent races of *Colletotrichum lindemuthianum.*[39] Under the conditions used, this probe also detected two other transcripts from related genes, corresponding to the HRGP2.13 and HRGP3.6 cDNAs.[15] In an incompatible interaction there was a massive induction of HRGP transcripts in the epidermal and cortical cells directly below the inoculation site, as well as significant accumulation in perivascular parenchyma tissue. In all three cell layers, transcript accumulation extended some distance from the site of inoculation, even though fungal hyphae, as shown by immunohistochemical staining, were confined to the epidermis. In the compatible interaction, HRGP transcripts accumulated only in the perivascular parenchyma. In this case, fungal hyphae had penetrated into the cortex but not into the perivascular parenchyma. Transcripts were also detected in the perivascular parenchyma 2 mm away from the inoculation site in sections that had not yet been penetrated by fungal hyphae. These results suggest the existence of some diffusible signal that can induce transcription at a distance from the invading fungus. However, the immunohistochemical studies cited above detected protein accumulation only in the walls of cells adjacent to the pathogen. This suggests that there may be some kind of posttranscriptional control. Perhaps there are two separate signals, a diffusible one that stimulates transcription, and one that requires cell-cell contact and stimulates translation of the accumulated transcripts.

### C. RESPONSE TO WOUNDING

Even before they were known to be affected by pathogen infection, cell wall HRGPs were found to accumulate in response to wounding.[40] In a number of different plants extensin mRNA has been shown to accumulate following wounding, and this accumulation has been assumed to be under transcriptional control.[15,38,41-46] It should be noted that this has been shown definitively only in one report where nuclear run-off experiments were used to show transcriptional activation following wounding, infection, or treatment with fungal elicitors.[29] Most of the reports of wound induction of HRGP expression have been in dicots. In barley and oats there was no appreciable change in HRGP levels 24 h after wounding by peeling away the lower epidermis.[47] In contrast, there was a rapid and dramatic, but transient, accumulation of HRGP mRNA in maize leaves in response to incision wounds.[43] The increase occurred within 2 h after wounding, with levels returning to normal by 24 h.

### D. SIGNALS FOR EXPRESSION OF CELL WALL HRGP GENES

The question of how infection and wounding lead to the increased synthesis of HRGPs is of great interest. What signals are released, and what are the signal transduction mechanisms that lead to changes in gene expression? Analysis of the signals involved in the regulation of the cell wall HRGP genes is complicated by the fact that, in all of the plants studied in detail, these genes exist as small multigene families with individual members responding differently to different conditions. In bean, for instance, transcripts of the HRGP2.13, HRGP3.6, and HRGP4.1 genes showed marked accumulation in response to wounding, elicitors, and infection.[15] Transcripts of the HRGP2.11 gene, on the other hand, rapidly disappeared after infection and accumulated only in response to wounding.[44] Each gene showed different kinetics of induction even to the single stimulus of wounding (Figure 1). Some HRGP genes even produce two different sized transcripts depending on the stimulus. Thus, the HRGP4.1 gene of bean produced a 2.5-kb transcript following infection or elicitation, but a 2.2-kb transcript after wounding.[15] In stored carrot roots, ethylene treatment increased the abundance of a 1.8-kb transcript and decreased the abundance of a 1.5-kb transcript, both of which are transcribed from a single gene.[42,48] Wounding, on the other hand, led to an increase in the 1.5-kb mRNA and a decrease in the 1.8-kb mRNA. These results suggest that wounding and ethylene are affecting different signal transduction pathways.

Elicitors of fungal and plant origin have been implicated as signal molecules in the induction of HRGP genes and in HRGP accumulation. Cell wall extracts from both *Colletotrichum lindemuthianum* and *Phytophthora megasperma* had a stimulatory effect on the incorporation of hydroxyproline into melon cell walls. A crude extract of melon cell walls was an even more potent elicitor of HRGP accumulation.[49] A crude elicitor prepared from autoclaved carrot roots also stimulated the accumulation of extensin mRNA in carrot cell cultures.[46] Experiments with bean cell cultures showed that fungal elicitors stimulated HRGP mRNA accumulation and that this was due to an activation of transcription.[15,29,32] Prolyl hydroxylase activity also increases in bean following elicitor treatment.[22]

Relatively little is known about the mechanism of plant recognition of viral infection. The coat protein of TMV may act as the signal molecule in some situations. Specific point mutations in the coat protein of TMV lead to a hypersensitive response (HR) on tobacco hosts that are not normally resistant.[50] However, in other tobacco hosts, which normally respond with necrotic local lesions, infectious mutants completely lacking the coat protein gene still induce a HR, implying the involvement of another viral factor. HRGPs do accumulate during the HR,[34] but were not specifically sought in this study.

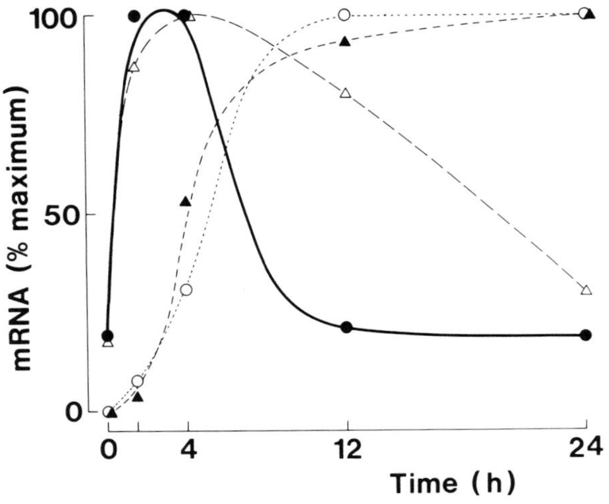


FIGURE 1. Kinetics of induction of specific HRGP transcripts in excision-wounded hypocotyl tissue: Hyp2.11 (●); Hyp3.6 (△); Hyp2.13 (▲); and Hyp4.1 (○). Total cellular RNA was isolated at different times after excision wounding, separated on 1% agarose gels, transferred to nitrocellulose, and hybridized to 32P-labeled sequences of the four different HRGP cDNA clones. Following autoradiography mRNA was quantitated by scanning densitometry. From Sauer, et al.[44]

Many plants respond to infection with enhanced ethylene production, and exogenously added ethylene has been found to activate certain biochemical defenses,[51] including the synthesis of cell wall HRGPs.[42] However, the role of ethylene as a transmissible signal in defense gene activation during wounding or infection is controversial. Investigations on the induction of phytoalexins,[52] and chitinase and glucanase[53] indicated that ethylene formation was more a symptom than a signal in the induction of these defenses. Furthermore, wounded carrot tissue produces virtually no ethylene, calling into doubt the physiological significance of the hormone's ability to induce the accumulation of HRGP mRNA transcripts in carrot. Nevertheless, ethylene has been implicated as a transmissible signal in the activation of HRGP genes in melon. AVG (aminoethoxy vinylglycine), an inhibitor of ethylene synthesis, inhibited both elicitor-induced ethylene and elicitor-induced HRGP synthesis, while ACC (1-aminocyclopropane-1-carboxylic acid), the direct precursor of ethylene, stimulated HRGP synthesis.[49] Such inhibitor studies must be viewed with caution, however, as the specificity of the inhibitor is not completely known.

Infection of melon with *Colletotrichum lagenarium* induced the appearance of 5 species of mRNA hybridizing to an HRGP probe: 4.5, 3.2, 2.15, 1.65, 1.4-kb, with the 1.65 and 1.4-kb species accumulating the most.[31] The relative abundance of hybridizable mRNAs was similar in leaves, roots, and stems, even though roots were not infected and showed no symptoms. This implies the existence of a transmissible signal which elicits transcription in tissues distant from the site of infection. After treatment with ethylene, maximum hybridization was reached within 2 days, with the 5 mRNA species in the same relative proportions, but somewhat less abundant than during infection.

It should be emphasized that there are most likely different signals and/or signal transduction pathways involved in the induction of different HRGP genes. In bean cell cultures, for instance, the transcripts corresponding to the HRGP4.1 and HRGP2.13 genes begin to accumulate in response to fungal elicitor only after a lag of ~2 h, and reach a maximum level of accumulation between 12 and 24 h. The HRGP3.6 and HRGP2.11 transcripts, on the other hand, begin to accumulate much more rapidly, peaking between 2 and 4 h and subsequently declining, much as the transcripts for the phenylpropanoid biosynthetic genes phenylalanine ammonia-lyase and chalcone synthase (Figure 1).[15,29,44]

## E. EXPRESSION DURING DEVELOPMENT

In addition to being stress-induced, many cell wall HRGPs are differentially expressed during development. As mentioned previously, extensin accumulates to high levels in the soybean seed coat.[54] It accumulates rapidly in the walls of the palisade epidermal cells and the underlying hourglass cells between 16 and 24 days after anthesis. Because of its location, the protein is thought to serve a protective function for the dormant seed.[25] A maize HRGP accumulates to high levels in the cell walls of maize pericarp during seed maturation.[17,55,56]

It has been known for some time that HRGPs are highly expressed in actively growing callus and cell cultures.[57,58] Although tissue culture is an artificial state, expression in tissue culture may reflect expression during normal stages of development when cells are actively dividing. Expression of a maize threonine-rich HRGP gene was particularly high in dividing cells, specifically in root tips and coleoptile nodes.[18] The steady-state level of this HRGP mRNA was shown to correlate with cell division by comparison with the level of histone H4 mRNA.[43] The amount of HRGP transcript dropped when actively dividing calli began to differentiate after being cultured on an auxin-depleted medium. However, HRGP gene expression decreased in maize coleoptiles stimulated to elongate by the synthetic auxin 2,4-D, but remained constant in control coleoptiles that did not elongate.[43] Thus, auxin can either stimulate or suppress HRGP accumulation, depending on the physiological state of the cells.

A novel HRGP gene was recently isolated from a tobacco genomic library. The presence of a hydrophobic leader sequence suggests that it is a cell wall protein. Based on gene fusion analysis, the gene is highly expressed only in the cells at the tip of emerging lateral roots, beginning in a subset of pericycle and endodermal cells, and continuing as the lateral root pushes its way through the cortex and epidermis of the main root.[59] Expression ceases soon after the lateral root emerges, suggesting that the protein has a very specific function, possibly in allowing the emerging lateral root to withstand the mechanical forces of breaking through the cortex and epidermis of the parental root.

Another extensin gene from tobacco also appears to be regulated in a tissue-specific manner. Although normally only expressed in roots and stems, it was approximately 100-fold more abundant in the leaves of transgenic plants containing an active T-DNA cytokinin gene than in normal leaves. Culture of normal shoots on cytokinin-containing medium led to an increase in expression comparable to that in the transgenic shoots.[60]

In transgenic tobacco plants containing 950 bp of 5′ flanking sequence from the bean HRGP4.1 gene fused to the reporter gene β-glucuronidase (GUS) (HRGP4.1-GUS), high levels of expression were found in root and shoot tips and in stem nodes.[61] The root tip expression was present in both primary and lateral roots at all stages of development, indicating that it is under a different control mechanism than the tobacco HRGP.

Recently tissue print northern and western blots were used to localize HRGP mRNA and protein in developing soybean and bean tissues.[62] In stem internodes, HRGP mRNA was most abundant in cambium cells, a few layers of cortical cells around the primary phloem, and in some parenchyma cells around primary xylem. Soluble HRGPs showed a similar pattern of accumulation. Soluble HRGPs were also abundant in young hypocotyl apical regions and in young root tips. *In situ* hybridization studies with the threonine-rich HRGP of maize found that mRNA accumulation was highest at sites of vascular differentiation in both the shoot, root tips and developing leaves.[63]

Cell wall HRGP genes may also be highly expressed in stigmas. A maize HRGP previously isolated from pericarps[17] accumulates to high levels in corn silks.[64] Evidence from HRGP4.1-GUS transgenic plants indicates that the promoter of this bean HRGP directs high levels of expression in stigmas.[61] Two clones isolated from a stigma-style specific cDNA library of tobacco have a deduced proline composition of 25% and 30%, and contain a few Ser(Pro)$_4$ repeats.[65] A cDNA clone that was expressed to a greater extent in tobacco TCL (thin cell layer) explants initiating floral meristems than in TCL explants initiating vegetative shoot meristems or possessing roots was identified as an extensin gene[66,67] Interestingly, while this extensin gene was expressed in TCL explants initiating vegetative shoots in the presence of kinetin, it was not expressed at all in explants initiating vegetative shoots in the presence of another cytokinin, zeatin.

### F. HRGP GENE REGULATORY ELEMENTS

Although some progress has been made toward identifying the molecules that affect HRGP expression, the ultimate signals have not yet been identified. Understanding how one particular HRGP gene, and not other genes in the same family, is regulated requires additional approaches. One such approach is functional dissection, in which the gene promoters (or deletions thereof) are placed in front of a reporter gene and expressed in transient or transgenic assays. Functional analysis of the bean HRGP4.1 gene in transgenic tobacco indicates that sequences responsible for root-tip, stem node, and wound-responsive expression are all located between −250 and −93 bp from the transcription start site, while sequences controlling expression in stigmas are located more than 700 bp upstream of this region.[68]

Another approach for analyzing the control of HRGP genes involves the use of mobility shift (gel retardation) assays and DNA footprinting. Recently a protein factor that bound to a fragment of a carrot extensin promoter was identified in nuclear extracts of carrot roots.[69] This "extensin gene-binding factor-1", EGBF-1, exhibited some degree of sequence-specificity (other plant genes, including the 35S promoter did not compete, but no mutagenesis was done on the wild-type promoter), and was organ specific (not present in leaves or stems). DNA footprint analysis revealed that the factor protected a region between −270 and −310 bp from the transcription start site of the 1.8-kb transcript.

While EGBF-1 activity in extracts from untreated roots remained steady over 48 h, the activity dropped sharply in wounded and ethylene treated roots between 12 and 36 h. Levels in wounded roots remained at undetectable levels while levels in ethylene-treated roots returned to pretreatment levels by 48 h. Because conditions which induce the accumulation of extensin mRNA (wounding, ethylene) lead to a reduction in EGBF-1, it was proposed that EGBF-1 is a repressor of extensin transcription and the binding site acts as a silencer. No functional data (deletions) exist to corroborate this yet. Since stem and leaf extracts contained no EGBF-1, there must be another mechanism responsible for the lack of expression in stems and leaves than simple repression. A second factor in wounded roots was found to inhibit EGBF-1 DNA-binding activity. The reaction between EGBF-1 and the inhibitor was saturable and reversible *in vitro*. The inhibitory factor was heat labile and larger than 12 kDa, suggesting that it is a protein.[70]

## III. HYDROXYPROLINE-PROLINE-RICH PROTEINS

### A. BIOCHEMISTRY, STRUCTURE, AND LOCALIZATION

At the same time that they cloned the first extensin cDNA, Chen and Varner[41] isolated a cDNA clone for a novel proline-rich protein that proved to be the first in a still-growing family of proteins involved in many aspects of development and plant-microbe interactions. At present this family is represented by three purified proteins and at least seven cDNA or genomic clones. The proteins, purified from soybean cell walls, contain about equal quantities of proline and hydroxyproline,[71,72] hence they have been termed hydroxyproline-proline-rich proteins (H/PRPs).[7] Like the cell wall HRGPs, the H/PRPs generally have highly repetitive sequences containing only a few different amino acids. They differ from the extensins in that their consensus repeats completely lack the Ser-$Pro_4$ motif (for a recent review, see Showalter and Rumeau[7]) The pentapeptides Pro-Pro-Val-Tyr-Lys and Pro-Pro-Val-Glu-Lys are common in the soybean cDNA clones SbPRP1, SbPRP2, and SbPRP3.[73,74] The sequence of the first 18 amino acids at the N-terminus of a H/PRP purified from soybean cell walls is identical to the predicted N-terminus for the protein product of the clone SbPRP2, with the second proline in the first three repeats modified to hydroxyproline.[72] Another soybean clone, 1A10, is 94% identical at the nucleotide level and 97.5% identical at the amino acid level to SbPRP1.[75] A fifth soybean clone, 1A10-2, is identical to SbPRP2 in the coding region, and up to nucleotide −534 from the transcription start site, beyond which it diverges.[76] The carrot clone p-33 contains several copies of the repeats Pro-Pro-Val(Ile)-His-Lys and Pro-Pro-Val-Tyr-Thr.[41] In addition, two of the H/PRP proteins that have been purified appear not to be glycosylated.[71]

A somewhat different repeating sequence is found in H/PRP genes expressed in early stages of nodule development in legumes during infection with *Rhizobium.* GmENOD2, a clone from soybean, contains several copies of the pentapeptide Pro-Pro-His-Glu-Lys and the hexapeptide Pro-Pro-Pro-Glu-Tyr-Gln.[77,78] Homologous genes with similar repeat sequences have been isolated from pea (PsENOD2)[79] and alfalfa (A2ENOD2).[80] Recently another gene was isolated from pea, ENOD12, which contains alternating pentapeptide repeats of Pro-Pro-Gln-Lys-Glu and Pro-Pro-Xaa-Xaa-Xaa.[81] The proteins corresponding to these early nodulin genes have not yet been purified, so it is not known what portion of their proline residues are hydroxylated. Nor is their subcellular localization known, but since they all encode putative signal peptides, it is presumed that they are localized in the cell wall.

## B. RESPONSE TO INFECTION AND WOUNDING

The p-33 cDNA of carrot was first isolated as a wound-responsive transcript.[41] The amount of extractable p33 protein was found to increase 5-fold in root tissue wounded for 24 h.[46] The amount of p33 mRNA in carrot roots increased within 1 h after wounding with a maximum increase of 50- to 100-fold within 2 h. An increase in extensin mRNA, by contrast, could not be detected until 8 h and continued to increase through 24 h. Ethylene treatment of unwounded tissue led to an increase in extensin mRNA, but not that of p33. Expression of p33 in wounded tissue appeared to be inhibited by ethylene. In addition, extensin mRNA was found to accumulate in cell cultures after 5 days of treatment with a crude carrot cell wall (endogenous elicitor) preparation, but p33 did not. All of the foregoing indicates that these two genes are under separate regulation.

There is no evidence as yet that either H/PRP transcripts or protein accumulate during pathogen infection, but several of them accumulate during nodule formation following infection of legume roots with *Rhizobium.* Cassab originally observed an increase in total hydroxyproline content of the nodule cortex during nodule development, but the identity of the proteins involved was not determined.[82] The first of the nodule-associated H/PRP clones to be isolated, pENOD2, was expressed in early stages of nodule development in soybean.[77] *Rhizobium fredii* USDA257 forms "empty" nodules on soybean. These nodules are void of bacteriods or infection threads, but ENOD2 mRNA can be found, suggesting that this protein functions in nodule morphogenesis and not the infection process.

*Rhizobium* strains cured of their Sym plasmid, but containing cosmids carrying all of the *nod* genes, are able to form ineffective nodules in *Vicia sativa.* Similar "empty" nodules are formed by an *Agrobacterium* strain cured of its Ti plasmid and carrying a Sym plasmid. In both of these unusual nodules, ENOD2 transcripts can also be found.[83] An *Agrobacterium* strain carrying the nod genes of *R. meliloti* had a similar effect on alfalfa and induced the alfalfa homolog A2ENOD2.[80] Additionally, an *Agrobacterium* without the nod genes did not elicit this nodulin. This suggests first, that the *Rhizobium nod* genes are alone responsible for ENOD2 induction and bacterial chromosomal genes are not required, and second, that ENOD2 induction is not simply a response to the presence of a potential pathogen.

The pea ENOD2 homolog occurs only in the inner cortex of the pea nodule. In soybean, the ENOD2 gene is expressed in the nodule inner cortex as well as in the tissue surrounding the vascular strand that connects the nodule with the central cylinder of the root. The very low oxygen concentration in the nodule central tissue is due in part to a diffusion barrier in the nodule inner cortex (also called the nodule parenchyma). This layer of cells has relatively few and small intercellular spaces, which may be the morphological trait required for the diffusion barrier, and the ENOD2 protein may play some role in creating this special morphology.[79]

Exopolysaccharide (*exo*) mutants of *R. meliloti,* which cannot penetrate alfalfa root tissue, nevertheless are able to induce ineffective nodules, in which the A2ENOD2 gene is one of the few nodulin genes expressed, suggesting that the *Rhizobium* mutants produce a signal that induces gene expression at a distance in the root cortex. The auxin transport inhibitors *N*(-1-naphthyl)-phthalamic acid (NPA) and 2,3,5-triiodobenzoic acid (TIBA) induce similar nodule-like structures on alfalfa, and also induce A2ENOD2 expression.[84] The nature of the bacterial signal is not yet known. Interestingly, A2ENOD2 expression was seen in nodules produced at low frequency on alfalfa in the absence of *Rhizobium.*[85]

ENOD12 is a H/PRP expressed transiently during nodule development and is present in root hairs as early as 24 h after inoculation. By *in situ* hybridization of root tissue 2 days after infection, it was shown that the ENOD12 mRNA is detected both in root hairs and in cortex cells containing infection threads. It is also present in root cortical cells in front of the infection thread that undergo morphological changes in "preparation" for infection thread passage, and in cells of the root inner cortex that form the nodule primordium. This expression in cells not yet penetrated by an infection thread implies the existence of a diffusible signal.

ENOD12 expression was elicited in root hairs by the culture medium from *Rhizobium* grown in the presence of the *nod* gene inducer naringenin, but not by culture filtrates from uninduced cells. Infection with strains carrying mutations in either *nodA* or *nodE* did not elicit ENOD12. This suggests that some diffusible product of the *nod* genes is acting as the signal for induction of ENOD12. The expression of the ENOD12 gene is not part of a general defense response, as it was not induced in response to infection by the fungus *Fusarium oxysporum* f.sp. *pisi.*[81] The nature of the signal that elicits the expression of ENOD12 is not yet known, but recent research suggests that a complex interchange of signals occurs between the plant and the bacterium during nodule initiation.[86,87] The *R. meliloti* host-range signal NodRm1 was identified as a sulfated β-1,4-tetrasaccharide of D-glucosamine in which three amino groups were acetylated and one was acylated with a $C_{16}$ *bis*-unsaturated fatty acid.[88] This factor induces root hair deformation on alfalfa plants, but its effect on nodule H/PRPs has not been tested.

## C. RAPID CELL WALL CHANGES

It now appears that plants may have a mechanism for responding to wounding or infection by making rapid changes in the cell wall that do not involve gene induction and synthesis of new proteins. A H/PRP similar, if not identical, to SbPRP2, is rapidly insolubilized in the walls of soybean following elicitation or wounding.

Bradley and Lamb[89] identified a protein of 33 kDa in extracts from cell walls of soybean suspension cultures that was no longer extractable following elicitation with glutathione (GSH) or fungal elicitor. The effect was rapid, with complete disappearance of the protein occurring within 5 min under optimal conditions (50-$\mu M$ GSH). Using an antibody raised against the purified protein, it was determined that the apparent disappearance was due to a rapid insolubilization in the cell wall. The insolubilization could be mimicked by adding

1 m*M* $H_2O_2$ and could be inhibited by addition of ascorbic acid (which can block $H_2O_2$ production) or catalase (which destroys $H_2O_2$). When isolated cell walls were used instead of live cells, only $H_2O_2$ was effective, indicating that plasma membrane or cytoplasmic components were necessary for the response to elicitor. N-terminal sequencing of the purified protein revealed a 35 aa sequence identical to the derived sequence of SbPRP2 after the putative signal peptide.[76]

In bean a cross-reacting 35-kDa protein is found in the upper elongating regions of etiolated hypocotyls and becomes progressively less extractable in more mature regions. Immunofluorescent staining, however, indicates that the protein is present in equivalent quantities in all sections. Wounding leads to a localized insolubilization within 2 min. $H_2O_2$ causes insolubilization in hypocotyl sections, while GSH and fungal elicitor has no reproducible effect.[89] A 100-kDa protein also disappears from the wall extracts, and preliminary biochemical evidence suggests it to be an extensin. It should be noted that wounding of barley and oat leaves led to a slight decrease of HRGP in the soluble fraction of the cell walls and a compensating increase in the insoluble fraction.[47] This is a phenomenon that merits further attention.

$H_2O_2$ produced by elicitation may be a second messenger in plants. Stimulation of cultured soybean cells with elicitors (crude fungal wall extract or oligogalacturonide) leads to a rapid (within 5 min) production of $H_2O_2$ and its use by extracellular peroxidases. Exogenously added $H_2O_2$ stimulated phytoalexin production, and elicitor-stimulated phytoalexin production was inhibited by the addition of catalase.[90] These results suggest that $H_2O_2$ may play an important role in inducing defense responses.

### D. EXPRESSION DURING DEVELOPMENT

The soybean H/PRPs SbPRP1, SbPRP2, and SbPRP3 are differentially expressed during development. PRP1 is the predominant form in mature hypocotyls, roots, and immature seed coats. PRP2 is the predominant form in the apical hypocotyl and young suspension cells. PRP3 is the major form in aerial parts of the plant and is not found in roots.[91] The 1A10 gene transcript, which shows a high sequence identity to SbPRP1, accumulates during early seedling growth in soybean, but is not found in the bottom 5 mm of the root tip.[75] All of this suggests that each protein serves a particular function during the normal development of the plant. A low level of transcription of ENOD12 was found in stems and flowers, suggesting that the ENOD12 gene product also serves some role in the plant outside of nodule initiation.[81]

## IV. CONCLUSIONS AND FUTURE PROSPECTS

A great deal has been learned in the last 6 years about the structure and expression of the HRGPs and H/PRPs with the advent of molecular biological approaches. Many genes have been cloned, and the application of promoter analysis is beginning to elucidate the molecular mechanisms regulating their expression. The discovery of $H_2O_2$ as a potential second messenger that stimulates both rapid cell wall changes and changes in gene expression opens up a new avenue of study.

Because of their repeating structure, rod-like appearance, and ability to cross-link, as well as their accumulation at points of stress in plants, HRGPs probably function in strengthening the wall. This may be important in certain cell types and at particular times during normal development, as well as specifically at sites of infection in order to impede pathogen growth. Is pathogen- or wound-induced HRGP expression due to induction of a completely separate signal pathway, or is it only "tapping into" the normal developmentally induced signal pathway?

The accumulation of the H/PRP ENOD2 in a tissue that is thought to act as a barrier to diffusion of oxygen into root nodules raises the possibility of an additional function for these proteins.[79] The nodule inner cortex has relatively few and small intercellular spaces, which may be the morphological trait required for the diffusion barrier. Interestingly, soybean nodules showed the ability to adapt to oxygen pressures above and below ambient levels by decreasing the number and size of cortical intercellular air-spaces with increasing oxygen pressure. This was the result of a change in cell structure and the deposition of an electron dense material within intercellular spaces.[92] Although no attempt was made to detect HRGPs or PRPs, this electron-dense material was similar in appearance to that seen filling intercellular spaces in necrotic tissue of tobacco infected by TMV.[34] This suggests an easily testable hypothesis: is the level of ENOD2 accumulation regulated by oxygen levels? Additionally, could the accumulation of HRGPs in intercellular spaces at the site of pathogen infection or wounding act as a barrier to oxygen diffusion, thus limiting pathogen growth or cellular oxidation damage?

A major question that remains to be answered is: what are the specific functions of each of the HRGPs and H/PRPs? Hypotheses have been based only on correlative evidence as there are no known mutants that lack any of the proteins characterized so far. However, because many of the genes have been cloned, and it is possible to transform and regenerate the source plants (tobacco, carrot, alfalfa), the tools are in hand to address the question directly. Expression of antisense or ribozyme constructs at the appropriate time and place to block translation of a HRGP or H/PRP, or overexpression or ectopic expression of the genes could provide more direct evidence about the function of these proteins.

## REFERENCES

1. **Dickman, M. B., Podilla, G. K., and Kolattukudy, P. E.,** Insertion of cutinase gene into a wound pathogen enables it to infect inact host, *Nature (London),* 352, 446, 1989.
2. **Cassab, G. I. and Varner, J. E.,** Cell wall proteins, *Annu. Rev. Plant. Physiol. Plant Mol. Biol.,* 39, 321, 1988.
3. **Cooper, J. B., Chen, J. A., Van-Holst, G. J., and Varner, J. E.,** Hydroxyproline-rich glycoproteins of plant cell walls, *Trends Biochem. Sci.,* 12, 24, 1987.
4. **Cooper, J. B.,** Cell wall extensin genes, in *Plant Gene Research: Basic Knowledge and Application: Temporal and Spatial Regulation of Plant Genes,* Verma, D. P. S. and Goldberg, R. B., Eds., Springer-Verlag, NY, 1988, 235.
5. **Jones, R. L. and Robinson, D. G.,** Protein secretion in plants, *New Phytol.,* 111, 567, 1989.
6. **Showalter, A. M. and Varner, J. E.,** Biology and molecular biology of plant hydroxyproline-rich glycoproteins, in *The Biochemistry of Plants: A Comprehensive Treatise,* Vol. 15, Marcus, A., Ed., Academic Press, San Diego, 1989, 485.
7. **Showalter, A. M. and Rumeau, D.,** Molecular biology of plant cell wall hydroxyproline-rich glycoproteins, in *Organization and Assembly of Plant and Animal Extracellular Matrix,* Adair, W. S. and Mecham, R. P., Eds., Academic Press, San Diego, 1990, 247.
8. **Tierney, M. L. and Varner, J. E.,** The extensins, *Plant Physiol.,* 84, 1, 1987.
9. **Varner, J. E. and Lin, L.-S.,** Plant cell wall architecture, *Cell,* 56, 231, 1989.
10. **Lamport, D. T. A. and Northcote, D. H.,** Hydroxyproline in primary cell walls of higher plants, *Nature (London),* 118, 665, 1960.
11. **Dougall, D. K. and Shimbayashi, K.,** Factors affecting growth of tobacco callus tissue and its incorporation of tyrosine, *Plant Physiol.,* 35, 396, 1960.
12. **Cooper, J. B. and Varner, J. E.,** Insolubilization of hydroxyproline-rich cell wall glycoprotein in aerated carrot root slices, *Biochem. Biophys. Res. Commun.,* 112, 161, 1983.
13. **Smith, J. J., Muldoon, E. P., and Lamport, D. T. A.,** Isolation of extensin precursors by direct elution of intact tomato cell suspension cultures, *Phytochemistry,* 23, 1233, 1984.
14. **Biggs, K. J. and Fry, S. C.,** Solubilization of covalently bound extensin from *Capsicum* cell walls, *Plant Physiol.,* 92, 197, 1990.

15. **Corbin, D. R., Sauer, N., and Lamb, C.,** Differential regulation of a hydroxyproline-rich glycoprotein gene family in wounded and infected plants, *Mol. Cell. Biol.*, 7, 4337, 1987.
16. **Kieliszewski, M. J., Leykam, J. F., and Lamport, D. T. A.,** Structure of the threonine-rich extensin from *Zea mays, Plant Physiol.*, 92, 316, 1990.
17. **Hood, E. E., Shen, Q. X., and Varner, J. E.,** A developmentally regulated hydroxyproline-rich glycoprotein in maize pericarp cell walls, *Plant Physiol.*, 87, 138, 1988.
18. **Stiefel, V., Péres-Grau, L., Albericio, F., Giralt, E., Ruiz-Avila, L., Dolors, L., and Puigdomènech, P.,** Molecular cloning of cDNAs encoding a putative cell wall protein from *Zea mays* and immunological identification of related polypeptides, *Plant Mol. Biol.*, 11, 483, 1988.
19. **Li, X., Kieliszewski, M., and Lamport, D. T. A.,** A chenopod extensin lacks repetitive tetrahydroxyproline blocks, *Plant Physiol.*, 92, 327, 1990.
20. **Stafstrom, J. P. and Staehelin, L. A.,** The role of carbohydrate in maintaining extensin in an extended conformation, *Plant Physiol.*, 81, 242, 1986.
21. **Everdeen, D. S., Kiefer, S., Willard, J. J., Muldoon, E. P., Dey, P. M., Li, X. B., and Lamport, D. T. A.,** Enzymic cross-linkage of monomeric extension precursors *in vitro, Plant Physiol.*, 87, 616, 1988.
22. **Bolwell, G. P., Robbins, M. P., and Dixon, R. A.,** Elicitor-induced prolyl hydroxylase EC-1.14.11.2 from french bean *Phaseolus vulgaris*. Localization, purification and properties, *Biochem. J.*, 229, 693, 1985.
23. **Stafstrom, J. P. and Staehelin, L. A.,** Antibody localization of extensin in cell walls of carrot storage roots, *Planta*, 174, 321, 1988.
24. **Evans, M. I., Gatehouse, L. N., Gatehouse, J. A., Yarwood, J. N., Boulter, D., and Croy, R. R. D.,** The extensin gene family in oilseed rape (*Brassica napus* L.): Characterization of sequences of representative members of the family, *Mol. Gen. Genet.*, 223, 273, 1990.
25. **Cassab, G. I. and Varner, J. E.,** Immunocytolocalization of extensin in developing soybean seed coats by immunogold-silver staining and by tissue printing on nitrocellulose paper, *J. Cell Biol.*, 105, 2581, 1987.
26. **Benhamou, N., Mazau, D., and Esquerré-Tugayé, M. T.,** Immunocytochemical localization of hydroxyproline-rich glycoproteins in tomato root cells infected by *Fusarium oxysporum* f.sp. *radicis-lycopersici:* study of a compatible interaction, *Phytopathology*, 80, 163, 1990.
27. **Esquerré-Tugayé, M-T. and Lamport, D. T. A.,** Cell surfaces in plant-microorganism interactions. I. A structural investigation of cell wall hydroxyproline-rich glycoproteins which accumulate in fungus-infected plants, *Plant Physiol.*, 64, 314, 1979.
28. **Hammerschmidt, R., Lamport, D. T. A., and Muldoon, E. P.,** Cell wall hydroxyproline enhancement and lignin deposition as an early event in the resistance of cucumber to *Cladosporium cucumerinum, Physiol. Plant Pathol.*, 24, 43, 1984.
29. **Lawton, M. A. and Lamb, C. J.,** Transcriptional activation of plant defense genes by fungal elicitor, wounding and infection, *Mol. Cell. Biol.*, 7, 335, 1987.
30. **O'Connell, R. J., Brown, I. R., Mansfield, J. W., Bailey, J. A., Mazau, D., Rumeau, D., and Esquerré-Tugayé, M-T.,** Immunocytochemical localization of hydroxyproline-rich glycoproteins accumulating in melon and bean at sites of resistance to bacteria and fungi, *Mol. Plant-Microbe Interact.*, 3, 33, 1990.
31. **Rumeau, D., Mazau, D., Panabieres, F., Delseny, M., and Esquerré-Tugayé, M-T.,** Accumulation of hydroxyproline-rich glycoprotein messenger RNA in infected or ethylene treated melon plants, *Physiol. Mol. Plant Pathol.*, 33, 419, 1988.
32. **Showalter, A. M., Bell, J. N., Cramer, C. L., Bailey, J. A., Varner, J. E., and Lamb, C. J.,** Accumulation of hydroxyproline-rich glycoprotein mRNAs in response to fungal elicitor and infection, *Proc. Natl. Acad. Sci. U.S.A.*, 82, 6551, 1985.
33. **Mazau, D. and Esquerré-Tugayé, M.-T.,** Hydroxyproline-rich glycoprotein accumulation in the cell walls of plants infected by various pathogens, *Physiol. Mol. Plant Pathol.*, 29, 147, 1986.
34. **Benhamou, N., Mazau, D., Esquerré-Tugayé, M-T., and Asselin, A.,** Immunogold localization of hydroxyproline-rich glycoproteins in necrotic tissue of *Nicotiana tabacum* L. cv. *Xanthi*-nc infected by tobacco mosaic virus, *Physiol. Mol. Plant Pathol.*, 36, 129, 1990.
35. **Wang, M. C., Lin, J. J., Duran-Vila, N., and Semancik, J. S.,** Alteration in cell wall composition and structure in viroid-infected cells, *Physiol. Mol. Plant Pathol.*, 28, 107, 1986.
36. **Esquerré-Tugayé, M-T., Lafitte, C., Mazau, D., Toppan, A., and Touzé, A.,** Cell surfaces in plant-microorganism interactions. II. Evidence for the accumulation of hydroxyproline-rich glycoproteins in the cell wall of diseased plants as a defense mechanism, *Plant Physiol.*, 64, 320, 1979.
37. **Stermer, B. A. and Hammerschmidt, R.,** Association of heat shock induced resistance to disease with increased accumulation of insoluble extensin and ethylene synthesis, *Physiol. Mol. Plant Pathol.*, 31, 453, 1987.

38. **Rumeau, D., Maher, E. A., Kelman, A., and Showalter, A. M.,** Extensin and phenylalanine ammonia-lyase gene expression altered in potato tubers in response to wounding, hypoxia, and *Erwinia carotovora* infection, *Plant Physiol.,* 93, 1134, 1990.
39. **Templeton, M. D., Dixon, R. A., Lamb, C. J., and Lawton, M. A.,** Hydroxyproline-rich glycoprotein transcripts exhibit differential spatial patterns of accumulation in compatible and incompatible interactions between *Phaseolus vulgaris* and *Colletotrichum lindemuthianum, Plant Physiol.,* 94, 1265, 1990.
40. **Chrispeels, M. J., Sadava, D., and Cho, Y. P.,** Enhancement of extensin biosynthesis in ageing disks of carrot storage tissue, *J. Exp. Bot.,* 25, 1157, 1974.
41. **Chen, J. and Varner, J. E.,** Isolation and characterization of complementary DNA clones for carrot extensin and a proline-rich 33-kilodalton protein, *Proc. Natl. Acad. Sci. U.S.A.,* 82, 4399, 1985.
42. **Ecker, J. R. and Davis, R. W.,** Plant defense genes are regulated by ethylene, *Proc. Natl. Acad. Sci. U.S.A.,* 84, 5202, 1987.
43. **Ludevid, M. D., Ruiz-Avila, L., Vallés, M. P., Steifel, V., Torrent, M., Torné, J. M., and Puigdomènech, P.,** Expression of genes for cell-wall proteins in dividing and wounded tissues of *Zea mays* L., *Planta,* 180, 524, 1990.
44. **Sauer, N., Corbin, D. R., Keller, B., and Lamb, C. J.,** Cloning and characterization of a wound-specific hydroxyproline-rich glycoprotein in *Phaseolus vulgaris, Plant Cell Environ.,* 13, 257, 1990.
45. **Showalter, A. M., Zhou, J., Rumeau, D., Worst, S. G., and Varner, J. E.,** Tomato extensin and extensin-like cDNAs: structure and expression in response to wounding, *Plant Mol. Biol.,* 16, 547, 1991.
46. **Tierney, M. L., Wiechert, J., and Pluymers, D.,** Analysis of the expression of extensin and P33-related cell wall proteins in carrot and soybean, *Mol. Gen. Genet.,* 211, 393, 1988.
47. **Li, Z.-C. and McClure, J. W.,** Soluble and bound hydroxyproline-rich glycoproteins in control and wounded oat and barley primary leaves, *Phytochemistry,* 29, 2811, 1990.
48. **Chen, J. and Varner, J. E.,** An extracellular matrix protein in plants. Characterization of a genomic clone for carrot extensin, *EMBO J.,* 4, 2145, 1985.
49. **Roby, D., Toppan, A., and Esquerré-Tugayé, M. T.,** Cell surfaces in plant-microorganism interactions VII. Elicitors of fungal and of plant origin trigger the synthesis of ethylene and of cell wall hydroxyproline-rich glycoprotein in plants, *Plant Physiol.,* 77, 700, 1985.
50. **Culver, J. N. and Dawson, W. O.,** Modifications of the coat protein gene of tobacco mosaic virus resulting in the induction of necrosis, in *Recognition and Response in Plant-Virus Interactions,* Vol. 41, Fraser, R. S. S., Ed., Springer-Verlag, Berlin, 1990, 337.
51. **Boller, T.,** Ethylene-induced biochemical defenses against pathogens, in *Plant Growth Substances,* Wareing, P. F., Ed., Academic Press, San Diego, 1982, 303.
52. **Paradies, I., Humme, B., Heitefuss, R., and Paxton, J.,** Ethylene: indicator but not inducer of phytoalexin synthesis in soybean, *Plant Physiol.,* 66, 1106, 1980.
53. **Mauch, F., Hadwiger, L. A., and Boller, T.,** Ethylene: symptom, not signal for the induction of chitinase and β-1,3 glucanase in pea pods by pathogens and elicitor, *Plant Physiol.,* 76, 607, 1984.
54. **Meyer, D. J., Afonso, C. L., and Galbraith, D. W.,** Isolation and characterization of monoclonal antibodies directed against plant plasma membrane and cell wall epitopes. Identification of a monoclonal antibody that recognizes extensin and analyis of the process of epitope biosynthesis in plant tissues and cell cultures, *J. Cell. Biol.,* 107, 163, 1988.
55. **Hood, E. E., Hood, K. R., and Fritz, S. E.,** Hydroxyproline-rich glycoproteins in cell walls of pericarp from maize, *Plant Sci.,* Vol. 76(in press), 1991.
56. **Fritz, S. E., Hood, K. R., and Hood, E. E.,** Localization of soluble and insoluble fractions of hydroxyproline-rich glycoproteins during maize kernel development, *J. Cell Sci.,* 98, 545, 1991.
57. **Lamport, D. T. A.,** The protein component of primary cell walls, *Adv. Bot. Res.,* 2, 151, 1965.
58. **Mazau, D., Rumeau, D., and Esquerré-Tugayé, M-T.,** Two different families of hydroxyproline-rich glycoproteins in melon cells, *Plant Physiol.,* 86, 540, 1988.
59. **Keller, B. and Lamb, C. J.,** Specific expression of a novel cell wall hydroxyproline-rich glycoprotein gene in lateral root initiation, *Genes Dev.,* 3, 1639, 1989.
60. **Memelink, J., Hoge, J. H. C., and Schilperoort, R. A.,** Cytokinin stress changes the developmental regulation of several defence-related genes in tobacco, *EMBO J.,* 6, 3579, 1987.
61. **Powell, P. A., Wycoff, K. L., Corbin, D. R., Hedrick, S. A., Dixon, R. A., and Lamb, C. J.,** Wound and infection activation of the promoter of a plant cell-wall hydroxyproline-rich glycoprotein gene superimposed on a novel pattern for tissue-specific developmental expression, *EMBO J.* (submitted), 1991.
62. **Ye, Z. H. and Varner, J. E.,** Tissue specific expression of cell wall proteins in developing soybean tissues, *Plant Cell,* 3, 23, 1990.
63. **Stiefel, V., Ruiz-Avila, L., Raz, R., Vallés, M. P., Gómez, J., Pagés, M., Martínez-Izquierdo, J. A., Ludevid, M. D., Langdale, J. A., Nelson, T., and Puigdomènech, P.,** Expression of a maize cell wall hydroxyproline-rich glycoprotein gene in early leaf and root vascular differentiation, *Plant Cell,* 2, 785, 1990.

64. **Hood, K. R., Baasiri, R. A., Fritz, S. E., and Hood, E. E.,** Biochemical and tissue-print analyses of hydroxyproline-rich glycoproteins in cell walls of sporophytic maize tissues, *Plant Physiol.*, 96, 1214, 1991.
65. **Goldman, M. H. de S.,** personal communication, 1991.
66. **Meeks-Wagner, D. R., Dennis, E. S., Tran Thanh Van, K., and Peacock, W. J.,** Tobacco genes expressed during in vitro floral initiation and their expression during normal plant development, *Plant Cell*, 1, 25, 1989.
67. **Neale, A. D., Wahleithner, J. A., Lund, M., Bonnett, H. T., Kelly, A., Meeks-Wagner, D. R., Peacock, W. J., and Dennis, E. S.,** Chitinase, β-1,3-glucanase, osmotin, and extensin are expressed in tobacco explants during flower formation, *Plant Cell*, 2, 673, 1990.
68. **Wycoff, K. L. Powell, P. A., Lamb, C. J., and Dixon, R. A.,** unpublished results, 1991.
69. **Holdsworth, M. J. and Laties, G. G.,** Site-specific binding of a nuclear factor to the carrot extensin gene is influenced by both ethylene and wounding, *Planta*, 179, 17, 1989.
70. **Holdsworth, M. J. and Laties, G. G.,** Identification of a wound-induced inhibitor of a nuclear factor that binds the carrot extensin gene, *Planta*, 180, 74, 1989.
71. **Datta, K., Schmidt, A., and Marcus, A.,** Characterization of two soybean repetitive proline-rich proteins and a cognate cDNA from germinated axes, *Plant Cell*, 1, 945, 1989.
72. **Kleis-San Francisco, S. M., and Tierney, M. L.,** Isolation and characterization of a proline-rich cell wall protein from soybean seedlings, *Plant Physiol.*, 94, 1897, 1990.
73. **Hong, J. C., Nagao, R. T., and Key, J. L.,** Characterization and sequence analysis of a developmentally regulated putative cell wall protein gene isolated from soybean, *J. Biol. Chem.*, 262, 8367, 1987.
74. **Hong, J. C., Nagao, R. T., and Key, J. L.,** Characterization of a proline-rich cell wall protein gene family of soybean, *J. Biol. Chem.*, 265, 2470, 1990.
75. **Averyhart-Fullard, V., Datta, K., and Marcus, A.,** A hydroxyproline-rich protein in the soybean cell wall, *Proc. Natl. Acad. Sci. U.S.A.*, 85, 1082, 1988.
76. **Datta, K. and Marcus, A.,** Nucleotide sequence of a gene encoding soybean repetitive proline-rich protein 3, *Plant Mol. Biol.*, 14, 285, 1990.
77. **Franssen, H. J., Nap, J.-P, Gloudemans, T., Stiekema, W., van Dam, H., Govers, F., Louwerse, J., von Kammen, A., and Bisseling, T.,** Characterization of cDNA for nodulin-75 of soybean: a gene product involved in early stages of root nodule development, *Proc. Natl. Acad. Sci. U.S.A.*, 84, 4495, 1987.
78. **Franssen, H. J., Thompson, D. V., Idler, K., Kormelink, R., Van, K. A., and Bisseling, T.,** Nucleotide sequence of two soybean ENOD2 early nodulin genes encoding Ngm-75, *Plant Mol. Biol.*, 14, 103, 1990.
79. **van de Wiel, C., Scheres, B., Franssen, H., van Lierop, M. J., van Lammeren, A., van Kammen, A., and Bisseling, T.,** The early nodulin transcript ENOD2 is located in the nodule parenchyma (inner cortex) of pea and soybean root nodules, *EMBO J.*, 9, 1, 1990.
80. **Dickstein, R., Bisseling, T., Reinhold, V. N., and Ausubel, F. M.,** Expression of nodule-specific genes in alfalfa root nodules blocked at an early stage of development, *Genes Dev.*, 2, 677, 1988.
81. **Scheres, B., van de Wiel, C., Zalensky, A., Horvath, B., Spaink, H., van Eck, H., Zwartkruis, F., Wolters, A. M., Gloudemans, T., van Kammen, A., and Bisseling, T.,** The ENOD12 gene product is involved in the infection process during the pea-*Rhizobium* interaction, *Cell*, 60, 281, 1990.
82. **Cassab, G. I.,** Arabinogalactan proteins during the development of soybean *glycine-max* root nodules, *Planta*, 168, 441, 1986.
83. **Moerman, M., Nap, J. P., Govers, F., Schilperoort, R., van Kammen, A., and Bisseling, T.,** *Rhizobium nod* genes are involved in the induction of two early nodulin genes in *Vicia sativa* root nodules, *Plant Mol. Biol.*, 9, 171, 1987.
84. **Hirsch, A. M., Bhuvaneswari, T. V., Torrey, J. G., and Bisseling, T.,** Early nodulin genes are induced in alfalfa root outgrowths elicited by auxin transport inhibitors, *Proc. Natl. Acad. Sci. U.S.A.*, 86, 1244, 1989.
85. **Truchet, G., Barker, D., Camut, S., de Billy, F., Vasse, J., and Huguet, T.,** Alfalfa nodulation in the absence of *Rhizobium*, *Mol. Gen. Genet.*, 219, 65, 1989.
86. **van Brussel, A. A. N., Recourt, K., Pees, E., Spaink, H. P., Tak, T., Wijffelman, C. A., Kijne, J. W., and Lugtenberg, J. J.,** A biovar-specific signal of *Rhizobium leguminosarum* bv. *viciae* induces increased nodulation gene-inducing activity in root exudate of *Vicia sativa* subsp. *nigra*, *J. Bacteriol.*, 172, 5394, 1990.
87. **Gloudemans, T. and Bisseling, T.,** Plant gene expression in early stages of *Rhizobium*-legume symbiosis, *Plant Sci.*, 65, 1, 1989.
88. **Lerouge, P., Roche, P., Faucher, C., Maillet, F., Truchet, G., Promé, J. C., and Dénarié, J.,** Symbiotic host-specificity of *Rhizobium meliloti* is determined by a sulphated and acylated glucosamine oligosaccharide signal, *Nature (London)*, 344, 781, 1990.
89. **Bradley, D. J. and Lamb, C. J.,** A rapid insolubilization of plant cell wall/extracellular proteins in response to elicitation, (in preparation), 1991.

90. **Apostol, I., Henstein, P. F., and Low, P. S.,** Rapid stimulation of an oxidative burst during elicitation of cultured plant cells, *Plant Physiol.*, 90, 109, 1989.
91. **Hong, J. C., Nagao, R. T., and Key, J. L.,** Developmentally regulated expression of soybean proline-rich cell wall protein genes, *Plant Cell,* 1, 937, 1989.
92. **Parsons, R. and Day, D. A.,** Mechanism of soybean nodule adaptation to different oxygen pressures, *Plant Cell Environ.*, 13, 501, 1990.

# MOLECULAR DETERMINANTS OF PLANT-VIRUS INTERACTION

**Steve Daubert**

## TABLE OF CONTENTS

# I. INTRODUCTION

Viruses communicate with their hosts from within. They associate intimately with their host's own information networks, parasitizing the host's gene expression systems for the expression of their own genetic potential. They then modify the host cell membrane so as to send themselves across the barriers that separate one cell from the next.

Some viral-coded proteins are specialized for the specific recognition of host proteins. The replicase complex of bacteriophage Q-β is an example. Only one of its components is a virus-coded protein; the other three proteins in the functional tetramer are host molecules.[1] Such specific intermolecular recognition events such as this are thought to be central to viral function in general, e.g., determining pathogenesis and host range in animal viruses[2] and in bacterial viruses.[3] Many proteins encoded by bacteriophage T4 are pathogenesis factors: some of them specifically bind host RNA polymerase to enhance the transcription of viral DNA while inhibiting host transcription.[4] Host transcription is also specifically inhibited by picornaviruses (e.g., by poliovirus infection).[5] Picornaviruses also express a specific inhibitor of translation, a protease that selectivity turns off host translation by recognizing and then destroying the host component which binds capped messenger RNA (to the benefit of the cap-independent viral RNA expression system).[6] And, not to overlook the host's potential for specificity in the host-virus interaction, pathogenesis can be mitigated by the targeted protein-protein recognition reactions arising in the immune system in higher animals.

Plant viruses and viroids, the noncellular parasites of plants, encompass a diverse set of structures and compositions (Table 1). As opposed to the viruses of animals and bacteria, a set of specific molecular interactions controlling cytopathology has not been described in the interactions of these pathogens with the cells of their plant hosts. Nonetheless, at levels of organization above that of the isolated cells, the replication of these microbes often alters host plant metabolism, with consequences that are readily apparent on the macroscopic level as disease symptoms. The appearance of symptoms reflects initiating molecular events, and the correlation of symptom induction with specific genetic determinants identifies the gene products that ultimately produce (or preclude) pathogenesis.

Specific molecular interactions appear to control the cell-to-cell and organ-to-organ spread of viral infection through the host plant. The failure of virus-coded determinants controlling systemic spread to interact productively with host components could block the establishment of systemic infection. This appears to be the case in nonhost plants, where virus fails to escape from initially infected cells. Host plants are defined as species in which virus can move cell-to-cell. In host plants, the appearance of symptoms, including the reactions of resistant hosts which limit the extent of infection, appear to reflect a response arising above the level of the single cell, e.g., at the organized-tissue level of cellular organization.

Specific, molecular-level host-virus recognition events are inferred when genetic determinants from both a particular host species and a particular viral strain interact to control susceptibility. Correlations between genetic analysis, physiological studies, and sequence data are beginning to reveal the classes of primary gene product-level events that eventually determine pathogenesis. These are discussed in this chapter in the order of the increasingly complex levels of cellular organization at which the interactions occur. This survey builds upon several previous reviews which cover determinants of symptom expression,[7] disease induction,[8] cell-cell movement,[9-11] and plasmodesmata.[12]

# II. NONSPECIFICITY OF THE INTERACTION ON THE INTRACELLULAR LEVEL

Initial viral entry into host cells is passive, aided mechanically by a vector organism. In the majority of cases analyzed to date, mechanically inoculated single plant cells non-

**TABLE 1**
**Plant Virus Characteristics**

| Group[a] (example) | Genome[b] | Particle morphology (length/diameter nm) |
|---|---|---|
| Tobamo-Tobacco mosaic | 1 RNA | Stiff rod (300) |
| Potex-Potato virus X | 1 RNA | Flexious rod (500) |
| Poty-Potato virus Y | 1 RNA | Flexious rod (750) |
| Luteo-Potato leafroll | 1 RNA | Sphere (24) |
| Como-Cowpea mosaic | 2 RNAs | Sphere (28) |
| Nepo-Tobacco ringspot | 2 RNAs (+ satellite RNA) | Sphere (28) |
| Cucumo-Cucumber mosaic | 3 RNAs (+ satellite RNA) | Sphere (30) |
| Bromo-Brome mosaic | 4 RNAs | Sphere (26) |
| Caulimo-Cauliflower mosaic | 1 DNA | Sphere (50) |
| Viroids[a] Potato spindle tuber | 1RNA | (No encapsidated particle) |

[a] Viroids are grouped separately from the viruses; their RNA genome is circular.
[b] Genomic RNAs are linear single strands; the caulimovirus genome is a circular, double-stranded DNA.

**TABLE 2**
**Replication of Viruses in Single Cells Derived from Plant Species Beyond the Usual Viral Host Range**

| Virus | Systemic host (family) | Nonhost source(s) of cells supporting replication |
|---|---|---|
| Cowpea mosaic | *Vigna unguiculata* (Leguminosae) | *Nicotiana tabacum* (Solanaceae) |
| Tobacco mosaic | *N. tabacum* (Solanaceae) | *Ipomoea batatas* (Convolvulaceae); *Hordeum vulgare* (Gramineae) |
| Brome mosaic | *H. vulgare* (Gramineae) | *Raphanus sativus* (Brassicaceae); *N. tabacum* (Solanaceae) |
| Cauliflower mosaic | *Brassica campestris* (Brassicaceae) | *Gossypium hursutis* (Malvaceae) |
| Cucumber mosaic | *Cucumis sativus* (Cucurbitaceae) | *Nicotiana tabacum* (Solanaceae) |

specifically support viral infection (Table 2). This does not correlate with the selectivity observed in viruses and their host plants in the establishment of systemic infections. Protoplasts from plants which are nonhosts for a given virus nonetheless support the replication of that virus. Examples include the replication of cowpea mosaic virus (CPMV) in protoplasts derived from tobacco, which is a nonhost plant for CPMV.[13] Sweet potato appears to be a nonhost for tobacco mosaic virus (TMV), yet viral replication is supported in up to 90% of protoplasts inoculated with TMV RNA by electroporation.[14] Systemic TMV infection is limited to dicotyledonous host plants, yet the virus infects barley protoplasts (Fig. 2a in Sacher et al.[15]). Brome mosaic virus (BMV) is generally limited in its systemic infection range to graminaceous hosts, but will propagate in protoplasts from nonhost plants in the Brassicaceae[16] or Solanaceae.[17] Cauliflower mosaic virus (CaMV), limited to brassicaceous and solanaceous host plants, infects cotton cells in culture.[18] Tobacco necrotic dwarf luteovirus is phloem-limited *in planta,* but will infect mesophyll protoplasts in culture.[19]

Another manifestation of this breakdown of host range barriers on the single-cell level is in the detection of very low levels of viral replication in the inoculated leaves of nonhosts. This phenomena, termed subliminal infection,[20] is exemplified in the characterization by Sulzinski and Zaitlin[21] of TMV infection of single cowpea or cotton leaf cells within inoculated leaves. Similarly, cucumber mosaic virus (CMV) will multiply subliminally in tobacco mesophyll cells[22] while failing to systemically infect that host. Other examples are given in Atabekov and Dorokhov.[9]

Specific reactions expressed by resistant plants are not recapitulated in protoplasts derived from them. For example, the cell death response is expressed in leaves of resistant *N/N* genotype tobacco plants against TMV, yet *in vitro*-inoculated tobacco protoplasts[23] derived from *N/N* plants support replication of TMV to the same levels as do protoplasts derived from susceptible *n/n* plants. Tomato protoplasts will asymptomatically support citrus exocortis viroid infection over many cell divisions[24] while intact plants would have shown obvious pathogenic symptoms. Thus, in general, host and tissue specificity, including specific resistance or pathogenesis reactions, is not evident at the intracellular or single-cell level.

## A. RESISTANCE AT THE INTRACELLULAR LEVEL

A few exceptions exist to the above generality. In these cases, resistance to systemic viral infection seen in specific host varieties was retained and expressed in isolated protoplasts. Thus, this resistance has a component that does not arise at points of cell-cell or higher-level tissue organization, but appears as an incompatibility between primary viral gene products and host components within single cells.

In an extensive analysis of the response of cowpea varieties and their protoplasts to CPMV, Beier et al.[25] reported that resistance most often appeared to function on the organized cell level. In "organized-cell resistance" (discussed below), hosts resistant at the whole plant level nonetheless give rise to susceptible protoplasts. However, Beier et al.[25] also found one host variety showing a rarer type of resistance, in which viral replication was inhibited in protoplasts as well as *in planta.* Viral accumulation in protoplasts from this host was reduced by 90%, in comparison with its accumulation in other varieties of cowpea,[26] and was undetectable *in planta.*

This cellular resistance has been characterized in molecular detail in another host-virus system, that of tomato and the tomato strain of TMV (TMV-L).[27] The host determinant controlling viral infectivity resides at the "*Tm-1*" genetic locus. TMV-L infects this variety of tomato to less than 5% of the titer it achieves in the corresponding +/+ genotype host, with similar or greater reductions seen in protoplasts derived from it.[28] However, the closely related TMV strain Lta1 is infectious in this host. The Lta1 determinant of infectivity has been identified as an alteration in the TMV-L replicase gene.

TMV strain Lta1 originated in a TMV-L infection and was selected by its growth/spread in tomato variety GCR 237. GCR 237 is genetically *Tm-1/Tm-1*; TMV-L inoculation of GCR 237 gives rise to no symptoms, and no virus can be detected postinoculation. TMV Lta1 is described as a "resistance-breaking" strain in this host.

Characterization of the Lta1 determinant revealed a 6-amino acid section in the viral replicase 126k domain (which is 1116 amino acids in length), in which Lta1 carries two

**TABLE 3**
**Determinants of Infectivity in the TMV Genome**

| Host (genotype) | Virus | Gene | Amino acid (position) | Host response |
|---|---|---|---|---|
| *Lycopersicon esculentum* (*Tm-1/Tm-1*) | TMV-L | Replicase | Gln(979),His(984) | Cellular resistance |
| | TMV Lta1 | | Glu(979),Tyr(984) | Susceptible |
| *Lycopersicon esculentum*(*Tm-2/Tm-2*) | TMV-L | m-Protein | Cys(68),Glu(133) | Non host |
| | TMV Ltb1 | | Phe(68),Lys(133) | Susceptible |
| *Nicotiana sylvestris* (*N'/N'*) | TMV U1 | Coat protein | Arg(46) | Susceptible |
| | TMV 46 | | Gly(46) | Organized cell resistance |

mutations relative to the parental strain (Table 3). Meshi et al.[27] further discovered that the single mutation at 984 in the replicase amino acid sequence was itself only weakly infectious in the *Tm-1/Tm-1* host, while the Gln-to-Glu single change at 979 was more infectious. Characterization of the virus recovered from the infection initiated by the 979 single mutant, however, revealed the reappearance of second-site mutations in the recovered isolates. These various secondary mutations were found in the same domain, between amino acid positions 975 and 992. This suggests that *Tm-1*-mediated resistance was initially overcome with the appearance of a mutation at position 979: this single-site mutant would have established limited replication in the restrictive host, giving rise to a viral population from which the subsequent mutation at position 984 most probably arose. The greater replication capacity of the double mutant would then have brought it to dominate the viral population. This same two-amino acid change was found in other independently isolated resistance-breaking strains.[27]

The TMV Lta1 characterization suggests that the viral replicase interacts with a host factor, perhaps at the cell membrane. In its simplest form, this scenario predicts that the *Tm-1* locus and its alleles would encode the host factor, and that the TMV-L-encoded replicase fails to productively interact with the *Tm-1*-encoded variant. *Tm-1*-mediated resistance to TMV-L should therefore be manifest within single cells, as was demonstrated by Watanabe et al.[28] in their description of TMV-L-resistant *Tm-1*-derived protoplasts. Extending this system for the analysis of viral replicase function, Okada et al.[29] have found that the Lta1 viral replicase will not support the replication of the nonmutant viral genome in *trans* when both are present in *Tm-1* protoplasts.

# III. SPECIFIC RESISTANCE DETERMINANTS AT THE CELL-CELL BARRIER

Systemic infection will follow only if virus replicating in initially infected cells encodes a determinant enabling it to escape and spread to other cells. In most plants, virus fails to escape. Resistance thus arises at an organizational level above that of the single cell, namely, at the intercellular communication channels that organize cells into tissues.[9] A block to viral escape from initially infected single cells occurs at these channels in nonhost plants.

Nonhost resistance arising from the blockade against viral cell-cell spread was initially deduced from observations on mixed viral infections. There is a vast catalog of modifications to symptoms arising from mixed infection, the archetype perhaps being the synergy of potato Y potyvirus (PVY) with potato X potexvirus (PVX).[30] In a few of these examples of interactive coinfection, the basis of the alteration to symptomatology can be traced to the release of a virus from a host-specific block to cell-cell movment, facilitated by a second

**TABLE 4**
**Viruses that Overcome Blocks to Systemic Infectivity in the Presence of a Second, Coinfecting Virus[a]**

| Virus | Resistance type | Plant | Coinfecting virus facilitating movement | Ref. |
|---|---|---|---|---|
| CMV | Nonhost: block to cell-cell spread | *Ipomoea batatas* | SPFMV | 31 |
| TMV | | *Hordeum vulgare* | BSMV | 32 |
| RCMV | | *Nicotiana tabacum* | TMV | 33 |
| TMV | | *Lycopersicon esculentum* (Tm-2) | PVX | 9 |
| SBMV-C | | *Phaseolus vulgaris* | TMV-1[c] | 34 |
| BGMV | *Phloem limitation* | *Phaseolus vulgaris* | TMV-1 | 35 |
| PLRV | | *Nicotiana clevelandii* | PVY | 36 |
| | | | PEBV | |
| | | | CMoV | |
| | | | NMV | |
| | | | TMV | |
| BWYV | | | PVY | |
| | | | PEVB | |
| | | | CMoV | |
| TSV | Organized cell resistance[b] | *Gossypium hirsutum* | Cotton anthocyanosis | 37 |
| TNV | | *Cucumis sativus* | CGMMV | 38 |

[a] Key to virus abbreviations: BGMV, bean golden mosaic geminivirus; BSMV, barley stripe mosaic hordeivirus; BWYV, beet western yellows luteovirus; CGMMV, cucumber green mottle mosaic tobamovirus; CMoV, carrot mottle (ungrouped) virus; CMV, cucumber mosaic cucumovirus; cotton anthocyanosis (ungrouped) virus; NMV, narcissus mosaic potexvirus; PEBV, pea early browning tobravirus; PLRV, potato leafroll luteovirus; PVX, potato X potexvirus; PVY, potato Y potyvirus; RCMV, red clover mottle comovirus; SBMV-C, southern bean mosaic sobemovirus (cowpea strain); SPFMV, sweet potato feathery mottle potyvirus; TMV, tobacco mosaic tobamovirus (legume strain); TNV, tobacco necrosis virus; TSV, tobacco streak ilarvirus.

[b] This resistance, defined in the text, involves a necrotic host response.

[c] In the SBMV-C/TMV-1 coinfection, SBMV escapes from nonhost resistance, then encounters organized cell resistance.

virus. For example, sweet potato appears to be a nonhost for CMV, but systemic CMV infection occurs when the plants are coinfected with sweet potato feathery mottle potyvirus (SPFMV).[31] The mixed infection is much more severe than infection by SPFMV alone. The observation suggested that sweet potato cells support CMV replication, but that CMV cannot move cell-to-cell until a second virus, one which is adept at systemic spread in sweet potato, coinfects.

TMV will infect barley protoplasts[15] but under field conditions will not systemically infect barley plants. However, systemic infection ensues when the plants are coinfected with barley stripe mosaic hordeivirus (BSMV).[32] In this case, even though the TMV concentration rises to 7 mg virus per milliliter of infected barley sap, the severity of the infection is no different from that of the BSMV infection alone. Other examples of viruses which facilitate the release of unrelated second viruses from the intercellular blockade against systemic infection are listed in Table 4.

## A. VIRUS-CODED MOVEMENT PROTEIN AS A DETERMINANT OF INFECTIVITY

The above examples suggest that a systemically infectious virus encodes a factor or factors that can interact with specific host components to (1) facilitate viral cell-to-cell spread, and (2) sometimes incidentally facilitate the spread of unrelated virus(es). In a nonhost, either a viral movement component is not expressed, or, if expressed, fails to stably interact with the host; the resulting failure of movement blocks infection of other cells.

The genetic basis of this viral block to infectivity at the intercellular barrier has been described in systems in which specific TMV strains fail to systemically infect host plants bearing specific determinants for resistance. In one such system, resistance to TMV-L is manifest in tomatoes bearing the *Tm-2* gene. Though TMV-L will normally infect protoplasts derived from this tomato variety,[39] the plants themselves do not support TMV infection. However, the virus can be released from the host intercellular barrier by coinfection with PVX.[9] The basis of TMV-L dysfunction in *Tm-2* tomato was described by the analysis of TMV Ltb1, a mutant which was selected for the ability to overcome *Tm-2*-mediated resistance. The mutation was mapped to the gene encoding the viral movement protein (m-protein; also called 30-K protein).

The TMV Ltb1 mutant originated in a TMV-L infection of a uniformly susceptible host, and was later selected by its transfer to the nonpermissive tomato variety, GCR 236[40] in which resistance is conferred by the *Tm-2* locus.[41] The resistance-breaking viral determinant was found to consist of two amino acid positions (Table 3). Further analysis of the separate mutations showed that virus carrying only the Glu-to-Lys change at amino acid position 133 in the m-protein induced either a necrotic reaction or no symptom at all in the *Tm-2/Tm-2* host; the Cys-to-Phe mutant (at position 68), however, established a mild systemic infection.[40] Thus, assuming that the two mutations did not occur simultaneously, the Cys-to-Phe single mutation likely occurred first. In the subsequent mild infection the second-site mutation would have arisen, resulting in a more aggressive mutant which would have replaced the single-site mutant in the population. Low level initial replication in the *Tm-2/Tm-2* genetic background is apparently also conferred by a single Glu-to-Lys mutation at position 52 in the m-protein amino acid sequence, which again fosters the Glu-to-Lys second-site change at position 68. This was observed in TMV C32, a mutant derived from TMV CH3 and characterized in the same study by Meshi et al.[40]

The TMV Ltb1 characterization suggested that an intermolecular interaction between the m-protein and a host factor determines the potential for viral movement. The *Tm-2* locus in tomato may encode an altered form of the host factor, with which the m-protein of TMV-L fails to productively interact. TMV-L can escape into the *Tm-2/+* host, causing a necrotic, partially restricted infection.[42]

In another case, a system with similar characteristics was developed to study the TMV LS-1 mutation, which is also in the m-protein gene.[43] Under nonpermissive conditions (temperatures above 32°C) the TMV LS-1 mutant infects protoplasts as well as the parental TMV-L strain does,[44] but its systemic spread *in planta* is completely restricted. Deom et al.[45] constructed transgenic tobacco plants in which the wild-type TMV-m protein is presumably synthesized in every cell. In these host plants, the capacity of TMV LS-1 to spread systemically is restored. Furthermore, aphid-inoculated potato leafroll luteovirus is seen to escape from its phloem-restricted tissue range and spread into all tissues of this transgenic host.[46] Thus, a compatible host-virus system has been artificially created: in unaltered plants the (as yet undescribed) host component with which the viral movement protein interacts serves as a resistance gene specific for TMV strain LS-1; the transgenic host carries a novel gene that renders it susceptible, permitting systemic infection by strain LS-1.

Watanabe et al.[47] described a determinant of TMV attenuation which appeared to reduce viral movement, yet mapped to the viral replicase gene. They concluded that the basis of attenuation was complex, arising ultimately from reduced synthesis of the m-protein RNA.

## B. CHARACTERIZATION OF THE CELL-CELL MOVEMENT FUNCTION

Tomenius et al.,[48] using immunogold microscopy, have reported that the m-protein is localized in host plasmodesmata. Wolf et al.[49] have analyzed the effective diameter of plasmodesmata in living, m-protein-containing cells. They have shown by analysis of dye exclusion limits that the maximum Stokes radius of particles passing through single plasmodesma is increased from approx. 1 nm in nontransgenic tobacco to 3 nm in transgenic cells expressing the m-protein. Deom et al.[50] have shown that this m-protein-catalyzed increase in plasmodesmatal exclusion limit is reduced in a tobacco variety in which resistance to TMV is mediated by the *N* gene. Moreover, this reduction is temperature sensitive: *N* gene resistance is not seen at temperatures above 28°C, and above that temperature the m-protein-increased plasmodesmatal exclusion limit returns to that seen in *n/n* tobacco cells.

This increase in maximum plasmodesmatal exclusion limit is still insufficient to allow the passage of TMV virions (18 nm minimum diameter), suggesting that some other viral form carries the infection cell-to-cell. Citovsky et al.[51] have shown that the m-protein of TMV is an RNA-binding protein *in vitro*. They have suggested that *in vivo* the protein may function in transplasmodesmatal movement of unincapsidated viral RNA.

Filamentous viruses are difficult to detect within plasmodesmatal[52] as particles, compared with the frequency of detection of isometric particles therein. In their description of uncommon micrographs of TMV and other filamentous virus particles within plasmodesmatal structures, Weintraub et al.[52] noted the potential for artifactual introduction of particles into plasmodesmatal during sample preparation, or in aged tissue. Elsewhere, the frequency of observation of TMV within plasmodesmatal has been described as "rarely, if ever,"[53] or, "no one has ever seen TMV in plasmodesmatal."[12] Encapsidated TMV would not necessarily be expected to be found between cells, since functional viral coat protein is not required for cell-to-cell movement.[11,54]

## C. NONSPECIFIC CELL-CELL MOVEMENT OF FILAMENTOUS VIRUSES

Table 2 lists examples of nonspecific facilitation of the movement of one virus by another. In these examples, the capacity to move systemically is conferred upon a virus which would itself be incapable of systemic infection if present alone. All of the viruses capable of this facilitation (Table 2, column 4) are morphologically filamentous (stiff or flexious rods). Thus, filamentous viruses appear nonexclusive in the facilitation of their own cell-cell movement, nonspecifically moving coinfecting secondary viruses as well.[33] This generality suggests that the filamentous viruses as a group may employ common aspects in their mechanism of transplasmodesmatal transport.

Nonspecific facilitation of movement is not usually seen, and coinfection does not often mobilize noninfectious viruses to overcome blocks to systemic infection (examples in Malyshenko et al.[33]). Once released from initially infected cells, the secondary virus must still overcome barriers to infectivity encountered by any potentially systemically infectious virus if systemic infection is to occur. These barriers include host resistance at higher levels of cellular organization, as well as potential interferences resulting from the presence of the primary infecting virus (reviewed by Ponz and Bruening[55]).

Citovsky et al.[51] have noted that the TMV movement protein is not sequence-specific in its binding to various RNAs *in vitro*. The protein even complexed with single-stranded DNAs. (Due to the limitations of their gel retardation assay system, these authors did not show that the m-protein interacts with TMV RNA.) Lack of m-protein specificity in RNA substrate binding is consistent with the nonspecific facilitation by TMV of the movement of unrelated coinfecting viruses. RNA transport through plasmodesmata may be a normal plant function which the filamentous viruses have adopted. It remains to be seen if viral perturbation of this function, or if viral-facilitated nonspecific intercellular movement of cellular RNAs correlates with symptom induction.

### D. SPECIFIC CELL-CELL MOVEMENT IN SPHERICAL VIRUSES

Polyhedral particles may employ a form of plasmodesmatal transport distinct from that of the filamentous particles. Their movement differs from that of the filamentous viruses in several ways: (1) while filamentous virus particles are rarely observed within plasmodesmata, spheriform viral particles are frequently observed physically within distended desmotubules (e.g., see Gibbs[12] and Kitajima and Lauritis[56]). The connections to the endoplasmic reticulum appear to have been disrupted in these virus-distended plasmodesmata, which also appear to have lost their central core proteins (reviewed by Robards and Lucas[53]). (2) Polyhedral viruses appear to show strict specificity in their intercellular transport. In contrast to the filamentous viruses, polyhedral viruses have not been reported to facilitate the movement of unrelated coinfecting viruses. For example, though cowpea chlorotic mottle bromovirus (CCMV) will propagate in barley protoplasts,[57] attempts fo demonstrate BMV-assisted systemic spread of CCMV through barley plants have been unsuccessful, as have reciprocal attempts at CCMV facilitation of BMV movement in cowpea plants.[58] The failure of BMV to facilitate the spread of CCMV in barley was first reported by Hamilton and Nichols;[59] these authors also noted that BMV also did not support the spread of alfalfa mosaic virus, southern bean mosaic sobemovirus, tobacco necrosis virus, or turnip yellow mosaic tymovirus in barley. In an exception to the exclusive specificity of polyhedral viruses for transport of homologous particles only, Hamilton and Nichols[59] observed a facilitation of TMV spread in barley by BMV. However, this was only seen in plants maintained continuously at temperatures of 30°C or above, while no *trans*-specific transport was found at lower temperatures. The 30°C growth regimen constitutes a stress condition for barley,[60] a condition under which TMV alone will systemically infect this plant to low titers.[60]

Barker[36] noted the failure of polyhedral viruses to facilitate an increase in titer of potato leafroll luteovirus (PLRV) or beet western yellows luteovirus in *Nicotiana* species. This was in contrast to the enhancement he observed in the titer of these luteoviruses in the presence of potex-, poty-, tobamo-, and tobraviruses (all filamentous morphologies), as well as by carrot mottle (ungrouped) virus (Table 2). Luteovirus infection is commonly restricted to phloem cells, and the increase in titer may reflect release from this phloem limitation by determinants of movement carried by the filamentous helper viruses. In this study, Barker[36] noted the failure of alfalfa mosaic virus (AlMV), CMV, broad bean mottle bromovirus, two nepoviruses, parsnip yellow fleck virus, and pea enation mosaic virus (all polyhedral particles) to stimulate the accumulation of the luteoviruses tested.

The observation of encapsidated polyhedral virions in plasmodesmata, coupled with the selectivity polyhedral viruses are seen to impose against nonspecific facilitation of the movement of unrelated viruses, suggests that polyhedral viruses may employ a second determinant complementing m-protein control over cell-cell spread. The coat protein may be that second factor through which polyhedral viruses maintain self-specificity in cell-cell movement. Wellink and van Kammen[61] have noted that there are two genes in CPMV, deletion mutations in either of which will block the systemic expression of CPMV symptoms. Allison et al.[62] have made a similar observation for BMV. In the examples of these BMV and CPMV mutants, one of the mutated genes encodes the putative movement protein, and the other gene essential for movement encodes the coat protein. In neither of these cases do the mutants express the limited, nonvascular, cell-to-cell only spread of infection that is typical of coat protein mutants of TMV.

Allison et al.[63] have separately exchanged each of the genes on bromovirus RNA 3, between CCMV and BMV. RNA 3 encodes the 3a protein which is the putative movement protein, as well as the capsid protein (3b). Recombinant viruses were made by the exchange of genes on RNA 3, followed by the reconstitution of the virus by the addition of RNAs 1 and 2. Although functional capsid protein is necessary for viral movement, capsid protein was not found to control host specificity, since recombinant viruses carrying a functional copy of the heterologous coat protein gene retained systemic infectivity in their original hosts. Host specificity in viral movement was inferred of the 3a protein, however, exchange of the 3a gene abolished systemic infectivity in the resulting recombinant forms of CCMV and BMV, even though such recombinants propagated successfully in isolated protoplasts.

## IV. BARRIERS TO LONG-RANGE VIRAL MOVEMENT

Virus spread *in planta* is of two types, short-range cell-to-cell movement, and long-range movement through the phloem (summarized in Gibbs[12]). Inhibition of movement can occur either at the cell-cell barrier or on a higher level at which long-range systemic movement is blocked even while localized cell-cell spread may still occur. Although there are plasmodesmata connecting phloem cells with the mesophyll, there is nevertheless a block to virus movement at the barrier surrounding phloem vasculature. Those determinants that overcome this blockade are expressed in the form of the relatively more severe infective phenotype produced when phloem limitations on viral infection are released, enabling the spread of infection through a greater extent of host tissue. This increased infectivity can result from viral release (1) into the phloem, or (2) out of the phloem. The two release points may well be functionally identical, operationally differentiated only as a consequence of the side of the barrier on which the virus initially occurs.

### A. VIRAL MOVEMENT INTO PHLOEM

Although coat protein is not required for the cell-to-cell movement of TMV, some TMV coat protein mutants produce abnormally slowly spreading symptoms in infected tissues. Similar observations have been made for coat protein mutants of another filamentous virus, beet necrotic yellow vein furvovirus.[64] This infection phenotype was interpreted for TMV by Siegal et al.[65] to suggest that though cell-cell spread still occurs in coat-protein mutated strains, movement across longer distances through the host vasculature is prevented. The requirement for coat protein function for long-distance spread is illustrated in the system reported by Osbourn et al.[66] in which movement-impaired, coat protein-defective TMV mutants appear to recover to a wild-type rate of systemic spread when inoculated to transgenic tobacco plants expressing wild-type TMV coat protein.

Specific alterations in the coat protein gene which simultaneously block both encapsidation and long-distance movement of TMV have been well characterized.[67,68] Dawson et al.[69] described in detail a series of coat protein mutations of TMV which produced the gradually spreading localized symptom. The mutations resulted in the production of truncated coat proteins which were unable to bind viral genomic RNA, giving rise to mutant strains incapable of assembling viral particles. A mutant producing no detectable coat protein also showed the slowly spreading localized symptom.[70] These various coat protein deletions each imparted distinctive symptomatic variations to the slowly spreading chlorotic infection phenotype. Culver and Dawson[70] characterized the delay in systemic spread of a coat proteinless mutant; movement became apparent in 9 to 12 weeks, a much longer time than wild-type virus requires for systemic spread. The virus appeared to spread through movement into already expanded leaves.

Saito et al.[68] have shown that the same slowly spreading phenotype is produced in an encapsidation mutant which carries an unaltered coat protein gene. The mutation in this case is in the origin-of-assembly sequence, the genomic locus at which coat protein recognition of a specific genomic RNA sequence initiates encapsidation. This locus is not in the coat protein gene in TMV-L, the strain used in this study; it is within the m-protein gene.

The above set of observations suggests that coat protein and origin-of-assembly loci function as determinants of long-distance spread by controlling the assembly of encapsidated TMV particles. Though mature viral particles have been seen in sieve elements,[71] the way they may have gotten there is currently unexplained. Protein synthesis likely does not take place in sieve tube cells. Thus, the capsid protein subunits coating TMV particles in the phloem are probably synthesized in other cells, then exported to the phloem, either in the form of encapsidated TMV particles, or as free subunits. Although mesophyll plasmodesmata appear to pass only the uncoated viral RNAs, phloem boundary plasmodesmata may have unique properties that account for the passage of intact filamentous-form viral particles. Alternatively, filamentous virus may appear in sieve elements after free capsid protein subunits pass into the phloem, there binding free viral RNA to assemble virion particles.

As with TMV, mutations in either the coat protein or in a second movement-associated gene can abolish systemic spread of spherical viruses.[61,62] However, as opposed to the observations with TMV, short-distance spread is also eliminated in these examples of coat protein mutants of isometric viruses. This difference complicates interpretations as to the cellular level at which the coat protein mutation effects transport of spherical particles.

Coat protein does not appear to be determinant of long-distance movement in BSMV. Pring[72] has suggested that BSMV infection in barley progresses through stages, with an initial severe episode involving the systemic spread of free viral RNA, followed by a milder phase during which encapsidated virions accumulate. Nonetheless, coat protein appears to be a determinant of severity of BSMV infection. Petty and Jackson[73] have produced coat proteinless mutants, e.g., by the alteration of the initiating AUG codon to AUC. This and similar such mutants are unaltered in their capabilities to initiate the initial systemic infection of barley, compared with the wild-type virus. However, they are even more virulent than the unaltered strain, appearing to be permanently arrested in the initial severe stage of infection. Petty and Jackson[73] have also observed that coat protein is not required for BSMV spread in dicotyledonous hosts.

Frame shift and deletion mutations in the coat protein gene of BSMV produce mutant strains that systemically infect barley, but are differentiated in *Chenopodium amaranticolor*. Some of the deletion strains are uninfectious in *C. amaranticolor,* while other coat proteinless mutants that retain the affected RNA sections (which are funtionally deleted, due to frame-shift mutations) retain infectivity. This suggests that, as opposed to the gene product, an intrinsic property of the BSMV coat protein gene itself controls infectivity in dicotyledonous hosts.

### B. VIRAL MOVEMENT OUT OF PHLOEM

The barrier to escape from the phloem is evident in classes of viruses that are phloem-restricted in their tissue range. This includes the beet western yellows luteovirus, which can be detected within plasmodesmatal structures, but only those between cells of phloem tissues.[74]

Egress from phloem can be conferred upon these viruses by separate, coinfecting viruses. The viruses reported to facilitate phloem escape (Table 4) all are filamentous. For example, PLRV infection in tobacco is restricted to phloem cells, but can be released to the mesophyll by coinfection with PVY.[75] The PLRV/PVY mixed infection shows a synergistic increase in symptomatic severity. Other examples from Barker[36] of the facilitation of luteovirus escape from phloem limitation are listed in Table 4. Viruses with filamentous morphologies have also been reported to facilitate the general cell-to-cell movement of coinfecting viruses (discussed above); if facilitation by filamentous viruses of cell-to-cell movement in, e.g., mesophyll tissue, is mediated through plasmodesmata, then facilitation of movement out of phloem may be through plasmodesmatal structures as well.

## V. RESISTANT HOST RESPONSE TO VIRAL INFECTION

Virus replication within and movement between the cells of a given plant defines that plant as a ''host'' for that virus. Host plants that can restrict the spread of virus infection are termed resistant. The genetics of this type of resistance has been extensively characterized.[76] Determinants of resistance are described in genetic terms, i.e., the viral determinants controlling the induction of the resistance reponse are termed "avirulence genes."

Vascular spread of the infection is blocked in resistant hosts. The appearance of virions is limited to a radially expanding disk of cells centered on the point of initial infection. A delayed response is elaborated behind this advancing boundary of infected cells. The response can take the form of starch lesions[77] or necrotic lesions.[78] In either case, virus-containing, but ultrastructurally normal cells are found just beyond the leading edge of the lesion. Lesion production is a secondary response[79] biochemically related to a wound reaction. It includes the elaboration of phenolic compounds, hydrolases, and phytoalexins[80,81] and is classically called "the hypersensitive response." As opposed to inoculation to nonhost plants, infectious virus can be recovered at the inoculation site from such resistant hosts.

The resistance reaction occurs in host cells which are integrated into the matrix of organized tissues. This "organized cell resistance" was studied by Takabe and colleagues, who analyzed the resistance reaction in TMV-inoculated protoplasts derived from tobacco plants in which resistance was mediated by the *N*-gene. They observed[23] that viral replication was not inhibited in protoplasts as it is *in planta,* but was identical to the infection supported by protoplasts derived from nonresistant (*n*/*n*) hosts; the cell death response did not occur in cells that had been removed from their organized tissue structure. These workers deduced that the resistance reactions were not initiated solely by intracellular events, but were dependent upon events involving intercellular communication.

Asymptomatic replication of avirulent virus in plant cells in culture, or in cells beyond the necrotic lesion edge *in planta,* suggests that the necrotic response may be determined by factors operating on an order of cellular organization above that of the single cell. However, the accumulation of virus in *N*/*N* protoplasts in culture can vary, depending on the phytohormonal constitution of the culture medium.[82] One interpretation of this result is that resistance responses may be fully induced in cell culture by the presence of auxin, and no further reaction is possible in response to virus inoculation. Nonetheless, these workers did not report protoplast culture conditions in which the cell death response seen *in planta* was duplicated *in vitro.* Still, the extrapolation from protoplasts, which have no plasmodesmata and, initially, no cell walls, to plant cells organized into the cellular matrix of leaf tissues, must be made with caution.

Organized cell resistance, wherein vascular viral spread is reduced *in planta* concomitantly with the formation of localized lesions, is very wide spread and agronomically significant.[83] Determinants of TMV inducing this response have been particularly well studied; viral determinants have been characterized on the nucleotide sequence level, while host determinants have only been characterized[83] as to their genetic segregation. The *N*-prime locus,[84] such as in *N. sylvestris,* may be allelic with the *N* locus from *N. glutinosa,*[85] but these loci are distinguished by a difference in corresponding inducing loci on the TMV genome; *N*/*N*-mediated resistance is induced by viral determinants located outside of the coat protein coding region,[67] while *N*-prime-mediated resistance is determined within the capsid protein gene.

The TMV coat protein gene encodes viral inducers of *N*-prime-mediated resistance,[86,87] while a coat-proteinless derivative of an inducing TMV strain, carrying the full length RNA with a mutation in the coat protein ATG initiation codon, no longer induces the response.[70] These viral determinants have been characterized on the coat protein amino acid sequence level. Mutagenesis of the common strain of TMV (TMV U1) can alter infection in *N. sylvestris* from systemic to necrotic/restricted spread, leaving the mutant strains still systemically infectious in, for example, *N. tabacum.* This was first recognized by Funatsu and Fraenkel-Conrat[88] and has been definitively characterized by W. Dawson and colleagues.[69] Mutations at a number of coat protein sites that render derivatives of TMV U1 avirulent in the *N'/N'* genetic background include Val[11]-to-Met, Pro[20]-to-Leu, Asn[25]-to-Ser, Arg[46]-to-Gly (Culver and Dawson[89]), and Ser[148]-to-Phe (Knorr and Dawson[90]). These mutations span most of the length of the gene, yet the amino acid changes are all located within the same domain of the coat protein — they all occur on the protein-protein interface between subunits in the assembled capsid structure.[91] Incompatibility in *N. sylvestris* could therefore arise from general deformation of virion structure and the resulting abnormalities of function; alternatively, a specific interaction with host components could occur at the ends of the virion, where alterations to the capsid subunit interfacial surfaces would be exposed. The mutation at TMV coat protein amino acid position 46, and others in this class of mutations, also affect the systemic symptoms induced in *N. tabacum* (*n/n*), the permissive host.[89] The number and nature of the steps lying between the initial molecular-level interaction of TMV capsid variants with host component(s), and the final elaboration of the resulting foliar necrosis, are currently completely undescribed.

Although mutations in the plasmodesmata-interactive m-protein do not appear to be involved in the (coat-protein mediated) induction of the response determined by the *N*-prime locus in the *N. sylvestris,* the m-protein gene may be involved with *N*-gene-mediated reactions in other *Nicotiana* species. The restriction of TMV spread by the *N* gene response appears to require cell-cell communication,[23] so plasmodesmatal physiology may well be involved. Deom et al.[50] have transformed *N/N*-genotype *N. tabacum* cv. *xanthi* with the TMV m-protein gene, and no necrotic reaction was observed in the m-protein-expressing plants. However, the inoculation of these transgeneic plants with a non-m-protein-expressing TMV mutant incited necrotic local lesions, similar to the general response of *N/N* tobacco to TMV. Since the non-m-protein-expressing TMV mutant by itself is not infectious and incites no responses in normal *N/N* tobacco plants, this experiment implicates the m-protein as a component of the interaction that induces necrotic lesion formation. The specific identity of the viral locus (loci) determining the response will be addressed by the characterization of the genome of TMV Ob-1, an *N*-gene resistance-breaking strain of TMV.[92] Host components involved will eventually be sought in *in vitro* reconstitutions of plasmodesmata from components isolated from plant cell membranes.

Mutations in CaMV gene VI can lead to the restriction of viral spread and concomitant necrotic lesion formation in one host but have no effect in a different host. Single base changes in the gene VI protein-coding region which have no effect in *Brassica campestris* (the permissive host), can result in viral strains which are restricted to necrotic lesions at the site of innoculation on *Datura stramonium,* the restrictive host.[93] One such strain was an acclimation mutant, derived from a CaMV strain which was originally systemically infectious in *D. stramonium.* After serial passage in *B. campestris,* a change arose in CaMV gene VI from Asp[488]-to-Asn; this had no effect in the permissive host, but resulted in restriction of the mutant to necrotic lesions in *D. stramonium.* The nature of the interaction of the product of CaMV gene VI with host plants is unknown.

Tobacco streak ilarvirus (TSV) is restricted in cotton to necrotic or chlorotic local lesions, but this barrier to systemic infection is overcome by coinfection with cotton anthocyanosis (ungrouped) virus.[37] Coinoculation results in the expression of "late mosaic," produced by systemic TSV infection.

Fuentes and Richardson (1988) have described a system which displays attributes of both nonhost resistance and resistance at the organized cell level. They have observed that the cowpea strain of southern bean mosaic sobemovirus (SBMV-C) will not move systemically in *Phaseolus* unless released from the barrier to cell-cell spread during coinfection with legume strain of TMV. When SBMV-C does spread, the host responds with a necrotic response which restricts the infection to the inoculated leaf.

One hypothesis for the induction of the host resistance response would predict that host receptors exist with the capacity to recognize specific viral determinants. The recognition event would then trigger a cellular defense response. The genetics of avirulence in bacterial and fungal pathogenesis has been interpreted in these terms (see chapter by Oku in this volume). Some plant-virus interactions can be similarly characterized as "gene-for-gene",[76] while other examples of plant-virus incompatibility do not conform to the "dominant avirulence" model.[8] Even though the plant response to viral infection may be traced ultimately to a specific viral gene, the host reaction to the resulting nonhomeostasis condition may nonetheless be nonspecific, involving the induction of the same general-purpose responses to trauma caused by other pathogens, herbivores, chemical, and physical stress (discussed by Carr and Klessig[80]). One possible inducer of a seemingly specific necrotic stress response is incorrectly folded protein. The accumulation of denatured protein can be cytotoxic (e.g., Poritz et al.[94]) and lead to the induction of stress reactions, including the heat shock response.[95] The replication of a mutant form (e.g., carrying an altered capsid protein or inclusion body protein) of an otherwise well-adapted high-titer virus could lead to the accumulation of incorrectly folded protein in host cell cytoplasm. An ensuing localized region of cell death, though not mediated by a specific host recognition reaction, would be difficult to distinguish from a specific reaction if analyzed at the phenotypic expression level.

## VI. MULTIPLE DETERMINANTS OF SYSTEMIC INFECTION

The viral determinants reviewed here that result in restriction of systemic viral infection have been precisely characterized at the gene sequence level. In the case of systemic infection of a susceptible host, however, evaluation of determinants is a much more difficult task, complicated by the complexity of the interaction. Systemic symptoms are revealed upon the expansion of plant organs already infected during their morphogenesis. The development of these symptoms involves the organizational stages of host tissue differentiation, and also invovles the integrated functioning of all the viral components necessary for systemic spread of the infection. Symptoms other than necroses appear to arise from morphogenetic disturbances.[96] Developmental hypoplasia of the plant vasculature during its morphogenetic differentiation would give rise to: (1) mosaics (chlorosis of leaf vasculature seen against the unaltered green lamina); (2) rugosity (failure of leaf vasculature to elongate in unison with the expansion of the leaf lamina, resulting in laminar sections bulging up above the restricting vascular net); (3) epinasty (decurvature of the mid-vein, resulting from asymmetric expansion of the upper and lower vein walls); and (4) leaf curvature (lateral asymmetry of mid-vein expansion). General foliar deformities may result from varying contributions from all of the above. Clearly, many host determinants are involved in tissue growth, and identification of the genetic elements involved in their infection-mediated dysfunction will await our better understanding of morphogenesis in general.

Similarly, all of the viral genes necessary for the propagation of the infection will have a determinant impact upon systemic symptomatology. For example, A. Jackson and coworkers have shown that systemic infection by BSMV can be determined by the effects of different viral regulatory factors which modulate levels of expression of primary viral functions. In one case,[97] a short open reading frame upstream from the replicase gene appears to control BSMV infection in *N. benthamiana* through its effect on expression of the replicase.

The interplay of viral genes in the evolution of disease symptoms has been seen in CaMV. The ease of application of recombinant DNA techniques to the DNA genome of this virus has facilitated the construction of genomic chimeras. Various hybrids derived from CaMV strains having distinct systemic infection phenotypes[98-100] have been seen to produce a continuum of gradations of symptom types. In these infections, specified infection traits and single viral genes could not be simply correlated, suggesting that systemic symptom induction is multigenic in origin.

In an attempt to single out the contribution of one of these genes, CaMV gene VI has been isolated for study. The viral gene VI DNA segment, under the control of the CaMV 19 S[101] or 35 S[102] promoter, has been expressed in transformed *N. tabacum*. Although gene expression should be systemically uniform in plants cloned from a single transformed cell, the expression of the gene VI product gave rise to nonuniform, viruslike symptoms in tobacco, described as flecked, sectored, or necrotic. These are difficult to relate to the symptoms incited by complete virions, as the strains of CaMV used as sources of gene VI do not cause the observed set of symptoms in *N. tabacum;* the viral strains used do not systemically infect this host. The symptomlike effects seen in the transgenic plants varied widely in type and degree, from necrosis and stunting to milder chlorosis. This could be due in part to protein dosage effects, which can have pathogenic consequences of their own in transgenic plants. Takahashi et al.[102] noted that the gene VI product was lethal when expressed at higher levels. This phenomenon may be more readily understood if repeated in transgenic plants which are systemic hosts for the virus.

## VII. SUBVIRAL PATHOGENS

This group of small, highly structured RNA agents includes the viroids and the viral satellites. The viroids are autonomously infectious, while the satellites depend upon "helper viruses" for their propagation. Satellite RNAs are small molecules transported within helper virus capsids, while satellite viruses, though still obligately associated with helper viruses, encode their own capsid proteins. The assessment of the basis of viroid- or satellite-induced modulations of pathogenicity is quite complex. However, this difficulty is balanced in small measure by the ease of identifying the determinants. Viroids and satellite RNAs are less than 400 bases in length, and complete sequences of sets of related isolates of the satellite RNAs from tobacco ringspot virus (sTobRV RNA) and CMV, and of two viroids, potato spindle tube (PSTVd) and citrus exocortis (CEVd), have been compared.

Satellite RNAs can alter the level of replication of their helper viruses,[103] as well as the severity of viral infection.[104,105] Effects on helper virus replication will, in turn, effect its interaction with its host plant, and the resulting symptomatology. Symptoms can vary depending on (1) host species or variety, (2) strain of helper virus, and (3) satellite isolate,[106] note to mention environmental conditions. A full understanding of the sequence determinants of satellite-mediated plant pathogenicity is thus complex and indirect and will not be easily grasped until we better understand details of the primary virus-satellite and host-virus interactions.

Since viroids do not encode proteins[107] the induction of pathological symptoms by viroids will not derive directly from protein-protein interactions. Similarly, the attenuation of the symptoms of tobacco ringspot virus by sTobRV RNA is apparently not protein-mediated, since sTobRV RNA also appears to encode no proteins.[108] Although many isolates of the satellite RNA of CMV appear capable of encoding a small protein from a reading frame between genomic positions 11-91 (see Hidaka et al.[109]), the sCMV RNA determinants which alter the severity and type of CMV infection symptoms lie beyond the protein coding regions (e.g., see Sleat and Palukaitis;[110] Masuta and Takanami[111]). Viroid effects not mediated through protein-protein interactions may involve interference with host RNA-protein interaction events;[112] this may also be the case for satellite effects.

Point mutations attenuating viroid aggressiveness or satellite impact on helper virus symptoms are often generally detrimental to these subviral agents, independent of host plant species. Since the characteristics of these agents appear to derive from their small and highly based-paired genomes, point mutations that disrupt structure will often simultaneously disrupt function in viroid or satellite RNAs. A case in point[113] is a stem in the autocatalytically self-cleaving domain of the sTobRV RNA. The base pairing in that domain forms a (inferred) helical stem which would be destroyed by transversion mutations in either strand in the region of position 240. Transversion mutations at that position in either of the two strands were found to result in satellite RNAs which were not replicated and did not attenuate the symptom of the supporting virus — the mutated sequences appear to render the satellite nonfunctional. However, the recombination of the two mutated domains to provide a correctly base-paired alternate structure restores function to the reconstituted double mutant. A similar observation correlating structure and function in these subviral pathogens has been made in the reconstruction of a viable double mutant of PSTVd[114] from a self-complementary pair of nonviable single-site mutants also located in a (inferred) base-paired structural position. No host-specific interaction is evident for this class of mutation.

Host-specific modification of CMV infection by one isolate of its satellite RNA was described by Waterworth et al.[104] as attenuating in most hosts with the exception of tomato, in which the opposite, lethal infection, occurred. Two general genomic domains of pathogenicity common to many strains of sCMV RNA have been delineated:[115,116] a domain controlling chlorotic symptom induction is within satellite positions 100-185 (Masuta and Takanami;[111] Palukaitis and Sleat[117]), a second domain at positions 276-300 controls necrotic symptom induction.[110,111] Jaegle et al.[118] have shown that in the chlorosis-determining domain of sCMV RNA Y, nucleotides 185-186 control chlorosis on tobacco. Palukaitis and Sleat[117] have made a similar observation, showing that host specificity resides at position 149 in sCMV RNA B5, a position analogous to the above locus in the Y-SAT. This nucleotide position determines in a mutually exclusive fashion whether the supporting CMV infection will induce chlorosis on tobacco or on tomato.

As opposed to sCMV RNA, only one domain appears to control virulence in the viroids. This domain has been delineated in PSTVd and CEVd.[119,120] As has been the case with the satellite RNAs, however, more detailed analysis of viroid structure have failed to resolve mutations affecting host response from mutations causing generalized structural dysfunction.[114] This is apparently due in part to the inseparability of viroid structure and function (discussed by Owens and Hammond, Riesner, and others in Diener[121]).

The relationship between latency and overt pathology is illustrated by an example of viroid pathogenesis. Columnea latent viroid (CLVd) was discovered by chance in asymptomatic association with *Columnea erythrophae,* a gesneriaceous tropical epiphyte. If transferred to potato or tomato, CLVd causes a severe PSTVd-like infection.[121] The inception of these symptoms resulted from an encounter, mediated by the investigator-as-vector, between two organisms which otherwise would never have come into contact. As discussed by Diener,[112] one analysis of this interaction would hold that coeval host-viroid relationships would tend to be nonpathogenic. Viroid diseases would then result from the unfortunate pairing of a parasite and a host in which there had been no previous coevolutionary optimization. Diener's thesis may be applicable beyond the viroids, extending to include viruses which may have moved from their coeval hosts and relocated to crop plants (discussed below).

## VIII. OUTLOOK

The symptomatology of infection has been useful in the identification of viral and viroid genetic loci which encode plant-interactive functions. However these descriptions have not provided an understanding of the selective advantage conferred, or of the molecular interactions underlying disease induction. This contrasts with the molecular basis of cytopathological symptoms caused by animal and bacterial viruses. These symptoms are readily correlated with recognition interactions between specific virus and host proteins: those interactions include the shut-off of host transcription, translation, or DNA replication.

Plant viruses and viroids are not generally as severely cytopathic as animal/bacterial viruses; plant cells in culture can support virus and viroid replication with little evidence of stress. Further, a growing list of asymptomatic viruses and viroids suggests that plants may have evolved a capacity for nonpathogenic acclimation with these obligate parasites. The appearance of pathogenic strains in crop plants may be an aberrant consequence of the impact of agricultural practices upon a previously established benign relationship of plants with their viruses.[122] In this scenario, adventitious pairings of plants and parasites, coupled with various disadaptive genetic manipulations suffered by host plants during their domestication, will have disturbed the potential for latent relationships. Multiple types of disease-inducing molecular interactions may have independently evolved from this disequilibrium.

Viral disease symptoms are generally not evident in single plant cells, but are expressed in differentiated host tissues. Symptoms are most noticeable in the vascular network through which host cells both communicate with each other and become exposed initially to the infection. The most common descriptor of plant viruses and viroids is the term ''mosaic,'' which indicates chlorosis of the cells in and near vascular channels. Mosaics arise in veinal networks in leaves infected early during their morphogenesis. The mechanism by which circulating virus penetrates developing plant organs through perivascular cells remains undescribed, but the investigation of plasmodesmatal channels may provide clues.

The modification of plasmodesmata by viruses escaping from the cells they were synthesized in may be the first example of plant-virus recognition to be understood in molecular detail. The protein-protein interface between viral movement proteins and plasmodesmatal components may well reveal specific recognition features that arose from the adaptation and coevolution of virus with host. This description of the molecular basis of host-virus interaction at the cell-to-cell junction may soon be realized in examples of the rod-shaped viruses. The more selective cell-cell movement systems of the polyhedral viruses may be more complex, requiring the description of interactions between multiple viral determinants prior to the characterization of the host-virus interaction.

# REFERENCES


1. **Blumenthal, T. and Carmichael, G.,** RNA replication: function and structure of Q-beta replicase, *Annu. Rev. Biochem.*, 48, 525, 1979.
2. **Lentz, T.,** The recognition event by virus and host cell receptor: a target for antiviral agents, *J. Gen. Virol.*, 71, 751, 1990.
3. **Lindberg, A.,** Bacteriophage receptors, *Annu. Rev. Microbiol.*, 27, 205, 1973.
4. **Orsini, G. and Brody, E.,** Phage T4 DNA codes for two distinct 10-kDa proteins which strongly bind to RNA polymerase, *Virology*, 162, 397, 1988.
5. **Rubenstein, S. and Dasgupta, A.,** Inhibition of rRNA synthesis by poliovirus: specific inactivation of transcription factors, *J. Virol.*, 63, 4689, 1989.
6. **Sonenberg, N.,** Regulation of translation by poliovirus, *Adv. Virus Res.*, 33, 174, 1987.
7. **Daubert, S.,** Sequence determinants of symptoms in the genomes of plant viruses, viroids, and satellites, *Mol. Plant-Microbe Interact.*, 1, 317, 1988.
8. **Van Loon, L.,** Disease induction by plant viruses, *Adv. Virus Res.*, 33, 205, 1987.
9. **Atabekov, J. and Dorokhov, Y.,** Plant virus-specific transport function and resistance of plants to viruses, *Adv. Virus Res.*, 29, 313, 1984.
10. **Hull, R.,** The movement of viruses in plants, *Annu. Rev. Phytopathol.*, 27, 213, 1989.
11. **Atabekov, J. and Talianski, M.,** Expression of a plant virus-coded transport function by different virus genomes, *Adv. Virus Res.*, 38, 201, 1990.
12. **Gibbs, A.,** Viruses and plasmodesmata, in *Intercellular Communication in Plants: Studies on Plasmodesmata*, Gunning, B. and Robards, A., Eds., Springer-Verlag, Berlin, 1976, 149.
13. **Huber, R., Rezelman, G., Hibi, T., and van Kammen, A.,** Cowpea mosaic virus infection of protoplasts from Samson tobacco leaves, *J. Gen. Virol.*, 34, 315, 1977.
14. **Nishiguchi, M. and Uehara, Y.,** Resistance of sweet potato to tobacco mosaic virus, Abstr. P81-009, 7th Int. Congr. Virol., Berlin, 1990.

15. **Sacher, R., French, R., and Ahlquist, P.,** Hybrid brome mosaic virus RNAs express and are packaged in tobacco mosaic virus coat protein *in vivo, Virology,* 167, 16, 1988.
16. **Furusawa, I. and Okuno, T.,** Infection with brome mosaic virus of mesophyll protoplasts isolated from five plant species, *J. Gen. Virol.,* 40, 489, 1978.
17. **Sakai, F., Dawson, J., and Watts, J.,** Interference in infections of tobacco protoplasts with two bromoviruses, *J. Gen. Virol.,* 64, 1347, 1983.
18. **Hussian, M., Melcher, U., Whittle, T., Williams, A., Brannan, C., and Mitchel, E.,** Replication of cauliflower mosaic virus in leaves and suspension culture protoplasts of cotton, *Plant Physiol.,* 83, 633, 1987.
19. **Kubo, S. and Takanami, Y.,** Infection of tobacco mesophyll protoplasts with tobacco necrotic dwarf virus, a phloem limited virus, *J. Gen. Virol.,* 42, 387, 1979.
20. **Cheo, P.,** Subliminal infection of cotton by tobacco mosaic virus, *Phytopathology,* 60, 41, 1970.
21. **Sulzinski, M. and Zaitlin, M.,** Tobacco mosaic virus replication in resistant and susceptible plants: in some resistant species virus is confined to a small number of initially infected cells, *Virology,* 121, 12, 1982.
22. **Hosokawa, D. and Mori, K.,** Investigation of CMV multiplication in tobacco epidermal cells by fluorescent antibody technique, *Ann. Phytopathol. Soc. Jpn.,* 40, 210, 1974.
23. **Otsuki, Y., Shimomura, T., and Takabe, I.,** Tobacco mosaic virus multiplication and expression of the N gene in necrotic responding tobacco varieties, *Virology,* 50, 45, 1972.
24. **Marton, L., Duran-Vila, N., Lin, J., and Semancik, J.,** Properties of cell cultures containing the citrus exocortis viroid, *Virology,* 122, 229, 1982.
25. **Beier, H., Siler, D., Russell, M., and Bruening, G.,** Survey of susceptibility to cowpea mosaic virus among protoplasts and intact plants from *Vigna sinesnis* lines, *Phytopathology,* 67, 917, 1977.
26. **Kiefer, M., Bruening, G., and Russell, M.,** RNA and capsid accumulation in cowpea protoplasts that are resistant to cowpea mosaic virus strain SB, *Virology,* 137, 71, 1984.
27. **Meshi, T., Motoyoshi, F., Adachi, A., Watanabe, H., Takamatsu, N., and Okada, Y.,** Two concomitant base substitutions in the putative replicase gene of TMV confer the ability to overcome the effects of a tomato resistance gene, Tm-1, *EMBO J.,* 7, 1575, 1988.
28. **Watanabe, Y., Kishibayashi, N., Motoyoshi, F., and Okada, Y.,** Characterization of Tm-1 gene action on replication of common isolates and a resistance-breaking isolate of TMV, *Virology,* 161, 527, 1987.
29. **Okada, Y.,** personal communication.
30. **Smith, K.,** On the composite nature of certain potato virus diseases of the mosaic group as revealed by the use of plant indicators and selective methods of transmission, *Proc. R. Soc. London Ser. B,* 109, 251, 1931.
31. **Cohen, J., Loebenstein, G., and Spiegel, S.,** Infection of sweet potato by cucumber mosaic virus depends on the presence of sweet potato feathery mottle virus, *Plant Dis.,* 72, 583, 1990.
32. **Dodds, A. and Hamilton, R.,** The influence of barley stripe mosaic virus on the replication of tobacco mosaic virus in *Hordeum* vulgare L., *Virology,* 50, 404, 1973.
33. **Malyshenko, S., Kondakova, O., Talinsky, M., and Atabekov, J.,** Plant virus transport function — Complementation by helper virus is non-specific, *J. Gen. Virol.,* 70, 2751, 1989.
34. **Fuentes, L. and Hamilton, R.,** Spread of the cowpea strain of southern bean mosaic virus in a non-permissive host is facilitated by infection with sunn-hemp mosaic virus. Abstr. P.1-2-114, *5th Int. Congr. Plant Pathol.,* Kyoto, Japan, 1988.
35. **Carr, R. and Kim, S.,** Evidence that bean golden mosaic virus invades non-phloem tissue in double infections with tobacco mosaic virus, *J. Gen. Virol.,* 64, 2489, 1983.
36. **Barker, H.,** Specificity of the effect of sap-transmissible viruses in increasing the accumulation of luteoviruses in co-infected plants, *Ann. Appl. Biol.,* 115, 71, 1989.
37. **Costa, A.,** Conditioning of the plant by one virus necessary for systemic invasion by another, *Phytopathol. Z.,* 65, 219, 1969.
38. **Thomas, B.,** Occurrence and epidemiology of the cucumber necrosis strain of tobacco necrosis virus in cucumber crops. Annual report glasshouse crops research institute 1982, Littlehampton, West Sussex, UK, 1984, p. 117.
39. **Motoyoshi, F. and Oshima, N.,** Expression of genetically controlled resistance to tobacco mosaic virus infection in isolated tomato leaf mesophyll protoplasts, *J. Gen. Virol.,* 34, 499, 1977.
40. **Meshi, T., Motoyoshi, F., Maeda, T., Yoshiwoka, S., Watanabe, H., and Okada, Y.,** Mutations in the tobacco mosaic virus 30-kd protein gene overcome Tm-2 resistance in tomato, *Plant Cell,* 1, 515, 1989.
41. **Hall, T.,** Resistance at the Tm-2 locus in the tomato to tomato mosaic virus, *Euphytica,* 29, 189, 1980.
42. **Pelham, J.,** Strain-genotype interaction of tobacco mosaic virus in tomato, *Ann. Appl. Biol.,* 71, 219, 1972.
43. **Meshi, T., Watanabe, H., Saito, T., Sugimoto, A., Maeda, T., and Okada, Y.,** Function of the 30-kd protein of TMV, *EMBO J.,* 6, 2556, 1987.

45. **Deom, C., Oliver, M., and Beachy, R.,** The 30-kilodalton gene product of TMV potentiates virus movement, *Science,* 237, 389, 1987.
46. **Beachy, R.,** personal communication.
47. **Watanabe, Y., Morita, N., Nishiguchi, M., and Okada, Y.,** Attenuated strains of tobacco mosaic virus. Reduced synthesis of a viral protein with cell-to-cell movement function, *J. Mol. Biol.,* 194, 699, 1987.
48. **Tomenius, K., Chapham, D., and Meshi, T.,** Localization by immunogold cytochemistry of the virus-coded 30-K protein in plasmodesmata of leaves infected with tobacco mosaic virus, *Virology,* 160, 363, 1987.
49. **Wolf, S., Deom, C., Beachy, R., and Lucas, W.,** Movement protein of tobacco mosaic virus modifies plasmodesmatal size exclusion limit, *Science,* 246, 377, 1989.
50. **Deom, C., Wolf, S., Holt, C., Lucas, W., and Beachy, R.,** Altered function of the tobacco mosaic virus movement protein in a hypersensitive host, *Virology,* 180, 257, 1991.
51. **Citovsky, V., Knorr, D., Schuster, G., and Zambryski, P.,** The P-30 movement protein of tobacco mosaic virus is a single-stranded nucleic acid binding protein, *Cell,* 60, 637, 1990.
52. **Weintraub, M., Ragetli, H., and Leung, E.,** Elongated virus particles in plasmodesmata, *J. Ultrastr. Res.,* 56, 351, 1976.
53. **Robards, A. and Lucas, W.,** Plasmodesmata, *Annu. Rev. Plant Physiol. Plant Mol. Biol.,* 41, 369, 1990.
54. **Deom, C., Schubert, K., Wolf, S., Holt, C., Lucas, W., and Beachy, R.,** Molecular characterization and biological function of the movement protein of tobacco mosaic virus in transgenic plants, *Proc. Natl. Acad. Sci. U.S.A.,* 87, 3284, 1990.
55. **Ponz, F. and Bruening, G.,** Mechanisms of resistance to plant viruses, *Annu. Rev. Phytopathol.,* 24, 355, 1986.
56. **Kitajima, E. and Lauritis, J.,** Plant virions in plasmodesmata, *Virology,* 37, 681, 1969.
57. **Allison, R., Janda, M., and Ahlquist, P.,** Infectious *in vitro* transcripts from cowpea chlorotic mottle virus cDNA clones and exchange of individual RNA components with brome mosaic virus, *J. Virol.,* 62, 3581, 1988.
58. **Allison, R.,** personal communication.
59. **Hamilton, R. and Nichols, C.,** Influence of bromegrass mosaic virus on the replication of tobacco mosaic virus in Hordeum vulgare, *Phytopathology,* 67, 4811, 1977.
60. **Hamilton, R. and Dodds, A.,** Infection of barley by tobacco mosaic virus in single and mixed infection, *Virology,* 4, 266, 1970.
61. **Wellink, J. and van Kammen, A.,** Cell-to-cell transport of cowpea mosaic virus requires both the 58K/48K proteins and the capsid proteins, *J. Gen. Virol.,* 70, 2279, 1989.
62. **Allison, R., Thompson, C., and Ahlquist, P.,** Regeneration of a functional RNA virus genome by recombination between deletion mutants and requirement for cowpea chlorotic mottle virus 3A and coat genes for systemic infection, *Proc. Natl. Acad. Sci. U.S.A.,* 87, 1820, 1990.
63. **Allison, R., Janda, M., Thompson, C., and Ahlquist, P.,** Coat genes of cowpea chlorotic mottle virus and brome mosaic virus can be exchanged without affecting host specificity of systemic movement. Abstr. 205, Am. Phytopathol. Soc. Meetings, Grand Rapids, Michigan, 1990.
64. **Quillet, L., Guilley, H., Jonard, G., and Richards, K.,** In vitro synthesis of biologically active beet necrotic yellow vein virus RNA, *Virology,* 172, 293, 1989.
65. **Siegal, A., Zaitlin, M., and Sehgal, O.,** The isolation of defective tobacco mosaic virus strains, *Proc. Natl. Acad. Sci. U.S.A.,* 48, 1845, 1962.
66. **Osbourn, J., Sarkar, S., and Wilson, T.,** Complementation of coat protein-defective TMV mutants in transgenic tobacco plants expressing TMV coat protein, *Virology,* 179, 921, 1990.
67. **Takamatsu, N., Ishikawa, M., Meshi, T., and Okada, Y.,** Expression of bacterial chloramphenicol acetyl transferase gene in tobacco plants mediated by TMV RNA, *EMJO J.,* 6, 307, 1987.
68. **Saito, T., Yamanaka, K., and Okada, Y.,** Long distance movement and viral assembly of tobacco mosaic virus mutants, *Virology,* 176, 329, 1990.
69. **Dawson, W., Bubrick, P., and Grantham, G.,** Modifications of the tobacco mosaic virus coat protein gene affecting replication, movement, and symptomatology, *Phytopathology,* 78, 783, 1988.
70. **Culver, J. and Dawson, W.,** Tobacco mosaic virus coat protein: an elicitor of the hypersensitive reaction but not required for the development of mosaic symptoms in *Nicotiana sylvestris, Virology,* 173, 755, 1989.
71. **Esau, K. and Cronshaw, J.,** Relation of tobacco mosaic virus to the host cells, *J. Cell. Biol.,* 33, 665, 1967.
72. **Pring, D.,** Viral and host RNA synthesis in BSMV-infected barley, *Virology,* 44, 54, 1971.
73. **Petty, I. and Jackson, A.,** Mutational analysis of barley stripe mosaic virus RNA-B, *Virology,* 179, in press, 1990.
74. **D'Arcy, C. and de Zoetin, G.,** Beet western yellows virus in phloem tissue of Thlaspi arvense, *Phytopathology,* 69, 1194, 1979.

75. **Barker, H.,** Invasion of non-phloem tissue in *Nicotiana clevelandii* by potato leafroll luteovirus is enhanced in plants also infected with potato Y potyvirus, *J. Gen. Virol.,* 68, 1223, 1987.
76. **Fraser, R.,** The genetics of resistance to plant viruses, *Annu. Rev. Phytopathol.,* 28, 179, 1990.
77. **Milne, R.,** Electron microscopy of tobacco mosaic virus in leaves of *Nicotiana glutinosa, Virology,* 28, 527, 1966.
78. **Cohen, J. and Loebenstein, G.,** An electron microscopic study of starch lesions in cucumber cotyledons infected with tobacco mosaic virus, *Phytopathology,* 65, 32, 1975.
79. **Fraser, R.,** *Biochemistry of Virus Infected Plants,* Research Studies Press/John Wiley and Sons, Chichester, 1987.
80. **Carr, J. and Klessig, D.,** The pathogenesis related proteins of plants, in *Genetic Engineering, Principles and Methods,* Vol. 11, Setlow, W., Ed., 1989, 65.
81. **Bol, J., Linthorst, H., and Cornelissen, J.,** Plant pathogenesis related proteins induced by virus infection, *Annu. Rev. Phytopathol.,* 28, 113, 1990.
82. **Loebenstein, G., Gera, A., Barnett, A., Shabtai, S., and Cohen, J.,** Effect of 2,4-dichlorophenoxyacetic acid on multiplication of tobacco mosaic virus in protoplasts from local-lesion and systemic-responding tobaccos, *Virology,* 100, 110, 1980.
83. **Fraser, R.,** Genes for resistance to plant viruses, *Crit. Rev. Plant Sci.,* 3, 257, 1986.
84. **Fraser, R.,** Varying effectiveness of the N′ gene for resistance to TMV in tobacco infected with virus strains differing in coat protein properties, *Physiol. Plant Pathol.,* 22, 109, 1983.
85. **Weber, P.,** Inheritance of a necrotic-lesion reaction to a mild strain of tobacco mosaic virus, *Phytopathology,* 41, 593, 1951.
86. **Saito, T., Meshi, T., Takamatsu, N., and Okada, Y.,** Coat protein gene sequence of tobacco mosaic virus encodes a host response determinant, *Proc. Natl. Acad. Sci. U.S.A.,* 84, 6074, 1987.
87. **Saito, T., Kimiko, Y., Watanabe, Y., Takamatsu, N., Meshi, T., and Okada, Y.,** Mutational analysis of the coat protein gene of tobacco mosaic virus in relation to hypersensitive response in tobacco plants with the N′ gene, *Virology,* 173, 11, 1989.
88. **Funatsu, G. and Fraenkel-Conrat, H.,** Location of amino acid exchanges in chemically evolved mutants of TMV, *Biochemistry,* 3, 1356, 1964.
89. **Culver, J. and Dawson, W.,** Point mutations in the coat protein gene of tobacco mosaic virus induce hypersensitivity in Nicotiana sylvestris, *Mol. Plant-Microbe Interact.,* 2, 209, 1989.
90. **Knorr, D. and Dawson, W.,** A point mutation in the TMV capsid protein gene induces hypersensitivity in N. sylvestris, *Proc. Natl. Acad. Sci. U.S.A.,* 85, 170, 1988.
91. **Culver, J. and Dawson, W.,** personal communication.
92. **Okada, Y.,** personal communication.
93. **Daubert, S. and Routh, G.,** Determinants of symptomatology in the DNA sequence of cauliflower mosaic virus gene VI, *Mol. Plant-Microbe Interact.,* 3, 341, 1990.
94. **Poritz, M., Bernstein, H., Strub, K., Zopf, D., Wilhelm, H., and Walter, P.,** An *E. coli* ribonucleoprotein containing 4.5S RNA resembles mammalian signal recognition particle, *Science,* 250, 1111, 1990.
95. **Goff, S. and Goldberg, A.,** Production of abnormal proteins in E. coli stimulates transcription of Ion and other heat shock genes, *Cell,* 41, 587, 1985.
96. **Esau, K.,** Anatomy of plant virus infection, *Annu. Rev. Phytopathol.,* 5, 47, 1967.
97. **Petty, I., Edwards, M., and Jackson, A.,** Systemic movement of an RNA plant virus is determined by a point mutation in a 5′ leader sequence, *Proc. Natl. Acad. Sci. U.S.A.,* 87, in press, 1990.
98. **Daubert, S., Schoelz, J., Li, D., and Shepherd, R.,** Expression of disease symptoms in cauliflower mosaic virus genomic hybrids, *J. Mol. Appl. Genet.,* 2, 537, 1984.
99. **Schoelz, J. and Shepherd, R.,** Host range of cauliflower mosaic virus, *Virology,* 162, 30, 1988.
100. **Stratford, R. and Covey, S.,** Segregation of cauliflower mosaic virus symptom genetic determinants, *Virology,* 172, 451, 1989.
101. **Baughman, G., Jacobs, J., and Howell, S.,** Cauliflower mosaic virus gene VI produces a symptomatic phenotype in transgenic tobacco plants, *Proc. Natl. Acad. Sci. U.S.A.,* 85, 733, 1988.
102. **Takahashi, H., Shimamoto, K., and Ehara, Y.,** Cauliflower mosaic virus gene VI causes growth suppression, development of necrotic spots, and expression of defense-related genes in transgenic tobacco plants, *Mol. Gen. Genet.,* 216, 188, 1989.
103. **Francki, R.,** Plant virus satellites, *Annu. Rev. Microbiol.,* 39, 151, 1985.
104. **Waterworth, H., Kaper, J., and Tousignant, M.,** CARNA 5, the small cucumber mosaic virus-dependent replicating RNA, regulates disease expression, *Science,* 204, 845, 1979.
105. **Kumar, I., Murant, A., and Robinson, D.,** A variant of the groundnut rosette virus satellite RNA that induces brilliant yellow blotch mosaic symptoms in Nicotiana benthamiana, Abstr. W86-004, 8th Int. Congr. Virol., Berlin, 1990.
106. **Sleat, D. and Palukaitis, P.,** Induction of tobacco chlorosis by certain cucumber mosaic virus satellite RNAs is specific to subgroup-II helper strains, *Virology,* 176, 292, 1990.

107. **Diener, T. and Owens, R.,** Viroids, in *The Biochemistry of Plants: A Comprehensive Treatise,* Vol. 15, Marcus, A., Ed., Academic Press, San Diego, 1989, 537.
108. **Owens, R. and Schneider, I.,** Satellite of tobacco ringspot virus RNA lacks detectable mRNA activity, *Virology,* 80, 222, 1977.
109. **Hidaka, S., Hanada, K., and Kiichi, I.,** *In vitro* messenger properties of a satellite RNA of cucumber mosaic virus, *J. Gen. Virol.,* 71, 439, 1990.
110. **Sleat, D. and Palukaitis, P.,** Site-directed mutagenesis of a plant viral satellite RNA changes its phenotype from ameliorative to necrogenic, *Proc. Natl. Acad. Sci. U.S.A.,* 87, 2946, 1990.
111. **Masuta, C. and Takanami, Y.,** Determinants of sequence and structural requirements for pathogenicity of a cucumber mosaic virus satellite RNA (Y-sat RNA), *Plant Cell,* 1, 1165, 1989.
112. **Diener, T.,** *Viroids and Viroid Diseases,* Interscience, NY, 1979.
113. **Van Tol, H., Buzayan, J., and Bruening, G.,** Evidence for spontaneous circle formation in the replication of the satellite RNA of tobacco ringspot virus, *Virology,* 180, 23, 1991.
114. **Hammond, M. and Owens, R.,** Mutational analysis of potato spindle tuber viroid reveals complex relationships between structure and infectivity, *Proc. Natl. Acad. Sci. U.S.A.,* 84, 3967, 1987.
115. **Devic, M., Jaegle, M., and Baulcombe, D.,** Symptom production on tobacco and tomato is determined by 2 distinct domains of the satellite RNA of cucumber mosaic virus (strain Y), *J. Gen. Virol.,* 70, 2675, 1989.
116. **Kurath, G. and Palukaitis, P.,** RNA sequence heterogeneity in natural populations of three satellite RNAs of cucumber mosaic virus, *Virology,* 173, 231, 1989.
117. **Palukaitis, P. and Sleat, D.,** Cucumber mosaic virus and its satellite RNAs: localization of sequences involved in pathogenicity, in Proc. Int. Meet. Biol., Fundacon Juan March, Genome Expression and Pathogenesis of Plant RNA Viruses. Madrid, 1990, p. 22.
118. **Jaegle, M., Devic, M., Longstaff, M., and Baulcombe, D.,** Cucumber mosaic virus satellite RNA (strain Y): analysis of sequences which affect yellow mosaic symptoms on tobacco, *J. Gen. Virol.,* 71, 1905, 1990.
119. **Schnolzer, M., Haas, B., Ramm, K., Hofmann, H., and Sanger, H.,** Correlation between structure and pathogenicity of potato spindle tuber viroid, *EMBO J.,* 4, 2181, 1985.
120. **Visvader, J. and Symons, R.,** Replication of in vitro constructed viroid mutants, *EMBO J.,* 5, 2051, 1986.
121. **Diener, T.,** *The Viroids,* Plenum Press, NY, 1987.
122. **Thresh, J.,** Cropping practices and virus spread, *Annu. Rev. Phytopathol.,* 20, 193, 1982.

# PLANTS AND ENDOMYCORRHIZAL FUNGI: THE CELLULAR AND MOLECULAR BASIS OF THEIR INTERACTION

**Paola Bonfante-Fasolo and Silvia Perotto**

## TABLE OF CONTENTS

# I. INTRODUCTION

Living together is a very common situation in the biological world, especially in the plant kingdom and in the underground environment. In addition to nitrogen-fixing symbiosis between leguminous plants and rhizobia,[1] the underground world harbors the most common type of mutualistic symbiosis, namely the mycorrhizae. Mycorrhizae are a tight association between the roots of higher green plants and soil fungi. Mycorrhizal associations involve the majority of plants and a large number of fungal species. In the British flora for example, 80% of the angiosperms, 100% of the gymnosperms and 70% of the pteridophytes are potentially mycorrhizal.[2]

The great success of mycorrhizal plants in all natural ecosystems raises the question of what features increase the fitness of mycorrhizal over nonmycorrhizal plants. The idea that both symbionts contribute to the nutrition of the partner with the establishment of a bi-directional flux of nutrients (carbon compounds moving from plant to fungus, mineral nutrients, mainly phosphate, in the opposite direction) dates back many years.[3] More recent studies suggest that the partners of a mycorrhizal association can influence each other during all levels of metabolic activity, from gene expression to morphogenesis. The result is a state of cellular and physiological compatibility that can be maintained for long periods of time (even for years) and is often recognized as a state of "mutual benefit."[2]

Mycorrhizae represent a composite world: the wide range of host plants is mirrored by a large variety of mycorrhizal fungi which can belong to all taxonomic divisions, from Zygomycetes to Ascomycetes, Basidiomycetes, and imperfect fungi.[2] The term mycorrhiza brings together types of associations which differ both morphologically and physiologically.[4] These different types of mycorrhizae can be often correlated to defined plant species and communities, so that their distribution changes from the Arctic regions to the tropics along with the distribution of the plants that harbor them.[5] Because of this wide scenario in physiology and morphology, this chapter will only focus on the most common type of mycorrhiza, namely the vesicular arbuscular mycorrhiza (VAM), which can occur in about 80% of land plants. VAM are endomycorrhizae, which means that the fungus forms intracellular structures in the host root. This feature makes them an interesting system to investigate the cellular and molecular basis of plant-fungal communication in symbiotic interactions. A general view on the structural and physiological aspects of this association can be found in some recent reviews.[6-9]

# II. THE LIFE CYCLE OF THE VAM FUNGUS DURING THE INFECTION PROCESS

VAM fungi belong to the order of Glomales (Zygomycetes) and include a limited number of genera (*Glomus, Acaulospora, Gigaspora, and Scutellospora*).[10] However, their taxonomy is still uncertain and is mainly based on the morphological features of the fungal spores. This is due, in part, to our inability to maintain these fungi in pure axenic culture and to obtain sexual structures. They are obligate symbionts and can only be grown successfully in the presence of a host plant. In a dual culture system, obtained by inoculating a host plant with fungal spores, the fungus completes its life cycle and undergoes a complex morphogenesis. During this process the spore germinates, producing a vegetative mycelium that contacts the host root surface and produces appressoria. The appressoria then originates hyphae which initiate the infection of the root and subsequently form intercellular hyphae, coils, vesicles and, most importantly, highly branched intracellular structures called arbuscules (Figure 1).

The recent use of an *in vitro* dual culture system with *Agrobacterium rhizogenes* T-DNA-transformed carrot roots and spores of *Gigaspora margarita* allows the sequence of

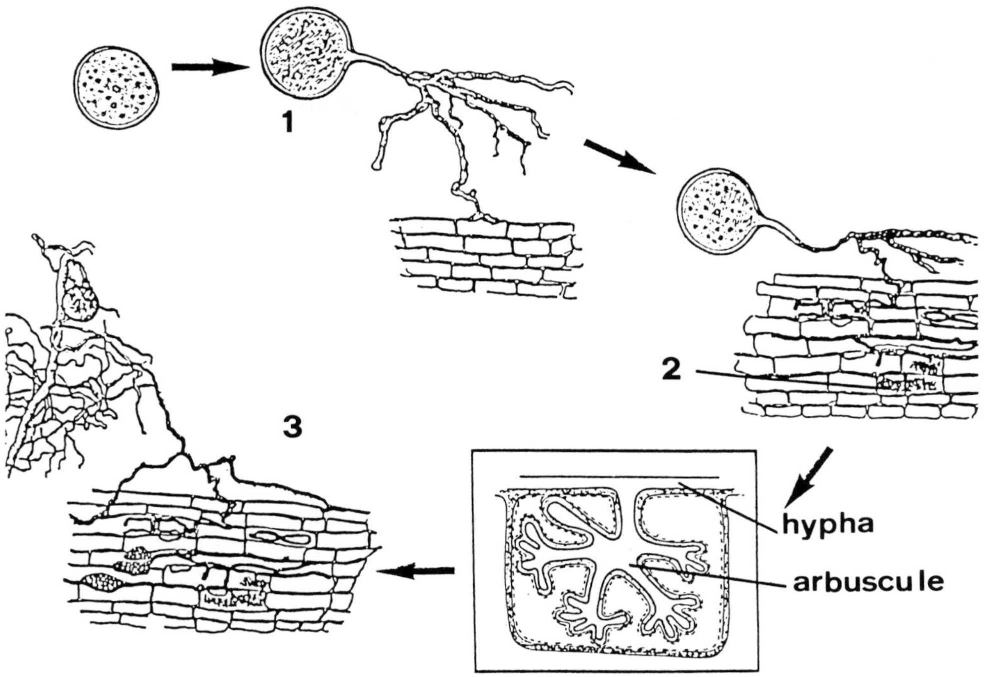


FIGURE 1. Scheme of the infection process of a vesicular arbuscular fungus (*Glomus* sp.) inside a host root. The process is described as a succession of space events: (1) extraradical phase, (2) intraradical phase with the development of intercellular hyphae and arbuscules, and (3) further production of new extraradical mycelium.

events in the early stages of the fungal life cycle to be defined.[11] In the absence of the carrot root, hyphae from a germinated spore elongate slowly and reach a length of 20 to 30 mm about 10 days after germination before growth ceasing. The introduction of carrot roots in the system, even after spore germination, induces a significant increase (20-fold) in hyphal growth. No physical contact with the root is required for this induction, suggesting that molecules are released by the root. Five to eight days after the introduction of the carrot roots, colonization by the fungus occurs, with extraradical hyphae that continue to grow.[11] On the basis of these observations, the authors suggest that the infection process consists of three different steps: (1) hyphal growth from a germinating spore, dependent initially upon its own nutritional supply; (2) stimulation of further fungal growth by root exudates and initiation of the infection process; and (3) fungal development of intracellular arbuscules which connect the fungus to the nutrient flux from the plant (Figure 2).

## A. WHY DOES THE FUNGUS NOT GROW SAPROPHYTICALLY IN PURE CULTURE?

The ability of many VAM spores to germinate in culture was demonstrated many years ago in the pioneering work of Mosse[12] and Hepper.[13] Unfortunately, no real progress has since been made in culturing VAM fungi separated from their host and sustaining their independent saprophytic phase.[14] Many different growth media and conditions have been unsuccessfully tried in order to grow VAM fungi for longer times in pure culture. Spores can germinate in poor media or even in deionized water with percentages of germination depending upon the fungal species (up to 80 to 100%).[14] Daily observations on the development of mycelium from germinated spores of *Scutellospora gregaria* show that regularly nucleated hyphae initially follow a linear growth pattern, with rare but regular secondary

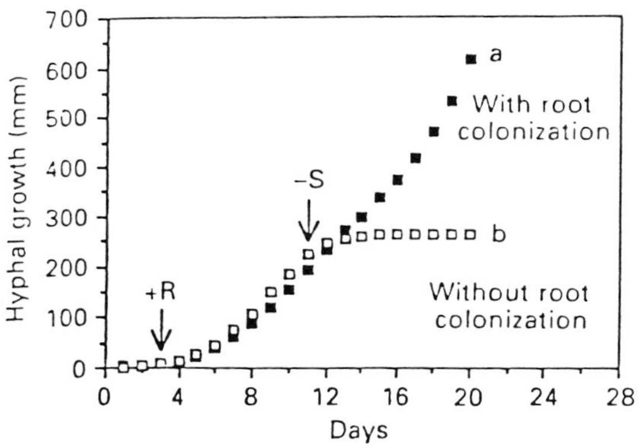


FIGURE 2. Time course of the infection process of a vesicular arbuscular fungus (*Gigaspora margarita*) in presence of the host root (+R). (a) With root colonization; (b), without root colonization both having the spore removed (-S) after 8 days of dual culture. (With permission by Becard, G. and Pichet, Y., *New Phytol.* 112, 77, 1989.)

branching (Figures 3 and 4).[15] Clusters of auxiliary cells are usually formed on the lateral branches. Abnormal features such as swollen or collapsed hyphae and frequent septation begin to appear after 15 to 20 days of culture, all leading to the arrest of hyphal growth.[15]

An interesting hypothesis to explain the limited growth of the fungus is the absence of nuclear DNA replication during spore germination.[14] A large number of nuclei, at least 20,000 for *G. margarita,* are easily detected in the chlamydospores, but the use of tritiated adenine indicates that there is no incorporation of this nucleotide in the nuclear DNA during or after spore germination. These data suggest that there is no nuclear division, at least *in vitro* and, therefore, that the nuclei present in the developing hyphae are derived from the germinating spore by migration. However, other authors suggest that mitotic processes may occur before and during the spore germination.[16]

The estimation of the amount of DNA during the germination of spores has not provided clear results in favor of one or the other hypothesis. Viera and Glenn[17] report values ranging from 13,700 to 32,400 pg DNA/spore by using a fluorometric assay on DNA extracted from *G. versiforme* spores. They estimate a value of 3.4 pg DNA/nucleus, by dividing the amount of nuclear DNA per spore by the estimated number of nuclei (about 5,000 per spore). Despite the high variation in the values obtained before and during germination, the authors conclude that DNA replication does not occur during germ tube development. Notwithstanding the tremendous variation in the fungal genome size, the value reported by Viera and Glenn is well above those reported for other fungi.[18] Our unpublished results using static and flow fluorimetry suggest a lower value of about 0.25 to 0.3 pg DNA/nucleus for an isolate of the same species *G. versiforme*.

Manipulation of possible secondary messengers, such as cAMP and calcium, and modification of the microtubular system did not increase the *in vitro* growth of VAM fungi or their morphology.[14] This is somehow different from what happens in pathogenic rust fungi, where the manipulation of secondary messengers affects cell differentiation and DNA replication.[19]

A speculative hypothesis to explain the metabolic deficiencies of the fungus during its saprophytic phase has been proposed by Fortin.[20] According to this hypothesis, genes involved in free-living growth or development have been lost by the VAM fungi during the long coevolution with the plant. Corresponding gene functions are provided somehow by the host during the initial steps of the symbiotic phase.[20] However, in order to obtain a

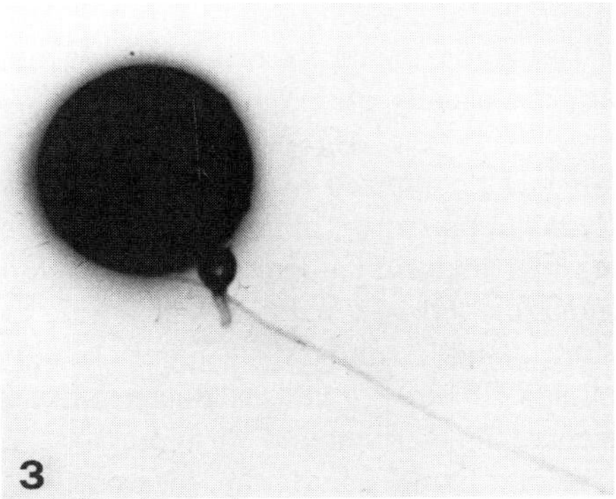


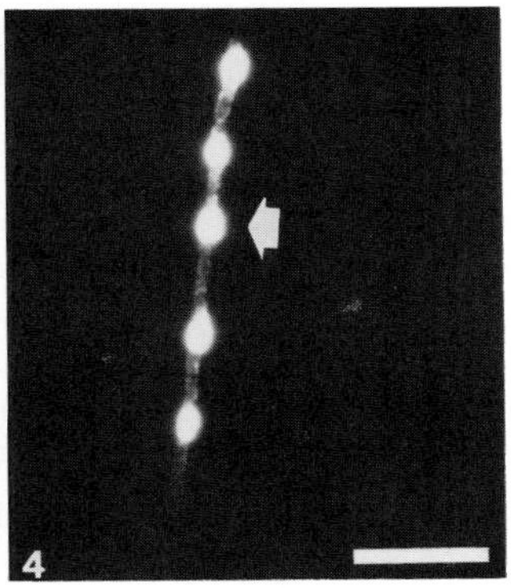


FIGURE 3. Germinating spore of *Scutellospora gregaria*. FIGURE 4. Regular distribution of nuclei (arrow) in the germinating spores of *S.gregaria* DAPI staining. (Bar: 50 μm.)

clearer picture on the ontogeny and genetic of VAM fungal spores, further studies must be made to clarify the details of the VAM fungi life cycle.

## B. THE PREINFECTION PHASE: DO SIGNAL MOLECULES EXIST?

The dual *in vitro* system for culturing vesicular-arbuscular mycorrhizal fungi has clearly demonstrated that the presence of the root is required for fungal growth. Soluble factors from the plant can significantly stimulate the growth of the fungus before physical contact with the root surface.[11] These data confirm previous reports that host roots produce factors which stimulate VAM fungal growth prior to infection,[13] while root exudates from nonhost plants do not show any effect.[21] If there are any signal molecules involved, the first questions concern their chemical nature and their range of specificity, considering the fact that 80% of land plants harbor VAM symbiosis. Reports derived from studies on interactions between plants and pathogenic or symbiotic bacteria have shown the importance of plant phenolic compounds in the early stages of their communication.[22] Wounded plant cells produce phenolic molecules that activate *vir* genes in *Agrobacterium,*[23] while roots exude flavonoids that act as activators of *nod* genes in *Rhizobium*.[1] These regulatory genes are necessary in both cases for infection of the host. When some flavonoids (naringenin, hesperitin, and apigenin) were added to VAM fungi at concentrations ranging from 0.15 to 1.5 μ*M*, hyphal growth of *G. margarita* was increased 2- to 10-fold over that obtained on water agar and there was also an increase in the rate of spore germination.[21] Flavonoids released from seeds of alfalfa during imbibition also increased mycelial branching in VAM fungi.[24]

These results demonstrate that plant-exuded flavonoids may enhance independent growth in VAM fungi. However, it is not clear whether they act as true signal molecules or simply as a nutritional factor. Also, a wider range of flavonoids should be tested to find out whether related molecules may have a different effect or some degree of host specificity, similar to that described for rhizobia.[25,26] However, in order to resolve this question and define the actual molecules responsible for induction of fungal growth in culture, activating crude extracts from plants should be fractionated and the individual compounds identified. Interestingly, flavonoid-like molecules are also phytoalexin, which are potent inhibitors of growth in invading pathogens.[27] If flavonoids act as signal molecules, the next step would be to identify the receptor in the fungus and to trace the regulation of gene activation.

## C. THE CONTACT PHASE

In a plant root-fungal culture, the fungal germ tubes take contact with the root, branching and growing on the root surface. Appressoria shaped as flat and swollen bodies, 20 to 40 μm long, are formed, adhering to the host surface as a result of the contact. It is not known whether topographical signals induce the formation of appressoria in VAM fungi as has been shown for rust fungi.[28] However, they are usually observed as lens shaped structures lying between adjacent epidermal cells.[29]

Detailed morphological observations of leeks colonized by *G.versiforme* revealed that two infection hyphae are usually formed by the appressoria. In this plant, they penetrate the underlying epidermal layer passing between two adjacent epidermal cells, developing either into short pegs or large hyphae. They contact the hypodermal cells and inside them, or in the underlying cortical cell, they loop to form coils.[29] Appressoria appear as multinucleate structures, slightly vacuolate and rich in lipids (Figures 5 and 6). Bacteria-like organisms directly immersed in the fungal cytoplasm are regularly found in appressoria and in the intraradical structures.[30] Hyphae which penetrate intracellularly are encased by a thick deposition of material continuous with the cell wall, and by the invaginated host plasma membrane. This type of surface interaction is essentially maintained along the whole infective process (see Section III.D).

Alternative patterns of fungal penetration across the outer root layer have also been observed, generally related to the type of host plant.[6] Instead of penetrating through tight junctions between two epidermal cells, the fungus can pass across the tangential epidermal cell wall or through root hairs. Roots of woody plants often develop a rhizodermis with thick and cutinized cell walls. In these plants the fungus runs under the sloughing cells, penetrating the first intact cortical cell layer.

All these different patterns of penetration seem to be influenced by the features of the host cell surface.[31] The outer layers of the root demonstrate a strong heterogeneity in cell wall composition. In *Ginkgo biloba* and leek there are abundant UV autofluorescent, cell wall-bound phenols associated with the layered and complex texture of the cellulosic wall.[32-33] All these features are comparable to those found in the cell wall of many epidermal tissues of shoots and leaves. Other species, such as pea, possess epidermal cell walls which, on the contrary, are weakly autofluorescent and which have an amorphous texture. Pectin and cellulose have been detected by using monoclonal antibodies against polygalacturonic acid, and a purified cellobiohydrolase binding to β-1-4-glucans.[34,35] Many factors make it difficult to quantify antigens based on the density of gold particles after gold immunolabeling. However, a dual labeling approach suggests that, in leek roots, cellulose is very abundant in the epidermal cell walls, while pectin antigens are rare on the same walls. In pea, walls in the root hairs and epidermal cells are on the contrary, strongly labeled by the probe for pectin (Figures 7 to 9). These data are consistent with chemical analysis of the pectin content between dicots and monocots, which indicate that monocots are poorer in pectin material.[36] It is known that pectins represent the more flexible wall component, which probably influences and facilitates the penetration of microorganisms.[36] The low content reported in monocots may result in a scarcely flexible wall, and this physicochemical feature has been proposed to explain the inability of *Agrobacterium* to penetrate Graminaceae.[37] Other authors demonstrated that pectin-enriched fraction from tomato cell walls can inhibit attachment of *Agrobacterium* to plant tissues, suggesting the presence of specific sites of binding for the bacterium on the pectin molecule.[38]

One can thus hypothesize that in VAM the preferential site for the fungal penetration (through junctions between the epidermal cells or directly through tangential walls or root hairs) is mainly controlled by the physicochemical features of the epidermal cell walls. More specific mechanisms, however, cannot be ruled out.

When the fungus penetrates the epidermal cells, a localized thickening of the plant wall in the epidermal and in the underlying cortical cell has been observed in leek and pea roots.[29,39] The nature of this thickening has not been further investigated, but it is morphologically comparable to the papillae described during plant-pathogen interactions,[40] although

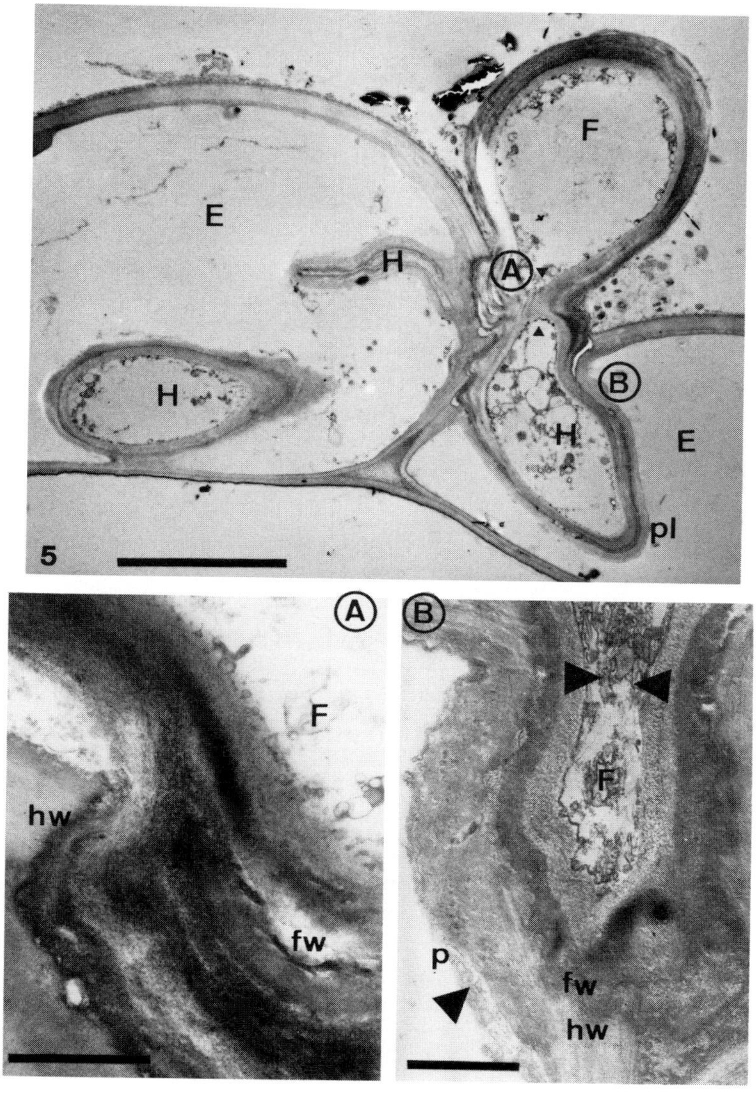


FIGURE 5. Transmission electron micrograph (TEM) of a leek root colonized by *Glomus versiforme*. The fungus (F) passes through two epidermal cells (E), forming two infecting hyphae (H). The fungus shows a constriction at the penetration point (arrow heads)and an important cell wall thickening (A, magnification of the detail). The host wall also shows a slight papilla-like thickening (p) (B, magnification of the detail). pl, host membrane; hw, host wall; fw, fungal wall. (Bars: 10 μm and 1 μm, in A and B, respectively.)

the size in mycorrhizal plants is very reduced (Figure 5B). A larger wall thickening, more similar to the papillae formed in response to pathogens, has been observed in some pea mutants when a VA mycorrhizal fungus attempts penetration.[41] These chemically induced mutants of *Pisum sativum* (called myc$^-$), identify the first genetic locus for resistance to mycorrhiza.[42] These mutants were derived from a larger group of mutants originally isolated as being unable to form root nodules (nod$^-$) with *Rhizobium*. In 21 out of 45 nod$^-$ mutants, the absence of nodulation is associated with the absence of mycorrhizal colonization. Analysis of the genetic behavior of the myc$^-$ pea mutants in diallele crosses has shown that myc$^-$

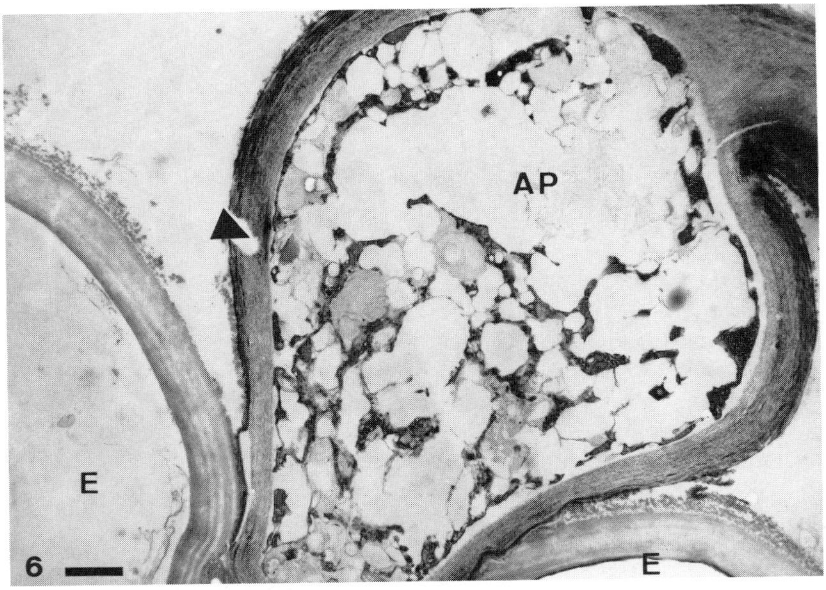


FIGURE 6. Ultrastructural aspects of a vacuolated appressorium (AP) of *Glomus versiforme* penetrating between two epidermal cells (E). Its cell wall is thick and fibrillar in texture (arrowhead). Note the layered texture of the epidermal cells of the host. (Bar: 1 μm.)

and nod$^-$ phenotype cannot be uncoupled, indicating that both phenotypes derive from the same genes. They also suggest that the myc$^-$ phenotype is under a recessive genetic control and that at least five separate genes are involved in VAM formation. Morphologically, myc$^-$ mutants are characterized by ill-defined appressoria which lead to aborted infections.[41] The formation of an abnormally thick papilla in the cell wall in contact with the fungus reinforces the hypothesis that one of the first mechanisms controlling the plant-fungal communication lies in molecules at the root cell surface. It would therefore be interesting to understand whether this nod$^-$/myc$^-$ phenotype is derived by changes in the regulatory control of cell wall synthesis, leading to an abnormal deposition of wall components that stop fungal penetration.

The characteristics of the plant cell surface may also influence the successive steps in the infection process, namely, the passage from the epidermal to the underlying cortical cells, which sometimes develop a well-defined layer called the hypodermis. Dimorphic hypodermis have been described and are characterized by deposits of suberin on the radial cell walls.[43] Suberification is a secondary modification of the cell wall that starts in the zone of differentiating cells. It is physiologically important because it can deeply influence the apoplastic radial transfer of solutes out of or into the root. In leek, suberified hypodermal cell walls are constitutive and occur both in mycorrhizal and nonmycorrhizal plants (Figure 10). However, during colonization of leek by a VAM fungus, the role of the dimorphic hypodermis is to channel fungal growth and development inside the root.[32] The invading fungus tends to penetrate and colonize those cells where there is little or no suberin deposited in the wall (Figure 11). Hypodermal cells where suberin is clearly detectable are avoided by the fungus and appear dead in contrast to the active and living colonized cells.[32] When the fungus finds itself surrounded by suberin-lined cells, it stops further colonization (Figures 12 and 13). Similar behavior has been observed in many other herbaceous woodland plants, where suberization of the hypodermis (referred to as exodermis) may occur before or after the passage of mycorrhizal fungi.[44]

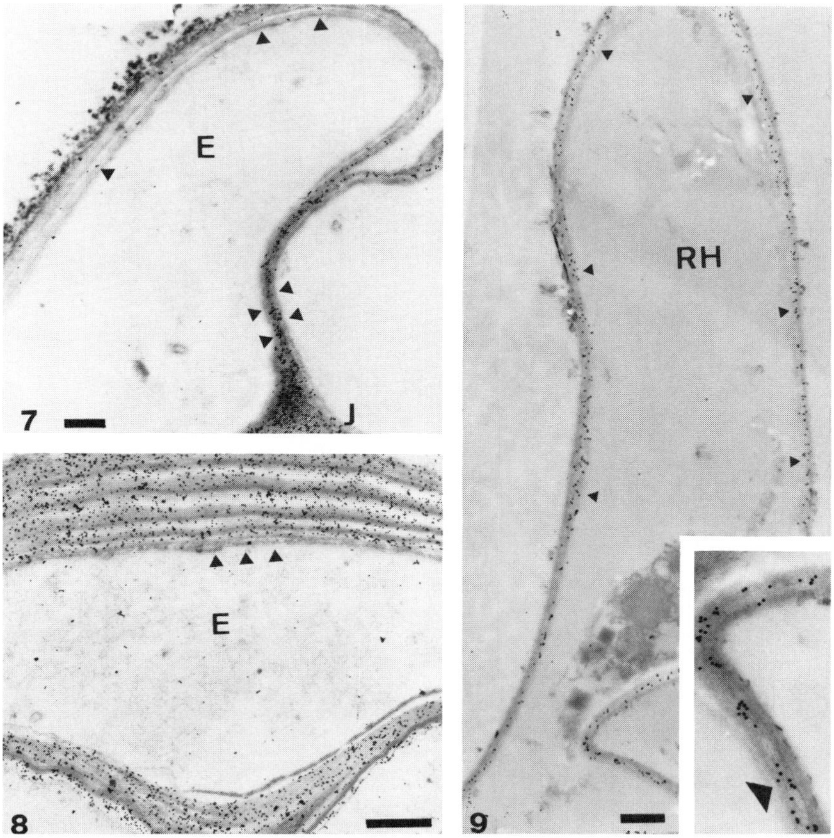


FIGURES 7-9. Comparison between the epidermal cell walls in leek and in pea. Figure 7: A monoclonal antibody (JIM 5) which binds to nonesterified pectins, reveals a limited number of antigens over the epidermal cell wall (E) of a leek root and a heavy labeling at the junctions (J) between the host cells (arrowheads). (Bar: 1 μm.) Figure 8: The complex cellobiohydrolase-gold which binds β-1-4 glucans, reveals a strong labeling over the epidermal cell wall of a leek root.(Bar: 1 μm.) Figure 9: Longitudinal section of a root hair (RH) in pea, treated as described in Figure 8. There is a regular distribution of gold granules (arrowheads) along the amorphous wall. (Bar: 1 μm.) Inset: magnification of the labeled wall. Figures 7 and 8 (from Bonfante-Fasolo, P. et al., *Planta,* 180, 537, 1990. With permission.)

Besides surface interactions, soluble shoot factors have also been suggested to influence the establishment of VAM symbiosis.[45] Reciprocal grafts have been obtained between shoots and roots of several cultivars of mycorrhizal and nonmycorrhizal legumes (pea and lupine, respectively). Only those of lupine shoots on pea roots survived, and the fact that mycelial growth and arbuscule development were inhibited in these chimeras suggests the existence of an inhibitory shoot factor in nonmycorrhizal plants.[45] Thus, genetic, morphologic, and cytochemical analysis point to the importance of the root cell surface in the early events of the formation of VAM symbiosis, being the primary physical site of contact between the plant and the fungus.

## D. THE SYMBIOTIC PHASE

Once the fungus overcomes the barrier represented by the outer root layers and reaches the parenchymatic tissues, it mostly develops hyphae running through the intercellular spaces. Small pegs are formed which grow into cortical cells and develop into arbuscules. The

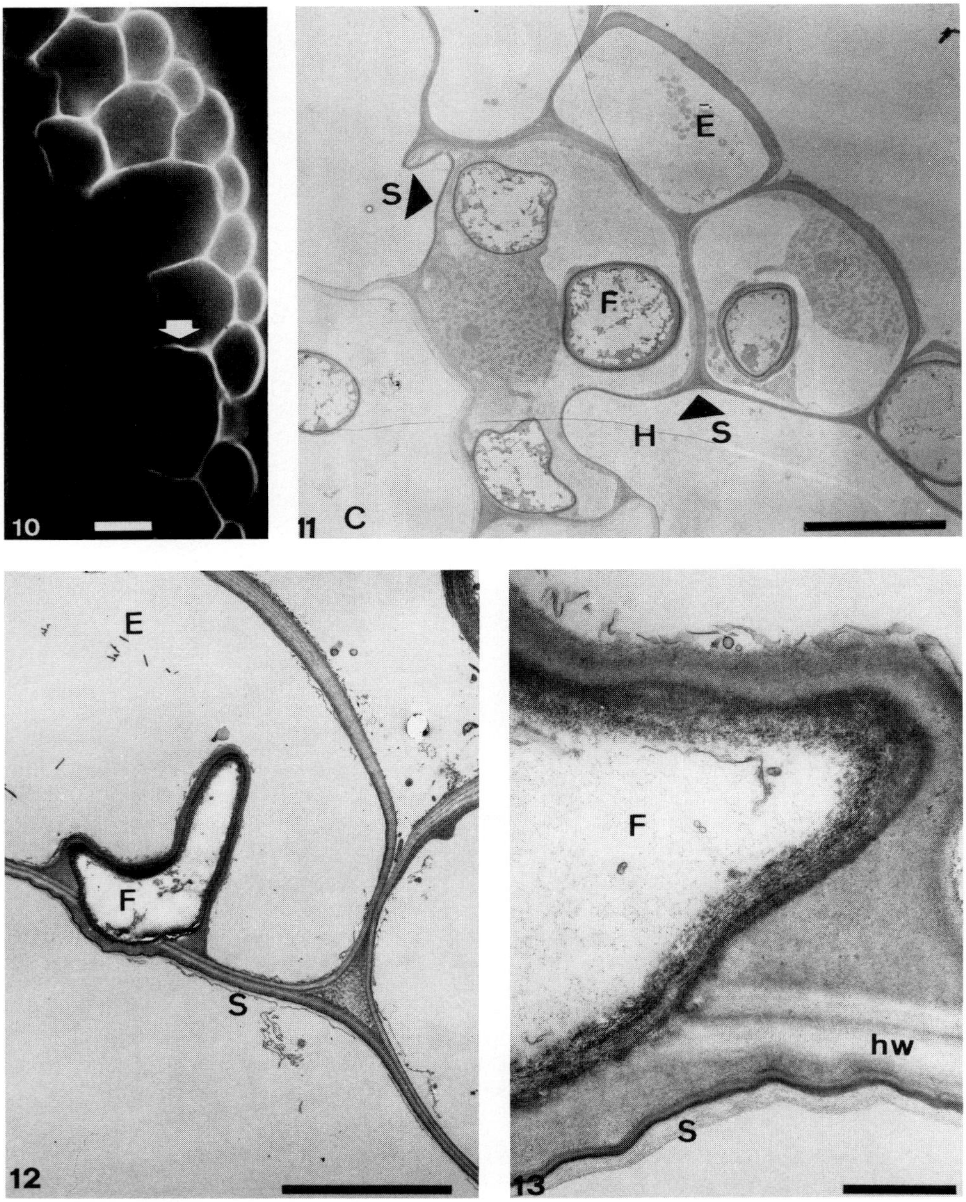


FIGURES 10-13. Cell wall features of epidermal and hypodermal cells in leek. Figure 10: Only the epidermal and hypodermal cell walls are autofluorescent under UV light. Suberin is responsible for the autofluorescence in the radial hypodermal cell walls (arrow). (Bar: 10 μm.) Figure 11: General view of the outer cortex showing two hypodermal (H) cells limited by suberin (S).They are dead and empty. The fungus (F) crosses living hypodermal cells, without suberin. C: cortical cells. E, epidermal cells. (Bar: 10 μm.) Figure 12: Unsuccessful colonization: the fungus is blocked inside an epidermal cell, probably due to the fact that a suberin layer (S) lines the whole underlying hypodermal cell. (Bar: 10 μm.) Figure 13: Detail of the suberin layer. hw, host cell wall (Bar: 1 μm). (Figures 10 and 11 from Bonfante-Fasolo, P. and Vian, B. *Ann. Bot. (Paris),* 10, 97, 1989.

formation of the arbuscule represents the most dramatic sign of the cellular integration between the two symbiotic eukaryotes. This highly branched structure plays an essential role in the transfer of nutrients (mostly sugars) from the root to the fungus and inorganic nutrients (mostly phosphate) from the arbuscule to the plant.[3]

Arbuscules start as a trunk with few branching hyphae and complete their development in 4 to 5 days.[46] The formation of a complete arbuscule induces further development of the extraradical mycelium, promoting new penetration points and new infection units (Figure 14). Arbuscules are ephemeral structures, existing for 5 to 6 days. They then rapidly collapse forming clumps, which are initially limited to the apical part of the arbuscular branches but quickly extend to the whole complex.[6]

The development of the infection is under host control and is largely dependent upon the root anatomy, as the same VA mycorrhizal fungal strain is reported to behave differently in different plants. In fact, it can form a high number of arbuscules in some plant species or run as very long intercellular hyphae with the formation of few arbuscules in others.[47]

The development as intercellular hyphae seems to be correlated to the presence of air channels and is the most common pattern of spreading of fungal infection, while coils are more common in wider roots where the intercellular spaces are filled with extracellular matrix or simply lack these channels.[44] Furthermore, some endomycorrhizal trees such as *Ginkgo biloba* or *Acer saccharum* determine a very unusual pattern in the arbuscule development of their endophytes.[48,49] These endophytes usually colonize angiosperms through the development of hyphae, each terminating in an arbuscule. On the contrary, on these endomycorrhizal trees the same endophytes show indeterminate growth, with single hyphae from which numerous arbuscules branch (Figure 15).

So far we do not know how fungal gene expression is regulated during the morphogenetic events occurring during the infection process. Ultrastructural observations reveal many cytological changes in the fungal hyphae during infection, including modifications to the cell wall, to the main storage components (lipids and glycogen), and to the nuclear organization. The fungal wall in the spore is about 12 μm, while in the coils it is about 1 μm and in the apical part of the arbuscular hyphae it is only 20 to 30 nm thick. This striking reduction in thickness is associated with a change in texture, from fibrillar in the intercellular hyphae to amorphous in the arbuscule.[31] Chitin is the most abundant skeletal molecule in VAM fungi and is normally polymerized into fibrils. The presence of *N*-acetylglucosamine residues in the cell wall of all fungal structures as revealed by lectins, suggests that chitin, or oligomers of chitin, are regularly synthesized during all stages of fungal development (Figure 16).[50] Therefore, the amorphous texture of the arbuscular wall is probably due to an inhibition of the polymerization of oligomers into a crystalline form of chitin during the arbuscule formation, rather then a block in synthesis.

In the thicker walls of intercellular hyphae, fibrillar chitin is embedded with other components that can be easily removed by mild alkali treatment. Solubility in alkali and reactivity with a lectin recognizing glucose/mannose (concanavalin A), indicates that they may be glucans. Labeling experiments with FITC-conjugated lectins on fresh tissue sections suggest that chitin is embedded by glucans mostly in the wall of coils and intercellular hyphae.[50] These glucans may protect the fungal wall from the action of plant hydrolytic enzymes during the early stages of infection.

The modulation of the fungal cell surface in the VAM fungus *Glomus versiforme* (thinning, chemical simplification, changes in texture) is comparable to what happens in some pathogenic fungi, where changes in the fungal wall characterize the different phases of infection.[51] These modifications can change the physicochemical properties of the fungal wall, from altering its permeability and resistance to turgor pressure, to influencing molecular exchanges between the two partners.[52] In addition, adjustments of the fungal surface could be important to avoid a host plant defense reaction. The embedding of the chitin fibrils by

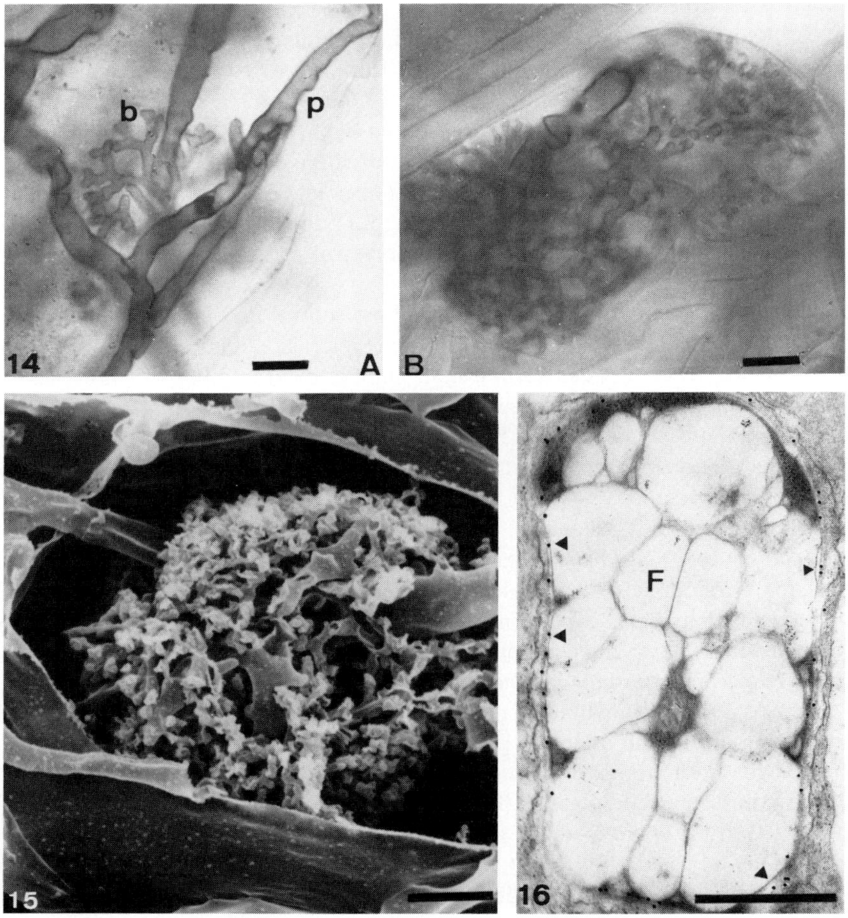


FIGURES 14-16. Arbuscule development. Figure 14: Arbuscule formation in leek after 12 days from the fungal spore inoculation (A) initial steps with peg formation (p) and first dichotomous branches (b). (B) Fully developed arbuscule. (Bar: 10 μm.) Figure 15: Scanning electron micrograph of a *Glomus* arbuscule inside a *Ginkgo* root. (Bar: 10 μm.) Figure 16: Transmission electron micrograph of an arbuscular branch of *Glomus versiforme,* The thin cell wall is regularly labeled by gold granules bound to wheat germ agglutinin (arrowheads). F, fungus, (Bar: 0.5 μm.)

other wall components may protect the fungal wall from the attack of lytic enzymes such as chitinases, which are produced by the VAM plant in response to the early fungal invasion (see Section III.E).[53] These data suggest that the modulation of synthesis and assembly of the fungal cell wall may be a key point in the establishment of cellular compatibility between the two partners.

## E. THE MOLECULAR BASIS OF THE INFECTION PROCESS: WHAT DOES THE FUNGUS COMMUNICATE TO THE PLANT?

Numerous signals are thought to be exchanged between the host-fungal partners during both compatible and incompatible interactions.[1,54] In plant-pathogen interactions, many of the signals produced during the early stages of infection induce a hypersensitive reaction in the plant. The extensive fungal colonization of VA mycorrhizal roots raises some questions as to how the fungi sustain a long-term infection without evoking or becoming susceptible

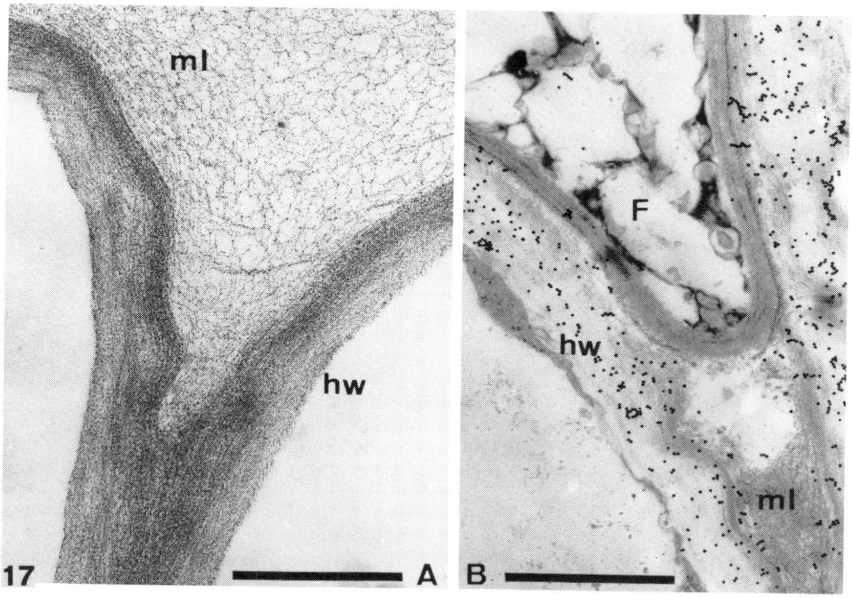


FIGURE 17. Texture of the middle lamella in leek roots. (A) The middle lamella (ml) consists of a regular net of fibrils. (B) In presence of the fungus (F), a clear empty zone can be seen. Host walls (hw) are labeled by the complex cellobiohydrolase-gold which binds β-(1-4-)glucan.(Bar: 1 μm.)

to a host defense response. Furthermore, how is the viability of host tissue maintained upon colonization?

The cell walls of both microbes and plants are considered to be one of the major structures from which signaling molecules are generated during interaction.[55] During the infection process, an important role in producing these signal molecules may be played by the hydrolytic enzymes produced by the fungus which degrade cell wall components.[56] To penetrate the plant tissue, invading microorganisms loosen the wall through the localized secretion of hydrolytic enzymes. At this stage, oligomeric fragments are released from the wall of the host plant which could regulate physiological events such as elicitation of the plant defense response. The plant may also produce enzymes that generate oligomers from the invading microorganism which have elicitor activity.[55] The role of wall-degrading enzymes produced by several pathogenic fungi in penetration and elicitation has been established through a number of correlations. Strong evidence is found in the correlation between the ability of the fungus to produce wall-degrading enzymes *in vitro* and its ability to penetrate the host walls, and the effects of purified enzymes on plants in generating elicitors. The detection of these enzymes and of their reaction products in infected tissues indicates that they are active *in planta*.[55-56]

In contrast, very little is known about mycorrhizal fungi enzymes. Since VAM fungi are intracellular symbionts, they must be able to penetrating the cell wall of root cells. Observations on the ultrastructure of plant cell walls in mycorrhizal plants indirectly supports the hypothesis that VAM fungi may penetrate by a localized production of hydrolytic enzymes. The VAM fungi develop between cortical cells which are joined by a middle lamella. Close to the fungal hyphae, this pectic lamella often shows a loosened texture (Figure 17). Mild treatment of the cell wall with chemical solvents, such as DMSO or EDTA resulted in the differential extraction of matrix components around the site of fungal penetration, revealing the cellulosic fibrillar texture.[32] As a comparison, it was observed that when some

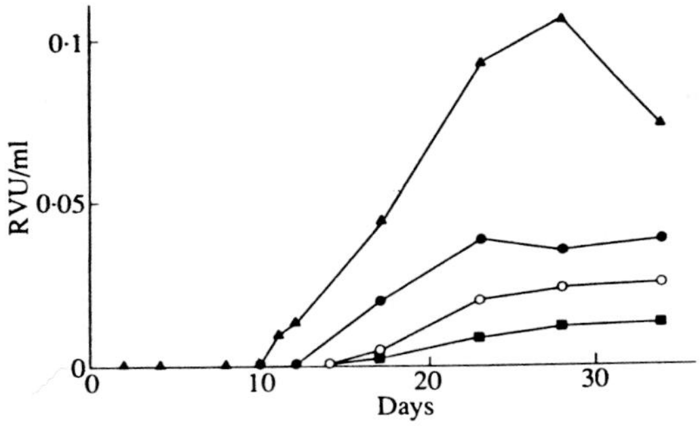


FIGURE 18. Time course of polygalacturonase synthesis of four ericoid mycorrhizal fungi grown on Czapeck-pectin. The strain which shows the highest activity possesses the lowest infection capability. (Reproduced with permission from S. Cervone et al., *Trans. Br. Mycol. Soc.* 91, 537, 1988. With permission).

pathogenic fungi (for example, the hemibiotrophic *Colletotrichum lindemuthianum*) penetrate the plant cell wall, the fibrillar texture of the host wall is revealed without need for solvent extractions.[57] Other pathogenic fungi, such as the necrotrophic *Rhizoctonia solani,* destroy all components of the host cell walls at the site of their penetration.[58] These results therefore suggest that VAM fungi may have a controlled and localized hydrolytic effect on wall matrix components, which may also be more selective compared to pathogenic fungi.

The inability to grow VAM fungi in pure culture makes it difficult to determine their enzymic activity *in vitro.* Insights on the significance of wall-degrading enzymes in plant-mycorrhizal fungus interaction had to come from observations on mycorrhizal fungi which are capable of growing *in vitro.* For example, ectomycorrhizal fungi, which do not penetrate normally into root cells are, nevertheless, able to grow within host cell walls.[59] Some ectomycorrhizal fungi such as *Laccaria laccata, Paxillus involutus,* and *Rhizopogon* reduce the viscosity of polygalacturonans, demonstrating a pectinolytic activity. Compared to most necrotrophic parasites, the activity detected in the culture medium of these ectomycorrhizal fungi is very low.[59] Similar results were obtained using *Hymenoschyphus ericae,* an endomycorrhizal ascomycetes which colonizes roots of Ericales.[60] Polygalacturonase activity was measurable in the culture medium at a much lower level compared to some pathogenic fungi such as *Fusarium* and *Rhizoctonia.* In addition, culture medium from strains with an increased mycorrhizal ability showed lower and delayed accumulation of polygalacturonase activity compared to poorly infective strains (Figure 18).[60] No biochemical data are available so far on the activity of these enzymes *in planta.*

The only biochemical data available on VAM are restricted to measurement of pectolytic enzymes in spore extracts and indirect observations on the ability of the fungus to penetrate roots in the presence of various pectic substrates.[61] Our preliminary results reveal polygalacturonase activity in the soluble fraction of root extracts of mycorrhizal plants, with a lower activity found in nonmycorrhizal roots. A polyclonal antibody raised against a fungal endopolygalacturonase,[62] was used to reveal the presence of the antigens in the mycorrhizal roots. Labeling was observed on the fungal cell wall of some extracellular hyphae at the root surface as well as at the tips of the arbuscular branches. The absence of labeling on fungal walls treated with saline buffers suggests that the enzyme, if any, would be weakly bound to the cell wall.[63]

Thus, different types of mycorrhizal fungi seem capable of producing limited quantities of polygalacturonases, whose limited activity may help to explain the absence of a vigorous hypersensitive reaction in the plant. These variations in fungal enzymes may be a direct response to signals from the host plant. A possible control of the host plant on the activity of fungal pectolytic enzymes is by the production of specific inhibitors. This mechanism is reinforced by the identification of proteins capable of inhibiting fungal polygalacturonases (PGIP) and produced in a variety of dicotyledonous plants.[64] By modulating the stability of the released active oligogalacturonides, the PGI protein from bean slows down the kinetics of the degrading enzyme.[64] Specific plant inhibitors of the fungal enzymes might be also produced in mycorrhizal symbiosis in response to the fungal colonization. However, the proposed function of this PGI protein on *in planta* has to be supported by cytological localization.

Plant signals may also be quite simply nutritional. For example, the production of pectolytic enzymes by many pathogenic fungi is repressed *in vitro* by preferred carbon sources.[56] This catabolite repression of the fungal enzymes has been suggested to lead to the different infective behavior of some hemibiotrophic fungi compared to necrotrophic fungi. In the initial stages of colonization by hemibiotrophs (e.g., *C. lindemuthianum*) very little symptoms and no pectolytic activity are visible in the plant.[65] It has been proposed that during this temporary phase, the fungal pectolytic enzymes are not produced due to catabolite repression, while in the subsequent pathogenic phase, similarly to what happens in necrotrophic fungi they somehow circumvent catabolite repression and produce large amounts of pectolytic enzymes.[65]

## III. HOST PLANT RESPONSE: ACTIVATION OF DEFENSE OR SYMBIOTIC GENES?

The colonization of the root system by VA mycorrhizal fungi and the establishment of an active bidirectional nutrient transport system induces modifications in the host plant that can be investigated at various levels.

### A. PLANT AND TISSUE RESPONSES

The growth effect shown by mycorrhizal plants is due to the high efficiency with which the fungal hyphae can explore the soil and adsorb water and salts.[3] These nutrients, particularly phosphate, are then delivered to the root cortical cells of the host plant. The nutrients are translocated through a net of extraradical mycelia, intraradical hyphae, and arbuscules, finally crossing the thin interface separating the fungal membrane from the plant membrane. Products of photosynthesis are transferred across the same interface, from the plant to the fungus and are stored in the fungal hyphae in the form of glycogen or lipids. However, the molecular basis of this bidirectional transport system is virtually unknown.[9]

### B. UNINFECTED TISSUE: LONG-DISTANCE EFFECTS

In the root, there are regions and tissues that are never colonized by the fungus at any stage of infection, such as the central cylinder and the meristematic apex. The inaccessibility of the central cylinder to fungal penetration is probably due to a physicochemical barrier associated with the root endodermis. The endodermis is a continuous layer of tightly connected cells surrounded by a thickened secondary wall rich in phenols, lignins, and suberins. As described previously, VAM fungi have a very low level of hydrolyzing enzymes, so that they may be incapable of degrading the encrusted and stiff endodermal wall. The lack of colonization of the root meristem, however, is far more difficult to explain on the basis of enzymic activity from the fungus. In fact, a very thin primary cell wall surrounds meristematic and early differentiating cells, and the VAM fungus might be considered potentially capable of penetrating this pectocellulosic layer. Some experiments demonstrate that the presence of the fungus influences the activity of the roots at the organ, tissue, and cellular level even without physical penetration.[66,67] Mycorrhizal roots show a higher number of apices and branching points compared to nonmycorrhizal plants, but the proportion of inactive or even necrotic root tips increases compared to control plants.[66] A comparison between the mitotic cycle of mycorrhizal and nonmycorrhizal root apices was made by feeding the tissues with tritiated thymidine.[67] The results indicated that mitosis in the apices of colonized roots is slowed down and that metaphase becomes longer with the progress of infection and fungal colonization. So far it is not known whether the VAM fungus directly modulates DNA transcription and the mitotic cycle of the root meristems through long distance signaling, or whether this effect is indirectly due to nutritional imbalance.[67] Both factors, hormonal and nutritional imbalance, are known to act and modify the mitotic cycle of root meristems in nonmycorrhizal plants, and they may well act in synergism. In particular, low sucrose concentration as well as high phosphate content have been reported to slow down the mitotic cycle.[67] Both conditions may be generated in the mycorrhizal root where the fungus is upstream with respect to the root tip meristem in regards to the flow of photosynthate and other components from the central cylinder. The root tip meristem that normally acts as a sink for the sugars derived from the photosynthesis, may be depleted of these components by the fungus, thus trapping sugars and perhaps other metabolites into the mycelial net. Other factors that limit the mitotic activity of root meristems, such as a high content of phosphate, have been known for a long time to occur in mycorrhizal roots [68] and could interact with other components affecting meristem development.

An alternative hypothesis must therefore be sought to explain what mechanisms prevent the VAM fungus from penetrating the meristematic tissue, and what kind of signals are sensed by the fungus. If the hyphae were chemotactically following a nutrient gradient (for example, sugars) inside the root, a possible explanation could be that meristems in mycorrhizal roots may be regions with low sugar content and therefore be avoided by the fungus. These kind of metabolites would be especially effective if they were present in the extracellular space. However, the presence of soluble sugars is difficult to determine precisely in the cell compartments.[69] A more general observation is that meristem is refractory of all microorganism infections including virus, bacteria, and fungi. However, the reason for this is not known.

## C. INFECTED TISSUES: TISSUE SPECIFICITY AND HOST RESPONSE

Concomitant with the branching of the fungal hypha, inside a cortical cell, there is also a dramatic modification of the host cell architecture: invagination of the plant plasmalemma, fragmentation of the vacuole, increase in the number of organelles, and unfolding of the host nuclear chromatin (Figure 19).[6,70] The penetration of the fungus in itself is not a sufficient stimulus in triggering these modifications either in the fungus or in the plant cell. Fungal coils are usually formed in the hypodermal root cells. These fungal structures do not branch like arbuscules do, and the plant cells harboring coils do not show any dramatic change in their cytology with their cytoplasm remaining peripheral (see Figure 11).

At a molecular level, growth and development of the intraradical fungus is probably modulated by the plant host through different mechanisms. A primary role, as it has been described earlier in this chapter (Section II.C), is played by the structure and composition of the plant cell surfaces that come in contact with the growing hyphae. In many cases, a specific distribution of secondary cell wall components seems to channel fungal growth inside the root, as described in leek, *Ginkgo,* and other species.[6,31,32,44,48] The cell wall, therefore, seems to play a primary role in allowing fungal penetration in certain cells and preventing it in others.

## D. THE HOST-FUNGAL INTERFACE AS A NEW APOPLASTIC COMPARTMENT

The wall material that is deposited by the plant around the hyphae during their intracellular growth is likely to play an important role in controlling fungal development. When-

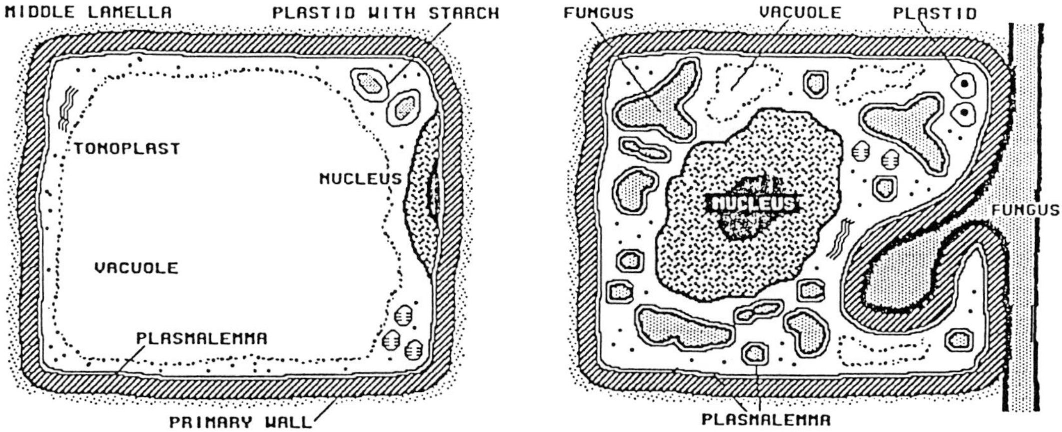


FIGURE 19. A comparison between an uninfected and an infected cortical cell.

ever the fungus penetrates the root cells, either to form coils or arbuscules, the host plasmalemma also invaginates and proliferates with the fungus. This newly synthesized membrane seems to retain the enzymic machinery involved in the deposition of cell wall material. It is also the site of targeting for vesicles carrying other common components of the primary cell wall derived from the Golgi, such as pectins. As a result of this synthetic activity, material is laid down at the interface between the symbionts, continuous with the plant cell wall but with a different texture.[31] In comparison to the peripheral cell wall, which seems to be a more constitutive structure opposed to the fungus, the interfacial material that is deposited around the intracellular hyphae derives from newly synthesized material. A detailed analysis of this interface may therefore provide clues to understand what kind of response the fungus is causing in the plant host.

During VAM fungal colonization there is apparently no callose production, which contrast with pathogen invasion.[31] Callose synthesis can be a very rapid reaction of the plant to invasion, as it seems not to involve any new synthesis of proteins, but only a switch in the activity of glucan synthetase on the plant membrane.[71] Callose has been suggested to occur in the papilla formed by myc$^-$ pea mutants (see Section II.C), and it may contribute to inhibiting VAM fungal colonization.[41]

Affinity probes such as enzymes, lectins, or antibodies conjugated to colloidal gold, have been used for the localization of specific molecules occurring on the interfacial material (Figure 20). β-1-4-Glucans, mostly cellulose, were localized around the intracellular fungus by using an enzyme, cellobiohydrolase I (CBH I, EC 3.2.1.g1), whose substrate is cellobiose. Nonesterified and rare methyl-esterified galacturonic acids were identified in the same compartment in pea, leek, and *Ginkgo* by using the monoclonal antibodies JIM5 and JIM7, respectively.[34,35,72] The presence of these wall polysaccharides indicated an amorphous texture of this interfacial layer which was very different from the peripheral cell wall, as if the components localized in this compartment were not assembled into a fully structured wall.[34]

Proteins were also localized on the interfacial material, in particular, molecules reacting with an antibody raised against an extensin-like melon hydroxyproline-rich glycoprotein (HRGP).[73] Extensins occur abundantly as structural components of primary cell walls, where they may act as a variable cross-bridging agent, changing the relative rigidity of the cell wall.[74] The role of extensin in hardening the cell wall would also explain the observation that it seems to increase in the presence of pathogens.[74] Extensins belong to a family of plant glycoproteins (HRGPs) whose members may have diverse roles. They are differentially expressed during tissue differentiation in plants and nodules.[75] The presence of extensin-

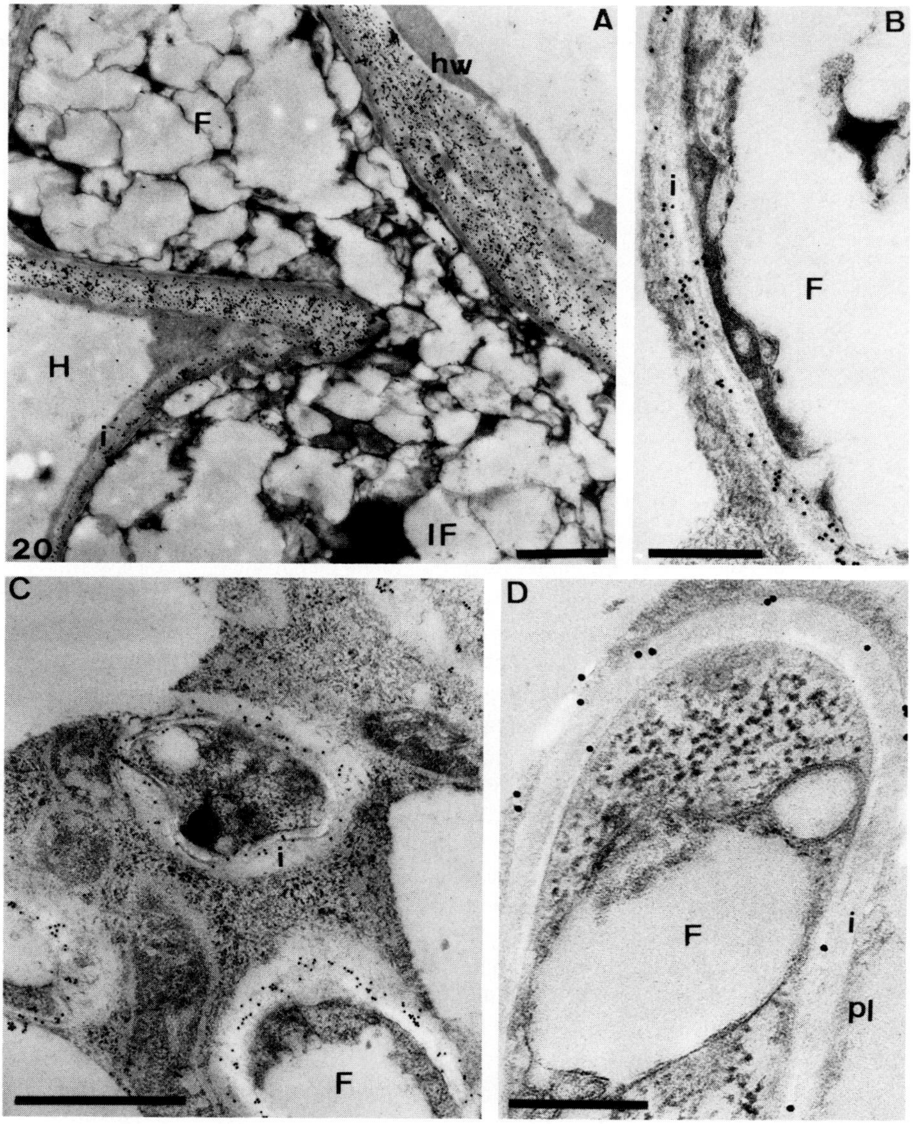


FIGURE 20. The cytochemical features of the interface zone in mycorrhizal roots of *Allium porrum*. (A) The complex cellobiohydrolase-gold reveals a labeling continuity between the peripheral cell wall (hw) and the interfacial material (i) around the intracellular fungus (IF) (Bar: 1 μm.) (B) A thin arbuscule branch (F) is surrounded by an interfacial material (i) rich in antigens which are revealed by the MAB JIM5 indicating the presence of pectins. (Bar: 0.5 μm.) (C) General view of thin arbuscule branches as in (B) (Bar: 1 μm) (D) Antigens which binds to an antibody against HRGP are revealed in the interfacial zone (i) around the fungus (F). pl = host membrane, (Bar: 0.25 μm.)

like molecules in the interfacial material, therefore, raises the interesting question about their particular structure and about their role in this extracellular compartment.

In certain aspects, the penetration of the VAM fungal hypha resembles the penetration of a root nodule primordium by an infection thread, both being described as apoplastic tunnels lined by a cell wall-like envelope containing cellulose and pectin.[34,35,76] Interesting differences occur however when the colonization leads to the formation of arbuscules in

VAM compared to the intracellular release of rhizobia. Both symbionts at this stage are surrounded by the host-derived membrane. The periarbuscular membrane in VAM retains synthetic activity with the deposition of β-1-4-glucans around the fungus, while the peribacteroid membrane surrounding rhizobia ceases synthetic activity and no glucans are found at the interface with the bacterium. Moreover, pectic material is targetted to the periarbuscular membrane but not to the peribacteroid membrane. It would, therefore, be interesting to know whether the different synthesis and targeting of wall components in these two types of interaction is also accompanied by a different distribution of membrane components.

The periarbuscular membrane is poorly characterized. The morphological continuity between the host plasmalemma and the invaginated membrane suggests their relatedness. However, important differences exist and concern the functional activities of the two membranes. Active membrane-bound ATPases have been described along the periarbuscular membrane but not on the peripheral plasma membrane.[77] Such cytochemical distribution is consistent with the double nutrient flow occurring in mycorrhizal symbiosis. Interestingly, in plant-pathogen interactions no active membrane-associated ATPases were found on the membrane surrounding haustoria. The absence of an active transport system on the plant perihaustorial membrane may explain the unidirectional nutrient flux toward the fungus observed in pathogenic interactions.[9]

The different distribution of ATPase can be correlated to difference in membrane potential: although in pathogenic associations a membrane depolarization is usually observed, VAMs show a stable hyperpolarization of the membrane.[78] Other experiments identify similarities between the periarbuscular membrane and plasma membrane. For example, monoclonal antibodies raised against membrane antigens in pea nodules, which recognize antigens both on the plasma membrane and on the plant membrane surrounding the bacteroid, target both to the plasma membrane and to the periarbuscular membrane in VAM pea plants.[79,80] Furthermore, recent studies have used *in vitro* translated mRNA from VAM soybean plants to identify components which immunologically resemble membrane-bound nodulins expressed by the same plant species in association with *Bradyrhizobium japonicum* (see next section).[81] It will be very exciting to show whether these components are expressed on the periarbuscular membrane.

## E. EXPRESSION OF NEW GENES IN THE SYMBIOTIC STATE

A clear sign that the cells harboring arbuscules are very active in the synthesis of new components is provided by studies on the nuclear chromatin structure. Nuclei of cortical cells harboring arbuscules have been described as hypertrophic by several authors.[6] By using a cytofluorimetric assay, Berta and co-workers showed that this increase in size is not due to a change in the ploidy status of mycorrhizal root cell nuclei compared to control plants, but rather to a decondensation of the chromatin.[70] The increase of decondensed chromatin has been related to an increase in transcriptional activity, as a response to infection.[70]

Great effort has been made in order to identify products of this new transcription activity that are specific for the mycorrhizal symbiosis, in the same way as nodulins have been identified as being specific to the nodule.

Analysis of soluble proteins from an ectomycorrhizal association, separated by two dimensional polyacrylamide gel electrophoresis, demonstrated a dramatic change in protein synthesis.[83] At least ten new polypeptides, mostly of low molecular weight, could be identified in the established association. They have been preliminary defined as ectomycorrhizins and have been suggested to be involved in the mycorrhizal morphogenesis.[82] The formation of ectomycorrhizae in fact, unlike VAM, produces extensive modifications in the morphology of both fungus and host plant. However, the characterization of these ectomycorrhizins as plant or fungal products awaits the study of purified mRNA populations and eventual cloning of the respective sequences.[84]

Dumas et al.[85] reported the appearance of new soluble proteins in mycorrhizal roots of tobacco and onion. Some of these polypeptides have been suggested to be of plant origin, as their mobility did not change when different endophytes were used as inoculum. Pakovsky[86] analyzed root soluble proteins from mycorrhizal and nonmycorrhizal soybean plants. In this experiment, the appearance of mycorrhiza-specific proteins was compared with the production of nodulins, the same plants of soybean being inoculated with or without *Bradyrhizobium*. At least five proteins were identified as specific to endomycorrhizal roots, independent from the presence of nodule-specific bands. There was only circumstantial evidence that the specific bands were plant-derived, as VAM spores do not contain detectable amount of molecules in this range of molecular size.[86]

These kinds of experiments raise the intriguing question of whether VA mycorrhizal infection triggers in the host a similar set of responses as does an invading endosymbiotic bacterium. Wyss et al.[81] demonstrated recently that there are similarities in the plant response to mycorrhizal fungi and bacteria. Polyadenylated mRNA was isolated from mycorrhizal and non-mycorrhizal soybean roots, and translated *in vitro*. Non-nodulating soybean mutants were used together with the wild type, to make sure that the expression of proteins corresponding to nodulins was not due to an undetected contamination by rhizobia. At least two polypeptides were expressed in mycorrhizal roots which could be immunoprecipitated with an antiserum against soluble nodulins, but which were absent in the nonmycorrhizal control. The size of these polypeptides did not correspond to any nodulin of known function. Also, three low-molecular weight polypeptides from the translation product of mycorrhizal roots were immunoprecipitated by an antiserum against membrane-bound nodulins. This is, so far, the first report of membrane-bound proteins present both in VA mycorrhizae and nodules, but the precise localization of these molecules in arbuscule-containing cells is required before any hypothesis can be made as to their function in transport or interactions with the endosymbiont.[81] One of these polypeptides revealed in soybean mycorrhizae shows a molecular weight corresponding to that of nodulin-26. This nodulin has been localized on the peribacteroid membrane[87] and shows high homology with a transmembrane protein described in *E. coli* and with a protein in the eye lens of cows, both probably involved in the membrane transport of small molecules.[88] However, an interesting discussion may arise on whether to call these proteins mycorrhizins or nodulins.

In the experiment described by Wyss et al.[81] two different mutants (which are unable to form nodules) were used along with the wild-type soybean. These mutants showed a normal VAM symbiosis as the wild type. These data complement those described by Duc et al.[42] in the pea system, where the inability to form nodules was in 50% of the plants accompanied by an inability to interact with mycorrhizal fungi.

The immunological cross-reaction of sera to proteins produced in mycorrhizal roots with those produced in nodules demonstrates a possible correlation between the response of a plant to VAM fungi and to endosymbiotic bacteria, with the expression of some common symbiotic genes. However, it is often observed in biological systems that delicate balances between organisms can shift rapidly from mutualistic to saprophytic or parasitic interactions. For example, even in some endomycorrhizae, such as ericoid and orchid mycorrhizae, the behavior of the fungal endophyte has been described in some stages as being saprophytic or necrotrophic.[89] Also, several mutants of *Bradyrhizobium japonicum*, that as a wild type normally nodulate soybean, have been reported to trigger a hypersensitive reaction in the plant.[90] It is, therefore, not surprising that attention has been paid to investigating the expression in mycorrhizae of some genes previously correlated to the response of plants to pathogens.[51,90]

The life strategy of endomycorrhizal fungi can be compared with that of fungal pathogens which spend part of their life as biotrophs on a host plant. These pathogens only activate defense genes in the plant during an incompatible reaction, eluding such defense mechanisms and inducing disease during compatible interactions. Investigations on the expression of pathogenesis-related proteins, such as chitinase and peroxidase, during mycorrhizal infection demonstrated, for both enzymes, a sharp peak of activity at the early stages of infection, when the fungus was still mostly intercellular.[53,92] Once the mycorrhiza was fully established and the fungus had fully colonized cortical cells (forming arbuscules), the level of activity for these enzymes was decreased and comparable to nonmycorrhizal plants.

The occurrence of phytoalexins was also examined during infection with mycorrhizal fungi. Some results suggested an accumulation of isoflavonoids in soybean,[93] while other detailed studies indicated that there is no clear accumulation in the early stages of infection or up to a stage of extensive root colonization.[94] The combined inoculation of the same roots with the necrotrophic fungus *R.solani* also showed that the plant can react with production of phytoalexins, demonstrating that the VAM fungus does not inhibit phytoalexin production, but simply avoids their induction. In this respect, compatible biotrophic fungi and VAM fungi behave in a similar way.[94]

## IV. THE PROMISES OF MOLECULAR BIOLOGY

A review of recent results obtained with the methods of cell biology shows that further basic knowledge on protein expression is needed by using biochemistry, immunology, and cytochemistry. However, in this aspect, molecular biology tools to dissect the complex events occurring in mycorrhizae, i.e., in interactions between two highly integrated eukaryotes, may prove to be more valuable.

Important advances have been made on the transformation of filamentous fungi, and some preliminary results are available for ectomycorrhizal fungi.[95] Protoplasts of the ectomycorrhizal fungus *Laccaria laccata* were transformed, using selection for hygromycin B on a vector containing a promoter region from *Aspergillus nidulans*. This is the first evidence for genetic transformation of a symbiotic mycorrhizal fungus and indicates the ability of promoters and termination signals of ascomycetous origin to function in taxonomically unrelated basidiomycetes.[95] These findings open important possibilities to improve symbiosis through transgenic manipulation of the fungal component through the introduction of genes which may prove beneficial to symbiosis. It would be interesting to see the effect on symbiosis caused by the introduction of genes coding, for example, for cell wall-degrading enzymes. In this case, the probes already isolated from pathogenic fungi could prove very useful, such as cutinases or endopolygalacturonases.[96-97] Such experiments, and the possibility to use promoters which may be regulated differently, would help to understand to what extent the production of hydrolytic enzymes play a role in the compatibility between plant and mycorrhizal fungi. These kinds of experiments will only be possible for VAM fungi when progress can be made in culturing the mycelium separated from its host. In the meanwhile, some understanding of the genetics of VAM fungi is being achieved with molecular biology techniques applied to taxonomy. Ongoing research has analyzed ribosomal RNA gene fragments from various VAM fungi using PCR, and has shown that previous taxonomic groupings based on morphology are generally consistent with molecular techniques.[98-99]

Improvement of DNA analysis in VAM fungi should allow us to begin to understand the developmental pattern of gene expression that correlates with changes in fungal morphogenesis. It also allows the monitoring and characterization of specific genes, such as those involved in penetration of the carbohydrate root layers. The possibility of comparing cDNA libraries obtained from germinating VAM fungal spores and mycorrhizal as well as nonmycorrhizal plants may provide clones encoding for plant or fungal polypeptides enhanced during the symbiotic phase. These clones may then be compared with those involved in other plant-microbe interactions.

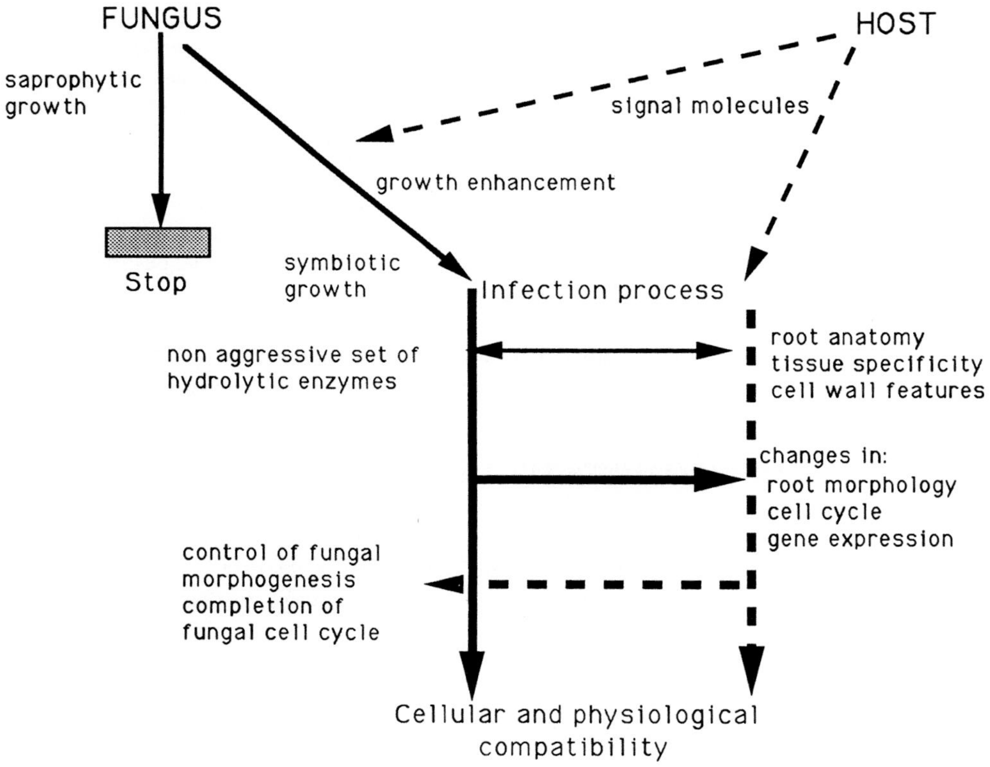


FIGURE 21. Summary of the different steps of the plant-fungal interactions at cellular level.

## V. CONCLUSIONS

The analysis of VAM mycorrhizae at the cellular level allows us to identify different stages of the plant-fungus interaction (Figure 21). VAM fungi are obligate biotrophs and the completion of their cell cycle depends upon the presence of their host plant. Signal molecules must be important in triggering the infection process. This process consists of a succession of events which seem to be under the control of the host and, in particular, on the physicochemical features of the plant root cell walls. The molecular basis of the penetration, as well as the large host range, could be explained by a nonaggressive set of fungal hydrolytic enzymes, which enable the fungus to penetrate differentiated tissues with nonencrusted cell walls. The fungus on the other hand, is capable of influencing the metabolic pathways of the infected plant, affecting both infected and uninfected tissues. These events correspond to a modification in gene expression which could be developmentally regulated in the fungus and tissue-specific in the host. However, many of the questions we have raised at the beginning of this chapter remain still unanswered and await application of molecular techniques capable of understanding the dialog that exists between VAM fungi and plants. As suggested by the above discussion, this may simply be a unique dialect in the common language of symbiotic partners.

# REFERENCES


1. **Long, S. R.,** Rhizobium-Legume Nodulation: life together in the underground, *Cell,* 56, 203, 1989.
2. **Harley, J.,** The significance of mycorrhiza, *Mycol. Res.,* 92, 129, 1989.
3. **Smith, S. E. and Gianinazzi-Pearson, V.,** Physiological interactions between symbionts in vesicular-arbuscular mycorrhizal plants, *Annu. Rev. Plant Physiol. Plant Mol. Biol.,* 39, 221, 1988.
4. **Scannerini, S. and Bonfante-Fasolo, P.,** Comparative ultrastructural analysis of mycorrhizal associations, *Can. J. Bot.,* 61, 917, 1983.
5. **Read, J. D.,** Mycorrhizas in ecosystems, *Experientia,* 47, 376, 1991.
6. **Bonfante-Fasolo, P.,** Anatomy and morphology of VA mycorrhizae, in *VA Mycorrhizas,* Powell, C. L. and Bagyaraj, D. J., Eds., CRC Press, Boca Raton, FL, 1984, 5.
7. **Bonfante-Fasolo, P.,** Vesicular-arbuscular mycorrhizae: fungus-plant interactions at the cellular level, *Symbiosis,* 3, 249, 1987.
8. **Gianinazzi-Pearson, V. and Gianinazzi, S.,** Morphological integration and functional compatibility between symbionts in vesicular arbuscular endomycorrhizal associations, in *Cell to Cell Signals in Plant, Animal, and Microbial Symbiosis,* NATO ASI Ser., Vol. 17, Scannerini, S., Smith, D. C., Bonfante-Fasolo, P., and Gianinazzi-Pearson, V., Eds., Springer-Verlag, Berlin, 1988, 73.
9. **Smith, S. E. and Smith, F. A.,** Structure and function of the interfaces in biotrophic symbioses as they relate to nutrient transport, *New Phytol.,* 114, 1, 1990.
10. **Morton, J. B. and Benny, G. L.,** Revised classification of arbuscular mycorrhizal fungi (Zygomycetes): a new order, Glomales, two new suborders, Glomineae and Gigasporineae, and two new families, Acaulosporinaceae and Gigasporaceae, with an emendation of Glomaceae, *Mycotaxon,* 37, 471, 1990.
11. **Becard, G. and Piché, Y.,** New aspects on the acquisition of biotrophic status by a vesicular-arbuscular mycorrhizal fungus *Gigaspora margarita, New Phytol.,* 112, 77, 1989.
12. **Mosse, B.,** The regular germination of resting spores and some observations on the growth requirements of an *Endogone* sp. causing vesicular-arbuscular mycorrhiza, *Trans. Br. Mycol. Soc.,* 42, 273, 1959.
13. **Hepper, M. C.,** Isolation and culture of VA mycorrhizal (VAM) fungi, in *VA Mycorrhizas,* Powell, C. L. and Bagyaraj, D. J., Eds., CRC Press, Boca Raton, FL, 1984, 95.
14. **Burggraaf, A. J. P. and Beringer, J. E.,** Absence of nuclear DNA synthesis in vesicular-arbuscular mycorrhizal fungi during *in vitro* development, *New Phytol.,* 111, 25, 1989.
15. **Bianciotto, V., Palazzo, D., and Bonfante-Fasolo, P.,** Germination process and hyphal growth of a vesicular-arbuscular mycorrhizal fungus, *Allionia,* 29, 17, 1989.
16. **Tommerup, I. C.,** Sexual and asexual spores of vesicular arbuscular mycorrhizal fungi: ontogeny, germination and genetics, paper presented at 4th Int. Mycol. Congr., Regensburg, 28 August-3rd September, 1990, 97.
17. **Viera, A. and Glenn, M. G.,** DNA content of vesicular-arbuscular mycorrhizal fungal spores, *Mycologia,* 82, 263, 1990.
18. **Cavalier-Smith, T.,** Eukaryotic gene number, non-coding DNA and genomic size, in *The Evolution of Genome Size,* Cavalier-Smith, T., Ed., John Wiley and Sons, NY, 1985, 69.
19. **Hoch, H. C. and Staples, R. C.,** Evidence that cAMP initiates nuclear division and infection structure formation in the bean rust fungus, *Uromyces phaseoli, Exp. Mycol.,* 8, 37, 1984.
20. **Fortin, J. A.,** New vistas on mycorrhizas research, paper presented at 8th N. Am. Conf. Mycorrhizae, Jackson, Wyoming, 5-8 September, 1990, 103.
21. **Gianinazzi-Pearson, V., Branzanti, B., and Gianinazzi, S.,** *In vitro* enhancement of spore germination and early hyphal growth of a vesicular-arbuscular mycorrhizal fungus by host root exudates and plant flavonoids, *Symbiosis,* 7, 243, 1989.
22. **Peters, N. K. and Verma, D. P. S.,** Phenolic compounds as regulators of gene expression in plant-microbe interactions, *Mol. Plant-Microbe Interact.,* 3, 48, 1990.
23. **Stachel, S. E., Messens, E., Van Montagu, M., and Zambrinsky, P.,** Identification of the signal molecules produced by wounded plant cell that activate T-DNA transfer in *Agrobacterium tumefaciens, Nature (London),* 318, 19, 1985.
24. **Tsai, S. M. and Philips, D. A.,** Alfalfa flavonoids affect vesicular-arbuscular mycorrhizae development *in vitro,* paper presented at 8th N. Am. Conf. *Mycorrhizae,* Jackson, Wyoming, 5-8 September, 1990, 290.
25. **Firmin, J. L., Wilson, K. E., Rossen, L., and Johnston, A. W. B.,** Flavonoid activation of nodulation genes in *Rhizobium* reversed by other compound present in plants, *Nature (London),* 324, 90, 1986.
26. **Zaat, S. A. J., Wijffelman, C. A., Spaink, H. P., Van Brussel, A. A. N., Okker, R. J. H., and Lugtenberg, G. J. J.,** Induction of the nodA promoter of *Rhizobium leguminosarum* Sym plasmid PRL1JI by plant flavanones and flavones, *J. Bacteriol.,* 16, 198, 1987.
27. **Darvill, A. and Albersheim, P.,** Phytoalexins and their elicitors — a defence against microbial infection in plants, *Annu. Rev. Plant Physiol.,* 35, 243, 1984.

28. **Wynn, W. K.,** Appressorium formation over stomates by the bean rust fungus: response to a surface contact stimulus, *Phytopathology,* 66, 136, 1976.
29. **Garriock, M. L., Peterson, R. L., and Ackerley, C. A.,** Early stages in colonization of *Allium porrum* (leek) roots by the VAM fungus, *Glomus versiforme, New Phytol.,* 112, 85, 1989.
30. **Scannerini, S. and Bonfante-Fasolo, P.,** Bacteria and bacteria-like objects in endomycorrhizal fungi (Glomaceae), in *Symbiosis as a Source of Evolutionary Innovation: Speciation and Morphogenesis,* Margulis, L. and Fester, R., Eds., MIT Press, Cambridge, 1991, 273.
31. **Bonfante-Fasolo, P.,** The role of the cell-wall as a signal in mycorrhizal associations, in *Cell to Cell Signals in Plant, Animal, and Microbial Symbiosis,* NATO ASI Ser., Vol. 17, Scannerini, S., Smith, D. C., Bonfante-Fasolo, P., and Gianinazzi-Pearson, V., Eds., Springer-Verlag, Berlin, 1988, 219.
32. **Bonfante-Fasolo, P. and Vian, B.,** Cell wall architecture in mycorrhizal roots of *Allium porrum, Ann. Bot. (Paris),* 10, 97, 1989.
33. **Codignola, A., Verotta, L., Maffei, M., Spanu, P., Scannerini, S., and Bonfante-Fasolo, P.,** Cell wall bound phenols in roots of vesicular-arbuscular mycorrhizal plants, *New Phytol.,* 112, 221, 1989.
34. **Bonfante-Fasolo, P., Vian, B., Perotto, S., Faccio, A., and Knox, J. P.,** Cellulose and pectin localization in roots of mycorrhizal *Allium porrum:* Labelling continuity between host cell wall and interfacial material, *Planta,* 180, 537, 1990.
35. **Perotto, S., Vandenbosch, K. A., Brewin, N. J., Faccio, A., Knox, J. P., and Bonfante-Fasolo, P.,** Modifications of the host cell wall during root colonization by Rhizobium and VAM Fungi, in *Endocytobiology IV,* Nardon P., Gianinazzi-Pearson, V., Grenier, A. M., Margulis, M., and Smith, D. C., Eds., INRA Press, Paris, 1990, 114.
36. **Jarvis, M. C., Forsyth, W., and Duncan, H. J.,** A survey of the pectic content of nonlignified monocot cell-walls, *Plant Physiol.,* 88, 309, 1988.
37. **Rao, S. S., Lippincott, B. B., and Lippincott, J. A.,** *Agrobacterium* adherence involves the pectic portion of the host cell wall and is sensitive to the degree of pectin methylation, *Physiol. Plant.,* 56, 374, 1982.
38. **Neff, N. T., Binns, A. N., and Brandt, C.,** Inhibitory effects of a pectin-enriched tomato cell wall fraction on *Agrobacterium tumefaciens* binding and tumor formation, *Plant Physiol.,* 83, 525, 1987.
39. **Bonfante-Fasolo, P., Vian, B., and Faccio, A.,** Texture of host cell walls in mycorrhizal leeks, *Agri. Ecosystems Environ.,* 29, 51, 1990.
40. **Aist, J. R.,** Papillae and related wound plugs of plant cells, *Annu. Rev. Phytopathol.,* 14, 145, 1976.
41. **Gianinazzi-Pearson, Gianinazzi, S., Guillemin, J. P., Trouvelot, A., and Duc, G.,** Genetic and cellular analysis of resistance to vesicular-arbuscular (VA) mycorrhizal fungi in pea mutants, in *Advances in Molecular Genetics of Plant-Microbe Interactions,* Hennecke, H. and Verma, P. S., Eds., Kluwer Academic Publishers, Netherlands, 336, 1991.
42. **Duc, G., Trouvelot, A., Gianinazzi-Pearson, V., and Gianinazzi, S.,** First report of non-mycorrhizal plant mutants ($Myc^-$) obtained in pea (*Pisum sativum* L.) and fababean (*Vicia faba* L.), *Plant Sci.,* 60, 215, 1989.
43. **Shishkoff, N.,** Distribution of the dimorphic hypodermis of roots in Angiosperm families, *Ann. Bot.,* 60, 1, 1987.
44. **Brundrett, M. and Kendrick, B.,** The roots and mycorrhizas of herbaceous woodland plants II. Structural aspects of morphology, *New Phytol.,* 114, 469, 1990.
45. **Gianinazzi-Pearson, V. and Gianinazzi, S.,** Cellular and genetical aspects of interactions between hosts and fungal symbionts in mycorrhizae, *Genome,* 31, 336, 1989.
46. **Brundrett, M. C., Piché, Y., and Peterson, R. L.,** A developmental study of the early stages in vesicular-arbuscular mycorrhiza formation, *Can. J. Bot.,* 63, 184, 1985.
47. **Jacquelinet-Jeanmougin, I., Gianinazzi-Pearson, V., and Gianinazzi, S.,** Endomycorrhizas in the Gentianaceae. II. Ultrastructural aspects of symbiont relationships in *Gentiana lutea* L., *Symbiosis,* 3, 269, 1987.
48. **Bonfante-Fasolo, P. and Fontana, A.,** VAM fungi in *Gingko biloba* roots. Their interactions at cellular level, *Symbiosis,* 1, 53, 1985.
49. **Yawney, W. J. and Schultz, R. C. S.,** Anatomy of a vesicular-arbuscular endomycorrhizal symbiosis between sugar maple (*Acer saccharum* Marsch) and *Glomus etunicatum* Becker & Gerdemann, *New Phytol.,* 114, 47, 1990.
50. **Bonfante-Fasolo, P., Faccio, A., Perotto, S., and Schubert, A.,** Correlation between chitin distribution and cell wall morphology in the mycorrhizal fungus *Glomus versiforme, Mycol. Res.,* 94, 157, 1990.
51. **Bonfante-Fasolo, P. and Perotto, S.,** Mycorrhizal and pathogenic fungi: do they share common features?, in *Electron Microscopy Applied in Plant Pathology,* Mendgen, K. and Lesemann, D. E., Eds., Springer-Verlag, Berlin, 1990, 265.
52. **Bonfante-Fasolo, P. and Scannerini, S.,** The cellular basis of plant-fungus interchanges in mycorrhizal associations, in *Functioning in Mycorrhizae,* Allen, M., Ed., Academic Press, San Diego, 1992, in press.
53. **Spanu, P., Boller, T., Ludwig, A., Wiemken, A., Faccio, A., and Bonfante-Fasolo, P.,** Chitinase in roots of mycorrhizal *Allium porrum:* regulation and localization, *Planta,* 177, 447, 1989.

54. **Lamb, C. J., Lawton, M. A., Dron, M., and Dixon, R. A.,** Signals and transduction mechanisms for activation of plant defenses against microbial attack, *Cell,* 56, 215, 1989.
55. **Hahan, M. G., Bucheli, P., Cervone, F., Doares, S., O'Neiil, R. A., Darvill, A., and Albersheim, P.,** The roles of cell wall constituents in plant-pathogen interactions, in *Plant-Microbe Interactions,* Vol. 3, Nester, E. and Kosuge, T., Eds., McGraw Hill, NY, 1989, 131.
56. **Cooper, R. M.,** The role of cell wall-degrading enzymes in infection and damage, in *Plant Diseases:* Infection, Damage and Loss, Wood, R. K. S. and Jellis, G. J., Eds., Blackwell Scientific, Oxford, 1984, 13.
57. **O'Connell, R. J. and Bailey, J. A.,** Cellular interactions between *Phaseolus vulgaris* and the hemibiotrophic fungus *Colletotrichum lindemuthianum,* in *Biology and Molecular Biology of Plant-Pathogen Interactions,* Bailey, J. A., Ed., NATO ASI Ser., Vol. H1, Springer-Verlag, Berlin, 1984, 39.
58. **Sneh, B., Ichielevich-Auster, and Shomer, I.,** Comparative anatomy of colonization of cotton hypocotyls and roots by virulent and hypovirulent isolates of *Rhizoctonia solani, Can. J. Bot.,* 67, 2142, 1989.
59. **Keon, J. P. R., Byrde, R. J. W., and Cooper, R. M.,** Some aspects of fungal enzymes that degrade plant cell walls, in *Fungal Infection of Plants,* Pegg, G. F. and Ayres, P. G., Eds., Cambridge University Press, Cambridge, 1987, 133.
60. **Cervone, S., Castoria, R., Spanu, P., and Bonfante-Fasolo, P.,** Pectinolytic activity in some ericoid mycorrhizal fungi, *Trans. Br. Mycol. Soc.,* 91, 537, 1988.
61. **Garcia-Romera, I., Garcia-Garrido, J. M., Martinez-Molina, E., and Ocampo, J. A.,** Possible influence of hydrolytic enzymes on vesicular-arbuscular mycorrhizal infection of alfalfa, *Soil. Biol. Biochem.,* 22, 149, 1990.
62. **De Lorenzo, G., Salvi, G., Degrà, L., D'Ovidio, R., and Cervone, F.,** Induction of extracellular polygalacturonase and its mRNA in the phytopathogenic fungus, *Fusarium moniliforme, J. Gen. Microbiol.,* 133, 3365, 1987.
63. **Peretto, R. and Bonfante-Fasolo, P.,** Endopolygalacturonase activity in vesicular-arbuscular mycorrhizal roots, paper presented at 4th Int. Mycol. Congr., Regensburg, 28 August-3rd September, 1990, 206.
64. **Cervone, S., Hahn, M. G., De Lorenzo, G., Darvill, A., and Albersheim, P.,** Host pathogen interactions. XXXIII. A plant protein converts a fungal pathogenesis factor into an elicitor of plant defense responses, *Plant Physiol.,* 90, 542, 1989.
65. **Wijesundera, R. L. C., Bailey, J. A., and Byrde, R. J. W.,** Production of pectin lyase by *Colletotrichum lindemuthianum* in culture and in infected bean *(Phaseolus vulgaris)* tissue, *J. Gen. Microbiol.,* 130, 285, 1984.
66. **Berta, G., Fusconi, A., Trotta, A., and Scannerini, S.,** Morphogenetic modifications induced by the mycorrhizal fungus *Glomus* strain $E_3$ on the root system of *Allium porrum* L., *New Phytol.,* 114, 207, 1990.
67. **Berta, G., Tagliasacchi, A. M., Fusconi, A., Gerlero, D., Trotta, A., and Scannerini, S.,** The mitotic cycle in root apical meristems of *Allium porrum* L. is controlled by the endomycorrhizal fungus *Glomus* sp. strain $E_3$, *Protoplasma,* 161, 12, 1991.
68. **Stribley, D. P., Tinker, P. B., and Rayner, J. H.,** Relation of internal phosphorus concentration and plant weight in plants infected by vesicular-arbuscular mycorrhizas, *New Phytol.,* 86, 261, 1980.
69. **Dick, P. S. and AP Rees, T.,** The pathway of sugar transport in roots of *Pisum sativum, J. Exp. Bot.,* 26, 305, 1975.
70. **Berta, G., Sgorbati, S., Soleri, V., Fusconi, A., Trotta, A., Citterio, M. G., Sparvoli, E., and Scannerini, S.,** Variations in chromatin structure in host nuclei of a vesicular arbuscular mycorrhiza, *New Phytol.,* 114, 199, 1990.
71. **Kauss, H.,** Callose biosynthesis as a $Ca^{2+}$ regulated process and possible relations to the induction of other metabolic changes, *J. Cell Sci. Suppl.* 2, 89, 1985.
72. **Bonfante-Fasolo, P., Perotto, S., and Peretto, R.,** Cell surface interactions in endomycorrhizal symbiosis, in *Perspectives in Plant Cell Recognition,* Callow, J. and Green, J., Eds., SEB Seminars, Cambridge University Press, in press, 1992.
73. **Mazau, D., Rumeau, D., and Esquerré-Tugayé, M. T.,** Two different families of hydroxyproline-rich glycoproteins in melon callus, *Plant Physiol.,* 86, 540, 1988.
74. **Varner, J. E. and Liang-Shiou, L.,** Plant cell wall architecture, *Cell,* 56, 231, 1989.
75. **Hong, J. C., Nagao, R. T., and Key, J. L.,** Developmentally regulated expression of soybean proline-rich cell wall protein genes, *Plant Cell,* 1, 937, 1989.
76. **VandenBosch, K. A., Bradley, D. J., Knox, J. P., Perotto, S., Butcher, G. W., and Brewin, J. B.,** Common components of the infection thread matrix and the intercellular space identified by immunocytochemical analysis of pea nodules and uninfected roots, *EMBO J.,* 8, 335, 1989.
77. **Gianinazzi-Pearson, V., Smith, S. E., Gianinazzi, S. and Smith, F. A.,** Enzymatic studies on the metabolism of vesicular-arbuscular mycorrhizas. V. Is $H^+$-ATPase a component of ATP-hydrolysing enzyme activities in plant-fungus interfaces? *New Phytologist,* 117, 61, 1991.

78. **Scannerini, S., Fieschi, M., Alloatti, G., Sacco, S., and Berta, G.,** Cell potential hyperpolarization in *Allium porrum* L. +*Glomus* sp. strain E3 VA mycorrhizae, paper presented at 8th N. Am. Conf. Mycorrhizae, Jackson, Wyoming, 5-8 September, 1990, 256.
79. **Gianinazzi-Pearson, V., Gianinazzi, S., and Brewin, N. J.,** Immunocytochemical localisation of antigenic sites in the perisymbiotic membrane of vesicular-arbuscular endomycorrhiza using monoclonal antibodies reacting against the peribacteroid membrane of nodules, in *Endocytobiology IV,* Nardon, P., Gianinazzi-Pearson, V., Grenier, A. M., Margulis, L., and Smith, D. C., Eds., INRA Press, Paris, 153, 1990.
80. **Perotto, S.,** unpublished data, 1990.
81. **Wyss, P., Mellor, R. B., and Wiemken, A.,** Vesicular-arbuscular mycorrhizas of wild type soybean and non-nodulating mutants with *Glomus mosseae* contain symbiosis specific polypeptides (mycorrhizins), immunologically cross-reactive with nodulins, *Planta,* 182, 22, 1990.
82. **Legocki, R. P. and Verma, D. P. S.,** Identification of "nodule specific" host proteins (nodulins) involved in the development of *Rhizobium*-legume symbiosis, *Cell,* 20, 153, 1980.
83. **Hilbert, J. L. and Martin, F. M.,** Regulation of gene expression in Ectomycorrhizae. I. Protein changes and the presence of ectomycorrhiza specific polypeptides in the *Pisolithus-Eucalyptus* symbiosis, *New Phytol.,* 110, 339, 1988.
84. **Martin, F. M., Hilbert, J. L., Henrion, B., and Costa, G.,** Biochemical and molecular changes during ectomycorrhiza formation, paper presented at 8th N. Am. Conf. Mycorrhizae, Jackson, Wyoming, 5-8 September, 1990, 200.
85. **Dumas, E., Tahiri-Alaoui, A., Gianinazzi, S., and Gianinazzi-Pearson, V.,** Observations on modifications in gene expression with VA endomycorrhiza development in tobacco: qualitative and quantitative changes in protein profiles, in *Endocytobiology IV,* Nardon, P., Gianinazzi-Pearson, V., Grenier, A. M., Margulis, L., and Smith, D. C., Eds., INRA Press, Paris, 1990, 153.
86. **Pacovsky, R. S.,** Carbohydrate, protein and amino acid status of Glycine-*Glomus-Bradyrhizobium* symbioses, *Physiol. Plant.,* 75, 346, 1989.
87. **Fortin, M. G., Morrison, N. A., and Verma, D. P. S.,** Nodulin-26, a peribacteroid membrane nodulin, is expressed independently of the development of the peribacteroid compartment, *Nucleic Acids Res.,* 15, 813, 1987.
88. **Baker, E. M. and Saier, M. H., Jr.,** A common ancestor for bovine lens fiber major intrinsic protein, soybean nodulin 26 protein, and *E. coli* glycerol facilitator, *Cell,* 60, 185, 1990.
89. **Bonfante-Fasolo, P. and Perotto, S.,** Ericoid mycorrhiza: new insights from ultrastructure allied to cytochemistry, in *Cell Interactions and Differentiation,* Ghiara, G., Ed., University of Naples, Naples, 1988, 27.
90. **Werner, D., Mellor, R. B., Hahn, M. G., and Grisebach, H.,** Glyceollin 1 accumulation in an ineffective type of soybean nodule with an early loss of peribacteroid membrane, *Z. Naturforsch.,* 40c, 179, 1985.
91. **Bonfante-Fasolo, P. and Spanu, P.,** Pathogenic and endomycorrhizal associations, in *Techniques in Mycorrhizae,* Varma, G. and Read, D. J., Eds., Academic Press, London, 1991.
92. **Spanu, P. and Bonfante-Fasolo, P.,** Cell-wall bound peroxidase activity in roots of mycorrhizal *Allium porrum, New Phytol.,* 109, 119, 1988.
93. **Morandi, D., Bailey, J. A., and Gianinazzi-Pearson, V.,** Isoflavonoid accumulation in soybean roots infected with vesicular-arbuscular mycorrhizal fungi, *Physiol. Plant Pathol.,* 24, 357, 1984.
94. **Wyss, P., Boller, T., and Wiemken, A.,** Phytoalexin response of soybean roots to separate or combined infection with mycorrhizal fungus and a pathogen, *Experientia,* 47, 395, 1991.
95. **Barrett, V., Shaw, J. J., and Lemke, P. A.,** Transformation of an ectomycorrhizal fungus, paper presented at 8th N. Am. Conf. *Mycorrhizae,* Jackson, Wyoming, 5-8 September, 1990, 17.
96. **Kolattukudy, P. E., Podila, G. K., Roberts, E., and Dickman, M. B.,** Gene expression resulting from the early signals in plant-fungus interaction, *Molecular Biology of Plant-Pathogen Interactions,* Alan R. Liss, 1989, 87.
97. **Dean, R. A. and Timberlake, W. E.,** Regulation of *Aspergillus nidulans* pectate lyase gene (pel/A), *Plant Cell,* 1, 275, 1989.
98. **Millner, P. D. and Meyer, R. J.,** Analysis of amplified ribosomal DNA fragments from VAM fungi, paper presented at 8th N. Am. Conf. Mycorrhizae, Jackson, Wyoming, 5-8 September, 1990, 210.
99. **Simon, L., Lalonde, M., and Bruns, T.,** Amplification and direct sequencing of ribosomal genes from VAM fungi, paper presented at 8th N. Am. Conf. Mycorrhizae, Jackson, Wyoming, 5-8 September, 1990, 265.

# A SEARCH FOR SIGNALS IN ENDOPHYTIC MICROORGANISMS

**Anton Quispel**

## TABLE OF CONTENTS

# I. INTRODUCTION

Endophytic microorganisms are organisms which live, at least during a considerable part of their lifecycle, inside the cells or tissues of a higher or lower plant. As such, they are comparable with endozoic organisms living inside cells or tissues of animals. They may be intercellular, still surrounded by the apparently unchanged cell walls of the host, or become intracellular after penetration or dissolving of the host cell walls. In all cases, the endophytes are surrounded by a cell membrane and often by a matrix of predominantly host origin.

These definitions need some further restrictions. Parasitic microorganisms which rapidly kill the host cells and then live in necrotic cells of tissues are no longer considered endophytes. In biotrophic parasites there may be a period in which the cells of the parasite are, for some time, in direct contact with living host cells and during this period we may consider them as endophytes. Although comparisons with such biotrophic parasites may be interesting, this chapter will be mainly concerned with endosymbionts.*

The definition of a symbiont or of the concept "symbiosis" is by no means easy. In all symbioses there are aspects of a definite parasitic character like infection and defense reactions. While in parasitic infections the host plant is more or less damaged and ecologically threatened, the most important aspect of all symbioses is an ecological benefit for the host from the presence of the symbiont. This especially applies to the endophytic state, during which essential nutrients or metabolites are transported from the endophyte to the host. As a consequence, the interface between endophyte and host forms the most essential region for our understanding of the symbiotic interactions. It has been emphasized that the most fundamental difference between parasitic and symbiotic interactions consists in the unidirectional flow of metabolites and nutrients from host to parasite and the bidirectional flow between host and symbiotic endophyte.[1]

Although this chapter must concentrate on the endophytic states during which the (endo)symbionts function to provide nutrition to their host, we must realize that this endophytic state is part of a dynamic process starting with infection and ending with senescence. Progress in our understanding of the signals during this dynamic process is restricted to the study of infection and subsequent development in parasitic organisms and in members of Rhizobiaceae. As yet it is impossible to give clear-cut examples of signals which play a role during establishment, preservation, and functioning of the endophytic states of any type of symbiosis. Hypotheses about such signals may be deduced from studies of the symbiotic systems, especially the structure and function of the interface between host and symbiont.

In Table 1 a survey is given of the main types of symbioses and parasitisms in which some form of endophytic state is observed. It is important to make comparisons between different types of endophytes and even include some studies on endozoic symbioses.[2-5] Even in the best and most extensively studied symbiosis, the root nodule symbiosis of legumes with *Rhizobium,* the symbiotic interactions and signals during the endophytic bacteroid phase are still insufficiently understood. Different symbioses may have special advantages for the study of particular aspects. A comparative study thus may lead us to a better understanding of the endophytic way of life and the possible signals which are involved in the different symbiotic interactions.

# II. STUDIES OF THE ENDOPHYTIC SYSTEMS

During their endophytic state, the symbionts live in a microhabitat protected against external environmental influences by the host tissues. In intercellular symbioses the endo-

* Although both organisms which participate in a symbiotic relationship may be called "symbionts" it is convenient to use this term for the organism which, in an endophytic form, is enclosed by a plant or animal host.

**TABLE 1**
**Survey of Endophytic and Endozoic Organisms**

| Host | Symbiont (or parasite) | Examples | Ref. |
|---|---|---|---|
| Bacteria | Bacteria | Bdellovibrio | |
| Fungi | Cyanobacteria | Some lichens as main or second phycobiont | 6 |
| | Green algae | Most lichens | 6 |
| | Fungi | Some parasites | |
| Green algae | Cyanobacteria | Cyanelles in some flagellates and diatoms | |
| | Fungi | Homoiomeric lichens | 6 |
| | Protists | Cryoptomonas in Mesodimium | |
| Plants | Eubacteria | Plant pathogens | |
| | | Associative symbioses | |
| | | Leaf nodules (with *Phyllobacter*) | |
| | | Stem nodules of legumes (with *Azorhizobium*) | |
| | | Root nodules of legumes (with *Bradyrhizobium)* | 7—9 |
| | Actinomycetes | Actinorhizal root nodules (with *Frankia*) | 10 |
| | Cyanobacteria | In mosses, liverworts, *Azolla, Cycadeae, Gunnera* (with *Nostoc* or *Anabaena*) | 11 |
| | Fungi | Plant pathogens | 12 |
| | | Ecto- and endomycorrhizae | 13,14 |
| | | Stem and leaf symbionts | 15 |
| Protozoa | Eubacteria | Kappa in Paramecium; respiratory bacteria in amoebae (*Pelomyxa*) | |
| | Green algae | Many Ciliates with *Chlorella* | 16 |
| | Protozoa | *Leptomonas* in macronucleus | |
| Metazoa | Eubacteria | Many parasites, intestinal bacteria, insect symbionts | |
| | Cyanobacteria | Porifera, Echiurida, Tunicata | |
| | Actinomycetes | Some parasites | |
| | Fungi and yeasts | Different parasites insect symbioses | |
| | Green algae | *Chlorella* and dinoflagellates in coelenterates | 16 |
| | Chloroplasts | Marine sponges and mollusks | 16 |
| | Protozoa | Different parasites | |

phytes are situated in the intercellular or in preformed cavities surrounded by the cellwalls of host cells. Since cell wall hydrolyzing enzymes play a role during infection, it is possible that these host cell walls around the endophytic symbionts are affected by such enzymes which may reduce their influence as diffusion barriers. In intracellular symbioses, the penetrating symbionts initially may be surrounded by a cell wall deposit as in the infection threads of the penetrating *Rhizobium* during the initiation of root nodules in legume plants. Such cell wall deposits may disappear at a later phase leading to bacterial release from the infection threads. In vesicular-arbuscular (VA) mycorrhizae as well, the surrounding cell walls are inconspicuous in the endophytic arbuscles. In other cases, like in the root nodules of Caesalpinoideae and the only nonleguminous host for *Rhizobium, Parasponia,* the bacteria remain surrounded by a cell wall like matrix.[17] This is observed in actinorhizal root nodules as well. An effective fixation of dinitrogen demonstrates the functioning of the symbiosis, and thus it must be concluded that the presence of a cell wall-like matrix does not lead to a considerable barrier against diffusion of metabolites.

All endophytic symbionts are surrounded by a cytoplasmic membrane of predominantly host origin. The function of these perisymbiont membranes is illustrated by the universal presence of an active $H^+$-ATPase,[18] which indicates the presence of transport systems. In the membranes around endophytic parasites, no $H^+$-ATPase is observed.

Advances in the studies of the endophytic cells and their surrounding perisymbiont membranes has been obtained by the isolation of endophytic cells with and without the perisymbiont membrane. Moreover, experiments with isolated endophytic states of symbionts allow comparisons with their cultivated forms.

Isolation of endosymbiotic cells appears to be relatively easy when they live in special organs, like the digestive cells of *Hydra,* or in special cavities like in *Azolla.* Even then, we must recognize the possibility that substances from the host still adhere to the surface of the isolated cells. For the isolation of intracellular endophytic cells, one applies the methods for the isolation of cell organelles and the same precautions must be taken. Endophytic cells must be prevented from damage during homogenization, namely, exposure to lytic enzymes, plant phenolics, or oxidizing conditions. For isolation and separation from cell organelles different methods are applied, such as differential centrifugation, filtration through sieves of different mesh width, and density gradients with sucrose or silica gel (Percol). From root nodules of *Alnus incana,* preparations of endophytic vesicle clusters could be obtained of which 98% consisted of vesicle clusters and only 0.4% consisted of host cell mitochondria.[19]

During isolation of intracellular endophytes a special problem arises: are the isolated cells still surrounded by the perisymbiont membrane? This aspect has been recently studied in the *Rhizobium*-legume root nodules. After careful and mild homogenization, peribacteroid units may be isolated in which the bacteroids are still surrounded by the peribacteroid membrane as peribacteroid units (PBU).[20] After application of, for example, mild osmotic shock, these peribacteroid membranes can be removed after which differential centrifugation yields fractions with the pure bacteroids and fractions of the peribacteroid membrane (PBM), while the supernatant can give some impression of the composition of the material in the peribacteroid space (PBS).

Comparisons between transport in PBUs and bacteroids enables us to study the role of the PBM as a selective permeability barrier which determines the direct environment around the bacteroids. Pure PBM fractions may be used for chemical analyses and for the preparation of monoclonal antibodies.[21] Such antibodies then form an important tool for further cytochemical studies of the structures around the endophytes in the root nodules. These studies have considerably increased our knowledge about the PBM and PBU. While the PBM originally has been considered as a continuation of the host cell plasmalemma, it appeared from studies with bacterial mutants that its composition is affected by bacterial genes.[22] They affect the fatty acid composition,[23] and induce some nodule specific proteins, like nodulin-26, which are inserted in the PBM.[24] Some of these nodulins are induced early in the nodulation process by as yet unidentified bacterial signals, while the induction of some other nodulins occurs later in the nodulation process after physical presence of bacteroids in the host cells. In other aspects the PBM more closely resembles the host endoplasmic reticulum (ER) or Golgi membranes. It thus appears that the PBM, although originally formed as an extension of the host plasmalemma, further develops as a complicated product from host ER and Golgi systems with many host cell proteins, which are partially induced under the influence of bacterial signals. The PBM thus must be considered to be a product of symbiotic interactions. This has far-reaching consequences for the microenvironment around the bacteroids. The presence of certain enzymes like $\alpha$-mannosidase, proteinase, and proteinase inhibitors indicate that the PBS has a vacuole content which resembles a typical plant vacuole with lysosomal characteristics.[25] In mutants, which fail to induce some of the later PBM nodulins, unstable situations arise leading to fusion of different PBMs and breakdown of the PBM and of the bacteroids.

# III. CULTIVATION OF THE ENDOPHYTIC ORGANISMS

It is by no means easy to ascertain that the cultures of endophytic organisms really belong to the same organism which formed the endophytic cells. There are several observations about a heterogenity of endophytic populations. Moreover, the viability of the functional endophytic cells may be disputed. In some cases of leguminous root nodules, the characteristic endophytic bacteroids could be cultivated; in other cases no cultures could be obtained.[26] In *Frankia* the vesicles may grow out and form new hyphae,[27] but it is not certain that this may be applied to all endophytic vesicles.

Identity of cultures with the endophytic organism is reasonably certain when the cultures are obtained from single-cell isolates. This is not always possible, e.g., when sufficiently dense inoculations are used to obtain growth. The necessity for a critical consideration of the identity of the cultivated strains is best illustrated for symbioses with cyanobacteria. In most hosts, the *Nostoc* or *Anabaena* is easily isolated and reinfections are possible, even with originally nonsymbiotic strains, showing a high, though not absolute, promiscuity of the isolated strains for their hosts. However, in *Azolla* the situation is quite different. Although many strains of *Anabaena azollae* have been isolated, reinfections of symbiont-free *Azolla* plantlets is impossible or, at best, ineffective. The enzymatic, antigenic and morphologic differences of the isolated and cultivated strains with isolated suspensions of the endophytic forms are so great that either the cultures were markedly changed during cultivation or belonged to a minor population, differing from the dominant endophytic form.[28] An even more complicated situation appears with the cultivation of *Frankia* strains from different actinorhizal root nodules. In 1978, the first cultures of *Frankia* became available. Since then many strains have been cultivated from nodule fragments after sufficient external disinfection and cultivation on suitable nutrient media. Most isolates infect and effectively nodulate roots of their original host species. However, strains which were isolated from nodules with abundant spore formation may give nodules without spores after reinfection and vice versa.[29,30] Strains which were isolated from fully effective nodules form noneffective nodules after reinfection on their original host species. They even may fail to infect their original hosts but form effective nodules on different species.[31] Again we may ask the question whether the isolated strains are not representative of the dominant endophytic population or are markedly changed during cultivation.

There are still many types of actinorhizal root nodules from which no *Frankia* strains could be cultivated. From some nodules, the outgrowth of only a few hyphae appear after several months of incubation, which makes the identity of the isolate rather suspicious. In root nodules from the spore-negative type of root nodules of *Alnus glutinosa,* such occasional hyphae were observed after 2 to 3 months of incubation on the usual nutrient media, although abundant outgrowth of hyphae was observed within 12 days if a small amount of a root lipid extract from alder roots had been added to the nutrient medium.[29] Only these hyphae can be considered to be representative of the dominant endophytic population. During further cultivation, the root lipid extract could be substituted by substances like Tween 80 or lecithin, most probably serving as suitable sources of fatty acids. These substances could not replace the root-lipid extract during primary isolation. During subsequent cultivation, no further additions were necessary, at least on media with propionic acid as the carbon and energy source. The specific factor in the root lipid extract which appeared to be essential for the primary isolation from the endophytic phase could be identified as the triterpene dipterocarpol[32] (Figure 1).

Since dipterocarpol is a natural constituent of the host cells of the endophyte the necessity of this compound during primary isolation and the intermediate function of less specific lipids indicates a gradual adaptation process from the endophytic to the free-living condition. It is tempting to relate this process to the controversial discussion around the viability of

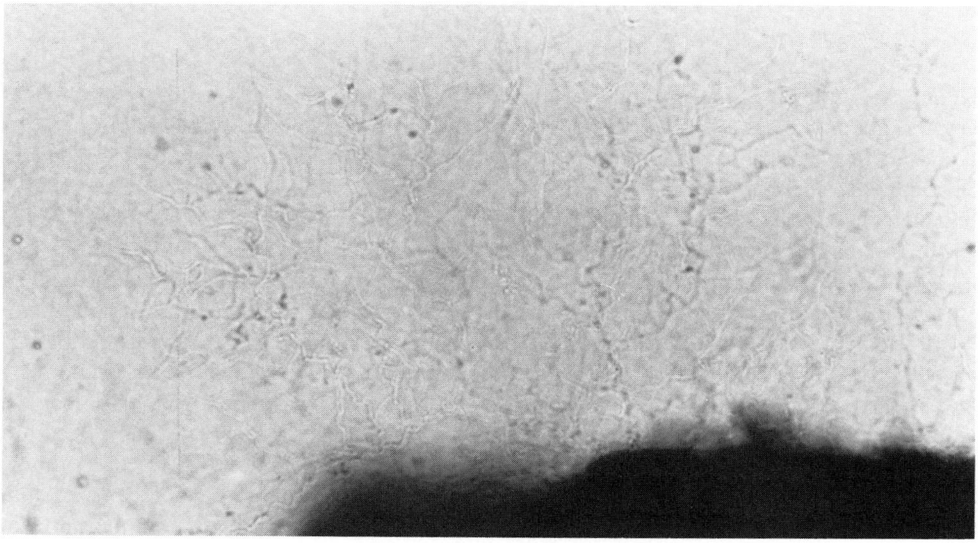

FIGURE 1. Outgrowth of hyphae of *Frankia* from a fragment of root nodules of *Alnus glutinosa* (spore-negative type) 12 days after inoculation in a nutrient medium with 20 g/ml dipterocarpol (further details[32]). Controls without dipterocarpol never showed any growth of hyphae within 2 months of incubation.

*Rhizobium* bacteroids.[26] Certainly, other factors play a role since dipterocarpol did not stimulate growth from spore-positive nodules of *Alnus glutinosa*. There are indications that such nodules contain inhibiting factors.[29]

In mycorrhizae, the fungus is only partially endophytic while a considerable part of the hyphae develops in the surrounding soil where fructifications and spores are formed. Cultures of many ectotrophic mycorrhizae have been obtained. Cultivation of the fungi from the VA mycorrhizae is still not possible. Spores can be collected which form hyphae in cultures and the growth of these hyphae may be stimulated by host cell products. However, after some time, this inevitably stops.[33] The spores contain abundant nuclei which do not divide but are distributed over the hyphal cells; no DNA synthesis takes place because the enzyme thymidylate synthase is inactive.[34] Of course, this enzyme has to be activated or formed in the symbiosis, but in cultures no activation was possible until recently.

## IV. COMPARISONS BETWEEN ENDOPHYTIC AND CULTIVATED FORMS

Comparisons between the freshly isolated endophytes and their cultivated strains may give valuable information about the characteristic aspects of the endophytic forms and may provide some indications about the factors which are responsible for these differences. However, such comparisons must be made in a highly critical way. The pitfalls of possible artefacts during the isolation procedure were already discussed. In general, it must be realized that endophytic forms represent a highly adapted form of a microorganism which is susceptible to sudden changes in the environment during isolation. When we compare them with cultivated cells, we must realize that these cells may be quite different in the exponential and in the stationary phase, since they will be markedly influenced by the nutrient conditions. On the other hand, systematic studies on the effects of different nutrient solutions in different growth phases may lead to possible conclusions about the conditions in the endophytic state.

## A. FORMS, STRUCTURE, AND COMPOSITION

In many symbionts there are no obvious differences between the endophytic and cultivated forms. Endophytic green algae may have somewhat larger dimensions and do not form spores, even though they are abundant in cultures.[35] They may have modified cell wall constituents as is evident from reaction with lectins or antigenic cross-reactions in algae from lichens.[36,37] Although there are no indications about the causes of such differences, the influence of fungal enzymes may be involved.

There are some reports about changes in pigment formation e.g., in lichens with *Trentepohlia* as green symbiont.[35] In symbioses with cyanobacteria, the photosynthetic pigments are usually still present, albeit at reduced level, in the endophytic cells even when they live entirely in a heterotrophic way.[38] In the symbiosis with *Cyanocyta korschikoffiana,* the synthesis of the typical prokaryotic phycobilins and chlorophyll is regulated by factors from the eukaryotic host *Cyanophora paradoxa.*[39]

Endophytic cyanobacteria have a high percentage of heterocysts, at least when their symbiotic function is nitrogen fixation.[40] Such an increase in heterocysts can be obtained in cultivated forms by nitrogen limitation.

In root nodule symbioses, the endophytic bacteroids of *Bradyrhizobium* consist mostly of swollen, globoid forms, while the bacteroids of *Rhizobium* species are characterized mostly by their branched and Y forms. Such forms may be observed in cultures during different "stress" conditions. It is by no means certain that bacteroids produced in culture are comparable with the endophytic forms. Since there are good indications that dicarboxylic acids are the natural carbon and energy sources in the endophytic situation, it is interesting to note that bacteroid-like cells may be induced by succinic acid in *B. japonicum.*[41]

The typical bacteroid forms may be the result of the absence of cell divisions with continued cell growth combined with a weakened cell wall. This weakening can be explained by changes in the lipopolysaccharides (LPS) in the outer membrane of the cell wall.[42,43] Immunochemical analyses of LPS of cultivated cells and bacteroids in *R. leguminosarum* showed that both types contained different LPS epitopes.[44] The expression of different cell-surface antigens appeared to be regulated by oxygen concentration and pH.[45,46] The suggestion that bacteroid formation is related to the induction of nitrogen fixation is unlikely since typical bacteroids have been described in the non-nitrogen fixing *Phyllobacterium* from leaf symbioses.[47]

The actinorhizal root nodules with *Frankia* are most suitable for the study of plant signals which affect the morphogenetic development of an endophyte. This actinomycete forms three types of structures: hyphae, vesicles, and sporangia with spores. All forms are observed in cultures and in nodules. The vesicles have been well established as the sites for nitrogen fixation. With the exception of some fix$^-$ mutants they are formed in all cultivated strains after nitrogen limitation and have the appearance of globules with characteristic septae, as seen by electron microscopic examination. Such globular vesicles are observed in the endophytic forms, e.g., in *Alnus* or *Hippophae,* or they are club-shaped like in the endophytes of *Myrica* and *Comptonia.* Since the cultivated strains all form normal globular vesicles, these club-shaped types must be the result of some plant interaction. In the Casuarinaceae, no vesicles are formed in nodules of *Casuarina,* though the same strains form vesicles in culture and in the nodules of *Gymnostoma,* again indicating remarkable plant effects.[48]

All cultivated *Frankia* strains form characteristic sporangia. In the nodules, spores may be abundant (spore-positive) or virtually absent (spore-negative). This difference is caused by genetic differences of the *Frankia* strains, as has been shown in cross-inoculation experiments.[49] However, strains from spore-negative nodules form abundant sporangia and spores in cultures. The formation of sporangia inside the nodules thus depends on different reactions to the host plant.[30]

In endomycorrhizae, no comparisons between endophytic and cultivated forms are possible. The typical endophytic formation of arbuscles is an indication for some morphogenetic effect of the surrounding root cells. It might be interesting to make comparisons with the formation of infective structures by pathogenic fungi under the influence of volatile factors.[50] In *Ustilago violaceae,* the inducer of infective structures has been identified as α-tocopherol.[51]

## B. CARBON METABOLISM

In all plant-microbe symbioses, at least one of the partners is a photosynthetic organism. In symbioses of higher plants as hosts, the symbionts depend on the host for the supply of organic substances which are formed during photosynthesis. In lichens and many animal symbioses, the endosymbiont is photosynthetic.

Photosynthetic autotrophic microorganisms as endosymbionts are either unicellular algae or cyanobacteria. Best studied are the symbioses of ciliates like *Paramecium* with *Chlorella,* and of coelenterates like *Chlorella* in *Hydra,* or dinoflagellates in many marine invertebrates and the lichen symbioses with the green alga *Trebouxia* and/or cyanobacteria. Most isolated algae from animal hosts show good photosynthesis in cultures. The question is whether in the symbiotic state the hosts are, in some way, able to stimulate photosynthesis. There are some claims that e.g., in marine corals, photosynthesis of the symbiotic algae is stimulated by some unknown factor[52] of host origin. A similar factor was claimed for the symbiosis of *Elysia viridis* with chloroplasts.[53]

However, it is not clear whether such effects are directly on photosynthesis or on a later biochemical reaction. In *Hydra,* no stimulation of homogenates on the photosynthesis of *Chlorella* cultures could be demonstrated.[54] Yet the cultivated *Chlorella* cells have a lower content of chlorophyll and less $CO_2$ assimilation. In lichens, with *Trebouxia* as the green symbiont, it has been generally observed that the algae need glucose or another sugar for good growth in culture. Yet, the inability of autotrophic growth in cultures is in contrast with the net $CO_2$ assimilation in the thallus. In so :e experiments with *Trebouxia* from different lichens, it appeared that autotrophic growth was possible after addition of ascorbic acid under a reducing ($H_2$) atmosphere.[55] The concentration of ascorbic acid was too low to act as an organic carbon source. In contrast with cultures on glucose the cells had far better chloroplasts with definite pyrenoids. Dioxymaleic acid had a similar effect and cysteine was inactive. We cannot yet conclude whether ascorbic acid acts on the formation of the chloroplasts or directly on photosynthesis or whether it is related to the establishment of reducing conditions.

On the other hand, there can be no doubt that at least in symbioses with cyanobacteria, photosynthesis can be inhibited. This only occurs in heterotrophic hosts when, like in the cephalodiae of lichens, the cyanobacteria are a second green symbiont together with green algae.[56] In lichens with *cyanobacteria* as the only symbionts there is a normal photosynthetic activity. When cyanobacteria are symbionts in green hosts, like *Azolla,* their photosynthesis is negligible. Action spectra of total photosynthesis of the *Azolla-Anabaena* system are identical with the spectra of *Azolla* alone.[57] Nothing is known about the signals which are responsible for this reduction of photosynthesis.

In all symbioses with green symbionts the hosts have a well-established effect on the further fate of the assimilated carbon dioxide. In green *Hydra,* freshly isolated algae excrete the $^{14}C$ after assimilation of $^{14}CO_2$ as maltose although in cultivated algae the $^{14}C$ is primarily found in sucrose. This change from sucrose to maltose appeared to be highly pH dependent and was explained from an effect on presumably membrane enzymes which form these sugars from hexosemonophosphate.[58]

While freshly isolated algae excrete maltose, this excretion gradually declines during further cultivation. In all other symbioses with algae, similar excretions of sugars have been observed. In marine corals, glycerol is excreted,[52] in lichens, the excreted sugar is fully determined by the algal species.[59] Excretion of sugar in lichen-algae has been elegantly demonstrated by the method of competitive infiltration by Drew and Smith.[60] After assimilation of $^{14}CO_2$, sufficient amounts of unlabeled sugars are added to the thallus. When a sugar is added, which competes with the labeled sugar which is excreted from the algae, this sugar competes for the transport system of the fungus so that more of the labeled sugar is found in the medium. A similar excretion of sugar into the medium is obtained when the transport system of the fungus is inhibited with digitonin.[61] The same sugars are excreted by freshly isolated algae but, as in *Hydra,* this excretion gradually declines after further cultivation. After decrease of sugar excretion from the algae, more $^{14}C$ from the $^{14}CO_2$ is found in the insoluble cell material. As in the case of the chlorellae from *Hydra, Trebouxia* and cells from Xanthoria, showed a stimulation of excretion under the influence of a low pH (Green cited by Richardson[59]).

Heterotrophic endosymbionts in plant roots depend for their carbon and energy source on the organic substances transported in the phloem in which sucrose is the dominant sugar. This sucrose is subjected to enzymic degradation in the root host cells while transport processes in the perisymbiont membranes select the substrates for immediate use by the endophytes. The differences in carbon metabolism between endophytes and their cultivated forms have been best studied in the *Rhizobium*-legume root nodules. For most *Rhizobium* cultures, sugars are excellent substrates, but the bacteroids generally can not use sugars as substrates for respiration and nitrogen fixation, although there are some exceptions. Since the enzymes of the Entner-Doudoroff and pentose phosphate pathway are all present in bacteroids, it is doubtful whether the inability to use sugars is caused by the absence of necessary enzymes or by deficiencies in transport systems, which will be discussed below. All experiments on the use of different substrates for respiration and nitrogen fixation in bacteroids agree that dicarboxylic acids like succinic acid or malic acid are the primary substrates. This conclusion is supported by different observations. Transport studies in isolated PBMs indicate that dicarboxylic acids must be present in the PBS.[20] The presence of malic enzyme in the bacteroids,[62] which in cultures can be induced by malic acid, is an indication for a similar induction in the endophytic situation; most convincing, is that nodules of mutants defective in the transport or metabolism of dicarboxylic acid are noneffective while most mutants in sugar transport or metabolism form normally effective root nodules.[63,64]

Many studies have been made on the use of different substrates for respiration and nitrogen fixation in isolated vesicle clusters and cultivated strains of *Frankia*. The results have been recently reviewed by Benson and Schultz and by Huss-Danell in the book by Schwintzer and Tjepkema.[10] There appear to be many differences between different strains. Growth of cultivated strains is possible on many nutrient media with different organic substrates like sugars, organic acids, casein hydrolyzate or peptones, and lipids. In many strains, excellent growth is obtained on fatty acids, especially propionic acid. It is doubtful whether this simple fatty acid is a natural substrate inside the root nodules. Enzymes of the Embden-Meyerhof-Parnas and pentose phosphate pathway have been demonstrated. Endophytic vesicles may use different sugars and organic acids and contain enzymes for hexose breakdown even on media with propionic acid. We cannot yet conclude that dicarboxylic acids play the same important role as in *Rhizobium* bacteroids. No conclusive experiments on vesicle clusters with and without their perisymbiont membranes have yet been carried out.

### C. NITROGEN FIXATION AND NITROGEN METABOLISM

Most diazotrophic endosymbionts are able to fix nitrogen in cultures and use the fixed N for nitrogen assimilation. In most diazotrophic cyanobacteria, the heterocysts are the sites of nitrogen fixation, although under microaerophilic conditions the vegetative cells may show some activity as well. In *Frankia,* the vesicles have a similar role. Here the activity of nitrogenase is remarkably insensitive to oxygen presumably because of the barriers against oxygen diffusion through the lipid layers of the thick cell walls. In both types of organisms nitrogen limitation in cultures as well as the endophytic situation lead to a considerable increase of heterocysts or vesicles and nitrogenase. In lichens this increase is only observed when the cyanobacterial symbionts occur as second ''phycobionts'' in cephalodia, where their symbiotic function is not photosynthesis but nitrogen fixation.

In *Rhizobium* symbioses the situation is more complicated. Strains of *Azorhizobium* form stem nodules on *Sesbania* and fix nitrogen in cultures, using the fixed nitrogen for their own growth. In *Bradyrhizobium,* nitrogenase activity may be induced on certain nutrient solutions under microaerophilic conditions but the activity of nitrogen fixation is uncoupled from nitrogen assimilation. As a consequence, the fixation product ammonium is excreted into the medium.[65] In cultures of *Rhizobium* strains, nitrogen fixation is absent or negligible. However, experiments with the *lacZ* fusion technique have demonstrated that under microaerophilic conditions the genes *nif H, D,* and *K,* which encode for the structural proteins of the enzyme nitrogenase, are activated.[66] The nitrogenase proteins thus formed may be active in the reduction of the alternative substrate CN but unable to reduce nitrogen.[67]

A regulated oxygen diffusion barrier is now recognized as being of fundamental importance for the functioning of the bacteroids.[68] A low oxygen concentration is essential for the regulation of the transcription of the *nif* genes,[69-71] for the protection of nitrogenase, and for the synthesis of the bacterial heme contribution to the synthesis of leghemoglobin.[72] Hemoglobin is important for enabling a sufficient respiratory activity in the bacteroids at this low oxygen level. Nitrogenase activity depends on the activation of the bacteroid cytoplasmic membrane potential[73], while there are indications that a $K^+$ pump is involved.[74,75]

During nodule development, the induction of nitrogenase activity is correlated with the development of the bacteroids.[76] However, development of bacteroids does not ensure nitrogen fixation, as is the case with the vesicles of *Frankia.*

In all diazotrophic symbionts, the fixation product, ammonium, is further assimilated by glutamine synthetase (GS) and glutamate synthase (GOGAT). *Rhizobium* and *Bradyrhizobium* strains, as well as *Frankia* strains,[77] are characterized by the presence of two, or even three different types of GS. The GS I is a typical prokaryotic-type enzyme, susceptible to end product inhibition and inactivation through adenylylation, while the GS II is more related to eukaryotic GS and is repressed by low oxygen concentration.[78] In all endophytic states, *Rhizobium* bacteroids, *Frankia* vesicles, and cyanobacterial symbionts, both types of GS are reduced or absent. The ammonium formed is thus not further assimilated by the endophyte, but is excreted. In the surrounding host cells, GS activity is considerably increased which can not simply be explained from the increased provision with ammonium.[79] If bacteroids excrete all of their ammonium to the host, the question arises whether they must rely for their, albeit small, nitrogen demand on amino acids recycled from their hosts.

Nitrogen recycling has been considered as an important aspect of the symbiotic relations in other symbioses, e.g., between animals and green algae. The algae secrete amino acids to the hosts where they are used for protein synthesis while some ammonium from protein breakdown or deamination is recycled to the endosymbiotic algae.[80] The release of maltose by the algae might decrease the rate of deamination in the host and thus lead to a better conservation of amino acid nitrogen.[81] Recycling of nitrogen might play a role in the symbiosis of grasses with endophytic fungi.[82]

## D. TRANSPORT PROCESSES

The bidirectional transport systems in the membrane interface between the host and endophyte must be considered as the most essential structure of mutualistic symbioses and one which deserves considerable attention.[18] Two membranes deserve special study: the perisymbiont membrane (e.g., the PBM around bacteroids in *Rhizobium* root nodules) and the cytoplasmic membranes of the endophytes. Cell wall and matrix layers between these two membranes are not considered as important barriers to diffusion. The cell walls of the prokaryotic endophytes may be important because binding proteins are situated between the cytoplasmic and the outer membrane. Changes in the LPS of the outer membrane during formation of the endophytic cells then might lead to loss of these binding proteins, but there is no evidence that in bacteroids, e.g., of *R. leguminosarum,* this is of great importance.[83]

Before considering the cytoplasmic transport systems in the endophytes, we must know how the perisymbiont membranes affect the direct environment of the endophytes. The only perisymbiont membrane which has been studied in this respect is the PBM around the bacteroids of *Rhizobium* or *Bradyrhizobium*. Comparisons between isolated PBUs (bacteroids with the surrounding PBM) and free bacteroids have indicated that the PBM contains good transport systems for dicarboxylic acids, but forms a barrier against transport of glucose, pyruvate, α-ketoglutarate, glutamate, and α-aminobutyrate.[20,84-87] The PBM contains a $H^+$-ATPase but does not contain an electron transport system.[20] It regulates the transport of metabolites, especially dicarboxylic acids to the bacteroids.

Conclusions about transport in the perisymbiont membranes of other symbioses are restricted to estimations of concentrations and fluxes of metabolites from host to endophytic symbionts.[88] They indicate considerable modifications and adaptations of the membranes as compared with the normal host cytoplasmic membranes which had to be expected after the earlier discussed evidence for the difference in constitution. Thus all perisymbiont membranes must be able to facilitate the efflux of metabolites from the host cells while the function of normal cytoplasmic membranes of root cells is based on different influx systems.

Studies of the transport systems of the endophytic cells have been mainly performed with bacteroids of *Rhizobium* and *Bradyrhizobium*. Comparisons between transport in cultivated bacteria and bacteroids are difficult as long as we are not absolutely certain that we are studying transport and not some subsequent metabolic reactions. Unfortunately, the use of isolated membrane vesicles has not been possible up until now.[89] Such studies using deoxyglucose show that bacteria have a good active transport system for glucose, while in the bacteroids this transport is nearby absent.[83] On the other hand, transport of dicarboxylic acids is increased in bacteroids as compared to the cultivated bacteria. Both effects may be explained from the accumulation of dicarboxylic acids in the vacuole around the bacteroids and their effects on the repression of the glucose transport system and on the induction of the dicarboxylic acid transport system.[83,62] The latter transport system has been extensively studied by using mutants in the genes for the dicarboxylic acid transport system, the *dct* genes. This gene system consists of the structural gene *dctA* and three regulatory genes *dct B, C,* and *D*. They operate in conjunction with the protein *Ntr* A and thus are influenced by the general status of the cell.[90] Comparisons of some mutants in *dctB* and *dctD* of *R. meliloti,* with $fix^+$ and $fix^-$ phenotypes on different substrates, indicated a possible regulatory link between the expression of the genes *nif H, D,* and *K* and the activity of the *dct* regulatory genes.[91]

Cultivated strains of *Rhizobium* and *Bradyrhizobium* accumulate ammonium by an active transport system as has been demonstrated by a selective electrode,[92] or by the ammonium analog $^{14}C$-methylammonium.[93] The transport system is repressed by atmospheric oxygen concentrations, but is highly active at 0.2% oxygen. It is induced by nitrogen-limiting conditions. In bacteroids, no activity can be observed. The excretion of ammonium by bacteroids can be explained by a simple diffusion process,[73] made possible by an ammonium concentration outside the bacteroids of nearby zero.[94]

Transport of aspartate and glutamate has been demonstrated in cultivated strains to be the same as in bacteroids, presumably through the dicarboxylate transport system. Since the PBM does not transport aspartate and there is a competitive inhibition by succinic acid, the biological function of aspartate for bacteroids has been doubted for *R. meliloti,* although it might be possible for *B. japonium.*[95] Both cultivated bacteria and bacteroids of *R. leguminosarum* have a transport system for the branched amino acids, leucine, isoleucine, and valine, but there appear to be marked differences with regard to the inhibition by other amino acids and the pH,[89] again demonstrating the differences in membrane transport systems between cultivated bacteria and bacteroids.

Far less data are available for other symbioses. For vesicle clusters of the endophytic *Frankia,* the existence of a malate-aspartate shuttle has been suggested.[96] Here, as in *Rhizobium,* the endophytic forms lack an active (methyl) ammonium transport system.[97]

A vast amount of data is available for the transport of phosphate and other ions from the soil through the mycorrhizal hyphae to the plant roots. Cytochemical studies demonstrate the presence of $H^+$-ATPase in the membranes of the host-fungus interface. At present, we can only speculate about the transport processes involved and the explanations for the reversed transport at the membranes of the intercellular arbuscles as compared with the fungus in the soil.

The only other type of symbioses for which direct studies on transport systems have been made is the *Hydra-Chlorella* symbiosis.[98-100] In *Chlorella,* two systems for transport of amino acids could be distinguished: a system for the basic amino acids and a less specific general amino acid transport system. Transport activity was higher in freshly isolated cells then after longer cultivation. The remarkable stimulation by arginine of the transport of other amino acids in freshly isolated cells might be explained by an interference with the $H^+/K^+$ system.[100] In cultures, the transport could be affected by the pH, by nitrogen starvation, and by a high $CO_2$ concentration.

Studies on the membranes in the host-symbiont interface and their transport systems are an underdeveloped field of study. This must be considered of the most fundamental importance for our understanding of host-symbiont interactions.

## E. REGULATION OF ENDOPHYTE POPULATIONS

In all mutualistic endosymbioses there exists a homoeostatic regulation which keeps the endophytic or endozoic population more or less constant in relation to the biomass of the hosts. Smith developed a regulation model for the biomass of symbionts consisting of three aspects: (1) restriction of the initiation and infection, (2) restriction of symbiont growth and cell divisions, and (3) expulsion of the surplus symbionts.[101] The first aspect starts with the infection if the symbiont has to enter the host from the environment. During this process the symbionts are transferred to the sites where they will perform their symbiotic functions. These aspects, though highly important for the establishment of a symbiosis and the development of the symbiont population, must fall outside the scope of this review. In symbioses with photosynthetic host plants, the success of this development is mainly determined by the efficiency of host photosynthesis. Moreover, it is affected by many developmental factors like plant hormones from plant and microbial origin.

Once the structures for the endosymbiosis have been established or the symbionts have been resorted to the preformed structures of the host, like in *Hydra* and *Azolla,* the further population of the symbionts is regulated by different factors in a way that is only partially understood. Most studies on this regulation have been performed with animal symbioses, like those of *Hydra* with *Chlorella,*[101] and with lichens and the *Azolla-Anabaena* symbiosis.[102] In root nodule symbioses, this highly important aspect has been relatively neglected and deserves much more attention.

During normal development of green *Hydra,* the population of the endosymbionts is regulated in coordination with the cell divisions and growth of the host. A distinction must be made between the factors which influence the upper limit of the endosymbiont population, like the feeding regime of the hosts or the light conditions, and the factors which influence cell divisions and cell growth. Douglas and Smith formulated the hypothesis that the division of algal symbionts is regulated by the pH within the perialgal vacuole — inhibited at a low pH and induced by a temporary increase within the range of permissible pH.[103]

If this hypothesis is true it must be concluded that the $H^+$ pumps in the perisymbiont membranes play an essential role in the regulation of the endosymbiont populations. This need not mean that the pH should be the only factor to explain the generally observed correlation between the cell divisions of the symbiotic algae and the cell divisions and growth of the host. Nutrient conditions certainly will be important. Growth of *Chlorella* cells in *Hydra* may be nitrogen limited and stimulated by the release of ammonium from the hosts.[104] In experiments in which the *Hydra* hosts were fed with $^{14}C$-leucine-labeled *Artemia*, MacAuley[99] observed a correlation between the uptake of $^{14}C$ by the symbiotic *Chlorella* cells and their cell divisions. After considering this correlation in relation to the effect of leucine on the transport systems for amino acids, he speculated that the often postulated "division factor" might consist of one or more amino acids.[99]

In *Azolla* and in lichens, cell divisions and growth of the endophytes are highly correlated with the growth of the host.[102] Hill suggested two controls: (1) the nutrient supply and (2) a constraint or stimulation of commitment to divide. As a consequence, cells of the endophytes which can not further divide but still obtain sufficient nutrients become larger in size. Increase of cell size as a consequence of the inhibition of cell division is a general phenomenon as exemplified by the formation of bacteroids in root- and leaf nodules.

It is now generally recognized that digestion or expulsion of endosymbionts does not play a major role in most normally functioning symbioses. However, it may be important under extreme conditions, e.g., during excessive feeding of hosts and at the end of a symbiotic development. Experiments with the introduction of foreign algae into the digestive glands of *Hydra* demonstrated the importance of the specific transfer of the homologous algae to the basic digestive cells. If newly introduced foreign algae are unable to reach the sites which are occupied by the already present endophytic algae, they are unable to prevent the fusion of the phagosomes, in which they are absorbed, with lysosomes.[105] In the normal symbiosis such fusion is prevented.

In root nodules with *Rhizobium*, the PBM may act as a barrier to prevent the attack by proteases from the host cells.[106] *Bradyrhizobium* mutants which fail to stimulate the plant to express the nodulin F gene result in the fusion of PBMs and the digestion of the bacteroids.[25] Here, as in the *Hydra-Chlorella* symbiosis, the symbiosis can only be functional as long as contact with lytic enzymes can be prevented. At later phases of the symbiosis, during senescence, fusion or collapse of the PBM leads to breakdown of the bacteroids system. Too early breakdown is the most general cause of ineffectivity.

Although we might expect that comparable processes are involved in the protection of *Frankia* vesicle clusters and the senescence in older cells of the root nodules in actinorhizal plants, nothing is known about their regulation. The absolute and specific necessity of dipterocarpol for stimulating growth from the endophytic form of *Frankia* in certain root nodules of *Alnus glutinosa* indicates one possibility for regulation of growth of this endophytic form. Although dipterocarpol is a natural constituent of the root nodule cells, we cannot yet draw conclusions about a possible role of this substance in the regulation of the development of *Frankia in planta*.[32]

## V. ANALYSIS OF THE SYMBIOTIC INTERACTIONS AND THE ROLE OF POSSIBLE SIGNALS

The preceding sections were meant to give a short overview of the main characteristic differences between endophytic organisms and their cultivated forms. These differences as well as conclusions based on the study of the different aspects of the endophytes inside the host plants (or animals) may lead to an analysis of the symbiotic interactions and suggestions about the possible signals involved. It must be emphasized that many, if not most, of the host-endophyte interactions are far from being understood to enable any suggestions about the regulating signals.

## A. THE MICROENVIRONMENT OF THE ENDOPHYTIC MICROORGANISMS

### 1. Microaerobic and Reducing Conditions

Since all endophytic organisms are surrounded by host tissues, their respiration will reduce the oxygen content. In the cortex of leguminous root nodules regulating barriers against oxygen diffusion have been demonstrated.[68] The synthesis of leghemoglobin by a combined biosynthesis of host and symbiont enables the provision of oxygen for the respiration of the symbiont at the prevailing low oxygen content. The effect of ascorbic acid in a reducing atmosphere on the autotrophic development of *Trebouxia* might be interpreted as indicating the importance of reducing conditions even in a lichen thallus. Many effects of reduced oxygen content have been demonstrated, most of them by studies of the leguminous root nodules. These effects may operate on the level of the regulation of transcription: the expression of the *nif* genes by regulation of the regulatory gene *nif*A,[71] the effects on the synthesis of many other bacteroid proteins,[107] the repression of the synthesis of GS II and the ammonium transport system, the increased synthesis of heme for leghemoglobin,[72] the formation of bacteroids and their characteristic cell wall LPS epitopes,[44] and the induction of DNA superhelicity.[108] Not all of these effects need be explained on the level of transcription. Direct inhibition by high oxygen concentrations may play a role as well, e.g., on the *Nif*A protein and the nitrogenase itself.

### 2. Regulation of the pH Around Endophytic Cells

The best evidence for the role of pH on metabolic and/or transport activities of an endosymbiotic organism has been obtained for the effect of a low pH on the excretion of maltose by *Chlorella* cells from *Hydra*.[58] Other pH effects have been described as well, e.g., the necessity of a pH of 5 to 6 as one of the factors for the formation of bacteroids in *Rhizobium* and the regulation of cell divisions in endosymbiotic *Chlorella*.[103] The universal presence of an active $H^+$-ATPase in perisymbiont membranes indicates that a proton pump must affect the pH of the vacuole around the endophytic cells. Moreover, the process of nitrogen fixation itself leads to considerable extrusion of $H^+$ions.[109]

### 3. The Concentration of Other Cations

The presence of an active proton pump in the perisymbiont membrane will have an effect on other transport systems as sym- or antiport. Changes in the concentration of cations in the environment of the endophytic cells will affect the operation of the transport systems in their cytoplasmic membranes. In the *Chlorella-Hydra* symbiosis, the stimulating effect of arginine on the transport of other amino acids by the *Chlorella* cells has been related to a possible interaction with their $H^+/K^+$ pump.[100] For other symbioses, only incidental observations about an effect of changes in the concentration of, for example, $K^+$ ions are available.[74,75,108]

In view of the complicated changes of transport systems in the endophytic state, more attention must be given to the functioning of cation pumps in the endophytic membranes.

## B. NUTRIENT EXCHANGE

The beneficial effects of symbiosis are mainly based on the availability of nutrients: the products of $CO_2$ assimilation, the products of nitrogen fixation, and the transfer of mineral nutrients. These effects are too obvious to discuss here. However, the provision of certain nutrients has further consequences. These nutrients may have regulatory effects which play an important role in the mutual adaptations. Certain metabolites affect existing enzymic reactions or transport processes by competitive or noncompetitive inhibitions, e.g., the effect of succinic acid in the peribacteroid space on the transport of aspartic acid,[95] or the inhibition of glutamine synthetase GS I by ammonium. More important are the effects at the transcription level, e.g., the repression of the transport system for glucose by cultivation of *R. leguminosarum* in the presence of succinic acid which must be present in sufficient concentrations around the *Rhizobium* bacteroids.[83] Nitrogen limitation or more generally the relation between carbon and nitrogen nutrition affects the transcription of the *ntr* genes. The protein NtrA, itself regulated by the sensor system NtrB/NtrC, has, in cooperation with other regulatory genes, manifold effects in the regulation of enzymic and transport processes like $NO_3$ reduction, nitrogen fixation, and transport of dicarboxylic acids.[90] As further examples of regulation under the influence of nitrogen limitation, the transport of ammonium[92,93] and the synthesis of glutamine synthetase GSI may be mentioned.[78,92,93]

## C. SPECIFIC SIGNALS FOR TRANSCRIPTION REGULATION

Studies on infection and nodulation by *Rhizobium* and crown gall formation by *Agrobacterium* have elucidated the important role of host plant flavonoids during the activation of specific bacterial genes. These aspects will be extensively discussed elsewhere in this volume. Here the question arises as to whether such flavonoids and the activated genes play a role during later phases of the development of root nodules and still function in the endophytic phase. While studies with *R. meliloti* originally indicated that the proteins NodA and NodC were present in high concentrations in the bacteroids,[110] later studies arrived at the conclusion that the expression of all 11 *nod* genes, as studied by the gusA fusion technique, disappeared during the later steps of nodule and bacteroid formation.[111] In this respect, it is important to know whether the same flavonoids are present inside the nodules as at the surface[8] or whether they are fully glucosylated.[112] There are indications for a function of other genes during the later steps of nodule and bacteroid formation and the induction of the late nodulins.[113,114]

A vast amount of literature is available about signals during infection processes either in the root nodule symbioses or in parasitic infection. Of specific importance are the role of lectins in recognition of cell surface constituents and elicitors for plant reactions like phytoalexin production and hypersensitive reactions. We refer for references to other parts of this volume. It is tempting to suggest that the very close contacts between host and symbionts during the endophytic phase must lead to similar reactions, but at this moment we can only speculate about such possibilities.

## D. DIRECT EFFECTS OF HOST ENZYMES

Notwithstanding the great importance of regulations of the endophytic genes at the level of transcription, we must not forget that other, more direct, host aspects are possible. The cells of an endophytic microorganism have to develop in a hostile environment which contains different host enzymes. Such enzymes may lead to the hydrolysis of the cell wall or even cytoplasmic constituents. In many symbioses, fusion with lysosomes leads to breakdown of the endophytic microorganisms and, although the vacuole around the *Rhizobium* bacteroids itself has been compared with a lysosome, the bacteroids must be protected, either by the PBM or by the presence of antiproteinases, against lysis by host plant enzymes.[25,106]

## E. FORMATION AND DEFORMATION OF CYTOPLASMIC MEMBRANES

Cell wall lysis, both of the host and the endophytic cells, leads to a more close, and sometimes even direct, contact between the cytoplasmic membranes of endophytic cells and their perisymbiont membranes.[115] Besides the changes in endophytic membrane proteins, we must envisage the possibility of an effect on the lipid constitution of the membranes. It is now well established that the constitution of membrane lipids not only influences their

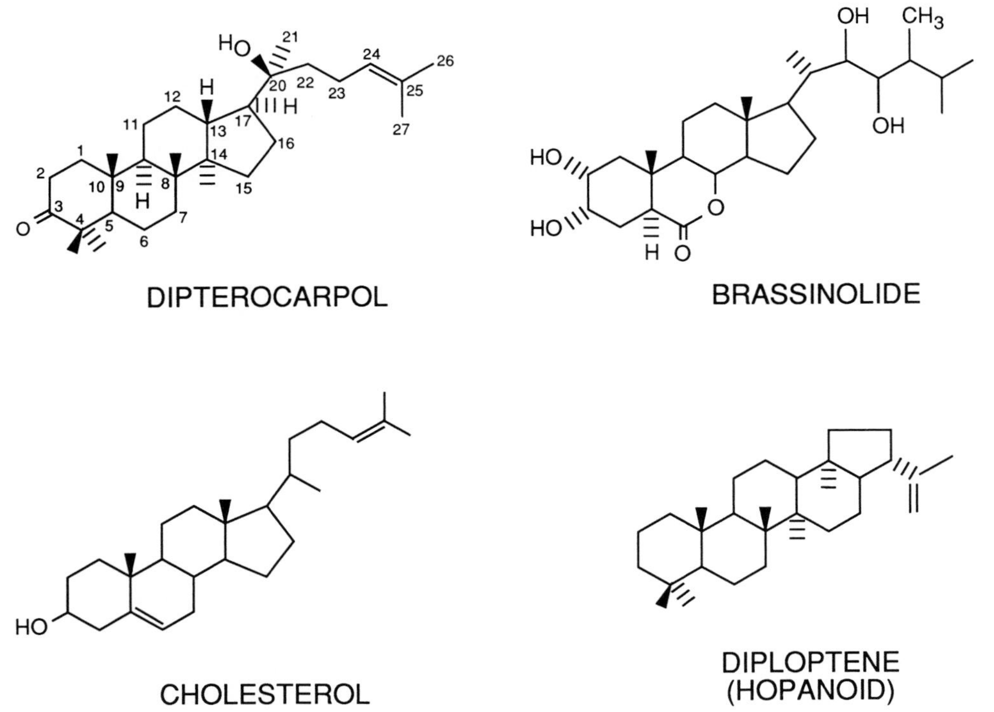


FIGURE 2. The chemical structure of dipterocarpol as compared with related physiologically active substances: the (plant) steroid hormone brassinolide and the membrane constituents cholesterol (in eukaryotic and prokaryotic membranes) and diploptene (belonging to the prokaryotic hopanoids).

physical properties, like the membrane fluidity, but the functioning of receptor sites and the activity of transport and enzyme systems as well.[116] Minor quantities of sterols, which as such can not directly influence membrane structure, can have specific effects on, for example, phospholipid synthesis,[117] and such effects are highly dependent on the composition of the acyl chains of the membrane fatty acids.[118] The only experiments with endophytic organisms, which could be related to this question, are the stimulation of hyphal outgrowth from vesicle clusters of *Frankia* in Dutch spore-negative *Alnus glutinosa* by the triterpenoid dipterocarpol,[32] and the necessity of α-tocopherol in the formation of infection structures in *Ustilago violaceae*.[51] The chemical structure of dipterocarpol resembles those of steroid hormones and many membrane sterols or their prokaryotic equivalents, the hopanoids (Figure 2). Triterpenoids may compete with sterols in eukaryotic membranes and then markedly affect their constitution and function.[119] Dipterocarpol is a consistuent of *Alnus* roots and root nodules and thus may be incorporated in the endophytic membranes. Loss of vitality in nutrient solutions may therefore depend on the absence of dipterocarpol. During a gradual adaptation process the specific necessity of dipterocarpol is replaced by far less specific effects of fatty acids, e.g., in the form of Tween or lecithin.[29] The observation that after full adaptation to nutrient solutions no further additions are necessary if propionic acid is the carbon and energy source while such additions are still needed on media with casamino acids might be an indication for the special requirements of *Frankia* with regard to their lipid metabolism.

It thus is possible that *Frankia* is especially suitable for the study of membrane constitution as an essential aspect of endophytic relations but there can be no doubt that in all endophytic symbioses studies on their membranes will appear to be most promising for our further understanding of the mutual interactions and the signals involved during the new life together.

# REFERENCES


1. **MacGee, P. A., Smith, S. E., and Smith, F. A.,** Research directions on the structure and function of the plant-microbe interface, *Aust., J. Plant Physiol.,* 16, 1, 1989.
2. **Henry, S. M.,** *Symbiosis,* Vol. 1 and 2, Academic Press, San Diego, London, 1966.
3. **Jennings, D. H. and Lee, D. L., Eds.,** Symbiosis, *Symp. Soc. Exp. Biol.,* 29, 1975.
4. **Wiersner, W. and Lorenzen, H., Eds.,** Intrazellulare und Interzellulare Erkennungs und Regulationsmechanismen in Algen und Symbiosen, *Ber. Deut. Bot. Ges.,* 94, 325, 1981.
5. **Linskens, H. F. and Heslop-Harrison, J., Eds.,** Cellular Interactions, *Encyclop. Plant Physiol.,* N.S., 17, 1984.
6. **Ahmadjian, V. and Hale, M. E., Eds.,** The Lichens, Academic Press, San Diego, 1973.
7. **Djordjevic, M. A., Gabriel, D. W., and Rolfe, B. G.,** *Rhizobium*-the refined parasite of legumes, *Annu. Rev. Plant Physiol.,* 25, 145, 1987.
8. **Quispel, A.,** Bacteria-plant interactions in symbiotic nitrogen fixation, *Physiol. Plant,* 74, 783, 1988.
9. **Greshoff, P. M., Newton, W. E., Roth, E. E., and Stacey, G.,** Eds., Nitrogen Fixation: Achievements and Objectives, Chapman and Hall, NY, 1990, in press.
10. **Schwintzer, Chr. R. and Tjepkema, J. D.,** *The Biology of Frankia and Actinorhizal Plants,* Academic Press, San Diego, 1990.
11. **Peters, G. A. and Meeks, J. C.,** The *Azolla-Anabaena* symbiosis: basic biology, *Annu. Rev. Plant Physiol. Plant Mol. Biol.,* 40, 193, 1989.
12. **Mayer, A. M.,** Plant-fungal interactions: a plant physiologists viewpoint, *Phytochemistry,* 28, 311, 1989.
13. **Harley, J. L. and Smith, S. E.,** *Mycorrhizal Symbiosis,* Acad. Press, San Diego, 1983.
14. **Smith, S. E. and Gianinazzi-Pearson, V.,** Physiological interactions between symbionts in vesicular-arbuscular mycorrhizal plants, *Annu. Rev. Plant Physiol. Plant Mol. Biol.,* 39, 221, 1988.
15. **Siegel, M. R., Latch, G. C. M., and Johnson, M. C.,** Fungal endophytes of grasses, *Annu. Rev. Phytopathol.,* 25, 293, 1987.
16. **Trench, R. K.,** The cell biology of plant-animal symbioses, *Annu. Rev. Plant Physiol.,* 30, 485, 1979.
17. **de Faria, S. M., MacInroy, S. G., and Sprent, J. I.,** The occurrence of infected cells with persistent infection threads in legume root nodules, *Can. J. Bot.,* 65, 553, 1987.
18. **Smith, S. E. and Smith, F. A.,** Structure and function of the interface in biotrophic symbiosis and their relation to nutrient transport, *New Phytol.,* 114, 1, 1990.
19. **Kacperska, A. and Huss-Danell, L.,** Purity of *Frankia* preparations from root nodules of *Alnus incana, Physiol. Plant,* 71, 489, 1987.
20. **Day, D. A., Price, G. D., and Udvardi, M. K.,** Membrane interface of the *Bradyrhizobium-Glycine max* symbiosis: peribacteroid units from soybean nodules, *Aust. J. Plant Physiol.,* 16, 69, 1989.
21. **Bradley, D. J., Wood, E. A., Larkins, A. P., Galfre, G., Butcher, G. W., and Brewin, N. J.,** Isolation of monoclonal antibodies reacting with peribacteroid membranes and other components of pea root nodules containing *Rhizobium leguminosarum, Planta,* 1973, 149, 1988.
22. **Werner, D., Mörschel, E., Garbers, Chr., Bassarab, S., and Mellor, R. B.,** Particle density and protein composition of the peribacteroid membrane from soybean root nodules is affected by mutation in the microsymbiont *Bradyrhizobium japonicum, Planta,* 174, 263, 1988.
23. **Bassarab, S., Schenk, S. U., and Werner, D.,** Fatty acid composition of the peribacteroid membrane and the ER in nodules of *Glycine max* varies after infection by different strains of the microsymbiont *Bradyrhizobium japonicum, Bot. Acta,* 101, 156, 1989.
24. **Fortin, M. G., Morrison, N. A., and Verma, D. P.,** Nodulin 26, a peribacteroid membrane nodulin, is expressed independently of the development of the peribacteroid compartment, *Nucleic Acids Res.,* 15, 813, 1987.
25. **Mellor, R. B.,** Bacteroids in the *Rhizobium*-legume symbiosis inhabit a plant internal lytic compartment, implication for other microbiobial endosymbionts, *J. Exp. Bot.,* 40, 831, 1989.
26. **MacDermott, T. R., Graham, P. H., and Brandwein, D. H.,** Viability of *Bradyrhizobium japonicum* bacteroids, *Arch. Microbiol.,* 148, 100, 1987.
27. **Schultz, N. A. and Benson, D. R.,** Developmental potential of *Frankia* vesicles, *J. Bacteriol.,* 171, 6873, 1989.
28. **Zimmerman, W. J., Rosen, B. H., and Lumpkin, Th. A.,** Enzymatic, lectin and morphological characterization and classification of presumptive Cyanobionts from *Azolla* Lam., *New Phytol.,* 113, 497, 1989.
29. **Burggraaf, A. J. P.,** Isolation cultivation and characterization of Frankia strains from actinorhizal root nodules, Ph.D. thesis, University of Leiden, Leiden.
30. **Torrey, J. G.,** Endophyte sporulation in root nodules of actinorhizal plants, *Physiol. Plant,* 70, 279, 1987.
31. **Nazaret, S., Simonet, P., Normand, Ph., and Bardin, R.,** Genetic diversity among *Frankia* isolated from *Casuarina* nodules, *Plant Soil,* 118, 241, 1989.

32. **Quispel, A., Baerheim Svendsen, A., Schripsema, J., Baas, W. J., Erkelens, C., and Lugtenburg, J.,** Identification of dipterocarpol as isolation factor for the induction of primary isolation of *Frankia* from root nodules of *Alnus glutinosa* (L.) Gaertn., *Mol. Plant-Microbe Interact.*, 2, 107, 1989.
33. **Paula, M. A. and Siqueira, J. O.,** Stimulation of hyphal growth of the VA mycorrhizal fungus *Gigaspora margarita* by suspension-cultured *Pueraria phaseoloides* cells and cell products, *New Phytol.*, 115, 69, 1990.
34. **Burggraaf, A. J. P. and Beringer, J. E.,** Absence of nuclear DNA synthesis in vesicular-arbuscular mycorrhizal fungi during *in vitro* development, *New Phytol.*, 111, 25, 1989.
35. **Galun, M. and Bubrick, P.,** Physiological interactions between the partners of the lichen symbiosis, *Encyclop. Plant Physiol.*, 362, 1984.
36. **Petit, P.,** Phytolectins from the nitrogen-fixing lichen *Peltigera horizontalis:* the binding pattern of primary protein extracts, *New Phytol.*, 91, 705, 1982.
37. **Bubrick, P., Galun, M., Ben-Yaacou, M., and Frensdorff, A.,** Antigenic similarities and differences between symbiotic and cultured phycobionts from the lichen *Xanthoria parietina, FEMS Microbiol. Lett.*, 13, 435, 1982.
38. **Tredici, M. R., Margheri, M. C., Giovanetti, L., de Philippis, R., and Vincenzini, M.,** Heterotrophic and diazotrophic growth of *Nostoc* sp. from *Cycas circinalis, Plant Soil,* 110, 199, 1988.
39. **Marten, S., Brandt, P., and Wiessner, W.,** On the developmental dependence between *Cyanophora paradoxa* and *Cyanocyta Korschikoffiana* in symbiosis, *Planta,* 155, 190, 1982.
40. **Stewart, W. D. P. and Rowell, P.,** Modifications of nitrogen-fixing algae in lichen symbioses, *Nature (London),* 265, 371, 1977.
41. **Reding, H. K. and Lepo, J. E.,** Physiological characteristics of dicarboxylate induced pleiomorphic forms of *Bradyrhizobium japonicum, Appl. Environ. Microbiol.*, 55, 666, 1989.
42. **van Brussel, A. A. N., Planque, K., and Quispel, A.,** The wall of *Rhizobium leguminosarum* in bacteroids and free-living forms, *J. Gen. Microbiol.*, 101, 51, 1977.
43. **Planque, K., van Nierop, J. J., Burgers, A., and Wilkinson, S. G.,** The lipopolysaccharide of free-living and bacteroid-forms of *Rhizobium leguminosarum, J. Gen. Microbiol.*, 110, 151, 1979.
44. **VandenBosch, K. A., Brewin, N. J., and Kannenberg, E. L.,** Developmental regulation of a *Rhizobium* cell-surface during growth of pea root nodules, *J. Bacteriol.*, 171, 4537, 1989.
45. **Kannenberg, E. L. and Brewin, N. J.,** Expression of cell surface antigen from *Rhizobium leguminosarum* 3841 is regulated by oxygen and pH, *J. Bacteriol.*, 171, 4543, 1989.
46. **Sindhu, S. S., Brewin, N. J., and Kannenberg, E. L.,** Immunological analysis of lipopolysaccharides from free-living and endosymbiotic forms of *Rhizobium leguminosarum, J. Bacteriol.*, 172, 1804, 1990.
47. **Miller, I. M. and Reporter, M.,** Bacterial leaf symbiosis in *Dioscorea sansibarensis:* morphology and ultrastructure of the acuminate leaf glands, *Plant Cell Environ.*, 10, 413, 1987.
48. **Racette, S. and Torrey, J. G.,** Root nodule initiation in *Gymnostoma* (Casuarinaceae) and *Shepherdia* (Eleagnaceae) induced by *Frankia* strain HFPG PI1, *Can. J. Bot.*, 67, 2873, 1989.
49. **van Dijk, C.,** Spore formation and endophyte diversity in root nodules of *Alnus glutinosa* (L.) Vill., *New Phytol.*, 81, 601, 1978.
50. **Grambow, H. J. and Riedel, S.,** The effect of morphologically active factors from host and non-host plants on the in vitro differentiation of infection structures of *Puccinia graminis* f. sp. *tritici, Physiol. Plant Pathol.*, 11, 213, 1977.
51. **Castle, A. J. and Day, A. W.,** Isolation and identification of α-tocopherol as an inducer of the parasitic phase of *Ustilago violaceae, Phytopathology,* 74, 1194, 1984.
52. **Muscatine, L., Pool, R. R., and Cernichiari, E.,** Some factors influencing soluble organic material by zooxanthellae from reef corals, *Marine Biol.*, 13, 298, 1972.
53. **Gallop, A.,** Evidence for the presence of a "factor" in *Elysia viridis* which stimulates photosynthate release from its symbiotic chloroplasts, *New Phytol.*, 73, 1111, 1974.
54. **Mews, L. K.,** The green *Hydra* symbiosis III. The biotrophic transport of carbohydrate from alga to animal, *Proc. R. Soc. London Ser. B,* 209, 377, 1980.
55. **Quispel, A.,** The mutual relations between algae and fungi in lichens, *Rec. Trav. Bot. Neer.*, 40, 413, 1943.
56. **Feige, G. B.,** Untersuchungen zur Physiologie der Cephalodien der Flechte *Peltigera aphtosa* (L.) Willd. II. Das photosynthetische $^{14}C$-Markierungsmuster und der Kohlenhydrattransfer zwischen Phycobiot und Mycobiot, *Z. Pflanzenphysiol.*, 80, 386, 1976.
57. **Ray, T. B., Mayne, B. C., Peters, G. A., and Toia, R. E.,** *Azolla-Anabaena* relationship. VIII. Photosynthetic characterization of the association and individual partners, *Plant Physiol.*, 64, 791, 1979.
58. **Cernichiari, E., Muscatine, L., and Smith, D. C.,** Maltose excretion by the symbiotic algae of *Hydra viridis, Proc. R. Soc. London Ser. B.*, 173, 557, 1969.
59. **Richardson, D. H. S.,** Photosynthesis and carbohydrate movement, in *The Lichens,* Ahmadjian, V. and Hale, M. E., Eds., Academic Press, San Diego, 1973, 249.

60. **Drew, E. A. and Smith, D. C.**, Studies in the physiology of lichens. VII. The physiology of the *Nostoc* symbiont of *Peltigera polydactyla* compared with cultures and free-living forms, *New Phytol.*, 66, 379, 1967.
61. **Chambers, S., Morris, M., and Smith, D. C.**, Lichen physiology. XV. The effect of digitonin and other treatments on biotrophic transport of glucose from alga to fungus in *Peltigera polydactyla, New Phytol.*, 76, 485, 1976.
62. **MacKay, I. A., Dilworth, M. J., and Glenn, A. R.**, $C_4$-dicarboxylate metabolism in free-living and bacteroid forms of *Rhizobium leguminosarum* MNF 3481, *J. Gen. Microbiol.*, 134, 1433, 1988.
63. **Finan, T. M., Wood, J. M., and Jordan, D. C.**, Symbiotic properties of $C_4$-dicarboxylic acid transport mutants of *Rhizobium leguminosarum, J. Bacteriol.*, 154, 1403, 1983.
64. **LaFontaine, P. J., LaFrenière, C., and Antour, H.**, Some properties of carbohydrate and $C_4$-dicarboxylic acid utilization negative mutants of *Rhizobium leguminosarum* biovar *phaseoli* strain P 121, *Plant Soil,* 120, 195, 1989.
65. **O'Gara, F. and Shanmugam, K. T.**, Regulation of nitrogen fixation by rhizobia, export of fixed $N_2$ as $NH_4^+$, *Biochim. Biophys. Acta,* 437, 313, 1976.
66. **Szeto, W. W., Nixon, B. T., Ronson, Cl. E., and Ausubel, F. M.**, Identification and characterization of the *Rhizobium meliloti ntr* C gene: *R. meliloti* has separate regulatory pathways of nitrogen fixation genes in free-living and symbiotic cells, *J. Bacteriol.*, 169, 1423, 1987.
67. **Stouthamer, A. H., Stam, H., de Vries, W., and van Vlerken, M.**, Some aspects of nitrogen fixation in free-living cultures of *Rhizobium,* in *Nitrogen Fixation: Hundred Years After,* Bothe, H., de Bruijn, F. J., and Newton, W. E., Eds., Gustav Fischer, Stuttgart, 1988, 257.
68. **Dakora, F. D. and Atkins, C. A.**, Diffusion of oxygen in relation to structure and function in legume root nodules, *Aust. J. Plant Physiol.*, 16, 131, 1989.
69. **Ditta, G., Virts, E., Palomares, A., and Kim, C. H.**, The *nif* A gene of *Rhizobium meliloti* is oxygen regulated, *J. Bacteriol.*, 169, 3217, 1987.
70. **Fischer, H. M. and Hennecke, H.**, Direct response of *Bradyrhizobium japonicum nif* A-mediated *nif* gene regulation to cellular oxygen status, *Mol. Gen. Genet.*, 209, 621, 1987.
71. **De Philip, P., Batut, J., and Boistard, P.**, *Rhizobium meliloti Fix* L is an oxygen sensor and regulates *R. meliloti nif* A and *fix* K genes differently in *Escherichia coli, J. Bacteriol.*, 172, 4255, 1990.
72. **Keithly, J. H. and Nadler, K. D.**, Protoporphyrin formation in *Rhizobium japonicum, J. Bacteriol.*, 154, 838, 1983.
73. **Laane, C., Krone, W., Konings, W. N., Haaker, H., and Veeger, C.**, The involvement of the membrane potential in nitrogen fixation by bacteroids, *FEBS Lett.*, 103, 328, 1980.
74. **Gober, J. W. and Kashket, E. R.**, $K^+$ regulates bacteroid associated functions of *Bradyrhizobium, Proc. Natl. Acad. Sci. U.S.A.*, 84, 4650, 1987.
75. **Kahn, D., David, M., Dommergue, O., Daveran, M. L., Ghai, J., Hirsch, P. R., and Batut, J.**, *Rhizobium meliloti fix* GHI sequence predicts involvement of a specific cation pump in symbiotic nitrogen fixation, *J. Bacteriol.*, 171, 929, 1989.
76. **Vasse, J., de Billy, F., Camut, S., and Truchet, G.**, Correlation between ultrastructural differentiation of bacteroids and nitrogen fixation in alfalfa nodules, *J. Bacteriol.*, 172, 4295, 1990.
77. **Tsai, Y. L. and Benson, D. R.**, Physiological characteristics of glutamine synthetase I and II of *Frankia* sp. strain CpI, *Arch. Microbiol.*, 152, 381, 1989.
78. **Ludwig, R. A. and de Vries, G. E.**, Biochemical physiology of *Rhizobium* dinitrogen fixation, in *Nitrogen Fixation,* Vol. 4, Broughton, W. J. and Pühler, A., Eds., Clarendon Press, Oxford, 1986, 50.
79. **Cock, J. M., Mould, R. M., Bennett, M. J., and Cullimore, J. V.**, Expression of glutamine synthetase genes in roots and nodules of *Phaseolus vulgaris* following changes in the ammonium supply and infection with various *Rhizobium* mutants, *Plant Mol. Biol.*, 14, 549, 1990.
80. **Rahav, O., Dubinsky, Z., Achituv, Y., and Falkowski, P. G.**, Ammonium metabolism in the zooxanthellate coral *Stylophora pistillata, Proc. R. Soc. London Ser. B,* 236, 325, 1989.
81. **Rees, T. A. V. and Ellard, F. M.**, Nitrogen conservation and the green *Hydra* symbiosis, *Proc. R. Soc. London Ser. B,* 236, 203, 1989.
82. **Lyons, Ph.C., Evans, J. J., and Bacon, Ch. W.**, Effects of the fungal endophyte *Acremonium coenophialum* on nitrogen accumulation and metabolism of tall fescue, *Plant Physiol.*, 92, 726, 1990.
83. **de Vries, G. E., van Brussel, A. A. N., and Quispel, A.**, Mechanism and regulation of glucose transport by *Rhizobium leguminosarum, J. Bacteriol.*, 149, 872, 1982.
84. **Price, G. D., Day, D. A., and Greshoff, P. M.**, Rapid isolation of intact peribacteroid envelopes from soybean nodules and demonstration of selective permeability to metabolites, *J. Plant Physiol.*, 130, 157, 1987.
85. **Udvardi, M. K., Price, G. D., Greshoff, P. M., and Day, D. A.**, A dicarboxylate transporter on the peribacteroid membrane of soybean nodules, *FEBS Lett.*, 231, 36, 1988.
86. **Udvardi, M. K., Salom, C. S., and Day, D. A.**, Transport of 1-glutamate across the bacteroid membrane but not the peribacteroid membrane from soybean root nodules, *Mol. Plant-Microbe Interact.*, 1, 250, 1988.

87. **Herrada, G., Puppo, A., and Rigaud, J.,** Uptake of metabolites by bacteroid-containing vesicles and by free bacteroids from French bean nodules, *J. Gen. Microbiol.,* 135, 3165, 1989.
88. **Patrick, J. W.,** Solute efflux from the host at plant-microbe interfaces, *Aust. J. Plant Physiol.,* 16, 53, 1989.
89. **Staal, H.,** Amino acid transport in *Rhizobium leguminosarum,* Ph.D. thesis, University of Leiden, Leiden, 1985.
90. **Ronson, C. W., Nixon, B. T., Albright, L. N., and Ausubel, F. M.,** *Rhizobium meliloti ntr* A (*rpo* N) gene is required for diverse metabolic functions, *J. Bacteriol,* 169, 2424, 1987.
91. **Birkenhead, K., Nooman, B., Reville, W. J., Boerten, B., Manian, S. S., and O'Gara, F.,** Carbon utilization and regulation of nitrogen fixation genes in *Rhizobium meliloti, Mol. Plant-Microbe Interact.,* 3, 167, 1990.
92. **O'Hara, G. W., Riley, G. W., Glenn, A. R., and Dilworth, M. J.,** The ammonium permease of *Rhizobium leguminosarum* MNF 3841, *J. Gen. Microbiol.,* 131, 757, 1985.
93. **Gober, J. W. and Kashket, E. R.,** Methylammonium uptake by *Rhizobium* sp. strain 32H1, *J. Bacteriol.,* 153, 1196, 1983.
94. **Streeter, J. G.,** Estimation of ammonium concentration in the cytosol of soybean nodules, *Plant Physiol.,* 90, 779, 1989.
95. **MacRae, D. G., Miller, R. W., Berndt, W. D., and Joy, K.,** Transport of $C_4$-dicarboxylates and amino acids by *Rhizobium meliloti* bacteroids, *Mol. Plant-Microbe Interact.,* 2, 273, 1989.
96. **Akkermans, A. D. L., Huss-Danell, K., and Roelofsen, W.,** Enzymes of the tricarboxylic acid cycle and the malate-aspartate shuttle in the $N_2$-fixing endophyte of *Alnus glutinosa, Physiol. Plant,* 53, 289, 1981.
97. **Mazzucco, C. and Benson, D. R.,** $^{14}$C-Methylammonium transport by *Frankia* sp. strain CpI1, *J. Bacteriol.,* 160, 636, 1989.
98. **MacAuley, P. J.,** Uptake of amino acids by cultured and freshly isolated symbiotic *Chlorella, New Phytol.,* 104, 415, 1986.
99. **MacAuley, P. J.,** Uptake of leucine by *Chlorella* symbionts of green *Hydra, Proc. R. Soc. London Ser. B,* 234, 319, 1988.
100. **MacAuley, P. J.,** The effect of arginine on rates of internalization of other amino acids by symbiotic *Chlorella* cells, *New Phytol.,* 112, 553, 1989.
101. **Smith, D.,** Regulation and change in symbiosis, *Ann. Bot.,* 60, 115, 1987.
102. **Hill, D. J.,** The control of the cell cycle in microbial symbionts, *New Phytol.,* 112, 175, 1989.
103. **Douglas, A. and Smith, D. C.,** The green *Hydra* symbiosis, VIII, Mechanisms in symbiont regulation, *Proc. R. Soc. London Ser. B,* 221, 291, 1984.
104. **Rees, T. A. V.,** The green *Hydra* symbiosis and ammonium, 1. the role of the host in ammonium assimilation and its possible regulatory significance, *Proc. R. Soc. London Ser. B,* 229, 299, 1986.
105. **MacAuley, P. J. and Smith, D. C.,** The green *Hydra* symbiosis. VII. Conservation of the host cell habitat by the symbiotic algae, *Proc. R. Soc. London Ser. B,* 217, 415, 1982.
106. **Pladys, D. and Rigaud, J.,** Lysis of bacteroids *in vitro* and during the senescence in *Phaseolus vulgaris* nodules, *Plant Physiol. Biochem.,* 26, 179, 1988.
107. **Allen, G. C. and Elkan, G. H.,** Growth, Respiration, and polypeptide patterns of *Bradyrhizobium* sp. *(Arachis)* strain 3G4b20 from succinate- or oxygen-limited continuous cultures, *Appl. Environ. Microbiol.,* 56, 1025, 1990.
108. **Gober, J. W. and Kashket, E. R.,** Role of DNA superhelicity in regulation of bacteroid-associated functions of *Bradyrhizobium* sp. strain 32H1, *Appl. Environ. Microbiol.,* 55, 1420, 1989.
109. **Raven, J. A., Franco, A. A., de Jesus, E. L., and Jacob-Neto, J.,** $H^+$ extrusion and organic acid synthesis in $N_2$-fixing symbioses involving vascular plants, *New Phytol.,* 114, 369, 1990.
110. **John, M., Schmidt, J., Wieneke, U., Krüssman, H. D., and Schell, J.,** Transmembrane orientation and receptor-like structure of the *Rhizobium meliloti* common nodulation protein *Nod* C., *EMBO J.,* 7, 583, 1988.
111. **Sharma, S. B. and Signer, E. R.,** Temporal and spatial regulation of the symbiotic genes of *Rhizobium meliloti* in planta revealed by transposon Tn5-gusA, *Genes Dev.,* 4, 344, 1990.
112. **Vickery, M. L. and Vickery, B.,** *Secondary Plant Metabolism,* MacMillan, NY, 1981, 183.
113. **Rossbach, S., Gloudemans, T., Bisseling, T., Stader, B., Kalluza, B., Ebeling, S., and Hennecke, H.,** Genetic and physiologic characterization of a *Bradyrhizobium japonicum* mutant defective in early bacteroid development, *Mol. Plant-Microbe Interact.,* 2, 233, 1989.
114. **Nap, J. P., van de Wiel, C., Spaink, H. P., Moorman, M., van den Heuvel, M., Djordjevic, M. A., van Lammeren, A. A. M., van Kammen, A., and Bisseling, T.,** The relationship between nodulin gene expression and the *Rhizobium nod* genes in *Vicia sativa* root nodule development, *Mol. Plant-Microbe Interact.,* 2, 56, 1989.
115. **Bradley, D. J., Butcher, G. W., Galfre, G. M., Wood, E. A., and Brewin, N. J.,** Physical association between the peribacteroid membrane and lipopolysaccharide from the bacteroid outer membrane in *Rhizobium*-infected pea root nodule cells, *J. Cell Sci.,* 85, 97, 1986.

116. **Yeagles, Ph. L.,** Lipid regulation of cell membrane structure and function, *FASEB J.*, 3, 1833, 1989.
117. **Kawasawki, S., Rampogal, M., Chin, J., and Bloch, K.,** Sterol control of the phosphatidylethanolamine-phosphatidylcholine choline conversion in the yeast mutant GL7, *Proc. Natl. Acad. Sci. U.S.A.*, 82, 5715, 1085.
118. **Rilfors, L., Wikander, G., and Wieslander, A.,** Lipid acyl chain-dependent effects of sterols in *Acheloplasma laidlawii* membranes, *J. Bacteriol.*, 169, 830, 1987.
119. **Nes, W. D. and Heftmann, E.,** A comparison of triterpenoids with steroids as membrane components, *J. Nat. Prod.*, 44, 377, 1981.

# SIGNALS AND REGULATION IN THE DEVELOPMENT OF *STRIGA* AND OTHER PARASITIC ANGIOSPERMS

**James L. Riopel and Michael P. Timko**

## TABLE OF CONTENTS

# I. INTRODUCTION

Parasitic angiosperms comprise an interesting group of plants. There are about 3000 known species, all of them dicots with the exception of the gymnosperm, *Parasitaxus ustus.*[1] Several reviews of these plants exist that provide details of their biology.[2-5] It is a highly diverse group linked by a single phenotypic expression, the development of the haustorium. This is a multicellular structure which is unlike its fungal counterpart in cell specialization and structural organization. It functions to convey both water and nutrients from its host.

Parasitic plants attach to their hosts either at foliar or root regions. This chapter reviews recent progress in understanding how the development of the root parasite is cued to specific chemical signals contained in the rhizosphere of neighboring plants. Exogenous signals have been identified for two events, seed germination and the initiation of the haustorium. Most of our information comes from recent studies of *Striga asiatica* (witchweed). Impetus for much of this work has come from the serious economic impact of *Striga* species on cereal crops in several regions of the world, especially in Third World countries. Witchweed is also present in a small area of the Carolinas where it has been contained by a vigorous U.S. Department of Agriculture eradication program.[6]

# II. GERMINATION

Seeds of most root parasites germinate when moistened or, if they are temperate species, after a vernalization period. For others, germination physiology is far more complex. A specific chemical signal is required and other conditions may also be necessary for germination to proceed.[4,7,8]

It is not surprising that obligate root parasite species, with small seeds and little food reserves, may have evolved mechanisms to utilize host signals to synchronize their germination and further development to the presence of host roots. The requirement for a germination stimulant was first noted by Fuller in 1900[9] and has been known for many years for species of *Alectra,*[10] *Aegenetia,*[11] *Orobanche,*[12] *Tozzia,*[13] and *Striga.*[14]

*Striga asiatica* seeds are 200 to 300 μm long with characteristic ridges on the surface (Figure 1). As many as 500,000 seeds are produced by a single plant and they remain viable in soil for periods up to 10 years or perhaps longer.[14,15] Several conditions are necessary for germination. Seeds of most *Striga* species require a period of afterripening of 1 to several months, followed by a preconditioning period under moist conditions lasting 1 to 2 weeks. Both of these processes can be accelerated by elevated temperatures.[16] Some of the metabolic changes that occur during these stages have recently been examined,[17,18] but the physiological basis of the afterripening and preconditioning requirements remain obscure.

The final requirement for germination is receipt of an exogenous stimulant. A broad range of compounds are known to stimulate germination. They include allylthiourea, coumarin, cytokinins, ethylene, scopeletin, and sodium hypochlorite.[8,19] Ethylene is sometimes used in the U.S. to promote suicidal germination as a control measure for *Striga.*[20] In *Orobanche,* the duration of the exposure to signal necessary to induce germination has been reported to be as brief as 30.[12] Germination and radicle extension then occur in 10 to 12 hr (Figure 1).

There are also naturally occurring germination stimulants present in root exudates. Strigol was isolated from exudates of cotton (*Gossypium hirsutum* L.).[21,22] It is active at concentrations as low as $10^{-12}$ *M* with germination rates of 70 to 90%. Methods of synthesis and the structure of strigol are now well established[23-27]. Strigol is a sesqueterpene containing four rings (labeled A to D, Figure 2A). There have been several comparative studies of strigol and compounds of related structure. Some representative compounds are shown in Figure 2. The results show how small modifications in structure influence germination promotion

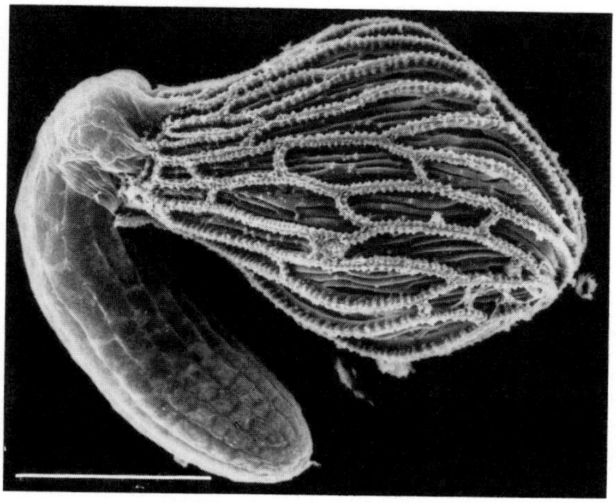

FIGURE 1. Scanning electron micrograph of germinating *Striga asiatica* at 12 h. Bar = 100 μm.)

but, as yet, do not give a structure-functional basis for activity. For example, alterations of the strigol configuration by the addition of a benzene ring to the butenolide side (ring *D*) typically decrease activity (Figure 2B), while promotion is retained by compounds with an added benzene on the A ring (Figure 2C, GR18; 2-D, GR24).[28]

At the USDA Southern Research Center in New Orleans, analogs of both sides of the strigol molecule, the aliphatic terpenoid A ring and the butenolide D ring, have been synthesized and tested.[29,30] Thirteen terpenoids were found to actively promote germination. Most have a trimethylcyclohexene structure (Figure 2 E through J). To generate more stability, the chemically active aldehyde group was replaced in some of the compounds with ester groups. At $10^{-4}$ *M* five of the compounds such as in Figure 2H,I promoted germination of *Striga* or *Orobanche ramosa* (broomrape). Again, a small change in structure can significantly alter function as shown in Figure 2I or J differing only at the C-4 position. Both *Orobanche* and *Striga* germinate with the compound shown in Figure 2I; only *Orobanche* with that in Figure 2J.

The butenolide D ring of strigol is a common constituent of most active compounds. With a few exceptions, however, D-ring analogs tested have low activity. Johnson et al.[25] reported no promotion of *Striga hermonthica* germination unless the butenolide was joined to several different lactone rings through a methyleneoxy bridge (Figure 2K). Pepperman et al.[29] tested several other monocyclic structures. Solubility problems requiring acetone or DMSO in bioassays and instability of the compounds caused problems for interpretation in these studies. Only the ethoxy ring structure proved to be active. Pepperman et al. concluded that the basic D-ring structure is without activity unless joined to other substituents (see Figure 2L).

Recently Fisher et al.[31,32] reported on a number of strigol-like compounds, all sesquiterpene lactones, that stimulate germination of *Striga*. Dihydroparthenolide, a natural product found in *Ambrosia,* stimulated 70% germination at $10^{-9}$ *M*. This level of stimulation is close to the activity of strigol. For the germacramolides, a class of sesquiterpenes, all compounds shown in Figure 3 induced 40 to 65% germination over a range from $10^{-4}$ to $10^{-9}$ *M*. The authors suggest that the conformation of the medium ring skeleton in conjunction with the lactone ring, and not a specific functional constituent, is important to this activity. Sesquiterpene lactones are abundant in plants. With literally thousands already known, these compounds offer a rich source of naturally occurring substances with potential for control measures of *Striga* and perhaps *Orobanche*[32].

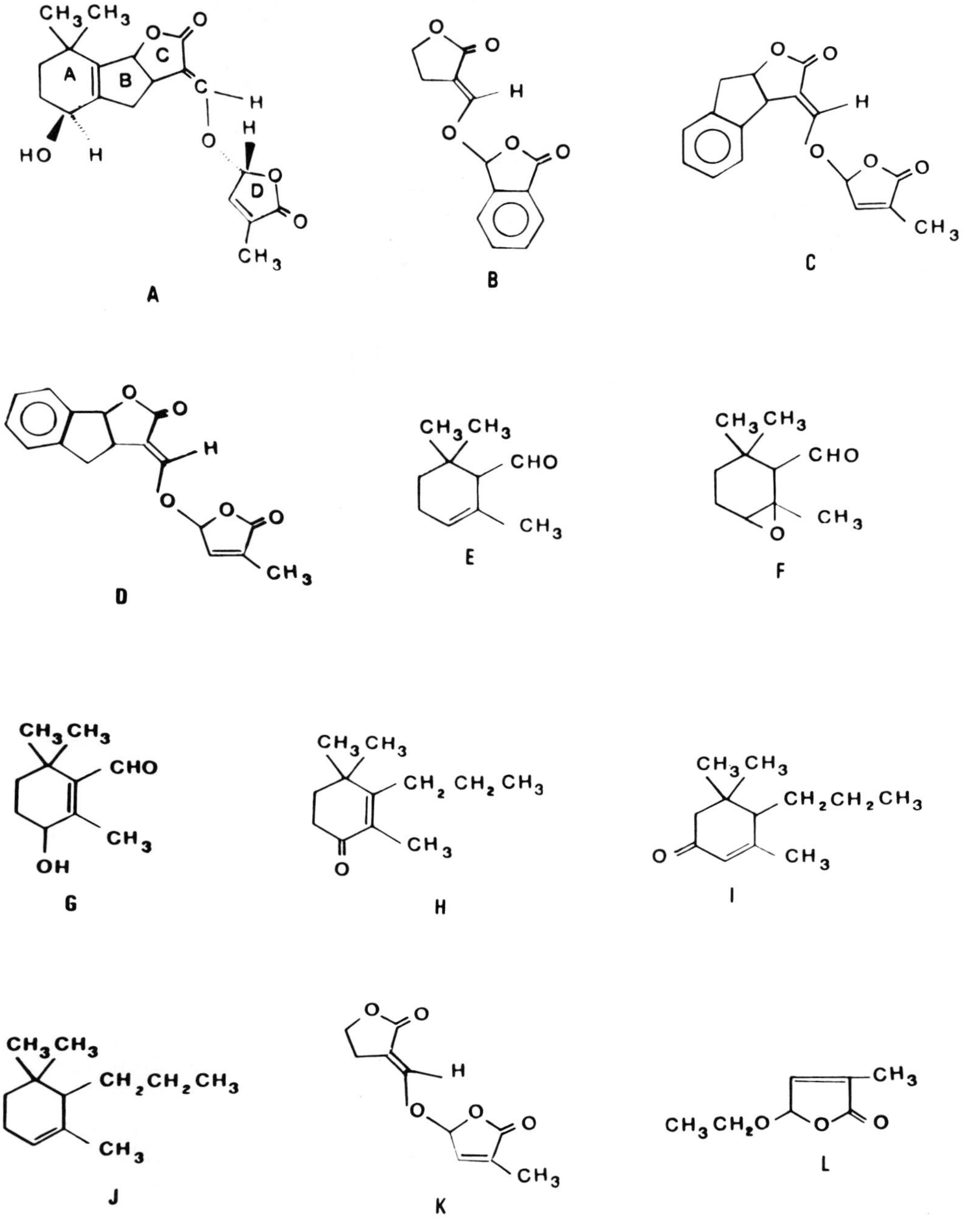

FIGURE 2. Structures of the germination stimulant strigol (A) and related synthesized compounds (B-L) tested for germination activity.

Recently, Netzly et al.,[33] isolated a compound that has now turned out to be the first host germination stimulant known for *Striga*. They noted prominent colored droplets associated with root hairs of sorghum seedlings growing on moistened filter paper in petri plates (Figure 4). Roots dipped in methylene chloride yielded a germination stimulant. Subsequent studies showed that these droplets consist of almost pure concentrations of four *p*-benzoquinones, termed sorgoleone, (Figure 5), which in the dihydroquinone configuration, are labile, but potent germination stimulants[34,35].

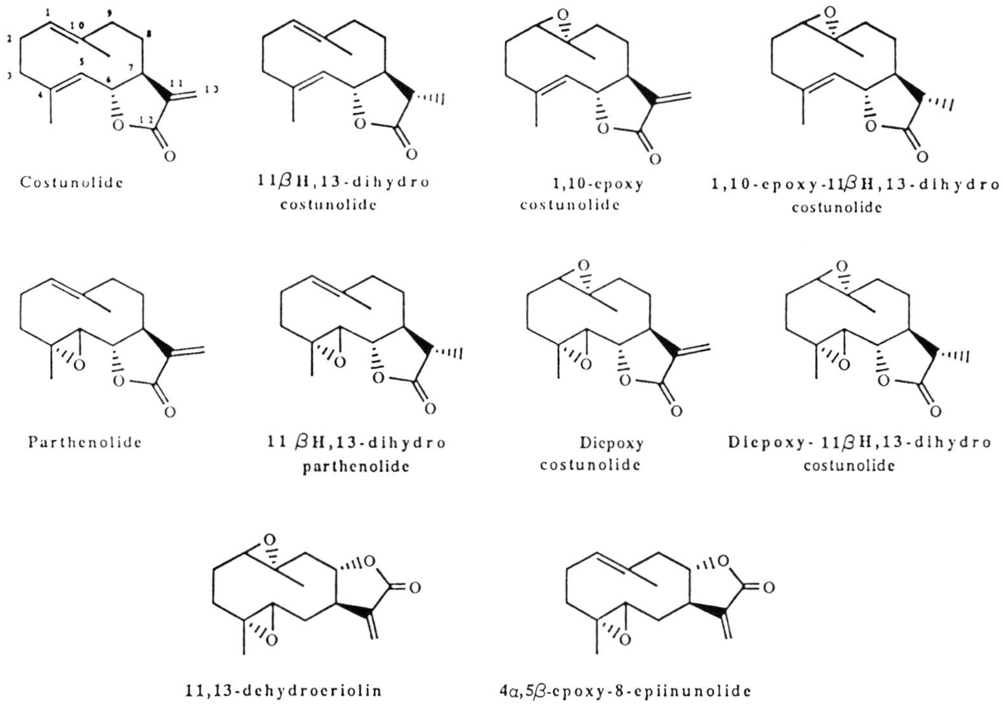


FIGURE 3. Structures of germacranolides active in *Striga* germination.

FIGURE 4. Root hairs of sorghum with droplets containing *Striga* germination factors, the sorgoleones. Bar = 100 μm.)

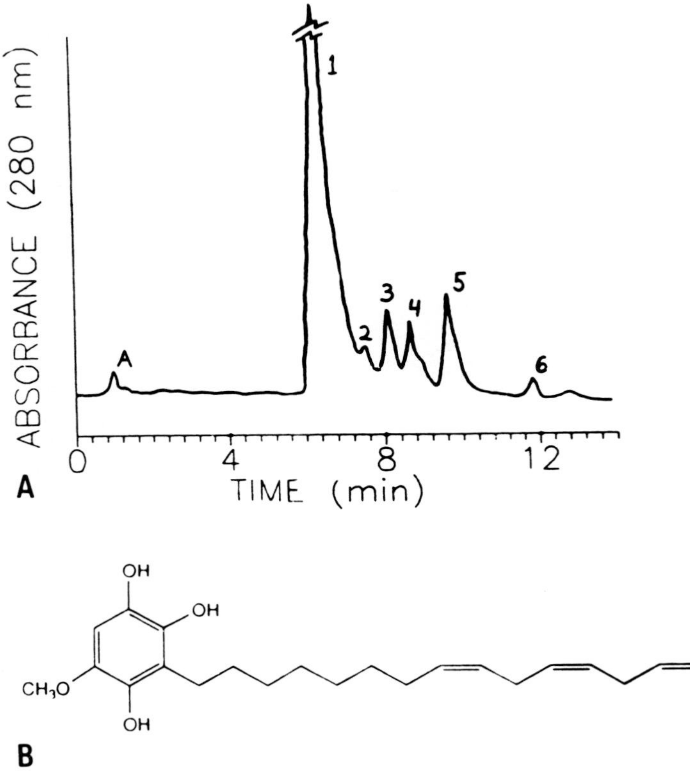


FIGURE 5. Isolation and structure of the sorgoleones. (A) shows an absorbance tracing from an HPLC separation of hydrophobic extract from sorghum root hairs. The major peaks listed represent (1) sorgoleone-358, (2) dihydroquinone of sorgoleone-358, (3) sorgoleone-360, (4) sorgoleone-386, (5) sorgoleone-362, (6) benzoquinone 6; and A, a labile component. (B) shows the molecular structure of the dihydroquinone of sorgoleone-358.

Further experiments were carried out to ascertain the function of sorgoleones for sorghum and corn grown under soil conditions. Soil containing sorghum roots yielded 160 to 230 mg of sorgoleones (equivalent to $10^{-4}$ to $10^{-5}$ mol/l in the soil), but soil containing corn roots showed none. Extractions of soil after roots had been removed yielded sorgoleones only from the sorghum soil. Since corn is a natural host for *Striga,* it is likely that the germination stimulus in corn is more labile or is unrelated chemically to the sorgoleones.

The work of Netzly et al.,is important not only in the discovery of the first host germination stimulant for these important parasites, but also in demonstrating the diffusion of hydrophobic exudate constituents into the rhizosphere. At the moment, the extent of the diffusion and the stability of these compounds in soil is uncertain. The rapid oxidation of the hydroquinone to the inactive quinone, along with its hydrophobic character, results in limited soil mobility and could generate a restricted rhizosphere promotion zone.[35] Such a strategy would be ideal as a host signal specific to *Striga* seeds and optimizing opportunities for host attachment. The discovery of the sorgoleones has provided the basis for a theory[36,37] that the germination zone for *Striga* is defined *exclusively* by a steady state concentration gradient of diminishing signal of the dihydroquinone from the host root surface. This, however, seems unlikely. The existence of several germination stimulants in root exudates including water-soluble fractions is well known [14,38-41] and suggests that quinones may well act in concert with other hydrophilic stimulants.

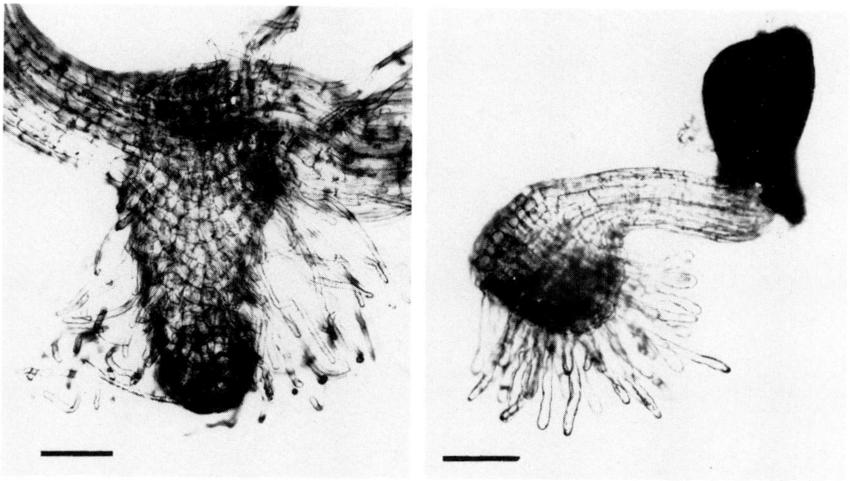

A B

FIGURE 6. (A) Three-week-old secondary haustorium of *Agalinis purpurea*, (B) Two-day-old haustorium of *Striga asiatica* induced by corn root. Bars = 100 μm.)

# III. HAUSTORIAL INDUCTION

For most root parasite species the haustorium is a prominent structure formed at the root tip (primary haustoria) or at lateral positions along the root (secondary haustoria) (Figure 6). The morphology of haustoria varies considerably, but before attachment and penetration of a host they typically appear as conical structures surrounded by abundant haustorial hairs. Development to the stage of attachment competency is one of the most rapid developmental events known for angiosperms. In *Striga* an established haustorium develops in 12 to 24 h.[42] In every case we have examined, the first visible cellular change occurs in the root cortex. Cells that normally elongate in the longitudinal dimension shift to expanding laterally. Thus, the first response is likely to involve localized effects on cell wall plasticity utilizing existing cell resources. Just hours later, haustorial hairs are initiated and cell divisions take place in the cortex and epidermis.

Haustoria are always present on field-collected specimens, but are reduced in frequency or absent on both hemi- and holoparasites grown in sterile culture without a host.[43,44] Observation of this phenomenon in cultures of *S. senegalensis* led Okonkwo[45] to suggest that this may be indirect evidence that a stimulus for the formation of haustoria comes from the host root. To test this idea, we have grown a variety of parasitic plants in culture without hosts. These included species of *Agalinis, Aureolaria, Buchnera, Castilleja,* and *Striga.* Although not extensive, our findings suggest that there is a species gradient in the requirement for the presence of another plant to induce haustoria. For example, *Aureolaria pedicularia* forms abundant haustoria without a host. Spontaneous haustoria occur in *Agalinis purpurea* in about 2% of the population in young cultures (3 to 6 weeks) and about 20 to 25% with older ones (2 to 4 months). In *Striga asiatica* the requirement for an exogenous signal seems absolute for induction of haustoria.

In our search for chemicals that induce haustoria, gum tragacanth, a foliar exudate of *Astragalus gummifer,* was found to be a potent inducer of haustoria in *Agalinis purpurea.*[46] Two phenylpropanoids, xenognosin A and B, were subsequently isolated by Steffens et al.[47] (Figure 7). Xenognosin A proved to be the more active constituent with 86% of the *Agalinis*

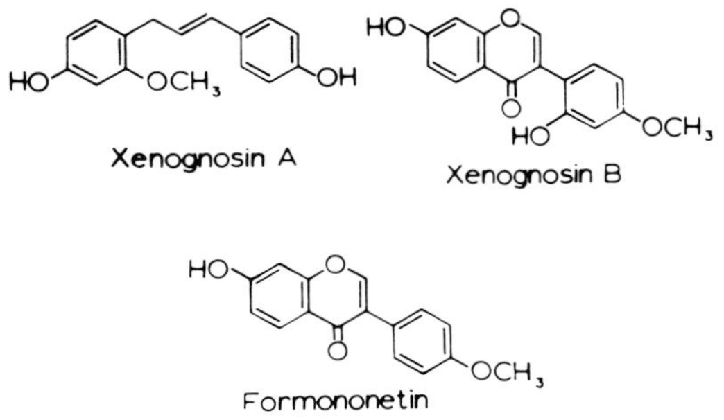


FIGURE 7. Chemical structures of *Agalinis* haustorial factors xenognosin A and B isolated from gum tragacanth and the inactive formononetin.

**TABLE 1**
**Active Substances in Promoting *Striga* Haustoria**

| Pheonlic compounds | Cytokinins |
|---|---|
| Arbutin | 6,Benzylaminopurine |
| 2,6-Dimethoxy-*p*-benzoquinone | 6-Benzylaminopurine riboside |
| Sinapic acid | 6-Benzylmercaptopurine |
| Syringic acid | 6-*N*-hexylaminopurine |
| Vanillin | Dihydrozeatin |
| Vanillic acid | Isopentenyl adenine |
| Isovanillin | Kinetin |
| *O*-Vanillin | Kinetin riboside |
| Vanillin acetate | Zeatin |
| Vanillin azine | |
| Umbelliferone | |
| Xenognosin A | |
| Gum tragacanth (contains xenognosin A and B) | |

plants producing haustoria at a concentration $10^{-4}$ *M* and 41% at $10^{-5}$ *M*. Although haustoria can be induced by numerous compounds (see Table 1), slight changes in chemical structure significantly alter promotive function. For example, formononetin initially was copurified with xenognosin B. It shows no activity, yet differs only by a single hydroxyl group from xenognosin B (Figure 7). The structural specificity of xenognosin A was further probed using a number of analogs. Several features were found to be related to activity. They were: (1) a methoxyl group, (2) a *meta* relationship of hydroxyl and methoxyl groups, and (3) an alkyl branching *ortho* to the methoxyl substituent.

In the *Agalinis* studies, Steffens also isolated a haustorial inducer from roots of *Lespedeza sericea*.[48,49] Surprisingly, it proved not to be a phenylpropanoid but was identified as a pentacyclic triterpene, soyasapogenol B (Figure 8). Soyasapogenol B is quite different in structure from xenognosin A and possesses lower promotive activity. Several constituents in addition to soyasapogenol B were required to reconstitute the full activity of the root extract. However, the presence of these components in root exudate was not established.

Part of the puzzle of understanding the mechanism of haustorial induction is the rather large number of compounds that are promotive.[42,43] Table 1 lists those found just for *Striga*. Also, even though host attachment is critical for survival, *Striga* initiates haustoria against all species of roots we have tried.

Recently 2,6-dimethoxy-*p*-benzoquinone (2,6-DMBQ) was identified as a haustorial inducer for *Striga*[50] (Figure 9). Significantly, this molecule is typically not found in sorghum

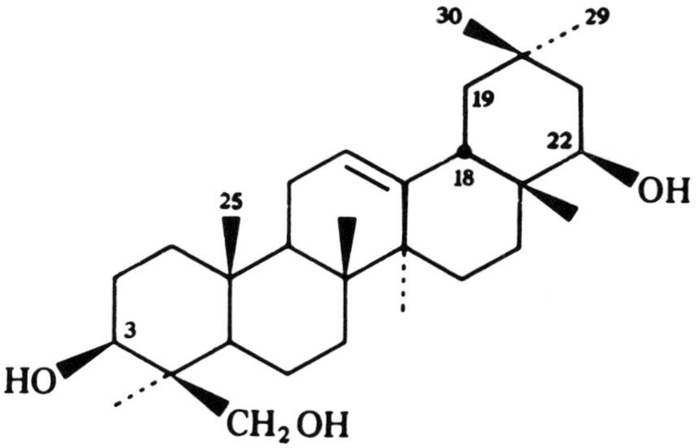


FIGURE 8. Chemical structure of *Agalinis* haustorial factor soyasapogenol B isolated from root extract of *Lespedeza*.

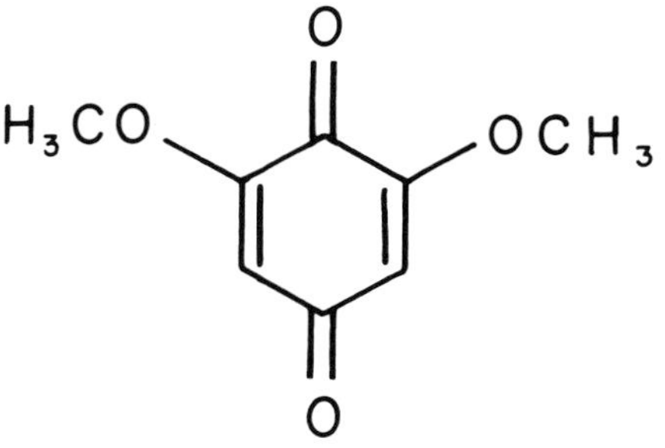


FIGURE 9. The molecular structure of 2,6-dimethoxy-*p*-benzoquinone.

root exudate. Active exudate is produced only when sorghum roots are sonicated or agitated in some way, prior to being used as an inducing substrate.

Using syringaldazine, Chang and Lynn[51] have detected laccase-type enzymes associated with the roots of *Striga asiatica* and *Agalinis*. They theorize that such enzymes released from the parasite root may cleave a quinone (2,6-DMBQ) from the cell wall complex of the host which then diffuses back to *Striga* functioning as host signal. In support of this theory, the authors show that host surface material is converted to the quinone in the presence of *Striga* seedlings. The conversion does not occur when *Striga* roots are washed immediately before the experiment. Lynn and Chang[52] have proposed that haustorial induction in *Striga* and possibly other root parasites results from the conversion of host root surface components, including flavanoids, to active quinones. They further conclude that zones for both *Striga* germination and haustorial induction are determined by a combination of factors including diffusion rates of promoters, the instability of active compounds, and critical threshold concentrations and exposure times required for induction.[53] As unifying concepts, these are attractive theories that will become clearer as more root parasites are examined and present questions are addressed.

In a recent study[54] we have looked at 25 strains of sorghum identified as potentially resistant to *Striga*. The procedure was to suspend *Striga* seeds in agar before it jelled, then introduce a sorghum seedling. Plates were fixed after either 4 or 12 days and germination and haustorial development along with other features of the interaction of host and parasite were examined. One resistant variety, P967083* (P96), was studied in detail in comparison

* Provided by Dr. Gebisa Ejita, Department of Agronomy, Purdue University, West Lafayette, IN.

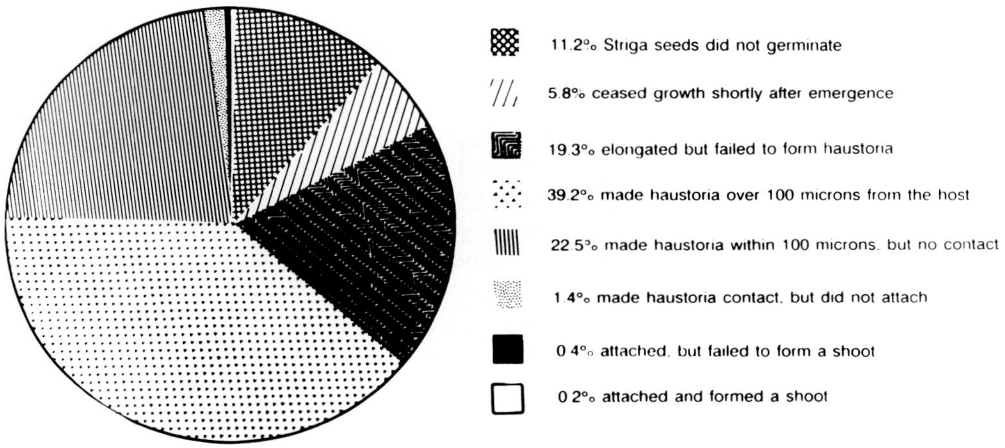


FIGURE 10. Fate profile of *Striga* responses in a 2-mm zone from roots of resistant sorghum cultivar P967083.

with Dabar, a control, susceptible variety. The findings of this work point to several important overall features of signal regulation in *Striga*. They are as follows:

1. Three zones of promotion are well defined. A germination zone extends to about 2 cm and, in some cases, further, with over 90% germination near the host root. A zone of positive chemotropism for *Striga* radicles extends to about 5- to 6-mm from the host root. The third zone for haustorial induction is the most narrow. No haustoria are observed beyond 3.4 mm for Dabar and 3.8-mm for P96. The average radicle length for Striga in these studies is 2.1 mm. Elongation is inhibited in *Striga* radicles closer than 3-mm from the host root. Radicles are occasionally 3- to 4-mm long outside of the 3-mm zone.
2. Within the haustorial zone there are significant differences between Dabar and P96. The haustorial zone in P96 extends further from the host root and there is a 33% increase in the number of haustoria. The increased frequency, however, is generated by the precocious development of haustoria positioned in the outer region with no opportunity to attach to the host surface. Formation of haustoria further from the host in P96 results in an average haustoria-host distance of 864 μm compared with 422 μm for Dabar. It is important to emphasize these differences exist only in the outer region of the promotion zone. Within 300 μm of the host surface, no significant differences were observed in haustorial frequency.
3. The most important finding of these studies relates to what happens to haustoria as they approach the host surface. A careful examination of haustoria within 300 μm of the host shows that in the resistant variety P96 many haustoria stop just short of the host root surface. In the susceptible variety Dabar, there were 16.4 contacts per petri plate for primary roots and only 4.2 contacts for P96. There is a futher reduction in success of attachment as measured by *Striga* shoot formation. This study shows an attachment success for Dabar of 1.7% compared with only 0.2% for P96. A profile of the overall fate of *Striga* seeds for a 2-mm zone next to host sorghum P96 is summarized in Figure 10.

At present, the basis for diminution in successful contacts is not known but possibly points to yet further dimensions of regulation operative just before contact is established. To what extent these signals are mitigated by the soil environment has not been determined. Obviously, there are many aspects of these plate experiments that do not simulate the physical and chemical environment of soil conditions. The absence of microbial degradation of metabolites and differences in diffusion rates would contribute in significant ways to the interphase chemistry of the host-parasite interactions. Yet the plate experiments discussed above are valuable to assess the interactive chemistry of these two organisms which might otherwise not be found, and to provide comparative assays between host varieties for several phases of parasite development cued to chemicals released from the host surface.

Finally, it is important to put in perspective the chemistry of recognition for these plants compared with other systems described in this volume. In the real world it is not likely that most parasitic angiosperms operate with precise host recognition chemistry. Parasitic plants are extremely successful, but for reasons that have little to do with host recognition. In the case of *Striga,* its reproductive strategy is to produce thousands of seeds, literally millions each year, in an infested area. The result is a soil seed bank of astronomical numbers of viable seeds. These seeds then germinate both on hosts and many nonhosts in a fairly wide range of plants that are known and undoubtedly many more that we simply do not know about. Thus, host specificity is not rended by the germination signal. Further, although considered by themselves, the instability and hydrophobic nature of quinones might be a mechanism to stimulate development only close to a potential host, the conclusion that they exclusively do so ignores reports of hydrophilic compounds present in *Striga* host exudate that are highly active with diffusion not limited by the constraints of labile, hydrophobic quinones. Significant reductions in *Striga* seeds in infested soil by the use of nonhost plants (trap crops) that promote germination, but can not sustain *Striga* growth is further testimony to the extensive diffusion of root exudate stimulants.[55]

All the same, *Striga* has evolved a seed physiology that makes it a worthy adversary for other plants. The required postripening period assures that seeds do not germinate at the end of the season. The subsequent water-conditioning period then arms seeds to respond to other plant exudates during the growing season when there has been sustained wet conditions. Finally, there is the germination signal itself, which like the other aspects of seed dormancy may be mostly important in conveying temporal rather than spacial information. In its native grassland, timing of germination concomitant to the vigorous growth of many grass species as well as other host crops assures that many potential host contacts will be made even if *Striga* stands still!

## IV. DISCUSSION

Parasitic angiosperms show unique examples of the use of exogenous chemical signals from other angiosperms that initiate developmental events. Although not true for all root parasite species, it is clear that in some species, such as *Striga,* specific signals elicit seed germination and later, the initiation of the haustorium. As we add information to the physiological profile of these parasites, it may be useful also to set forth what is not yet known. Again, *Striga* can serve as a useful model. Indeed it appears we know more about this species than any other. It will prove very interesting to determine how mechanisms present and active for *Striga* also hold true for other parasite species.

At this point we know that in agar plates the sorghum rhizosphere holds three zones. The widest is the germination zone extending at least 2 cm from the host surface. Next, is a positive chemotropic region for *Striga* radicles extending out about 5 to 6 mm. The last is the zone of haustorial formation which is 2 to 3 mm wide. In all cases, the responses diminish outward from the host source. The chemistry of these regions is undoubtedly very complex. We have learned that *Striga* may have simplified it somewhat, using quinones, though different ones, for germination and haustorial promotion. There is as yet no definitive study for the basis of chemotropism. At this stage, it may be correct to say that quinones can be thought of as *primary* signals. However, both germination and haustorial development are promoted by a number of other chemicals, some clearly understood to be contained in root exudates (cytokinins, for example)[56] raising the possibility that other secondary signals may function here as well.

The comparative analysis of susceptible and resistant cultivars of sorghum by Lesny[54] has been informative. Two significant points emerge. First, angiosperm root parasitism functions with low efficiency. Over 90% of the seeds germinate too far away to ever achieve host contact. In addition, of those radicles that form haustoria, between 20 to 30% do so before they are close enough to attach. For others, lying within 100 μm of the host, another approximately 20% for both varieties form haustoria that fail to attach. Clearly, the start signals for these plants provide basically large windows for host contact, but factors as yet not understood result in very narrow levels of parasitism.

The second point is that although promotive signals have been defined, we have observed increasing inhibitory effects on radicle and haustorial morphogenesis near the sorghum surface. Some of these effects are: (1) inhibition of elongation, (2) diminished haustorial hair frequency, (3) irregular swelling and enlargement of haustoria, (4) multiple haustorial initiations, and (5) aborted radicle development after germination. These effects were common to both sorghum cultivars and became especially apparent as the host surface is approached by *Striga*.

We believe an important difference between strains and possibly a hint for additional chemistry important to parasitism is the reduction in contacts of haustoria for the resistant cultivar followed later by a lower success rate of shoot development by the parasite. Are there other signals at the attachment stage? Perhaps, or it seems just as possible, the inhibitory effects noted above are amplified at the contact stage for the resistant strain of sorghum. What we can be sure of is that both specific signals and a series of as yet unknown other factors play important roles in achieving vascular hook-up of *Striga* to another angiosperm.

As the influence of the host root environment is studied, there are also questions on how these signals are processed by the plant cell. The molecular specificity of these signals argues for the existence of specific receptors either at the cell surface or located within the cytosol. Recent studies in our laboratories have demonstrated that both haustorial induction by 2,6-DMBQ and the completion of haustorial development require continued protein synthesis.[57] Interruption of the normal developmental pattern by inhibiting protein synthesis results, in most cases, in an irreversible arrest at an intermediate stage. Treatment of elongating radicles with protein synthesis inhibitors also results in a decrease in the ability to respond to 2,6-DMBQ. There are a number of interpretations possible for these results including the turnover of one or more proteins that are involved in either perception of the inducer or that are necessary in haustorial maturation. Obviously, continued protein synthesis is necessary for development beyond radicle swelling.

We also demonstrated that a different biosynthetic program is initiated during the transition from radicle to haustorium. Comparison of the two-dimensional-PAGE profiles of 2-day germinated radicles versus radicles of equivalent age, but induced to form haustoria by 2,6-DMBQ clearly show that compositional differences (both qualitative and quantitative) are present. Changes in protein synthetic activities are observed as early as 3 h following induction by 2,6-DMBQ and many polypeptides accumulated in 24-h developed haustoria. Studies are currently underway to identify the products of genes differentially expressed at several stages of haustorial development. It is hoped that this approach will provide more insight into the transduction pathway that leads from perception of the exogenous signal to the dramatic alterations in cell division, growth and differentiation that lead to haustorial formation.

Advancements in the last decade have provided new insights on how parasitic angiosperms function. The recent discovery of quinones as signals leaves questions both for how chemicals like these are processed at the cell surface and, indeed, if there are other types of recognition molecules. Are the zones generated by the chemical parameters of quinone oxidation and diffusion, or also by the diffusion of other hydrophilic molecules of different character? For now, the possibility that other signals may be discovered seems of secondary importance. For the first time the identity of quinone signaling provides a chemical basis for at least one way parasitic angiosperms communicate in the milieu of complex chemistry that is the root rhizosphere.

## ACKNOWLEDGMENTS

The authors thank Tamela Davis for help in preparing this manuscript, Dr. Wm. V. Baird, Dept. of Horticulture, Clemson University for use of the SEM photograph, and Michael Hood for his photograph of sorghum root hairs. We also thank Christa Florea for her thoughtful comments. This study was supported in part by grants from the Thomas F. and Kate Miller Jeffress Memorial Trust (MPT) and the National Science Foundation DCB 8905106 (JLR, MPT). MH was supported by funds from the Research Experiences for Undergraduates Program (NSF).

## DEDICATION

We dedicate this paper to Dr. Johann Visser who passed away January 15, 1990. He was 59. To all those who met Johann he was a friend. He was the consummate professor, a dedicated teacher, active researcher, and always the enthusiastic student of parasitic plants. In this day of specialized disciplines, Johann was the exception who was at home in the laboratory or on field excursions. His breadth of knowledge is shown by two of his recent published works, *South African Parasitic Flowering Plants* (1981)[58] and one of his last publications entitled, Germination Requirements of Some Root-Parasitic Flowering Plants.[59]

## REFERENCES

1. **Musselman, L. J.,** The biology of *Striga, Orobanche* and other root–parasitic weeds, *Annu. Rev. Phytopathol.*, p. 463, 1980.
2. **Kuijt, Job,** *The Biology of Parasitic Flowering Plants,* University of California Press, Chicago, 1969.
3. **Musselman, L. J., Ed.,** *Parasitic Weed in Agriculture,* Vol. I. *Striga,* CRC Press, Boca Raton, FL, 1987.
4. **Sahai, A. and Shivanna, K. R.,** Seed germination and seedling morphogenesis in parasitic angiosperms of the families Scrophulariaceae and Orbanchaceae, *Seed Sci. Technol.*, 10, 565, 1982.
5. **Stewart, G. R. and Press, M. C.,** The physiology and biochemistry of parasitic angiosperms, *Annu. Rev. Plant Physiol. Plant Mol. Biol.*, p. 127, 1990.
6. **Eplee, R. E. and Morris, R. S.,** Chemical control of *Striga,* in *Parasitic Weeds in Agriculture,* Vol. I, *Striga,* Musselman, L. J., Ed., CRC Press, Boca Raton, FL, 1987, 174.
7. **Riopel, J. L.,** The biology of parasitic plants: physiological aspects, in *Vegetative Compatibility,* Moore, R., Ed., Academic Press, San Diego, 1983, 13.
8. **Worsham, A. D.,** Germination of witchweed seeds, in *Parasitic Weeds in Agriculture,* Vol. I, Musselman, L. J., Ed., CRC Press, Boca Raton, FL, 1987, 45.
9. **Fuller, C.,** Root-bloom or witchweed, First Rept. Gov. Eatom, Natal, 1899-1900, p. 20, 1900.
10. **Botha, P. J.,** The germination of the seeds of angiospermous root parasites. I. The nature of the changes occurring during pre-exposure of the seed to *Alectra vogelii.* Benth., *J. S. Afr. Bot.*, 14, 63, 1948.
11. **Kusano, S.,** Further studies on *Aeginetia indica,* Beih., *Bot. Zentralb.*, 24(1), 286, 1908.
12. **Chabrolin, C.,** Contribution a l'etude de la germination des graines de l'Orobanche de la feve, *An. Serv. Bot. Agron. Tunis,* 14, 91, 1938.
13. **Heinricher, E.,** Die grünen Halbachmar-otzer. III. *Bartschia* und *Tozzia* nebst Bemerkungen zur Frage nach der assimilatorischen Luslungsfahigkeit der grünen Halbachmarotzer, *Jahrb. Wiss. Bot.*, 36, 665, 1901.

14. **Saunders, A. R.,** Studies in phanerogamic parasitism with particular reference to *Striga lutea* Lour., Sci. Bull., No. 128, Department of Agriculture, South Africa, 1933.
15. **Bebawi, F. F., Eplee, R. E., Harris, C. E., and Norris, R.,** Longevity of witchweed *(Striga asiatica)* seed, *Weed Sci.*, 32, 494, 1984.
16. **Hsiao, A. I., Worsham, A. D., and Moreland, D. E.,** Effects of temperature and dl-strigol on seed conditioning and germination of witchweed *(Striga asiatica) Ann. Bot.*, 61, 65, 1988.
17. **Menetrez, M. L., Fites, R. C., and Wilson, R. F.,** Lipid changes during pregermination and germination of *Striga asiatica* seeds, *J. Am. Oil Chem. Soc.*, 65, 634, 1988.
18. **Bkrathalakshmi, B.,** Studies on the root parasite– *Striga asiatica* (L.) Kuntze, Ph.D. thesis, Bangalore University, Bangalore, 1982.
19. **Visser, J. H.,** Germination requirements of some root-parasitic flowering plants, *Naturwissenschaften*, 76, 253, 1989.
20. **Eplee, R. E.,** Effectiveness of ethylene in control of witchweed, *Proc. South. Weed. Sci. Soc.*, 28, 300, 1975.
21. **Cook, C. E., Whichard, L. P., Turner, B., Wall, M. E., and Egley, G. H.,** Germination of witchweed *(Striga lutea* Lour.): Isolation and properties of a potent stimulant, *Science*, 154, 1189, 1966.
22. **Cook, C. E., Wichard, L. P., Wall, M. E., Egley, G. H., Coggon, P., Luhan, P. A., and McPhail, A. T.,** Germination stimulants. II. The structure of strigol — a potent seed germination stimulant for witchweed *(Striga lutea* Lour.), *J. Am. Chem. Soc.*, 94, 6198, 1972.
23. **Heather, J. B., Mittal, R. S. D., and Sih, C. J.,** Synthesis of the witchweed seed germination stimulant (±)strigol, *J. Am. Chem. Soc.*, 98, 3661, 1976.
24. **Brooks, D. W., Bevinakatti, H. S., and Powell, D. R.,** The absolute structure of (±)strigol, *J. Org. Chem.*, 50, 3779, 1985.
25. **Johnson, A. W., Rosebery, G., and Parker, C.,** A novel approach to *Striga* and *Orobanche* control using synthetic germination stimulants, *Weed Res.*, 16, 223, 1976.
26. **Dailey, O. D. and Vail, S. L.,** An improved partial synthesis of (±) strigol, in *The Chemistry of Allebopathy: Biochemical Interactions Among Plants, ACS Symp. Ser. 268,* Thompson, A. C., Ed., Amer. Chem. Soc., Washington, DC, 1985, 427.
27. **Dailey, O. D.,** A new synthetic route to (±) strigol, *J. Org. Chem.*, 52, 1984, 1987.
28. **Hassanoli, A.,** Strigol analogues: synthetic achievements and prospects, in *Striga Biology and Its Control,* Ayensu, E. S., Doggett, H., Keynes, R. D., Marton-Lefevre, J., Musselman, L. J., Parker, C., and Bickering, A., Eds., International Council of Scientific Unions, Paris, 1984, 125.
29. **Pepperman, A. B., Connick, W. J., Jr., Vail, S. L., Worsham, A. D., Pavlista, A. D., and Moreland, D. E.,** Education of precursors and analogs of strigol as witchweed *(Striga asiatica)* seed germination stimulants, *Weed Sci.*, 30, 561, 1982.
30. **Vail, S. L., Dailey, O. D., Blanchard, E. J., Pepperman, A. B., and Riopel, J. L.,** Terpenoid precursors of strigol as seed germination stimulants of broomrape *(Orobanche ramosa)* and witchweed *(Striga asiatica), J. Plant Growth Regulation,* 9(2), 77, 1990.
31. **Fisher, N. H., Weidenhamer, J. D., and Bradow, J. M.,** Dihydroparthenolide and other sesquiterpene lactones stimulate witchweed germination, *Phytochemistry,* 28(9), 2315, 1989.
32. **Fisher, N. H., Weidenhamer, J. D., Riopel, J. L., Quijano, L., and Minelaou, M. A.,** Stimulation of witchweed germination by sesquiterpene lactones: a structure-activity study, *Phytochemistry,* 29(8), 2479, 1990.
33. **Netzly, D. H. and Butler, L. G.,** Roots of *Sorghum bicolor* exude hydrophobic droplets that contain biologically active materials, *Crop. Sci.*, 26, 775, 1986.
34. **Chang, M., Netzly, D. H., Butler, L. G., and Lynn, D. G.,** Chemical regulation of distance: characterization of the first natural host germination stimulant for *Striga asiatica, J. Am. Chem. Soc.*, 108, 7858, 1986.
35. **Netzly, D. H., Riopel, J. L., Ejeta, G., and Butler, L. G.,** Germination stimulants of witchweed *(Striga asiatica)* from hydrophobic root exudate of sorghum *(Sorghum bicolor), Weed Sci.*, 36, 441, 1988.
36. **Smith, C. E., Orr, J. D., and Lynn, D. G.,** Chemical communication and the control of development, in *Natural Products Isolation,* Cooper, R. and Wagman, G. H., Eds., Elsevier, NY, 1989, 562.
37. **Fate, G., Chang, M., and Lynn, D. G.,** Control of germination in *Striga asiatica:* chemistry of spatial definition, *Plant Physiol.*, 93, 201, 1990.
38. **Brown, R. and Edwards, M.,** The germination of the seed of *Striga lutea*. I. Host influence and the progress of germination. *Ann. Bot.* [*NS*], 8, 131, 1944.
39. **Sunderland, N.,** The production of the *Striga* and *Orobanche* germination stimulants by maize roots. I. The number and variety of stimulants. *J. Exp. Bot.*, 11, 236, 1960.
40. **Worsham, A. D., Moreland, D. E., and Kling, G. C.,** Characterization of the *Striga asiatica* (Witchweed) germination stimulant from *Zea mays* L., *J. Exp. Bot.*, 15, 556, 1964.
41. **Dale, J. E. and Egley, G. H.,** Stimulation of witchweed germination by run-off water and plant tissues, *Weed Sci.*, 19, 678, 1971.

42. **Riopel, J. L. and Baird, W. V.,** Morphogenesis of the early development of primary haustoria in *Striga asiatica, in Parasite Weeds in Agriculture,* Vol. I. *Striga*, Musselman, L. J., Ed., CRC Press, Boca Raton, FL, 1987, 107.
43. **Riopel, J. L.,** Experimental studies on induction of haustoria in *Agalinis Purpurea, Proc. 2nd Int. Symp. Parasitic Weeds,* N. Carolina State University, Raleigh, NC, p. 165, 1979.
44. **Riopel, J. L. and Musselman, L. J.,** Experimental initiation of haustoria in *Agalinis purpurea* (Scrophu-
45. **Okonkwo, S. N. C.,** Studies on *Striga senegalensis.* III. In vitro culture of seedlings. Establishment of cultures, *Am. J. Bot.,* 53, 687, 1966.
46. **Lynn, D. G., Steffens, J. C., Kamat, V. S., Graden, D. W., Shabanowitz, J., and Riopel, J. L.,** Isolation and characterization of the first host recognition substances for parasitic angiosperms, *J. Am. Chem. Soc.,* 103, 1868, 1981.
47. **Steffens, J. C., Lynn, D. G., Kamat, V., and Riopel, J. L.,** Molecular specificity of haustorial induction in *Agalinis purpurea* (L) Raf. (Scrophulariaceae), *Ann. Bot.,* 50, 1, 1982.
48. **Steffens, J. C.,** Biochemistry of host recognition in *Agalinis purpurea,* Ph.D. dissertation, University of Virginia, Charlottesville, VA, 1979.
49. **Steffens, J. C., Lynn, D. G., and Riopel, J. L.,** An haustorial inducer for the root parasite *Agalinis purpurea, Phytochemistry,* 25, 2291, 1986.
50. **Chang, M.,** Isolation and characterization of semiochemicals involved in host recognition in *Striga asiatica,* Ph.D. dissertation, Dept. of Chemistry, University of Chicago, Chicago, 1986.
51. **Chang, M. and Lynn, D. G.,** Haustoria and the chemistry of host recognition in parasitic angiosperms, *J. Chem. Ecol.,* 12, 561, 1986.
52. **Lynn, D. G. and Chang, M.,** Phenolic signals in cohabitation: Implications for plant development, *Annu. Rev. Plant Physiol. Plant Mol. Biol.,* 41, 497, 1990.
53. **Smith, C. E., Dudley, M. W., and Lynn, D. G.,** Vegetative/parasitic transition: control and plasticity in *Striga* development, *Plant Physiol.,* 93, 208, 1990.
54. **Lesny, M. E.,** *Striga asiatica* chemotropism and haustoria formation in the presence of resistant and susceptible sorghum cultivars. M.S. thesis, University of Virginia, Charlottesville, VA, 1990.
55. **Bebawi, F. F.,** Cultural practices in witchweed management, in *Parasite Weeds in Agriculture, Vol. I. Striga,* Musselman, L. J., Ed., CRC Press, Boca Raton, FL, 1987, 159.
56. **Van Staden, J. and Smith, A. R.,** The synthesis of cytokinins in excised roots of maize and tomato under aseptic conditions, *Ann. Bot.,* 42, 751, 1978.
57. **Timko, M. P., Florea, C. S., and Riopel, J. L.,** Control of germination and early development in parasitic angiosperms, in *Recent Advances in the Development of Seeds,* Taylorson, R. B., Ed., Plenum Press, NY, 1990, 225.
58. **Visser, J.,** *South African Parasitic Flowering Plants,* Juta and Co., (P.O. Box 123, Kenwyn, S. Africa 7790), 1981.
59. **Visser, J.,** Germination requirements of some root-parasitic flowering plants, *Naturwissenchaften,* 76, 253, 1989.

# *Index*

# INDEX

## A

## B



## C

## G



## H

## I



## K



## L

## O



## P

## Q



## R

## S

## T



## U



## V

**W**



**X**